W9-CGP-881

STATISTICS
for the social sciences
SECOND EDITION

William L. Hays
University of Georgia

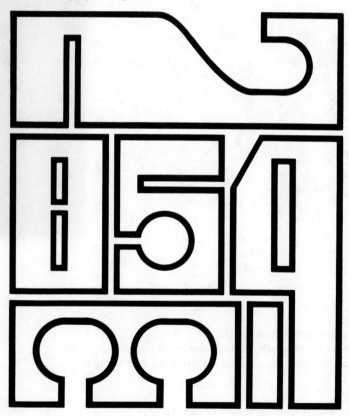

HOLT, RINEHART AND WINSTON, INC.
New York   Chicago   San Francisco   Atlanta
Dallas   Montreal   Toronto   London   Sydney

© 1973 by Holt, Rinehart and Winston, Inc.
First edition published under the title *Statistics for Psychologists,* © 1963
by Holt, Rinehart and Winston, Inc., and later reprinted under the title
of *Statistics.*

All Rights Reserved
Library of Congress Catalog Card Number: 72-85484

ISBN: 0-03-077945-6

Printed in the United States of America
7890 006 9876

Preface: To the Teacher

The first edition of this book was addressed to students in psychology, and most of the examples and explanations were designed with that audience in mind. Happily, the book has also appealed to a wider group, and has had a favorable reception by students and teachers in most of the other social and behavioral sciences. In view of this reception, I have been persuaded by the publisher to give the new edition a title which suggests its usefulness beyond the immediate psychological community.

On the other hand, I have not taken the further step of trying to rewrite the book so as to appeal to every conceivable user, since I felt this would destroy its essential unity of approach. Hence, the book often reads as though it were still addressed to students in psychology. I do not feel this to be any great barrier to its use by students in other fields, however. The examples cited are by no means exotic in their psychological content, and most of the issues discussed and the methods presented occur in all fields of social and behavioral research.

The aim of the book remains precisely the same as outlined in the preface to the first edition: to give the elements of modern statistics in a relatively nonmathematical form, but in somewhat more detail than is customary in such texts, and with considerably more emphasis on the theoretical than upon the applied and computational aspects of the methods. Indeed, I feel that the utility of such an approach may be even greater now than it was ten years ago. A large proportion of statistical analysis is now done on the computer, and the research worker relies considerably on a library of ready-made statistical programs. The problem is no longer so much "how to do it" as it is how to select an appropriate technique which will give results in the format needed. Nevertheless, the interpretation of a result remains just as big a problem as before. It now seems increasingly important for the student to understand the background of the large number of statistical methods available, and to know how and when to carry his problem to a statistician for advice and assistance. The author continues to hope that this book will help the student to learn to do these things.

The content of the book has been expanded in several ways. A good bit of additional space has been given to elementary distribution theory, in order to provide the student with somewhat more groundwork in this area. Several additional families of distributions, such as the Pascal, the gamma, and the beta, are discussed, both because of their intrinsic importance and because they provide some useful techniques. Exercises have been included at the ends of the chapters, along with solutions to odd-numbered problems. Hopefully, these will be useful to student and teacher alike, although teachers will likely wish to expand both the range and the content with additional exercises.

A major change from the first edition is the inclusion of a fairly long chapter on simple Bayesian methods. I believe that this approach and these methods are now too important for an introductory text to ignore. The placement of this chapter at the end of the book reflects my own opinion that a student should have a pretty fair grounding in the classical methods before the Bayesian approach is attempted.

It may be that much of the material on set and function theory is now superfluous for students brought up on the new mathematics. If so, well and good, let these students start with Chapter 2. However, if the new math is retained like the old, then much of this exposure may have worn off by the middle or late university years, and a quick review may be welcomed by many students.

Finally, in addition to all of those whose help I acknowledged in the preface to the first edition, I wish to express my thanks to Professor William Kruskal of the University of Chicago, who contributed many helpful suggestions toward this revision, and especially to Professor J. E. Keith Smith of the University of Michigan, whose careful review of the first edition had much to do with shaping the present version. I wish also to thank Professor E. S. Pearson and the trustees of *Biometrika* for their permission to reproduce selections from the *Biometrika Tables for Statisticians*, Vol. I (3d ed.), and to Professor R. S. Burington and the McGraw-Hill Company for permission to reproduce the table of binomial probabilities from R. S. Burington and D. C. May, *Handbook of Probability and Statistics with Tables* (2d ed.). Most of all, I wish to thank all of the many students and teachers who have so thoughtfully contributed corrections and suggestions over the years. I hope that my work is worthy of you.

W.L.H.

Ann Arbor
November 1972

Preface: To the Student

In its original version, this book was designed for students in experimental psychology. Therefore, many of the examples and much of the discussion deal with issues in that field. Most of these same problems occur in almost identical form in the other social and behavioral sciences, however, and you as a student, whatever your field, should not feel uncomfortable in translating psychological examples and issues directly into your own content area. The psychological examples are simple and easily understood, and every issue having to do with psychological research applies equally well to the other social and behavioral sciences.

In writing this book I assumed that its readers would be serious students just beginning their late undergraduate or early graduate studies. I have tried not to oversimplify or to write down to such students, and I believe that the kinds of students I have in mind will not be dismayed by some tough issues, by some algebraic manipulations, or by the prospect of learning more mathematics. In fact, many students find it interesting, and often exciting, to follow a logical argument and to try to anticipate what the next step will be. I have provided a great deal more explanation and discussion than is customary in statistics texts, largely because I believe that if a student is serious enough to contemplate a research career, he should be also serious enough to want to understand his research tools as fully as possible. I am not so naive as to believe that this will be true of all students, and some of you are going to find this book long-winded, complicated, and *deadly*. To you my sympathy! On the other hand, I am happy to say that many students find the content interesting, and even downright fascinating. These are the students I have in mind as I write. They are the "you" in this book.

Anyone who has had any exposure at all to the social and behavioral sciences does not need to be told that statistics is an important tool in these fields. Statistics serves in at least two capacities. First, it gives methods for organizing, summarizing, and communicating data. Second, it provides methods for making inferences beyond the observations actually made to statements about large classes of *potential* observations. The set of methods serving the first of these functions is generally called **descriptive statistics,** the body of techniques for effective organization and communication of data. When the man on the street speaks of "statistics" he usually means data organized by these methods. However, the major emphasis in this book is on **inferential statistics,** the body of methods for arriving at conclusions extending beyond the immediate data. A large part of the mathematical theory of statistics is concerned with the problem of inference, and with the development of inferential methods. Furthermore, most of the interesting and important applications of mathematical statistics to the sciences concern problems of inference. This book attempts to lay some of the groundwork for an understanding of the origins of inferential methods and their applications to data.

You will soon discover that the main concern in this book is with the mathematical theory underlying inferential methods, rather than with a detailed exposition of all the different methods psychologists and others find useful. The author had no intention of writing a "cookbook" that would equip the student to meet every possible situation he might encounter in his work. Many methods will be introduced, it is true, and we will, in fact, discuss most of the elementary techniques for statistical inference currently in use. However, in the past few years the concerns of the social scientist have begun to grow increasingly complicated. Psychological theory is growing, psychologists are turning their attention to new problems, and techniques for experimentation are becoming much more sophisticated than in the past. The same thing is happening in the other social and behavioral sciences. The statistical analyses required in many such experiments are simply not in the "cookbooks." From all indications, this trend will continue, and by the time that you, the student, are in the midst of your professional career it may well be the case that entirely new statistical methods will be required, replacing many of the methods currently found useful.

As social and behavioral research becomes more sophisticated and mathematical statistics produces more and more methods appropriate to particular situations, a point is rapidly being reached where the research worker simply cannot be familiar with all the statistical methods appropriate to his work. It seems unfair to demand that each competent researcher must also be a competent mathematical statistician as well, although a few gifted individuals (*not* including the author) have somehow found time and brain-power to be both. Furthermore, the advent of electronic computers has opened up new avenues of data-analysis, making it possible to answer questions that were formerly unanswerable because of the sheer arithmetical complexity of the analysis involved. In short, the days when each researcher was his own "do it yourself" statistician, relying on his handy cookbook, are about over.

What, then, is the research scientist to do? He wants the answers that statistical analysis can give him, but he may not know the range of methods open to him within theoretical statistics itself. The answer is very simple: when in doubt, ask a statistician, a man whose principal training and commitment is in mathematical statistics and the development of such methods. A large part of the work of most applied statisticians consists of consultation on problems of design and analysis of experiments, and many are available for such consultation on a professional basis. The statistician can usually provide answers to the research worker's questions, *provided that the statistician is asked about the problem before the data are collected, and can participate in the efficient and logical planning of the experiment.* It is most unreasonable to expect the statistician to reach in his hat and pull out a method that will extract meaning from a poorly designed or executed study.

In order to use the resources of mathematical statistics and statisticians the research scientist must know something about the nature of mathematical statistics. He should be able at least to talk to the statistician in terms they both understand. The statistician does not expect the scientist to know all about theoretical statistics, nor does the scientist expect the statistician to know all about his particular problem. But to work together effectively, each must have some idea of the basic concepts the other uses. This is the reason for the theoretical emphasis in this book. At the very outset, the student needs to know something about the nature of theoretical statistics if he is to appreciate the resources of statistics and not become lost in the complexities of using statistical methods effectively.

This book is not, nor does it pretend to be, a first course in mathematical statistics. Ideally, the serious student in the social or behavioral sciences should take at least one such course. However, there are two practical difficulties: The content and the organization of courses in mathematical statistics are framed for the training of statisticians, not behavioral scientists and the peculiar problems of these research areas are not emphasized in such courses: This is as it should be. In the second place, to become a really good researcher is a full-time job, and the student may not have the time to devote to the mathematical statistics courses and their prerequisites in order to gain the essential background he needs.

Thus, this book contains some of the concepts, results, and theoretical arguments that come from mathematical statistics, but these results and arguments are given at a far more intuitive and informal level than would be the case for a student in mathematics. Only very seldom will the level of mathematics used rise above the high school level, although the mathematical concepts used will occasionally be unfamiliar to most students. In particular we will use some results coming from the application of the calculus, especially results having to do with the idea of a "limit"; these ideas really cannot be treated adequately at an elementary level. From a mathematician's technical point of view, many of our statements are incomplete, poorly framed, or imprecise. On the other hand, many of these ideas can be grasped intuitively by the serious student, and the author feels that this intuitive understanding is better than no understanding at all, *provided* that the student understands the limitations of a presentation such as this.

A number of topics have been included that have little or no direct application to social or behavioral science at this time, because the author feels that these topics do help to clarify some theoretical point. On the other hand, a few topics ordinarily included in elementary statistics books have been omitted, largely because they have rather minor importance and the author preferred to devote space to other matters. Finally, some techniques are included simply because the author feels that you, as a research worker, might want to know these methods. These are techniques that are useful even in elementary experimentation, although they are usually given extensive coverage only in more advanced texts.

The student also should understand that the examples in this book are hypothetical. This, admittedly, goes against current practice in such texts. On the other hand, the author feels that it is more important to have a fairly simple and plausible problem that the beginner in research can understand and that illustrates the method clearly, than to try to provoke the student into exclaiming, "Gee, they really *do* use statistics in research!" Presumably, a student who has had an adequate introductory course in his field knows this already.

A glance at the table of contents reveals the topics covered, and there is little point in a detailed listing here. However, it should be pointed out that the chapters in this book fall roughly into two sections: Chapters 1 through 8 deal very largely with the essential ideas of probability and of distributions, the two central notions of theoretical statistics. The first chapter lays a foundation for these topics by introducing three very fundamental mathematical concepts: set, relation, and function. A clear idea of these concepts can do a great deal to clarify the remainder of the book. Chapters 2 through 8 are very closely related in the topics covered, and each succeeding chapter builds on the concepts introduced in the preceding ones.

Chapter 9 develops some of the issues connected with the actual use of results from theoretical statistics, particularly the problem of making up one's mind from data. Chapters 10 through 18 discuss particular methods for various kinds of inferences to be made in simple experimental situations. Finally, Chapter 19 gives some of the basic ideas of Bayesian statistics, an alternate approach.

A theme that runs throughout this book is the search for relationships through experimentation. A statistical relation will be said to exist when knowledge of one property of an object or event *reduces our uncertainty* about another property that object or event will show. A statistical relation occurs when things tend to "go together" in a systematic way. This theme will recur *ad nauseam* in the chapters to follow, but it is an important one.

Very many mathematical expressions occur throughout this text. These are of three kinds: algebraic equivalences serving as steps in some derivation, actual definitions or principles stated mathematically, and computational formulas useful in some method. Some of the mathematical expressions are numbered; ordinarily this occurs when some reference will be made to that expression at a later point. If the number for any expression is followed by an asterisk (*), then this is an important definition or relationship that is worthy of your special attention. If an expression is primarily a computing formula, then this will be given a dagger (†) following the number.

A few words must also be said about the symbols we will use. Generally, when a new symbol is introduced, it will be given an "on the spot" definition. However, there are a few symbols in such widespread use that the author may omit their definition; or you may have forgotten what the symbol meant on its first introduction. In either case you will find the glossary of symbols in the back of the book helpful. Furthermore, Appendixes A and B, rules for the manipulation of summations and of expectations, are very important, since we will use these rules to considerable extent in our simple derivations of results.

So far we have talked at length about the author's expectations about the student and the reasons underlying this book, but we have failed to say much about the topic itself. Next we will take an overview of what inferential statistics is about. In addition, some ideas about formal systems and mathematical models will be given, which may help the student understand how "statistics" can mean both a body of applied methods and a mathematical theory.

Applications of statistics occur in virtually all fields of research endeavor—the physical sciences, the biological sciences, the social sciences, engineering, market and consumer research, quality control in industry, and so on, almost without end. Although the actual methods differ somewhat in the different fields, the applications all rest on the same general theory of statistics. By examining what the fields have in common in their applications of statistics we can gain a picture of the basic problem studied in mathematical statistics. *The major applications of statistics in any field all rest on the possibility of repeated observations or experiments made under essentially the same conditions.* That is, either the researcher actually can observe the same process repeated many times, as in industrial quality control, or there is the *conceptual* possibility of repeated observation, as in a scientific experiment that might, in principle, be repeated under identical conditions. However, in any circumstance where repeated observations are made, even though every precaution is taken to make conditions exactly the same the results of observations will vary, or tend to be different, from trial to trial. The researcher has control over some, but not all, of the factors that make outcomes of observations tend to differ from each other.

In some areas of research, objects or phenomena viewed under the same conditions will vary only to a small extent. This is certainly true in some branches of physical science, where observations made under carefully controlled conditions give virtually identical results. On the other hand, in the biological, and especially the social, sciences, even though the experimenter exerts almost superhuman efforts to observe repeatedly under precisely the same conditions, some differences among his observations will occur, and these differences are ordinarily *not* negligible.

When observations are made under the same conditions in one or more respects, but they give outcomes differing in other ways, then there is some *uncertainty* connected with observation of any given object or phenomenon. Even though some things are known to be true about that object in advance of the observation, the experimenter cannot predict with complete certainty what its other characteristics will be. Given enough repeated observations of the same object or kind of object the experimenter may be able to formulate a good bet about what the other characteristics are likely to be, but he cannot be completely sure of the status of any given object.

This fact leads us to the central problem of theoretical statistics: *in one sense, mathematical statistics is a theory about uncertainty, the tendency of outcomes to vary when repeated observations are made under identical conditions.* Granted that certain conditions are fulfilled, theoretical statistics permits deductions about the *likelihood* of the various possible outcomes of observation. The essential concepts in statistics derive from the theory of probability, and the deductions made within the theory of statistics are, by and large, statements about the probability of particular kinds of outcomes, given that initial, mathematical, conditions are met.

Mathematical statistics is a formal mathematical system. Any mathematical system consists of these basic parts:

1. A collection of undefined "**things**" or "**elements,**" considered only as abstract entities;
2. A set of undefined **operations,** or possible relations among the abstract elements;
3. A set of **postulates** and **definitions,** each asserting that some specific relation holds among the various elements, the various operations, or both.

In any mathematical system the application of logic to combinations of the postulates and definitions leads to *new* statements, or theorems, about the undefined elements of the system. *Given* that the original postulates and definitions are true, then the new statements *must* be true. Mathematical systems are purely abstract, and essentially undefined, **deductive** structures. In the first chapter, the theory of sets will be used as an example of an abstract system of this sort, and the theory of probability also has this character, as we shall see.

Mathematical systems are not really "about" anything in particular. They are systems of statements about "things" having the formal properties given by the postulates. The mathematician does not, in fact, have to commit himself about what he really has in mind to call these abstract elements; indeed, he may have absolutely nothing in mind that exists in the real world of experience, and his sole concern may be in what he can derive about the other necessary relations among abstract elements given particular sets of postulates. It is perfectly true, of course, that many mathematical systems originated from attempts to describe real objects or phenomena and their interrelationships: historically, the abstract systems of geometry, school algebra, and the calculus grew out of problems where something very concrete was in the back of the mathematician's mind. However, as *mathematics* these systems deal with completely abstract entities.

When a mathematical system is interpreted in terms of real objects or events, then the system is said to be a **mathematical model** for those objects or events. Somewhat more precisely, the undefined terms in the mathematical system are identified with particular, relevant, **properties** of objects or events; thus, in applications of arithmetic, the number symbols are identified with magnitudes or amounts of some particular property that objects possess, such as weight, or extent, or numerosity. The system of arithmetic need not apply to other characteristics of the same objects, as, for example, their colors. Once this identification can be made between the mathematical system and the relevant properties of objects, then anything that is a logical consequence in the system is a true statement about objects in the model, *provided*, of course, *that the formal characteristics of the system actually parallel the real characteristics of objects in terms of the particular properties considered.* In short, in order to be useful as a mathematical model, a mathematical system must have a formal structure that "fits" at least one aspect of a real situation.

Probability theory and statistics are each both mathematical systems *and* mathematical models. Probability theory deals with elements called "events," which are completely abstract. Furthermore, these abstract things are paired with numbers called "probabilities." The theory itself is the system of logical relations among these essentially undefined things. The experimenter uses this abstract system as a mathematical model: his experiment produces a real outcome, which he calls an event, and he uses the model to find a probability, which he interprets as the relative frequency of occurrence for that outcome. If the requirements of the model are met, this is a true, and perhaps useful result. If his experiment really does not fit the requirements of probability theory as a system, then the statement he makes about his actual result need not be true. (This point must not be overstressed, however. We will find that often a statistical method can yield practically useful results even when its requirements are not fully satisfied. Much of the art in applying statistical methods lies in understanding when and how this is true.)

Mathematical systems such as probability theory and the theory of statistics are, by their very nature, **deductive.** That is, formal assertions are postulated as true, and then by logical argument true conclusions are reached. All well-developed theories have this formal, logico-deductive character.

On the other hand, the problem of the empirical scientist is essentially different from that of the logician or mathematician. Scientists search for general relations among events; these general relations are those which can be expected to hold whenever the appropriate set of circumstances exists. The very name "empirical science" asserts that these laws shall be discovered and verified by the actual observation of what happens in the real world of experience. However, no mortal scientist ever observes all the phenomena about which he would like to make a generalization. He must always draw his conclusions about what would happen for *all* of a certain class of phenomena by observing a very few particular cases of that phenomenon.

The student acquainted with logic will recognize that this is a problem of **induction.** The rules of logical *deduction* are rules for arriving at true consequences from true premises. Scientific theories are, for the most part, systems of deductions from basic principles held to be true. If the basic principles are true, then the deductions must be true. However, how does one go about arriving at and checking the truth of the initial propositions? The answer is, for an empirical science, observation and inductive generalization—going from what is true of some observations to a statement that this is true for *all possible* observations made under the same conditions. Any empirical science begins with observation and generalization.

Furthermore, even after deductive theories exist in a science, experimentation is used to check on the truth of these theories. Observations that contradict deductions made within the theory are prima-facie evidence against the truth of the theory itself. Yet, how does the scientist know that his results are not an accident, the product of some chance variation in procedure or conditions over which he has no control? Would his result be the same in the long run if the experiment could be repeated many times?

It takes only a little imagination to see that this process of going from the specific to the general is a very risky one. Each observation the scientist makes is different in some way from the next. Innumerable influences are at work altering—sometimes minutely, sometimes radically—the similarities and differences the scientist observes among events. Controlled experimentation in any science is an attempt to minimize at least part of the accidental variation or "error" in observation. Precise techniques of measurement are aids to the scientist in sharpening his own rather dull powers of observation and comparison among events. So-called "exact sciences," such as physics and chemistry, have thus been able to remove a substantial amount of the unwanted variation among observations from time to time, place to place, observer to observer, and hence are often able to make general statements about physical phenomena with great assurance from the observation of quite limited numbers of events. Observations in these sciences can often be made in such a way that the generality of conclusions is not a major point at issue.

In the biological and social sciences, however, the situation is radically different. In these sciences the variations between observations are not subject to the precise experimental controls that are possible in the physical sciences. Refined measurement techniques have not reached the stage of development that they have attained in physics and chemistry. Consequently, the drawing of general conclusions is a much more dangerous business in these fields, where the sources of variability among living things are extremely difficult to identify, measure, and control. And yet the aim of the social or biological scientist is precisely the same as that of the physical scientist—arriving at general statements about the phenomena under study.

Faced as he is with only a limited number of observations or with an experiment that he can conduct only once, the scientist can reach general conclusions only in the form of a "bet" about what the true, long run, situation actually is like. Given only sample evidence, the scientist is always unsure of the "goodness" of any assertion he makes about the true state of affairs. The theory of statistics provides ways to assess this uncertainty and to calculate the probability that he will be wrong if he decides in a particular way. *Provided that the experimenter can make some assumptions about what is true, then the deductive theory of statistics tells him how likely he is to observe particular results.* Armed with this information, the experimenter is in better position to decide, if he must, what he will say about the true situation. Regardless of what he decides from his evidence, he *could* be wrong; but using deductive statistical theory he can at least determine the probabilities of error in a particular decision.

In recent years, a branch of mathematics has been developed around this problem of decision-making under uncertain conditions. This is sometimes called "statistical decision theory." One of the main problems treated in decision theory is the choice of a decision rule, or "deciding how to decide" from evidence. Decision theory evaluates rules for deciding from evidence in the light of what the decision-maker wants to accomplish. As we shall see in later chapters, mathematics can tell us wise ways to decide how to decide under some circumstances, but it can never tell the experimenter how he *must* decide in any particular situation. The theory of statistics supplies one very important piece of information to the experimenter: the probability of sample results *given* certain conditions. Decision theory supplies another: optimal ways of using this and other information to accomplish certain ends. Nevertheless, neither theory tells the experimenter *exactly* how to decide—how to make the inductive leap from what he observes to what is true in general. This is the experimenter's problem, and he must seek the answer outside of deductive mathematics, and in the light of what he is trying to do.

These, then, are a few of the reasons for studying inferential statistics. The rest of this book will go into the background and details of how these methods are developed and used. I hope that you enjoy learning about them as much as I have enjoyed trying to explain them for you.

W.L.H.

Ann Arbor
November 1972

Contents

13. THE ANALYSIS OF VARIANCE: MODELS II AND III, RANDOM EFFECTS AND MIXED MODELS   524

14. INDIVIDUAL COMPARISONS AMONG MEANS   581

17. COMPARING ENTIRE DISTRIBUTIONS: CHI-SQUARE TESTS   717

18. SOME ORDER STATISTICS   760

19. SOME ELEMENTARY BAYESIAN METHODS   809

APPENDIX A. RULES OF SUMMATION   861

APPENDIX B. THE ALGEBRA OF EXPECTATIONS   871

APPENDIX C. TABLES   878

REFERENCES   908

GLOSSARY OF SYMBOLS   913

SOLUTIONS TO SELECTED PROBLEMS   922

INDEX   948

1 Sets and Functions

It may seem surprising that a book about statistics starts off with a discussion of sets. Although the study of sets and functions is not usually a part of a course in statistics, these topics actually provide the most fundamental concepts we will use. Set theory will be the basis for our discussion of probability theory, to be introduced in the next chapter. The idea of a function pervades virtually all mathematics and is a key concept in modern scientific work as well. Even at the high-school level most of us are exposed to the notion of a mathematical function, and we know that saying that "Y is a function of X" expresses something about a relation between two things. However, unless the student has a very good background in mathematics, he is usually somewhat vague about the precise meaning of the word "function" used in mathematical or scientific writing. One of the primary purposes of this chapter is to give a very concrete and restricted meaning to this term. It will be shown that the idea of a function is a very simple one, which grows out of the notion of a set.

1.1 SETS

The concept of a set is the starting point for all of modern mathematics, and yet this idea could hardly be more simple:

Any well-defined collection of objects is a set.

The individual objects making up the set are known as the "elements" or "members" of the set. The set itself is the aggregate or totality of its members. If a given object is in the set, then one says that the object is an **element** of, or a **member** of, or **belongs** to, the set.

For example, all students enrolled at a given university in a given year is a set, and any particular student is an element of that set. All living men whose last name is "Jones" is another set. All of the whole numbers between 10 and 10,000 is a set, all animals of the species *Rattus norvegicus* living at this particular moment make up a set, all possible neural pathways in John Doe's brain form a set, and so on, ad infinitum.

It is important to note that in the definition of a set the qualification "well-defined" occurs. This means that *it must be possible, at least in principle, to specify the set so that one can decide whether any given object does or does not belong.* This does not mean that sets can be discussed only if their members actually exist: it is perfectly possible to speak of the set of all women presidents of the United States, for example, even though there are not any such objects at this writing. What *is* required is some procedure or rule for deciding whether or not an object is or is not in the set; given an object, one can decide if it meets the qualifications of a female president of the United States, and thus the set is well-defined.

The word "object" in the definition can also be interpreted quite liberally. Often, sets are discussed having members that are not objects in the usual sense but rather are "phenomena," or "happenings," or possible outcomes of observation. We will have occasion to use sets of "logical possibilities," all the different ways something might happen, where each distinct possibility is thought of as one member of the set. Very often the members of a set will be sets themselves, or the names of sets. For example, a set of numbers can be thought of as a set of sets, each number ultimately signifying a set of objects each showing the same property of "numerosity." A set of nouns can also be regarded as a set of sets, where each noun designates some set of objects.

1.2   WAYS OF SPECIFYING SETS

In discussing sets, we will follow the practice of letting a capital letter, such as A, symbolize the set itself, with a small letter, such as a, used to indicate a particular member of the set. The symbol "ϵ" (small Greek epsilon) is often used to indicate "is a member of"; thus,

$$a \; \epsilon \; A$$

is read "a is a member of A."

There are two different ways of specifying a set. The first way is by *listing* all of the members. For example,

$$A = \{1,2,3,4,5\}$$

is a complete specification of the set A, saying that it consists of the numbers 1, 2, 3, 4, and 5. The braces around the listing are used to indicate that the list makes up a set. Another set specified in a similar way is

$$B = \{\text{orange, grapefruit, lemon, lime, tangerine, kumquat, citron}\}$$

and another is

$$C = \{\text{Roosevelt, Truman, Eisenhower, Kennedy, Johnson, Nixon}\}.$$

In each instance the respective elements of the set are simply listed. These sets are thought of as unordered, since the order of the listing is completely irrelevant so long as each member is included once and only once in the list.

The other way of specifying a set is to give a *rule* that lets one decide whether or not an object is a member of the set in question. Thus, set A may be specified by

$$A = \{a | a \text{ is an integer and } 1 \leq a \leq 5\}.$$

(The symbol | is read as "such that," so that the expression above is, in words, "the set of all elements a such that a is an integer between 1 and 5 inclusive.") Given this rule, and any potential element of the set A, we can decide immediately whether or not the object actually does belong to the set.

The rule for set B would be

$$B = \{b | b \text{ is the name of a kind of citrus fruit}\}$$

and for set C

$$C = \{c | c \text{ is a United States President elected between 1932 and 1968, inclusive}\}.$$

It is quite possible to specify sets in terms of other sets. For example, we might specify a set D by

$$D = \{d | d = a + 15, \text{ for all } a \in A\}.$$

When listed, the elements of D would be

$$D = \{16, 17, 18, 19, 20\}.$$

It is usually far more convenient to specify a set by its rule than by listing the elements. For example, sets such as the following would be awkward or impossible to list:

$$G = \{g | g \text{ is a human born in 1955}\}$$
$$X = \{x | x \text{ is a positive number}\}$$
$$Y = \{y | y \in X \text{ and } y \text{ is an integer}\}.$$

1.3   UNIVERSAL SETS

There are many instances in set theory when it becomes convenient to consider only objects belonging to some "large" set. Then, given that all objects to be discussed belong to this "universal" set, we proceed to talk of particular groupings of elements. The introduction of a universal set acts to set the stage for the kinds of sets that will be introduced. For example, we may wish to deal only with U.S. students enrolled in college in the year 1970, and so we begin by specifying a universal set:

$$W = \{w | w \text{ was a U.S. student enrolled in college in 1970}\}.$$

Then particular subsets of W are introduced:

$$A = \{a|a \,\epsilon\, W \text{ and } a \text{ was a student at Harvard}\}$$
$$B = \{b|b \,\epsilon\, W \text{ and } b \text{ was classified as a sophomore}\}$$

and so on. Or perhaps the universal set is

$$W = \{w|w \text{ is a symbol for a sound}\}$$
$$A = \{a|a \,\epsilon\, W \text{ and } a \text{ is a letter in the Cyrillic alphabet}\}$$
$$B = \{b|b \,\epsilon\, W \text{ and } b \text{ is a letter in the Greek alphabet}\}$$

and so on.

Up until this point, the individual member of a set has been denoted by a small letter, a, b, w, and so on, depending on the capital letter used for the set itself. This is not strictly necessary, however, as any symbol that serves as a "place holder" in the rule specifying the set would do as well. In future sections it will sometimes prove convenient to use the same symbol for a member of several sets, so that the neutral symbol x will be used, indicating any member of the universal set under discussion. Furthermore, it is redundant to state that $x \,\epsilon\, W$ after the universal set W has been specified, and so this statement will be omitted from the rule for particular subsets of the universal set. Nevertheless, it is always understood that any member of a set is automatically a member of some universe W. In the example just given, the sets are adequately specified by

$$W = \{x|x \text{ is a symbol for a sound}\}$$
$$A = \{x|x \text{ is a letter in the Cyrillic alphabet}\}$$
$$B = \{x|x \text{ is a letter in the Greek alphabet}\}.$$

1.4   SUBSETS

Suppose that there were some set A and another set B, so that any element which is in B is also in A. Then B *is a subset of* A. This is symbolized by

$$B \subseteq A,$$

read as "B is a subset of A." More formally,

The set B is a subset of the set A if and only if
for each $x \,\epsilon\, B$ then $x \,\epsilon\, A$.

For example, consider the set A of all male citizens of the United States at this moment. Now let set B be the set of all living U.S. Army generals; the set B is a subset of the set A. Or, let A be the set of all numbers, and let B indicate the set of all whole numbers. Once again, B is a subset of A.

If there is at least one element of A not in B, then B is a **proper subset** of A, symbolized by

$$B \subset A.$$

In both the examples just given the set B is a proper subset of A, since some U.S.

men are not generals, and some numbers are not whole. On the other hand, if every single element in B is in A, *and* every single element in A is in B, so that $B \subseteq A$ and $A \subseteq B$, then

$$A = B,$$

the two sets are equivalent or equal. Two sets are equal only if they each contain exactly the same elements.

Note the similarity of the symbols "$\subseteq$," for "is a subset of," and "$\subset$," for "is a proper subset of," to the symbols "$\leq$," for "is less than or equal to," and "$<$," for "is less than." The difference in "$\subseteq$" and "$\subset$" reflects the fact that one set can be a subset of another even though the two sets actually are identical or equal element by element; a set can be a proper subset of another only if the two sets are not equal. It should also be noted that the statements "$A \supseteq B$" and "$B \subseteq A$" are equivalent, as are the statements "$A \supset B$" and "$B \subset A$." In each case the set at the rounded end of the "horseshoe" is the subset or proper subset, and the set at the open end is the larger set. Observe that this is similar to the correspondence between "$x \geq y$" and "$y \leq x$" in the case of numbers.

Given the universal set W for some problem, then every set A to be discussed is a subset of the universal set,

$$A \subseteq W, \text{ for every } A.$$

1.5   FINITE AND INFINITE SETS

In future work it will be necessary to distinguish between finite and infinite sets. For our purposes, a finite set has members equal in number to some specifiable positive integer or to zero. The members of a finite set can thus be counted exactly, at least in principle. The set of all names in the Manhattan telephone book, the set of all books printed in the nineteenth century, the set of all houses in Rhode Island, the set of all living mammals, are all finite sets, even though the number of members each includes is very large.

Sets with infinite numbers of members fall into two general classifications. Some sets are said to be **countably infinite.** This means that, in principle, the members of the set can be put into 1-to-1 correspondence with the set of **natural** or **counting numbers,** $\{1,2,3,4,5, \cdots \}$, so that each member of the set would be associated with one and only one number. However, in such a countably infinite set, one would never run out of numbers, nor of members to assign them to. The members of the set, though countable in the sense of 1-to-1 correspondence with the counting numbers, exceed any exact number that one could possibly name. In some conceptions the entire set of stars in the cosmos make up such a countably infinite set. In principle each separate star could be assigned one of the counting numbers, but one might go on forever with such a count. A more familiar example is the set of even integers. How would one show that this set is also countably infinite?

Still another kind of set is **uncountably infinite.** Here the elements of the set are so "dense" that even the counting numbers would fail to exhaust all of them. One example of such an infinite set consists of all possible points on a straight line. Regardless of how thickly counting numbers are assigned to such points on a line, there will always be points left over. Similarly, the set of all points lying on a circle, the set of all points in a plane, the set of all real numbers (including all rational and irrational, positive and negative numbers), the set of all intervals into which a straight line may be divided, the set of all possible intervals within a given span of time, all are examples of such uncountably infinite sets.

In our discussion in the next few sections, all sets will be treated as though they were finite. The general ideas apply to infinite sets as well, but certain qualifications have to be made in some of the definitions, and we will have to overlook these.

1.6   Venn diagrams

A scheme that is very useful for illustrating sets and for showing relations among them is the Venn diagram (named after the logician J. Venn, 1834–1923). These are sometimes referred to as Euler diagrams (after the mathematician L. Euler, 1707–1783; both men made important contributions to the theory of sets). A Venn diagram pictures a set as all points contained within a circle, square, or other closed geometrical figure.

Since there is an infinite number of possible points within any such figure, Venn diagrams actually represent infinite sets, but they are useful for

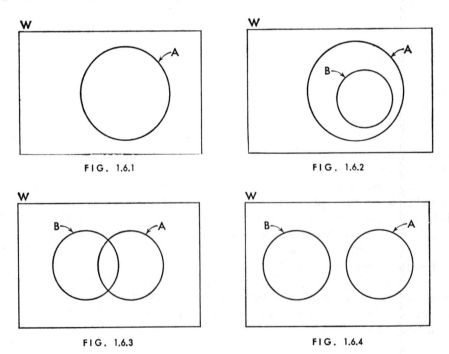

FIG. 1.6.1   FIG. 1.6.2

FIG. 1.6.3   FIG. 1.6.4

representing any set. For example, a Venn diagram picturing the universal set W and the subset A would look like Figure 1.6.1. Since A is a subset of W, the area of the circle A is completely included in the area of the rectangle W.

Now consider another set $B \subseteq W$. If $B \subset A$, the Venn diagram would be as shown by Figure 1.6.2. On the other hand, if B were not a subset of A, then Figure 1.6.3 would be the Venn diagram if A and B shared members in common. Figure 1.6.4, would be the Venn diagram if the two sets had no members in common.

1.7 THE EMPTY SET

Just as the number "zero" is important in ordinary arithmetic, so the theory of sets requires the notion of the "empty set." **Any set that contains no members is called the empty set.** The empty set is usually identified by the symbol $\emptyset$, a zero with a diagonal slash. We have already had one example of the empty set, the set of all women presidents of the United States. It is not hard to dream up other examples: the set of all months containing 38 days, the set of all people with 8 legs, the set of all numbers not divisible by 1, and so on.

This set has a very special property: **the empty set is a subset of every set.** Notice that the definition of a subset as given in Section 1.4 does not rule out the empty set; since the empty set has no members, the definition of subset is not contradicted, and so $\emptyset \subseteq A$ for every A.

1.8 NEW SETS FROM OLD: OPERATIONS WITH SETS

Given some universal set W, and a set of its subsets, A, B, C, and so on, it is possible to "operate" on sets to form new sets, each of which is also a subset of W.

The first of these operations is the **union** of two or more sets. Given W and the two subsets A and B, then the union of A and B is written $A \cup B$ (a useful mnemonic device for the symbol $\cup$ is the u in *u*nion). By definition,

$A \cup B$ **is the set of all elements that are members of A, or of B, or of both:**

$$A \cup B = \{x | x \,\epsilon\, A \text{ or } x \,\epsilon\, B, \text{ or both}\}.$$

For example, let

$$W = \{x | x \text{ is a living American war veteran}\}$$
$$A = \{x | x \text{ is a veteran of World War II}\}$$
$$B = \{x | x \text{ is a veteran of the Korean War}\}.$$

Then

$$A \cup B = \{x | x \text{ is a veteran of World War II, or}$$
$$x \text{ is a veteran of the Korean War, or}$$
$$x \text{ is a veteran of both}\}.$$

In the Venn diagram (Fig. 1.8.1), the shaded portion shows the union of A and B. Notice that we include the possibility that an element of the union could be a

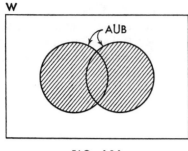

FIG. 1.8.1

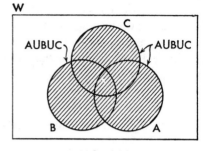

FIG. 1.8.2

member of *both* A and B. Note further that in this example, $A \cup B \subset W$, the union is a *proper* subset of the universal set, since there are some living American war veterans who belong neither to A nor B.

The idea of the union may be extended to more than two sets. For instance, given W and three sets A, B, and C, then

$$A \cup B \cup C = \{x | x \, \epsilon \, A, \text{ or } x \, \epsilon \, B, \text{ or } x \, \epsilon \, C\},$$

as shown in Figure 1.8.2. In general, the union of K sets, A_1, A_2, $\cdots$, A_K is defined as follows:

$$\bigcup_{i=1}^{K} A_i = \{x | x \, \epsilon \, A_1, \text{ or } x \, \epsilon \, A_2, \text{ or } x \, \epsilon \, A_3, \cdots, \text{ or } x \, \epsilon \, A_K\}$$

$$= \{x | x \text{ is a member of at least one of the } K \text{ sets } A_1, A_2, \cdots, A_K\}.$$

where $i = 1, 2, \cdots, K$.

(Here, the counting numbers $\{1, 2, \cdots, K\}$ are being used as an "index set" to distinguish the various sets A from each other. The symbol i refers to any one of the index numbers; the symbol A_i refers to any one of the sets, with index number i. The symbol $\bigcup_{i=1}^{K} A_i$ is a convenient shorthand for $A_1 \cup A_2 \cup \cdots \cup A_K$. The idea of an index set will be used repeatedly in chapters to follow.)

As an example of the union of three sets, let

$$W = \{x | x \text{ is a positive integer, } 1 \angle x \angle 10\}$$
$$A = \{x | x \text{ is a perfect square}\} = \{1, 4, 9\}$$
$$B = \{x | x \text{ is divisible by 3}\} = \{3, 6, 9\}$$
$$C = \{x | x \text{ is divisible by 7}\} = \{7\}.$$

Then

$$A \cup B \cup C = \{1, 3, 4, 6, 7, 9\}.$$

The union of any set with a proper subset of itself is simply the larger set: given $B \subset A$, then $A \cup B = B \cup A = A$. It follows that

$$A \cup \emptyset = A$$

and
$$W \cup A = W.$$

1.9 THE INTERSECTION OF SETS

The verbal rule for the union of two sets always involves the word "or"; the union $A \cup B$ is the set made by finding the elements that are members of A *or* of B or of both. However, what of the set of elements in *both* A and B? This set is included in the union, but we are most often interested in this set by itself as the **intersection** of A and B:

The intersection of sets A and B, written $A \cap B$, is the set of all members belonging to both A and B: $A \cap B = \{x | x \in A \text{ and } x \in B\}$.

For example, let
$$W = \{x | x \text{ is a chemical compound}\}$$
$$A = \{x | x \text{ contains chlorine}\}$$
$$B = \{x | x \text{ contains oxygen}\}.$$
Then
$$A \cap B = \{x | x \text{ contains chlorine and oxygen}\}.$$

As another example, consider the sets of numbers
$$W = \{x | x \text{ is a positive integer}\}$$
$$A = \{1,2,3,4,5,6,7,8,9,10\}$$
$$B = \{8,9,10,11,12,13\}.$$
Then
$$A \cap B = \{8,9,10\}$$
since only these elements appear in both A and B.

The intersection of two sets A and B is always a subset of their union:
$$A \cap B \subseteq A \cup B.$$

If the intersection of two sets is empty, $A \cap B = \emptyset$, then the sets are said to be **disjoint** or **mutually exclusive**.

The intersection of two sets presented in a Venn diagram appears as Figure 1.9.1.

Whenever B is a subset of A, then the intersection $A \cap B$ or $B \cap A$ is equal to the *smaller* of the two sets, or B. Thus,
$$A \cap \emptyset = \emptyset,$$
and
$$A \cap W = A,$$
since $\emptyset$ is a subset of A, and A is a subset of W.

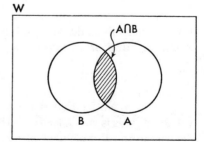

FIG. 1.9.1

The intersection may be defined for any number of sets taken together. For example, the intersection of three sets A, B, and C is

$$A \cap B \cap C = \{x | x \, \epsilon \, A \text{ and } x \, \epsilon \, B \text{ and } x \, \epsilon \, C\}.$$

Thus, if A were all women, B were all librarians, and C were all persons with blue eyes, then $A \cap B \cap C$ would be the set of all women librarians having blue eyes. In general, the intersection of K sets, A_1, A_2, $\cdots$, A_K, is defined as follows:

$$\bigcap_{i=1}^{K} A_i = A_1 \cap A_2 \cap \cdots \cap A_K$$
$$= \{x | x \, \epsilon \, A_1 \text{ and } x \, \epsilon \, A_2 \text{ and } \cdots x \, \epsilon \, A_K\}$$
$$= \{x | x \text{ is a member of all of the sets } A_1, \, A_2, \, \cdots, \, A_K\}.$$

where $i = 1, 2, \cdots, K$.

1.10   THE COMPLEMENT OF A SET

Another operation on sets is called taking the **complement** of a set. Given W and a subset A, then the complement of A is made up of all members of W that are *not* in A. In symbols,

$$\text{the complement of } A = \bar{A} = \{x | x \, \notin \, A\}$$

where the symbol $\notin$ is read as "not a member of" the set following. For example, if

$$W = \{x | x \text{ is a name in the 1972 Detroit, Michigan, telephone directory}\}$$

and

$$A = \{x | x \text{ begins with the letter "S"}\},$$

then

$$\bar{A} = \{x | x \text{ does not begin with the letter "S"}\}.$$

The intersection of any set and its complement is always empty,

$$A \cap \bar{A} = \emptyset$$

since no element could be simultaneously a member of both A and $\bar{A}$. The union of A and $\bar{A}$, however, always equals the universal set W

$$A \cup \bar{A} = W.$$

Notice that the complement of any set is always relative to the universal set; this is why specifying the universal set is so important, since one cannot determine the complement of a set without doing so. For example, suppose that we specified the following set:

$$A = \{x|x \text{ has blue eyes}\}.$$

What is $\bar{A}$? That depends on what we assumed W to be. If

$$W = \{x|x \text{ is a woman living in } \\ \text{the United States}\},$$

then $\bar{A}$ consists of all non-blue-eyed women living in the U.S. However, if

$$W = \{x|x \text{ is a person living in the U.S.}\},$$

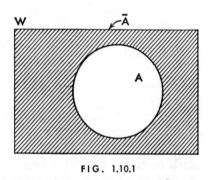

FIG. 1.10.1

the $\bar{A}$ is quite a different set, including non-blue-eyed men and children as well. If

$$W = \{x|x \text{ is an organism}\},$$

then $\bar{A}$ would include all non-blue-eyed organisms, among them some that have no eyes at all. The complement of a set simply has no meaning without a universal set for reference.

The complement of a set is shown in Figure 1.10.1 as the shaded area in the Venn diagram.

1.11   THE PARTITION OF A SET

As we have seen, the two sets A and B are mutually exclusive, or disjoint, when $A \cap B = \emptyset$. This concept may be extended to three or more sets. The sets A, B, and C are mutually exclusive if $A \cap B = \emptyset$, $A \cap C = \emptyset$, and $B \cap C = \emptyset$. That is, for three sets to be mutually exclusive, each *pair* of sets must be mutually exclusive. Bear in mind that it is not sufficient simply that $A \cap B \cap C = \emptyset$ for the three sets to be mutually exclusive. On the other hand, it does follow from the definition of three mutually exclusive sets that $A \cap B \cap C$ will be equal to the empty set when A, B, and C are mutually exclusive; since $A \cap B \cap C \subseteq A \cap B$, then when $A \cap B = \emptyset$, $A \cap B \cap C = \emptyset$. The same argument holds for $A \cap C$ and $B \cap C$ as well.

Now consider a set of some K sets, $A_1, A_2, \cdots, A_K$, where each is a subset of some W. Then these K sets are mutually exclusive if $A_i \cap A_j = \emptyset$ for each pair of sets A_i and A_j, where $i \neq j$. Once again, the sets are all mutually exclusive only if each possible pair is mutually exclusive.

We will also have occasion to use the concept of an **exhaustive** set of sets. Given some universal set W and two subsets A and B, then A and B are said to be exhaustive if $A \cup B = W$. That is, between them the sets A and B literally "exhaust" the elements of W, since every element of W must fall into A or into B or into both, if $A \cup B = W$. Similarly, three sets A, B, and C are exhaustive if $A \cup B \cup C = W$. In general, given K sets $A_1, A_2, \cdots, A_i$, $\cdots, A_K$, where each set $A_i \subseteq W$, then the set of sets is exhaustive if

$$A_1 \cup A_2 \cup A_3 \cup \cdots \cup A_i \cup \cdots \cup A_K = W$$

or

$$\bigcup_{i=1}^{K} A_i = W.$$

Note that for any $A \subseteq W$, the sets A and $\bar{A}$ are exhaustive, since it must be true that $A \cup \bar{A} = W$.

Finally consider once again a set of K sets $A_1, A_2, \cdots, A_K$, where $A_i \subseteq W$ for any of the sets A_i. Then the set of sets is said to be a **partition** of W if the set of sets are all mutually exclusive, and if the set of sets is exhaustive. Each member of W must be a member of exactly one and only one of $A_1, \cdots, A_K$ if that set of sets is a partition of W.

For example, the sets "men," "women," and "children" may be considered a partition of the universal set "human being," since the three sets are mutually exclusive and exhaustive. If the 365 days of the year are elements of the universal set W, then the 12 months make up a partition, since each day falls into one and only one month. In many courses the letter grades "A," "B," "C," and so on, form a partition of the universal set "possible course grades," since a student must receive one and only one of these grades. We will find this concept of a partition useful in later sections when we come to discuss probability and the treatment of sets of data.

1.12 THE DIFFERENCE BETWEEN TWO SETS

The difference between two sets is closely allied to the idea of a complement. Given the universal set, the **difference** between sets A and B is

$$A - B = A \cap \bar{B} = \{x | x \, \epsilon \, A \text{ and } x \, \epsilon \, \bar{B}\}.$$

In other words, the difference contains all elements that are members of A *but not* members of B.

Although it makes no difference which set we write first in the symbols for union and intersection,

$$A \cup B = B \cup A$$

and

$$A \cap B = B \cap A,$$

the order is very important for the difference between two sets. The difference $A - B$ is *not* the same as the difference $B - A$.

As an example of the two differences, $A - B$ and $B - A$, let

$W = \{x | x \text{ was an elected official of the U.S. government in 1970}\}$
$A = \{x | x \text{ was a member of the U.S. Congress}\}$
$B = \{x | x \text{ was a lawyer}\}.$

Then

$A - B = \{x | x \text{ was a member of the U.S. Congress who was not a lawyer}\}$

and

$B - A = \{x|x$ was a lawyer who was not a member of the U.S. Congress$\}$.

It is easy to see that these two differences are quite different sets: A congressman who is not a lawyer is not the same as a lawyer who is not a congressman.

The Venn diagram for a difference $A - B$ is the shaded area in Figure 1.12.1.

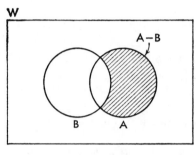

FIG. 1.12.1

Still another concept of some importance in set theory is the *symmetric difference* between two sets, symbolized by $|A - B|$. The symmetric difference is defined by

$$|A - B| = |B - A| = (A - B) \cup (B - A).$$

1.13   THE ALGEBRA OF SETS

In the algebra that everyone studies in high school, there are certain rules which the operations symbolized by "$+$," "$\cdot$," and so on must obey. The study of algebra is largely the study of these basic rules and of the mathematical consequences of the application of these operations to numbers. In the same way, an algebra of sets exists, consisting of a system for forming and manipulating sets by operations of union, intersection, complement, and so on, according to a specific set of rules or postulates. An algebra of sets has a great many similarities to, as well as some important differences from, ordinary school algebra. In a course in modern mathematics one learns to use set algebra (often called a Boolean algebra, after the nineteenth-century British mathematician, G. Boole, who first developed it). Although there is little point in dwelling upon set algebra at length in a course in statistics, it may be instructive to list the basic "rules" for an algebra of sets and to give examples of how simple theorems about sets may be proved using those rules. Just as in any formal mathematical system, the basic postulates are used to derive new conclusions by logical argument. Since we are going to be discussing a closely related system in the next chapter, a view of the foundations of this simplest of all mathematical systems may give the reader a somewhat better idea of how postulates are used.

These postulates refer to **abstract sets,** any collections of undefined elements. These statements about abstract sets are assumed true without question; they are simply "postulated." In order for set theory to be a model of sets of *real* objects, it must be true that the conditions specified by the postulates are met by the real sets. For most practical purposes these postulates *are* true of real sets. Every statement that one can make about new sets formed from old by the operations "∪," "∩," and complement can be deduced from these postulates.

1.14   POSTULATES AND THEOREMS ABOUT SETS

The postulates of set theory can be stated as follows:

1. Closure laws:
   (a) if A and B are sets, then $A \cup B$ is a set;
   (b) if A and B are sets, then $A \cap B$ is a set.
2. Identity laws:
   (a) there is one and only one set $\emptyset$, such that $A \cup \emptyset = A$ for any A;
   (b) there is one and only one set W, such that $A \cap W = A$ for any set A.
3. Commutative laws:
   (a) $A \cup B = B \cup A$;
   (b) $A \cap B = B \cap A$.
4. Associative laws:
   (a) $(A \cup B) \cup C = A \cup (B \cup C)$;
   (b) $(A \cap B) \cap C = A \cap (B \cap C)$.
5. Distributive laws:
   (a) $A \cup (B \cap C) = (A \cup B) \cap (A \cup C)$;
   (b) $A \cap (B \cup C) = (A \cap B) \cup (A \cap C)$.
6. For every set A there is one and only one set $\bar{A}$ such that
$$A \cup \bar{A} = W \text{ and } A \cap \bar{A} = \emptyset.$$
7. (a) $A = B$ and $C = D$ implies that $A \cup C = B \cup D$;
   (b) $A = B$ and $C = D$ implies that $A \cap C = B \cap D$;
   (c) $A = B$ implies that $\bar{A} = \bar{B}$.
8. There are at least two distinct sets.

Upon these statements about sets and the operations "∪," "∩," and "complement," an elaborate mathematical structure can be erected. It is very important to note that nowhere in these postulates is there an explicit definition of what is meant by a "set," by "∪," and by "∩." The postulates deal with equivalences among *undefined operations* carried out on *undefined things.* In this sense, the set of postulates is said to be "formal."

It is interesting to note that some of these postulates are true if we happen to be talking about the ordinary algebra of numbers rather than sets, and if "$\cup$" is replaced by "$+$" and "$\cap$" is replaced by "$\cdot$." For instance, postulate 1(a) corresponds to a similar postulate about numbers: if x and y are numbers, then $x + y$ is a number. Similarly, postulates 3, 4, and 5(b) have equivalent statements in ordinary algebra:

Commutative laws: $x + y = y + x$ and $x \cdot y = y \cdot x$.

Associative laws:

$$(x \cdot y) \cdot z = x \cdot (y \cdot z) \text{ and } x + (y + z) = (x + y) + z.$$

Distributive law: $x \cdot (y + z) = x \cdot y + x \cdot z.$

Notice, however, that a statement corresponding to postulate 5(a) is *not*, in general, true of numbers:

$$x + (y \cdot z) \text{ is not equal to } (x + y) \cdot (x + z).$$

Thus the algebra of sets, though similar to ordinary school algebra, does differ from it in important respects.

Now an example will be given of how these postulates are used to arrive at new statements, or theorems, not specifically among these original statements. As with any mathematical system, a logical argument is used to arrive at true conclusions *given* that the postulates are true. First of all, a simple but very important theorem will be proved:

THEOREM I: For any set A, $A \cap A = A$.

$$
\begin{aligned}
A \cap A &= (A \cap A) \cup \emptyset & &\text{by Postulate 2(a)} \\
&= (A \cap A) \cup (A \cap \bar{A}) & &\text{Postulate 6} \\
&= A \cap (A \cup \bar{A}) & &\text{Postulate 5(b)} \\
&= A \cap W & &\text{Postulate 6} \\
&= A & &\text{Postulate 2(b).}
\end{aligned}
$$

By a series of equivalent statements (the "substitutions" familiar from school algebra), each justified by a postulate, we have arrived at a new statement, $A \cap A = A$, which we know must be true whenever the postulates are true.

As a slightly more complicated example of how these postulates are used, together with theorems already proved, consider the following:

THEOREM II: If $A \cup B = B$, then $A \cap B = A$.

$$
\begin{aligned}
B &= A \cup B & &\text{Given.} \\
A \cap B &= A \cap (A \cup B) & &\text{Postulate 7(b)} \\
&= (A \cap A) \cup (A \cap B) & &\text{Postulate 5(b)} \\
&= A \cup (A \cap B) & &\text{Theorem I} \\
&= (A \cap W) \cup (A \cap B) & &\text{Postulate 2(b)} \\
&= A \cap (W \cup B) & &\text{Postulate 5(b).}
\end{aligned}
$$

However,

$$
\begin{aligned}
(B \cup W) &= (B \cup W) \cap (B \cup \bar{B}) & &\text{Postulates 2(b), and 6} \\
&= B \cup (W \cap \bar{B}) & &\text{Postulate 5(a)}
\end{aligned}
$$

$$= B \cup \bar{B} \qquad\qquad \text{Postulate 2(b)}$$
$$= W \qquad\qquad \text{Postulate 6}$$

so that

$$A \cap B = A \cap W \qquad\qquad \text{Substitution}$$
$$A \cap B = A \qquad\qquad \text{Postulate 2(b).}$$

Here we have proved a much less obvious statement to be true. Anyone who has studied high-school geometry will appreciate the fact that with an accumulation of such theorems, plus the original postulates, there is virtually no end to the number of new theorems that can be proved in this way. One extremely powerful aid in the proof of theorems exists within set theory: the so-called **duality principle**. This principle states that for every true statement about sets there is another true statement with union and intersection reversed, and with W and $\emptyset$ reversed. Thus for every true theorem there is another true theorem, its dual theorem, and the proof of the second theorem is established from the first simply by invoking the duality principle. In the postulates given above, notice how the pairs of postulates under each heading illustrate this principle. Furthermore, we may apply the principle to the theorem just proved: thus, by the duality principle, if $A \cap B = B$, then $A \cup B = A$.

The interesting and important thing about the theorems derived from set theory is that each of them will ordinarily be true if one takes sets of real objects and combines and recombines them in the ways represented by "$\cup$" and "$\cap$." Remember that the algebra of sets is not about any particular set of sets, but about "sets" defined abstractly. Propositions about abstract sets will be true of real sets satisfying the postulates exactly as an expression in school algebra will be true when the symbols are turned into numbers. Just as school algebra is a mathematical model for solving problems about real quantities, and one can find characteristics of real figures using the model of Euclidean geometry, so does the algebra of sets become a useful mathematical model in some real situations. One striking example is the theory underlying the large electronic computers, although very many other instances could be given. Set theory is providing a useful model for areas in psychology as well, especially in problems of learning, perception, and cognition.

Although this very hurried sketch of set theory cannot possibly give you any real facility with the set-language, perhaps it will help when the essential ideas about sets recur in later chapters. However, the real motive that underlies the discussion of sets is that it permits us to go on to a topic of great importance, both in statistics and in scientific enterprise in general: the study of mathematical relations.

1.15   SET PRODUCTS AND RELATIONS

The business of science or of any other field of knowledge is to discover and describe relationships among things. Everybody knows what is meant by a relationship or connection among objects or phenomena; in order to use language itself we must group our experiences into classes or sets (using the nouns and adjectives) and then state relationships linking one kind of thing with another (using the verbs). However, it is very important that we settle on a formal definition of what makes a **mathematical relation** before turning to other matters. It will be seen that the idea of a mathematical relation is nothing more than an extension of the ideas of set and subset.

Just as a set is defined as a collection of objects, it is also possible to define a special kind of set made up of *pairs of objects*. Let a pair of objects be symbolized by (a,b), where a is a member of some set A, and b is a member of some set B. Each pair of objects is thought of as **ordered,** meaning that the way the objects are listed in a pair is important. Every day we encounter such ordered pairs of objects, where the two members are distinguished from each other by the role they play: husband-wife pairs, right and left hands, first and second movie in a double-feature, and so on. For an ordered pair (a,b), the order of listing is significant for the role played by each element of the pair, so that the pair (a,b) is not necessarily the same as the pair (b,a).

Now suppose that we have two sets, A and B. *All possible* pairs (a,b) are found, each pair associating a member of A with a member B. This set of all possible pairings of an $a \, \epsilon \, A$ with a $b \, \epsilon \, B$ is called the **Cartesian product,** or the **set product,** of A and B:

$$A \times B = \{(a,b) | a \, \epsilon \, A, \, b \, \epsilon \, B\}.$$

The product $A \times B$ is only a set, but this time the elements are the possible pairs, as symbolized by (a,b). The product $B \times A$ would be different set, with members (b,a).

This idea of a set of ordered pairs originated with the mathematician and philosopher Descartes (1596–1650; Latin: Cartesius), who used it as the foundation of analytic geometry; hence the name *Cartesian* product. The "product" part of the name is attributable to the fact that the total number of possible pairs is always the number of members of A *times* the number of members of B.

As a simple example of a set product, consider these two sets:

$$A = \{\text{Mary, Susan, Tom, Bill}\}$$
$$B = \{\text{Smith, Jones, Brown, Adams}\}.$$

The product of the two sets, $A \times B$, is nothing more than the set of all possible pairings of one of the first names with one of the last names:

$A \times B = \{$(Mary, Smith) (Mary, Jones) (Mary, Brown) (Mary, Adams)
(Susan, Smith) (Susan, Jones) (Susan, Brown) (Susan, Adams)
(Tom, Smith) (Tom, Jones) (Tom, Brown) (Tom, Adams)
(Bill, Smith) (Bill, Jones) (Bill, Brown) (Bill, Adams)$\}$.

Notice that a pair (b,a) such as (Jones, Mary) is not a member of the product, since the first element must be a member of A (first names) and the second a mem-

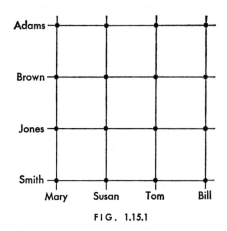

FIG. 1.15.1

ber of B (last names). This illustrates that the ordering of a pair is important in determining which pairs are included.

The pairs making up $A \times B$ can also be shown graphically, as in Figure 1.15.1. In this graph, each point formed by the intersection of a vertical line with one of the horizontal lines symbolizes one pair in the set $A \times B$.

It is also possible to talk about the product of a set with itself. That is, if we have the set A, then $A \times A$ consists of all possible pairings of a member of A with a member of A. For the example just given,

$$A \times A = \{(\text{Mary, Mary}) \ (\text{Mary, Susan}) \ (\text{Mary, Tom}) \ (\text{Mary, Bill})$$
$$(\text{Susan, Mary}) \ (\text{Susan, Susan}) \ (\text{Susan, Tom}) \ (\text{Susan, Bill})$$
$$(\text{Tom, Mary}) \ (\text{Tom, Susan}) \ (\text{Tom, Tom}) \ (\text{Tom, Bill})$$
$$(\text{Bill, Mary}) \ (\text{Bill, Susan}) \ (\text{Bill, Tom}) \ (\text{Bill, Bill})\}.$$

Again, it *does* make a difference which member comes first in a pair; thus the pair (Mary, Susan) is not considered the same as (Susan, Mary).

One of the most important examples of a Cartesian product is known to anyone who has taken high-school algebra. Suppose that we have the set R, which is the set of all possible **real numbers** (that is, the set of all numbers that can be put into exact one-to-one correspondence with points on a straight line). We take the product $R \times R$, which is the set of all pairings of *two* real numbers. Any given pair might be symbolized (x,y), where x is a number called the value on the X-coordinate or "abscissa," and y is a number standing for value on the Y-coordinate, or "ordinate." This should be a familiar idea, since the product $R \times R$ is used any time one plots points on a graph or a curve. The first set R is the "X-axis," and the second set R is the "Y-axis" of the "Cartesian coordinates."

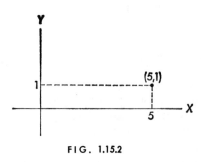

FIG. 1.15.2

When a pair such as (5,1) is used to locate a point on a graph, it means that the number 5 is the element from the first set of numbers, and 1 from the second set. The entire set $R \times R$ consists of all points that one *could* locate in the plane where the two axes of the graph lie. Pictorially, the product $R \times R$ and the pair (5,1) might be shown as in Figure 1.15.2. Regardless of whether you are dealing with pairs of objects, as in the first example, or pairs of numbers, as in the second, the idea of a Cartesian product is the same: the set of all ordered pairs, the first being a member of one specified set, and the second a member of another.

Once again, be very careful to notice that in the example just given, the pair (5,1) is *not* the same as the pair (1,5). These would be represented by very different points on the graph. In a Cartesian product, the order of the members of any pair is of the greatest importance.

The sets in a Cartesian product need not contain the same kinds of elements. To illustrate a product of a set of objects with a set of numbers, let

$$A = \{a | a \text{ is a man living in the United States}\}$$
$$G = \{g | g \text{ is a weight in pounds}\}.$$

Then

$$A \times G = \{(a,g) | a \in A, g \in G\},$$

the set of all possible pairings of a man with a weight. If each man were represented by a point along a horizontal axis, and each weight by a point along a vertical axis, then each intersection of a vertical with a horizontal line stands for a possible man-weight pair, such as (John Jones, 165 pounds).

Given the idea of a product of two sets, then it is finally possible to state what we mean by a **mathematical relation:**

A mathematical relation on two sets A and B is a subset of $A \times B$.

In other words, any mathematical relation is a subset of a Cartesian product, *some specific set of pairs out of all possible pairs.* At first blush, this seems to be an extremely trivial idea, but it actually is a remarkably subtle and elegant way to approach a difficult problem.

We often speak loosely of the husband-wife relation, the hand-fits-glove relation, the pitcher-catcher relation, the height-weight relation, the relation of the side of a square to its area, and so on. In each case, the fact of the relation implies that *some* pairs out of all possible pairs make a statement "*a* plays such and such a role relative to *b*" a true one. For any "husband" not all women qualify as "wife"; for any hand only certain gloves fit; for any side length of a square, only a certain number can represent its area.

As we have seen, all points in a plane, as represented in a graph, can be thought of as a set of ordered pairs of numerical coordinates. Now consider any line in that plane. The line includes some, but not all, of the points in the plane. Such a line serves to restrict, or reduce, or delimit the possible points that qualify in this subset of the plane. Since the line stands for a subset of the full set of points making up the Cartesian product, the line is a mathematical relation.

On the other hand, a few words of warning are in order before this topic is pushed further. It is important not to confuse the idea of a mathematical relation with the idea of a *true relationship* among objects. All relationships that are true in our experience can be represented as mathematical relations, but it is *not necessarily true* that every mathematical relation we might invent must correspond to a real relationship among things. Furthermore, several different relationships, meaning quite different things in our ordinary experience, may show up as the *same* mathematical relation, in that exactly the same pairs qualify for the relation. For example, it *might* happen that for some set of men and some set of women, the relationship "*a* is married to *b*" would involve exactly the same pairs as "*a* files a joint income tax return with *b*." The two *relations* would be identical, but the *relationship* represented means something quite different in each instance.

1.16 SPECIFYING RELATIONS

Being a set, a mathematical relation can be specified in either of the two ways used for any set. In the first place, a relation may be specified by a listing of all pairs that qualify for the relation. In the example of Section 1.15, the set *A* was four first names, and the set *B* was four last names. In this circumstance the relation *L might* be

$$L = \{(\text{Mary, Jones}) \, (\text{Susan, Brown}) \, (\text{Tom, Adams}) \, (\text{Bill, Smith})\}.$$

This relation could also be displayed graphically (Fig. 1.16.1), the filled dots in the diagram showing the pairs in the relation *L*; all the remaining pairs in $A \times B$ but not in *L* are shown unfilled.

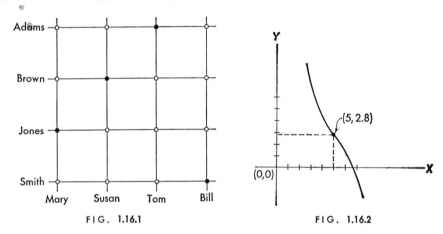

FIG. 1.16.1   FIG. 1.16.2

Quite often, the graph itself serves as a kind of symbolic listing when the pairs are too numerous to list explicitly. For example, for the product $R \times R$, the curve in Figure 1.16.2 stands in place of a listing of (x,y) pairs. In this instance, the number-pair represented by any point falling *on* the curve is *in* the relation. The point's two coordinates denote the particular pair of numbers. This specifies the relation unambiguously, and really serves the purpose of a list. The test of membership for any pair of numbers is whether or not the corresponding point falls exactly on the curve.

Plots of other relations may show up as areas or sectors rather than curves, as for example, in Figure 1.16.3. Here, any pair of numbers (x,y), such as $(7,5)$, qualifies for the relation if the corresponding point falls into the shaded sector of the graph.

The more usual way of specifying a relation is by a statement of the

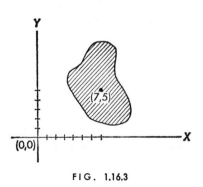

FIG. 1.16.3

rule by which a pair qualifies. For example, if A is the set of all women, and B is the set of all men, the relation S might be specified by

$$S = \{(a,b) \in (A \times B) | a \text{ is married to } b\}.$$

If A is the set of all men in the United States, and G is the set of all weights in

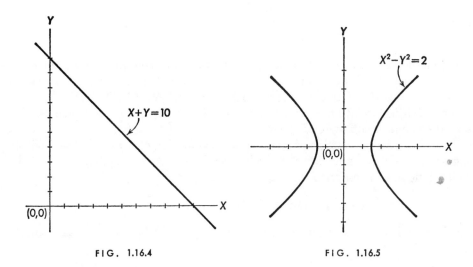

FIG. 1.16.4   FIG. 1.16.5

pounds, a relation T might be specified by

$$T = \{(a,g) \in A \times G | \text{the weight of } a \text{ is } g\}.$$

If the Cartesian product is $R \times R$, then the most common way to specify a relation is by a mathematical expression giving the qualifications a number-pair must meet: As examples, consider

$$F = \{(x,y) | x + y = 10\}$$
$$V = \{(x,y) | x^2 - y^2 = 2\}$$
$$Z = \{(x,y) | 3 \angle x \angle 5, \text{ and } -2 \angle y \angle -1\}$$

which are but a few of the countless ways of specifying a relation by a mathematical rule. For these numerical relations, the corresponding graphs are shown in

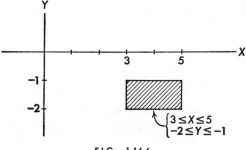

FIG. 1.16.6

Figures 1.16.4–6. Notice that the relation Z is actually the intersection of two other relations: That is, we could just as well have written

$$G = \{(x,y)|3 \leqq x \leqq 5\}$$

and

$$H = \{(x,y)|-2 \leqq y \leqq -1\}$$

so that

$$Z = G \cap H.$$

Since mathematical relations are simply subsets of a Cartesian product, the notation and rules of set theory apply to relations as well. Thus, for the universal set $R \times R$ we might wish to consider such relations as $\bar{Z}, G \cup V, F \cap (\bar{V} \cup Z)$, and so on. Each new relation is simply a subset of $R \times R$, since it is formed by set operations out of subsets of $R \times R$.

Any discussion of sets includes a definition of a universal set and the empty set, and the same is true for relations. The universal relation is called the "all-relation." The all-relation is simply $A \times B$,

$$\text{the all-relation} = \{(a,b) \; \epsilon \; A \times B\} = A \times B.$$

In the all-relation, every single member of a is paired with every single member of b.

On the other hand, the empty relation includes *no* pairs (a,b):

$$\emptyset = \{(a,b) \; \epsilon \; A \times B | (a,b) \notin A \times B\}.$$

1.17   THE IMPORTANCE OF MATHEMATICAL RELATIONS

Scientific knowledge is a body of statements about presumably true relationships. Putting the matter very crudely, you might say that science is concerned with questions of "what goes with what." Given a particular set of phenomena, some, but not all, of another set of phenomena may occur, and there exists a relationship that can, in principle, be described as a mathematical relation. For example, given a set of varieties of animals, each variety can breed with some, but not all, of the others; this relation in part gives rise to notions of species and evolutionary connections. The fact that some chemical elements will displace other elements in a compound can be stated as a relation, giving rise to many important ideas in chemistry. In any text in elementary psychology there are many examples of well-established relations, such as that between the spacing of trials in a given learning task and the rate of forgetting of materials once learned, to give only one example.

Any relation that corresponds to a true relationship among things is, quite literally, a statement of "what goes with what." If there is some set of phenomena A, and another set of phenomena B, and if the *all* relation is true for A and B, *anything can go with anything;* specifying an element of A does not narrow down the possibilities for elements of B at all. On the other hand, suppose that a true relation exists between these two sets of phenomena, so that only a certain *proper subset* of B can be paired with any given element of A. Now if you give me the element a, I will be able to narrow down the possibilities for b more than if I knew nothing at all about the element a. One might say that when there is a true relation A provides *information* about B, and that knowledge of an element a increases my ability to *predict* the element of B. One goal of scientific knowledge is the ability to predict one thing more or less *exactly* from knowledge of another, and the discovery of a true relation is the first step in this direction.

Given the discovery of some true relation, scientific enterprise does not stop, however. Ideally, the scientist hopes to study and understand the relation well enough so that finally he may be able to state a rule, or principle, that summarizes and generalizes the original finding. He also tries to account for the existence of the relation by appealing to other known or theoretical relations; this is one role of a scientific theory, which is a relatively small set of postulated relations from which it is possible to deduce and account for a great many other relations observable among things.

When a relation among sets of phenomena has been found and studied enough that the scientist knows *exactly* the conditions under which the relation may be observed, he can sometimes give the relation a *rule* completely specifying it. True relations that can be given mathematical rules are sometimes called **laws**. The laws of physical science are mathematical specifications of relations that will always be true under the appropriate circumstances. For example, Ohm's law for direct current can be stated as follows: for a given voltage v and a given resistance r, then the current c is $c = v/r$. Notice how this is a way of specifying a relation: let the voltage, for example, be fixed at some number. Then given the resistance r, what is the current c that is paired with r in this relation? The only number pairs that qualify are those for which $c = v/r$, or $cr = v$, so that this physical law is a mathematical specification of a relation. This law specifies that given v, not only are some *possible* (r,c) pairs disqualified, but also that *only one c* can be paired with each r.

We might also interpret Ohm's law as a relation in another way. Think of the set of all (r,c) pairs as itself a set with elements a. Then the law says that the only elements v that can be associated with a given pair a are those for which $cr = v$. Thus $(a,v) = ((10,5),2)$ does not qualify for the relation, although $(a,v) = ((2,5),10)$ does so.

Laws in the physical and other sciences are, obviously, often more complicated than the simple relation just given. Fortunately, the mathematical idea of a relation is extremely general and can be extended to more than just two sets. For three sets, one may define the Cartesian product $A \times B \times C$ as the set of all ordered triples, (a,b,c). A relation on the three sets is *some subset* of $A \times B \times C$. If there are some N sets, A, B, $\cdots$, T, the product is $A \times B \times \cdots \times T$, consisting of all *n-tuples* of ordered elements $(a, b, \cdots , t)$. The more complicated scientific relations are described in terms of these "larger" set-products. However, the principle is the same: a mathematical relation is some subset of a Cartesian product, and is specified by a listing procedure or by a statement of the rule that qualifies the n-tuple for the relation.

1.18 THE DOMAIN AND RANGE OF A RELATION

In discussing relations, we need some way to distinguish between the different sets of elements that play different roles. In a relation that is a subset of $A \times B$, it may be true that only *some* elements a enter into one or more (a,b) pairs; what we want is a way to discuss those elements of A that actually *are* paired with b elements in the relation itself. This subset of elements of A actually figuring in the relation is called the **domain,** and it is defined as follows:

Given some relation S, which is a subset of $A \times B$, then

$$\text{domain of } S = \{a \in A \,|\, (a,b) \in S, \text{ for at least one } b \in B\}.$$

Notice that **the domain of the relation S is always a subset of** A. The domain includes each element in A that actually plays a role in the relation, vis-à-vis some b.

For example, let A be the set of all men, and B the set of all women. Let the relation $S \subseteq A \times B$ be "a is married to b." Then the domain of S is "all married men," since only the subset of married men in A can figure in at least one pair (a,b) in the relation.

As another example, suppose that the product is $R \times R$, and the relation is given by the rule

$$C = \{(x,y) \,|\, x^2 + y^2 = 4\}.$$

What is the domain of C? The very largest number that x could be is 2 for any (x,y) pair in C, and the very smallest number is -2. Thus, x can be any number between -2 and 2 inclusive, but all other numbers in R are excluded. Hence,

$$\text{domain of } C = \{x \in R \,|\, -2 \leqq x \leqq 2\}.$$

The idea of the **range** of a relation is very similar to that of the domain, except that it applies to members of the set B, the second members of pairs (a,b).

The set of all elements b in B paired with at least one a in A in the relation S is called the range of S:

$$\text{range of } S = \{b \in B \,|\, (a,b) \in S, \text{ for at least one } a \in A\}.$$

In the example of the relation "a is married to b," the range is the set of all married women, a subset of B. In the example of the relation C, for y to be a real number the value of y must lie between -2 and 2, and so the range is

$$\text{range of } C = \{y \in R \,|\, -2 \leqq y \leqq 2\}.$$

It is entirely possible for the domain to be equal to A, and for the range to be equal to B, in some examples. In others, both the range and the domain will be proper subsets of A and B, respectively. On the other hand, neither the range nor the domain may be empty unless the relation itself is empty.

1.19 Functional relations

We come now to one of the most important mathematical concepts, from the points of view both of mathematics itself and of its applications in science. This is the idea of a **functional relation, or function.** The definition we will use is:

A relation is said to be a *functional relation* or a *function* if each member of the domain is paired with *one and only one* member of the range.

That is, in a functional relation each a entering into the relation has *exactly one pair-mate b.* Thus a function is just a special kind of relation. In some mathematical writing this is called a "single-valued function," and a relation is a "multiple-valued function." However, the author feels that it is useful to call only the former a "function."

This idea will become clearer if we inspect some relations that are, and that are not, functions:

Given A as the set of all men and B the set of all women in some society, the relation

$$\{(a,b) \; \epsilon \; A \times B | \text{``}a \text{ is married to } b\text{''}\}$$

is a function *if the society is monogamous,* so that each man may have only one wife. Here, each member of the domain, a married man, has one and only one wife, a member of the range. If, on the other hand, the society is polygamous so that a man may have two or more wives, then the relation is not necessarily a function.

Consider the relation defined on pairs of real numbers:

$$\{(x,y) | x^2 = y\}.$$

This relation is a function; corresponding to each x, there is one and only one y, which is the same as the square of x. Thus, $x = 2$ can be paired only with $y = 4$, $x = 5$ only with $y = 25$, and so on. Contrast this with the relation

$$\{(x,y) | x^2 = y^2\}.$$

In this case, the pair $(2,2)$ qualifies, but so does the pair $(2, -2)$. Each value of x can be associated with *two* values of y by this rule, and so the relation is not a function. Figures 1.19.1–2 show how these two relations can be plotted. Notice how in the first example a vertical line drawn from any point on the X axis will intercept the curve in at most *one* place, while in the second example *two* interceptions are possible. The same criterion may be applied to the plot of any relation between two numbers in order to determine if it is a function.

There is nothing in the definition of a function that limits the number of times an element b *in the range* may appear in a pair. Thus, in the marriage example, the relation is still a function even though the same woman is married to several men, and in the relation with rule $x^2 = y$, the same y value is associated with two x values (for example, $y = 4$ is paired both with $x = 2$ and $x = -2$). It follows that a given rule relating x and y may specify a function for (x,y) pairs, but may specify a nonfunctional relation for (y,x) pairs.

When a situation exists in which the set of (x,y) pairs make up a function, and the associated set of (y,x) pairs make up a function as well, the function is said to be 1-to-1. For example, the function defined by the rule $y = 3x + 6$ is a 1-to-1 function, since for each (x,y) pair such as $(1,9)$ qualifying under the rule, there is exactly one (y,x) pair such as $(9,1)$ fitting the function rule $x = (y/3) - 2$. On the other hand, a function such as $y = x^2$ is not 1-to-1: for any x value there is exactly one possible value of y, but for any y there are two possible values of x.

More generally, consider the Cartesian product $A \times B$, and let F be a function on $A \times B$. Now consider the Cartesian product $B \times A$, and let G be a relation on $B \times A$. The function F will consist of pairs (a,b) and the relation G of pairs (b,a). Then G is said to be the *inverse* of the function F if every pair $(b,a) \in G$ implies and is implied by $(a,b) \in F$. F and G consist of exactly the same pairs, but with the order reversed, if G is the inverse of F. As we have just seen, the inverse of a function may or may not be a function. If the inverse G of a function F is itself a function, then the function F is said to be 1-to-1.

Functions are relations, and hence are **sets of pairs,** and the set itself should be distinguished from the rule specifying the set. The distinction is

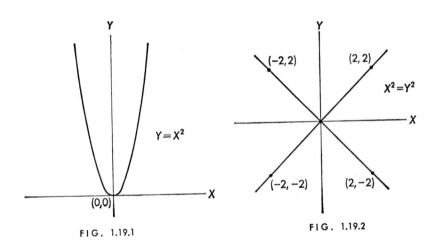

FIG. 1.19.1   FIG. 1.19.2

not always clear in mathematical writing, where commonly both the set and the rule are called functions. However, if you remember that the function itself is the set of pairs, you should have little difficulty in deciding which is meant in a discussion using this word.

A very large part of mathematics deals with functions and the properties of their rules as specifications of what the set itself, or certain subsets, must be like. When a mathematician describes the "curve" or plot of a numerical function on the basis of its mathematical rule, he is really saying something about how the set is delimited by the rule. Any curve that can be plotted for a function can be regarded as a symbolic listing of pairs of numbers.

The function idea, of course, applies to products of *any* sets. As we have seen, it is quite possible to describe a function where the elements are pairs of people. Other examples can be given where a variety of other things are paired. However, for mathematicians the most interesting functions involve numbers, either as the range of the function, or as both range and domain.

An important kind of function is one having numbers as the range and a **set of sets** as the domain. This is called a **set function.** One very important type of set function will be discussed in the section to follow.

Nevertheless, most of the interesting and useful functions relate numbers to numbers, since their rules can be given quite precise algebraic or symbolic form. Some of the most highly developed parts of mathematics are devoted, essentially, to the study of such functions. Numerical functions are also extremely important in science, where the mathematical function rules form precise statements of the relations between measured quantities. Indeed, the emphasis on functions in mathematics is partly a reflection of their historical importance in physical science.

1.20   WEIGHT FUNCTIONS AND MEASURE FUNCTIONS ON SETS

In sections to come we will often use a special kind of function in which the domain is a set of sets and the range is a set of nonnegative numbers. Because of their special importance in the theory of statistics, such **measure functions on sets** deserve some special comment here.

Consider a set A consisting of elements $a \in A$. Then let the set U consist of the positive numbers and zero: $U = \{x | 0 \leq x\}$. The Cartesian product $A \times U$ consists of all pairs (a, x). Let a function H be defined that associates with each element $a \in A$ an element from U. We will designate by the symbol $h(a)$ the number x that is associated with a particular element $a \in A$. Such a function pairing a nonnegative number with each element $a \in A$, is often called a **weight function.**

Consider now a set of sets, α, consisting of some T sets $A_i \in \alpha$. On the Cartesian product $\alpha \times U$ we define a function M, which pairs a number $x = m(A)$ with each possible $A \in \alpha$. The function M is called a **measure function** if the following conditions hold:

1. $m(A) = 0$ if and only if $A = \emptyset$,
2. if $A_i \subseteq A_j$, then $m(A_i) \leq m(A_j)$, for any A_i, A_j,
3. $m(A_i \cup A_j) = m(A_i) + m(A_j)$ if and only if $A_i \cap A_j = \emptyset$.

It follows that for a measure function, if α is a partition of some set W, then

$$m(W) = m(A_1) + \cdots + m(A_T) = \sum_i m(A_i).$$

(The symbol "Σ" means simply "the sum of.")

If the weight of an element is $h(a)$, then the measure of the set A must be

$$m(A) = \sum_{a \epsilon A} h(a).$$

The measure of a set is the sum of the weights of all of the elements making up that set. Furthermore, if the set α is a partition of the set W, then the measure of W is both the sum of the measures of the sets making up the partition, and the sum of the weights of all of the elements contained in W:

$$m(W) = \sum_{i} m(A_i) = \sum_{a \epsilon W} h(a).$$

Examples of such measure functions on sets are very easy to find. Consider all of the undergraduates enrolled at some **school** on November 20,1971. Call this total set of students W. Then let the set of sets α consist of the five sets {Freshman, Sophomore, Junior, Senior, Special Student}. The weight given to any student falling into the set "Freshman" is 1.00, and the same weight is assigned to the students belonging to the other possible sets. Then the measure given to the set "Freshman" is the sum of the weights given to all of the students who are members of that set. In other words, the measure of the set "Freshman" is simply the number of students in that classification. The same thing is true of the other sets, and the measure of W is thus the total number of students.

In many of the important applications of the idea of a measure function on sets, the weight given to each element is either unity or some other constant number, and the measures of the various sets correspond or are proportional to the number of elements each contains. On the other hand, examples can be found in which the weights given to the elements differ, and the measure function corresponds to something other than simple counts of elements. For example, let

$W = \{a|a$ is a new automobile sold in the United States during the calendar year 1972}
$U = \{x|0 \leq x$ and x is the price in dollars charged for an automobile}.

Then, the function H assigns to each $a \epsilon W$ a weight $h(a) = x$; that is, paired with each new automobile sold is the price charged.

Now let the set α consist of the sets $\alpha = \{A_1, A_2, \cdots, A_{12}\}$, where

$$A_1 = \{a \epsilon W|a \text{ is an automobile sold in January}\},$$
$$A_2 = \{a \epsilon W|a \text{ is an automobile sold in February}\},$$

and so on through A_{12}. The set α constitutes a partition of W. Then, the measure function M assigns a value $m(A)$ to each set A. The measure of A_1 is the sum of the measures (weights) of the elements in A_1. Here the value $m(A_1)$ is not the number of automobiles sold in January, but rather the total of the prices charged for new automobiles in January. Similarly, the values of $m(A_2)$, $m(A_3)$, and the rest are total prices for new automobiles in the other months. The measure of W is the total price charged for new automobiles over the entire year. Notice that although this function satisfies all of the requirements of a measure function, the values attached to each set really have nothing to do with the number of elements that each contains (with two important exceptions $\cdots$ what are they?)

We will encounter many examples of measure functions in the sections to follow. The most important application of this idea will occur when we come to a discussion of the theory of probability in the next chapter.

1.21   Variables and functional notation

Because of the importance of functional relations both in mathematics and science a special notation has been developed for discussing functions, and we will find this useful in our study of statistics. Before we deal with this notation, the notion of a "variable" must be clarified:

A variable is a symbol that can be replaced by any one of the elements of some specified set. The particular set is called the range of the variable.

Note carefully that the variable is only a stand-in or placeholder, which can *always* be replaced by a particular element from some set of possibilities. Thus, if x is a variable, wherever x appears in a mathematical expression it can be replaced by *one* element from some specified set. This idea should be a familiar one, since school algebra deals largely with variables ranging over the real numbers. Throughout our discussion of set theory, a variable such as X symbolized a set, which could be any member of some *set of sets;* the variable x stood for *some particular member* of the set X.

Variables may range over any well-defined set. For example, the variable C might symbolize any of a set of colors; the variable g might range over the 365 different dates in the year; the variable y might be any one of the infinite set of possible temperatures in Farenheit degrees. When someone says, "You are a no-good blankety-blank!" he is using "blankety-blank" much like a variable, standing for any one of a set of expressions; the range is left to your imagination.

Despite its name, a variable is not something that varies, or wiggles, or scurries around while you work with it. If a variable is used as a symbol in a given mathematical discussion, then replacing the variable by some specific element of its range at one place requires you to replace it with the same element *wherever* the variable appears. The same variable may appear many times in the same discussion, and it retains its identity throughout; if x is replaced by some number in one place, it is replaced by the same number in all other appearances. The name "variable" actually comes from the fact that a symbol represents "various" values. This distinguishes such symbols from mathematical *constants*. Unlike a variable, which can stand for a variety of things, a symbol that can be replaced by *one and only one* number in a given expression is a constant for the expression. For example, consider the well-known mathematical expression

$$c = 2\pi r.$$

The symbols c and r here are variables; each stands for any one of an infinite set of *positive numbers*. However, the symbol π is a constant, which can be replaced only with a *particular* number. Regardless of the values assumed by r and c, π is the same. Note also, in this instance, that c and r are *functionally related* variables: the value that c can assume is dictated by the value of r (and vice versa).

Most of the variables that will concern us here represent sets of numbers, so that given some range of numbers, x stands for any specific member of the range. For instance, the statement, "let X be a variable ranging over the real numbers" means that X is a stand-in for any one of the numbers that can be represented as points on a line. The X-axis in a graph is a representation of the range of this variable. When one number, say 10, is selected to replace X, then we say "X assumes the value 10." Quite often we want to discuss several variables X, Y, Z, and so on, simultaneously. Each of these symbolizes one of a range of numbers, but *not necessarily the same number, or even the same range of numbers*. For example, the X-axis and the Y-axis in a graph each symbolize ranges of possible numbers; when X assumes the particular value x, and Y the particular value y, we have the pair of numbers (x,y).

Sometimes the range of a numerical variable is specified as other than all real numbers. For example, one might have the variable X, followed by the statement $3 \angle X \angle 5$, meaning that the only values that X symbolizes are those lying between 3 and 5 inclusive (X *ranges* between 3 and 5 inclusive). Or perhaps X is a variable that is integral, meaning that the only numbers that can replace the symbol X are whole numbers.

The mathematical use of the term "variable" here is slightly different from its use in some scientific writing. A "variable" in science is often a name for a *factor* or *influence* that enters into our observation or measurement of a phenomenon; it is something that makes a difference in something else. However, as you can see, the mathematical use of the term "variable" is much more restricted than this; it is simply a symbol, usually for any one of some set of numerical values. There is no connotation that the variable means something or does something. Nevertheless, mathematical variables are useful devices for science, since they enter into the mathematical rules for true relations. Here they are stand-ins for the measured values of real "variables," which do mean something.

Variables are the basis for the notation most commonly used for functions. Given the variables X and Y, each with specified range, a relation between the variables is a subset of all possible (x,y) pairs. If this relation is a function, then this fact is symbolized by any of the statements such as

$$y = f(x)$$
$$y = G(x)$$
$$y = \varphi(x)$$

and so on. In words, these expressions all say:

> "There is a rule that pairs each possible value
> of X with at most one value of Y."

The different Roman and Greek letters that precede the symbol (x) simply indicate different rules, giving different functions or subsets of all (x,y) pairs. *This notation does nothing more than assert that the function rule exists.* It says "Give me the rule, and the value of X, and I'll give you Y."

Notice that a symbol $f(x)$ actually stands for a value of Y: this is the value of Y that is paired with x by the function rule. The symbol $f(x)$ is *not* the function, but an indicator that the function rule exists, turning a value x into some value $f(x)$, or y.

Sometimes the function rule itself is stated:

$$y = f(x) = x^3 - 2x^2 + 4$$
$$y = G(x) = k \log (x), \qquad x > 0$$
$$y = \varphi(x) = \frac{1}{\sqrt{2\pi}\sigma} \exp\left[-\frac{(x-\mu)^2}{2\sigma^2} \right]$$

are but a few of the countless function rules that might be stated for the variables X and Y, both ranging over the real numbers. The terms such as k, e, σ, and μ in these rules are simply constants having the same value in a given function rule regardless of the value that x symbolizes. In each example, the function rule stated permits the association of some y value with a given value of x. *Be sure to notice that these expressions are the rules, not the functions.*

Occasionally students are troubled by the use of the word "range" both as an expression for the possible values that may be substituted for a variable, and as an expression of the possible y values that may be paired with the x values in the (x, y) pairs of a function. However, this really need cause no difficulty. The range of a function can be thought of as the range of the variable Y or the variable $f(x)$ when $y = f(x)$. The value of this variable $f(x)$ depends upon the value x, and the possible values that $f(x)$ can assume is the range of the function.

Functional notation has been discussed here only for the case in which both X and Y range over numbers, largely because this is the most common situation in mathematics. However, the same form of notation may be extended to other situations as well. In the previous section we employed the symbol $m(A)$ to stand for the numerical value paired with a set A in a measure function on sets, and the same general form can be used for other functions. Suppose that the variable a stands for any one of a set of men, and the variable y stands for any one of the set of positive numbers that are possible scores on an intelligence test. Granted that each man must have one and only one score, then a function exists relating a and y. This could be symbolized by

$$y = f(a)$$

where a designates a particular man. We would be hard put, however, to state the rule for this function, and so we cannot go beyond the simple assertion that a function exists, unless we undertook to list each man paired with his score. In general, such functions are a pain in the neck to specify, and this accounts for some of the elaborate attempts made in science to describe relationships as numerical functions which *can* be given mathematical rules.

Once again, you are warned that the word "function" is often used to refer *both* to the set of pairs, which is actually the function, and the function rule, which specifies the set. In particular, when one says, "*Y* is a function of *X*," he is making the assertion symbolized by $y = f(x)$, that some rule exists pairing no more than one *y* with each *x*. Then he may go ahead to specify the function rule. On the other hand, when he talks about *the* function relating *X* and *Y*, he is usually referring to the set of pairings of an *x* with a *y* given by that rule.

1.22 INDEPENDENT AND DEPENDENT VARIABLES

It is customary to talk about the symbol for values in the **domain** of the function, values that come first in (x,y) pairs, as the **independent variable,** or the **argument** of the function rule. In practice, this is the *X* value known first, or given, if one is to find the $f(x)$ or *y* value. The variable *X* is "independent" because its value can be anything in the domain of the function. However, given the value of *X*, and the rule, the value of *Y* is completely determined, making *Y* the **dependent variable.**

These terms have carried over into experimental practice: the feature in the experiment that one manipulates at will is often called the independent variable. Then the measured value standing for some other aspect of the phenomena, presumably determined by the initial manipulation, is called the dependent variable.

1.23 CONTINUOUS VARIABLES AND FUNCTIONS

In statistics it is often necessary to specify that a variable or a function is continuous. Unfortunately, an accurate definition of "continuous" requires more mathematics than many students can command at this point. However, the following way of thinking about continuous variables and functions, though not really adequate as a definition, will serve our purposes.

Consider a variable *X*. If there are two numbers *u* and *v* such that the range of *X* is $u \angle x \angle v$, *all real numbers lying between u and v*, then *X* is said to be continuous in the interval $\langle u,v \rangle$. If the variable *x* is continuous over all possible intervals defined by pairs of real numbers *u* and *v*, then *X* is said to be continuous over the real numbers.

The idea of a continuous function is best conveyed by a picture of a curve without "gaps" or "breaks" in terms of *y* values. For a function $y = f(x)$ to be continuous, several things must be true: The variable *X* must itself be continuous, and for each value of *X* there must be some defined value of *Y* or $f(x)$. Most important, in a continuous function if the value of *X* is set "near" some value *a*, then $f(x)$ must be "near" the value $f(a)$.

This last condition may be put somewhat more formally. Consider some function $y = f(x)$, and two values of *X*, say *a* and *b*, where $b - a = \Delta$, or $b = a + \Delta$. That is, let Δ be some positive number that is the difference between *a* and *b*. The function with these points indicated is shown in Figure 1.23.1. Now in the function $y = f(x)$, the y values corresponding to *a* and $a + \Delta$ must be $f(a)$ and $f(a + \Delta)$, and their difference must be $f(a + \Delta) - f(a)$. These points are also shown in Figure 1.23.1. Now let ϵ (epsilon) stand for some arbitrarily small positive number. Then for a continuous function, if Δ is allowed to approach 0, with a corresponding change in the value of $f(a + \Delta)$, the absolute difference $|f(a + \Delta) - f(a)|$ will be less than ϵ, regardless of how small this arbitrary positive number ϵ is taken to be. That is, for a continuous function,

$$\lim_{\Delta \to 0} |f(a + \Delta) - f(a)| < \epsilon \qquad \text{for any positive } \epsilon.$$

This is simply a formal way of saying that for a continuous function, an x value very close to some value *a* must be associated with a y value very close to $f(a)$.

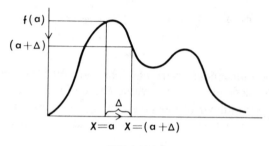

FIG. 1.23.1

A function may be continuous only in some particular interval of values such as $u \leq x \leq v$, or the function may be continuous over all of the real numbers. Thus, for example, the function given by the rule $y = 3x^2 - 5$ is continuous over all of the real numbers. On the other hand, a function with the rule $y = 1/(x + 1)$ is discontinuous at $x = -1$, since for this value of x the value of y is indeterminate. Even a function such as $y = 3x^2 - 5$ can be made discontinuous if special conditions are set down in its rule. Suppose that the rule for the function were "$y = 3x^2 - 5$ except that when $x = 0$, $y = 7$, and when $x = 2$, $y = -5$." This is a perfectly legitimate function rule, but such a function would no longer be continuous. (Can you see how the principle given above would be violated in this case? As x comes close to 0, does y come close to 7?)

1.24 FUNCTIONS WITH SEVERAL VARIABLES

The idea of a numerical function can be extended without difficulty to the case with three or more variables. Here the domain of the function is a set of *pairs* of numbers. Given variables X, Y, and Z, each ranging over the real numbers, the statement

$$y = f(x,z)$$

means that a rule exists such that given any pair of numbers x and z in the domain, then the value of y is known. For example, it may be that

$$y = f(x,z) = 3x^2 - 2z^2 + 4,$$

so that given this rule, and letting x and z assume some pair of values (x,z), we have completely determined the value of y. Or, it may be that a different rule applies, such as

$$y = g(x,z) = 2(x)^z,$$

making this second function a very different subset from among all (x,z,y) triples.

In general, if there is some function with three or more variables, and we fix one or more variables at a constant value but leave the other variables free to take on different values, then this also results in a function. For example, consider the function having the following rule:

$$y = xz.$$

Suppose that x were fixed at the value 3, while z can take on any value. Then we have the function

$$y = 3z,$$

which has only one independent variable. The particular mathematical rule for $y = f(z)$ in this case depends upon the value chosen for x. The same idea applies to functions with any number of variables: fixing the value of one or more variables results in a new function, where the fixed values appear as constants in the mathematical rule.

1.25   FUNCTIONS AND PRECISE PREDICTION

In the scientist's use of mathematics, numerical variables symbolize real quantities or magnitudes that can be found (potentially) from measurements of phenomena. Real relationships are summarized as mathematical relations, having some mathematical expression as the specifying rule. When the scientific fact is a true functional relation between quantities, then this fact is communicated in its most concise and elegant form by a statement of the function rule.

One of the goals of any science is prediction: given some set of known circumstances, what else can we expect to be true? If the scientific relation is functional in form, then *precise* prediction is possible; given the true value of X in the domain, precisely one value of Y will be observed, all other things being constant. However, in this qualification, "all other things being constant," lies one of the central problems of any science.

Each of us, scientist or layman, has learned from infancy that there are relationships in the world about us; things go together, and some things lead to other things. On the other hand, everyone knows that precise prediction of the world about him is virtually impossible. Even given some information about an object or event, we never know *exactly* what its other properties will be. True enough, providing us with information may let us restrict the scope of things we *expect* to observe, but the possibility of exact prediction is almost unknown in the everyday world. Nevertheless, the more advanced sciences are very successful in describing relationships *as though* they were mathematical functions. How has this come about? What does the scientist do that makes it possible for him to make precise predictions from known situations, when the world of the man on the street is such a disorderly and unpredictable affair?

It seems safe to say that the world is full of marvelously complicated relationships and that any event we experience must have its character determined by a vast number of influences. Some of these may exert major forces on what we observe, and others may be quite minor in shaping a given event. At one time most scientists subscribed to the idea that, ultimately, if the values of all of the independent variables were known—all of the influences that go into the determination of an event—together with the rule that relates these variables to the event itself, then precise prediction would be possible. Modern science tends to be more dubious of the possibility of such ultimate precision; and even if it were possible, it would do the scientist little good, because he is never given all of the information. On the other hand, if Y is, ultimately, a function of many variables, $X, Z, W, \cdots$, and if it really were possible to hold everything almost constant except X and Y, then we should begin to approximate a functional relationship between X and Y. The scientist, who searches for more or less precise functional relations among all of this real-world confusion, must restrict the kinds of things that he studies and the situations in which he studies them. Only through controlled experimentation and observation is one able to demonstrate something like precise, functional relations. Ideally, a controlled experiment is a device for showing the consequences of the variation of some things on the measured value of another thing, with all other factors held constant. The ability to make relatively precise predictions is bought by paying the price of careful experimental controls.

This is all well and good, but the experiment has never been done where all of the "other" things that might make a difference in the result have been controlled. Even though the experiment is a marvel of ingenuity, it is never possible to make several observations under truly identical conditions. Even the act of observation itself may influence the outcome. Fortunately, however, the ideal of a controlled experiment can at least be approximated. In the first place, some features attending different observations of the same phenomena are more or less irrelevant: the relation emerging from the data is more or less the same whether or not these features are controlled.

Nonetheless, the discovery of the irrelevances attending any relation, the things that safely can be ignored, is not the smallest problem of a science. It took a man of Galileo's intellectual stature to realize that some of the really obvious differences between a feather and a stone are quite irrelevant if we want to know how quickly each will fall from a given height. How many centuries it must have taken man to learn what to ignore before even the simplest physical relationships were determined!

Of course, other variables are quite relevant and do obscure the presence of functional relations unless they are controlled (taken at constant value) or otherwise represented systematically in the experiment. For example, we might, as Torricelli did, observe a relation between altitude and the height of a column of mercury in a glass tube (a simple barometer). We might also infer that this relation gives an indication of the air pressure at any given point. Suppose, however, that in discovering the relation we took the readings of our barometer for various places *and* times; once on a mountain top in a period of clear sunny weather, and again at sea-level just before a storm. A relation of sorts would still emerge between altitude and the barometer reading, but we would probably not approach the true functional relation. Things beside altitude influence the barometer reading; in particular, the weather conditions. Only when various altitudes are studied under exactly the same weather conditions will the true functional relation begin to show up. Then, once the function relating altitude and barometric reading is known for a *given* weather condition, the *change* in this function for different weather conditions can be studied. Finally, we might find the function showing how *both* altitude and atmospheric conditions determine the mercury barometer's height, other things being equal. The two methods suggested here, control and systematic variation, are the time-honored strategies for finding relations and refining them into functional relations.

Even primitive man must have noticed some of the connections that eventually became physical laws, although he had neither the accumulated experience nor the symbols to let him express and refine his observations. Nevertheless, the irrelevancies were gradually weeded away and the factors that affected the observed relation were discovered. In the centuries that led up to modern physics the point was finally reached where mathematical expressions relating X and Y could be stated, giving at least a good estimate of the Y value that goes with any X, under the appropriate conditions. The first triumphs of modern physics were the laws that gave a precise, functional statement of these relations.

However, are the functions specified by physical laws really true? The answer is both yes and no, depending on what you mean by "really." If you mean, "Is the prediction of the actual value of an observation as precise as the mathematical rule suggests?" then the answer is no. As we have already noted, such laws are modified by the phrases "other things being equal" or "holding other relevant variables constant." To observe phenomena in the relations that the laws embody, you must arrange things in a particular way. If it is possible to achieve this ideal condition for observation, then the phenomena should behave almost *exactly* in the predicted way. But this is a very large "if." It is well-nigh impossible to hold all the relevant variables constant in a real situation, and the best that the scientist can do is to restrict the range of these "nuisance" variables to the point where the relations that *are* observed are *almost* functions. This means that given x only a relatively small range of y values can occur. The value of y observed is seldom exactly the value given by the function rule, but the discrepancy can be made so slight as to be negligible. Physics and related sciences are precise in this sense: function rules exist that permit predictions from known x values to exact y values which differ only negligibly from actual observation.

There is another important aspect to the fact that much of physical science deals with functional relations: the ability to describe relations mathematically gives great power and elegance to the scientists' results. The meaningless or misleading "gimcracks and gingerbread" that ornament ordinary verbal communication are stripped away, so that the scientist not only knows something exact, he can say it exactly, with a minimum of effort. Imagine what a precise scientific statement would be like without mathematical language! For example, consider Ohm's law once again. Without the ability to embody this function mathematically, the scientist could communicate only by giving all pairings of values, something like this: "Under fixed conditions, and with voltage equal to 3, then if resistance is equal to 1, current is equal to 3; resistance equal to 2, current equal to 1.5; resistance equal to 3, current equal to 1; . . . " and so on ad infinitum. Historians of science tell us that this way of transmitting scientific knowledge occurred in ancient times: for example, the Egyptians had observed a relationship among the sides of right triangles, but this was part of the lore of builders and surveyors and was known only for certain proportions among the sides, such as $3:4:5$ or $7:24:25$ or $12:35:37$. It remained for the Greeks to specify the relation mathematically in the famous Pythagorean theorem (the square of the hypotenuse is equal to the sum of the squares of the remaining sides), summarizing a fact applicable to every idealized right triangle in a plane.

Once the relations among a number of interdependent variables are put into precise mathematical form, then mathematical logic comes into play. Well-established principles serve as axioms for deducing new mathematical relations that should be true. In modern physics, for example, mathematics ranks alongside actual experimentation as a device for seeking knowledge about the physical world. Experimentation is, of course, the proof of the pudding, but the basic relationships among the things studied are so well known, and can be specified so exactly in mathematical form, that the physicist has great confidence that deductions from these basic principles will also be found true.

In summary, the laws of a quantitative science are usually function rules, giving the mathematical expressions that real measurements will satisfy to a high degree of precision. These laws are idealizations, of course, and always specify that the function specified will be observed *if* certain conditions are true. Nevertheless, within these limitations they are almost exact, and the ability to specify the relation by a mathematical function rule reflects this exactitude. Furthermore, once the interrelations among many kinds of measurements are put into mathematical form, then mathematical logic alone can take over the job of seeking out new relations for study.

1.26   STATISTICS IN THE SEARCH FOR RELATIONS

The other sciences, and notably the behavioral sciences, are not so far along in this use of mathematics to specify real relations. This does not mean that these scientists do not rejoice in known mathematical functions describing behavioral phenomena when these are found. And found they sometimes are; a large body of established relations, many of known mathematical forms, already exists in psychology. In order to appreciate this one has only to consider the 1362 pages of relations, many of them quite solidly established, outlined in the *Handbook of Experimental Psychology* (Stevens, 1951). Psychologists already know a great deal on which to base a full-fledged quantitative science of behavior, and it is not too much to expect that psychology will continue to evolve in this direction.

Nevertheless, the precision and control in experimentation often possible in the physical sciences is not attainable in most psychological experiments at this time, largely because the psychologist cannot specify exactly how variables relevant to his experimental result influence the things he actually observes. Among the foremost of the missing ingredients are numerical measurements for many important features we believe to underlie behavior. Measurement procedures exist for some things, it is true, but many of the really key concepts in behavioral science are captured only crudely, if at all, in numerical measurements at this time. We should not, of course, discount the possibility of establishing empirical relations using only qualitative or nonnumerical features of behavior; there is certainly nothing in the nature of scientific investigation that limits the interesting and important relations to those based on numerical measurements. We have just seen that it is perfectly possible to discuss a mathematical relation or even a function purely in terms of qualitatively defined sets, and there are unlimited possibilities for development of nonnumerical mathematical models appropriate for analyzing and describing certain kinds of behavior.

However, if we hope ever to develop powerful and general psychological theories following the example of the physical sciences, theories that permit far-reaching deductions from a few central concepts, then the problem of precise numerical measurement becomes an important one. A satisfactory scientific theory must permit deductions to be made that can then be subjected to empirical test, and for this to be possible the constructs entering into the theory must have *specified* relations with each other and with observable phenomena. The scientist's ability to make deductions within a theory actually depends on the fact that constructs are related to each other in known ways. Furthermore, his ability to check the deductions arrived at from the theory depends on known relations between what he observes and the various constructs entering into the theory. If either the relations among the constructs or, most important, the relation between the constructs and observables, cannot be made precise, then the problem of assessing the adequacy of the theory from obtained data is a most difficult one.

In short, the development of **powerful** and **testable** scientific theories depends on our ability to frame known relations in relatively precise terms, and the greatest economy and exactitude of statement in mathematical form is possible when the variables involved represent numerical measurements. On the other hand, for a developing science like psychology, we cannot be sure of the relations of our constructs to each other, or of the relation of behavior to construct, which makes the experimental test of almost any theoretical proposition ambiguous. Do the data fail to agree with theoretical prediction because the theory is inadequate, or because of a nonfunctional relation between the measurements and one or more constructs in the theory? If we knew how to measure the relevant factors more precisely, and especially if we knew how to measure, represent, and control other factors which we wish to rule out of the experiment, would the result show more or less agreement with some theoretical prediction? The use of statistics can handle some, though not all, of the problems arising from our imperfect knowledge of how to measure important psychological factors. The psychologist often uses statistical methods on his experimental data to help him judge agreement with some theoretical prediction in spite of his inability to measure or to specify relations among factors he believes to be relevant to his result.

A second barrier to functional principles in a behavioral science is the lack of knowledge about the features in an observed relationship which *are* relevant. The psychologist is still trying to find the variables that are irrelevant and those that must either be manipulated or controlled if functional relations or their close approximations are to emerge from his data. He does this by studying the factors that seem to make a difference in (that is, are related to) behavior. However, in an actual experiment, only a very few factors can be studied at one time. If behavior is, ultimately, the consequence of a vast number of influences, how does he know that what he observes in an experiment is the result of what he did, rather than of something completely out of his control? How does he know that his failure to find a certain kind of relation is not due to extraneous circumstances masking the true relationship?

In a field such as psychology, where the full range of relevant factors is quite unknown, statistics becomes a most useful tool. The theory of inferential statistics deals with the problem of "error," the failure of an observation to agree with a true situation. Error is the product of "chance," the influence of the innumerable uncontrolled factors determining an experimental result. The concept of "probability" is introduced to evaluate the likelihood that a given observation will disagree to a certain extent with a true value. Instead of being certain or nearly certain about the conclusions reached from observations, as the physical scientist often is, the scientist using statistics draws his conclusion much like a bet.

The gambler placing a bet is faced with a situation he knows very little about; heaven alone knows the real reasons why a coin comes up heads on a given toss or a particular card comes up on the third draw of a poker game. However, given that some things are true, it can be deduced that some things are more or less *likely* to occur, and still other things *unlikely*, regardless of the reasons why. The gambler knows only that there is a given probability of his being right, and also of his being wrong, in a given decision, and the wise gambler places his bets accordingly. In the same general way, the scientist faced with data containing error uses deductions from the theory of probability in deciding how to interpret his results. However, his decisions usually deal with relationships among things, and he "wins" if he infers the true relation, and loses otherwise.

In succeeding sections we will continue to compare the task of the scientist to that of the gambler, and to show how statistical theory aids the scientist in making his own sorts of bets. In order to do this, we need some of the ideas of probability theory, to be introduced in the next chapter.

In summary, the discovery of fundamental relations that can be put into mathematical form, with formal mathematical rules like those of physics, depends upon our ability to observe the *right things* in the *right ways* under the *right circumstances*. Presumably, as we learn more and more about the relation of behavior to a variety of factors, we will begin to construct more formal and deductively powerful theories based on a number of fundamental constructs and their relations. However, we are still limited by our ignorance of the factors that are relevant in accounting for particular kinds of behavior, and particularly by our lack of knowledge of the relations of those factors to each other and to the behavior we observe. Given these limitations, statistical methods are particularly useful in drawing conclusions from psychological data, and undoubtedly will continue to be so for many years to come. Experimental evidence is accumulating at a rapid rate in psychology, and efforts at constructing psychological theories with mathematical deductive power are constantly being made. However, it seems safe to say that it will be some time before there are psychological laws and theories on a par with those of physics. The absence of a general theory does **not** imply that known relations are missing or unimportant in psychology; the discovery and specification of relations is the process by which theories are built. Conversely, our present inability to give exact mathematical expression to many known relations does not mean that we have no theories or that they are irrelevant to our experimental work. Our theories, however, do not yet have the powerful, formal, deductive character we hope they eventually will have. We do not yet know how to measure and control most of the factors responsible for variability in our data. Until psychological knowledge reaches this level, the theory of statistics will continue to be useful to the psychologist both in his task of hunting for fundamental relations and in testing his current theories against evidence.

EXERCISES

1. Specify the following sets by first listing all of their members and then by stating a formal rule:
    (a) The set of all positive integers less than 17.
    (b) The set of all even integers less than or equal to 17 and greater than or equal to 4.
    (c) The set of all odd positive integers greater than 3 and less than or equal to 13.
    (d) What is the relation of the sets in part (b) and (c) to the set in part (a)?
2. Characterize each of the following sets according to whether it is most appropriately considered to be finite, countably infinite, or uncountably infinite:
    (a) The set of all species of animals belonging to the phylum Chordata.
    (b) The set of all women married in New York City between 1900 and 1931.
    (c) The set of all possible distances along a circle measured between the two hands on a clock-face.
    (d) The possible numbers of passengers that an airline might carry before having its first fatal accident.
    (e) The true weights, accurate to any number of decimal places, of a specific group of 100 American men.
    (f) The possible true height in inches, accurate to any number of decimal places, of an American woman.
    (g) The possible number of bridge hands that might be dealt before one of the players receives 13 spades.
3. Write the following in symbolic notation:
    (a) The union of sets A and B.

(b) The intersection of A and the complement of B.

(c) The union of the empty set and the complement of B.

(d) The intersection of set A and the set formed by the union of sets B and C.

(e) The difference between the set C and the intersection of sets A and B.

4. Write out in symbols the set expressions represented by the shaded portions of the following Venn diagrams:

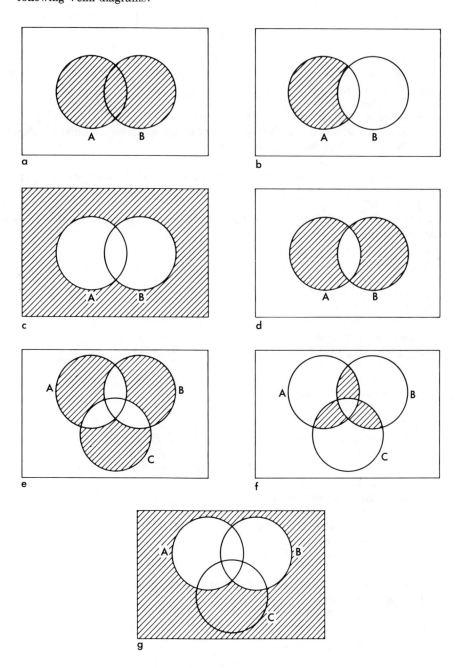

5. How many different sets can be formed from:
   - (a) Exactly one object?
   - (b) Two objects?
   - (c) Three objects?
   - (d) What is the principle for finding the number of sets that can be formed from some number N objects, where N is any positive whole number?

6. Given that A and B are each subsets of W, which of the following sets of sets must necessarily constitute a partition of the set W?
   - (a) $\{A,W\}$
   - (b) $\{A,\emptyset\}$
   - (c) $\{A,\bar{A}\}$
   - (d) $\{W,\emptyset\}$
   - (e) $\{A - B, B - A\}$
   - (f) $\{A - B, A \cap B, B \cap \bar{A}, W - (A \cup B)\}$
   - (g) $\{A, B - A, \bar{A} \cap \bar{B}\}$

7. Draw Venn diagrams showing:
   - (a) Three mutually exclusive sets.
   - (b) Three sets where $A \cap B \cap C = \emptyset$ but where the sets are not mutually exclusive.
   - (c) Three sets which form a partition of W.

   (Hint: shapes other than circles may be useful here.)

8. Show (using a Venn diagram if necessary) that even though the three sets A, B, and C do not form a partition of W, subsets exist which, along with $\bar{A} \cap \bar{B} \cap \bar{C}$, do form a partition of W.

9. Let the set $A = \{0,1,2,3,4,5,6\}$. List the elements in the following sets that are also members of the set A:
   - (a) $\{x | x/2 \text{ is a whole number or } 0\}$
   - (b) $\{x | x + 2 = 6\}$
   - (c) $\{x | x^2 - 2x + 6 = 5\}$
   - (d) $\{x | x^2 - 6x + 8 = 0\}$
   - (e) $\{x | x - 2 > 0\}$
   - (f) $\{x | 2x^2 + 3x - 1 = 0\}$

10. Consider the following sets:
    $$A = \{-3,-2,-1,0,1,2,3\}$$
    $$B = \{-4,-3,-1,1,3,4\}$$
    $$C = \{1,2,3,4,5\}$$
    List the members of the following sets:
    - (a) $A \cup B \cup C = W$
    - (b) $A \cap B$
    - (c) $A - B$
    - (d) $(B - A) \cup (A - B)$
    - (e) $(A \cap B) \cup C$
    - (f) $(A \cup B) \cap C$
    - (g) $\bar{A} \cap \bar{B} \cup C$
    - (h) $(A \cap B) - C$

11. Using the sets A and C of problem 10 above, let B be a set of real numbers variously defined as follows:
    - (a) $B = \{x | x^2 - 3x + 2 = 0\}$
    - (b) $B = \{x | x + 1 > 0\}$
    - (c) $B = \{x | 6 \geq 2x^2 - 5 \geq -2\}$
    - (d) $B = \{x | 4x^3 + 10 \geq -7\}$

    For each definition of set B, find $A \cap B$, $B \cap C$, $A \cap B \cap C$, $B \cup (A - C)$, $B \cap (C - A)$.

12. Let the universal set W be the set of all real numbers, and let:
    $$A = \{x | 4 \geq x^2 \geq 0\}$$
    $$B = \{x | x^2 + x - 2 = 0\}$$
    $$C = \{x | 5 \geq x + 5\}$$

Find:

   (a) $A \cup B$

   (b) $A \cap B$

   (c) $A \cup C$

   (d) $A \cap C$

   (e) $A - C$

   (f) $(A - B) \cup (C - B)$

13. Let the set W consist of all members of the House of Representatives of the 91st Congress of the United States. Then let:

   A = set of all Representatives who are above sixty years of age at the opening of the Congress

   B = set of all women Representatives

   C = set of all Representatives from urban districts

   D = set of all Democratic Representatives

Describe the following sets verbally:

   (a) $A \cup (B \cap C)$ (b) $(A \cap B) \cup (C \cap D)$

   (c) $(\bar{A} \cap B) \cup (\bar{C} \cap \bar{D})$ (d) $(C - D) \cap B$

   (e) $\overline{(A \cup C)} \cap (\bar{B} - D)$ (f) $(B \cap \bar{D}) \cup (B \cap C)$

14. In a class in esthetics the teacher asked each student for his preference among three pictures: a Van Gogh, a Renoir, and a Matisse. He recorded the choices for the students according to their class levels: Freshman, Sophomore, Junior, and Senior. The results were as follows (the numerals indicate the number of students voting for each picture) according to class:

CLASS	PICTURES		
	VAN GOGH (V)	RENOIR (R)	MATISSE (M)
Freshman (I)	13	4	3
Sophomore (II)	16	8	11
Junior (III)	12	21	17
Senior (IV)	2	3	10

Counting each student as an element of a set, how many students are in each of the following:

   (a) $V \cup R = \{x | x$ prefers the Van Gogh or the Matisse$\}$

   (b) $\bar{R} \cup \bar{M}$

   (c) $(\text{III} \cup \text{IV}) \cap R$

   (d) $(\text{I} - V) \cup (\text{IV} - M)$

   (e) $(\text{II} \cup \bar{V}) \cap (\text{II} \cup \bar{R})$

15. In a study of an industrial organization, a psychologist studied the verbal communication patterns of a group of four workers for a period of one week. In particular, he was interested in the verbal communication between the supervisor and each of the other workers. Let:

   A = set of all instances of verbal communication from the supervisor to worker I during the period observed and similarly, set B for worker 2 and set C for worker 3

These three sets and their intersections contained the following number of elements:

A 30 $A \cap B$ 15
B 34 $A \cap C$ 12
C 21 $B \cap C$ 7
                        $A \cap B \cap C$ 5

How many communications from the supervisor were as follows:

(a) To worker 1, but not to 2 or 3?
(b) To workers 1 and 2, or to workers 1 and 3, but not to all three?
(c) To exactly one of the three workers?
(d) To two or more of the workers?
(e) How many verbal communications in all did the supervisor make to the workers during the week observed?

16. Consider the set $A = \{2,3,4,5,6\}$. Graph the following relations, each of which is a subset of $A \times A$:

$G = \{(2,2), (3,2), (4,2), (5,2), (6,2)\}$
$H = \{(2,2), (2,3), (2,4), (2,5), (2,6)\}$
$I = \{(2,6), (3,5), (4,4), (5,3), (6,2)\}$
$J = \{(2,6), (3,5), (4,3), (5,5), (6,6)\}$
$K = \{(2,2), (3,6), (4,3), (5,5), (6,4)\}$

$G \cap H$ $J \cup K$
$G \cup H$ $H \cap I \cap J$
$I \cup J$ $H \cap \bar{I} \cap J \cap \bar{K}$

(Define $A \times A = W$.)

17. For the relations of problem 16, state the domain and range of each. Which of the relations are functions?

18. In each of the following, list the elements of the relation and give its domain and range, where $U = \{1,2,3,4,5,6,7,8,9,10\}$ and (x,y) is an element of $U \times U$:

(a) $\{(x,y)|x = 4\}$ (b) $\{(x,y)|y = 7\}$
(c) $\{(x,y)|y < x\}$ (d) $\{(x,y)|y = 2x\}$
(e) $\{(x,y)|y = 1/2x\}$ (f) $\{(x,y)|y \neq x + 1\}$

19. Given the set $A = \{2,3,4,5,6\}$ and defining $(x,y) \, \epsilon \, A \times A$, list the sets of (x,y) pairs that are members of the following relations:

$G = \{(x,y)|y$ is a multiple of $x\}$ (include $y = x$ here)
$H = \{(x,y)|y$ is less than or equal to $x\}$
$I = \{(x,y)|y$ is greater than $x\}$
$J = \{(x,y)|y = x + 1\}$
$K = \{(x,y)|y = x - 2\}$

$G \cap I$ $J \cap K$
$G - I$ $\bar{G}$
$I \cup K$ $I - J$

20. Give the domain and range of each of the relations in problem 19. Which of these relations are functions?

21. Given the set of real numbers $A = \{x|-5 \leq x \leq 5\}$, and letting $(x,y) \, \epsilon \, A \times A$, draw the graph of the following relations, each of which is a subset of $A \times A$:

(a) $\{(x,y)|x = 3\}$ (b) $\{(x,y)|y = 4\}$
(c) $\{(x,y)|y = 2x + 1\}$ (d) $\{(x,y)|y = x^2 - 1\}$
(e) $\{(x,y)|xy = 2\}$ (f) $\{(x,y)|y \geq x + 1\}$

22. In exercise 21, give the domain and range of each relation. Which of the relations are functions? If A had been defined as $A = \{x|0 \leq x \leq 5\}$, which of the relations would be functions?

23. A used car dealer has seven models of automobile for sale on his lot represented as
$A = \{a,b,c,d,e,f,g\}$. Each car has a mileage drawn from the set $M = \{m|m$ is the mileage
of a car$\}$, and a price drawn from $P = \{p|p$ is the price of a car$\}$.
The members of these sets are as follows:

CAR	MILEAGE	PRICE
a	10,500	$950
b	41,000	$500
c	32,000	$550
d	8,100	$1000
e	19,200	$4000
f	31,000	$400
g	25,000	$700

(The mileage and price of each car are shown on the line beside it.)
  (a) What are the elements of $A \times A$ falling into the relation $\{(x,y)|x$ has more
      mileage than $y\}$?
  (b) What are the elements of $A \times A$ falling into the relation $\{(x,y)|x$ is more expen-
      sive than $y\}$?
  (c) What elements belong to the intersection of the relations in (1) and (2) above?
  (d) What are the elements (x,y) of $M \times P$ falling into the relation $\{(x,y)|y$ is the
      price of a car with mileage $x\}$?
  (e) If each car has a life expectancy of, say, 60,000 miles before repairs, which
      elements of $M \times P$ fall into the relation $\{(x,y)|$cost per mile before repairs ≤ 2
      cents$\}$ or $\{(x,y)|[y/(60,000 - x)] \leq .02\}$?
24. If each line of the table in problem 23 is thought of as an *ordered triple*, how would you
    symbolize the Cartesian product of which it is an element? How might one characterize
    the table in problem 23 as a relation on some Cartesian product? (Hint: can a set of
    pairs serve either as the domain or the range of a relation?)
25. Each and every child born has exactly two biological parents. How would you make a
    formal description of the relation "x has parents y and z"? Is this relation a function?
26. A die (one of a pair of dice) has spots on its six sides numbering from 1 to 6. The
    possible results of tossing a single die can be represented by the set $D = \{1,2,3,4,5,6\}$.
    How many elements of $D \times D$ fall into the following sets:
  (a) $\{(x,y) \in D \times D|x + y = 7\}$
  (b) $\{(x,y)|x + y = 11\}$
  (c) $\{(x,y)|x + y \leq 4\}$
  (d) $\{(x,y)|x + y > 8\}$
  (e) $\{(x,y)|x + y = 2\} \cup \{(x,y)|x + y = 12\}$
What fraction of the total number of elements in $D \times D$ does each of these sets of
elements represent? Can you think of any interpretation that might be given to these
proportions?
27. Let $\mathcal{C}$ be a set of sets, $\mathcal{C} = \{A_1, A_2, \cdots, A_N\}$ and let M be a measure function on
    $\mathcal{C} \times U$, where $U = \{x|x \geq 0\}$. Given the definition of a measure function, prove the
    following:
  (a) $m(A_i - A_j) = m(A_i) - m(A_i \cap A_j)$
  (b) if $A_i \subset A_j$, then $m(A_i) < m(A_j)$

(c) if $A_i \cap A_j \neq \emptyset$, then $m(A_i \cup A_j) = m(A_i) + m(A_j) - m(A_i \cap A_j)$

28. Using the elements of the Cartesian product $D \times D$ as defined in problem 26, now let $Z = \{z | z = x + y, \ (x,y) \in D \times D\}$ and let $U = \{u | u \geq 0\}$. List the elements of $Z \times U$ falling into the relation $\{(z,u) \in (Z \times U) | u$ is the number of elements in $z\}$. Demonstrate that this is or is not a measure function.

29. In a certain municipal election, three bonding proposals were on the ballot. A voter could vote for any or all of the propositions. Of the total votes cast, 42 percent voted for proposition A, 51 percent for proposition B, and 32 percent for proposition C. In addition, it is known that:

   13 percent voted for A and not B and not C.
   12 percent voted for A and for B but not for C.
   26 percent voted for B but not for A or C.
   10 percent voted for C but not for A or B.

   (a) What percentage voted for all three propositions?
   (b) What percent voted for A and C but not for B?
   (c) What percent voted for at least one proposition?
   (Hint: construct a Venn diagram.)

2 Elementary Probability Theory

Statistical inference involves statements about probability. Everyone has the words "probable" or "likely" in his vocabulary, and most of us have some notion of the meaning of statements such as "The probability is one-half that the coin will come up heads." However, in order to understand the methods of statistical inference and use them correctly, one must have some grasp of probability theory. This is a mathematical system, closely related to the algebra of sets described in the last chapter. Like many mathematical systems, probability theory becomes a useful model when its elements are identified with particular things, in this case the *outcomes of real or conceptual experiments*. Then the theory lets us deduce propositions about the likelihood of various outcomes, if certain conditions are true.

Originally, of course, the theory of probability was developed to serve as a model of games of chance—in this case the "experiment" in question was rolling a die, or spinning a roulette wheel, or dealing a hand from a deck of playing cards. As this theory developed, it became apparent that it could also serve as a model for many other kinds of things having little obvious connection with games, such as results in the sciences. One feature is common to most applications of this theory, however: the observer is uncertain about what the outcome of some observation will be, and he must eventually infer or guess what will happen. In particular, he needs to know what will happen *in the long run* if observations could be made indefinitely.

In this respect the scientist is like the gambler keeping track of the numbers coming up on a roulette wheel. At any given opportunity he can observe only the tiniest part of the total set of things he would like to know about. Given his human frailty as an observer, he must fall back on logic; *given* that certain things are true, he can make deductions about what *should* be true in the long run. The logical machinery of "in the long run" is formalized in the theory of probability. The scientist's statements about *all* observations of such and such phenomena are on a par with the gambler's; both are deductions about what should be true, if the initial conditions are met. Furthermore, the gambler, the business man, the engineer, the scientist, and, indeed, the man on the street must make decisions based on incomplete evidence. Each does so in the face of risk about how good those decisions will turn out to be. Probability theory alone does not tell any of these people how they should decide, but it does give ways of evaluating the degree of risk one takes for some decisions.

The theory of probability is very closely tied to the theory of sets. Indeed, the main undefined terms in the theory, the "events," are simply sets of possibilities. Before we develop this idea further, we need to talk about the ways that events are made to happen: simple experiments.

2.1   SIMPLE EXPERIMENTS

We shall mean nothing very fancy by the term "simple experiment," and there is no implication that a simple experiment need be anything even remotely resembling a laboratory experiment. **A simple experiment is some well-defined act or process that leads to a single well-defined outcome.** Some simple experiments are tossing a coin to see whether heads or tails comes up, cutting a deck of cards and noting the particular card that appears on the bottom of the cut, opening a book at random and noting the first word on the right-hand page, running a rat through a T maze and noting whether he turns to the right or the left, lifting a telephone receiver and recording the time until the dial-tone is heard, asking a person his political preference, giving a person an intelligence test and computing his score, counting the number of colonies of bacteria seen through a microscope, and so on, literally without end. The simple experiment may be real (actually carried out) or conceptual (completely idealized), but it must be capable of being described and be repeatable (at least in principle). We also require that each performance of the simple experiment have one and only one outcome, that we can know when it occurs, and that the set of all possible outcomes can be specified. Any single performance, or trial, of the experiment must eventuate in one of these possibilities.

Obviously, this concept of a simple experiment is a very broad one, and almost any describable act or process can be called a simple experiment. There is no implication that the act or process even be initiated by the experimenter; he need function only in the role of an observer. On the other hand, it is essential that the outcome, whatever it is, be unambiguous and capable of being categorized among all possibilities.

2.2   EVENTS

The basic elements of probability theory are the possible distinct outcomes of some idealized simple experiment. The set of all possible distinct outcomes for a simple experiment will be called the **sample space** for that experiment. Any member of the sample space is called a **sample point,** or an **elementary event.** Every separate and "thinkable" result of an experiment is represented by *one and only one* sample point or elementary event. Any elementary event is *one possible result* of a single trial of the experiment.

For example, we have a standard deck of playing cards. The simple experiment is drawing out one card, haphazardly, from the shuffled deck. The sample space consists of the fifty-two separate and distinct cards that we might draw. If the experiment is stopping a person on the street, then the sample space consists of all the different individuals we might possibly stop. If the experiment is reading a thermometer under particular conditions, then the sample space is all the different numerical readings that the thermometer might show.

Seldom, however, is the *particular* elementary event that occurs on a trial of any special interest. Rather the actual outcome takes on importance only in relation to all possible outcomes. Ordinarily we are interested in the kind or class of outcome an observation represents. The outcome of an experiment is *measured,* at least by allotting it to some qualitative class. For this reason, the main concern of probability theory is with *sets* of elementary events. **Any set of elementary events is known simply as an event, or an event-class.**

Imagine, once again, that the experiment is carried out by drawing playing cards from a standard pack. The fifty-two sample points (the distinct cards) can be grouped into sets in a variety of ways. The suits of the cards make up four sets of thirteen cards each. The event "spades" is the set of all card possibilities showing this suit, the event "hearts" is another set, and so on. The event "ace" consists of four different elementary events, as does the event "king," and so on.

If the experiment involves stopping a person on the street and noting his eye color, the experimenter may designate seven or eight different eye-color classes. Each such set is an event. The event "blue eyes" is said to "occur" when we encounter a person who is a member of the class "blue eyes." If, on the other hand, we find the weight of each person we stop, then there may be a vast number of "weight events," different weight numbers standing for classes into which people may fall. If a person is stopped who weighs exactly 168 pounds, then the event "168 pounds" is said to occur.

In short, **events are sets, or classes, having the elementary events as members.** The elementary events are the raw materials that make up event-classes. The occurrence of any member of event-class A makes us say that event A has occurred. Since *any* subset of the sample space is an event, then some event *must* occur on each and every trial of the experiment.

2.3 Events as sets

The symbol $\S$ will be used to stand for the *sample space,* the set of all elementary events that are possible outcomes of some simple experiment. Then capital letters A, B, and so on will represent events, each of which is some subset of $\mathcal{S}$. Notice how the sample space $\mathcal{S}$ is used like a universal set; for a given simple experiment, *every event discussed must be some subset of $\mathcal{S}$.*

The set $\mathcal{S}$ may contain either a finite or an infinite number of elements. In this elementary discussion the number of elements assumed in $\mathcal{S}$ will always be finite. Nevertheless, in many of the most important instances in the theory the sample space is infinitely large.

Since any event A is a subset of $\mathcal{S}$, the operations and postulates of set theory carry over directly to operations on events to define other events.

First of all, the set $\mathcal{S}$ is an event: $\mathcal{S}$ is the "sure event," since it is *bound* to occur (one of its members must turn up on every trial). The event $\emptyset$ is called the "impossible event," since it cannot occur, having no members. It is important to remember that the empty or impossible event $\emptyset$ is, nevertheless, an *event* according to our definition, since the empty set is a subset of every set, and thus $\emptyset \subset \mathcal{S}$.

By our definition of an event as a subset of $\mathcal{S}$, then if A and B are both events, the union $A \cup B$ is also an event. The event $A \cup B$ occurs if we observe a member of A, or a member of B, or a member of both.

In the same way, $A \cap B$ is an event. The event $A \cap B$ requires an outcome that is *both* an A and a B in order to occur. Notice that any occurrence of $A \cap B$ is automatically an occurrence of $A \cup B$, but that the reverse is not true.

If A is an event, then so is its complement $\bar{A}$, and both A and $\bar{A}$ cannot occur since $A \cap \bar{A} = \emptyset$. On the other hand, either A or $\bar{A}$ must occur, since $A \cup \bar{A} = \mathcal{S}$. The difference between two events, $A - B$, is likewise an event. **In short, each and every subset formed from the elementary events in $\mathcal{S}$ is an event.**

Since events are sets, other concepts from set theory apply to events as well. Thus, two events are said to be mutually exclusive if $A \cap B = \emptyset$—that is, if $A \cap B$ cannot occur. Any set of N events is said to be mutually exclusive or disjoint if all possible pairs of events selected from among the N events are mutually exclusive. Thus, three events A, B, and C, are mutually exclusive if $A \cap B = \emptyset$, $A \cap C = \emptyset$, and $B \cap C = \emptyset$.

Two or more events are exhaustive if their union must occur. Thus A and $\bar{A}$ are exhaustive, since $A \cup \bar{A} = \mathcal{S}$. If any set of N events, A_1, A_2, $\cdots$, A_N are mutually exclusive and exhaustive, we say that they form an N-fold partition of the sample space $\mathcal{S}$. Note that A and $\bar{A}$ form a twofold partition of $\mathcal{S}$.

Let us take a concrete example of some events. Suppose that we had a list containing the name of every living person in the United States. We close our eyes, point a finger at some spot on the list, and choose one person to observe. The elementary event is the actual person we see as the result of that choice, and the set $\mathcal{S}$ is the total set of possible persons. Suppose that the event A is the set "female," and B is the set "red headed," among this total set of persons. If our chosen person turns out to be female, then event A occurs; if not, event $\bar{A}$. If he turns out to be red headed, this is an occurrence of event B; if not, event $\bar{B}$. If the observation shows up as both female and red headed, $A \cap B$ occurs; if either female or red headed, then event $A \cup B$ occurs. If the person is female but not red headed, this is an occurrence of the event $A - B$. Only one thing is sure; we will observe a person living in the United States, and the event $\mathcal{S}$ must occur.

We can continue the example above by specifying that the event C is the set "under 12 years of age," the event D is "between the ages of 12 and 50 years inclusive," and the event E is "over 50 years of age." The events C, D, and E are mutually exclusive, since no two can occur simultaneously. These events are also exhaustive, since one *must* occur. Thus, C, D, and E form a threefold partition of the sample space.

2.4 Families of events

In the example just given, the choice of the five events A, B, C, D, and E was completely arbitrary, and the example could have been given just as well with any other scheme for arranging elementary events into event classes. Ordinarily, there is some restricted set of events of interest, but any scheme for arranging elementary events into event classes can be used.

In the next section we are going to define probability in terms of events. However, we want to do this in some way so that probability can be discussed regardless of how the sample space is broken into subsets. This is accomplished by considering all possible subsets of sample space $\mathcal{S}$. Given N elementary events in $\mathcal{S}$, there are exactly 2^N event classes that can be formed, including $\emptyset$ and $\mathcal{S}$. *The set $\mathcal{C}$ consisting of all possible subsets of the sample space $\mathcal{S}$ will be called the family of events* in $\mathcal{S}$. This set of subsets $\mathcal{C}$ will be the basis for the definition of probability to follow.

2.5 Probability functions

We are now ready for a formal definition of what is meant by probability. This definition will be given in terms of a probability function, a pairing of each event with a positive real number (or zero), its probability. The question of how this probability is to be interpreted will be discussed a little later.

DEFINITION: Given the sample space $\mathcal{S}$, and the family $\mathcal{C}$ of events in $\mathcal{S}$, a probability function associates with each event A in $\mathcal{C}$ a real number $p(A)$, the *probability* of event A, such that the following axioms are true:
1. $p(A) \geqq 0$ for every event A.
2. $p(\mathcal{S}) = 1$.
3. If there exists some countable set of events, $\{A_1, A_2, \cdots, A_N\}$, and if these events are all mutually exclusive, then $p(A_1 \cup A_2 \cup \cdots \cup A_N) = p(A_1) + p(A_2) + \cdots + p(A_N)$ (the probability of the union of mutually exclusive events is the sum of their separate probabilities).

(Remember that in this connection, the word "countable" means simply that it is possible to associate each and every distinct event with one and only one of the "natural" or counting numbers. Thus the set of events $\{A_1, A_2, \cdots, A_N\}$ considered here may be either finite or countably infinite. Later, when we deal with uncountably infinite sets of events, the definition will be extended accordingly.)

In essence, this definition states that paired with each event A is a nonnegative number, probability $p(A)$, and that the probability of the sure event S, or $p(\mathsf{S})$, is always 1.00. Furthermore, if A and B are any two *mutually exclusive* events in the sample space, the probability of their union, $p(A \cup B)$, is simply the sum of their two probabilities, $p(A) + p(B)$.

Still another view of this definition is supplied by application of the terminology introduced in the previous chapter. Consider the family of events $\mathcal{C}$, and a set of positive numbers and zero, $U = \{x | 0 \leq x \leq 1\}$. Then the probability function defined above is a particular measure function P on $\mathcal{C} \times U$. That is, P is a measure function that assigns to each $A \in \mathcal{C}$ a value $p(A) \in U$, such that $0 \leq p(\mathsf{S}) \leq 1$ and $p(\mathsf{S}) = 1$. A probability function is thus a special case of a measure function on sets. Note that what we have actually defined is *a* probability function, and not *the* probability function. For any family of events there are very many ways in which numbers between 0 and 1 might be assigned to the events so as to represent the various probabilities and satisfy this definition.

It is important to remember at this stage that this is a purely formal definition of probability in terms of a function assigning numbers to sets. *Events* have probabilities, and in order to discuss probability we must always discuss the events to which these probabilities belong. When we speak of the probability that a person in the United States has red hair, we are speaking of the probability number assigned to the event "red hair" in the sample space "all persons in the United States." Similarly, when we say that the probability that a coin will come up heads is .50, we are saying that the number .50 is assigned in the probability function to the event "heads" in the sample space "all possible results of tossing a particular coin."

This may seem to be an extremely unmotivated and arbitrary way to discuss probability. We all know that the word "probability" means more than a mere number assigned to a set, and we shall certainly give these probability numbers meaning in the sections to follow. For the nonce, however, let us simply accept this purely formal definition at face value.

Given our definition of a probability function, we can begin to deduce other features that probabilities of events must have. Now we shall proceed to derive some consequences of the formal axioms included in our definition above; not only will this demonstrate that deductions do, in fact, follow directly from the axioms of this formal mathematical system, but also we will find the elementary rules of probability to be derived extremely useful when we begin to calculate probabilities.

First of all, we will give an informal proof of the following statement:

$$p(\bar{A}) = 1 - p(A), \qquad\qquad [2.5.1*]$$

for any event A, the probability of the complementary event "not-A" is 1 minus the probability of A. We proceed as follows: we know from the algebra of sets that A and $\bar{A}$ are mutually exclusive $(A \cap \bar{A} = \emptyset)$, and that $A \cup \bar{A} = \mathcal{S}$. Then, by axiom 2 above

$$p(A \cup \bar{A}) = p(\mathcal{S}) = 1.$$

Furthermore, by axiom 3, since A and $\bar{A}$ are mutually exclusive,

$$p(A \cup \bar{A}) = p(A) + p(\bar{A}).$$

Then it follows that

$$p(A) + p(\bar{A}) = 1$$

so that

$$p(\bar{A}) = 1 - p(A)$$

and

$$p(A) = 1 - p(\bar{A}).$$

In this simple way we have proved an elementary theorem in the formal theory of probability.

Next, we will show that

$$0 \leq p(A) \leq 1, \qquad\qquad [2.5.2*]$$

for any event A in the family of events $\mathcal{C}$. That is, **the probability of any event must lie between zero and one inclusive.** Suppose that some event A could be found where $p(A) > 1$; would this lead to a contradiction of one or more of our axioms? If so, then under these axioms no event can have a probability greater than 1. From the theorem just proved above,

$$p(A) + p(\bar{A}) = 1$$

If $p(A)$ should be greater than 1, then it must be true that $p(\bar{A})$ is less than zero; however, axiom 1 dictates that *any* event must have a probability greater than or equal to zero. Thus, if $p(A)$ were greater than 1 a contradiction is generated, and this means than any event must have a probability lying between zero and 1 inclusive.

The third theorem we shall prove states that

$$p(\emptyset) = 0, \qquad\qquad [2.5.3*]$$

the probability of the empty or "impossible" event is zero. To show this we recall that

$$\bar{\mathcal{S}} = \emptyset,$$

the set of all elementary events not in $\mathcal{S}$ is empty. Also, we know that

$$\mathcal{S} \cup \bar{\mathcal{S}} = \mathcal{S} \cup \emptyset = \mathcal{S}.$$

Then, by **2.5.1,**

$$p(\emptyset) = 1 - p(\mathcal{S})$$
$$= 0,$$

proving the theorem.

Still another important elementary theorem of probability goes as follows:

If $A \subseteq B$, where A and B are two events in α, then

$$p(A) \leq p(B). \qquad\qquad [2.5.4^*]$$

In other words, if the occurrence of an event A implies that an event B occurs, so that the event-class A is a subset of the event-class B, then the probability of A is less than or equal to the probability of B. Here, we start by using a set-theoretic result: if $A \subseteq B$, then

$$A \cap B = A$$
$$A \cup B = B.$$

Now the set B can be thought of as the union of *two* mutually exclusive sets:

$$B = (A \cap B) \cup (\bar{A} \cap B)$$

or, since A is a subset of B,

$$B = A \cup (\bar{A} \cap B).$$

Then, by axiom 3,

$$p(B) = p(A) + p(\bar{A} \cap B)$$

and

$$p(B) - p(A) = p(\bar{A} \cap B).$$

Since $p(\bar{A} \cap B)$ cannot be negative, $p(B)$ must be greater than or equal to $p(A)$.

Two immediate and useful consequences of the theorem just proved are the following statements:

For any pair of events, A and B, $p(A \cap B) \leq p(A \cup B)$.   [2.5.5*]
If $A \subseteq B$, then $p(B - A) = p(B) - p(A)$.   [2.5.6*]

The last theorem to be proved here is actually a most important one for all sorts of probability calculations:

Given two events A and B, then

$$p(A \cup B) = p(A) + p(B) - p(A \cap B). \qquad\qquad [2.5.7^*]$$

In order to see why this is true, recall that

$$A = (A \cap B) \cup (A \cap \bar{B})$$

so that

$$p(A) = p(A \cap B) + p(A \cap \bar{B}).$$

In the same way,

$$p(B) = p(A \cap B) + p(\bar{A} \cap B).$$

On the other hand, if we break the *union* of A and B into the union of mutually exclusive sets, we find that

$$(A \cup B) = (A \cap \bar{B}) \cup (A \cap B) \cup (\bar{A} \cap B)$$

so that, by axiom 3,

$$p(A \cup B) = p(A \cap \bar{B}) + p(A \cap B) + p(\bar{A} \cap B).$$

If we added $p(A \cap B)$ to both sides of this equation, we would have

$$p(A \cup B) + p(A \cap B) = p(A \cap \bar{B}) + p(A \cap B) + p(\bar{A} \cap B) + p(A \cap B)$$

so that on substituting $p(A)$ and $p(B)$ for their equivalents found above, we have

$$p(A \cup B) + p(A \cap B) = p(A) + p(B)$$

or

$$p(A \cup B) = p(A) + p(B) - p(A \cap B).$$

In other words, **to find the probability that** A **or** B **(or both) occurs, we must know the probability of** A**, the probability of** B**, and also the probability that** A **and** B **both occur.** Be sure to notice, however, that if A and B are *mutually exclusive* events, so that $p(A \cap B) = 0$, then

$$p(A \cup B) = p(A) + p(B),$$

just as provided by axiom 3.

Finally, we will prove a useful theorem about an N-fold partition of a sample space S. *Let the set of events* $\{A_1, A_2, \cdots, A_N\}$ *be an N-fold partition of the sample space* S. *Then the sum of the probabilities over the events in the partition is equal to 1.00:*

$$p(A_1) + p(A_2) + \cdots + p(A_N) = p(S) = 1.00. \qquad [2.5.8*]$$

By the definition of a partition as a set of mutually exclusive and exhaustive events, it must be true that

$$A_1 \cup A_2 \cup \cdots \cup A_N = S,$$

so that

$$p(A_1 \cup A_2 \cup \cdots \cup A_N) = p(S).$$

However, by Axiom 3, for these mutually exclusive events

$$p(A_1 \cup A_2 \cup \cdots \cup A_N) = p(A_1) + p(A_2) + \cdots + p(A_N).$$

Furthermore, by Axiom 2

$$p(S) = 1.00.$$

Thus, substituting in the expression above, we find

$$p(A_1) + p(A_2) + \cdots + p(A_N) = p(S) = 1.00.$$

It should be obvious that there is no end to the number of new deductions we could make using the original three axioms and the accumulated theorems such as those just proved. Enormous volumes have been filled with such probability theorems, all deduced in essentially the same way from the same axioms we have been using. Naturally, only some of the most elementary results have been shown here, and these particular results have been chosen because of their simple proofs and because they will be useful for us to have in future sections. The idea to be conveyed, however, is that one can deduce all sorts of consequences which *must* be true of these numbers we call probabilities, if these numbers obey the *formal* definition set forth at the beginning of this section. Not once has it been necessary for us to say what probability "really" means in order to deduce all sorts of rules that probabilities must obey. In short, probability theory can perfectly well be studied strictly as an abstract mathematical system without any interpretation at all.

The simple rules we have deduced and the three basic axioms will all be useful to us in learning to calculate probabilities. For the while, however, we must turn to the interpretation of these numbers called probabilities. What do we usually mean by this number when we assign it to a possible outcome of an experiment we actually carry out?

2.6   A SPECIAL CASE: EQUALLY PROBABLE ELEMENTARY EVENTS

Most simple probability calculations actually rest on the "addition" principle embodied in axiom 3 of Section 2.5. The elementary events making up the sample space S are conceived as the set of all possible *distinctly different* outcomes of the particular simple experiment. Each such elementary event is an event in its own right, since each constitutes a one-element subset of S. Furthermore, the elementary events are mutually exclusive; if one distinct elementary outcome occurs, no other elementary event may occur. Now under this view, if we somehow knew the probability for each and every elementary event making up the sample space S, then we would also be able to compute the probability of any event in α. Since the elementary events can be regarded as themselves mutually exclusive events, **the probability of any event in α is simply the sum of the probabilities of the elementary events qualifying for that event class;** this is an immediate consequence of axiom 3. Thus, the probability of any event corresponding to a subset of S can be found provided that the probability associated with each and every elementary event is known.

However, the practical difficulty in actually computing probabilities is in knowing the probabilities of the elementary events themselves. Fortunately, a great many of the simple experiments for which we need to calculate probabilities, and particularly games of chance, have a feature that does away with this problem. A great many simple experiments are conducted in such a way that it is reasonable to assume that each and every distinct elementary event has the *same* probability. When tickets are drawn in a lottery, for example, great pains are taken to have the tickets well shaken-up in a tumbler before each one is drawn; this mixing operation makes it reasonable to believe that any particular ticket has the same chance of being drawn as any other. Cards are thoroughly shuffled and cut, perfectly balanced dice are thrown in a dice-cage, and, in fact, almost all gambling situations have some feature which makes this equal-chances assumption reasonable. As we shall see presently, even experiments that are not games of chance also are carried in such a way that each and every elementary event should have the same probability.

When there is some finite number of elementary events in a sample space, where each elementary event has exactly the same probability, then the probability of any event is particularly easy to compute. (Observe that this is the same as treating the probability function as a measure function in which each element has the same weight). Imagine a sample space containing N distinct elementary events. For any event A, let the number of elementary events that are members of A be denoted by $n(A)$. The following principle applies:

If all elementary events in the sample space S have exactly the same probability, then the probability of any event A is given by

$$p(A) = \frac{\text{number of elementary events in } A}{\text{total number of elementary events in } S}$$

$$= \frac{n(A)}{N}. \qquad\qquad [2.6.1*]$$

It is easy to see that this rule must hold for the probability of any event provided that the various elementary events are equally probable. Suppose that there were N elements in S. Now consider a subset A consisting of exactly two elementary events, $A = \{a_1, a_2\}$. What is the probability of A? Since all such events are equally likely, $p(a_1) = 1/N$ and $p(a_2) = 1/N$. The set A is the union $a_1 \cup a_2$ of two mutually exclusive sets; thus, by axiom 3, Section 2.5, $p(A) = 1/N + 1/N$ or $2/N$. Proceeding in this way, suppose the set A contained n events. Here $p(A) = 1/N + 1/N + \cdots + 1/N$, or $1/N$ summed n times, so that for any n, $0 \leq n \leq N$,

$$p(A) = \frac{\text{number of elements in } A}{\text{total number of elements}} = \frac{n(A)}{N}.$$

For equally likely elementary events, the probability of an event A is its relative frequency in the sample space.

As an illustration, suppose that a box contains ten marbles. Five of these marbles are white, three are red, and two are black. We perform the simple experiment of drawing a marble out of the box (without looking) in such a way that the marbles are well mixed up, and that there is no reason for any given marble to be favored in our drawing. Now we can identify our experiment with the model of probability we have been discussing. An outcome is the result of our drawing a marble and looking at it. An elementary event is a particular marble, in this instance, and there are exactly ten distinct marbles; hence there are ten elementary events making up the sample space S. The marble-events are mutually exclusive, and each has probability $1/10$.

We are concerned with the three events: "white," "red," and "black." Notice that these three events are subsets of S, containing five, three, and two elementary events (marbles), respectively. What is the probability of our drawing a red marble? The answer is given by

$$p(\text{red}) = \frac{\text{number of red marbles}}{\text{total number of marbles}}$$

$$= \frac{3}{10}$$

so that we may say correctly that the probability of the event "red" in this experiment is .30. In the same way, one can find

$$p(\text{white}) = .50, \text{ and } p(\text{black}) = .20.$$

Furthermore, it is easy to see, by applying the rules of Section 2.5, that

$p(\text{no color}) = .00$
$p(\text{red or white}) = .30 + .50 = .80$
$p(\text{red and white}) = .00$
$p(\text{red or black}) = .30 + .20 = .50$
$p(\text{red or white or black}) = .30 + .20 + .50 = 1.00$

and so on, for any other event.

Consider another example. A teacher has thirty children in a class-room. In some completely "random" and unsystematic way, the teacher chooses a child to tell his father's occupation. The children are each equally likely to be chosen by the teacher. In this example the elementary events are the different children who might be chosen: there are thirty such elementary events possible, making up the sample space $\mathcal{S}$. Now suppose that there are only four classes of occupation represented in the room: professional, white-collar, skilled labor, and unskilled labor. These four classes will be labeled events A, B, C, and D. If three children represent group A, fifteen, group B, ten, group C, and two, group D, what is the probability that a child chosen will fall into a given group? In other words, what are the probabilities of the four different events? Once again, these probabilities are given by the relative frequencies:

$$p(A) = \frac{\text{number of elementary events in } A}{\text{total number of elementary events}}$$

$$= \frac{3}{30} = .10.$$

In the same way we find:

$$p(B) = \frac{15}{30} = .50$$

$$p(C) = \frac{10}{30} = .33$$

$$p(D) = \frac{2}{30} = .07.$$

Finally, suppose that the experiment in question were that of throwing a fair (that is, unloaded) die, marked with spots from 1 to 6 on its sides. The die is constructed and thrown so that each of the sides is equally likely to appear. There are six distinct outcomes, so that there are six elementary events making up $\mathcal{S}$. The probability of the event "1" is thus 1/6, the event "2" has probability 1/6, and so on. What is the probability of the event "odd number"? An odd number occurs when the die comes up with one, three, or five spots. That is, the desired probability is

$$p(\text{one or three or five spots}) = p(\text{odd number}).$$

Since these events are mutually exclusive, the desired probability is found from axiom 3 to be

$$p(\text{one or three or five spots}) = \frac{1}{6} + \frac{1}{6} + \frac{1}{6} = .50.$$

2.7 "IN THE LONG RUN"

Each of these examples illustrates how the outcomes of an experiment are identified with the elementary events in a sample space, and how probabilities may be computed as relative frequencies. However, we have still not really answered the question of what these probabilities mean. How are probabilities reflected in what we observe?

In the first example, if one were to keep drawing marbles out of the box, and after each draw were to replace the marble in the box before drawing again, then in the *long run*, after very many observations, *what proportion of the marbles drawn should be red?* It should take only a very little thought to arrive at the conclusion that one should *expect* about 3 in 10 such observations to be red. In the same way, in the second example, suppose that the teacher keeps choosing children in this unsystematic way, and any given child can be chosen over and over again. If each child is equally likely to be chosen at any time, then it seems reasonable that in the long run, over a large number of such observations, about .10 or one in ten of the children chosen should be of occupation group A.

Finally, in the die example, about half the time in the long run one should expect an odd number to turn up on the die if it were thrown repeatedly. In each example, **the probability of an event denotes the relative frequency of occurrence of that event to be expected in the long run.**

Simple examples of this sort all involve equally likely elementary events, and in many applications of probability theory, especially to games of chance, this assumption is made. On the other hand, the same idea applies even when elementary events are *not* equally likely: *in the long run, the relative frequency of occurrence approaches the probability for any event.*

Thus, the idea of relative frequency is connected with probability in two ways:

1. If elementary events are equally probable, the probability of an event is its relative frequency in the sample space.
2. The *long-run* relative frequency of occurrences of event A over trials of the experiment *should approach* $p(A)$. This is true regardless of whether elementary events are equally probable or not, provided that observations are made "independently" and "at random."

2.8   BERNOULLI'S THEOREM

The connection between probability and long-run relative frequency is both a simple and appealing one, and it does form a tie between the purely formal notion, "the probability of an event," and something we can actually observe, a relative frequency of occurrence. A statement of probability tells us what to *expect* about the relative frequency of occurrence of an event given that enough sample observations are made at random: In the long run, the relative frequency of occurrence of event X should approach the probability of this event, if independent trials are made at random over an indefinitely long sequence. As it stands, this idea is a familiar and reasonable one, but it will be useful to have a more exact statement of this principle for use in future work. The principle was first formulated and proved by James Bernoulli in the early eighteenth century, and it often goes by the name "Bernoulli's theorem." A more or less precise statement of Bernoulli's theorem goes as follows:

If the probability of occurrence of the event X is $p(X)$, and if N trials are made, independently and under exactly the same conditions, then the probability that the relative frequency of occurrence of X differs from $p(X)$ by any amount, however small, approaches zero as the number of trials grows indefinitely large.

The theorem may also be represented by the following mathematical expression:

$$\text{prob}\left(\left|\frac{f}{N} - p\right| \geqq \epsilon\right) \to 0, \text{ as } N \to \infty,$$

where f stands for the frequency of occurrence of event X among the N trials, p is the probability of X, ϵ is some arbitrarily small positive number, and $N \to \infty$ indicates that the number of trials increases until it approaches an infinite number. The mathematical statistician says that the proportion f/N approaches the value p "in probability" as N becomes indefinitely large.

In effect, the theorem says this: Imagine some N trials made at random and in such a way that the outcome on any one trial cannot possibly influence the probability of outcomes on any other trial; that is, N independent trials are made. This act of taking N independent trials of the same simple experiment may itself be regarded as another experiment in which the possible outcomes are the numbers of times event X occurs out of N trials. Each possible such outcome has a probability. Then the most probable result of carrying out N trials is a proportion of occurrences of X that is very nearly the value of $p(X)$ for one trial (equal to $p(X)$ when that proportion can occur; see Section 5.15). Although this is the most probable outcome of N trials of the same experiment, this does not mean, however, that the proportion of X occurrences among any N trials *must* be $p(X)$; the proportion actually observed might be any number between 0 and 1. Nevertheless, given more and more trials, the relative frequency of X occurrences may be expected to come closer and closer to $p(X)$.

It may be helpful to use a little of the machinery of Chapter 1 in interpreting Bernoulli's theorem. Let $\mathcal{C}$ symbolize the family of events that can occur on any single trial of a simple experiment. Now we deal with a series of N trials, labeled $\{1,2, \cdots ,i, \cdots ,N\}$. The family of events possible on trial 1 will be designated by $\mathcal{C}_1$, the family of events possible on trial 2 by $\mathcal{C}_2$, and so on. For any given trial i, the family of events will be symbolized by $\mathcal{C}_i$. The outcome of any series of N trials is thus a point in the sample space $\mathcal{C}_1 \times \mathcal{C}_2 \times \cdots \times \mathcal{C}_i \times \cdots \times \mathcal{C}_N$. That is, the results of a series of N trials can be thought of as an N-tuple, consisting of the outcome of the first trial, the outcome of the second trial, and so on through the outcome of the Nth trial. Each elementary event in this sample space of N-tuples is a specific sequence of outcomes of N trials. For example, if a coin is tossed, on any trial the family of events is {Head,Tail}. For any series of five tosses, the sample space is the set of quintuples $\mathcal{C}_1 \times \mathcal{C}_2 \times \mathcal{C}_3 \times \mathcal{C}_4 \times \mathcal{C}_5$ with elements such as (Head,Tail,Tail,Head,Head), (Head,Head,Tail,Head,Head), and so on.

After we have chosen some event in the family $\mathcal{C}$ to be the event X, we can identify in any sequence the number of times that X occurred. This number of occurrences of X in a sequence is the value of f for that sequence. Furthermore, on this sample space of N-tuples we can define the event $|(f/N) - p| \geqq \epsilon$. Such an event will have a probability, symbolized by $p(|f/N - p| \geqq \epsilon)$. Then the substance of the theorem is that these probabilities $p(|f/N - p| \geqq \epsilon)$ tend to grow smaller the larger that N is taken to be, regardless of the positive number chosen for ϵ.

Although Bernoulli's theorem works, it does so not because of any necessary compensation for early "misses" by more "hits" later on. Rather, the theorem holds because departures from what we expect to occur are simply *swamped out* as the number of trials becomes large. This is fairly easy to demonstrate, even for small numbers of trials. Suppose that a fair coin is being tossed, so that $p(\text{Head}) = 1/2$ (that is, $p = 1/2$). This means that on any series of N tosses we should expect one-half of the outcomes to be Heads. Now imagine that the coin is tossed 10 times, and Heads occur only 3 times. This means that f/N, the relative frequency of Heads, is only 3/10. The difference $|(f/N) - p|$ is thus $|.3 - .5|$ or .2. Now suppose that 10 more tosses of the coin are carried out and that this time we do get what we expect: five Heads and five Tails. For the entire sequence of twenty tosses $f/N = (3 + 5)/(10 + 10)$ or 8/20, and thus the difference is now $|.4 - .5|$ or .10. Even though there was no excess of Heads later on to make up for too few Heads on the first ten trials, the total sequence nevertheless shows a relative frequency of Heads which has come closer to p than did the first, shorter, sequence. This has happened because the first trials are swamped by the outcomes of the longer sequence. If we carried out ten more trials, and if, as we expect, 5 Heads and 5 Tails occurred once again, then the difference $|(f/N) - p|$ would be even smaller: $|13/30 - .5|$ or .067. Any fixed set of trials have less weight in a series of thirty than they do in a series of twenty.

In one way, this principle is what people have in mind when they refer to "the law of averages." Most of us have the intuition that the relative number of times an event occurs over repeated opportunities reflects its probability.

Given enough experimental trials or sample observations from the sample space, we expect the relative frequency of occurrence to work out to equal the probability.

However, *do not fall into the error of thinking that an event is ever due to occur on any given trial.* If you toss a coin 100 times, you expect about 50 percent of the tosses to show the event "heads," since the theoretical probability of that event for a fair coin is .50. This does not mean, however, that for any given one-hundred trials (or one-thousand, or million, or billion trials) the coin must show 50 percent heads. This need not be true at all. Every one of your 100 tosses *could* result in the event heads, or none of them might result in this event—the coin does not ever have to come up heads in any finite number of tosses. Only in an infinite number of tosses must the relative frequency equal .50. The only thing which we can say with assurance is that .50 is the relative frequency of heads we should expect to observe in any given number N of tosses, and that it is increasingly probable that we observe close to 50 percent heads as the N grows larger. But on any finite number of tosses of the coin, the relative frequency of heads observed can be anything. This same is true for the occurrence of any event in any simple experiment in which observations are made at random and in which the probability $p(X)$ of the event X is other than 1 or 0.

Although it is true that the relative frequency of occurrence of any event must exactly equal the probability only for an infinite number of independent trials, this point must not be overstressed. Even with relatively small numbers of trials we have very good reason to expect the observed relative frequency to be quite close to the probability. The rate of convergence of the relative frequency to the probability is very rapid, even at the lower levels of sample size, although the probability of a small discrepancy between relative frequency and probability is much smaller for extremely large than for extremely small samples. A probability is not a curiosity requiring unattainable conditions, but rather a value that can be estimated with considerable accuracy from a sample. Our best bet about the probability of an event is the actual relative frequency we have observed from some N trials, and the larger N is, the better the bet.

In a later section we will actually prove the main features of Bernoulli's theorem, but this must wait until Chapter 6. As yet, we lack the vocabulary to give a sufficiently precise meaning to terms such as "at random" and "independent events," that figure in this theorem. Nevertheless, accepting this principle at face value for the moment, we can show a simple example of its use.

2.9   AN EXAMPLE OF SIMPLE STATISTICAL INFERENCE

Imagine that we were faced with the following problem: We are given a coin and asked to decide if the coin is fair (that is, if the probability of heads is .50). This is to be decided through the repeated simple experiment of tossing the coin. If the coin actually is fair, then in an infinite number of tosses the relative frequency of the event heads must occur with probability of exactly .50. However, on any given number of tosses, we should only expect a relative frequency of .50 to occur. The larger the number of tosses, the more nearly should we expect the relative frequency to approach .50.

Now we toss the coin four times; suppose that four heads occur. This would cast no special doubt upon the fairness of the coin, since it is relatively likely that the relative frequency of heads will depart widely from .50 on such a small number of trials. However, if after 500 tosses, the number of heads were 500, we would begin to have very serious doubts about the fairness of the coin; although this *could* happen, it is far from what one *expects*. Nevertheless, we still could not be sure. If we tossed the coin 500 million times, still with 100 percent heads, we could be practically certain that the probability of a head for this coin is not .50— *practically* certain, that is, but still not *completely* certain. Only if we tossed the coin an infinite number of times would we be absolutely certain that the coin was not fair, given that the long-run relative frequency differed from .50 in any way.

This hypothetical example exhibits one of the basic features of statistical inference as applied to many real problems. One starts out with a theoretical state of affairs that can be represented by a statement of the probability for one or more events. He wishes to use empirical evidence to check the truth of that theoretical situation. Therefore he conducts an experiment in which the event or events in question are possible outcomes, and he takes some N observations at random. The theoretical probabilities tell him what to expect with regard to the relative frequencies of the events in question, and he compares the obtained relative frequencies against those expected. To the extent that the obtained relative frequencies of events depart widely from the expected, he has some evidence that the theory is not true. Moreover, given any degree of departure of observed from the expected relative frequency, he is more certain that the theory is not true the larger the sample N. However, *he can never be completely sure in his judgment that the theory is false unless he has made an infinite number of observations.*

By the same token, he can never be completely sure that the theory is true. An extremely biased coin (short of a coin which must come up heads with probability 1 or tails with probability 1) *can* give exactly 50 percent heads on any set of N trials. Even though the evidence appears to agree well with the theory, this in no way implies that the theory must be true. Only after an infinite number of trials, when the relative frequency of occurrence of events matches the theoretical probability exactly, can one assert with complete confidence that the theory is true. Increased numbers of trials lend confidence to our judgments about the true state of affairs, but we can always be wrong, short of an infinite number of observations.

The principle embodied in Bernoulli's theorem can be used not only to compare empirical results with theoretical probabilities, but also to let us *estimate* true probabilities of events by observing their relative frequencies over some limited number of trials. For example, suppose that once again we had a box containing marbles of different solid colors. We do not know how many marbles there are in the box, and we do not know how many different colors, but our task is to find out the probability of each color. We conduct the simple experiment of mixing up the marbles well, drawing one out, observing its color, and putting it back (random sampling with replacement). Now suppose that the first marble is white.

The observed relative frequencies of colors are white 1.00, other .00. We wish to decide about the probabilities in such a way that our estimate would, if correct, have made the occurrence of this particular sample result have the greatest prior probability. What probability of the event "white" would make drawing a white marble most likely? The answer is $p(\text{white}) = 1.00$, and $p(\text{other}) = .00$, so that this is our first estimate of the probabilities of the colors of the marbles.

However, we now draw a second marble at random with replacement. This marble turns out to be red, and our best guess about the probabilities is now white .50, red .50, other .00, as these probabilities would make the occurrence of the actual sample of two marbles most likely. If we kept on drawing, observing, and replacing marbles for 100 trials, shaking the box so that the marbles (elementary events) are equally likely to be chosen on a trial, we might begin to get a reasonably clear picture of the probabilities of the colors:

Color	Relative frequency
white	.24
red	.50
blue	.26

By the time we had made 10,000 observations, we could be even more confident that the probabilities are close to:

Color	Relative frequency
white	.24
red	.50
blue	.24
green	.02

and so on. Given an infinite number of trials, we could specify the relative frequencies (that is, the probabilities) of the marbles in the box precisely. The larger the number of observations, the less do we expect to "miss" in our estimates of what the box contains. However, for fewer than an infinite number of observations, the observed relative frequencies *need not* reflect the probabilities of the colors exactly, though they may.

It is important to note that since we are drawing samples *with replacement* in this situation the fact that there may be only some finite numbers of distinct elementary events (marbles in this case) in no way prevents us from taking a very large or even an infinite number of sample observations with replacement after each trial. Thus Bernoulli's theorem applies perfectly well to situations in which there may be a very small number of distinct elementary events possible, so long as one can draw an unlimited number of samples or make an unlimited number of trials of the experiment (as in coin tossing or die rolling).

The scientist making observations is doing something quite similar to drawing marbles from a box. He cannot see "into the box" and observe all such phenomena about which he wishes to generalize, but he can observe samples drawn from among all such elementary events, and generalize from what he actually observes. His generalization is much like a bet, and how good the bet is depends to great extent on how many sample observations he makes. He can never be completely sure that his generalization is the correct one, but he can make the risk of being wrong very small by making a sufficient number of observations.

2.10 Other interpretations of probability

Long-run relative frequency is but one interpretation that can be given to the formal notion of probability. It is important to remember that this is an *interpretation* of the abstract model. The model per se is a system of relations among and rules for calculating with numbers that happen to be called probabilities. There is no "true meaning" of probability, any more than there is a true meaning of the symbol "x" as used in school algebra. Probability is an abstract mathematical concept that takes on meaning in the ordinary sense only when it is identified with something in our experience, such as relative frequency of real events.

The probability concept acquired its interpretation as relative frequency because it was originally developed to describe certain games of chance where plays (such as spinning a roulette wheel or tossing dice or dealing cards) are indeed repeated for very many trials. Similarly, there are situations in which the scientist makes many observations under the same conditions, and so the mathematical theory of probability is given a relative frequency interpretation here as well.

On the other hand, there are some students of the matter who object to this as the exclusive interpretation of probability, and who have shown that quite different interpretations can be advanced that do not identify formal probability exclusively with relative frequency. Indeed, an everyday use of the probability concept does not have this relative frequency connotation at all. We say "It will probably rain tomorrow," or "The Yankees will probably win the pennant," or "I am unlikely to pass this test"; our hearers have no difficulty in understanding what we mean in each instance, but it is very difficult to see how these statements describe long-run relative frequencies of outcomes of simple experiments repeated over and over again. Each such statement describes the speaker's *certainty* or *degree of belief* about an event that will occur once and once only. Our inclination to use naive notions of probability in this way is one of the reasons theorists have sought other interpretations.

For many years some theorists of probability and statistics have studied the implications of a subjective, or personalistic, approach to the interpretation of probability. Such an interpretation is not so much an alternative to, as an extension of, the usual relative frequency interpretation of probability. Under this subjective approach, a probability is seen as a measure of individual degree of belief, the quantified judgment of a particular individual with respect to the occurrence of a particular event. An individual is assumed to hold, with some measurable degree of confidence, the belief that a particular event will or will not occur. Then the measure of that degree of belief about the event is represented as a probability. Under this conception, it is perfectly reasonable to assign a probability value to an event that can occur only once. The event "rain tomorrow" for a particular day is such a nonrepetitive event. That particular "tomorrow" will occur once and only once, and it will or will not rain at the location in question. When an individual says that "the probability is 1/5 that it will rain at this place tomorrow," he is stating a value reflecting the strength of his belief that the event in question will occur. It is very difficult to interpret such a statement in terms of relative frequency, but it does make intuitive sense when viewed as an individual's subjective assessment of a situation. We make similar statements all the time without a moment's second thought.

It is also quite possible to apply the subjective interpretation of probability to events that are repetitive, and which thus lend themselves to a relative frequency interpretation as well. Again consider a game of dice. A player might very well have a definite degree of belief that the dice will turn up "7" on the next play. Given certain assumptions, such that each die is "fair," that the method of tossing does not influence the outcome, and so on, an intelligent player might well behave as though the chances were 1/6 that the dice would total 7 on the next throw. This is the same value that simple probability calculations dictate under these assumptions. After all, remember that the probability computations are absolutely neutral with respect to the interpretation to be placed on the probabilities. The difference between the subjective and the relative frequency approaches lies only in the way the probability values are interpreted, not in the way they are treated mathematically. In one view, probability calculations can be thought of as ways of settling on an appropriate degree of belief, given certain assumptions and information about the circumstances of the event in question.

Critics of this approach are quick to point out that because they stand for degrees of belief, subjective probabilities may vary from individual to individual for the same event. Since individuals vary in their backgrounds, knowledge, and available information, it is quite reasonable to suppose that they will vary in the degree of belief they hold for the occurrence of a given event. If this were not the case, professional gamblers, as well as thousands of perfectly respectable businessmen, would soon be out of work. However, such individual differences in degrees of belief do not unduly trouble advocates of the subjective approach, provided that differing probability values accurately reflect differences in judgment among individuals. As we shall see, students of the subjective approach to probability are especially interested in the choice behavior of an individual based on his own subjective probabilities. Furthermore, they are concerned with changes in degree of belief by individuals, as new information is gained and as information is shared. Thus, the individuality of subjective probabilities is not an overwhelming obstacle to the most important uses of this concept.

Still another objection to the subjective approach is that people do not consciously carry probability values around in their heads. It is claimed that there is no really objective way to measure degree of belief and convert these values into probabilities. It is entirely true, of course, that we seldom think directly in terms of degree of belief or of probability. Indeed, one is seldom conscious of such concerns. On the other hand, we are continually in the process of making choices and decisions, usually in the face of uncertainty about events that may or may not occur. Degree of belief, and thus subjective probability, can be inferred from the choice behavior of an individual. It is not from what an individual says, but rather from the choices that he makes, that a measure of subjective probability can be gained. It is true that, in order to find these measures, we must make certain assumptions about behavior. Certain reasonable "axioms of choice" or "axioms of coherence" must be satisfied if the values inferred from individual behavior are to be treated as probabilities satisfying the mathematical properties given above. In this rapid overview we cannot take the time to go into these axioms. Suffice it to say that if an individual's behavior is consistent in the ways specified by these axioms, then probabilities inferred from his behavior *can* be treated by the full machinery of probability theory. Next we will show a few of the ways that have been devised for measuring degree of belief and converting these measures into probabilities. Since some of these methods depend upon individual choices among "betting odds" and "lotteries," a brief discussion of these topics follows.

2.11 PROBABILITIES, BETTING ODDS, AND LOTTERIES

One common way of expressing the probabilities of two mutually exclusive events is in terms of betting odds. The formal connection between betting odds and probabilities is as follows:

If the probability of an event is p, then the odds in favor of the event are p to $(1 - p)$.

Thus, if some event A has a probability $p(A) = 3/4$, then the odds in favor of A are $3/4$ to $1/4$, or 3 to 1. If $p(A) = 1/8$, then the odds in favor of A are only $1/8$ to $7/8$, or 1 to 7. When two fair dice are tossed, the probability of the event "7" is $1/6$, so that the odds in favor of a "7" are 1 to 5. The odds against the occurrence of a "7" are 5 to 1. As another example, suppose that a class contains 15 girls and 11 boys. A student is selected at random from the class. What are the odds in favor of the selection of a girl? Since the probability of the event "girl" is $15/26$, the odds in favor of the selection of a girl are 15 to 11.

It is equally simple to convert statements of odds into statements of probability:

If the odds in favor of some event A are x to y, then the probability of that event is given by $p(A) = x/(x + y)$.

If the odds for some event are 9 to 2, then the probability of that event is $9/(9 + 2)$ or $9/11$. If the odds for some event are 1 to 1, then the probability of that event must be $1/2$.

In betting situations individuals often give or accept odds. This is true not only in games of chance but also in situations where the "objective" probability of the event in question would be difficult to determine. In such situations, a statement of acceptable odds can sometimes be taken as a reflection of subjective probability. When someone says that he believes the odds are 2 to 1 that the Detroit Lions will beat the Los Angeles Rams in their impending game, he is saying that the subjective probability is, for him, $p(\text{Lions beat Rams}) = 2/3$. When the weatherman says that the odds are 5 to 2 against rain tomorrow, he is saying that the subjective probability $p(\text{no rain tomorrow}) = 5/7$, or $p(\text{rain tomorrow}) = 2/7$.

Odds are often stated in monetary terms as well. Suppose that bettor I and bettor II agree that the odds are 2 to 1 in favor of the Detroit Lions against the Los Angeles Rams (there will be a sudden-death playoff, so that no tie is possible in this game). Bettor I picks the Lions and II picks the Rams. How much should each put up in order to make this a fair bet? The first bettor has chosen the event he believes to be the more likely, and thus he should stand to gain less if he wins than bettor II, who has the less likely event. The bet becomes fair when bettor I agrees to put up \$2 to \$1 for bettor II. In this way bettor I gets only \$1 if he wins, whereas bettor II gets \$2. In general, a bet is fair when the ratio of the moneys put up is the same as the odds. In this case, the bettor on the more likely event puts up \$2 for each \$1 on the less likely event, since the odds are 2 to 1. Looked at in another way, *a bet is fair if the following relationship holds:*

$$(\text{amount won if } A \text{ occurs})[p(A)] = (\text{amount lost if } \bar{A} \text{ occurs})[1 - p(A)].$$

When a bet is fair, it is also true that

$$\frac{(\text{amount bet on event } A)}{(\text{amount bet against event } A)} = \text{odds in favor of } A.$$

The amounts bet in a fair bet thus give the odds, and consequently the probabilities involved. When someone says "Five dollars will get you ten dollars that such and such will occur," he is actually saying that he wants to bet on an event with odds of 2 to 1, so that he is willing to put up \$2 for every \$1 you put up. This is another way of saying that the probability is, in his judgment, $2/3$.

Strictly speaking, in a bet such as we have been describing, the two bettors agree only that true odds are *at least* as favorable as those accepted by the person who bets on the event, and *at most* as favorable as those accepted by the person who bets against it. If bettor I accepts odds of 2 to 1 in the example given above, but really believes that the odds are, say, 4 to 1, he should also believe that the bet is biased in his favor. On the other hand, if bettor II believes that the odds are less in favor of the event, say 1 to 1 rather than 2 to 1, and he nevertheless accepts the 2-to-1 bet, he should also believe the bet to be biased in his favor. In short, when two people agree on a bet with odds of x to y, we can say only that the subjective probability of the person who bets on event A is $p(A) \gneq x/(x + y)$, and that the subjective probability of the person who bets against event A is $p(A) \lneq x/(x + y)$. Even so, the fact that two individuals can agree on a bet gives information on the degree of belief each holds about the occurrence of an event.

Another way of assessing subjective probabilities is based upon the preference of any individual among several betting situations, or lotteries. A lottery is a real or hypothetical betting situation of the following form:

$$\begin{cases} \text{You win \$} X \text{ if event } A \text{ occurs} \\ \text{You win (or lose) \$} Y \text{ if event } A \text{ does not occur} \end{cases}$$

Sometimes the lottery takes an alternate form:

$$\begin{cases} \text{You win \$} X \text{ with probability } p(A) \\ \text{You win (lose) \$} Y \text{ with probability } 1 - p(A) \end{cases}$$

For example, two lotteries might be:

Lottery I: $\begin{cases} \text{You win \$50 if it rains tomorrow} \\ \text{You lose \$30 otherwise} \end{cases}$

and

Lottery II: $\begin{cases} \text{You win \$.10 with probability 3/4} \\ \text{You win \$.00 with probability 1/4} \end{cases}$

Still another lottery might be

Lottery III: $\begin{cases} \text{You win a trip to Paris with probability 1/8} \\ \text{You win a trip to Peoria with probability 7/8} \end{cases}$

Each lottery thus consists of two mutually exclusive events, to which a probability either is or may be assigned. Each event is accompanied with an outcome, which may be monetary or of some other value for the individual.

In an experimental situation, an individual is asked to state his preferences from among two or more lotteries. Under the proper conditions, a series of such preference choices gives information about his subjective probabilities, or, in some instances, about the value he attaches to certain outcomes.

When an individual is asked his preference between two lotteries, how should he decide? The usual assumption is that the individual prefers the lottery which he expects to yield the higher value for him. That is, consider the following lotteries:

L_1: $\begin{cases} \text{You get value } v(A) \text{ with probability } p(A) \\ \text{You get value } v(\bar{A}) \text{ with probability } 1 - p(A) \end{cases}$

L_2: $\begin{cases} \text{You get value } v(B) \text{ with probability } p(B) \\ \text{You get value } v(\bar{B}) \text{ with probability } 1 - p(B) \end{cases}$

We define the value of the first lottery as follows:

$$\text{value of lottery } L_1 = v(A)p(A) + v(\bar{A})[1 - p(A)].$$

Similarly, the value of the second lottery is given by

$$\text{value of lottery } L_2 = v(B)p(B) + v(\bar{B})[1 - p(B)].$$

Then we assume that an individual prefers lottery L_1 to lottery L_2 if and only if value $L_1 >$ value L_2. If the two lotteries are exactly equal in value, the individual should be indifferent between them.

Thus, for the two lotteries

$$L_1: \quad \begin{cases} \text{You win \$16 with probability 1/4} \\ \text{You lose \$4 with probability 3/4} \end{cases}$$

and

$$L_2: \quad \begin{cases} \text{You win \$9 with probability 2/5} \\ \text{You lose \$4 with probability 3/5} \end{cases}$$

the values can be found by letting

$$v(A) = \$16, \quad p(A) = 1/4, \quad v(\bar{A}) = -\$4$$

and

$$v(B) = \$9, \quad p(B) = 2/5, \quad v(\bar{B}) = -\$4,$$

so that

$$\text{value of } L_1 = \$16(1/4) - \$4(3/4)$$
$$= \$1$$

and

$$\text{value of } L_2 = \$9(2/5) - \$4(3/5)$$
$$= \$1.20.$$

Here an individual should prefer lottery L_2 to lottery L_1, since the value of the second lottery for him is higher than that of the first. On the other hand, the following lottery

$$L_3: \quad \begin{cases} \text{You win \$5 with probability 2/3} \\ \text{You lose \$4 with probability 1/3} \end{cases}$$

should be preferred to both the first and the second lotteries, since it has a value of \$2.

Given the assumption that preferences among lotteries are determined by their values, it is clear that two things go into the determination of preferences: the values of the various outcomes for the individual, and the probabilities of the events producing those outcomes. Given some information about the values of the various outcomes for an individual, then his preference behavior in choosing among lotteries gives information about his subjective probabilities. For example, consider the following lotteries involving monetary values:

$$L_1: \quad \begin{cases} \text{You win \$.50 if the weather is fair tomorrow} \\ \text{You lose \$.10 if the weather is not fair tomorrow} \end{cases}$$

and

$$L_2: \quad \begin{cases} \text{You win \$.50 with probability 1/2} \\ \text{You lose \$.10 with probability 1/2} \end{cases}$$

If the first lottery is preferred to the second, then the value of the first is assumed to be greater than the value of the second. Thus, if the first is preferred,

$$\$.50p(A) - \$.10[1 - p(A)] > \$.50(1/2) - \$.10(1/2)$$

so that $p(A) > 1/2$. On the other hand, if lottery L_2 is preferred, similar reasoning shows that $p(A) < 1/2$. The preference behavior between these two lotteries has thus given us one "fix" on the value of the subjective probability, $p(\text{fair tomorrow})$.

Now suppose that a third lottery is introduced:

$$L_3: \begin{cases} \text{You win } \$.50 \text{ with probability } 5/8 \\ \text{You lose } \$.10 \text{ with probability } 3/8 \end{cases}$$

and suppose that although the individual prefers L_1 to L_2, he also prefers L_3 to both L_1 and L_2. This means that the value of L_3 is greater than the value of L_1, so that

$$\$.50(5/8) - \$.10(3/8) > \$.50p(A) - \$.10[1 - p(A)]$$

and $5/8 > p(A)$. This gives a new limitation on the value of $p(A)$, and since L_3 is preferred to L_1, which is in turn preferred to L_2, we can state that

$$5/8 > p(A) > 1/2.$$

In principle, one can keep on comparing a lottery such as L_1, with its unknown value $p(A)$, to other lotteries with known probability values. This is done until the individual encounters some lottery that he finds equal in attractiveness to the lottery in question, so that he is indifferent between the first and the new lottery. For example, suppose that the individual is indifferent between L_1 and the following lottery:

$$\begin{cases} \text{You win } \$.80 \text{ with probability } .55 \\ \text{You lose } \$.40 \text{ with probability } .45 \end{cases}$$

The value of this lottery is given by

$$.80(.55) - .40(.45) = .26.$$

This means that the value of L_1 must be $\$.26$ as well, since the individual subject is indifferent between L_1 and this lottery. Hence we can solve exactly for the value of $p(A)$ by substituting .26 for the value of L_1:

$$\begin{aligned} \text{value of } L_1 = \$.50[p(A)] - \$.10[1 - p(A)] &= \$.26, \\ \$.50p(A) + \$.10p(A) - \$.10 &= \$.26, \\ \$.60p(A) &= .36, \\ p(A) &= .60. \end{aligned}$$

In this way, provided that all our assumptions are true, we have found that the subjective or personal probability of the individual for the event A is $p(A) = .60$.

Subjective probabilities of individuals for events can be determined by exploration of each individual's pattern of preference behaviors among lotteries. Of course, at best this method might be laborious, and some definite practical problems are involved. Furthermore, there are some assumptions that one must make. Not only is it necessary to assume that preferences among lotteries are determined by their values, but also that the individual behaves in accordance with the axioms of coherence and transitivity of choice referred to above. In effect these require that an individual behave in a consistent and "orderly" way during a series of choices. Although we are not going into these axioms in detail, it is important to note that unless the individual behaves in such a way as to satisfy them, there is no guarantee that the values obtained as subjective probabilities will actually obey the laws of probability.

The important point to carry away from this discussion is that subjective probabilities can be assessed from individual preference behavior, and in this sense they are just as "objective" as probabilities interpreted in other ways. The practical difficulty in discovering these probability values is somewhat beside the point. Occasionally, probabilities interpreted as relative frequencies are also extremely difficult to evaluate, and must be based upon assumptions fully as stringent as those underlying the measurement of subjective probability. The behavioral basis for the measurement of subjective probabilities should itself be appealing to behavioral scientists. As we shall see, many of the important ramifications of probability extend to the problem of individual choice behavior and decision-making, core concerns of many of the behavioral sciences.

In an elementary discussion such as this, probability can usually be interpreted as relative frequency. Most applications of statistics deal with situations where sampling can be repeated many times, at least in theory, and in these situations the relative-frequency interpretation does make sense. However, when we deal with decision-making based upon statistical information, we will have occasion to discuss subjective probabilities once again. In Chapter 19, the problem of statistical inference will be examined from this point of view.

Generally, in the material to follow, we will work with probabilities without specifying whether these are to be interpreted in one way or the other. We can do this because, essentially, probability is an abstract mathematical concept, for which certain consequences must follow from certain premises, regardless of what the "real" interpretation of the concept may be.

2.12   RANDOM SAMPLING

Given some simple experiment and the sample space of elementary events, the set of outcomes of N separate trials is a **sample**. When the sample is drawn *with replacement*, the same elementary event can occur more than once. When the sample is drawn *without replacement*, the same elementary event can occur no more than once in a given sample. Ordinarily, probability and statistics deal with **random samples**. The word "random" has been used rather loosely in the foregoing discussion, and now the time has come to give it a more restricted meaning in connection with random samples.

It has already been suggested that probability calculations can be made quite simple when the elementary events in a sample space have equal probabilities. The theory of statistics deals with samples of size N from a specified sample space, and here, too, great simplification is introduced if each distinct sample of a particular size can be assumed to have equal probability of selection. For this reason, the elementary theory of statistics is based on the idea of simple random sampling:

A method of drawing samples such that each and every distinct sample of the same size N has exactly the same probability of being selected is called simple random sampling.

In other words, our discussion will be confined to the situation where all possible samples of the same size have exactly the same probability. For us, sampling "at random" will always mean simple random sampling, as defined above. This does not mean, however, that the theory of statistics does not apply to situations where samples have unequal probabilities of occurrence; in more advanced work the theory and methods can be extended to any sampling scheme where the probabilities of the various samples are *known*, even though they are unequal.

An alternative definition of simple random sampling in terms of elementary events can also be given: "Simple random sampling is a process of selecting elementary events for observation in such a way that each and every elementary event has precisely the same probability of being included in any sample of N observations." In *random sampling with replacement*, each elementary event has exactly the same probability of occurring on each trial. In random sampling *without replacement*, the composition of sample space $\mathcal{S}$ changes with each trial since an elementary event can occur only once in N trials. However, we shall assume that among the elementary events available for selection on a given trial, the probabilities are equal.

We shall also have many occasions to require **independent random sampling.** You may recall that independence was specified in Bernoulli's theorem. Although an adequate definition of independence of events cannot be given until later (Chapter 4), we can, for the moment, use the term as follows: elementary events are sampled independently when the occurrence of one elementary event has *absolutely no connection* with the probability of occurrence of the same or another elementary event on a subsequent trial. A series of tosses of a coin can be thought of as a random and independent sampling of events: what happens on one toss has no conceivable connection with what happens on any other. Each trial of a simple experiment is a sampling of elementary events, and the trials are independent when no connection exists between the particular outcomes of different trials.

Take care to notice that our assuming samples of size N thus to be equally probable does not imply that *events* must be equally probable. Thus, in the marble example above, samples of N marbles were assumed equally probable, but the probabilities of the colors of the marbles were not equal. The assumption of equal probabilities is simply a way of saying that any elementary event has just as good an opportunity as any other to serve as a sample on any given experimental trial.

2.13 RANDOM NUMBER TABLES

In practical situations, there are a number of schemes for sampling the elementary events randomly and independently. One of the most common is by use of a table of random numbers. Random number tables consist of many pages filled with digits, from 0 through 9. These tables have been composed in such a way that each digit is approximately equally likely to occur in any spot in the table and there is no systematic connection between the occurrence of any digit in the sequence and any other. Most books on the design of experiments contain pages of random numbers, and very extensive tables of random numbers may be found in a book prepared by the RAND Corporation (1955).

In order to use these tables for random sampling, one must list the possible elementary events (distinct units that might be observed). Each unit is then given a different number. The table of random numbers is entered by choosing some starting point at random and reading to the right or left, up or down. If the sample space contains only ten or fewer individual points, then the digits 0 through 9 themselves are used; if there are one-hundred or fewer members of the sample space, digits are grouped by pairs, 00 through 99, and so on. Then the first N nonrepeated random numbers or groups of numbers are chosen, where N is the required sample size. The individual units corresponding to these numbers comprise the sample, which can be regarded as chosen at random.

This method is subject to two drawbacks. In the first place, unless one has an infinitely large table of random numbers, the one he uses must contain either some differences in probability or some dependence among the digits. This makes a sample chosen with any given table only approximately random. Second, and much more important, is the fact that a *listing* of the members of the sample space (all potential observation-units) must be possible. Except in some very restricted situations this is exceedingly difficult, or even impossible, to do. What usually happens in psychology is that selection is made from some sample space much smaller than the one the experimenter would really like to use. For example, he might like to make statements about all people of a certain age, but the only group he can really sample randomly is college sophomores in a certain locale. Then one of two things must occur; either the experimenter assumes that the sample space actually used is itself a random sample from the larger space, or he restricts his inferences to the group that he can sample randomly. Regardless of which course he takes, if he is going to make probability statements the sample must be drawn at random from *some* specific sample space of potential observations.

Tables of random numbers are used in the same way to achieve randomization in experiments. For example, an experimenter is going to administer three different treatments to a sample of rats, with one third of the rats getting each treatment. He lists the members of his total sample, and then assigns them to the different groups by selecting random numbers. This randomization is an extremely important part of experimental procedure, as we shall see, even though the sample itself is a random selection from some sample space.

Quite apart from their utility for drawing random samples, random number tables are interesting as another instance of the notion of "randomness" that is idealized in the theory of probability. A random process is one in which only chance factors determine the particular outcome of any single trial of the experimental operation. The possible outcomes are known in advance, but not the exact outcome to be realized on any given trial. Nevertheless, built into the process is some regularity, so that each class of outcome (or event) can reasonably be assigned a probability, representing its long-run relative frequency. Perhaps the simplest example of a random phenomenon is the result of tossing a coin. Only chance factors determine whether heads or tails will come up on a particular toss, but a fair coin is so constructed that there is just as much physical reason for heads to appear as for tails, and so we have the justification we need in order to say that, in the long run, heads will appear just as often as tails. This property of the coin is idealized when we say that the probability of heads is .50. A fair die is constructed in such a way that each of the six sides should have the same physical opportunity to appear on a given throw, although unknown factors dictate which side actually does appear on that throw; we idealize this property of the die when we say that the probability of, say, three spots is 1/6. One can construct a random number table by use of a ten-sided die, each side labeled with a number from 0 through 9. Then any sequence of tosses gives a sequence of random numbers. In practice, of course, random numbers are generated electronically, but the general idea is the same: the physical process is such that no single number is favored for any particular outcome, and chance alone actually dictates the number that does occur in any place in a sequence. When we use random numbers to select samples, we are using this property of randomness, which was inherent in the process by which the numbers themselves were generated. The point is that the ideas of randomness and of probability are not just misty abstractions; we know how to create processes and devise operations that are approximately "random" and we *do* know how to manufacture events with particular probabilities, at least with probabilities that are approximately known.

In summary, the calculation of probabilities requires that the basic outcomes of observation, or elementary events, occur with probabilities that are known. Random sampling methods are ways of observing sequences or sets of N such elementary events so that the probabilities associated with the various possible sample results actually *are* known. In simple random sampling, each potential observation-unit or elementary event has the same probability of appearing in a sample of N observations.

The student may be a bit puzzled as we proceed by the insistence on random sampling with *replacement*, when it is perfectly obvious that in most sampling N different subjects or observation units are chosen, and each unit is *not* replaced before the next is sampled. This is done in experimental psychology and related fields because the number of potential observations in the sample space is extremely large. The selection and nonreplacement of a unit does virtually nothing to the probabilities of events. In such situations, sampling without replacement can be treated exactly like sampling with replacement. On the other hand, as we shall see in Section 7.18, when the number of elementary events in the sample space is relatively small, sampling without replacement requires special procedures, since the probabilities of events are altered appreciably with each new observation made. This does no real violence to the notion of simple random sampling, however; instead of requiring that each elementary event be equally likely to occur on each trial, we require that *all samples of N possible outcomes be equally likely to occur*. This will be illustrated in Chapter 5 when we deal with sequences of events. However, in virtually all elementary applications in experimental psychology sampling is thought of as being with replacement.

EXERCISES

1. A simple experiment consists of drawing exactly one playing card from a well-shuffled standard deck. The suit and value of the card is noted. How many elements qualify for the following events?
   (a) Ace
   (b) Red suit
   (c) A face card (i.e. king, queen, jack)
   (d) An even-valued card, not a face card
   (e) A spade or a diamond
   (f) A 10 in a red suit
2. A person is selected at random from a list of registered voters in some community. Make up a list of ten events that might be the outcome of this simple experiment.
3. Two dice are rolled simultaneously. One die is white with black spots, and the other is black with white spots. Let an elementary event be a pair (x,y) where x is the number appearing on the white die, and y is the number appearing on the black die. List the elements of the sample space $\mathcal{S}$. How many elements does it contain? Find the numbers of elements in the following events:
   (a) $x + y \geq 7$
   (b) $x = y$
   (c) $x > y$
   (d) x is odd and y is even
   (e) x is even or y is odd
   (f) $xy \geq 8$
   (g) $(x = 4) \cup (y = 2)$
   (h) $[(x = 6) \cap (y = 4)] \cup [(x = 5) \cap (y \neq 2)]$
4. If the elementary events of problem 3 are equally probable, find the probability of each of the events listed.
5. The numbers on the jerseys of two football teams are as follows:

POSITION	OFFENSE	DEFENSE
Right end	82	84
Right tackle	77	70
Right guard	60	68
Center	55	57
Left guard	63	62
Left tackle	72	73
Left end	86	81
Quarterback	10	15
Right halfback	21	30
Left halfback	33	26
Fullback	41	45

One of the players is selected at random. If each player is equally likely to be selected, find the probability of the following events:
   (a) The player is on the offensive team and has an even-numbered jersey.
   (b) The player is a tackle or a guard.

   (c) The player is on the defensive team and has a higher number than the player of the same position on the offensive team.

   (d) The player is one of the backs with a number higher than 25.

   (e) The player is not a center and is not an end.

   (f) The number on the player's jersey is greater than 40 or is an even number.

   (g) The player is not a back or has a number between 40 and 75.

6. Suppose that from the football teams of problem 5, two players are selected at random, one from the offensive team and one from the defensive team. Describe the sample space. How many elements would it include? Find the probability of the following events:

   (a) The two players chosen are both backs.

   (b) The two players chosen play the same position.

   (c) The first player chosen from the offense is either a tackle or a guard and the second player is an end.

   (d) The numbers on the jerseys of both players are less than 40.

   (e) The first player chosen is a center and the second player chosen is not a center.

7. A letter is chosen at random from the English alphabet. If all letters are equally likely to be chosen, what is the probability of the following events:

   (a) The letter is a vowel (include y as a vowel).

   (b) The letter is a consonant, other than s or t.

   (c) The letter is a consonant from the first half of the alphabet.

   (d) The letter chosen appears in the word PSYCHOLOGIST.

   (e) The letter appears either in the word BEGGAR or the word BENIGN or both.

   (f) Both the event (d) and the event (e) occur.

8. Suppose that a letter is drawn at random from the word SENSATION. Then that letter is replaced and a second letter drawn. Find the probabilities of the following events. (Hints: treat the two occurrences of the letter S in SENSATION as two elementary events, and the same for N. A simple graphic plot of all of the elements of the sample space will be helpful.)

   (a) The two letters drawn are the same.

   (b) The letters drawn are both consonants.

   (c) The letters drawn are both vowels.

   (d) The letters drawn are two different consonants.

   (e) The letters drawn are two different vowels.

   (f) The letters drawn are a vowel followed by a consonant.

   (g) The letters drawn in order form any of the words SO, AS, or IS.

   (h) The letters form any of the words IN, ON, AT, or AN.

   (i) The two letters drawn form *any* common English word.

9. A box contains fifty marbles. Some of the marbles are large (the set L) and others are small (the set $\bar{L}$). Some are blue (the set B) and others are red ($\bar{B}$). The simple experiment is performed of drawing one marble at random from the box. Specify the complete family of events for this experiment. If there are 23 small marbles, 37 marbles that are either large or blue, and 15 marbles that are large but not blue, find the following probabilities:

   (a) $p(L)$

   (b) $p(B)$

   (c) $p(\bar{L} \cap B)$

   (d) $p(\bar{L} \cap \bar{B})$

   (e) $p(L \cap B)$

   (f) $p(L \cup \bar{B})$

10. In example 9, what should one expect as the result of the simple experiment in terms of the color of the marble? the size of the marble? the color and size of the marble? If you drew marbles at random with replacement from the box for 150 trials, how many times should you expect the following events to occur:
    (a) A blue marble?
    (b) A small marble?
    (c) A small, red marble?
    (d) A small red or a large, blue marble?
    (e) A small blue or a large, red marble?

11. A box contains marbles that are either blue (B) or red $(\bar{B})$, either large (L) or small $(\bar{L})$, and either clear glass (C) or opaque glass $(\bar{C})$. One hundred marbles were drown from the box at random with replacement, and the following events occurred.

EVENT	NUMBER OF OCCURRENCES
$(L \cap \bar{B} \cap \bar{C})$	5
$(L \cap \bar{B} \cap C)$	15
$(L \cap B \cap \bar{C})$	30
$(L \cap B \cap C)$	0
$(\bar{L} \cap B \cap \bar{C})$	10
$(\bar{L} \cap B \cap C)$	6
$(\bar{L} \cap \bar{B} \cap C)$	9
$(\bar{L} \cap \bar{B} \cap \bar{C})$	25

Using these results estimate the probabilities of the following events:
    (a) $p(L)$
    (b) $p(B)$
    (c) $p(C)$
    (d) $p(L \cap B)$
    (e) $p(L \cap \bar{C})$
    (f) $p(B \cap C)$
    (g) $p(L \cup \bar{B})$
    (h) $p(L \cup B \cup C)$

12. For each of the events corresponding to the sets found in problem 15, Chapter I, calculate the probability. Assume that each elementary event is equally probable.

13. In problem 26 of Chapter I, let each element of $D \times D$ be an elementary event so that each relation on $D \times D$ is an event. If the elementary events are equally probable, find the probability for each of the five events listed.

14. Given that it is equally probable that a person chosen at random was born on any of the 365 days of the year, find the probabilities of the following events:
    (a) The person was born between September 1 and December 31 inclusive.
    (b) The person was born in a month with 30 days.
    (c) The person was born in June or July, or was born in January, February, or March.
    (d) The person was born between August 1 and November 30, or between September 1 and December 31.
    (e) The person was born between February 12 and March 15 inclusive, or between March 1 and April 2 inclusive.
    (Hint: for these calculations ignore the possibility of a leap year.)

15. A child's game has a metal spinner on a card. This card consists of a circular area divided into six colored sectors of equal area: black, white, red, yellow, green, and blue. The spinner is attached to the center of the card, and is free to come to rest at any point along the circle. If the spinner is equally likely to come to rest in any sector then what is the probability of the following:
    (a) Black or white?
    (b) Yellow?
    (c) Neither white nor yellow?
    (d) Either green or not blue?
    (e) Blue or neither red nor black?
16. Suppose that the card described in the problem above were such that black, white, and yellow sectors were each 1.5 times the size of the individual sectors devoted to red, blue, and green. Calculate the probabilities of the events listed above under these circumstances, assuming that the pointer has equal probability of stopping at any point on the circle.
17. Suppose that the following simple experiment is carried out: An individual repeatedly draws a single card at random and with replacement from a standard deck, shuffling the cards thoroughly after each draw. He keeps doing this until he has drawn one ace. The outcome of this experiment is thus the trial number on which the first ace appears. Discuss the sample space of outcomes for this experiment.
18. A married couple is drawn at random from a set of such couples. Let a be the month of birth of the husband, and let b be the month of the birth of the wife. If all (a,b) pairs have the same probability of occurrence, what is the probability of the following:
    (a) The husband and wife were born in the same month.
    (b) The husband and wife were born in immediately adjacent months, such as May-June, August-September.
    (c) The husband was born in a month with 30 days, and the wife was born in a month with 31 days.
19. State the probabilities corresponding to the following odds and bets:
    (a) 5 to 3 in favor of event a.
    (b) 2 to 9 against event a.
    (c) 6 to 4 in favor of event a.
    (d) 17 to 15 in favor of a.
    (e) \$5 to \$2 bet on a.
    (f) \$9 to \$4 bet against a.
    (g) \$1 to \$49 bet on a.
20. The following events can occur only once. If an individual regards the bets listed as fair bets, estimate his subjective probability for these events:
    (a) A bet of \$.50 to \$.10 that it will rain tomorrow.
    (b) A bet of \$3 to \$7 that the Big Ten representative will win or tie in the Rose Bowl game next New Year's Day.
    (c) A bet of \$1 to \$1 that this individual will fail his Psychology course this term.
21. State the odds corresponding to the probabilities of the events listed in problem **3**.
22. Give dollar values corresponding to fair bets about the events listed in problem **3**.
23. An individual is presented with the following statements about Lottery A:

> You win \$18 if player Z wins the Heisman Trophy this year.
> You lose \$2 if player Z does not win the Heisman Trophy.

What can you say about the subjective probability of the event for this individual under each of the following conditions:

(a) He prefers Lottery A to the lottery whose terms state:

{ You win $6 with probability 1/3.
  You lose $1 with probability 2/3.

(b) He prefers the following lottery to Lottery A:

{ You win $8 with probability 3/4.
  You lose $2 with probability 1/4.

(c) He is indifferent between the following lottery and Lottery A:

{ You win $9 with probability 2/3.
  You lose $3 with probability 1/3.

24. A voter believes that his candidate has a 5/8 chance of winning an election. A fair bet on his candidate thus corresponds to the following lottery:

{ He wins $3 if the candidate wins the election.
  He loses $5 if the candidate loses the election.

In each of the following lotteries, what is the value of $p(a)$ that would make the voter indifferent between his election bet and the lottery:

(a) { He wins $18 with probability $p(a)$.
      He loses $15 with probability $1 - p(a)$.

(b) { He wins $130 with probability $p(a)$.
      He loses $1.20 with probability $1 - p(a)$.

(c) { He wins $1.95 with probability $p(a)$.
      He loses $.39 with probability $1 - p(a)$.

3 Frequency and Probability Distributions

Given the definitions of "event" and of "probability" in the last chapter, we are now ready to take up the first major set of concepts in statistics. First of all, the idea of a frequency distribution for sets of observations will be introduced, together with some of the mechanics for constructing distributions of data. Then this idea of a frequency distribution will be paralleled by that of a probability distribution. The important special case of numerical events will be introduced under the heading "random variables."

Prior to any concerns he may have with summarizing his data, or comparing it with some theoretical state of affairs, the experimenter must first measure the phenomena he studies. It therefore seems appropriate to mention the process of measurement at this point, and to indicate the role measurement considerations will play in the remainder of this text. The author will not even attempt to undertake a thorough discussion of this topic, but he will try to suggest its relevance to statistics.

3.1 MEASUREMENT SCALES

Whenever the scientist makes observations of any kind, he employs some scheme for classifying and recording what he observes. Any phenomenon or "thing" will have many distinguishable characteristics or attributes, but *the scientist must first single out those properties relevant to the question being studied.* For example, a scientist tries out a new serum on laboratory rats infected with a disease. Any single rat differs from any other rat in innumerable ways: his coat markings, his heart rate, the length of his tail, his exact age, and so on. The scientist is interested, however, in only one thing: did the rat recover or did he die of the disease? Or, perhaps, the intelligence level of a group of boys is being compared with that of a group of girls. Individual boys and girls differ and same-sex groups differ among themselves in ways as diverse as body temperature, size of head, weight, color of hair, father's income, name, and so on ad infinitum. All these properties are ignored by the experimenter as immaterial to his immediate purpose, which is giving each child a number representing his intelligence.

The classifying scheme the scientist uses is based upon differences in some *particular* property or attribute that objects of observation exhibit. The scientist simply cannot pay attention to all the ways that things differ from each other. Many of these differences obviously are irrelevant to his purpose, and other potential differences are controlled by the scientist in making his observations: only rats (not all kinds of animals) are given the serum, and each child is given the same intelligence test in the same way. Still other differences that are germane to his conclusions, but not specifically controlled, are treated as statistical "error."

When the scientist has singled out the property or properties he wants to study, he applies his classifying scheme to each observation. Such a scheme is essential if he wants to record, summarize, and communicate what he observes. At its very simplest, *this scheme is a rule for arranging observations into equivalence classes, so that observations falling into the same set are thought of as qualitatively the same and those in different classes as qualitatively different in some respect.* In general, *each observation is placed in one and only one class, making the classes mutually exclusive and exhaustive.*

The process of grouping individual observations into qualitative classes is measurement at its most primitive level. Sometimes this is called **categorical or nominal scaling.** The set of equivalence classes itself is called a **nominal scale.** There are many areas in science where the best one can do is to group observations into classes, each given a distinguishing symbol or name. For example, taxonomy in biology consists of grouping living things into phyla, genera, species, and so on, which are simply sets or classes. Psychiatric nosology contains classes such as schizophrenic, manic-depressive, paretic, and so on; individual patients are classified on the basis of symptoms and the history of their disease.

Notice that even this kind of measurement can be thought of in terms of a function: the domain of this function is a set of individual observations, and the range is the set of equivalence classes (the nominal scale). The measurement procedure is like a function rule that permits the assignment of each individual observation to one and only one qualitative class. The class paired with an individual informs us of some qualitative property that he exhibits.

We may also observe that the set of equivalence classes is a partition of some universal set W, since each object that is a member of the set W is assigned to one and only one class from among the partition $\{A_1, A_2, \cdots, A_i, \cdots, A_N\}$ by the process of measurement. In nominal measurement, any other partition of W consisting of N classes, such as $\{B_1, B_2, \cdots, B_j, B_N\}$ could just as well have been used, *provided* that a 1-to-1 function exists pairing each class A_i with exactly one class B_j in such a way that if an object $x \in A_i$, then $x \in B_j$ and no other class B. In other words, in nominal measurement, any set of equivalence classes may be transformed into another set of equivalence classes, provided that common class membership, or equivalence, is strictly preserved. In effect, the membership of the sets stay the same, and only the names of the sets are changed. Thus, consider a set of seniors at a given University in a given year. They might be classified as "liberal arts major," "engineering major," "education major," "business major," and "other major." However, they could just as well be classified as "group alpha," "group beta," "group gamma," "group delta," and "group omega" so far as the measurement process is concerned, provided that all members of a group in the first instance wind up as members of the same group in the second instance. Only the names of the classifications are changed. Any such equivalence-preserving transformation is permissible in nominal measurement.

The word "measurement" is usually reserved, however, for the situation where each individual is assigned a *number;* this number reflects a magnitude of some *quantitative* property. A function exists associating one and only one number with each individual, and the measurement procedure fills the role of a rule for this function.

There are at least three kinds of numerical measurement that can be distinguished: these are often called ordinal scaling interval scaling, and ratio scaling. A really accurate description of each kind of measurement is beyond the scope of this book, but we can gain some idea of these distinctions by using the function terminology once again.

Imagine a set of objects O, and any pair of objects in that set, say o_1 and o_2. Now suppose that there is some property that all objects in the set show, such as temperature, or weight, or length, or intelligence, or motivation. Furthermore, let us suppose that each object o has a certain amount or degree of that property. In principle we could assign a number, $t(o)$, to any object standing for the amount that o actually "has" of that characteristic.

Ideally, in order to measure an object o, we would like to determine this number $t(o)$ directly. However, this is not always (or even usually) possible, and so what we do is to devise a procedure for pairing each object with another number $m(o)$, that can be called its *numerical measurement*. The measurement procedure we use constitutes a function rule, telling how to give an object o its $m(o)$ value. But just any measurement procedure will not do; *we want the various $m(o)$ values at least to reflect the different $t(o)$ values for different degrees of the property*. A measurement rule would be nonsense if it gave numbers having no connection at all with the amounts of some property different objects possess. Even though we may never be able to determine the $t(o)$ values exactly, we would like to find $m(o)$ numbers that will at least be related to them in a systematic way. The measurement numbers must be good reflections of the true quantities, so that information about magnitudes actually is contained in our numerical measurements.

Measurement operations or procedures differ in the information that the numerical measurements themselves provide about the true magnitudes. Some ways of measuring permit us to make very strong statements about what the differences or ratios among the true magnitudes must be, and thus about the actual differences in, or proportional amounts of, some property that different objects possess. On the other hand, some measurement operations permit only the roughest inferences to be made about true magnitudes from the measurement numbers themselves.

Suppose that we had a measurement procedure or rule for assigning a number $m(o)$ to each object o, and suppose that the following statements were true:

1. For any pair of objects o_1 and o_2,
   $$m(o_1) \neq m(o_2) \text{ only if } t(o_1) \neq t(o_2).$$
2. $m(o_1) > m(o_2)$ only if $t(o_1) > t(o_2)$.

In other words, by this rule we can at least say that if two measurements are unequal the true magnitudes are unequal, and if one measurement is larger than another one magnitude exceeds another. Any measurement procedure for which both statements 1 and 2 are true is an example of **ordinal scaling,** or measurement at the **ordinal level.**

For example, suppose that the objects in question were minerals of various kinds. Each mineral has a certain degree of *hardness*, represented by the quantity $t(o)$. We have no way to know these quantities directly, and so we devise the following measurement rule: take each pair of minerals and find if one *scratches* the other. Presumably, the harder mineral will scratch the softer in each case. When this has been done for each pair of minerals, give the mineral scratched by everything some number, the mineral scratched by all but the first a higher number, and so on, until the mineral scratching all others but scratched by nothing else gets the highest number of all. In each pair the "scratcher" gets a higher number than the "scratchee." (Here we are assuming tacitly that "*a* scratches *b*" and "*b* scratches *c*" implies that "*a* scratches *c*," and that we will get a *simple ordering* of "what scratches what.")

This measurement procedure gives an example of *ordinal* scaling. The possible numerical measurements themselves might be some set of numbers, such as $(1, 2, 3, 4, \cdots)$ or $(10, 17, 24, \cdots)$ or even other symbols having some conventional order, as $(A, B, C, \cdots)$. In any case, the assignment of numbers or symbols to objects is a form of ranking, showing which is "more" something. In the example, if one mineral gets a higher number than another then we can say that the first mineral is harder than the second. Notice, however, that this is really all we can say about the degree or amount of hardness each possesses. If we find two objects where $m(o_1) = 3$ and $m(o_2) = 2$ by this procedure, then it is perfectly true that the difference between the *numbers* is $3 - 2 = 1$, but we have no basis at all for saying that $t(o_1) - t(o_2) = 1$. The difference in hardness between the two minerals could be any positive number. In short, *although the numbers standing for ordinal measurements may be manipulated by arithmetic, the answer cannot necessarily be interpreted as a statement about the true magnitudes of objects, nor about the true amounts of some property.*

The paragraph above also illustrates that it is possible to change the symbols, numerical or otherwise, involved in ordinal measurement without losing the essential ordinal quality of the measurement scale. However, an ordinal scale may be transformed into another ordinal scale only if order is strictly preserved in the transformation. Thus, let numbers such as $m(o_1)$ and $m(o_2)$ be transformed into numbers such as $g(o_1)$ and $g(o_2)$ for objects such as o_1 and o_2. Then, for an ordinal scale, this transformation is permissible if the following conditions hold:

1. $g(o_1) \neq g(o_2)$ if and only if $m(o_1) \neq m(o_2)$.
2. $g(o_1) > g(o_2)$ if and only if $m(o_1) > m(o_2)$.

In short, only order-preserving transformations are permissible with an ordinal scale.

Finally, it is worth noting that the definition given above does allow for grossness of measurement. That is, the measuring process itself may not be sufficiently sensitive to reflect true differences in magnitudes as differences in order between objects. Thus, it may be that $m(o_1) = m(o_2)$ even though $t(o_1) > t(o_2)$. Even with this degree of insensitivity, a measuring procedure can still provide a useful ordinal scale.

Other measurement procedures give functions pairing objects with numbers where much stronger statements can be made about the true magnitudes from the numerical measurements. Suppose that the following statement, in addition to statements 1 and 2, were true:

3. $t(o) = x$ if and only if $m(o) = ax + b$, where $a > 0$.

That is, the measurement number $m(o)$ is some *linear function* of the true magnitude x (the rule for a linear function is x multiplied by some constant a and added to some constant b). When the statements 1, 2, and 3 are all true, the measurement operation is called **interval scaling,** or measurement at the **interval-scale level.**

Much stronger inferences about magnitudes can be made from interval-scale measurements than from ordinal measurements. In particular, we can say something precise about *differences* in objects in terms of magnitude. Consider two objects, o_1 and o_2, once again. Then if we find

$$m(o_1) - m(o_2) = 4,$$

we can conclude that

$$t(o_1) - t(o_2) = 4/a,$$

the difference between the magnitudes of the two objects is 4 units, where a is simply some constant changing measurement units into "real" units, whatever they may be.

For example, finding temperature in Fahrenheit units is measurement on an interval scale. If object o_1 has a reading of 180° and o_2 has 140°, the difference $(180 - 160)$ times a constant actually *is* the difference in "temperature-magnitude" between the objects. It is perfectly meaningful to say that the first object has twenty units more temperature than the second. However, even stronger statements can be made. If $m(o_3) = 140°$, so that

$$m(o_1) - m(o_3) = 40°$$
$$m(o_1) - m(o_2) = 20°$$

then

$$\frac{40°}{20°} = \frac{t(o_1) - t(o_3)}{t(o_1) - t(o_2)} = 2.$$

One can say that in true amount of "temperature," object o_3 is twice as different from o_1 as o_2 is from o_1.

Given some measurement operation yielding an interval scale of some property, then *any other measurement operation for the same property also gives an interval scale, provided that the second way of measuring yields numbers that are a linear function of the first:* thus, let some second measurement operation give numbers $g(o)$ to each object o, and let it be true that

$$g(o) = c[m(o)] + d, \qquad c > 0,$$

where c and d are constants. However, by statement 3,

$$m(o) = ax + b$$

so that

$$g(o) = c[ax + b] + d = cax + bc + d.$$

The number ca and the number $bc + d$ are constants, and so the $g(o)$ values satisfy statement 3 as well, showing that this second measurement procedure gives an interval scale. A familiar example of this principle is temperature measurement in Fahrenheit and in Centigrade degrees; each is an interval-scale measurement procedure, and the reading on one scale is a function of the reading on the other. If the $m(o)$ are Fahrenheit readings, and the $g(o)$ are Centigrade, then it will always be true that for any three objects

$$\frac{m(o_1) - m(o_3)}{m(o_1) - m(o_2)} = \frac{g(o_1) - g(o_3)}{g(o_1) - g(o_2)}.$$

Hence, if we conclude from a Fahrenheit thermometer that o_3 is twice as different from o_1 as o_2 is from o_1, we would reach exactly the same conclusion from a Centigrade thermometer.

When measurement is at the interval-scale level, any of the ordinary operations of arithmetic may be applied to the differences between numerical measurements, and the result interpreted as a statement about magnitudes of the underlying property. The important part is this interpretation of a numerical result as a quantitative statement about the property shown by the objects. This is not generally possible for ordinal-scale measurement numbers, but it *can* be done for differences between interval-scale numbers. In very simple language: one can do arithmetic to his heart's content on any set of numbers, but his results are not necessarily true statements about amounts of some property objects possess unless interval-scale requirements are met by the procedure for obtaining those numbers.

Interval scaling is about the best one can do in most scientific work, and even this level of measurement is all too rare in psychology. However, especially in physical science, it is sometimes possible to find measurement operations making the following statement true:

4. $t(o) = x$ if and only if $m(o) = ax$, where $a > 0$.

When the measurement operation defines a function such that statements 1 through 4 are all true, then measurement is said to be at the **ratio-scale level**. For such scales, *ratios* of numerical measurements can be interpreted directly as ratios of magnitudes of objects:

$$\frac{m(o_1)}{m(o_2)} = \frac{t(o_1)}{t(o_2)}.$$

For example, the usual procedure for finding the length of objects provides a ratio scale. If one object has a measurement value of 10 feet, and another a value of 20 feet, then it is quite legitimate to say that the second object has twice as much length as the first. Notice that this is not a statement one ordinarily makes about the *temperatures* of objects (on an interval scale): if the first object has a temperature reading of 10° and the second 20°, we do not ordinarily say that the second has twice the temperature of the first. Only when scaling is at the ratio level can the full force of ordinary arithmetic be applied directly to the measurements themselves, and the results reinterpreted as statements about magnitudes of objects.

By now, an important connection between interval scaling and ratio scaling should become clear. **When objects are measured on an interval scale, then differences between objects are measured on a ratio scale.** The concept of zero difference, or zero "distance," does have a fixed and nonarbitrary definition, and differences between objects can be treated by any of the methods available for ratio-scale values, provided that the original objects were measured on an interval scale. This accounts, in part, for the considerable preoccupation with differences among measurements that one encounters not only in the physical sciences but also in statistics.

Only one sort of transformation is permissible for values measured on a ratio scale. This is multiplication of each value by the same positive constant. Thus, if $m(o)$ is the ratio-scale measurement of object o on some property, then only a transformation of the type $g(o) = km(o)$, where $k > 0$, alone is permissible, if the new values are, in turn, to be a ratio scale. In other words, the only permissible change is in the unit of measurement employed.

There are any number of examples that could be adduced to illustrate the differences among these levels of measurement, and other possible intermediate levels as well. The student who is interested in this topic enough to explore it further is urged to look into the initial chapters of the books by Stevens (1951), by Thrall, Coombs, and Davis (1954), and by Torgerson (1958); each gives a slightly different perspective on psychological measurement. Our immediate concern, however, is to see the implications of different levels of measurement for probability and statistics.

3.2   MEASUREMENT LEVELS AND STATISTICAL METHODS

Any scientific investigation uses measurement, if only at the nominal-scale level, and certainly it behooves the investigator to think very carefully about the level of measurement he is using. Can he take the numbers he assigns to objects and deduce true statements about the property measured? Do mathematical manipulations of those numbers *necessarily* yield answers that are themselves interpretable as statements about magnitudes of some characteristic? It is quite easy to show a situation in psychology where it is not possible to go from numerical scores to statements about quantities. For example, in intelligence testing, a score called a "mental age" is sometimes the end product of giving a person a test. It is reasonable that the more intelligence a person really has the greater this mental age should be, other things being equal. *As numbers*, mental ages can be subjected to any arithmetic operation we choose, but the rub comes in translating the resultant numbers back into statements about amounts of intelligence. Thus, suppose that one five-year-old child has a mental age of 4 years, and another a mental age of 8 years. As numbers, it is quite true that 4 goes into 8 exactly 2 times; but can we assert that the second child is twice as intelligent as the first? Not without considerably more justification than we have at present. There is no solid theoretical or empirical reason for thinking that intelligence scores must relate to "real" amounts of intelligence in the way specified by the definitions of ratio, or even interval, scaling. On the other hand, we do feel justified in thinking of intelligence tests as giving at least ordinal scales, if not more. The point is that one cannot simply assume that a certain level of scaling is or is not reached without some theoretical or empirical basis, showing how the quantities we hope to measure actually are reflected in the numbers we get.

The problem of measurement, and especially of attaining interval scales, is an extremely serious one for psychology. It is unfortunate but true that in their search for quantitative methods psychologists sometimes overlook the question of level of measurement and tend to read quite unjustified meanings into their results. This has brought about a reaction in some quarters, where people insist that psychologists are not justified in using most of the machinery of mathematics on their generally low-level measurements. Others have argued that psychologists are too concerned with level of measurement, and that ordinary mathematical techniques apply to virtually any numerical measurements. There has been considerable controversy about the applicability of statistical methods according to the scale requirements met by the data. Indeed, some statistics texts have even been organized on the basis of the different levels of measurement.

It seems to this author that statistics qua statistics is quite neutral on this issue. In developing procedures, mathematical statisticians have assumed that techniques involving numerical scores, orderings, or categorization are to be applied where these numbers or classes are appropriate and meaningful within the experimenter's problem. If the statistical method involves the procedures of arithmetic used on numerical scores, then the numerical answer is formally correct. Even if the numbers are the purest nonsense, having no relation to real magnitudes or the properties of real things, the answers are still right *as numbers*. The difficulty comes with the interpretation of these numbers back into statements about the real world. If nonsense is put into the mathematical system, nonsense is sure to come out.

On the other hand, there are many instances where the experimenter is interested in numerical scores *as scores*. For example, he may want to compare one group of five-year-old children with another on the basis of mental age scores, and so he averages the numerical scores for each group. The fact that he cannot say that one group has an average of 65 units of intelligence and the other 60 units does not alter the practical advantage he may gain by finding average scores. There are very many instances where the level of measurement may not reach the level supposed by the statistical technique, and yet the method itself may be quite adequate for showing what the experimenter wants to know. Any statistical method designed for numerical scores can be used, provided that the purely mathematical-statistical requirements of the method are met, and provided that the experimenter wants to reach a conclusion about such measurements. However, he is not necessarily entitled to translate the statistical result back into a direct numerical statement about quantities or magnitudes of some property, unless measurement is at the interval- or ratio-scale level. Even so, although the statistical result does not translate into a conclusion about the properties of objects, it is nonetheless a true conclusion about *measurement numbers of this sort.* In some instances, this may be enough for the experimenter's immediate purpose.

The problem of level of measurement really lies outside the province of mathematics and statistics. Statistics deals with numbers, and statistical methods yield conclusions based on numbers, but there is absolutely nothing in the mathematical machinery that whistles and waves a flag to show that the numbers supplied are not really interval-scale measurements of some property of objects or human beings. The machinery works the same way and gives the same result regardless of whether the numbers are made up from the whole cloth or are the product of the most refined measurement procedures imaginable. Only the users of the statistical result, the experimenter and his readers, can judge the reinterpretability of the numerical result into a valid statement about properties of things. Sometimes the psychologist regards the numerical conclusion as a statement about *scores*, and does not go beyond this statement to a conclusion about the real magnitudes of some property. In other instances the whole investigation makes no sense at all unless the numerical conclusion can refer directly to magnitudes of some property. Surely the individual psychologist knows his problem and his measurement procedures better than a stranger dogmatizing about hypothetical situations in a textbook! The experimenting psychologist must face the problem of the interpretation of statistical results *within psychology* and on *extramathematical* grounds.

For this reason, very little more will be said about scales of measurement in this book. Everything depends upon what the psychologist is measuring, how he goes about it, what he is trying to find out, and, most of all, what he wants to say about the "real properties" underlying his numerical measurements.

Most of the statistical techniques introduced in this book are designed for numerical scores, presumably arising from application of measurement techniques at the interval-scale level. This does not mean that these techniques cannot be applied to numerical data where the interval-scale level of measurement is not met; these techniques can be applied to *any* numerical data where the purely *statistical* assumptions are met. It does mean, however, that caution must be exercised in the interpretation of these statistical results in terms of some property the experimenter intended to measure. The statistical conclusion may be quite valid for these scores *as scores*, but the experimenter must think quite seriously about the further interpretations he gives to the result.

Of course, it is completely obvious that interval scales simply cannot be attained for some things. It may be that an ordering of objects on some basis is the very best the experimenter can do. In this case, ordinal methods like those in

Chapter 18 are helpful; each of these techniques is designed specifically for ordinal data. Even when measurement is only nominal, statistical methods can be applied; some of these are discussed in Chapter 17. The most common situation is, however, one where numerical measurements are made, but where the interval-scale assumptions are hard to justify. In these cases, the user of the statistical techniques should understand the limitations imposed by his measurement device (not by statistics): the road from objects to numbers may be easy, but the return trip from numbers to properties of objects is not.

3.3　FREQUENCY DISTRIBUTIONS

As we have just seen, even in the simplest instances of measurement the experimenter makes some N observations and classifies them into a set of qualitative measurement classes. These qualitative classes are mutually exclusive and exhaustive, so that each observation falls into one and only one class.

As a summarization of his observations, the experimenter often reports the various possible classes, together with the number of observations falling into each. This may be done by a simple listing of classes, each paired with its frequency number, the number of cases observed in that category. The same information may be displayed as a graph, perhaps with the different classes represented by points or segments of a horizontal axis, the frequency shown by a point or a vertical bar above each class. Regardless of how this information is displayed, such a listing of classes and their frequencies is called a *frequency distribution. Any representation of the relation between a set of mutually exclusive and exhaustive measurement classes and the frequency of each is a frequency distribution.*

Notice that a frequency distribution is a funct on: each of a set of classes is paired with a number, its frequency. Thus, in principle, a frequency distribution of real or theoretical data can be shown by any of the three ways one specifies any function: an explicit listing of class and frequency pairs, graphically, or by the statement of a rule for pairing a class of observations with its frequency. In describing actual data, the first two methods are almost always used alone, but there are circumstances where the experimenter wants to describe some theoretical frequency distribution, and in this case the mathematical rule for the function sometimes is stated.

The set of measurement classes may correspond to a nominal, an ordinal, an interval, or a ratio scale. Although the various possible classes will be qualitative in some instances and number classes in others, a frequency distribution can always be constructed, provided that each and every observation goes into one and only one class. Thus, for example, suppose that some N native United States citizens are observed, and the state in which each was born noted. This is like nominal measurement, where the measurement classes consist of the fifty states (and the District of Columbia) into which subjects could be categorized. On the other hand, the subjects might be students in a course and graded according to A, B, C, and so on. Here the frequency distribution would show how many got A, B, and so on. Notice that in this case the measurement actually is ordinal: A is better than B, B better than C. Nevertheless, the frequency distribution is constructed in the same way, except that the various classes are displayed in their proper order.

Even when measurement is at the interval- or ratio-scale level, one actually reports frequency distributions in this general way. Suppose that N college students were each weighed, and the weights noted to the nearest pound. The set of weight classes into which students are placed would perhaps consist of fifty or sixty different numbers; a weight class is the set of students getting the same weight number. There may be only one, or even no students, in a particular weight class. Nevertheless, the frequency distribution would show the pairing of each possible class with some number from zero to N, its frequency.

Perhaps it will be useful to recall where we have already encountered ideas very similar to these. In Chapter 1 a measure function was introduced that assigns a positive numerical value, or zero, to each of a set of sets. A frequency distribution is just such a measure function, with a domain consisting of measurement-classes, or numerical value-classes, which form a partition of the total set of objects measured. The range of this measure function consists of all the possible frequencies that are paired with the respective classes. Thus a frequency distribution is still another example of a measure function on sets, and is closely related to another example we have discussed, a probability function.

As we shall see, when the number of possible classes is very large, or even potentially infinite, the task of constructing a frequency distribution is made easier by a process of combining classes. The principle is, however, always the same: a display of the function relating classes and their frequencies.

3.4 FREQUENCY DISTRIBUTIONS WITH A SMALL NUMBER OF MEASUREMENT CLASSES

The reasons for dealing with frequency distributions rather than raw data are not hard to see. Raw data almost always take the form of some sort of listing of pairs, each consisting of an object and its measurement class or number. Thus, if we were noting the hair color of eleven persons, our raw data might be something like this:

John Jones	black
Mary Smith	blonde
Jim Hardy	brown
Horace Goodman	brown
Alice Adams	blonde
Ann Wilk	red
William Thomas	brown
Bert Fox	red
Homer Giddens	black
Shirley Snider	brown
Richard Rowan	blonde

This set of pairs does contain all the relevant information, but if there are a great many objects being measured such a listing is not only laborious but also very confusing to anyone who is trying to get a picture of the set of observations as a whole. If we are interested only in the "pattern" of hair color in the group the names of the persons are irrelevant, and all that is necessary is the number of individuals having each hair color.

There are only four measurement classes in this example that show one or more cases, and so the frequency distribution can be displayed in this way:

Hair color	Frequency
black	2
red	2
blonde	3
brown	4
	11 = N

It is possible to illustrate this idea of a frequency distribution in any number of ways: consider a study done on a group of twenty-five males, in order to determine their blood types. The subjects were classified variously as having blood types "A," "B," "AB," or "O." Table 3.4.1 lists the measurements of these

Table 3.4.1

Man number	Blood type
1	A
2	B
3	A
4	A
5	A
6	AB
7	O
8	A
9	A
10	A
11	O
12	B
13	O
14	B
15	A
16	B
17	O
18	B
19	O
20	A
21	B
22	B
23	A
24	A
25	O

twenty-five men. These data condense into the following frequency distribution:

Class	f
A	11
B	7
AB	1
O	6
	$\overline{}$
	$25 = N$

In another study, 2000 families in some city were interviewed about their preferences in art, music, literature, recreation, and so forth. After each interview, the family was characterized as having "highbrow," "middlebrow," or "lowbrow" tastes. This we can consider as a case of ordinal-scale measurement, since these three "taste" categories do seem to be ordered: "highbrow" and "lowbrow" should be more different from each other than "middlebrow" is from either. A listing of families in terms of these classes would be confusing, if not grounds for libel suits. However, just as before, we can construct a frequency distribution summarizing these data:

Class	f
highbrow	50
middlebrow	990
lowbrow	960
	$\overline{}$
	$2000 = N$

Notice how much more clearly the characteristics of the group as a whole emerge from a frequency distribution such as this than they possibly could from a listing of 2000 names each paired with a rating.

It is appropriate to point out that a frequency distribution provides clarity at the expense of some information in the data. It is not possible to know from the frequency distribution alone whether the family of John Jones at 2193 Spring Street is high-, middle-, or low-browed. Such information about *particular* objects is sacrificed in a frequency distribution to gain a picture of the group of measurements *as a whole*. This is true of all descriptive statistics; we want clear pictures of large numbers of measurements, and we can do this only by losing detail about particular objects. The process of weeding out particular qualities of the object that happen to be irrelevant to our purpose, begun whenever we measure, is continued when we summarize a set of measurements.

3.5 GROUPED DISTRIBUTIONS

In the distributions shown above, there were a few "natural" categories into which all the data fit. It often happens, however, that data are

measured in some way giving a great many categories into which a given observation might fall. Indeed, the number of potential categories of description often
vastly exceeds the number of cases observed, so that little or no economy of
description would be gained by constructing a distribution showing each *possible*
class. The most usual such situation occurs when the data are measured in numerical terms. For instance, in the measurement of height or weight, there are, in
principle, an infinite number of numerical classes into which an observation might
fall: all the positive real numbers. Even when the measurements result in fewer
than an infinite number of possibilities, it is still quite common for the number of
available categories to be very large.

For this reason, it is necessary to form **grouped** frequency distributions. Here the domain of the frequency function is not each possible measurement
category, but rather intervals or groupings of categories.

For example, consider a number of people who have been given an
intelligence test. The data of interest are the numerical IQ scores. Immediately,
we run into a problem of procedure. It could very well turn out that for, say 150
individuals, the IQ score would be different for each case, and thus a frequency
distribution using each different IQ number as a measurement class would not
condense our data any more than a simple listing.

The solution to this problem lies in grouping the possible measurement classes into new classes, called **class intervals,** each including a range of
score possibilities. Proceeding in this way, if the IQs 105, 104, 100, 101, and 102
should turn up in our data, instead of listing each in a different class, we might put
them all into a single class, 100–105, along with any other IQ measurements that
fall between these limits. Similarly, we group other sets of numbers into class
intervals. On our doing this, one way that the frequency distribution might look
for a group of 150 persons is shown in Table 3.5.1.

Table 3.5.1

Class interval	Midpoint	f
124–129	126.5	8
118–123	120.5	0
112–117	114.5	10
106–111	108.5	20
100–105	102.5	65
94– 99	96.5	22
88– 93	90.5	23
82– 87	84.5	2
		$150 = N$

Forming class intervals has enabled us to condense the data so that a simple
statement of the frequency distribution can be made in terms of only a few
classes. In this distribution, the various class intervals are shown in order in the

first column on the left. The extreme right column lists the frequencies, each paired with the class in that row of the table. The sum of the frequencies must be N, the total number of observations. The middle column contains the midpoint of each class interval; these will be discussed in Section 3.8.

3.6   CLASS INTERVAL SIZE AND CLASS LIMITS

The first problem in constructing a grouped frequency distribution is deciding upon a class interval size. Just how large shall the range of numbers included in a class be? The class interval size for a grouped distribution is the difference between the largest and the smallest number that may go into any class interval. We will let the symbol i denote this interval size.

In the example above, as in most examples to follow, the size i is the same for each class interval. This is the accepted convention for most work in psychology. (Some exceptions will be mentioned in Section 3.10, however.) For the example, $i = 6$, which means that the largest and the smallest score going into a class interval differ by six units. Take the class interval labelled 100–105. It may seem that the smallest number going into this interval is 100, and the largest 105; this is not, however, true. Actually, the numbers 99.5, 99.6, 99.7 would also go in this interval, should they occur in the data. Also, the numbers 105.1, 105.2 are included. On the other hand, 99.2 or 105.8 would be excluded. The interval actually includes any number *greater than or equal to* 99.5 *and less than* 105.5. These are called the **real limits** of the interval, in contrast to those actually listed, which are the **apparent limits.** Thus, the real limits of the interval 100–105 are

$$99.5 \text{ to } 105.5$$

and

$$i = 105.5 - 99.5 = 6.$$

In general,

real lower limit = apparent lower limit minus .5(unit difference)
real upper limit = apparent upper limit plus .5(unit difference).

The term "unit difference" demands some explanation. In measuring something we usually find some limitation on the accuracy of the measurement, and seldom can one measure with *any* desired degree of accuracy. For this reason, measurement in numerical terms is always rounded, either during the measurement operation itself, or after the measurement has taken place. In measuring weight, for example, we obtain accuracy only within the nearest pound, or the nearest one tenth of a pound, or one hundredth of a pound, and so on. If one were constructing a frequency distribution where weight had been rounded to the nearest pound, then a unit difference is one pound, and the real limits of an interval such as 150–190 would be 149.5–190.5. The i here would be 41 pounds.

On the other hand, suppose that weights were accurate to the nearest one tenth of a pound; then the unit difference would be .1 pound, and the real class interval limits would be

$$149.05-190.05.$$

The way that rounding has been carried out will have a real bearing on how the successive class intervals will be constructed. Suppose that we wanted to construct the distribution of annual income for a group of American men, where the income figures have been rounded to the nearest thousand dollars. We have decided that the apparent upper limit of the top interval shall be 25,000 dollars, and that i should equal 5,000. Then the classes would have the following limits:

Real	Apparent
20,500–25,500	21,000–25,000
15,500–20,500	16,000–20,000
10,500–15,500	11,000–15,000
5,500–10,500	6,000–10,000
500– 5,500	1,000– 5,000

Here, one half of a unit difference is 500 dollars.

Now suppose that each man's income has been rounded to the nearest one hundred dollars. Once again, with $i = \$5,000$ and with the top apparent limit \$25,000, we form class intervals, but this time the class limits are:

Real	Apparent
20,050–25,050	20,100–25,000
15,050–20,050	15,100–20,000
10,050–15,050	10,100–15,000
5,050–10,050	5,100–10,000
50– 5,050	100– 5,000

This difference in unit for the two distributions could make a difference in the "picture" we get of the data. For example, a man making exactly \$20,100 would fall into the top interval in the second distribution, but into the second from top interval in the first. It is important to decide upon the accuracy represented in the data before the real limits for the class intervals are chosen. Is every digit recorded in the raw data regarded as significant, or will the data be rounded to the "nearest" unit?

3.7 INTERVAL SIZE AND THE NUMBER OF CLASS INTERVALS

One can use any number for i in setting up a distribution. However, it should be obvious that there is very little point in choosing i smaller than the unit difference (for example, letting $i = .1$ pound when the data are in nearest whole pounds), or in choosing i so large that all observations fall into the same class interval. Furthermore, i is usually chosen to be a *whole number* of units, whatever the unit may be. Even within these restrictions, there is considerable flexibility of choice. For example, the data of Section 3.5 could have been put into a distribution with $i = 3$, as shown in Table 3.7.1. Notice that this distribution with $i = 3$ is somewhat different from the distribution with $i = 6$, even though they are based on the same set of data. For one thing, here there are sixteen class intervals, whereas there were eight before. This second distribution also gives somewhat more detail than the first about the original set of data. We now know, for example, that there was no one in the group of persons who got an IQ score between the real limits of 83.5 and 86.5, whereas we could not have told this from the first distribution. We can also tell that more cases showed IQs between 101.5 and 104.5 than between 98.5 and 101.5; conceivably this could be a fact of some importance. On the other hand, the first distribution gives a simpler and, in a sense, a neater picture of the group than does the second.

Table 3.7.1

Class interval	Midpoint	f
126–128	127	5
123–125	124	3
120–122	121	0
117–119	118	0
114–116	115	7
111–113	112	3
108–110	109	8
105–107	106	13
102–104	103	44
99–101	100	20
96– 98	97	10
93– 95	94	12
90– 92	91	14
87– 89	88	9
84– 86	85	0
81– 83	82	2
		$150 = N$

The decision facing the maker of frequency distributions is, "Shall I use a small class-interval size and thus get more detail, or shall I use a large class-interval size and get more condensation of the data?" There is no fixed answer to this question; it depends on the kind of data and the uses to which they will be put. However, a convention does exist which says that for most purposes *ten to twenty class intervals* give a good balance between condensation and necessary detail, and in practice one usually chooses his class-interval size to make about that number of intervals.

A handy rule of thumb for deciding about the size of class intervals is given by

$$i = \frac{\text{highest score in data} - \text{lowest score}}{\text{number of class intervals}}.$$

After deciding on some convenient number of class intervals, you divide this number into the difference between the highest and lowest scores, or the **range** of scores, and find the size that i will have to be. In practice, this may give an i that is not a whole number, in which case you simply use the nearest whole number for i. In the example just given, the data showed 128 as the highest and 82 as the lowest scores, with a range of 46. To find i giving about 16 class intervals we would divide the range by 16:

$$\frac{128 - 82}{16} = 2.87.$$

Since this is a decimal number, we choose the nearest whole number, or 3.

Note that in the last distribution some intervals show a frequency of 0; it is not absolutely necessary that such intervals be listed. However, intervals with zero frequency are usually listed when they fall between other intervals that do not have a zero frequency. We could have had an interval 129–131 having a zero frequency in the last distribution; however, we did not list this interval because, unlike the interval 120–122, it did not fall between other nonzero intervals. In addition, it is important to take the total N into account when deciding upon the number and size of class intervals. It is really not very interesting to look at a distribution with fewer than five or so observations per class interval on the average, although in some exceptional situations this might be allowable. Usually, then, when fairly small numbers of observations are involved, it will pay to examine the average number of cases per class interval. This can be found from N/C or from iN/range. If this number comes out to be as small as 5 or less, consideration should be given to a larger interval size, or perhaps even to an unequal interval size over some intervals of values.

Two other conventions are useful in making distributions from data. The first is that the class intervals are usually listed starting with low numbers at the bottom of the list and with the high numbers at the top, as shown in the example. Another convention followed here is to start figuring the class intervals by listing the *highest score* in the data as the *apparent upper limit of the top interval;* then it is a simple matter to find the other apparent upper limits by subtracting the class-interval size successively from this number. The apparent *lower* limit of each interval is then *one unit more* than the *upper* apparent limit of the interval below it. For example, suppose 128 were the top score in the data, and that i were 3. Taking 128 as the highest apparent limit, the upper limit of the next interval would be $128 - 3$, or 125, next would be $125 - 3$ or 122, and so on, until all upper limits were found. Then, given an interval with apparent upper limit of 122, the lower limit of the next interval up would be 123, and given the interval with upper limit 125, the lower limit of the next interval up would be 126, and so on.

3.8   Midpoints of class intervals

There is one more feature of frequency distributions that has not yet been discussed. When one puts a set of measurements into a frequency distribution, he loses the power to say exactly what the original numbers were. For the example of Section 3.7, if a person in the group of 150 cases has an IQ of 102, he is simply counted as one of the 44 individuals who make up the frequency for the interval 102–104. You cannot tell from the distribution exactly what the IQs of those 44 individuals were, but only that they fell within the limits 102–104. What do we call the IQs of these 44 individuals? We call them all 103, the midpoint of the interval. **The midpoint of any class interval is that number which would fall exactly halfway among the possible numbers included in the interval.** A moment's thought will convince you that 103 falls halfway between the real limits of 101.5 and 104.5. In a like fashion, all 5 cases falling into the class interval 126–128 will be called by the midpoint 127; all 3 cases in the interval 123–125 will be called 124, and so on for each of the other class intervals.

The real limits can also be defined in terms of the midpoint:

$$\text{real limits} = \text{midpoint} \pm .5i.$$

Thus, given only the midpoints of the distribution, one can find the class limits.

In computations using grouped distributions the midpoint is used to substitute for each raw score in the interval. For this reason, it is convenient to choose an odd number for i whenever possible; this makes the midpoint a whole number of units and simplifies computations.

3.9   Another example of a grouped frequency distribution

Suppose that seventy-five students in high school had been used as subjects in an experiment on verbal memory. Each student was given a list containing forty-eight pairs of words and allowed to study the list as a whole for ten minutes. The first member of each pair was then shown in order to the individual student, and his task was to recall the second word. The raw data are shown in Table 3.9.1. The largest number of words recalled by any student was thirty-five, and the smallest number was ten. The range was thus $35 - 10$ or 25.

Now we want to make a frequency distribution having about ten class intervals. According to our rule of thumb, given above:

$$i = \frac{\text{range}}{\text{number of intervals}} = \frac{25}{10} = 2.5.$$

Thus $i = 3$ should provide us with a convenient class-interval size.

We start with the largest number, 35, and make it the upper apparent limit of the highest class interval. That interval must then have real limits of 32.5 and 35.5 with apparent limits of 33–35. The real limits of the second from

Table 3.9.1

SCORES OF SEVENTY-FIVE STUDENTS IN ONE TRIAL OF
PAIRED-ASSOCIATE LEARNING

Student	Score	Student	Score	Student	Score
1	19	26	18	51	15
2	16	27	14	52	15
3	11	28	26	53	14
4	13	29	16	54	11
5	12	30	19	55	23
6	20	31	21	56	13
7	11	32	17	57	19
8	24	33	20	58	17
9	19	34	12	59	21
10	16	35	22	60	15
11	12	36	11	61	20
12	24	37	13	62	16
13	19	38	16	63	12
14	17	39	10	64	10
15	18	40	17	65	10
16	25	41	17	66	15
17	16	42	19	67	18
18	20	43	11	68	10
19	17	44	10	69	19
20	19	45	15	70	14
21	15	46	10	71	12
22	35	47	13	72	18
23	16	48	13	73	14
24	11	49	14	74	15
25	26	50	10	75	20

highest interval must be 29.5–32.5, and so on. Each time, the difference between the real limits to an interval must be i, or 3, and the differences between the successive lower apparent limits (and also successive upper limits) must also be 3. Proceeding in this way, we find the class intervals shown in Table 3.9.2.

Table 3.9.2 is completed by inspecting the raw data to find the number of cases that fall into each class interval. Only one case falls between the real limits 32.5–35.5, and so the frequency for the highest interval is 1. The frequency for the lowest interval is 13, since exactly thirteen individuals showed scores between the real limits 8.5 and 11.5. Just as in the preceding examples, the midpoint for each interval stands exactly midway between the two real limits.

What can the experimenter tell from looking at this distribution that would not have been obvious from the raw data? First of all the "typical" range of performance is really a fairly short interval of scores; the large majority of students scored in the range of 11 points from 9 through 20. The most "popular" score interval is 15–17. Note that this concentration of cases lies at the low end of

Table 3.9.2

NUMBER OF PAIRED ASSOCIATES
RECALLED ON THE FIRST TRIAL
BY A SAMPLE OF SEVENTY-FIVE
HIGH SCHOOL STUDENTS

Class interval	Midpoint	f
33–35	34	1
30–32	31	0
27–29	28	0
24–26	25	5
21–23	22	4
18–20	19	17
15–17	16	20
12–14	13	15
9–11	10	13
		75 = N

the *conceivable range* of scores 0 through 48. If the experimenter thinks of a score as reflecting the student's ability to memorize paired associates, then the task is, by and large, a hard one for the students. The single individual scoring in the interval 33–35 is really most atypical. His score falls very far from the bulk of the cases in the distribution.

Naturally, there are other, more precise statements that the experimenter can make about what this distribution shows. We will discuss these summary indices in Chapter 6. For the moment, however, it should be clear that a distribution does communicate information about a set of observations *as a whole* even without further analysis of the data.

3.10   FREQUENCY DISTRIBUTIONS WITH OPEN OR UNEQUAL CLASS INTERVALS

Quite often the data are such that it is not possible to make a frequency distribution with intervals of a constant size. This most commonly occurs in psychology when exact scores are not known for some of the cases. For example, in a study of the trials that it takes an animal to learn a discrimination problem the experimenter found that out of 100 animals, 5 could not learn the problem in sixty trials. He felt that some of the animals would never learn the problem at all, and he ceased running these animals at the sixtieth trial. This means, however, that the 5 animals could not be given an exact score, and could only be put in the top class of the score distribution, in an interval called "sixty or more." This interval is **open,** since there is no way to determine its upper real limit. Furthermore, there is no way to give such an open interval a midpoint.

It is perfectly correct to show open intervals at either end of a frequency distribution. However, no computations involving the midpoints can be carried out from this distribution, unless the cases in the open intervals are dropped from the sample.

In other instances, it may be that there are extreme scores in a distribution that are widely separated from the bulk of the cases. When this is true, a class-interval size that is small enough to show "detail" in the more concentrated part of the distribution will eventuate in very many classes with zero frequency before the extreme scores are included. Enlarging the class-interval size will reduce the number of unnecessary intervals, but will also sacrifice detail. When this situation arises, it is often wise to have a varying class-interval size, so that the class intervals are narrow where the detail is desired, but rather broad toward the extreme or extremes of the distribution. In this case, class-interval limits and midpoints can be found in the usual way, provided that one takes care to notice where the interval size changes. Furthermore, midpoints of these intervals can be used for further computations using the methods given in Chapter 6.

3.11   GRAPHS OF DISTRIBUTIONS: HISTOGRAMS

Now we direct our attention for a time to the problem of putting a frequency distribution into graphic form. Often a graph shows that the old bromide, "a picture is worth a thousand words," is really true. If the purpose is to provide an easily grasped summary of data, nothing is so effective as a graph of a distribution.

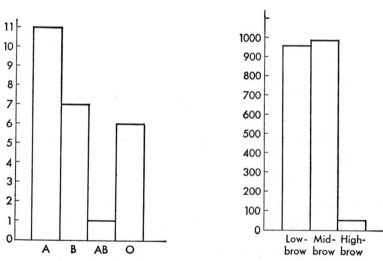

FIG. 3.11.1 (*left*). Blood types of twenty-five American males. FIG. 3.11.2 (*right*). Taste ratings for two thousand American families.

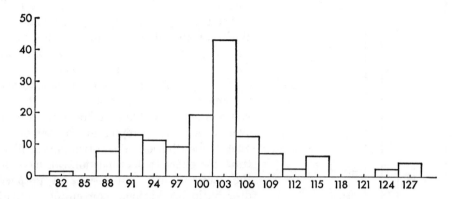

FIG. 3.11.3. Intelligence quotients of a sample of one hundred and fifty subjects.

There are undoubtedly all sorts of ways any frequency distribution might be graphed. A glance at any national news magazine will show many examples of graphs which have been made striking by some ingenious artist. However, we shall deal with only three "garden varieties": the histogram, the frequency polygon, and the cumulative frequency polygon.

The histogram is really a version of the familiar "bar graph." In a histogram, each class or class interval is represented by a portion of the horizontal axis. Then over each class interval is erected a bar or rectangle equal in height to

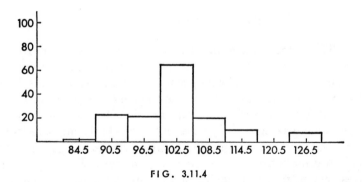

FIG. 3.11.4

the frequency for that particular class or class interval. The histograms representing the frequency distributions in Sections 3.4 and 3.7 are shown in Figures 3.11.1–3. It is customary to label both the horizontal and the vertical axes as shown in these figures, and to give a label to the graph as a whole. In the case where class intervals are employed to group numerical measurements, a saving of time and space may

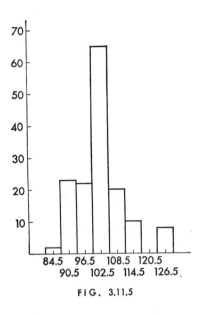

84.5   96.5   108.5   120.5
   90.5   102.5   114.5   126.5

FIG. 3.11.5

be accomplished by labeling the class intervals by their midpoints (as in Figure 3.11.3) rather than by their apparent or real limits. With categorical measurements, this problem does not arise, of course, since each class can only be given its name or symbol.

If you will look at the two histograms shown in Figures 3.11.4 and 3.11.5, you will get very different first impressions, even though they represent the same frequency distribution. These two figures illustrate the effect that proportion can have on the viewer's first impressions of graphed distributions. Various conventions about graphs of distributions exist in different fields, of course. In fact, those who make a business of using statistics for persuasion often choose a proportion designed to give a certain impression. However, in psychology, it is most usual to find graphs arranged so that the vertical axis is about three fourths as long as the horizontal. This usually will give a clear and esthetically pleasing picture of the distribution. (An amusing and instructive recital of ways to "adjust" graphs and other statistical presentations so as to get a desired effect is given in the little book by Huff, *How To Lie with Statistics*.)

3.12 FREQUENCY POLYGONS

The histogram is a useful way to picture any sort of frequency distribution, regardless of the scale of measurement of the data. However, a second kind of graph is often used to show frequency distributions, particularly those that are based upon numerical data: this is the frequency polygon. In order to construct the frequency polygon from a frequency distribution one proceeds exactly as though one were making a histogram; that is, the horizontal axis is marked off into class intervals and the vertical into numbers representing frequencies. However, instead of using a bar to show the frequency for each class, a point on the graph corresponding to the midpoint of the interval and the frequency of the interval is found. These points are then joined by straight lines, each being connected to the point immediately succeeding and the point immediately following, as in Figure 3.12.1.

The frequency polygon is especially useful when there are a great many potential class intervals, and it thus finds its chief use with numerical data that could, in principle, represent a continuous variable. For a distribution based on a very large number of numerical measurements on a potentially continuous

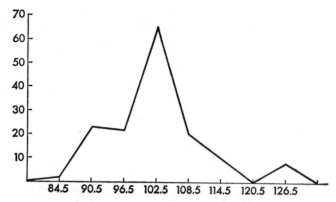

FIG. 3.12.1. Intelligence quotients of a sample of one hundred and fifty subjects.

scale, if we maintained the same proportions in our graph, but employed a much greater number of class intervals, the frequency polygon would provide a picture more like a smooth curve, as shown in Figures 3.12.2–4, based on several thousand cases. Although the function rule describing the frequency polygon may be extremely complicated to state, the function rule for a smooth curve approximately the same as that of the distribution may be relatively easy to find. For this reason, a frequency polygon based upon interval-or ratio-scale measurement with a relatively large number of classes is sometimes "smoothed" by creating a curve that approximates the shape of the frequency polygon. This new curve may have a rule that is simple to state and that may serve as an approximation of the rule

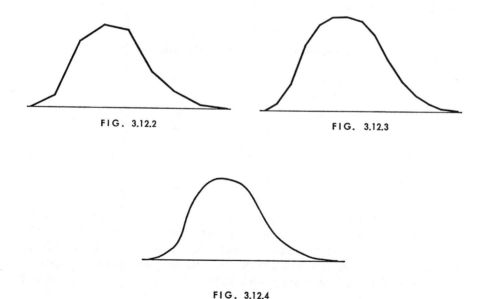

FIG. 3.12.2

FIG. 3.12.3

FIG. 3.12.4

describing the frequency distribution itself. While the procedures for smoothing such a frequency polygon and finding a function rule that comes close to describing the distribution are outside the scope of this book, they may be found in many advanced texts on the analysis of experimental data. Suffice it to say, for the moment, that when we come to a discussion of the "shapes" of distributions, we shall often refer to smooth curves as examples of types of distributions, when actually the frequency distributions themselves could only be represented by "jagged" polygons composed of straight lines. Nevertheless we may use the smooth curves as good approximations to the picture that the frequency polygon presents of the distribution.

3.13   CUMULATIVE FREQUENCY DISTRIBUTIONS

For some purposes it is convenient to make a different arrangement of data into a distribution, called a cumulative frequency distribution. Instead of showing the relation that exists between a class interval and its frequency, **a cumulative distribution shows the relation between a class interval and the frequency at or below its real upper limit.** In other words, the cumulative frequency shows how many cases fall into *this interval or below*. Thus, in the distribution in Section 3.5, the lowest interval shows two cases, and its cumulative frequency is 2. Now the next class interval has a frequency of 23, so that its cumulative frequency is 23 + 2, or 25. The third interval has a frequency of 22, and its cumulative frequency is 22 + 23 + 2, or 47, and so on. The cumulative frequency of the very top interval must always be N, since all cases must lie either in the top interval or below. The cumulative frequency distribution for the example of Section 3.5 is given below.

Class interval	Cumulative frequency
124–129	150
118–123	142
112–117	142
106–111	132
100–105	112
94– 99	47
88– 93	25
82– 87	2

Cumulative frequency distributions are often graphed as polygons. The axes of the graph are laid out just as for a frequency polygon, except that the class intervals are labeled by their *real upper limits* rather than by their midpoints. The numbers on the vertical axis are now cumulative frequencies, of course. The graph of the example from Section 3.5 is shown in Figure 3.13.1. Cumulative fre-

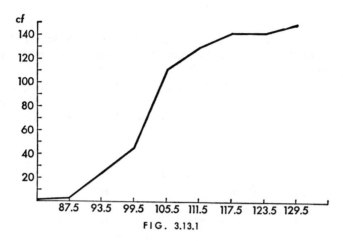

FIG. 3.13.1

quency distributions often have more or less this characteristic **S** shape, although the slope and the size of the "tails" on the **S** will vary greatly from distribution to distribution. Like a frequency polygon, a cumulative frequency polygon is sometimes smoothed into a curve as similar as possible to the polygon. You sometimes encounter a reference to such a smoothed cumulative distribution curve under the term **ogive**.

3.14 PROBABILITY DISTRIBUTIONS

Suppose that there were some finite set of objects, and that we sampled these objects at random with replacement. Each object sampled is measured in some way. The measurement may be at the nominal-, ordinal-, interval-, or ratio-scale level, so long as each object is measured by the same procedure. *Now notice that the sampling of each object is an elementary event, and that the actual measurement class or number assigned each object is an event.* Since many different elementary events can be allotted the same number or classification, each measurement class is an event class. These measurement events are mutually exclusive and exhaustive, each object being given one and only one number or category.

Furthermore, since the different possible measurement classes define events, there is a probability that can be assigned to each; there is, for example, some probability that the score of an object will be 10, or that the object will be assigned to the class "high need for achievement." In short, given the sample space S and a set of measurement events, we can think of a probability function pairing some probability with each possible measurement class.

Any statement of a probability function having a set of mutually exclusive and exhaustive events for its domain is a probability distribution.

Notice how the idea of probability distribution exactly parallels that of frequency distribution: each is a function on a set of mutually exclusive and exhaustive measurement classes. In a frequency distribution, each measurement class is paired with a frequency, and in a probability distribution each is paired with its probability. Just as the sum of the frequencies must be N in a distribution, so must the sum of the probabilities be 1.00.

As a simple example of a probability distribution, imagine a sample space of all boys of high-school age. A boy is selected at random and classified as "right-handed," "left-handed," or "ambidextrous." The probability distribution might be:

Class	p
right	.60
left	.30
ambidextrous	.10
	1.00

Or, perhaps, the height of a boy drawn at random is measured. Some seven class intervals are used to record the height of any boy, and the probability distribution might be

Height in inches	p
78–82	.002
73–77	.021
68–72	.136
63–67	.682
58–62	.136
53–57	.021
48–52	.002
	1.000

Here, each class interval is an event.

Corresponding to any frequency distribution of N cases there is a probability distribution, if single cases are selected from the total group of N cases at random with replacement. Thus, imagine a set of 50 persons, whose scores on some test formed the following distribution:

Scores	f
90–99	3
80–89	8
70–79	15
60–69	14
50–59	10
	50

One person is drawn at random from among this set. What are the probabilities associated with the various intervals he might fall into? If the individuals are all assumed equally likely to be observed, then for any class interval A,

$$p(A) = \frac{\text{frequency of } A}{\text{total frequency}}.$$

Thus, the class interval 90–99 has a probability of 3/50, or .06, and the other probabilities can be found in the same way, giving

Scores	p
90–99	.06
80–89	.16
70–79	.30
60–69	.28
50–59	.20
	1.00

In short, any group of cases summarized in a frequency distribution can also be regarded as a sample space of *potential* observations. If a single case is sampled at random, then the probabilities of the various classes are the same as the relative frequencies of those classes.

Furthermore, given some theoretical probability distribution, and N observations made independently and at random with replacement, then there is a *theoretical frequency distribution* as well, where for any event class A

$$\text{theoretical frequency} = Np(A).$$

This is the frequency distribution we *expect* to observe, given the theoretical probability distribution. For example, suppose that the probability distribution of heights of boys given above were true for some specific population of boys. What is the theoretical frequency distribution of heights for 1000 boys sampled independently and at random? The answer is

Height in inches	$f = Np(A)$
78–82	2
73–77	21
68–72	136
63–67	682
58–62	136
53–57	21
48–52	2
	1000

Given that the probability distribution is correct, then for a sample of 1000 boys we should expect our obtained frequency distribution to show 682 boys in the interval 63–67 inches, only 2 boys in the interval 78–82, and so on. This distribution might never actually occur in practice, but it is our best guess about what will occur if the probability distribution is a correct statement. Whereas the frequencies in real distributions must always be whole numbers, in theoretical distributions it is possible to have fractional frequencies; if only 100 boys were observed in this example, then the frequencies for the top and bottom intervals would be, in theory, .2 each.

Like frequency distributions, a probability distribution may be specified either by a listing such as we have shown here, or in graphic form. However, unlike actual obtained distributions of data, the most important probability distributions are theoretical ones that can be given a mathematical rule. In order to discuss such distributions we need the concept of a "random variable."

3.15 RANDOM VARIABLES

A special terminology is useful for discussing probability distributions of numerical scores. Imagine that a single number could be assigned to each and every possible elementary event in some S. Various elementary events are paired with various values of a variable. Thus, an elementary event may be a person, with some height in inches; or the elementary event may be the result of tossing a pair of dice, with the assigned number being the total of the spots that come up; or the elementary event may be a rat, with the number standing for the trials taken to learn a maze. Each and every elementary event is thought of as getting one and only one such number.

Let X represent a function that associates a real number with each and every elementary event in some sample space S. Then X is called a random variable on the sample space S.

Bear in mind the distinction between the elementary events themselves, which may or may not be numbers, and the values associated with the various elementary events, which are the values of the random variable. Thus, for example, consider the set of American males 21 years or older. An individual is drawn at random from this set. The sample space here consists of the entire set of American males, and each such male is an elementary event in this sample space. Now we can define a function X on this sample space that associates with each elementary event a real value, the income of the man during the current calendar year. The values that the random variable X can thus assume are the income values associated with the men. The value x occurs when a man is chosen who has income x.

On the other hand, suppose that a box were filled with slips of paper, each inscribed with a real number. Here the elementary events are the slips bearing the numbers. We might define X to be the function assigning numbers to particular slips, the value x being associated with a particular slip, so that x occurs when the particular slip is drawn at random. However, any number of other functions might be chosen for X. Thus, we might define X as the square of the number on a slip, so that $X = 4$ occurs when a slip is drawn bearing the number 2. The point is that a random variable represents values that are associated in some way with elementary events, so that particular values of X occur when the appropriate elementary events occur. The way in which the numerical values get associated with the elementary events is open to the widest latitude of definition. The numerical values really need have nothing in common with the elementary events at all except by arbitrary definition. Thus, we might toss a penny and assign the number 335 to Heads and -62 to Tails. Then X, which can take on the values 335 and -62, is a random variable on the sample space consisting of the two elementary events, Heads and Tails.

Although the term "random variable" is rather awkward, it will be used here because of its popularity in statistical writing. The terms "chance variable" or "stochastic variable" are sometimes used. These all mean precisely the same thing, however; a symbol for number events each having a probability.

Given the random variable X, other events may be defined. Thus, given two numbers a and b,

$$(a \leq X \leq b)$$

is the event of some value of X lying between the numbers a and b (inclusive). This event has some probability $p(a \leq X \leq b)$. Furthermore, there is the event

$$(a \leq X),$$

a value greater than or equal to a; this has probability $p(a \leq X)$.

This leads us to an alternate, and somewhat more modern, definition of a random variable:

A quantity symbolized by X is said to be a random variable if for every real number a there exists a probability $p(X \leq a)$ that X takes on a value that is less than or equal to a.

As we shall see, this definition has its advantages when the variable X is continuous, ranging over all the real numbers, although there is nothing in this definition that says that the random variable *must* be continuous.

As another example of a random variable, suppose that X symbolizes the height of an American man, measured to the nearest inch. Here, there is some probability that $X \leq 60$, or $p(X \leq 60)$, since there is presumably some set of American men with heights less than or equal to 60 inches. Furthermore, there is some probability $p(70 \leq X \leq 72)$, since there is a set of American men having heights between five feet ten inches and six feet, inclusive. Here, X symbolizes the numerical value (height in inches) assigned to an American man (an elementary event). This symbol X represents any one of many different such values, and for any arbitrary number a there is some probability that the particular value x is less than or equal to the value of a.

Although the notation employed for random variables varies considerably among various authors, we will find it convenient to use capital letters, such as X, Y, Z, to denote random variables, and lower-case letters, such as x, y, z and a, b, c, to denote particular values that the random variable in question can take on. (Occasionally, this convention becomes awkward, and it will be violated from time to time in future sections. However, the context will usually make the random variable notation clear in these instances.) The expression $p(X = x)$ symbolizes the probability that the random variable X takes on the particular value x. Often, for convenience, this will be written simply as $p(x)$. In the same way, $p(a \leq X)$ stands for the probability that X takes on some particular value x greater than or equal to the value symbolized by a, $p(a \leq X \leq b)$ symbolizes the event in which X takes on some value lying between the two values symbolized by a and b, and so on.

3.16 Discrete random variables

In very many situations a random variable X can assume only a finite, or a countably infinite, set of values. Speaking rather crudely, the range of such a random variable has "holes" in it, so that some values have zero probability even though there are values to either side that have probability greater than zero. The set of possible values may be either finite in extent, or the possible values themselves, even though infinite in extent, can be put into 1-to-1 correspondence with the counting numbers 1, 2, 3, $\cdot \cdot \cdot$ and so on. As an example of a finite set of values that a random variable can assume, suppose that a volume of the *Encyclopedia Britannica* is opened at random, and the page number noted. In this instance, the values the random variable could assume would be all whole numbers between 1 and, say, 1009 (with a few Roman numerals thrown in, perhaps). Certainly, the number of values is finite. Numbers such as 34.6 and 1.1103 would not occur, of course.

On the other hand, some random variables can assume a countably infinite number of values. Suppose that the simple experiment consists of drawing cards with replacement from a well-shuffled deck until the queen of hearts first appears. Let the random variable X be the trial number on which this card turns up for the first time. It could happen on the first trial, or the second, or the third, or the 9069th or the nine-millionth, or never, short of an infinite number of trials. Then the values of X here are countably infinite in extent. Certainly the trials can be counted, just as the pages can be counted in the previous example. The difference is that in principle we might never run out of *possible* trial numbers, or values of X.

This leads us to our definition of a discrete random variable: **If a random variable X can assume only a particular finite or countably infinite** range of values, it is said to be a discrete random variable. As we shall see, not all random variables are discrete, but a large number of random variables of practical and theoretical interest to us will have this property.

Given a discrete random variable, which can take on only the K different values $a_1, a_2, \cdots, a_K$, then it must be true that

$$p(X = a_1) + p(X = a_2) + \cdots + p(X = a_K) = 1.00.$$

Naturally, it is also true that $p(X = a_i) \geqq 0$, for $i = 1, 2, \cdots, K$, since no probability may have a value that is less than zero. Both of these conditions follow directly from the basic definition of a probability function in Chapter 2.

Quite often one is interested in finding the probability that the obtained value of X will fall *between* two particular values, or in some interval. For example, the number of spots coming up on a pair of dice is a discrete random variable taking on only the whole number values 2 through 12 (the probabilities of 0 and of 1 are both zero, of course). What is the probability that X, the number of spots, is between 3 and 5 inclusive?

In order to find such probabilities, we rely once again on rule 3 of Section 2.5. The various possible values of X are mutually exclusive events, and so

$$p(3 \leqq X \leqq 5) = p(3 \cup 4 \cup 5) = p(3) + p(4) + p(5).$$

If X is the number of spots on a pair of dice, then the random variable can be shown to have the distribution of Table 3.16.1.

Table 3.16.1

x	$p(x)$
12	1/36
11	2/36
10	3/36
9	4/36
8	5/36
7	6/36
6	5/36
5	4/36
4	3/36
3	2/36
2	1/36
	36/36

Using these probabilities, we find

$$p(3 \leqq X \leqq 5) = p(X = 3) + p(X = 4) + p(X = 5)$$
$$= \frac{2 + 3 + 4}{36} = \frac{1}{4}.$$

Similarly,

$$p(9 \leq X \leq 10) = p(X = 9) + p(X = 10) = \frac{7}{36}.$$

In general, the probability that X falls in the interval between any two numbers a and b, inclusive, is found by the sum of probabilities for X over *all possible* values between a and b, inclusive:

$$p(a \leq X \leq b) = \text{sum of } p(X = c) \text{ for all } c \text{ such that } a \leq c \leq b.$$

By the same argument,

$$p(a \leq X) = \text{sum of } p(X = c) \text{ for all } c \text{ such that } a \leq c.$$

Furthermore,

$$p(X < a) = 1 - p(a \leq X).$$

For this example,

$$p(5 \leq X) = p(X = 5) + \cdots + p(X = 12) = \frac{4}{36} + \cdots + \frac{1}{36}$$

$$= \frac{30}{36} = \frac{5}{6},$$

$$p(X < 5) = p(X = 4) + p(X = 3) + p(X = 2)$$

$$= \frac{3}{36} + \frac{2}{36} + \frac{1}{36} = \frac{1}{6}.$$

If cases are sampled from any frequency distribution that has been grouped into class intervals, then the random variable X is usually regarded as symbolizing the midpoints of the intervals. For a finite number of class intervals, such an X is a discrete random variable. Also, when measurements have been rounded to the *nearest* unit, as in height to the nearest inch, this is like forming class intervals, and the random variable standing for such a rounded measurement is often thought of as discrete.

3.17   GRAPHS OF PROBABILITY DISTRIBUTIONS

The same general procedure is used for graphing either a frequency distribution or the probability distribution of a discrete random variable. The two most common forms are histograms and probability mass functions.

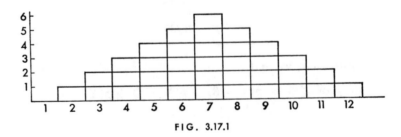

FIG. 3.17.1

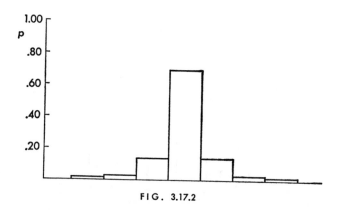

FIG. 3.17.2

First, let us deal with the histogram. When the random variable is discrete and can take on only a few values, it is customary to show each value as an interval on the X axis, with the height of the bar above that interval indicating the probability. For example, the distribution in Section 3.16 gives the histogram of Figure 3.17.1. Notice that each of the blocks making up the column above a value of X has exactly the same area, and the number of blocks divided by the total number gives the probability. In other words, if each block had area of 1/36, the total area would be 1.00, and the area of any column would be the probability of that value of X.

This idea extends to the histogram of any probability distribution: the total area covered by all columns is regarded as 1.00, and the area of any single column representing an interval or class is thought of as a probability for that interval or class. Another example is given by Fig. 3.17.2.

The use of a histogram for the distribution of a discrete random variable emphasizes the analogy between probabilities and relative areas, with total probability corresponding to total area in the graph. The values on the X axis to either side of any midpoint of an interval are treated as though they could occur with the same probability as the midpoint itself. This is, of course not generally the case with a discrete random variable, although this convenient fiction is adopted in order that the connection between areas and probabilities can be seen.

An alternate method of graphing the probability distribution for a discrete random variable emphasizes the discreteness of the random variable at the expense of the analogy between probability and area. This alternate form of graph is called a **probability mass function**. An example of a probability mass function is shown in Figure 3.17.3. (In much statistical writing, probability mass function, or PMF, refers both to the rule for the probability function of a discrete random variable and to the graphic representation of such a function. Since the function rule and the graph are only different ways of specifying the same thing, the probability distribution associated with a given discrete random variable, this should cause no undue confusion.)

Note that in the probability mass function graphed in Figure 3.17.3 the same basic features exist as in a histogram. The X axis of the graph is marked

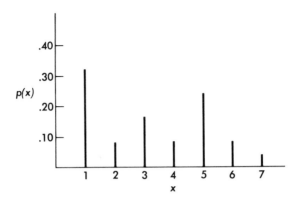

F I G . 3.17.3. Probability mass function for a discrete random variable

off with the values that the random variable can actually assume. The Y axis, which may or may not be explicitly shown, is a measure of units of probability. Then above each possible value x of X a vertical line is raised to the height corresponding to $p(X = x)$. The discreteness of the random variable is emphasized by the gaps between the vertical lines standing for the various possible values of X. A value corresponding to a point in one of these gaps has zero probability.

The probabilities, or masses, in any such probability mass function must be nonnegative, of course, and the sum of the probabilities over all possible values of X must be equal to 1.00.

3.18   FUNCTION RULES FOR DISCRETE RANDOM VARIABLES

In many instances, particularly in theoretical statistics, it will be much more convenient to specify the distribution of a discrete random variable by the rule for its probability function, rather than by a simple listing or by a graph. Three very simple examples will show the form that such function rules often take. Later we will deal with much more complicated function rules, but the general ideas presented here will still obtain.

Suppose that we are interested in a discrete random variable X, which can take on only one of K different values on any occurrence, the integers $1, 2, \cdots, K$. Furthermore, the probability for the occurrence of any particular value from among these K possible values is exactly the same as for any other. How would one symbolize the rule for this random variable? The answer is

$$p(x) = \begin{cases} 1/K, & \text{if } x = 1, 2, 3, \cdots, K, \\ 0, & \text{otherwise.} \end{cases}$$

You can check for yourself to see that the requirements of a probability function

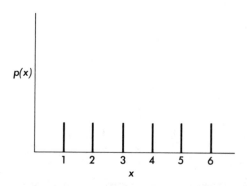

FIG. 3.18.1. Probability mass function for a discrete random variable, where p(x) = 1/6, for x = 1,2, . . . 6

are met for this random variable: the probability of any value is nonnegative, and the sum of probabilities over all possible values is 1.00. Figure 3.18.1 shows the probability mass function for the case in which $K = 6$, so that the possible values of X are 1, 2, 3, 4, 5, 6.

As another example, consider the discrete random variable X, which, once again, can assume only the values 1, 2, 3, 4, 5, and 6. This time, however, the probability function for the random variable obeys the following rule:

$$p(x) = \begin{cases} x/G, & \text{if } x = 1, 2, 3, 4, 5, 6, \\ 0, & \text{otherwise.} \end{cases}$$

This time there is a little problem to be solved before the exact probability values can be found. We must determine the value of the constant G. This is easy enough to do if we recall that the sum of all of the probabilities must be equal to 1.00, if this is to be a legitimate probability function. Thus, take the following

$$\text{sum of } p(x) \text{ over } 1, 2, 3, 4, 5, 6 = \frac{1 + 2 + 3 + 4 + 5 + 6}{G}$$

or

$$\frac{21}{G} = 1.00 \quad \text{so that} \quad G = 21.$$

Thus $p(X = 1) = 1/21$, $p(X = 2) = 2/21$, and so on. The probability mass function is graphed as Figure 3.18.2.

Finally, once again consider a random variable X that can take on the values 1, 2, 3, 4, 5, and 6. This time, however, let the function rule be as follows:

$$p(x) = \begin{cases} x/G, & \text{if } x = 1, 2, 3, \\ (7 - x)/G, & \text{if } x = 4, 5, 6, \\ 0, & \text{otherwise.} \end{cases}$$

Again, we need to find the value of the constant G, so that the sum of all of the

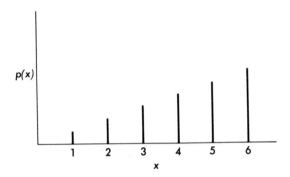

F I G . 3.18.2. Probability mass function where p(x) = x/21 for x = 1,2,3,4,5,6

probabilities will be equal to 1.00. Thus:

$$\frac{1}{G} + \frac{2}{G} + \frac{3}{G} + \frac{7-4}{G} + \frac{7-5}{G} + \frac{7-6}{G} = 1.00$$

so that

$$G = 12.$$

This probability mass function is shown as Figure 3.18.3. Note how this rule produces a very different picture from those given by the first two rules.

These are among the simplest of all function rules for discrete random variables, of course. Those of ultimate interest for us will tend to be more elaborate. Nevertheless, even though the possible function rules for discrete random variables are limitless in their variety, they all show the same basic features. These are rules for pairing with each possible value of the random variable, $X = x$, a probability $p(X = x)$ that the random variable takes on that value. Each probability provided by the rule is nonnegative, and the sum of the probabilities over all possible values of X must be equal to 1.00.

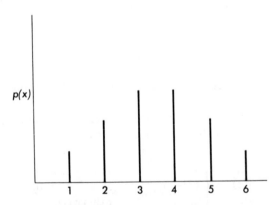

FIG. 3.18.3. Probability mass function where p(x) = x/12 for x = 1,2,3 and p(x) = (7 − x)/12 for x = 4,5,6

3.19   CONTINUOUS RANDOM VARIABLES

Just as an intuitive picture of a continuous function consists of a smooth curve, without gaps, breaks, or sudden jumps, so the idea of a continuous random variable should also suggest a smooth curve. However, the distribution of a continuous random variable has a slightly different interpretation from that of a discrete variable, and thus demands a somewhat different terminology.

We can set the stage for the discussion of continuous random variables by considering a variable X whose values are grouped into intervals. Consider once again the examples of heights from Section 3.14. In Figure 3.19.1, each interval of values is shown as the corresponding interval on the X-axis. If the total area of the histogram is equal to 1.00, as it should be, then the area of any column is the probability of the corresponding interval.

Now suppose that we were able to measure height to any degree of precision, regardless of how many decimal places this might involve. In other words, suppose that our measurements, and the population being measured, were such that we could choose any class interval size i for the random variable X, and still have a nonzero probability that values in the interval would occur, so long as the interval lay between two extreme values, say u and v. In this instance we could arrange measurements of height according to a scheme of class intervals of any size, however small, and still have a chance of observing a case in any interval.

For an interval size smaller than the $i = 5$ used above, say $i = 1$, our histogram might look like that shown in Figure 3.19.2. Here the number of intervals is larger and the histogram area for each is smaller. Suppose that the class interval size were made *extremely* small, $i = \Delta X$, still under the assumption that any class interval within the limits u and v will have a probability greater

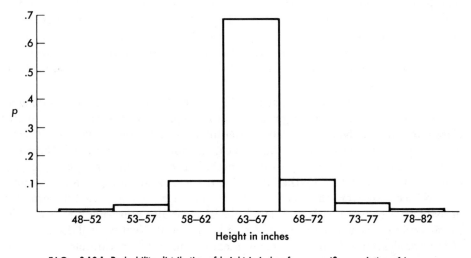

FIG. 3.19.1. Probability distribution of height in inches for a specific population of boys

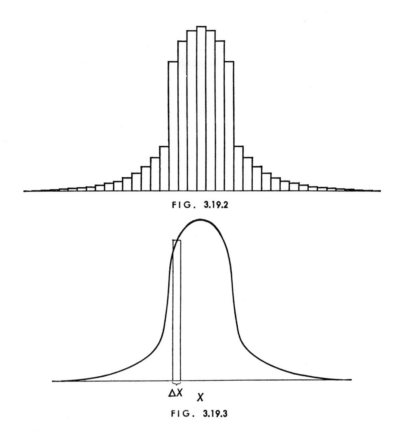

FIG. 3.19.2

FIG. 3.19.3

than zero. In this circumstance each class interval will have an extremely small corresponding area, and the area for the histogram as a whole would be very nearly the same as enclosed in a smooth curve (Fig. 3.19.3). The width of the bar shown on the graph is a representative class interval. Notice that the area cut off by the interval with width ΔX under the curve is almost the same as the area of the histogram bar itself; that is, the area under the curve for that interval is almost the same as the probability.

Finally, imagine that the class-interval size is made *very nearly zero*. As the class-interval size ΔX approaches zero, the probability associated with any class interval and the area cut off by the interval under the curve should agree increasingly well. A random variable that can be represented in terms of arbitrarily small class intervals of size ΔX, with the probability of any interval corresponding *exactly* to the area cut off by the interval under a smooth curve, is called **continuous.** Notice, however, that as the class-interval size ΔX approaches zero, the probability associated with any class interval must also approach zero, since the corresponding area under the curve is steadily reduced. Next we will discuss some of the ramifications of this property of a continuous random variable.

In Section 3.18 it was possible for us to discuss discrete random variables in terms of probabilities $p(x)$, the probability that the random variable takes on exactly some value. However, the fact we have just illustrated, that the area in any interval under a smooth curve approaches zero when interval size goes toward zero, makes it necessary to discuss continuous random variables in a special way.

Consider the example of an infinite set of men weighed with any degree of accuracy. What is the probability that a man will be observed at random weighing *exactly* 160 pounds? This event is an interval of values smaller than 159.5–160.5, or 159.95–160.05, or any other interval $160 \pm .5i$, where i is not exactly zero. Since the smaller the interval, the smaller the probability, the probability of exactly 160 pounds is, in effect, zero. **In general, for continuous random variables, the occurrence of any exact value of X may be regarded as having zero probability.**

For this reason, one does not usually discuss the probability *per se* for a value of a continuous random variable; probabilities are discussed only for *intervals* of X values in a continuous distribution. Instead of the probability that X takes on some value a, we deal with the so-called **probability density** of X at a, symbolized by

$$f(a) = \text{probability density of } X \text{ at } a.$$

A fairly "rough" definition of a probability density can be given as follows: for any distribution it is proper to speak of the probability associated with an interval. Imagine an interval with limits a and b. Then the probability of that interval is

$$p(a \leq X \leq b) = p(X \leq b) - p(X < a).$$

Now suppose that we let the size of the interval be denoted by

$$\Delta X = b - a$$

so that

$$b = a + \Delta X.$$

Then the probability of the interval *relative* to ΔX is just

$$\frac{p(a \leq X \leq b)}{\Delta X} = \frac{p(X \leq a + \Delta X) - p(X < a)}{\Delta X}.$$

Now suppose that we fix a, but allow ΔX to vary; in fact, let ΔX approach zero in size. What happens to the probability of this interval relative to the interval size? Both numbers will change, and the ratio will change, but this ratio will approach some **limiting value** as ΔX comes close to zero. That is, we can speak of the limit of $p(a \leq X \leq b)/\Delta X$ as ΔX approaches zero. *This limit gives the probability density of the variable X at value a.* Loosely speaking, one can say that the probability density at a is the *rate of change* in the probability of an interval with lower limit a, for minute changes in the size of the interval. This rate of change will depend on two things: the function rule assigning probabilities to intervals such as $(X \leq a)$ and the particular "region" of X values we happen to be talking about. Rather than talk about probabilities of X values per se, for continuous random variables it is mathematically far more convenient to discuss the probability density and reserve the idea of probability only for the discussion of *intervals* of X values. This need not trouble us especially, as we will be interested only in intervals of values in the first place. We will even continue to speak loosely of the probabilities for different X values. However, distribution functions plotted as smooth curves are really plots of probability *densities*, and the only probabilities represented are the areas cut off by nonzero intervals under these curves. Furthermore, when we come to look at a function rule for a continuous random variable, we will actually be looking at the rule for a density function.

If the random variable is discrete, then we can say that

$$p(X = a) = f(a) = \text{the probability density of } X \text{ at value } a.$$

The terms "probability at value x" and "density at value x" can be used as though they were synonomous in the case of discrete random variables. On the other hand, these terms are not interchangeable in the case of continuous random variables. For a continuous random variable probabilities are defined only for intervals of values: e.g. $p(X \leq a)$.

Following this introduction, we are now able to give a somewhat more formal definition of a continuous random variable.

A random variable X is said to be continuous if its density function $f(x)$ is continuous.

This means that if each possible value $X = x$ for the random variable is plotted against its probability density $f(x)$, the resultant function will be continuous. (See Section 1.23.) Graphically, the plot of the density function will be a smooth curve, without breaks, undefined points, or any other marks of discontinuity.

Although probability densities such as $f(x)$ are not probabilities, **intervals of values can always be assigned probabilities,** regardless of whether the variable is discrete or continuous.

We have already seen that for discrete variables, the probability of an interval, say $a \leq X \leq b$, is simply the sum of the probabilities for all values of X such that $a \leq X$ and $X \leq b$. For continuous random variables, the probability of any interval depends on the probability density associated with each value in the interval. The probability of any continuous interval is given by

$$p(a \leq X \leq b) = \int_a^b f(x) \, dx.$$

The mathematically unsophisticated reader need not worry over the symbolism used here; **it suffices to say that the probability of an interval is the same as the area cut off by that interval under the curve for the probability densities, when the random variable is continuous, and the total area is equal to 1.00.** The expression $f(x) \, dx$ can be thought of as the area of a minute interval with midpoint

$X = x$, somewhere between a and b. When the number of such intervals approaches an infinite number and their size approaches zero, the sum of all these areas is the *entire* area cut off by the limits a and b. Since there is an infinite number of such intervals, this sum is expressed by the definite integral sign $\int_a^b$. This agrees in general form with the definition of the probability of an interval for a discrete variable, which also is a sum, though of probabilities rather than probability densities. Furthermore, in a histogram for a discrete variable, the area in an interval is a sum of areas, and this too is like the summing of areas under a smooth curve, yielding the probability of an interval for a continuous variable.

3.20 CUMULATIVE DISTRIBUTION FUNCTIONS

As we have seen, the distribution of a discrete random variable may be represented by a histogram or by a probability mass function, and that of a continuous random variable by a probability density function. However, there is another important way to describe the distribution of a random variable. This is through use of the so-called "cumulative distribution function." Any frequency distribution can be converted into a cumulative frequency distribution, which shows the relation between a class interval and the frequency of cases falling at or below the interval's upper limit. In the same way, a probability or density function can be converted into a cumulative probability distribution. A cumulative probability distribution shows the relation between the possible values a of a random variable X, and the probability that the value of X is less than or equal to a. That is, the cumulative probability distribution is a function relating all possible values a to the probabilities $p(X \leq a)$.

The probability that a random variable X takes on a value less than or equal to some particular value a is often written as $F(a)$:

$$F(a) = p(X \leq a).$$

The symbol $F(a)$ denotes the particular probability for the interval $X \leq a$; the general symbol $F(X)$ is sometimes used to represent the existence of the function relating the various values of X to the corresponding *cumulative* probabilities.

The probability that a continuous random variable takes on **any** value between limits a and b can be found from

$$p(a \leq X \leq b) = F(b) - F(a).$$

This is seen easily if it is recalled that $F(b)$ is the probability that X takes on value b or below, $F(a)$ is the probability that X takes on value a or below; their difference must be the probability of a value between a and b.

In the inequalities such as $a \leq X \leq b$, the "less than or equal to" sign is used. These could just as well be written with "less than" signs alone, however, and the statements would still be true for a *continuous* distribution. The reason is that for a continuous random variable the probability that X equals any exact number is regarded as zero, and thus the probabilities remain the same whether or not a or b or both are considered inside or outside of the interval. However, for discrete variables, $<$ and $\leq$ signs may lead to different probabilities.

The symbol $F(a)$ can be used to represent the cumulative probability that X is less than or equal to a either for a continuous or a discrete random variable. All random variables have cumulative distribution functions. However, remember that symbols such as $f(a)$ are read differently for continuous and discrete variables. For a continuous variable, $f(a)$ means the *probability density* for X at the exact value a; this is not a probability. On the other hand, if the symbol $f(a)$ is applied to a discrete random variable, this *can* be read as the probability for $X = a$. To avoid possible confusion the density notation $f(a)$ will be reserved for *continuous* variables in all of the following, and $p(a)$ will be used both for discrete variables and for *intervals* of continuous variables, as in $p(a \leq X \leq b)$. Occasionally in mathematical statistics the terms "distribution function" or "probability function" are used to refer only to the *cumulative* distribution of a random variable, and "density function" is used where we have used "probability distribution." However, the author believes the terms used here are simpler for the beginning student to learn and use.

Any cumulative distribution function must satisfy certain mathematical requirements, among which are the following:

1. $0 \leq F(x) \leq 1$,
2. if $a < b$, then $F(a) \leq F(b)$,
3. $\lim\limits_{x \to \infty} F(x) = 1.00, \quad \lim\limits_{x \to -\infty} F(x) = 0.$

The first of these requirements follows simply from the fact that $F(x)$ is a probability, $p(X \leq x)$. That the second requirement must be met can be shown quite simply: If $a < b$, then the event $X \leq b$ is the same as the event

$$(X \leq a) \cup (a < X \leq b).$$

However, since these two events are mutually exclusive,

$$p(X \leq b) = p(X \leq a) + p(a < X \leq b).$$

Then since $p(a < X \leq b) \geq 0$, $p(X \leq a)$ cannot be greater than $p(X \leq b)$.

The third statement says that as a is made indefinitely large, the probability that $X \leq a$ must approach 1.00. This is quite reasonable, since as a is set indefinitely larger, it becomes increasingly certain that any value of X that might occur must be less than or equal to a. Conversely, as a is made indefinitely small, it becomes more and more unlikely that a value of X can occur that is less than a: hence, $F(x) = 0$.
$$\scriptstyle x \to -\infty$$

Given the probability mass function of a discrete variable, its cumulative distribution function is uniquely determined. Since X is discrete, $F(a)$ is just the sum of all the probabilities $p(X = x)$ for which $x \leq a$. That is,

$$F(a) = p(X \leq a) = \text{sum of all } p(x) \text{ such that } x \leq a.$$

Thus, given the probabilities associated with the possible values of a discrete random variable, it is a matter of addition to find the $F(a)$ values.

As an example of the determination of a cumulative distribution function from a probability mass function, consider a discrete random variable with the following rule:

$$p(x) = \begin{cases} 1/4 & \text{for } x = 0, \\ 1/2 & x = 1, \\ 1/4 & x = 2, \\ 0 & \text{otherwise.} \end{cases}$$

Now, to determine the cumulative distribution function, we need only the values $F(0)$, $F(1)$ and $F(2)$, since these are the only values of X with nonzero probabilities. For $X < 0$, the probability is zero, of course, and for $X = 0$, the probability is 1/4. Hence

$$F(0) = 0 + 1/4 = 1/4.$$

For $F(1)$ we have

$$F(1) = F(0) + p(1) = 1/4 + 1/2 = 3/4.$$

For $F(2)$, we find

$$F(2) = F(1) + p(2) = 3/4 + 1/4 = 1.00.$$

For any value of X above 2, the value of $F(x) = 1.00$.

The cumulative distribution function for this random variable is now fully determined. We can plot this function as in Figure 3.20.1. Note that in this plot the same $F(x)$ value occurs for all points in the interval $0 \le X < 1$, the same value holds for all points in the interval $1 \le X < 2$, and so on. This gives the graph the distinctive appearance of a set of plateaus, or "steps," and for this reason such a graph is called a "step function." The cumulative distribution function of any such discrete variable will produce such a step function. Although we might have proceeded just as for the cumulative distribution of frequencies, and connected the points standing for the $(x, F(x))$ pairs to produce a polygon,

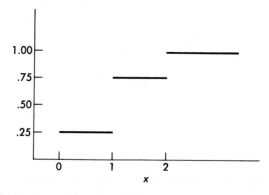

FIG. 3.20.1. A step function, showing the cumulative probabilities for values of a discrete random variable

the step function does have the advantage of emphasizing the discrete nature of the variable X. In the light of the previous discussion of continuous functions, it should be obvious that any step function must be discontinuous.

3.21   GRAPHIC REPRESENTATIONS OF CONTINUOUS DISTRIBUTIONS

A continuous probability distribution is always represented as a smooth curve erected over the horizontal axis, which itself represents the possible values of the random variable X. A curve for some hypothetical distribution is

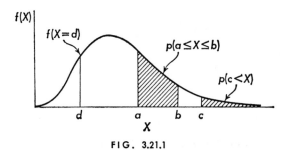

FIG. 3.21.1

shown in Figure 3.21.1. The two points a and b marked off on the horizontal axis represent limits to some interval. The shaded portion between a and b shows

$$p(a \leq X \leq b),$$

the probability that X takes on a value between the limits a and b. Remember that this probability corresponds to an *area* under the curve: in any continuous distribution, the probability of an interval can be represented by an area under the distribution curve. The total area under the curve represents 1.00, the probability that X takes on *some* value.

In Figure 3.21.1, the shaded portion of the curve to the right of the point c is the probability $p(c < X)$. This is found by first taking

$$F(c) = p(X \leq c).$$

Since the total probability is 1.00, and since $(X \leq c)$ and $(c < X)$ are mutually exclusive events, then

$$p(c < X) = 1.00 - F(c).$$

Cumulative distribution curves often have more or less the characteristic **S** shape shown in Figure 3.21.2. Here, the horizontal axis shows possible

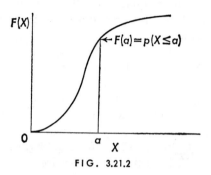

FIG. 3.21.2

values of X; for any point on the axis, a, the *height* of the curve, $F(a)$, is the probability that X is less than or equal to a.

Many of the features of continuous random variables can be demonstrated by an examination of a few of the very simplest probability density functions. Perhaps the simplest type is given by a rule such as

$$f(x) = \begin{cases} k & \text{for } a \leq x \leq b, \text{ where } 0 < k, \\ 0 & \text{otherwise.} \end{cases}$$

Here the density is a positive constant k, for any x lying between two values a and b inclusive; otherwise, the density is zero. This is known as a **uniform density function**.

However, the constant k cannot be just any value; it must be such that the total area under the curve described by this density function is equal to 1.00, so that $p(a \leq X \leq b) = 1.00$. In mathematical terms this means that k must be chosen in such a way that

$$\int_a^b k \, dx = 1.00.$$

Now it is easy enough to see that since the density will always be the same, or k, for any value of x, $a \leq x \leq b$, the curve for this function will simply be a horizontal straight line (see Figure 3.21.3). Furthermore, the area under this curve

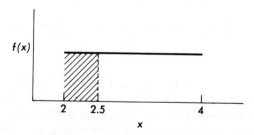

FIG. 3.21.3. A uniform density function for $2 \leq x \leq 4$

must be a rectangle, with a base equal in length to $b - a$, and a height equal to k. The area of such a rectangle is, of course, $k(b - a)$. In view of this, we can solve for the necessary value of k:

$$k(b - a) = 1.00, \qquad k = \frac{1}{b - a}.$$

For example, consider a random variable of this type for which $a = 2$. $b = 4$. Then $k = 1/(4 - 2)$ or $1/2$. The rule is thus:

$$f(x) = \begin{cases} 1/2 & \text{for } 2 \leq x \leq 4, \\ 0 & \text{otherwise.} \end{cases}$$

This density function is plotted as Figure 3.21.3. The total area under the curve

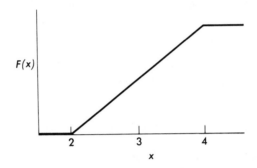

FIG. 3.21.4. Cumulative distribution function for the random variable of Figure 3.21.3

(in this case a straight line) is equal to 1.00. This is a continuous function between the values a and b, or 2 and 4.

Now to find the cumulative distribution function for this random variable, let us first find the probability of an interval such as $2 \leq X \leq 2.5$. This interval is shown as shaded in Figure 3.21.3. Note that this shaded area, which represents the desired probability, is also a rectangle. Thus, its area must be equal to its base times its height, or $k(2.5 - 2)$, which is $1/4$. Thus

$$p(2 \leq X \leq 2.5) = 1/4.$$

It follows that since $p(X < 2) = 0$, and $p(2 \leq X \leq 2.5) = 1/4$, then

$$F(2.5) = 1/4.$$

In exactly the same way, we find $F(3) = 1/2$, and $F(3.5) = 3/4$, and so on for $F(x)$ for any value x we might be interested in. On the basis of these F values, the cumulative distribution function can be plotted, as in Figure 3.21.4. Here the X axis shows possible values of the random variable, and the Y axis

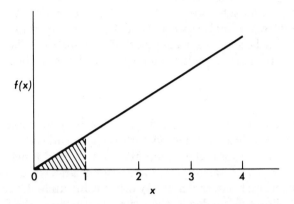

FIG. 3.21.5. Density function where f(x) = x/8, 0 $\leq$ x $\leq$ 4

shows the possible $F(x)$ value, or cumulative probability, associated with any given x value.

Generally, for any uniform distribution, the cumulative probability at any value c is given by $F(c) = (c - a)/(b - a)$, where a and b are the limits given in the rule above, and $a \leq c \leq b$.

Another very simple type of density function follows a rule of the following form:

$$f(x) = \begin{cases} kx & \text{for } 0 \leq x \leq a, \text{ where } 0 < k, \\ 0 & \text{otherwise.} \end{cases}$$

The graph of one example of such a function is shown in Figure 3.21.5. Again, in order to specify a particular random variable, we must define the values of k and of a. Suppose that we investigate the random variable for which a is equal to 4, so that we can find the density values $f(x)$ for $0 \leq X \leq 4$. Next, of course, we must determine the required value of k. Notice that in the graph of such a density function, the area between any given x value and zero is that of a right triangle, with

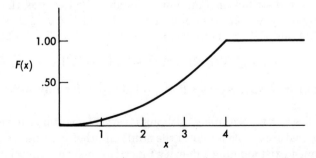

FIG. 3.21.6. Cumulative probability function for the density function of Figure 3.21.5

a height equal to $f(x)$, and a base equal to x. In particular, for $x = 4$, the height must be $f(4)$ or $4k$, and the base equal to 4. We also know that the entire area under the curve (straight line) between 0 and 4 must be equal to 1.00, or the total probability. Thus, it becomes easy enough to determine the value of k. Recalling that the area of any right triangle is equal to 1/2 the base times the height, we take:

$$(1/2)(4)(4k) = 1.00, \qquad k = 1/8.$$

Thus the density for any value x, $0 \angle X \angle 4$, is simply $x/8$. Figure 3.21.5 gives a plot of the various values of X paired with their densities.

In order to form the cumulative distribution function for this random variable, we rely once again on the fact that the area cut off under the curve of the density function, between 0 and any other value up to 4, is a right triangle. One such area is shown as shaded in the figure. Let us find, for example, $F(1)$, or $p(X \angle 1)$ for this random variable. The probability that $0 \angle X \angle 1$ is given by the area of the triangle with base equal to 1 and height equal to $f(1)$, or 1/8. Then the area is equal to $(1/2)(1)(1/8)$ or 1/16. Thus,

$$F(1) = p(0 \angle X \angle 1) + p(X < 0) = 1/16 + 0 = 1/16.$$

In the same way we find that $p(0 \angle X \angle 2) = (1/2)(2)(2/8)$ or 1/4, so that

$$F(2) = p(0 \angle X \angle 2) + p(x < 0) = 1/4 + 0 = 1/4.$$

Proceeding in this way, we find enough values of $F(x)$ to sketch the curve (a real curve this time) of the cumulative distribution function as shown in Figure 3.21.6. This cumulative distribution function follows the rule

$$F(x) = \frac{x^2}{16}, \qquad 0 \angle x \angle 4.$$

Suppose that we wished to find the probability that a value of this random variable occurs falling in the interval $1 \angle X \angle 2$. This is the same as the difference between the two cumulative probabilities $F(2) - F(1)$, which appear in the cumulative distribution function. It is also the same as the area cut off under the curve of the density function by the interval $1 \angle x \angle 2$. Since we know that $F(2) = 1/4$ and $F(1) = 1/16$, then

$$p(1 \angle X \angle 2) = 1/4 - 1/16 = 3/16.$$

In exactly the same way, we can find probabilities for any other intervals of values.

These two examples, although very elementary, nevertheless embody all of the features to be found in the study of other continuous random variables. A rule must exist defining a density function, such that associated with each possible value x of a random variable X is a nonnegative number $f(x)$, the probability density at that value. The total area under the curve described by the density function must be equal to 1.00. The area cut off under the curve by any interval of values is then the probability that X takes on a value falling in that interval. Furthermore, the cumulative distribution function shows the relationship between possible values x and the probability that X takes on a value less than or equal to x.

Just as the density function uniquely defines the cumulative distribution function, so the cumulative distribution function in turn uniquely specifies the density function. Basically, the probability density for the random variable X at any given value a is the rate of change in $F(x)$ for any minute change in the neighborhood of a. That is,

$$f(a) = \operatorname*{limit}_{\Delta X \to 0} \frac{F(a + \Delta X) - F(a)}{\Delta X}.$$

In the language of the calculus, $f(a)$ is simply the derivative of $F(x)$ with respect to X, evaluated at the value a. In the example of the uniform distribution above,

$$\operatorname*{limit}_{\Delta X \to 0} \frac{F(a + \Delta X) - F(a)}{\Delta X} = \operatorname*{limit}_{\Delta X \to 0} \left(\frac{a + \Delta X}{2} - \frac{a}{2} \right) \Big/ \Delta X$$

$$= \operatorname*{limit}_{\Delta X \to 0} \frac{\Delta X}{2\,\Delta X}$$

$$= \operatorname*{limit}_{\Delta X \to 0} \frac{1}{2} = \frac{1}{2},$$

so that

$$f(a) = \frac{1}{2},$$

just as specified in the original density-function rule. In a similar way, the original density-function rule can be recovered from the rule for the cumulative distribution function for any continuous random variable (at least in principle). The practical consequence is that random variables may be specified either by their density functions or by their cumulative distribution functions, whichever is the more convenient in any given use.

Tables of theoretical probability distributions are most often given in terms of the cumulative distribution, or $F(x)$, for a random variable. Given these cumulative probabilities for various values of X it is easy to find the probability for any interval by subtraction. When we come to use the table of the so-called "normal" distribution, we will find that cumulative probabilities are shown.

This short discussion cannot, of course, even begin to suggest the mathematical complexity of the matter. Some of the mathematical rules specifying theoretical density or cumulative functions are extremely complicated. It may be very difficult even to arrive at the mathematical specification of some random variable of interest. Furthermore, not all random variables can be categorized neatly as discrete or continuous. Theoretically, at least, there can be mixed types, which are continuous in some intervals of values and discrete in others. Nevertheless, these basic ideas will underlie all of our future discussions of random variables and distributions, and the basic set of terms employed will always be the same.

3.22   CONTINUOUS RANDOM VARIABLES AS AN IDEALIZATION

A continuous distribution of a random variable is a theoretical state of affairs. Continuous distributions can never be observed, but are only idealizations. In the first place, no set of potential observations is actually infinite in number; the mortal scientist can deal only with finite numbers of real things, and so obtained data distributions are always discrete. Regardless of the size of his sample, it will never be large enough to permit each of the possible real numbers in any interval to be observed as values of the random variable. In the second place, measurements are imprecise. Not even in the most precise work known can accuracy be obtained to *any* number of decimals. This puts a limit to the actual possibility of encountering a continuous distribution in practice.

Why, then, do statisticians so often deal with these idealizations? The answer is that, mathematically, continuous distributions are far more tractable than discrete distributions. The function rules for continuous distributions are relatively easy to state and to study using the full power of mathematical analysis. This is not usually true for discrete distributions. On the other hand, continuous distributions are very good approximations to many truly discrete distributions. This fact makes it possible to organize statistical theory about a few such idealized distributions and find methods that are good approximations to results for the more complicated discrete situations. Nevertheless, the student should realize that these continuous distributions are mathematical abstractions that happen to be quite useful; they do not necessarily describe "real" situations.

3.23   FREQUENCY AND PROBABILITY DISTRIBUTIONS IN USE

Both frequency and probability distributions exhibit the same basic features: each is a statement of a relation between the various possible measurement classes into which observations may fall and a number attached to each class. In the case of frequency distributions, this number is the frequency of cases actually observed in a particular class or class interval. For probability distributions, each class or interval is accompanied by its probability. In future chapters we will see that the same language is applied in the further summarization of either kind of distribution.

However, there is an important difference between frequency distributions of data and probability distributions. The latter almost invariably represent *hypothetical* or *theoretical* distributions. A probability distribution is some idealization of the way things might be, could we but have all of the information. The frequency distribution represents what we have actually seen to be true from some limited number of observations.

The connection between frequency and probability distributions should be obvious by now: the probability distribution dictates what we should expect to observe in a frequency distribution, if the given state of affairs is true. Thus, if theoretically the random variable X has probability of .30 of taking on a value in the interval between 100 and 102, then given a frequency distribution of 500 sample observations we should expect the class interval 100–102 to contain (500)(.30) or 150 cases. This does not mean that this will be true for any given sample; we *might* observe any frequency between 0 and 500. Nevertheless, in the long run, if we sample indefinitely, 30 percent of all cases sampled should show scores in that interval, provided that our hypothetical probability distribution is right.

This is the reason for the parallel development of frequency and probability distributions carried on here. Each form of distribution plays a role in inference: the probability distribution specifies what to expect when the data are put into distribution form, and the obtained frequency distribution supplies us with our best evidence about the probability distribution (the *relative* frequencies of occurrence being estimates of the true probabilities). This parallel discussion will be continued in other chapters. However, the immediate goal is to show how theoretical probability distributions can be constructed. In order to do this we must pick up a few more rudiments of probability theory. These will be presented in the next chapter.

EXERCISES

1. Given data of the following types, state the scale of measurement that each type seems most clearly to represent:
   (a) The nationality of an individual's male parent.
   (b) Hand pressure as applied to a bulb (i.e. on a dynamometer).
   (c) Memory ability as measured by the number of words recalled from an initially-memorized list.

(d) The excellence of baseball teams as determined from their won-lost records at the end of the season.

(e) The time to clotting of samples of human blood.

(f) Distance by air between New York and other cities in the United States.

(g) Time as measured before and after the discovery of America by Columbus in 1492.

(h) The tensile strength (force required to break) for a piece of wire.

(i) U.S. Department of Agriculture classifications of fresh meats ("choice", etc.).

(j) Gains and losses in value of stocks during a day of trading on the New York Stock Exchange.

(k) Magnitudes of stars, as judged from their apparent brightness to the naked eye.

2. A well-known magazine gives evaluative reports to consumers on the quality of various products. In a report on television sets of several brands, the ratings were as follows:

BRAND	RATING
A	Acceptable
B	Good
C	Good
D	Acceptable
E	Poor
F	Acceptable
G	Acceptable
H	Poor
I	Very Good
J	Good
K	Acceptable
L	Acceptable
M	Poor
N	Very Good
O	Acceptable
P	Poor

Make a frequency distribution table displaying these data.

3. In a particular year, the manufacturer's suggested retail prices (in dollars) for a group of 25 new automobiles were as follows:

2669	2627	2809	2963	2797
2808	2752	2887	2856	2900
2755	2333	3010	2874	2950
2478	2337	3213	2963	3053
2593	2475	2767	2750	3131

Make a frequency distribution table based upon these data.

4. A student of the history of the English language noted the dates of the first recorded appearance in written English of 1000 words of foreign origin. The dates were as follows:

DATES	FREQUENCY	DATES	FREQUENCY
Before 1050	2	1451–1500	76
1051–1100	2	1501–1550	84
1101–1150	1	1551–1600	91
1151–1200	15	1601–1650	69
1201–1250	64	1651–1700	34
1251–1300	127	1701–1750	24
1301–1350	120	1751–1800	16
1351–1400	180	1801–1850	23
1401–1450	70	1851–1900	2

Plot the cumulative frequency distribution for these data.

5. Suppose that a word were drawn at random from among the 1000 words given in the distribution of problem 4 above. What is the probability that the following events occur:
   (a) The word drawn is one which appeared in English before 1301?
   (b) The word drawn appeared in English after 1601?
   (c) The word drawn appeared in English between 1501 and 1700 inclusive?
   (d) The word drawn appeared in English either before 1201 or after 1801?
6. For the data of problem 3, construct a histogram with $i = 100$.
7. For the data of problem 4, construct a relative frequency polygon with $i = 150$.
8. On a test of mechanical aptitude, the score is the number of seconds required in order to finish a certain task. A group of 52 students in the second grade received the following scores:

76.3	90.1	57.5	57.8
66.0	92.0	55.0	59.0
80.0	87.4	61.9	55.0
82.0	91.0	59.2	61.1
76.2	94.8	60.0	59.9
80.0	95.0	61.3	60.0
84.7	91.9	74.0	61.7
86.0	94.5	90.4	59.0
80.6	93.2	48.0	52.0
86.5	57.1	57.2	53.0
85.0	53.6	53.5	44.0
87.0	55.7	55.6	51.5
74.0	56.0	56.0	45.0

Construct a frequency distribution table for these data.

9. For the data of problem 8, construct a cumulative relative frequency distribution. From this distribution, estimate those score values which cut off the *middle* 50 percent of cases in this distribution. Estimate the value above which exactly 50 percent of cases fall, and below which exactly 50 percent of cases fall.

10. The following data represent weights, to the nearest pound, of 84 boys in junior high school. Arrange these data into a frequency polygon with $i = 5$.

122	122	110	118	120	111
117	122	107	127	146	113
116	119	108	118	127	116
114	118	153	125	138	126
110	117	148	119	133	113
109	112	134	125	134	106
107	108	140	119	128	118
105	117	108	124	144	115
103	117	126	120	132	119
102	126	103	121	137	133
134	123	136	128	136	148
126	118	112	116	146	137
123	113	118	127	152	124
123	112	130	141	135	145

11. Construct a cumulative relative frequency polygon for the data of problem 10. Based upon this graph, what is the relative frequency of each of the following intervals (where x represents a weight value)?

   (a) $(111 < x)$
   (c) $(109 < x \leq 135)$
   (e) $(129 < x)$

   (b) $(x \leq 124)$
   (d) $(140 < x \leq 150)$

12. A random variable has a probability function defined as follows:

$$f(x) = \begin{cases} \dfrac{1}{10} & -3 \leq x \leq -1 \\[2mm] \dfrac{2}{5} & -1 < x \leq 0 \\[2mm] \dfrac{1}{5} & 0 < x \leq 1 \\[2mm] \dfrac{3}{10} & 1 < x \leq 3 \\[2mm] 0 & \text{elsewhere} \end{cases}$$

   (a) Plot this probability function.
   (b) Plot the cumulative distribution function.
   (c) Is this a continuous or a discrete distribution? How can one tell?

13. Suppose that a card is drawn at random from a well-shuffled deck. Let v represent the value of the card, from 1 through 10, and let a face card (king, queen, jack) be assigned

the value 11. If the suit is black, let $x = v$, and if the suit is red, let $x = 2v$. Show the probability distribution of the random variable x.

14. Suppose that two cards are drawn at random with replacement from a well-shuffled deck, and let v_1 be the value of the first card drawn, and v_2 the value of the second card. Both v_1 and v_2 are defined as in problem 13 above. Plot the probability mass function and the cumulative distribution function for each of the following random variables:

   (a) $x = -1$ if $v_1 < v_2$
       $x = 0$ if $v_1 = v_2$
       $x = 1$ if $v_1 > v_2$

   (b) $y = -2$ if v_1 and v_2 are both even numbers
       $y = -1$ if v_1 is even and v_2 is odd
       $y = 1$ if v_1 is odd and v_2 is even
       $y = 2$ if v_1 and v_2 are both odd numbers

15. Consider the following cumulative distribution function for the random variable X:

a	$P(X \leq a)$
20	1.00
18	.98
16	.83
14	.74
12	.63
10	.52
8	.37
6	.21
4	.10
2	.03
0	0

Suppose that 48 observations were made, independently and at random, from the process generating this random variable. Construct the frequency distribution of observations which one should *expect* to obtain.

16. Using the following definition of a probability density,

$$f(a) = \lim_{\Delta x \to 0} \frac{p(x \leq a + \Delta x) - p(x \leq a)}{\Delta x}$$

see if you can find the expression for the probability density function associated with each of the following cumulative density functions:

   (a) $p(x \leq a) = a$, $0 \leq a \leq 1$
   (b) $p(x \leq a) = a^2/4$, $0 \leq a \leq 2$

17. If density function for a random variable follows the rule

$$f(x) = \begin{cases} k & -2 \leq x \leq 2 \\ 0 & \text{elsewhere} \end{cases}$$

and is to be a proper density function, what is the required value of k? (Hints: what form will the plot of the curve show? What must the area under the curve equal?) Having found k, plot the density function.

18. For the density function of problem 17, plot the cumulative distribution function, and find the following probabilities:

(a) $p(x \leq -1)$ (b) $p(1/3 \leq x \leq 2/3)$

(c) $p(-1/5 \leq x \leq 1)$ (d) $p(1/4 < x)$

(Hint: what must be true of the shape of this plot? In finding the probabilities, remember that the area of a right triangle is $1/2$ the base times the altitude.)

19. Graph the following density function, and then try to sketch the corresponding cumulative distribution function:

$$f(x) = \begin{cases} 1/2 & 0 \leq x \leq 1 \\ 1/2 & x = 2 \\ 0 & \text{elsewhere} \end{cases}$$

Comment on the continuity of this random variable.

20. Sketch the curve of the density function:

$$f(x) = \begin{cases} \dfrac{k}{1+x} & 0 \leq x \leq 6 \\ 0 & \text{elsewhere} \end{cases}$$

by arbitrarily letting $k = 1$. Then construct a histogram, with columns equal in height to the midpoints of the respective intervals $0 - 1$, $1 - 2$, etc. By means of the areas of this histogram, show that the required value of k is actually in the neighborhood of .5. (Remember that for a proper density function, the area defined under the curve should be equal to 1.00.) How could one refine this estimate of the required value of k?

21. Given the density function as plotted in problem 20 above, and given $k = .514$, estimate the following probabilities as areas under the curve:

(a) $p(4.5 \leq x)$

(b) $p(x \leq .8)$

(c) $p(1.7 \leq x \leq 3.2)$

(d) $p(x \leq .4 \text{ or } 4.6 \leq x)$

(Hint: plot the cumulative distribution function, and "smooth" it into a curve.)

22. Consider the following function:

$$f(x) = \begin{cases} 1 + x & \text{for } -1 \leq x \leq 0 \\ 1 - x & \text{for } 0 < x \leq 1 \\ 0 & \text{otherwise} \end{cases}$$

By means of a plot of this function, show that it is a proper probability density function. (Hint: use the principle which states that the area of a right triangle is one half the base times the altitude).

23. For the probability density function of problem 22, estimate the following probabilities from areas cut off by the curve:

(a) $p(0 \leq x \leq 1/2)$ (b) $p(x \leq -1/2 \text{ or } x \geq 1/2)$

(c) $p(x \leq -1/5$ or $x \leq 3/5)$ (d) $p(1/2 \leq x \leq 1)$

(e) $p(x \leq -1/3$ or $x \geq 2/3)$

24. A special kind of die is shaped like a triangular pyramid. Such a die has four equal faces, marked with the numbers 1, 2, 3, and 4. Now suppose that the four-sided die is thrown twice, and the number of the side on which the die comes to rest is noted for each trial. Let the number occurring on the first trial be designated x_1 and that for the second trial x_2. If each (x_1,x_2) pair is equiprobable, find the distribution of the following random variables:

(a) $Y = X_1 - X_2$

(b) $Y = 2X_1 + X_2$

(Hint: list each (x_1,x_2) possibility.)

4 Joint Events and Independence

The idea of independent events has already figured in our discussion, and we have dealt only with events sampled independently. Now this concept of independence will be elaborated, both for qualitative events and for random variables. However, the idea of joint events must first be introduced.

4.1 JOINT EVENTS

As we have seen, the elementary events making up the sample space may be grouped into events in a variety of ways. For example, if the sample space is a set of outcomes of drawing one card from a deck of playing cards, we might find the probability of the event "red suit," or the event "picture card," or the event "ten." However, notice that there is nothing to keep any given card from qualifying for two or more events, provided those events are not mutually exclusive. For example, an observed card might be both a red card and a picture card. When we draw such a card, the joint event "red *and* picture card" occurs.

The same idea applies to any sample space: **any event that is the intersection of two or more events is a joint event.** Thus, given the event A and the event B, the joint event is $A \cap B$. The event $A \cap B$ occurs when any elementary event belonging to this set is actually observed.

Since a member of $A \cap B$ must be a member of set A, and also of set B, both A and B events occur when $A \cap B$ occurs. Furthermore, a member of $A \cap B$ is a member of $A \cup B$, and so whenever the joint event occurs, $A \cup B$ occurs as well.

Provided that the elements of S are all equally likely to occur, the probability of the joint event is found in the usual way:

140

$$p(A \cap B) = \frac{\text{number of elementary events in } A \cap B}{\text{total number of elementary events}}.$$ [4.1.1]

With playing cards, the joint event "red and three" is the set of all cards having a red suit and three spots. There are exactly two cards, the three of diamonds and the three of hearts, meeting this qualification, and so

$$p(\text{red and three}) = \frac{2}{52} = \frac{1}{26}.$$

As another example, imagine a simple experiment consisting of rolling a die *and* tossing a penny. Each elementary event consists of the number that comes up on the die and the side that comes up on the penny. The sample space is shown below:

SIDE OF PENNY

$$\begin{array}{c} \text{H} \quad \bullet \quad \bullet \quad \bullet \quad \bullet \quad \bullet \quad \bullet \\ \text{T} \quad \bullet \quad \bullet \quad \bullet \quad \bullet \quad \bullet \quad \bullet \\ \quad 1 \quad 2 \quad 3 \quad 4 \quad 5 \quad 6 \end{array}$$

NUMBER ON DIE

Each point in the diagram is one elementary event, a possible outcome of this experiment, and there are exactly $(2)(6) = 12$ such outcomes. It is also useful to notice that in this example we have a sample space $S = P \times D$, where P is the set of all "penny" outcomes, and D is the set of all "die" outcomes.

Consider the joint event "coin comes up heads, and die comes up with an odd number." There are exactly three elementary events that qualify: (heads, 1) (heads, 3), and (heads, 5). Since there are twelve elementary events in all,

$$p(\text{H and odd}) = \frac{3}{12} = \frac{1}{4}.$$

This simple experiment illustrates that an elementary event can also be a joint event: the event (head, 1) is the intersection of the event "heads" and the event "1," and so on. This is very often the case in experiments where all events of interest are not mutually exclusive; each elementary event is the intersection of two or more events. Referring to the example of cards once again, we see that each distinct card represents the intersection of some suit-event with some number or picture-event.

Many situations give simple examples of joint events: imagine that someone samples students on a college campus at random and asks each his age, his sex (if necessary), and his field of concentration. Let the event A be that the student is 21 years or older, the event B that he is male, and C that he is an English major. The set of all students on the campus who are 21 or over and male English majors is $A \cap B \cap C$. The probability that a student sampled at random represents this joint event is

$$p(A \cap B \cap C) = \frac{\text{number of students in } A \cap B \cap C}{\text{total number of students}}. \qquad \text{[4.1.2]}$$

4.2 COMBINING PROBABILITIES OF JOINT EVENTS

In Chapter 2 it was pointed out that for any event A there is also an event $\bar{A}$, consisting of all of the elementary events not in the set A. In the preceding chapter it was pointed out that

$$p(A) + P(\bar{A}) = 1.$$

Furthermore, along with $A \cap B$ we can discuss the other joint events:

$A \cap \bar{B}$, consisting of all elementary events in A and not in B;
$\bar{A} \cap B$, consisting of all elementary events not in A but in B;
$\bar{A} \cap \bar{B}$, consisting of all elementary events neither in A nor in B.

These four events are all mutually exclusive, so that

$$
\begin{aligned}
p(A \cap B) + p(A \cap \bar{B}) &= p(A) \qquad \text{[4.2.1]}\\
p(A \cap B) + p(\bar{A} \cap B) &= p(B)\\
p(\bar{A} \cap B) + p(\bar{A} \cap \bar{B}) &= p(\bar{A})\\
p(A \cap \bar{B}) + p(\bar{A} \cap \bar{B}) &= p(\bar{B}).
\end{aligned}
$$

Another principle for combining the probabilities of joint events also comes from rule **2.5.7** given in Section 2.5. Suppose that there are two events A and B, and that we wish to determine the probability of the occurrence of the event A or B or both. That is, we wish to determine the probability of the event-class $A \cup B$. As we saw in Section 2.5, this probability is given by

$$p(A \cup B) = p(A) + p(B) - p(A \cap B). \qquad \text{[4.2.2]}$$

In other words, the probability of the occurrence of an A event or a B event or an elementary event falling into *both* classes is given by the probability of A plus the probability of B *minus* the probability of the compound event A and B. Recall that this is true because

$$p(A) + p(B) = p(A \cap \bar{B}) + p(\bar{A} \cap B) + 2p(A \cap B),$$

and

$$p(A \cup B) = p(A \cap \bar{B}) + p(\bar{A} \cap B) + p(A \cap B).$$

Thus, $p(A \cap B)$ must be subtracted from $p(A) + p(B)$ to find the probability of the union $A \cup B$. This rule for finding the probability of the union of two event classes is sometimes called the "or" rule, because $p(A \cup B)$ can be read as the probability of A *or* B *or* both.

The most important case of the *or* rule is that in which the two events A and B are *mutually exclusive*, so that the probability of the joint event $A \cap B$ is zero. Here, axiom 3 of Section 2.5 applies directly. In this case, for mutually exclusive events

$$p(A \cup B) = p(A) + p(B). \qquad\qquad [4.2.3]$$

As an illustration of the use of the *or* rule for combining probabilities, think of a city school system, where we are going to sample one pupil at random. In this school system, we know that 35 percent of the pupils are left-handed, so that we know also that the probability is .35 for the event "left-handed." In the same way we know that the probability is .51 of our observing a girl, and that the probability is .10 of observing a girl who is left-handed. What is the probability of observing *either* a left-handed student *or* a girl (or both)? The answer according to the *or* rule is

$$p(\text{left-handed or girl}) = .35 + 51 - .10 = .76.$$

Now suppose that .25 is the probability of observing a left-handed boy, and .10 is the probability of observing a left-handed girl. What is the probability of observing either a left-handed boy or a left-handed girl? It is not possible for an elementary event to fall both into the set of left-handed boys and the set of left-handed girls: these two events are mutually exclusive. Thus the desired probability is

$$p(\text{left-handed boy or left-handed girl}) = .25 + .10 = .35,$$

which is just the probability of "left-handed" alone.

The *or* rule for events may be generalized to more than two events. Consider three events A, B, and C. Then

$$p(A \cup B \cup C) = p(A) + p(B) + p(C) - p(A \cap B) - p(A \cap C)$$
$$- p(B \cap C) + p(A \cap B \cap C). \quad [4.2.4]$$

Similar expressions exist for the union of more than three events, but we will not have occasion to use them in our elementary work. If three or more events are mutually exclusive, then the union is just the sum of the separate probabilities:

$$p(A \cup B \cup C) = p(A) + p(B) + p(C), \qquad\qquad [4.2.5]$$

as given in axiom 3, Section 2.5.

4.3 CONDITIONAL PROBABILITY

Suppose that in sampling children in a school system, we restrict our observations only to girls. Here, there is a new sample space including only part of the elementary events in the original set. In this new sample space, consisting only of girls, what is the probability of observing a left-handed person? If all girls in the school system are equally likely to be observed, it is obvious that

$$p(\text{left-handed, among the girls}) = \frac{\text{number left-handed girls}}{\text{total number of girls}}.$$

This probability can be found from the *original* probabilities for the total sample space: first of all we know that

$$p(\text{left-handed and girl}) = \frac{\text{number left-handed girls}}{\text{total number of pupils}} = .10$$

and that

$$p(\text{girl}) = \frac{\text{total number of girls}}{\text{total number of pupils}} = .51.$$

It follows that

$$p(\text{left-handed, among the girls}) = \frac{p(\text{left-handed and girl})}{p(\text{girl})} = \frac{.10}{.51} = .196$$

since the two probabilities put into ratio each have the same denominator.

The probability just found, based on only a part of the total sample space, is called a conditional probability. A more formal definition follows:

Let A and B be events in a sample space made up of a finite number of elementary events. Then the conditional probability of B given A, denoted by $p(B|A)$, is

$$p(B|A) = \frac{p(A \cap B)}{p(A)}, \qquad\qquad [4.3.1*]$$

provided that $p(A)$ is not zero.

Notice that the conditional probability symbol, $p(B|A)$, is read as "the probability of B *given* A." In the example it was given that the observation would be a girl; the value desired was the probability of "left-handed," given that a girl were observed.

For any two events A and B, there are two conditional probabilities that may be calculated:

$$p(B|A) = \frac{p(A \cap B)}{p(A)}$$

and

$$p(A|B) = \frac{p(A \cap B)}{p(B)}.$$

We might find the probability of left-handed, given girl $(.10/.51 = .19)$ or the conditional probability of girl, given left-handed $(.10/.35 = .29)$. In general, these two conditional probabilities will not be equal, since they represent quite different sets of elementary events.

As another example, take the probabilities found in the example for a coin and a die in Section 4.1. Here, the probability of a tail coming up on the coin is .50, and the probability of an odd number on the die *and* a tail on the coin is .25. The probability that the die comes up with an odd number, *given* that the coin comes up tails is

$$p(\text{odd}|\text{tails}) = \frac{.25}{.50} = .50.$$

The probability that the coin comes up tails and the die comes up with a 1 or a 2 can be found from the example to be 1/6 or .167. If we wish to know the probability that the die comes up with a 1 or a 2 given that the coin comes up tails, the probability is

$$p(1 \text{ or } 2|\text{tails}) = \frac{.167}{.50} = .33.$$

In a sense all probabilities are conditional probabilities. Whenever we specify the circumstances surrounding, and the assumptions underlying, a given simple experiment, we are laying down certain conditions. These conditions could, if we wished, be represented through the use of conditional probabilities. For example, many times in preceding sections the experiment of drawing one card from a well-shuffled standard deck of playing cards has been used. In view of the stated condition "well-shuffled deck of standard playing cards," an event such as "king of spades" actually has a conditional probability given by

p(king of spades|well-shuffled standard deck of playing cards).

The sample space is limited by the condition "well-shuffled deck . . . ." To see that this is true, imagine the experiment to be broadened, so that there are several decks, some of which are well-shuffled and others not, and some of which are standard decks and others pinochle decks. One deck is then selected at random, and from this deck a card is drawn. Here the basic sample space would be the Cartesian product {decks} × {card suits and values}. In such a situation it is quite meaningful to think of the probability of the suit and value of the card as depending upon the deck from which it is drawn. Usually, however, we simply assume that a well-shuffled standard deck has been chosen, and then proceed to ignore this condition in the remainder of the discussion.

Most problems are stated in terms of probability rather than conditional probability for two reasons. Largely, this is just a way of saving time and symbols and avoiding tiresome repetition. Also, it turns out that conditional probabilities are rather trivial in these circumstances. When we assume that some condition A holds, this is the same as saying that for event A, $p(A) = 1.00$, and that $P(A \cap B) = p(B)$ for any event B. Then it follows that $p(B|A) = p(B)$ as well. In most circumstances it really makes no difference whether or not we keep repeating that the probability for an event B is dependent upon the condition A.

Nevertheless, there will be circumstances in which alternate or competing assumptions might underlie a given simple experiment, and in such cases we will find it convenient to use the language of conditional probability. The important thing to remember at this stage is that all probabilities are conditional upon some set of circumstances assumed to be true, even when these factors are not explicitly noted in the probability statements themselves.

4.4 RELATIONS AMONG CONDITIONAL PROBABILITIES

The conditional probability $p(B|A)$ can either be larger or smaller than $p(B)$; there is no necessary relation between the size of $p(B)$ and the size of $p(B|A)$, other than that they both must lie between zero and one.

For any event and its complement,

$$p(B|A) + p(\bar{B}|A) = 1. \qquad [4.4.1^*]$$

This is easily seen to be true if one remembers that

$$p(A \cap B) + p(A \cap \bar{B}) = p(A),$$

so that

$$\frac{p(A \cap B)}{p(A)} + \frac{p(A \cap \bar{B})}{p(A)} = 1.$$

If there are three events, A, B, and C, then

$$p(B|A) + p(C|A) - p(B \cap C|A) = p(B \cup C|A), \qquad [4.4.2]$$

since *within the set* A it is also true that

$$p(A \cap B) + p(A \cap C) - p(A \cap B \cap C) = p[(A \cap B) \cup (A \cap C)]$$

so that

$$\frac{p(A \cap B)}{p(A)} + \frac{p(A \cap C)}{p(A)} - \frac{p(A \cap B \cap C)}{p(A)} = \frac{p[A \cap (B \cup C)]}{p(A)}.$$

If we know the conditional probability $p(B|A)$ and also $p(A)$, then we can also find the joint probability $p(A \cap B)$:

$$p(A)p(B|A) = p(A \cap B) \qquad [4.4.3^*]$$

and

$$p(B)p(A|B) = p(A \cap B),$$

since, by definition,

$$p(B|A) = \frac{p(A \cap B)}{p(A)}$$

and

$$p(A|B) = \frac{p(A \cap B)}{p(B)}.$$

It follows that

$$p(A)p(B|A) = p(B)p(A|B) \qquad [4.4.4]$$

and that

$$\frac{p(B|A)}{p(A|B)} = \frac{p(B)}{p(A)}.$$

4.5 BAYES' THEOREM

The relation among various conditional probabilities are embodied in the theorem named for Bayes, an English clergyman who did early work in probability and decision theory. The simplest version of this theorem follows: for two events A and B, the following relation must hold:

$$p(A|B) = \frac{p(A \cap B)}{p(B)}$$

$$= \frac{p(B|A)p(A)}{p(B|A)p(A) + p(B|\bar{A})p(\bar{A})}.$$ [4.5.1*]

For example, suppose you were given

$$p(A) = .40 \text{ and } p(\bar{A}) = .60$$
$$p(B|A) = .80$$
$$p(B|\bar{A}) = .30.$$

You want to find the conditional probability of A given B, or $p(A|B)$. This must be

$$p(A|B) = \frac{(.80)(.40)}{(.80)(.40) + (.30)(.60)}$$
$$= \frac{.32}{.50} = .64.$$

To illustrate Bayes' theorem, suppose you awake in the middle of the night with a headache and stumble into the bathroom without turning on the light. In the dark you grab one of three bottles containing pills and take one. An hour later you become very ill, and the thought strikes you that one of the three bottles contains a poison, the other two, aspirin. With a few assumptions this can be made into a problem in conditional probability. (Before you jump to any unflattering conclusions about a professor's thoughts as he lies dying, be assured that a doctor most certainly would be called. If you will, this is an exercise to fill the time until the doctor comes.)

Let's assume that you have a medical text handy, and that it shows that 80 percent of normal individuals show your symptoms after taking the poison, and that only 5 percent have these symptoms from taking aspirin. Let B stand for the event "symptoms," and let A stand for the event "took poison," with $\bar{A}$ for the event "took aspirin." In effect, the sample space is restricted to those individuals who took *either* aspirin or poison. According to your information, you assume that

$$p(B|A) = .80$$
$$p(B|\bar{A}) = .05.$$

If each bottle had equal probability of being chosen in the dark, then

$$p(A) = 1/3$$
$$p(\bar{A}) = 2/3.$$

Given these figures, what is the probability that someone in this situation has taken poison, given that he has the symptoms? This is

$$p(A|B) = \frac{(.80)(.33)}{(.80)(.33) + (.05)(.67)} = .89.$$

If you consider yourself a random sample from among all normal persons who might be subjected to this "simple experiment," then it is a very good thing that the doctor *was* called! The conditional probability of "took poison, given the symptoms" is high.

Although this example is dramatic, it hardly illuminates the real uses to which this idea can be put, provided that one knows the necessary initial probabilities. But even so, this little example is not too far removed from the continuing everyday preoccupation that people must have with conditional probabilities. We must constantly make choices and decisions, from the most trivial all the way to some that are of tremendous importance. These decisions are all predicated upon things that may or may not be true, on things that may or may not happen. The very best information we have to go on is usually no more than a probability. These probabilities are conditional, since virtually all of our information is of an "if-then" character. "If so and so is true, then the probability of this event must be such and such." It is not surprising that a concern with changes in conditional probability, and with decisions based upon such probabilities, has been an important part of probability theory since its very beginning.

Consider this additional example: at a university it is decided to try out a new placement test for admitting students to a special mathematics class. Experience has shown that in general only 60 percent of students applying for admission actually can pass this course. Heretofore, each student applying for the course was admitted. Of the students who passed the course, some 80 percent passed the placement test beforehand, whereas only 40 percent of those who failed the course could pass the placement test initially. If this test is to be used for placement, and only students who pass the test are to be admitted to the course, what is the probability that such a student will pass the course?

First of all, event A is defined to be "passes the course," and it is assumed that $p(A) = .60$. The event B is "passes the test." The conditional probabilities are taken to be

$$p(B|A) = .80 \qquad p(\bar{B}|A) = .20$$
$$p(B|\bar{A}) = .40 \qquad p(\bar{B}|\bar{A}) = .60.$$

However, we want to know the conditional probability, $p(A|B)$, that a student will pass the course, *given* that he has passed the test. Bayes' theorem shows

$$p(A|B) = \frac{(.80)(.60)}{(.80)(.60) + (.40)(.40)}$$

$$= \frac{.48}{.48 + .16} = .75.$$

Thus, the probability is .75 for a student's passing the course *given* that he passed the test. If we wanted to find other conditional probabilities, we could also apply the theorem. For example,

$$p(\bar{A}|\bar{B}) = \frac{(.60)(.40)}{(.60)(.40) + (.20)(.60)}$$

$$= \frac{.24}{.24 + .12} = .67.$$

Notice that the probability of a student's passing the course given that he passes the test is greater than the probability in general of passing the course. Any student is a better bet to pass the course if he has passed the placement test.

What is the probability of being *right* if students are admitted or refused the course strictly on the basis of this test? The probability of a correct decision can be found by

$$p(A \cap B) + p(\bar{A} \cap \bar{B}) = p(\text{correct}).$$

By equation **4.4.1** this is

$$p(\text{correct}) = p(B|A)p(A) + p(\bar{B}|\bar{A})p(\bar{A})$$
$$= (.80)(.60) + (.60)(.40) = .72.$$

In other words, the administrator will have a probability of .72 of being right about a student's proper placement if he uses the test. If he does not use the test and simply admits all students to the course, the probability of being right (of the student passing the course) is only .60. Thus, the test does something for the administrator; its use allows him to increase the probability of being right about a given student selected at random.

Bayes' theorem and other calculations with conditional probability are often used in this way, especially in questions of selection or diagnosis of subjects where initial probabilities are known. Good selection or diagnostic procedures are those permitting an increase in the probability of being right about an individual given some prior information, and such *conditional* probabilities can often be calculated by Bayes' theorem.

Bayes' theorem can be put into much more general form: If $\{A_1, A_2, \cdots, A_J\}$ represents a set of J mutually exclusive events, and if $B \subset A_1 \cup A_2 \cup \cdots \cup A_J$, so that $p(B) < p(A_1) + p(A_2) + \cdots + p(A_J)$, then

$$p(A_j|B) = \frac{p(B|A_j)p(A_j)}{p(B|A_1)p(A_1) + \cdots + p(B|A_j)p(A_j) + \cdots + p(B|A_J)p(A_J)}$$
$$[4.5.2^*]$$

for any event A_j in the set. (Notice that A_j and A_J do not necessarily symbolize the same event.)

As a mathematical result, Bayes' theorem is necessarily true for conditional probabilities satisfying the basic axioms of probability theory. In and of itself Bayes' theorem is in no sense controversial. However, the question of its appropriate use has, in years past, been a focal point in the controversy between those who favor a strict relative-frequency interpretation of probability and those who would admit a subjective interpretation as well. The issue emerges quite clearly when some of the probabilities figuring in Bayes' theorem are associated with "states of nature" or with nonrepetitive events. As we have seen, it is usually quite difficult to give meaningful relative-frequency interpretations to probabilities for such states or events. The difficulty was compounded by the fact that in some past applications, Bayes' theorem yielded rather ridiculous results. This had the effect of casting into disrepute most applications of the theorem. The history of the development of statistics in the late nineteenth and early twentieth centuries demonstrates some rather elaborate attempts to ignore or to "finesse" the problems of prior information and subjective probability.

In recent years, some theorists have shown that subjective probabilities, including those for nonrepetitive events, can be given an axiomatic status on a par with probabilities subject only to relative-frequency interpretations. Furthermore, as we have already suggested, a basis exists for the experimental determination of subjective probabilities. Consequently, there has been a renewed interest in the application of Bayes' theorem. Not all statisticians agree that some of these applications are proper, just as not all agree that a subjective interpretation of probability is meaningful. Nevertheless, the use of Bayes' theorem is becoming a feature of a considerable portion of the modern theory of statistics and of decision-making, as these theories begin to allow once again for the subjective interpretation of probabilities. We will examine this approach more closely in Chapter 9 and particularly in Chapter 19.

4.6   INDEPENDENCE

Now we are ready to take up the topic of independence. The general idea used heretofore is that independent events are those having nothing to do with each other: the occurrence of one event in no wise affects the probability of the other event. But how does one know if two events A and B are independent? If the occurrence of the event A has nothing whatever to do with the occurrence of event B, then we should expect the conditional probability of B given A to be exactly the same as the probability of B, $p(B|A) = p(B)$. Likewise, the conditional probability of A given B should be equal to the probability of A, or $p(A|B) = p(A)$. The information that one event has occurred does not affect the probability of the other event, when the events are independent.

The condition of independence of two events may also be stated in another form: if $p(B|A) = p(B)$, then

$$\frac{p(A \cap B)}{p(A)} = p(B) \qquad\qquad [4.6.1^*]$$

so that

$$p(A \cap B) = p(A)p(B).$$

In the same way,

$$p(A|B) = p(A) \qquad\qquad [4.6.2^*]$$

leads to the statement that

$$p(A \cap B) = p(A)p(B).$$

This fact leads to the usual definition of independence:

If events A and B are independent, then the joint probability $p(A \cap B)$ is equal to the probability of A times the probability of B,

$$p(A \cap B) = p(A)p(B). \qquad [4.6.3^*]$$

For example, suppose that you go into a library and at random select one book (each book having an equal likelihood of being drawn). The sample space consists of all distinct books that a person might select. Suppose that the proportion of books then on the shelves and classified as "fiction" is exactly .15, so that the probability of selecting such a book is also .15. Furthermore, suppose that the proportion of books having red covers is exactly .30. If the event "fiction" is independent of the event "red cover," the probability of the joint event is found very easily:

$$p(\text{fiction and red cover}) = p(\text{fiction})p(\text{red cover}) = (.15)(.30) = .045.$$

You should come up with a red-covered piece of fiction in about forty-five out of one thousand random selections.

The example in which a coin is tossed and a die rolled also illustrates independent events. In this example, the probability that the coin comes up heads is .50, and the probability that the die comes up with an odd number is also .50. If we assume that the two events are independent, then the probability of the event "heads and odd number" is given by

$$p(\text{heads and odd number}) = p(\text{heads})p(\text{odd number}) = (.50)(.50) = .25.$$

If you look at the sample space for this experiment, pictured in Section 4.1, you will find that this number .25 agrees with the probability for the joint event "heads and odd number" calculated directly. When all the elementary events in this sample space are equally likely, these two events actually are independent.

It is certainly not true that all events must be independent. It is very easy to give examples where the joint probability of two events is not equal to the product of their separate probabilities. For example, suppose that children were selected at random. A child is observed and his hair color noted, and also whether or not he is freckled. For our purposes, the event A consists of the set of all children having red hair, and the event B is the set of all children who are freckled. Is it reasonable that the event A will be independent of the event B? The answer is no; everyone knows that among red-heads freckles are much more common than among children in general. One would expect in this case $p(B|A)$ to be *greater* than $p(B)$, so that it should *not* be true that $p(A \cap B) = p(A)p(B)$. The events "red hair" and "freckled" *do* tend to occur together and are not ordinarily regarded as independent.

This example suggests one of the uses of the concept of independent events. The definition of independence permits us to decide whether or not events *are associated* or *dependent* in some way:

If, for two events A and B, $p(A \cap B)$ is not equal to $p(A)p(B)$, then A and B are said to be associated or dependent.

For example, consider a sample space with elementary events consisting of all the adult male persons in the United States. We wish to answer the question, "Is making over twenty-thousand dollars a year independent of having a college education?" Let us call event A "makes over twenty-thousand dollars a year," and B "has a college education." Now suppose that the proportion of adult males in the United States who make more than twenty-thousand a year is .06; if each adult male is equally likely to be selected, the probability of event A is thus .06. Suppose also that the proportion of adult males with a college education is .30, so that $p(B) = .30$. Finally, the probability of an event $A \cap B$, our observing an adult male making over twenty-thousand a year and who has a college education is .04. If the events were independent, this probability $p(A \cap B)$ *should* be $(.06)(.30) = .018$; the actual probability is .04. This leads immediately to the conclusion that these two events are *not* independent (or *are* associated). The probability that event B will occur given event A is much greater than it should be if the events were independent: $p(B|A) = .04/.06 = .67$, whereas if the events were independent, it should be true that

$$p(B|A) = p(B) = .30.$$

It is also helpful to think of independence in terms of the proportion of elements belonging to events A and B in the sample space $\mathcal{S}$. Let $n(A)$ be the number of elements in A, $n(B)$ the number in B, and $n(\mathcal{S})$ the total number of elements. Then, if A and B are independent, the proportion of elements of A that are also in B is just the same as the proportion of B elements in the total $\mathcal{S}$. That is, when A and B are independent

$$\frac{n(A \cap B)}{n(A)} = \frac{n(B)}{n(\mathcal{S})}.$$

Then, for equally likely elementary events it must be true that

$$p(A \cap B) = p(A)p(B).$$

Incidentally, some students tend to confuse "independent" with "mutually exclusive" events. These are by no means the same! In fact, **two events A and B that are mutually exclusive cannot be independent unless one or both events have zero probability.** This is easy to show: Consider two mutually exclusive events, A and B. Since they are mutually exclusive,

$$A \cap B = \emptyset,$$

so that $P(A \cap B) = 0$. Now if A and B were independent, it would be true that $p(A \cap B) = p(A)p(B)$, but since $p(A \cap B) = 0$, this cannot be unless $p(A)$ or $p(B)$ or both are zero. Even at a common-sense level the difference between mutually exclusive and independent events is easy to show. Suppose that all men are either members of the class "balding" or the class "full head of hair." These two classes are mutually exclusive. Then if the events "balding" and "full head of hair" were independent, among those who are balding there would be the same proportion of men with a full head of hair, as the proportion of men with full heads of hair generally [i.e., $p(B|A) = p(B)$]. You must admit that this is a little hard to visualize.

The idea of independence is very important to many of the statistical techniques to be discussed in later chapters. The question of association or dependence of events is central to the scientific question of the relationships among the phenomena we observe (literally, the question of "what goes with what"). The techniques for studying the presence and degree of relatedness among observations, particularly those of Chapter 17, will depend directly upon this idea of comparing probabilities for joint events with the probabilities when events are independent.

The concept of association of events must not be confused with the idea of causation. Association, as used in statistics, means simply that the events are not independent, and that a certain correspondence between their joint and separate probabilities does not hold. When events A and B are associated, it need not mean that A causes B or that B causes A, but only that the events occur together with probability different from the product of their separate probabilities. This warning carries force whenever we talk of the association of events; association may be a consequence of causation, but this need not be true.

4.7 REPRESENTING JOINT EVENTS IN TABLES

Sometimes it is convenient to list joint events by means of a table, with each cell representing one joint possibility. For example, consider a sample space consisting of all individual students at a particular college campus. Each student is either male or female, of course, and when asked his opinion on some question, a student will respond either "yes" or "no." The possible joint events are as follows:

Male	(male and yes)	(male and no)
Female	(female and yes)	(female and no)
	Yes	No

It might be that on this particular campus, the probability of our observing a male student is .55 and of observing a female .45. Furthermore, suppose that the probability of obtaining a "yes" answer is .40, and .60 for obtaining a "no." Then the table of probabilities follows this general pattern:

p(Male) = .55	p(Male and Yes)	p(Male and No)
p(Female) = .45	p(Female and Yes)	p(Female and No)
	p(Yes) = .40	p(No) = .60

The values of p(Male), p(Female), p(Yes), and p(No) are called the **marginal**

probabilities (they get the name from the obvious circumstance of appearing in the *margin* of the table). Each marginal probability is a sum of all the joint probabilities in some particular row or column of the table:

$$p(\text{Male}) = p(\text{Male and Yes}) + p(\text{Male and No});$$
$$p(\text{Female}) = p(\text{Female and Yes}) + p(\text{Female and No});$$
$$p(\text{Yes}) = p(\text{Male and Yes}) + p(\text{Female and Yes});$$
$$p(\text{No}) = p(\text{Male and No}) + p(\text{Female and No}).$$

Ordinarily the event classes that appear together along any margin of the table are *mutually exclusive and exhaustive*. That is, the set of event classes forms a partition of the total sample space. Thus, the events "Yes" and "No" are mutually exclusive and exhaustive and so are the events "Male" and "Female." Any set of mutually exclusive and exhaustive event classes that make up one margin of the table can be called an **attribute** or a **dimension**. This table embodies two attributes, "the sex of the student" and "the response of the student," respectively.

Suppose that the two attributes of the table were independent: this means that any event along one margin must be independent of every event along the other margin. Here, this is tantamount to saying that the sex of the student has absolutely nothing to do with how he answers the question. **If independence exists, the probability of each joint event** (as given in a cell of the table) **must be equal to the product of the probabilities of the corresponding marginal events.**

For example, if the two attributes are independent, then the probability $p(\text{Male and Yes})$ must be equal to the product of the two marginal probabilities,

$$p(\text{Male and Yes}) = p(\text{Male})p(\text{Yes}) = (.55)(.40) = .22.$$

In the same way the joint probabilities for each of the other cells may be found from products of marginal probabilities, and the following table should be correct:

$p(\text{Male})$ = .55	$p(\text{Male})p(\text{Yes})$ = (.55)(.40) = .22	$p(\text{Male})p(\text{No})$ = (.55)(.60) = .33
$p(\text{Female})$ = (.45)	$p(\text{Female})p(\text{Yes})$ = (.45)(.40) = .18	$p(\text{Female})p(\text{No})$ = (.45)(.60) = .27
	$p(\text{Yes})$ = .40	$p(\text{No})$ = .60

It is important to remember, however, that these will be the correct joint probabilities only if the attributes *are* independent. It might be that the following table is the true one:

$p(\text{Male})$ = .55	$p(\text{Male and Yes})$ = .10	$p(\text{Male and No})$ = .45
$p(\text{Female})$ = .45	$p(\text{Female and Yes})$ = .30	$p(\text{Female and No})$ = .15
	$p(\text{Yes})$ = .40	$p(\text{No}) \neq .60$

When this is true it is safe to assert that for this sample space the sex of the student *is* associated with how he will answer the question. Look at the conditional probabilities: here, the probability of the student's answering "yes" if he is a male is

$$p(\text{Yes}|\text{Male}) = \frac{.10}{.55} = .18$$

and the probability of a "yes" answer if the student is a female is

$$p(\text{Yes}|\text{Female}) = \frac{.30}{.45} = .67.$$

Thus, a female is much more likely to answer "yes" to this question than is a male. On the other hand, had the two attributes been independent, these two conditional probabilities *should* have been the same:

$$p(\text{Yes}|\text{Male}) = \frac{.22}{.55} = .40$$

$$p(\text{Yes}|\text{Female}) = \frac{.18}{.45} = .40,$$

indicating that sex gives no information about how a person tends to respond to the question.

This idea of representing probabilities in tabular form can be extended to any number of attributes, each consisting of any number of event classes. For example, consider four mutually exclusive and exhaustive classes labeled $\{A_1, A_2, A_3, A_4\}$, which we may call the attribute A, and another set of classes $\{B_1, B_2, B_3\}$, which is the attribute B. Then the joint events consisting of the intersection of an event class from A with an event class from B may be represented by the table

B_1 $(A_1 \cap B_1)$	$(A_2 \cap B_1)$	$(A_3 \cap B_1)$	$(A_4 \cap B_1)$
B_2 $(A_1 \cap B_2)$	$(A_2 \cap B_2)$	$(A_3 \cap B_2)$	$(A_4 \cap B_2)$
B_3 $(A_1 \cap B_3)$	$(A_2 \cap B_3)$	$(A_3 \cap B_3)$	$(A_4 \cap B_3)$
A_1	A_2	A_3	A_4

(Notice the similarity of this formulation of joint events to the idea of a Cartesian product introduced in Section 1.14; the set of all joint-event pairs is nothing more than the Cartesian product $A \times B$.)

Associated with any joint-event class there is a probability, such as $p(A_1 \cap B_1)$ or $p(A_1 \cap B_2)$. Each of the marginal probabilities is a sum of the probabilities in a row or a column of the table: thus,

$$p(B_1) = p(A_1 \cap B_1) + p(A_2 \cap B_1) + p(A_3 \cap B_1) + p(A_4 \cap B_1),$$

and so on for the other marginal probabilities.

Given that the attributes are independent, then the joint probabilities are the products of the marginal probabilities, and the table becomes:

$$
\begin{array}{ccccc}
p(B_1) & p(A_1)p(B_1) & p(A_2)p(B_1) & p(A_3)p(B_1) & p(A_4)p(B_1) \\
p(B_2) & p(A_1)p(B_2) & p(A_2)p(B_2) & p(A_3)p(B_2) & p(A_4)p(B_2) \\
p(B_3) & p(A_1)p(B_3) & p(A_2)p(B_3) & p(A_3)p(B_3) & p(A_4)p(B_3) \\
 & p(A_1) & p(A_2) & p(A_3) & p(A_4)
\end{array}
$$

Again it must be emphasized, however, that the joint probabilities are equal to the products of the marginals *only* when the attributes represented by the margins are independent.

It is of interest to note that *some* pairs of event classes in such a table may fit the requirements for independence, even though all such classes do not do so, and hence the attributes themselves are not independent. For example, consider the following table, which gives the probabilities with two attributes, as well as the intersections of their event classes:

$$
\begin{array}{ccccc}
B_1 & .2 & .04 & .06 & .10 \\
B_2 & .6 & .14 & .40 & .06 \\
B_3 & .2 & .02 & .14 & .04 \\
 & & .2 & .6 & .2 \\
 & & A_1 & A_2 & A_3
\end{array}
$$

Now notice that for the events A_1 and B_1, $p(A_1 \cap B_1) = p(A_1)p(B_1)$, so that taken alone, these two events can be regarded as independent. In addition, the events A_3 and B_3 are independent, as shown in this table. On the other hand, no other pair of A and B events show this property. Thus, even though some pairs of events seem to show independence, the two attributes involved are not independent, since this would have required *all pairs* of A_i and B_j events to satisfy the requirement $p(A_i \cap B_j) = P(A_i)p(B_j)$.

4.8   INDEPENDENCE OF RANDOM VARIABLES

Suppose that we have two discrete random variables:

X, ranging over the values $\{a_1, a_2, \cdots, a_J\}$

and

Y, ranging over the values $\{b_1, b_2, \cdots, b_K\}$.

We can discuss the joint event that X takes on some value a *and* that Y takes on some value b: this is the event $(X = a, Y = b)$, with probability $p(X = a, Y = b)$.

Furthermore, the conditional probability that $X = a$, given that $Y = b$ is

$$
p(X = a | Y = b) = \frac{p(X = a, Y = b)}{p(Y = b)}.
\qquad [4.8.1]
$$

In the same way, the conditional probability of $Y = b$ given $X = a$ is found from

$$p(Y = b|X = a) = \frac{p(X = a,\ Y = b)}{p(X = a)}.$$

[4.8.2]

For example, consider a sample space of American men, once again. Let the random variable X be the hat size of a man, and let Y be his shoe size. Imagine that we know

$$p(X = 7) = .30$$
$$p(Y = 9) = .20.$$

Furthermore,

$$p(X = 7,\ Y = 9) = .10.$$

Then the probability that a man wears hat size 7 given that his shoe size is 9 is found to be

$$p(X = 7|Y = 9) = \frac{.10}{.20} = .50.$$

Two discrete random variables are independent if and only if

$$p(X = a,\ Y = b) = p(X = a)p(Y = b)$$

[4.8.3*]

for all pairs of values a and b.

Thus, in the example just given, if shoe size were independent of hat size, then it would be true that

$$p(X = 7,\ Y = 9) = p(X = 7)p(Y = 9) = (.30)(.20) = .06.$$

If the joint probability .10 is the true one, then X and Y are *not* independent.

The same idea applies to continuous random variables as well, except that the definition is stated in terms of probability densities:

A random variable X with density $g(a)$ at value a, and a random variable Y with density $h(b)$ at value b are independent if and only if for all (a,b)

$$f(a,b) = g(a)h(b)$$

[4.8.4*]

where $f(a,b)$ is the *joint* density for the event (a,b).

Just as for conditional probability, it is also possible to define **conditional density:**

$$w(b|a) = \frac{f(a,b)}{g(a)}$$

[4.8.5]

where $w(b|a)$ symbolizes the density for $Y = b$ given some value a of X. *The conditional distribution of Y given X will be exactly the same as the distribution of Y with X left unspecified when X and Y are independent:* for all (a,b),

$$w(b|a) = h(b), \qquad X \text{ and } Y \text{ independent.}$$

In the same way,

$$w(a|b) = g(a)$$

when the variables are independent.

For any random variables, one often has occasion to consider **joint intervals,** the event

$$[(a \leq X \leq b) \text{ and } (c \leq Y \leq d)],$$

some value of X lying between a and b, and some value of Y between c and d. For independent random variables:

$$p(a \leq X \leq b \text{ and } c \leq Y \leq d) = p(a \leq X \leq b)p(c \leq Y \leq d) \quad [4.8.5^*]$$

just as for any independent events.

The idea of independent random variables permeates all statistical theory. In some instances, X and Y are different kinds of measurements, and one is interested in the association between the attributes they represent. However, equally important is the situation where a series of observations are made, each observation producing one value of the *same* random variable X. The series of outcomes can be regarded as a set of values of random variables X_1, X_2, X_3 and so on, each having exactly the same distribution. The subscripts refer only to the order of sampling in the sequence, the value observed first, second, and so on. For example we might be measuring the same individual over time, or drawing several cases from the same group of people measured in the same way. If these observations are independent, then for, say, X_1 and X_2,

$$f(x_2|x_1) = f(x_2)$$

and

$$f(x_1|x_2) = f(x_1),$$

the distribution of X is the same regardless of which observation in the sequence or in the total sample we are considering. This amounts to saying that the chances for different intervals of values of X to occur stay the same regardless of where we are in the sampling process.

4.9 INDEPENDENCE OF FUNCTIONS OF RANDOM VARIABLES

The idea of independence of random variables can be extended to functions of random variables as well. That is, suppose we have a random variable X with a density $g(x)$, and an *independent* random variable Y with density $h(y)$, as before.

Now imagine some function of X: associated with each possible x value is a new number $v = t(x)$. For example, it might be that

$$v = ax + b$$

so that V is some linear function of X. Or perhaps,

$$v = 3x^2 + 2x,$$

and so on for any other function rule. Furthermore, let there be some function of Y:

$$w = s(y),$$

so that a new number w is associated with each possible value of Y. *Then V and W are independent random variables, provided that X and Y are independent.* This principle applies both to continuous and to discrete random variables, and to all the ordinary functions studied in elementary mathematical analysis. We will make extensive use of this principle in later sections on sampling random variables.

It is especially interesting to apply this principle to measurements as functions of underlying magnitudes of properties. Imagine that the true magnitudes on two properties for some sample space of objects are represented by the random variables X and Y, and that the numerical measurements assigned to these objects are represented by V and W. Furthermore, suppose that V is some function of X, and W is some function of Y. Then if the underlying variables X and Y are independent, the measurements V and W must be independent as well. Conversely, if V and W are not independent, X and Y are not independent.

4.10   REPRESENTATION OF JOINT RANDOM VARIABLES

The joint distribution of two discrete random variables may also be displayed in a table. This shows the possible pairs of values X and Y, each pair associated with a joint probability. The different X values are assigned to the columns of the table, with Y values to the rows, so that any cell is an (x,y) pair. For example, the joint distribution of two random variables is shown in Table 4.10.1.

Table 4.10.1

y	$p(y)$				
10	.0950	.0875	.0075	.0000	.0000
9	.1300	.0975	.0300	.0025	.0000
8	.1150	.0475	.0575	.0100	.0000
7	.1150	.0150	.0675	.0300	.0025
6	.1250	.0025	.0500	.0525	.0200
5	.1275	.0000	.0275	.0625	.0375
4	.1225	.0000	.0100	.0525	.0600
3	.0825	.0000	.0000	.0300	.0525
2	.0600	.0000	.0000	.0100	.0500
1	.0200	.0000	.0000	.0000	.0200
0	.0075	.0000	.0000	.0000	.0075
	1.0000	.2500	.2500	.2500	.2500	$p(x)$
		8	16	24	36	x

The sample space for this joint distribution might be conceived in this way: some specific group of human subjects is sampled at random and independently with replacement. Each subject sampled is also allotted *at random* to one of four experimental treatments. In one treatment he is deprived of sleep for eight hours, in the second he goes for sixteen hours without sleep, with twenty-four, and thirty-six hours sleep-deprivation being given as the other treatments. The elements in the sample space in this example can be thought of as the possible outcomes of giving each of a set of persons each of this set of treatments.

Since each and every person sampled is assigned some particular treatment on a trial, any treatment-class is an event, and the number of hours of sleep-deprivation are values of a discrete random variable X. By the nature of the assignment of observations to treatments, the probability of each possible value of X is .25, as shown in the marginal distribution for columns.

In the experiment proper, each person is given a motor-skills performance test after losing the specified amount of sleep. The scores on this performance are values of a discrete random variable Y, with probabilities given by the marginal distribution of the rows. Then the joint probability of any value of X paired with any value of Y is shown in the body of the table.

If one knew these probabilities, could he say that motor performance scores are independent of hours of sleep deprivation? Look at any cell of the table, say, for $(X = 24, Y = 5)$. If the two random variables are independent, then it must be true that

$$p(24,5) = p(24)p(5) = (.25)(.1275) = .0319.$$

This is different from the probability for the event $(X = 24, Y = 5)$ in the joint distribution, which is .0625. Since these are the actual probabilities of occurrence, this discrepancy alone is sufficient to let us say that the variables X and Y are not independent, and *are* associated. For X and Y to be independent, the probability of every event (x,y) must be the product of the two marginal probabilities, $p(x)p(y)$.

The association between X and Y can also be seen from a conditional distribution, showing $p(y|x)$ for some fixed x and over the different y values. Consider the column $X = 8$. The conditional distribution is shown in Table 4.10.2. This is rather different from the *marginal* distribution of Y. For instance, a Y score of 10 has a probability of .35 for $X = 8$, but only .095 if X is not specified. This also demonstrates that X and Y are not independent, since any conditional distribution must be identical to the corresponding marginal distribution when variables are independent. We *can* say that performance scores on this test are associated with hours of sleep-deprivation.

This same idea can be employed for either discrete random variables or continuous random variables grouped into class intervals. After such grouping, any variable can be treated as discrete, and independence or association can be judged from the joint probabilities of *intervals* of X and Y relative to the product of the marginal probabilities of those intervals.

As another example, consider observations of rats in a laboratory

Table 4.10.2

y	$p(y\mid X = 8) = \dfrac{p[(X = 8)\cap y]}{p(X = 8)}$		
10	.35	=	.0875/.25
9	.39	=	.0975/.25
8	.19	=	.0475/.25
7	.06	=	.0150/.25
6	.01	=	.0025/.25
5	.00	=	.0000/.25
4	.00	=	.0000/.25
3	.00	=	.0000/.25
2	.00	=	.0000/.25
1	.00	=	.0000/.25
0	.00	=	.0000/.25
	1.00		

situation. Each rat has an age in months (the variable X) at the time of the experiment. In the experiment proper, an animal is confined alone in a cage containing an activity wheel, which records how active the given animal is in any period of time. Let the variable Y be the number of revolutions in the first twenty-four hours. This set of elementary events can be thought of as the collection of all rats actually given this treatment in some laboratory, so that the probabilities are relative frequencies of actual rats. It is also possible to think of this set of rats as hypothetical, or as all rats of a specified kind who *might* be given this treatment; in this case the probabilities are purely hypothetical. Nevertheless, in either case, we are discussing probabilities in a specified sample space.

In principle, these two variables X and Y might either be continuous or discrete. However, in either circumstance, if the marginal distributions are arranged into class intervals, then the various *joint intervals* might be represented as cells in a table:

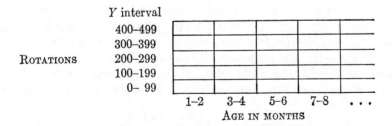

Each column or row will give a conditional distribution, based on the joint probabilities. If X and Y are independent, so that age is unrelated to activity in this experimental situation, each conditional probability distribution must be identical to the corresponding marginal distribution. *If we but knew these*

probability distributions, we could say without hesitation that there is or is not an association between these two variables in the specified sample space.

4.11   Joint frequency distributions and inferences about association

It should be apparent that actual data can be put into the form of a *joint frequency distribution* in exactly the same way that events are shown in a joint probability distribution. The only difference is that the observed frequency is shown for each joint event or joint interval. When the events of both attributes are qualitative, then the joint distribution is often called a **contingency table.** For numerical scores, it is often convenient to show occurrence of a joint event by a point on a graph, where the X and Y axes symbolize the two variables. However, for the same reason that numerical distributions are grouped into class intervals, it may be more convenient to break each axis into class intervals and show the frequency of occurrence of the joint intervals in the data. Such bivariate frequency distributions are often called **scatter plots** or **scatter diagrams.**

The principle embodied in Bernoulli's theorem holds for such bivariate (or even multivariate) frequency distributions: given random samples from a specific joint or bivariate distribution of events, the obtained relative frequencies of the various joint events must, in the long run, equal the probabilities for such joint events. Thus, any sample's joint distribution provides an estimate of the joint probability distribution, just as for single variate distributions.

Can the independence of events be judged from a sample joint distribution? Suppose that in one of the experiments of the section above some N cases were sampled in all. A joint frequency distribution similar to the joint distributions above could be constructed. In any cell the obtained relative frequency of a joint event $x \cap y$, or

$$\text{relative frequency } (x \cap y) = \frac{\text{freq.}(x \cap y)}{N}, \qquad [4.11.1]$$

is our best estimate of $p(x \cap y)$. Because of sampling error, this relative frequency in the sample need not be equal exactly to $p(x \cap y)$. Nevertheless, in the long run, as the sample size grows indefinitely large, the probability approaches 1.00 that the joint relative frequency will agree very closely with the true joint probability. Similarly, the probability is large that the relative frequency associated with any X will agree closely with $p(X)$, and the same is true of Y. **If the variables are independent, then we should expect that in the long run**

$$\frac{\text{freq.}(x \cap y)}{N} = \left(\frac{\text{freq.}(x)}{N}\right)\left(\frac{\text{freq.}(y)}{N}\right),$$

the relative frequency of any joint event $(x \cap y)$ **should equal the relative frequency of** x **times the relative frequency of** y **for those particular values of** X **and** Y.

Provided that X and Y are independent, the best estimate we can make of what to *expect* in a cell, given only the marginal frequencies, is

$$\text{expected frequency } (x \cap y) = \frac{\text{freq.}(x) \text{ times freq.}(y)}{N}. \qquad [4.11.2]$$

Now suppose that the experimenter draws a sample and actually constructs the table of joint frequencies. By the rule just given, he can estimate what the expected frequency of each cell should be, provided that X and Y were independent. Any departure of any cell's frequency from this expected number is a small piece of evidence that X and Y are not independent, and he may even say that X and Y are *apparently associated*. However, for any sample, it is quite possible for X and Y apparently to be associated, even though the variables are really independent. Conversely, finding relative frequencies that *exactly* conform to expectation about independent variables does not necessarily mean that X and Y are independent. Unequivocal statements about independence or association can never be made from sample results alone; these statements can be made with complete confidence only if the probabilities are known. The experimenter *never* has enough evidence to say for sure, unless his set of observations includes the entire sample space.

On the other hand, the more the joint frequencies depart from the frequencies to be expected for independent X and Y, the better is the bet that X and Y are associated. Big departures from expectation should be unusual for relatively large sample size and for truly independent random variables. However, we still cannot say for sure, even for very large departures, that X and Y are associated.

4.12   STATISTICAL RELATIONS AND ASSOCIATION

Most applications of statistics to experimental data are concerned with association among two or more variables. Regardless of what the scientist studies, he can be sure that *for any observation showing a given value on one variable, there will be a whole distribution of values that might occur on some other variable.* That is, given the value of X, there is some conditional distribution of Y. Almost never will it be possible for one and only one Y value to occur for each possible X; instead there will be any number of different Y values that one might observe, each with some probability.

This is what we mean when we say that any set of data shows *variability*. No matter how we restrict the sample space, making it true that all observations represent the same event when looked at in one way, the individuals sampled will "scatter" over many events when regarded in some other way. As we said in Chapter 1, in some physical sciences this variability can be made so small as to be negligible; given X, different possibilities for Y still exist, but different observations give Y values so nearly alike as to be practically identical.

This is seldom the case with observations of living things. Human monozygotic twins, who are nearly identical in every physical respect, and who have been reared in virtually the same environment, still differ in important psychological characteristics. Even observations of the same individual taken over time show variability, and most of all on measures reflecting psychological processes.

For this reason, it is useful to think of each value of X as paired with some distribution of Y values, and to discuss this pairing of X values with Y distributions as a *statistical relation*. When each and every X value possible is paired with exactly the same Y distribution, then the variables X and Y are statistically independent, and one says that there is *no* statistical relationship. On the other hand, when different conditional distributions of Y are paired with different values of X, then one says that a statistical relationship *does* exist. For example, in Section 4.10 each event corresponding to a number of hours of sleep deprivation, or a value of X, is paired with a whole conditional distribution of Y, or values representing performance on a motor task. In this example, the conditional distributions were different, and so we could say that a statistical relationship exists, or that X and Y are *associated*. On the other hand, had every conditional distribution of Y given X been the same, the variables would have been independent.

Psychologists using statistics are very interested in comparing distributions to see if they are identical or different in some way; the reason for this interest becomes clear when differences among distributions are viewed as suggesting a statistical relationship or association between variables. Perhaps the psychologist arranges an experiment, and both samples subjects at random and allots them at random to experimental groups. The assignment of a particular subject at random to a particular experimental group is an occurrence of an event X; this may actually be a number, such as the amount of some drug administered, or a qualitative class, such as a particular kind of visual stimulation given the subject. In either instance, the experimenter actually *makes* some experimental events occur in his sample. Then he looks at the distribution of Y for each different X possibility. Are these distributions different? If so, he has some evidence for association between X and Y. The more that specifying the value of X reduces the uncertainty about Y, the stronger is the evidence for association.

In other problems, nature or society has already defined the X events that may occur. Such events as "male," "15 years old," "college sophomore," "rat," "brain-injured," "Democrat," are well-defined classes of observations that may occur in samples. The experimenter actually samples members of several of these classes, and then looks at the distribution of Y within each sample. Or he may sample individuals at random, and look at the distribution of (x,y) pairs. Once again, association is inferred from differences among conditional distributions.

In short, the comparison of obtained Y distributions, conditional to specific X values, or different qualitative categories, gives clues to association between variables. Nevertheless, even though differences among frequency distributions may suggest association, the goodness of this evidence must be evaluated in terms of probability. In particular, the experimenter knows that sample distributions need not be like the true probability distributions, and, unfortunately, association is defined in terms of the true state of affairs for all such observations. The theory of statistical inference gives the ingredients for this judgment about association from samples.

Several things must be considered before we can discuss how this is done. In the first place, more must be said about probability distributions and random variables, and especially about how a purely theoretical distribution can be constructed. Second, it is awkward to compare entire distributions. In most instances distributions can be contrasted much more efficiently in terms of some single number, such as an average. We need to take up the summary character-istics of distributions, ways of compressing the information they contain into much more compact form. Finally, we need to know the rules for statistical infer-ence; how do you decide *how to decide* on the basis of probability?

The first of these issues will be developed in the next chapter. Chap-ter 6 will introduce the common summary measures, both for frequency and probability distributions. The succeeding chapters build toward an answer to the last question.

Exercises

1. In a certain game three decks of cards are used. These are labelled d_1, d_2, and d_3. One deck is selected at random, and then a card is drawn from the deck and its suit (spades, hearts, diamonds, clubs) noted. Make a table of the joint events of the form (deck, suit of card) making up the sample space of possible outcomes.

2. If each of the joint events in the sample space found in problem 1 is equally probable, find the following conditional probabilities:
   (a) $p(\text{spade}|d_1)$
   (b) $p(\text{red suit}|d_3)$
   (c) $p(\text{spade or heart}|d_1 \cup d_2)$
   (d) $p(d_3|\text{diamond})$
   (e) $p(d_1 \cup d_3|\text{club or diamond})$

   What do these conditional probabilities imply about the independence or non-inde-pendence of the "deck" and the "suit" events?

3. In a certain state legislature, there are thirty-six elected legislators. Let A be the set of legislators who are Republican, B the set of legislators who are above 45 years of age, and C those who represent rural districts. For an individual drawn at random from this group, the following probabilities obtain:

$$p(A) = 1/6 \qquad p(B) = 1/3 \qquad p(C) = 4/9$$
$$p(A \cap \bar{B} \cap \bar{C}) = 1/18 \quad p(\bar{A} \cap B \cap \bar{C}) = 7/36 \quad p(\bar{A} \cap \bar{B} \cap C) = 5/18$$
$$p(A \cap B \cap C) = 1/36 \quad p(\bar{A} \cap \bar{B} \cap \bar{C}) = 5/18$$

   Find the probabilities of the following events:
   (a) $(A \cap B)$
   (b) $(A \cap C)$
   (c) $(B \cup C)$
   (d) $(A \cup B) \cap (A \cup C)$
   (e) $(A \cap B) \cup C$
   (f) $(A \cup B \cup C)$
   (g) $\bar{A} \cap (B \cup C)$
   (h) $(\bar{A} \cap \bar{B}) \cup C$

4. For the data of problem 3, find the following conditional probabilities:
   (a) $p(A|B)$
   (b) $p(C|A)$
   (c) $p(B|C)$
   (d) $p(\bar{A}|B)$

(e) $p(\bar{A}|\bar{C})$ (f) $p(A \cap B|C)$
(g) $p(A \cup B|\bar{C})$ (h) $p(B|A \cap C)$
(i) $p(A|C \cup B)$ (j) $p(A \cup B|A \cup C)$

5. Using the probabilities found in problem 3, check the truth of the following statements:
   (a) $p(\bar{A} \cup A|B) = 1.00$
   (b) $p(B|A \cup \bar{A}) = p(B)$
   (c) $p(A \cap C|A \cup C) = \dfrac{p(A \cap C)}{p(A) + p(C) - p(A \cap C)}$
   (d) $p(A \cap B \cap C)p(C|B) = p(C|A \cap B)p(B|C)p(A|B)$
   (e) $p(A \cap C|B)/p(B \cap C|A) = P(A)/p(B)$
   (f) $p[(A \cup B)|C] = p(A|C) + p(B|C) - p(A \cap B|C)$

6. In a psychological experiment, three experimental conditions were used. Let three conditions be symbolized by A_1, A_2, and A_3. Subjects in this experiment were scored on the number of errors they made in performing a task under each of these conditions. The errors were classified as 0, 1, and 2 or more. Suppose that the joint probabilities for numbers of errors and conditions were given by the following table:

ERRORS	CONDITION		
	A_1	A_2	A_3
0	.05	.02	.13
1	.08	.17	.10
2 or more	.20	.15	.10

That is, the joint probability for condition A_1 and 0 errors was .05, and so on. Can one say that experimental condition and number of errors are independent? How should this table have appeared if number of errors had been independent of the experimental conditions?

7. Two random variables, X and Y, have the following joint distribution

X

Y	1	2	3	4
1	$\dfrac{1}{48}$	$\dfrac{1}{16}$	$\dfrac{1}{16}$	$\dfrac{1}{48}$
2	$\dfrac{1}{24}$	$\dfrac{1}{8}$	$\dfrac{1}{8}$	$\dfrac{1}{24}$
3	$\dfrac{1}{24}$	$\dfrac{1}{8}$	$\dfrac{1}{8}$	$\dfrac{1}{24}$
4	$\dfrac{1}{48}$	$\dfrac{1}{16}$	$\dfrac{1}{16}$	$\dfrac{1}{48}$

Show that the conditional distribution within any column (i.e. $p(y|x)$ is the same as the marginal distribution for Y (i.e. $p(y)$). What does this establish about X and Y?

8. Given X and Y as in problem 7, define new random variables as follows

$$Z = 4 - 3X$$
$$W = Y^2$$

Show the joint distribution of Z and W. Are Z and W independent? (Hint: the probability for $W = 4$ is the same as the probability for $Y = 2$, since each corresponds uniquely to the same event.)

9. Using the joint distribution of problem 7 again, this time let

$$Z = X + Y$$
$$W = X - Y$$

Now find the joint distribution of Z and W. (Proceed by writing down all x, y pairs and the corresponding (z,w) pairs; since each (z,w) corresponds to only one (x,y) event, $p(z,w) = p(x,y)$). Are Z and W independent? How can you tell?

10. For Z and W as defined in problem 9, find the marginal (i.e. unconditional) distribution for each of these random variables. Then find
    (a) $p(z = 6|w = 0)$
    (b) $p(4 \angle z \angle 7|w < 0)$
    (c) $p(|w| = 3|z = 5)$

11. In an experiment, two boxes of marbles are used. In Box I, 70 percent of the marbles are black and 30 percent are red. In Box II, 40 percent are black and 60 percent are red. One of the two boxes is drawn at random, and then a marble is drawn at random from the box selected. If the two boxes are equally likely to be selected, and given that a red marble is drawn, what is the probability that it came from Box I? If the odds are 2 to 1 that Box I will be selected, and a black marble is drawn, what is the probability that it came from Box I?

12. Given that $p(A) = .50$, $p(B|A) = .80$, and $p(B|\bar{A}) = .40$, find $p(A|B)$, $p(\bar{A}|B$, $p(A|\bar{B})$, and $p(\bar{A}|\bar{B})$.

13. An archaeologist believes that three different cultures (C_1, C_2, C_3) existed at various times on a given site. Two different kinds of artifacts, A and B, are found on this site. He has reason to believe that the probability of an artifact of type A arising from culture C_1, or $p(A|C_1)$ is .65. In addition he believes that $p(A|C_2) = .40$ and that $p(A|C_3) = .15$. Furthermore, it appears reasonable to assume that the three cultures produced about the same total numbers of surviving artifacts, and that very large numbers of these artifacts exist, all of which are now thoroughly mixed up at the site. Suppose then that an artifact of type A is discovered. If artifacts are equally likely to come from any culture, what is the probability that this artifact represents C_1? C_2? C_3?

14. A passage in a certain historical document could have been the work of any of three writers, W_1, W_2, or W_3. In the known works of writer W_1, the word "thee" appears in place of the word "you" with probability of .90. For writer W_2 this probability is .51, and for W_3 the probability is only .15. If we assume that each writer had equal prior probability of having written any passage, what is the probability that a passage containing the word "thee" was written by W_1? By W_2? By W_3?

15. As a defensive football player who has studied the films of opposing players carefully, you have noticed that when the opposing quarterback in today's game is going to.pass, the probability is about .80 that he wipes his hands just before he calls the play. However, he wipes his hands with probability of only .30 when he does not plan to pass. It is now third down and 4 yards to go, a situation in which the probability of a pass play should be about .60. However, the quarterback does *not* wipe his hands. What is the probability that he will pass on this play?

16. Let A be an attribute consisting of the partition $\{A_1, A_2, A_3, A_4\}$ and let the attribute B consist of the partition $\{B_1, B_2, B_3\}$. Then $A \times B$ consists of the joint events of the form (A_i, B_j). Show that if each joint event has the same probability, the two attributes A and B must be independent. (Hint: begin by finding the probabilities $p(A_i)$ and $p(B_j)$ for each A_i and B_j.)

17. For example 16, show that if all A_i are equally probable and all B_j are equally probable, and if A and B are independent, then all (A_i, B_j) must be equally probable.

18. Generalize the principle of example 11 to any two attributes $A = \{A_1, \ldots, A_i, \ldots, A_N\}$ and $B = \{B_1, \ldots, B_j, \ldots, B_M\}$.

19. Consider an attribute A consisting of the partition $\{A_1, \ldots, A_i, A_j, \ldots, A_N\}$. Show that any pair of events such as A_i and A_j, where $p(A_i) \neq 0$ and $p(A_j) \neq 0$, cannot be independent.

20. Complete the following tables, and indicate whether or not the two attributes are independent in each case:

$p(B_1)$	1/4		1/8
$p(B_2) = 1/2$			
	$p(A_1)$	$p(A_2) = 1/3$	$p(A_3) = 1/6$

$p(B_1)$			2/15
$p(B_2) = 2/3$	4/15	2/15	
$p(B_3) = 1/6$		1/30	0
	$p(A_1)$	$p(A_2) = 1/5$	$p(A_3)$

What does the occurrence of one or more 0 cells in a contingency table say about the attributes involved?

5 Some Discrete Random Variables: The Binomial, Pascal, and Poisson Distributions

The main burden of this chapter is to show how the theoretical distribution of a discrete random variable can be constructed. Principally, we will develop the distribution known as the "binomial," which will play a very important role in much that follows. Then we will also discuss some other discrete random variables that are closely related to the binomial, and likewise have practical applications. However, first of all we will need to show some elementary probability calculations and to introduce some "counting rules" that underlie the development to follow.

5.1 COMPUTING PROBABILITIES

Almost any simple problem in probability can be reduced to a problem in counting. Especially for a sample space containing equally probable elementary events, the computation of probability involves two quantities, both of which are *counts of possibilities:* the total number of elementary events, and the number qualifying for a particular event class. The key to solving probability problems is to learn to ask: "How many distinctly different ways can this event happen?" It may be possible simply to list the number of different elementary events that make up the event-class in question, but it is often much more convenient to use a rule for finding this number. Once the probabilities of events are found, the rules of Chapters 2 and 4 permit one to deduce other probabilities.

There is really no way to become expert in probability computations except by practice in the application of these various counting procedures. Naturally, problems differ widely in the particular principles they involve, but most problems should be approached by these steps:

      1. Determine exactly the sample space of elementary events with which this problem deals.

      2. Find out how many elementary events make up the sample space. What are all the distinct outcomes that might conceivably occur? If the elementary events are equally probable, then the probability of any single elementary event is one over the total number of elementary events.

      3. Decide on the particular events for which probabilities are to be found. How many elementary events qualify for each event class? If no other counting method is available, *list* the elements of these different events and count them. Be sure to remember that if event A can occur in $n(A)$ ways and event B in $n(B)$ ways, then event $(A \cap B)$ might occur in $n(A)n(B)$ ways.

      4. For equally probable elementary events the probability of any event A is simply the ratio of the number of members of A to the total number of elementary events. Even though elementary events are not equally probable, we can still use the fact that elementary events are mutually exclusive to find the probability of any event A by taking the *sum* of the probabilities of all members of that set.

      A game of chance such as roulette shows how probabilities may be computed simply by listing elementary events. A standard roulette wheel has 37 equally spaced slots into which a ball may come to rest after the wheel is spun. These slots are numbered from 0 through 36. One half the numbers 1 through 36 are red, the others are black, and the zero is generally green.

      In playing roulette, you may bet on any single number, on certain groups of numbers, on colors, etc. One such bet might be on the odd numbers (excluding zero of course). The only elementary events in the sample space are the numbers 0 through 36 with their respective colors. Which numbers qualify for the event "odd"? It is easy to count these numbers: they form the set

$$\{1,3,5,7,9,11,13,15,17,19,21,23,25,27,29,31,33,35\}.$$

Since there are exactly 18 elementary events qualifying as odd numbers, if each slot on the roulette wheel is equally likely to receive the ball, the probability of "odd" is 18/37, slightly less than 1/2.

      As a more complicated example that can be solved by listing, let us find the probability that the number contains a "3" as one of the digits. The elementary events qualifying are:

$$\{3,13,23,30,31,32,33,34,35,36\}.$$

Here the probability is 10/37, if the 37 different elementary events are equally likely.

      As a final example, consider the probability of a number that is "even and red" on the roulette wheel. These are

$$\{12,14,16,18,30,32,34,36\}$$

and the probability is 8/37.

There is no end to the examples that might be brought in at this point to show this counting principle in operation, but there is little point in spending more time with it. Just remember that almost all simple problems in probability can be solved in this same way. The important thing to understand is, "When in doubt, make a list." This will usually work, although, as we shall see, other counting methods are often more efficient.

5.2 Sequences of events

In many problems, an elementary event may be *the set of outcomes of a series or sequence of observations*. Such sequences can also be considered joint events, but it is convenient to regard an entire sequence as an elementary event. Suppose that any trial of some simple experiment must result in one of K mutually exclusive and exhaustive events, $\{A_1, \cdots, A_K\}$. Now the experiment is *repeated* N times. This leads to a sequence of events: the outcome of the first trial, the outcome of the second, and so on in order through the outcome of the Nth trial. The outcome of the whole series of trials might be the sequence $(A_3, A_1, A_2, \cdots, A_3)$. This denotes that the event A_3 occurred on the first trial, A_1 on the second trial, A_2 on the third, and so on. The place in order gives the trial on which the event occurred, and the symbol occupying that place shows the event occurring for that trial. For example, it might be that the simple experiment is drawing one marble at a time from a box, with replacement after each drawing. The marbles observed may be red, white, or black. For four drawings a sequence of outcomes might be (R,W,B,W), or perhaps (W,W,B,B).

The notion of a Cartesian product can be useful when we are thinking of sequences of events. If one defines the set of possible outcomes of one draw in the simple experiment just described as $S = \{R,B,W\}$, then the outcome of two successive draws is an elementary event in the new sample space $S' = S \times S$. Each elementary event in this new sample space consists of a *pair* of outcomes (outcome of first draw, outcome of second draw). We can further visualize the set of sequences resulting from three draws as $S'' = S \times S \times S$, the set of triples of outcomes (outcome of first draw, outcome of second draw, outcome of third draw). The sample space for four draws can be represented as $S''' = S \times S \times S \times S$, and so on.

The idea of a sequence also extends to a series of *different* simple experiments. Here the place in the sequence corresponds to the particular experiment being performed. For example, suppose that on the first trial, the experimenter draws a marble from a box, on the second he flips a coin, and on the third he tosses a die. Then the sequence of events observed might be (R,Heads,6) for example, where the position describes the particular simple experiment, and the symbol in that position tells the observed event. In this instance, we might symbolize the sample space for the first experiment by $\mathfrak{I} = \{R,B,W\}$, for the second experiment by $\mathfrak{U} = \{Heads,Tails\}$, and for the third experiment by

$$\mathfrak{V} = \{1,2,3,4,5,6\}.$$

Then the new sample space consisting of sequences of outcomes would be the Cartesian product $\mathcal{S} = \mathcal{I} \times \mathcal{U} \times \mathcal{V}$, composed of triples of outcomes such as (R,Heads,6).

Remember that each possible sequence can be thought of in two ways: it is a joint event from the standpoint of a series of simple experiments, but it is also an elementary event if one thinks of the *experiment itself* as the series of trials with a sample space consisting of *n*-tuples, or possible sequences of outcomes. Each *whole* or *compound experiment* has as its outcome one and only one sequence. Such experiments producing sequences as outcomes are very important to statistics, and so we will begin our study of counting rules by finding how many different sequences a given series of trials could produce.

5.3   NUMBER OF POSSIBLE SEQUENCES FOR N TRIALS: COUNTING RULE 1

Suppose that a series of N trials were carried out, and that on each trial any of K events might occur. Then the following rule holds:

Counting rule 1:   If any one of K mutually exclusive and exhaustive events can occur on each of N trials, then there are K^N different sequences that may result from a set of trials.

As an example of this rule, consider a coin's being tossed. Each toss can result only in a H or a T event ($K = 2$). Now the coin is tossed five times, so that $N = 5$. The total number of *possible* results of tossing the coin five times is $K^N = 2^5 = 32$ sequences. Exactly the same number is obtained for the possible outcomes of tossing five coins simultaneously, if the coins are thought of as numbered and a sequence describes what happens to coin 1, to coin 2, and so on.

As another example, the outcome of tossing two dice is a sequence: what number comes up on the first die, and what number comes up on the second. Here $K = 6$ (six different numbers per die), and $N = 2$. There are exactly $6^2 = 36$ different sequences possible as results of this experiment.

5.4   COUNTING RULE 2

Sometimes the number of possible events in the first trial of a series is different from the number possible in the second, the second different from the third, and so on. It is obvious that if there are different numbers $K_1, K_2, \cdots, K_N$ of events possible on the respective trials, then the total number of sequences will not be given by rule 1. Instead, the following rule holds:

Counting rule 2:   If $K_1, \cdots, K_N$ are the numbers of distinct events that can occur on trials 1, $\cdots$, N in a series, then the number of different sequences of N events that can occur is $(K_1)(K_2) \cdots (K_N)$.

For example, suppose that for the first trial you toss a coin (two possible outcomes) and for the second you roll a die (six possible outcomes). Then the total number of different sequences would be $(2)(6) = 12$.

Notice that counting rule 1 is actually a special case of rule 2. If the same number K of events can occur on any trial, then the total number of sequences is K multiplied by itself N times, or K^N.

Both counting rules 1 and 2 follow as a simple consequence if the sample space of sequences is viewed as a Cartesian product. Suppose that on any trial the sample space S consists of K possible events. Then, if there are two trials, the new sample space $S' = S \times S$ must consist of $K \times K$ or K^2 pairs of events. For three trials, $S'' = S \times S \times S$ will consist of K^3 triples of events, and so on, so that for N trials there must be K^N possible n-tuples of events.

Arguing in the same way, if the sample space S_1 for the first trial contains K_1 possibilities, and sample space S_2 contains K_2 possibilities, the total set of sequences of two trials must contain $K_1 \times K_2$ possibilities. Three trials will provide a sample space $S = S_1 \times S_2 \times S_3$ that consists of $K_1 \times K_2 \times K_3$ possibilities, and so on.

5.5   COUNTING RULE 3: PERMUTATIONS

A rule of extreme importance in probability computations concerns the number of ways that objects may be arranged in order. The rule will be given for arrangements of objects, but it is equally applicable to sequences of events:

Counting rule 3:   The number of different ways that N distinct things may be arranged in order is $N! = (1)(2)(3) \cdots (N-1)(N)$, (where $0! = 1$). An arrangement in order is called a permutation, so that the total number of permutations of N objects is $N!$ The symbol $N!$ is called "N factorial."

As an illustration of this rule, suppose that a classroom contained exactly 10 seats for 10 students. How many ways could the students be assigned to the chairs? Any of the students could be put into the first chair, making 10 possibilities for chair 1. But, given the occupancy of chair 1, there are only 9 students for chair 2; the total number of ways chair 1 and chair 2 could be filled is $(10)(9) = 90$ ways. Now consider chair 3. With chairs 1 and 2 occupied, 8 students remain, so that there are $(10)(9)(8) = 720$ ways to fill chairs 1, 2, and 3. Finally, when 9 chairs have been filled there remains only 1 student to fill the remaining place, so that there are

$$(10)(9)(8)(7)(6)(5)(4)(3)(2)(1) = (10)! = 3628800$$

ways of arranging the 10 students into the 10 chairs.

Now suppose that there are only N different events that can be observed in N trials. Imagine that the occurrence of any given event "uses up" that event for the sequence, so that each event may occur *once and only once* in the sequence. In this case there must be $N!$ different orders in which these events might occur in sequence. Each sequence is a permutation of the N possible events.

To take a homely example, suppose that a man is observed dressing. At each point in our observation there are three articles of clothing he might put on: his shoes, his pants, or his shirt. However each article can be put on only once. We might observe the following sequences:

(shoes, pants, shirt)
(shoes, shirt, pants)
(shirt, shoes, pants)
(shirt, pants, shoes)
(pants, shoes, shirt)
(pants, shirt, shoes)

The man's putting on any one of these articles of clothing literally uses up that outcome for any sequence of observations, so that there are $N! = (1)(2)(3) = 6$ possible permutation sequences.

The procedure of counting sequences as permutations in order is especially important for sampling without replacement. For example, suppose that a teacher has the names of five students in a hat. She draws the names out at random one at a time, *without* replacement. Then there are exactly 5! or 120 different sequences of names that she might observe. If all sequences of names are equally likely to be drawn, then the probability for any one sequence is 1/120.

What is the probability that the name of any given child will be drawn first? Given that a certain child is drawn first, the order of the remainder of the sequence is still unspecified. There are $(N - 1)! = 4! = 24$ different orders in which the other children can appear. Thus the probability of any given child being first in sequence is $24/120 = .20$ or $1/5$.

Suppose that exactly *two* of the names in the hat are girls'. What is the probability that the first two names drawn belong to the girls? Given the first two names are girls', there are still $3! = 6$ ways three boys may be arranged. The girls' names may themselves be ordered in two ways. Thus, the probability of the girls' names being drawn first and second is $(2)(6)/120 = 1/10$.

5.6 COUNTING RULE 4: ORDERED COMBINATIONS

Sometimes it is necessary to count the number of ways that r objects might be selected from among some N objects in all, $(r \leq N)$. Furthermore, each different *arrangement* of the r objects is considered separately. Then the following rule applies:

Counting rule 4: The number of ways of selecting and arranging r objects from among N distinct objects is $\dfrac{N!}{(N - r)!}$.

The reasoning underlying this rule becomes clear if a simple example is taken: Consider a classroom teacher, once again, who has 10 students to be assigned to seats. This time, however, imagine that there are only 5 seats. How many different ways could the teacher select 5 students and arrange them into the available seats? Notice that there are 10 ways that the first seat might be filled, 9 ways for the second, and so on, until seat 5 could be filled in 6 ways. Thus there are

$$(10)(9)(8)(7)(6)$$

ways to select students to fill the 5 seats. This number is equivalent to

$$\frac{10!}{5!} = \frac{N!}{(N-r)!},$$

the number of ways that 10 students out of 10 may be selected and arranged, but divided by the number of arrangements of the 5 *unselected* students in the 5 *missing* seats.

As an example of the use of this principle in probability calculations, consider the following: In lotteries it is usual for the first person whose name is drawn to receive a large amount, the second some smaller amount, and so on, until some r prizes are awarded. This means that some r names are drawn in all, and the *order* in which those names are drawn determines the size of the prizes awarded to individuals. Suppose that in a rather small lottery 40 tickets had been sold, each to a different person, and only 3 were to be drawn for first, second, and third prizes. Here $N = 40$ and $r = 3$. How many different assignments of prizes to persons could there be? The answer, by counting principle 4, is

$$\frac{40!}{(40-3)!} = \frac{40!}{37!} = (38)(39)(40) = 59,280.$$

On how many of these possible sequences of winners would a given person, John Doe, appear as first, second, *or* third prize winner? If he were drawn first, the number of possible selections for second and third prize would be $39!/37! = 1482$. Similarly, there would be 1482 sequences in which he could appear second, and a like number where he could appear third. Thus, the probability that he appears in a sequence of three drawings, winning first, second, or third prize is (by axiom 3, Section 2.5)

$$p(\text{first, second, or third}) = \frac{3(1482)}{59,280} = \frac{3}{40}.$$

5.7 COUNTING RULE 5: COMBINATIONS

In a very large class of probability problems, we are not interested in the *order* of events, but only in the number of ways that r things could be selected from among N things, *irrespective of order*. We have just seen that the total number of ways of selecting r things from N and ordering them is $\dfrac{N!}{(N-r)!}$, by rule 4. Each set of r objects has $r!$ possible orderings, by rule 3. A combination of these two facts gives us

Counting rule 5:   The total number of ways of selecting r distinct combinations of N objects, irrespective of order, is

$$\frac{N!}{r!\,(N-r)!} = \binom{N}{r}.$$

The symbol $\binom{N}{r}$ is *not* a symbol for a fraction, but instead denotes the number of combinations of N things, taken r at a time. Sometimes the number of combinations is known as a "binomial coefficient," and occasionally $\binom{N}{r}$ is replaced by the symbols $^{N}C_r$ or $_{N}C_r$. However, the name and symbol introduced in rule 5 will be used here.

It is helpful to note that

$$\binom{N}{r} = \binom{N}{N-r}.$$

Thus, $\binom{10}{3} = \binom{10}{7}$, $\binom{50}{49} = \binom{50}{1}$, and so on.

As an example of the use of this rule, suppose that a total of 33 men were candidates for the board of supervisors in some community. Three supervisors are to be elected at large; how many ways could 3 men be selected from among these 33 candidates? Here, $r = 3$, $N = 33$, so that

$$\binom{N}{r} = \frac{(33)!}{3!\,(30)!} = \frac{(1 \cdot 2 \cdot \ \cdots \ \cdot 32 \cdot 33)}{(1 \cdot 2 \cdot 3)(1 \cdot 2 \cdot \ \cdots \ \cdot 29 \cdot 30)}.$$

Cancelling in numerator and denominator, we get

$$\binom{33}{3} = \frac{(31)(32)(33)}{6} = (31)(16)(11) = 5456.$$

If all sets of three men are equally likely to be chosen, the probability for any given set is $1/5456$.

Because of their utility in probability calculations, a table of $\binom{N}{r}$ values for various values of N and r is included in Appendix C. Although this table shows values of N only up to 20, and values of r up to 10, other values can be found by the relation given above, and also by the relation known as Pascal's Rule:

$$\binom{N}{r} = \binom{N-1}{r-1} + \binom{N-1}{r}.$$

Still other values may be worked out from the table of factorials also included in the Appendix C, in terms of rule 5.

5.8 SOME EXAMPLES: POKER HANDS

These five rules provide the basis for many calculations of probability when the number of elementary events is finite, especially when sampling is without replacement. One of the easiest examples of their application is the calculation of probabilities for poker hands.

The game of poker as discussed here will be highly simplified and not very exciting to play: the player simply deals five cards to himself from a well-shuffled standard deck of fifty-two cards. Nevertheless, the probabilities of the various hands are interesting, and can be computed quite easily.

The particular hands we will examine are the following:

    a. one pair, with three different remaining cards
    b. full house (three of a kind, and one pair)
    c. flush (all cards of the same suit).

First of all we need to know how many different *hands* may be drawn. A given hand of five cards will be thought of as an event; notice that the order in which the cards appear in a hand is immaterial, so that a number of sequences of cards can correspond to a given hand. By rule 5, since there are 52 different cards in all and only 5 are selected, then there are

$$\binom{52}{5} = \frac{52!}{5!\,47!}$$

different hands that might be drawn. If all hands are equally likely, then the probability of any given one is $1 \Big/ \binom{52}{5}$.

There are 13 numbers that a pair of cards may show (counting the picture cards), and each member of a pair must have a suit. Thus by rule 5 there are $13\binom{4}{2}$ different pairs that might be observed. The remaining cards must show three of the twelve remaining numbers and each of the three cards may be of any suit. By rules 1 and 5 there are $\binom{12}{3}(4)(4)(4)$ ways of filling out the hand in this way. Finally, we find that there are

$$13\binom{4}{2}\binom{12}{3}4^3$$

different ways for this event to occur. Thus,

$$p(\text{one pair}) = \frac{13\binom{4}{2}\binom{12}{3}4^3}{\binom{52}{5}}.$$

This number can be worked out by writing out the factorials, cancelling in numerator and denominator, and dividing. It is approximately equal to .42. The chances are roughly four in ten of drawing a single pair in five cards, if all possible hands are equally likely to be drawn.

This same scheme can be followed to find the probability of a full house. There are thirteen numbers that the three of a kind may have, and then twelve numbers possible for the pair. The three of a kind must represent three of four suits, and the pair two of four suits. This gives

$$13 \binom{4}{3} 12 \binom{4}{2}$$

different ways to get a full house. The probability is

$$p(\text{full house}) = \frac{13 \binom{4}{3} 12 \binom{4}{2}}{\binom{52}{5}}$$

which is about .0014.

Finally, a flush is a hand of five cards all of which are in the same suit. There are exactly four suit possibilities, and a selection of five out of thirteen numbers that the cards may show. Hence, the number of different flushes is $4 \binom{13}{5}$, and

$$p(\text{flush of five cards}) = \frac{4 \binom{13}{5}}{\binom{52}{5}}$$

or about .002.

The probabilities of the various other hands can be worked out in a similar way. The point of this illustration is that it typifies the use of these counting rules for actually figuring probabilities of complicated events, such as particular poker hands. Naturally, a great deal of practice is usually necessary before one can visualize and carry out probability calculations from "scratch" with any facility. Nevertheless, such probability calculations usually depend upon counting how many ways events of a certain kind can occur, if the elementary events are finite in number.

Now we will turn to a special use of these counting rules in finding the distribution of a discrete random variable.

5.9   Bernoulli trials

The very simplest probability distribution is one with only two event classes. For example, a coin is tossed and one of two events, heads or tails must occur, each with some probability. Or a normal human being is selected at random and his sex noted: the outcome can be only Male or Female. Such an experiment or process that can eventuate in only one of two outcomes is usually called a Bernoulli trial, and we will call the two event classes and their associated probabilities a Bernoulli process.

In general, one of the two events is called a "success" and the other a "failure" or "nonsuccess." These names serve only to tell the events apart, and are not meant to bear any connotation of "goodness" of the event. In the discussion to follow, the symbol p will stand for the probability of a success, and $q = 1 - p$ for the probability of a failure. Thus, in tossing a fair coin, let a head be a success. Then $p = 1/2$, $q = 1 - p = 1/2$. If the coin is biased, so that heads are twice as likely to come up as tails, then $p = 2/3$, $q = 1/3$. In the following, take care to distinguish between p, standing for the probability of a success, and $p(x)$, standing for the probability of some value of a random variable.

5.10   SAMPLING FROM A BERNOULLI PROCESS

Suppose that some sample space exists fitting a Bernoulli process. Furthermore, suppose that either we sample independently with replacement, or that an infinite number of elementary events exist, so that for each sample observation out of N trials, p and q are unchanged. When the outcome is generated by the same process on every trial, so that p and q remain constant over the trials, the process is said to be **stationary**. If the p and q values change from trial to trial, as may well be the case in some practical situations, then the Bernoulli process is said to be nonstationary. For the time being, we will confine our attention to stationary processes and to independent trials made from such a process.

Now we proceed to make N independent observations. Let N be, say, 5. How many different sequences of five outcomes could be observed? The answer by rule 1, is $2^5 = 32$. *However, here it is not necessarily true that all sequences will be equally probable.* The probability of a given sequence depends upon p and q, the probabilities of the two events. Fortunately, since trials are independent one can compute the probability of any sequence by the application of rule **4.6.3.**

We want to find the probability of the particular sequence of events:

$$(S, S, F, F, S)$$

where S stands for a success and F for failure. The probability of first observing an S is p. If the second observation is independent of the first, then by rule **4.6.3,**

probability of $(S, S) = p \cdot p = p^2$.

The probability of an F on the third trial is q, so that the probability of (S, S) followed by F is p^2q. In the same way the probability of $(S, S, F, F) = p^2q^2$, and that of the entire sequence is $p^2q^2p = p^3q^2$.

The same argument shows that the probability of the sequence (S, F, F, F, F) is pq^4, that of (S, S, S, S, S) is p^5, of (F, S, S, S, F) is p^3q^2, and so on.

Now if we write out all of the possible sequences and their probabilities, an interesting fact emerges: **the probability of any given sequence of N independent Bernoulli trials depends only on the number of successes and p,**

the probability of a success. That is, regardless of the *order* in which successes and failures occur in a sequence, the probability is

$$p^r q^{N-r} \qquad\qquad [5.10.1]$$

where r is the number of successes, and $N - r$ is the number of failures. Suppose that in a sequence of 10 trials, exactly 4 successes occur. Then the probability of that particular sequence is

$$p^4 q^6.$$

If $p = 2/3$, then the probability can be worked out from

$$\left(\frac{2}{3}\right)^4 \left(\frac{1}{3}\right)^6.$$

The same procedure would be followed for any r successes out of N trials for any p.

For example, if we toss a fair coin ($p = 1/2$) six times, what is the probability of observing three heads followed in order by three tails? The answer is

$$p^3 q^3 = \left(\frac{1}{2}\right)^3 \left(\frac{1}{2}\right)^3 = \frac{1}{64}.$$

This is also the probability of the sequence (H, T, H, T, H, T), of the sequence (H, T, T, T, H, H), and of any other sequence containing exactly three successes or heads.

The probabilities just found are for *particular sequences*, arrangements of r successes and $N - r$ failures in a certain order. What we have found is that if we want to know the probability of a particular sequence of outcomes of independent Bernoulli trials, that sequence will have the same probability as any other sequence with exactly the same number of successes, given N and p.

In most instances, however, one is not especially interested in particular sequences *in order*. We would like to know probabilities of given numbers of successes *regardless* of order in which they occur. For example, when a coin is tossed five times, there are several sequences of outcomes where exactly two heads occur:

$$(H, H, T, T, T)$$
$$(H, T, H, T, T)$$
$$(H, T, T, H, T)$$
$$(H, T, T, T, H)$$
$$(T, H, T, T, H)$$
$$(T, H, H, T, T)$$
$$(T, H, T, H, T)$$
$$(T, T, H, H, T)$$
$$(T, T, H, T, H)$$
$$(T, T, T, H, H).$$

Each and every one of these different sequences must have the same probability, $p^2 q^3$, since each shows exactly two successes and three failures. Notice that there are $\binom{N}{r} = \binom{5}{2} = 10$ different such sequences, exactly as counting rule 5 gives for the number of ways 5 things can be taken 2 at a time.

What we want now is the probability that $r = 2$ successes will occur *regardless* of order. This could be paraphrased as "the probability of the sequence (H, H, T, T, T) *or* the sequence (H, T, H, T, T) *or* any other sequence showing exactly 2 successes in 5 trials." Such "or" statements about mutually exclusive events recall axiom 3 in Section 2.5: if A and B are mutually exclusive events, then $p(A \cup B) = p(A) + p(B)$. Thus, the probability of 2 successes in any sequence of 5 trials is

$$p(2 \text{ successes in 5 trials}) = p^2 q^3 + p^2 q^3 + \cdots + p^2 q^3 = \binom{5}{2} p^2 q^3$$

since each of these sequences has the same probability and there are $\binom{5}{2}$ of them.

Generalizing this idea for any r, N, and p, we have the following principle: **in sampling from a stationary Bernoulli process, with the probability of a success equal to** p, **the probability of observing exactly** r **successes in** N **independent trials is**

$$p(r \text{ successes};N,p) = \binom{N}{r} p^r q^{N-r}. \qquad \text{[5.10.1*]}$$

In understanding the basis for this rule, the thing to keep in mind is that $p^r q^{N-r}$ is the probability of *any* of the events consisting of a specific sequence showing exactly r successes out of N trials. Then $\binom{N}{r}$ is the number of such sequence events that qualify for the event "exactly r successes in N trials." It is important to notice that there is an exact correspondence between the binomial coefficients $\binom{N}{r}$, and the number of sequences possible in which exactly r successes occur out of N trials.

An experiment carried out in such a way that N independent trials are made from a stationary Bernoulli process is known as **binomial sampling.** In binomial sampling the value of N is predetermined, and it is the value of r, the number of successes, that is left to chance.

For example, imagine that in some very large population of animals 80 percent of the individuals have normal coloration, and only 20 percent are albino (no skin and hair pigmentation). This may be regarded as a Bernoulli process, with "albino" being a success and "normal" a failure. Suppose that the probabilities are $.20 = p$ and $.80 = q$. A biologist manages to sample this population at random, catching three animals. What is the probability that he will catch one albino? Here, $N = 3$, $r = 1$, so that

$$p(1 \text{ albino in 3 animals}) = \binom{3}{1}(.20)^1(.80)^2$$
$$= .384.$$

If the sampling is random, and if the population is so large that sampling without replacement still permits one to regard the results of the successive trials as independent, then the biologist has about 38 chances in 100 of observing exactly 1 albino in his sample of 3 animals.

As a more complex example, consider a psychological experiment in which an animal has the task of running through a maze. The maze has eight choice points, where the animal has to go either to the right or to the left. If the animal is equally likely to choose the right or the wrong path on any choice point, and if successive choices are independent, what is the probability that he will make exactly four errors in traversing the maze? Here the sample space is the set of all possible behaviors for the animal at choice points: these are grouped into only two classes, "right" or "wrong." The eight points in the maze are regarded as a random sample of such choice behaviors, and each occurrence of a choice behavior is assumed to be independent of every other. Then,

$$p(4 \text{ errors in 8 trials}) = \binom{8}{4} (.5)^4 (.5)^4$$

$$= \frac{(8)(7)(6)(5)}{(1)(2)(3)(4)} (.5)^8 = .27.$$

This event of 4 errors in 8 trials should occur with relative frequency of about 27 in 100 independent repetitions of the experiment.

5.11   NUMBER OF SUCCESSES AS A RANDOM VARIABLE: THE BINOMIAL DISTRIBUTION

When samples of N trials are taken from a Bernoulli process, the number of successes is a discrete random variable. Since the various values are counts of successes out of N observations the random variable can take on only the whole values from 0 through N. We have just seen how the probability for any given number of successes can be found. Now we will discuss the distribution pairing each possible number of successes with its probability. This distribution of number of successes in N trials is called the **binomial distribution**. This is the first distribution we have studied that can easily be described not only by listing or graphic methods but also by its mathematical rule.

A binomial distribution can be illustrated most simply as follows: Consider a simple experiment repeated independently five times. Each trial must result in only one of two outcomes and the result of five trials is a sequence of outcomes like those in the preceding sections. However, we are not at all interested in the order of the outcomes, but only in the number of successes in the set of trials.

In order to find the probability for each value of the discrete random variable, given N and p, let us begin with the largest value, $X = 5$. By counting rule 5, there must be $\binom{5}{5} = 1$ possible sequence where all of the outcomes are successes. The probability of this sequence is $p^5 q^0 = p^5$, so that we have

$$p(X = 5; N = 5, p) = \binom{5}{5} p^5 = p^5.$$

For four successes, we find that by counting rule 5,

$$\binom{5}{4} = \frac{5!}{4!\,1!} = \frac{(1)(2)(3)(4)(5)}{(1)(2)(3)(4)(1)} = 5,$$

so that four successes can appear in five different sequences. Each sequence has probability $p^4 q^1$. Thus,

$$p(X = 4; N = 5, p) = \binom{5}{4} p^4 q^1 = 5p^4 q^1.$$

Going on in this way, we find

$$p(X = 3; N = 5, p) = \binom{5}{3} p^3 q^2 = 10p^3 q^2$$

$$p(X = 2; N = 5, p) = \binom{5}{2} p^2 q^3 = 10p^2 q^3$$

$$p(X = 1; N = 5, p) = \binom{5}{1} pq^4 = 5pq^4$$

$$p(X = 0; N = 5, p) = \binom{5}{0} q^5 = q^5.$$

(Note that $\binom{5}{0} = \frac{5!}{0!\,5!} = 1$, since 0! is 1 by definition.)

To take a concrete instance of this binomial distribution, let the experiment be that of tossing a fair coin five times. Then $p = 1/2$, and the binomial distribution for the number of heads is

x	$p(x)$	
5	$(1/2)^5$	$= 1/32$
4	$5(1/2)^4(1/2)$	$= 5/32$
3	$10(1/2)^3(1/2)^2$	$= 10/32$
2	$10(1/2)^2(1/2)^3$	$= 10/32$
1	$5(1/2)(1/2)^4$	$= 5/32$
0	$(1/2)^5$	$= 1/32$
		$32/32$

Notice that the probabilities over all values of X sum to 1.00, just as they must for any probability distribution.

Now consider another example that is formally identical to this last one, but provides different probability values. Suppose that among American male college students who are undergraduates, only one in ten is married. A sample of five male students is drawn at random. Let X be the number of married students observed. (We will assume the total set of students to be large enough that sampling can be without replacement without affecting the probabilities and that observations are independent.) Here, $p = .10$ and the distribution is

x	$p(x)$	
5	$(1/10)^5$ =	1/100,000
4	$5(1/10)^4(9/10)$ =	45/100,000
3	$10(1/10)^3(9/10)^2$ =	810/100,000
2	$10(1/10)^2(9/10)^3$ =	7,290/100,000
1	$5(1/10)(9/10)^4$ =	32,805/100,000
0	$(9/10)^5$ =	59,049/100,000
		100,000/100,000

Contrast this distribution with the preceding one: When p was $1/2$ the distribution showed the greatest probability for $X = 2$ and $X = 3$, with the probabilities diminishing gradually both toward $X = 0$ and toward $X = 5$. On the other hand, in the second distribution, the most probable value of X is 0, with a steady decrease in probability for the values 1 through 5. The distribution over such values of X is very different in these two situations, even though the probabilities are found by exactly the same *formal* rule. This illustrates that **the binomial is actually a family of theoretical distributions, each following the same mathematical rule for associating probabilities with values of the random variable, but differing in particular probabilities depending upon N and p.**

The general definition of the binomial distribution can be stated as follows:

Any random variable X with probability function given by

$$p(X = r;N,p) = \binom{N}{r} p^r q^{N-r}, \qquad 0 \le X \le N,$$

is said to have a binomial distribution with parameters N and p.

The unspecified mathematical constants such as N and p that enter into the rules for probability or density functions indicate **parameters.** Families of distributions share the same mathematical rule for assigning probabilities or probability densities to values of X; in these rules the parameters are simply symbolized as constants. Actually assigning values to the parameters, such as we did for p and N in the distributions above, gives some particular distribution belonging to the family. Thus, *the* binomial distribution usually refers to the family of distributions having the same rule, and *a* binomial distribution is a particular one of this family found by fixing N and p.

Almost all theoretical distributions of interest in statistics can be specified by stating the function rule. The way this simplifies the discussion of distributions will be obvious as we go along; indeed, continuous distributions cannot really be discussed at all except in terms of their function rule.

5.12 THE BINOMIAL DISTRIBUTION AND THE BINOMIAL EXPANSION

In school algebra you were very likely taught how to expand an expression such as $(a + b)^n$ by the following rule:

$$(a + b)^n = a^n + \frac{n!}{(n-1)!\,1!} a^{n-1}b + \frac{n!}{(n-2)!\,2!} a^{n-2}b^2 + \cdots$$
$$+ \frac{n!}{1!\,(n-1)!} ab^{n-1} + b^n.$$

For example, $(a + b)^3 = a^3 + 3a^2b + 3ab^2 + b^3$ according to this rule. This is the familiar "binomial theorem" for expanding a sum of two terms raised to a power.

Notice that the various probabilities in the binomial distribution are simply terms in such a binomial expansion. Thus, if we take $a = p$, $b = q$, and $n = N$,

$$(p + q)^N = p^N + \binom{N}{N-1} p^{N-1}q + \binom{N}{N-2} p^{N-2}q^2 + \cdots + q^N.$$

Since $p + q$ must equal 1.00, then $(p + q)^N = 1.00$, and the sum of all of the probabilities in a binomial distribution is 1.00.

5.13 PROBABILITIES OF INTERVALS IN THE BINOMIAL DISTRIBUTION

In Chapter 3 we saw how to find a probability that a value of a random variable lies in an interval, such as $p(2 \le X \le 8)$, the probability that the random variable X takes on some value between 2 and 8 inclusive. This idea is easy to extend to a binomial variable.

Consider the binomial distribution shown in Table 5.13.1 (see p.187), $p = .3$ and $N = 10$.

First of all we will find the probability $p(1 \le X \le 7)$, that X lies between the values 1 and 7 inclusive. This is given by the sum

$p(X = 1)$	.12106
$+p(X = 2)$	.23347
$+p(X = 3)$	.26683
$+p(X = 4)$	.20012
$+p(X = 5)$	.10292
$+p(X = 6)$	.03676
$+p(X = 7)$	.00900
	.97016 $= p(1 \le X \le 7)$.

The probability is about .97 that an observed value of X will lie between 1 and 7 inclusive. Thus, if we were drawing random samples of 10 observations with replacement from a Bernoulli process where the probability of a success were .3, we should be very likely to observe a number of successes between 1 and 7 inclusive.

By the same token, we can find

$$p(8 \angle X) = p(X = 8) + p(X = 9) + p(X = 10) = .00160.$$

The chances are less than two in a thousand of observing eight or more successes, *if p* = .30.

Notice that

$$p(X = 0, \text{ or } 8 \angle X) = .02824 + .00160 = .02984,$$

which is the same as

$$1 - p(1 \angle X \angle 7) = 1 - .97016 = .02984,$$

so that this is also that probability that X falls *outside* the interval bounded by 1 and 7.

Binomial distributions can also be put into cumulative form. Thus, the probability that X falls at or below a certain value a is the probability of the interval $X \angle a$. For this particular distribution, the corresponding cumulative distribution is

r	$p(X \angle r)$
10	1.00000
9	.99999
8	.99985
7	.99840
6	.98940
5	.95264
4	.84972
3	.64960
2	.38277
1	.14930
0	.02824

In this distribution, we see that about 65 percent of samples of 10 should show 3 or fewer successes, about 85 percent should have 4 or fewer, 99.8 percent should have 7 or fewer. Every sample (100 percent) must have ten or fewer successes, of course, since $N = 10$.

It must be re-emphasized that the binomial distribution is *theoretical*. It shows the probabilities for various numbers of successes out of N trials *if* independent random samplings are carried out from a stationary Bernoulli process and *if p* is the probability of a success. Given a different value of p or of N (or of both) the probabilities will be different. Nevertheless, regardless of N or p, the probabilities are found by the same, binomial, rule.

Table 5.13.1

r	$\binom{N}{r} p^r q^{N-r} = p(X = r)$
10	.00001
9	.00014
8	.00145
7	.00900
6	.03676
5	.10292
4	.20012
3	.26683
2	.23347
1	.12106
0	.02824
	1.00000

Table II of Appendix C gives binomial probabilities for $0 \leq r \leq N$ and $1 \leq N \leq 20$, for selected values of p. For p values which are greater than .50, one simply deals with $q = 1 - p$ and with the number of failures, or $N - r$. Thus, for $p = .70$ and $N = 10$, the probability of 6 successes is the same as the probability of 4 failures given $q = .30$. In the table under $N = 10, r = 4$, and $p = .30$, this is .2001. Cumulative probabilities such as $p(X \leq r;N, p)$ are found from the sum of the values of $p(x)$ for $x = 0, 1, \cdots, r$, given N and p. Probabilities for larger values of N can be found directly by use of the table of factorials provided in Appendix C.

5.14   THE BINOMIAL DISTRIBUTION OF PROPORTIONS

Quite often researchers are interested not in the number of successes that occur for some N trial observations, but rather in the *proportion* of successes, r/N. The proportion of successes is also a random variable, taking on fractional (or decimal) values between 0 and 1.00. Such sample proportions will be designated by the capital letter P, to distinguish them from p, the probability of a success.

The probability of any given proportion P of successes among N cases sampled from a given Bernoulli process is exactly the same as the probability of the number of successes; that is

$$p\left(P = \frac{r}{N}\right) = p(X = r).$$ [5.14.1]

For instance, if $N = 6$ and $r = 4$

$$p(X = 4) = p\left(P = \frac{4}{6}\right) = \binom{6}{4} p^4 q^2.$$

The distribution of sample proportions is therefore given by the binomial distribution, the only difference being that each possible value of X becomes a value of $P = X/N$, so that prob.$(P = a) = $ prob.$(X = Na)$ for any particular value a. In the further discussion of the binomial distribution in Chapter 8, care will be taken to specify whether the random variable is regarded as X or P, since the arithmetic is slightly different in the two situations. Nevertheless, the probability of any given P is the same as for the corresponding X, given the sample size N.

5.15   The form of a binomial distribution

Although the mathematical rule for a binomial distribution is the same regardless of the particular values of N and p entering into the expression, the "shape" of a histogram or other representation of a binomial distribution will depend upon N and p. In general, the probabilities increase for increasing values of X until some maximum point is reached, and then for X values beyond this point the probabilities decrease once again. This makes the picture given by the distribution show a single "hump" somewhere between $X = 0$ and $X = N$, with probabilities gradually decreasing to either side of this maximum point. The exact location of this highest probability relative to X depends, of course, on N and p.

It can be shown that when Np is an integer, the probability associated with $X = Np$ is greater than that for any other value of X; in this situation, the maximum point in a graph of the binomial distribution occurs at $X = Np$. Therefore, one finds that in this special situation the value $P = p$ is the most probable relative frequency of occurrence of an event with probability p.

On the other hand, it may be that X cannot attain the exact value Np, since Np need not be an integer. In this situation the value $X = r$ with the highest probability is *no more* than p or q away from the value of Np; that is

$$-q \leq r - Np \leq p,$$

where r is the value of X having highest probability in the binomial distribution. Actually, the statement interpreting Bernoulli's theorem in Section 2.8 should be read as meaning "the most probable proportion of the event's occurrences to the total number of trials is also p, whenever that value of X/N can actually occur." Nevertheless, the most probable relative frequency is always a number *close* to the true value of p, even though this exact value may not occur for some values of N.

5.16   The binomial as a sampling distribution

A binomial distribution is one example of a very important kind of theoretical distribution encountered again and again in statistics. These are sampling distributions, where a random variable X denotes a possible value of some measure *summarizing* the outcomes of N distinct observations. In the binomial case, this random variable is either the number of successes out of N

trials, or, perhaps, P, the proportion of successes out of the N trials. In other contexts the random variable may be some other number summarizing a sample result, such as an average, or some index of variability. There are many different kinds of sampling distributions, depending on the basic sample space and the way that samples are summarized. Nevertheless, the essential idea is the same: a sample of N observations is drawn at random, so that each and every distinct sample possible has an equal probability of occurrence. A number is attached to the sample, in some way summarizing the N outcomes observed. In principle, different values will be found for different samples, and each sample-value may be regarded as a value of a random variable. The sampling distribution is a theoretical statement of the probability of observing various intervals of values of such a random variable over all possible samples of the same size N drawn at random.

For example, imagine that all United States adults could be classified into two categories:

$$\text{success} = \text{prefers foreign movies}$$
$$\text{failure} = \text{prefers U.S. movies.}$$

We will assume that each person has one of these two preferences, and that there is an almost infinite number of such persons. The outcomes of observations drawn from this population could be regarded as a Bernoulli process, pairing each of the two classes with its proportion, p or q.

We sample United States adults at random, asking each his preference. Suppose that our random sample consists of exactly 10 adults. The number of successes observed could be any number from 0 through 10, and there is some probability that can be associated with each possible number, given the true p. The pairing of different sample values one might get and the probability of each is the sampling distribution. Note well, however, that the sampling distribution depends upon how the sampling was done, how we attached the value to a sample, and most of all, what the *true* situation is. If only .30 of the population of adults prefer foreign movies, then the sampling distribution for number of successes is given by Table 5.13.1. This tells us that if we observe 10 successes out of 10 cases, then we have encountered an event that should occur only once in 100,000 samples of 10 people, *if* p is .30. On the other hand, for $p = .30$, we should expect 3 out of 10 to occur relatively often; about 26 in 100 samples should give us this exact value. About 70 in 100 samples should give us 2, 3, or 4 people preferring foreign movies, and so on. In short, a sampling distribution allows one to judge the probability of a particular kind of sample result, given that something is true in general about the population being sampled. Before one can find the sampling distribution, however, he must postulate something true about the general situation: in this case, we postulated that exactly 30 percent of United States adults prefer foreign movies. Given that this is true, the sampling distribution specifies the probabilities of various sample numbers who show this preference.

In the next section a little preview of statistical inference will be given, in which the binomial rule provides a sampling distribution. This example cannot be complete at this time simply because all the necessary ingredients have not been introduced. On the other hand, the author feels it important that the student get some feel for the use of theoretical distributions as early as possible, to provide a basis for the more technical development to follow.

5.17 A PREVIEW OF A USE OF THE BINOMIAL DISTRIBUTION

It is now possible to point ahead to an important use of the binomial distribution. This example will deal with an experiment in psychology where we must use the data to decide whether or not a particular hypothesis actually seems to fit the data we obtain.

The context of the experiment is this: a psychophysical threshold or limen is that value on some physical measurement of a stimulus object at which a human subject is just capable of responding—in somewhat inexact terms, if the stimulus is a point of light in a darkened room, the threshold might be the physical intensity the light would have to have so that the subject would just be able to "see" the stimulus. It has, however, been suspected for some time that subjects may be able to respond in certain ways to stimuli which are actually below their known threshold of awareness. Such stimuli are said to be subliminal; the subject may not really be conscious that he "sees" the light—nevertheless, he may be able to respond as though he were capable of seeing the light.

Think of a hypothetical study of this question: "If a human is subjected to a stimulus below his threshold of conscious awareness, can his behavior somehow still be influenced by the presence of the stimulus?" The experimental task is as follows: the subject is seated in a room in front of a square screen divided into four equal parts. He is instructed that his task is to guess in which part of the screen a small, very faint, spot of light is thrown. He is to do this for many trials, and is told the light will be projected on the screen in a completely "haphazard," "random" manner over the trials. The light projected is made to be so faint that the subject cannot in any sense actually "see" the light. However, unknown to the subject, the spot is always projected into the same one of the four parts of the screen over the various trials. For our computational convenience, suppose that only 10 trials are taken for this one subject.

Our hypothesis goes like this: if the subject really is in no way being influenced by the small "invisible" spot of light, then his guesses should be random, haphazard affairs themselves, so that he should be right in his guess only 1/4 of the time by accident. Thus, under this hypothesis of "only guessing," the sample space of the subject's response to this situation should be distributed in this way:

Class	p
right	1/4
wrong	3/4
	4/4 = 1.00

The explicit assumption made is that the various trials for the subject are independent of each other. Now what sampling distribution holds for 10 trials for this subject if this hypothesis is true? His number of correct guesses *could* range from 10 right to 0 right, and the distribution would be a binomial distribution with p(right) $= 1/4$, $N = 10$. If we apply the rule for the binomial, we find the distribution that appears in Table 5.17.1. This binomial distribution gives all the possible numbers of correct guesses that this subject *could* make in the ten trials, and the probability for each, *if* he truly were guessing in a haphazard manner.

Given this theoretical distribution of possible outcomes, we turn to the actual result of the experiment. It is found that the subject guessed correctly on 7 out of the 10 trials. What then is the probability that exactly this result *should* have come up by chance? The binomial probability for 7 correct is 3,240 out of 1,048,576, or about .0031. Thus the probability of his getting exactly this number correct by random guessing alone is about 31 chances in 10,000. In other words, if we repeated the experiment, giving the subject 10,000 independent sets of ten trials, about 31 of these repetitions should give us exactly 7 correct guesses.

Table 5.17.1

x	$p(x)$
10 right	$\binom{10}{10}(1/4)^{10}(3/4)^0 = .0000(+)$
9	$\binom{10}{9}(1/4)^9(3/4)^1 = .0000(+)$
8	$\binom{10}{8}(1/4)^8(3/4)^2 = .0004$
7	$\binom{10}{7}(1/4)^7(3/4)^3 = .0031$
6	$\binom{10}{6}(1/4)^6(3/4)^4 = .0162$
5	$\binom{10}{5}(1/4)^5(3/4)^5 = .0584$
4	$\binom{10}{4}(1/4)^4(3/4)^6 = .1460$
3	$\binom{10}{3}(1/4)^3(3/4)^7 = .2503$
2	$\binom{10}{2}(1/4)^2(3/4)^8 = .2816$
1	$\binom{10}{1}(1/4)^1(3/4)^9 = .1877$
0	$\binom{10}{0}(1/4)^0(3/4)^{10} = .0563$

The exact sample result obtained is *not* a very likely thing to occur if the hypothesis is correct.

However, we should be interested in the probability not only of his getting exactly 7 correct, but rather the probability of getting *this many or more* correct, since we are really interested in this result as evidence of whether or not he is guessing, or doing something which will make him be more often correct than should simple, haphazard guessing. What is the probability that he should get *7 or more* correct trials? The answer is readily seen if we remember that this is asking for the probability of an interval:

$$p(7 \leq X) = p(7) + p(8) + p(9) + p(10) = .0035(+).$$

Seven or more correct guesses should occur only about 35 times in 10,000 independent replications of this experiment, if guessing alone is responsible for the subject's behavior. Does this unlikely result cast any doubt on the theory that the stimulus has no effect on guessing behavior? The answer is yes. For a theory to be "good," it should forecast results that agree with what we actually obtain. If the subject had come up with 2, 3, or 4 correct responses, then we would have little reason to doubt the "guessing" hypothesis, since these results fall among those that are quite probable according to the sampling distribution. However, the results obtained are quite unlikely to occur if the hypothesis is true, and so the evidence does not seem to favor this hypothesis.

On the other hand, are we completely safe in inferring that the subject was not just guessing? The answer is, of course, no. Even though this many or more correct responses obtained is improbable if the hypothesis is true, it is still not impossible that the subject might have been only guessing. We need some way to state just how much of a chance we are taking in saying that the hypothesis is not true, and that the subject is in some way being influenced by the stimulus. *The best measure of the amount of risk we run by abandoning this hypothesis on the evidence is given by the probability of sample results as extreme or more extreme than those actually obtained, if the hypothesis were true.* The evidence apparently diverges considerably from what we should expect (about 1/4 correct) if the subject were guessing, and the probability of such a divergent sample is only about 35 in 10,000. The probability of our being *wrong* in rejecting the guessing-hypothesis equals the probability of the divergent result, or .0035. We *could* be wrong in rejecting the hypothesis that the subject is guessing, but the probability of our being wrong is not very great. The more our sample result departs from what we expect given a hypothetical situation, and the more improbable that departure is, the less credence is given the hypothesis. In a later chapter we shall discuss in detail the rules for (and precautions in) evaluating hypotheses in the light of obtained results. However, this illustration exhibits the general logic underlying all "tests" of hypotheses: from the hypothetical population distribution one obtains a theoretical sampling distribution. Then the obtained results are compared with the sampling distribution probabilities. If the probability of samples such as the one obtained is high, the hypothesis is regarded as tenable. On the other hand, if the probability of such a sample (or one in more extreme disagreement with what is expected) is quite small, then doubt is cast on the hypothesis.

Although this little example should not be taken as a model of sophisticated scientific practice, it does suggest the use of a theoretical distribution of sample results as an aid in making inferences about a hypothetical situation. The binomial distribution is only one of a number of such sampling distributions we shall employ in this general way.

Since we have been rather heavily theoretical in our discussion up to this point, the next section will deal with two methods based on the binomial distribution that *do* have some practical use in experimental situations. Although not all of the qualifications involved in the use of these methods can be explained at this time, perhaps they will prove interesting and useful nevertheless. Then, some close relatives of the binomial distribution will be introduced, along with some indications of how these distributions are used as well.

5.18 THE SIGN AND THE MEDIAN TESTS FOR TWO GROUPS

We have already seen that the binomial distribution can be used to test a hypothesis about a proportion on the basis of a random sample of N independent observations. This is but one of a very large number of uses of the idea of a binomial variable in applied statistics. Now, two methods based on the binomial distribution will be introduced, each of which is designed for the comparison of two groups of observations.

The first of these simple methods based on the binomial distribution is the so-called "sign test." The sign test is used in situations in which N *pairs* of *matched* observations are made. The first member of each pair is an observation of some type A, and the second an observation of some type B. Thus, husband-wife pairs may be observed, in which each husband (type A) is paired with a wife (type B). Now each A individual has a numerical score X_A, and this is paired with a numerical score for the B individual, X_B. The question to be answered is, "Is the distribution of X_A scores identical to the distribution of X_B scores in the long run, if all possible pairs could be observed?" We approach this question by noting the difference $(x_A - x_B)$ for each of the N pairs. If $(x_A - x_B)$ is positive in value, then a "+" or a "success" is recorded. If $(x_A - x_B)$ is negative, the a "−" or "failure" is recorded. When $(x_A - x_B) = 0$, a fair coin is tossed, and the result is noted as a success or a failure depending on the outcome of the coin.

If the distribution of X_A values and the distribution of the X_B values really are identical, in the long run for all possible pairs, then it should be true that $p(+) = p(-) = 1/2$. Furthermore, the occurrence of successes and failures should correspond to a Bernoulli process, with N equal to the number of pairs observed, and $p = 1/2$. In order to reach a conclusion about the tenability of the hypothesis that $p = 1/2$, one calculates the probability that results as extreme, or more extreme, should occur, given that $p = 1/2$, and given the value of N that is actually used. If this probability is sufficiently small (say, .05 or less), then the hypothesis that $p = 1/2$ is rejected, meaning that the hypothesis of equal distributions is rejected as well. If a number of successes or failures equally or more extreme than the number actually obtained does not have a sufficiently small probability, then the hypothesis is not rejected.

For example, suppose that the basic question is, "Are men better drivers than their wives?" In order to shed light on this question, we take 20 couples at random, and separately give each wife and each husband a driving test (we assume here that all subjects know how to drive). If a husband scores higher than his wife, then this is a "success" or "+". If a wife scores higher than her husband, then this is a "failure" or "−". We decide that we will reject the hypothesis that there is no difference in driving ability between husbands and wives only if the number of successes is such that this number should be equaled or exceeded with a probability no larger than .05. That is, if y is the number of successes actually obtained, then we will reject the hypothesis only if $y \geq z$, where z is some number such that $p(y \geq z;20, 1/2) = .05$, approximately.

The required number z can be found quite simply from Table II in Appendix C. Here $N = 20$ and $p = 1/2$, so that we find the column labeled $p = .50$ and the section for $N = 20$. The table shows that 14 or more successes out of 20 will occur with probability of approximately .05, given that $p = 1/2$. If 14 or more successes occur, then the hypothesis of equal ability will be rejected. If fewer than 14 successes occur, the hypothesis will not be rejected.

On the other hand, suppose that we had been concerned with the hypothesis that husbands and wives are simply different in driving ability, without specifying how they differ. This is opposed to the hypothesis that husbands and wives are equal in this respect. Then we would reject the hypothesis of equal ability if the number of successes fell either below some number z_1 or above some number z_2, where

$$p(z_1 < y < z_2;20, 1/2) = .95, \quad \text{approximately.}$$

If $z_1 = 5$ and $z_2 = 15$, then the probability that $y \leq z_1$ or $y \geq z_2$ is just under .05. Thus, the hypothesis of equal driving ability would be rejected if the number of successes were either 5 or fewer, or 15 or more; otherwise, the evidence would be viewed as insufficient to permit rejection of the hypothesis of equal driving ability.

The sign test can be quite useful in a variety of situations. It can also be extended to the case in which the sample size is too large to make the calculation of binomial probabilities, or the use of binomial tables, practicable. This extension is discussed in Chapter 18.

Still another simple statistical technique based directly upon the binomial distribution is the so-called "median test" for two samples. The median test can be applied when there are two *unmatched* or *unpaired* samples, with N_1 observations drawn at random in the first sample, and N_2 observations drawn at random in the second. The value of some random variable, say Y, is associated with each member of the first sample, and a value of the same random variable with each member of the second sample. The question to be asked is, "Does Y have the same distribution among the group represented by sample I as it does in the group represented by sample II?" That is, if we had made all possible observations in the total group from which sample I was drawn, and all possible observations in the group from which sample II was drawn, would the distribution of Y values in total group I have been the same as the distribution of Y values in the total group II?

The median test is carried out as follows. First, samples I and II are pooled into a single sample, and the Y values are arranged in order of magnitude. Then the *middle* Y value is found: this value is known as **the median.** If the total of $N_1 + N_2$ is an even number, then the median is taken to be the average of the two middle values, where these two middle values have exactly $(N_1 + N_2 - 2)/2$ sample values below them, and the same number of sample values above them. If the total $N_1 + N_2$ is an odd number, then there will be one middle value, and this is the median. The median will in this instance have exactly $(N_1 + N_2 - 1)/2$ values below it, and the same number of values above. For the moment, let us symbolize the median value by Md.

Now we separate the samples once again. We look at sample I and note how many cases have Y values that fall *above* the median, Md. Call the number of cases falling above Md in sample I, a_1. Then the number falling at or below the value Md in sample I must be $N_1 - a_1$. Similarly, in sample II, find the number of cases falling above Md, and symbolize this number of cases by a_2. Then the number of cases falling at or below Md in sample II is $N_2 - a_2$.

If we let a "success" be the event "case falls above the grand median Md," then the probability of a_1 successes among the N_1 possibilities in sample I is the binomial probability

$$\binom{N_1}{a_1} p^{a_1} q^{N_1 - a_1}.$$

Similarly, the probability of a_2 successes among the N_2 possibilities in sample II is the binomial probability

$$\binom{N_2}{a_2} p^{a_2} q^{N_2 - a_2}.$$

Since the samples are regarded as independent, the joint probability of a_1 successes and a_2 successes is the product:

$$\binom{N_1}{a_1} p^{a_1} q^{N_1 - a_1} \binom{N_2}{a_2} p^{a_2} q^{N_2 - a_2}.$$

However, we want the conditional probability, given $a_1 + a_2$ successes out of a possible $N_1 + N_2$ trials. This is

$$
\begin{aligned}
p(a_1, a_2 | a_1 + a_2, N) &= \frac{\binom{N_1}{a_1} p^{a_1} q^{N_1 - a_1} \binom{N_2}{a_2} p^{a_2} q^{N_2 - a_2}}{\binom{N}{a_1 + a_2} p^{a_1 + a_2} q^{N - a_1 - a_2}} \\[2ex]
&= \frac{\binom{N_1}{a_1}\binom{N_2}{a_2}}{\binom{N}{a_1 + a_2}},
\end{aligned}
$$

where $N = N_1 + N_2$.

If the hypothesis is that the two groups I and II are identical with respect to their Y values, as opposed to the hypothesis that group I tends to have higher Y values, then one looks for values of a_1 that are so extremely large as to make their occurrence quite unlikely when groups I and II are actually identical. If such extremely large a_1 values occur, one rejects the hypothesis of identity in favor of the hypothesis of higher Y values in group I. Likewise, if the alternative to the hypothesis of identical groups is the hypothesis of higher values in group II, one looks for extremely high values of a_2. Finally, if the alternative to the hypothesis of identical groups is the hypothesis that they are simply not identical, without respect to which has the higher Y values, then one looks for high values either of a_1 or a_2 as the basis for rejecting the hypothesis that the two groups are identical.

For fairly small samples (say, $N_1 + N_2 \leq 20$) this test is not too laborious to carry out. For example, suppose once again that we are interested in driving ability. However, this time we are not going to deal with husband-and-wife pairs, but rather with a group of 9 men and 11 women selected at random. Each is given a driving test and his score (Y) recorded. Suppose that the results are as follows:

Y scores for men: 24, 28, 29, 29, 34, 36, 40, 41, 60,
Y scores for women: 21, 31, 34, 37, 38, 39, 42, 43, 44, 50, 51.

Combining the groups, we have

21, 24, 28, 29, 29, 31, 34, 34, 36, 37, 38, 39, 40, 41, 42, 43, 44, 50, 51, 60.

The middle score, or median, must lie between the two middle scores 37 and 38, since the total number, $N_1 + N_2$, is even. Thus, we take

$$Md = \frac{37 + 38}{2} = 37.5.$$

Returning to the sample of men, we find that 3 men had scores above 37.5, the median. On the other hand, $9 - 3 = 6$ men had scores below the median. Among the sample of women, some 7 women had scores above the median, and 4 had scores below it.

Now suppose we are interested in testing the hypothesis that men and women have the same distribution of Y scores, as opposed to the hypothesis that women tend to have higher scores. What value of a_2, or number of women above the median, would be required if that value or a higher value can occur with a probability of only .05 or less? The highest possible value a_2 would be 10, since no more than ten scores out of 20 may fall above the median. We find the probability that $a_2 = 10$ by taking

$$\frac{\binom{N_1}{a_1}\binom{N_2}{a_2}}{\binom{N}{a_1 + a_2}} = \frac{N_1! N_2! (N - a_1 - a_2)! (a_1 + a_2)!}{a_1! a_2! (N_1 - a_1)! (N_2 - a_2)! N!}$$

$$= \frac{9!11!(20-10)!(10)!}{0!10!(9-0)!(11-10)!20!}$$

$$= \frac{10!(11)!}{20!}$$

$$= \frac{(3.63 \times 10^6)(3.99 \times 10^7)}{2.43 \times 10^{18}}$$

$$= \frac{5.9}{10^5} \quad \text{or} \quad \frac{5.9}{100000},$$

which is a very small probability indeed!

Continuing in the same way, we find that the probability for $a_2 = 9$, $a_1 = 1$ is $265.5/100000$, so that the probability that $a_2 \geqq 9$ is $(5.9 + 265.5)/100000$, or $.002714$.

The probability for $a_2 = 8$, $a_1 = 2$ is given by

$$\frac{\binom{9}{2}\binom{11}{8}}{\binom{20}{10}} = .03186,$$

so that the probability that a_2 is equal to or greater than 8 is

$$.00271 + .03186 = .03457$$

approximately.

When we calculate the probability for $a_2 = 7$ and $a_1 = 3$ by taking

$$\frac{\binom{9}{3}\binom{11}{7}}{\binom{20}{10}}$$

we obtain $.14860$. This means that the probability of an a_2 value greater than or equal to 7 is $.14860 + .03457 = .18317$, or $.18$ approximately. This means that if we want to find some value b such that $p(a_2 \geqq b | a_1 + a_2, N) \leqq .05$, then that value must be 8, since 7 or more has a probability of $.18$ while 8 or more has a probability of only about $.035$.

Now to return to our sample results. Here we found $a_2 = 7$, so that 7 out of our sample of 11 women showed scores above the median. Furthermore, we had decided to reject the hypothesis of identical distributions for men and women only if the results showed a number of women above the median such that a number this extreme or more could occur with probability of only $.05$ or less. The value of a_2 would have had to be 8 for us to be able to reject the hypothesis of no difference in the two distributions, and so we do not reject the hypothesis on the basis of these results.

It should be obvious that a fair amount of labor is involved in using the median test for larger samples. Fortunately, there is an alternative method for applying the median test for large samples; this will be discussed in Chapter 18.

5.19   THE PASCAL AND GEOMETRIC DISTRIBUTIONS

A close relative of the binomial distribution is the Pascal distribution, named for the French mathematician-philosopher Blaise Pascal (1623–1662). Whereas a random variable following the binomial rule corresponds to the number of successes r out of a fixed number of trials N, the random variable following the Pascal rule has a different interpretation. Here we are interested in the number of trials, N, necessary in order to achieve a given number of successes r. Thus N is the random variable and r is a constant in the Pascal distribution. Unlike binomial sampling, in Pascal sampling r is fixed and N is left to chance.

Consider a stationary and independent Bernoulli process in which the probability of a success on any trial is p. Then for any sequence of trials we might ask, "What is the probability that the first success occurs on the first trial?" or "What is the probability that the first success occurs on the sixth trial?" and so on for a trial of any number. Since the first success never *has* to occur at all, short of an indefinitely large number of trials, then the possible trials on which the first success *might* occur are countably infinite in number. The distribution of N, the trial number on which the first success occurs, given trials from a stable Bernoulli process, is known as the **geometric distribution.** In a geometric distribution, the random variable N can take on any value that is one of the counting numbers 1, 2, 3, $\cdots$. This means that, unlike a binomial variable, a geometric variable takes on a countably infinite set of values. However, such a geometric variable is still discrete, since it can take on only whole-number values.

Now, the probability that the first success occurs on the first trial is

$$p(N = 1;p) = p.$$

The probability that the first success occurs on the second trial must be

$$p(N = 2;p) = (1 - p)p,$$

since the first trial must be a failure if the first success occurs on the second trial. In the same way, we can show that

$$p(N = 3;p) = (1 - p)^2 p$$

and so on, until for the probability that the first success occurs on trial $N = n$ we have

$$p(N = n;p) = (1 - p)^{n-1} p.$$

This is the rule for the geometric distribution, in which the random variable is the trial number on which the first success occurs, in trials from a stable, independent, Bernoulli process with probability p.

The Pascal distribution can be thought of as a generalization of the geometric distribution. That is, the random variable in the Pascal distribution is the trial number on which the rth success occurs, where r can be any whole number $r = 1, 2, \cdots$, and where $r \leq N$. The geometric distribution is a Pascal distribution in which $r = 1$. For a Pascal distribution, the probability that N equals a given value n depends on the fixed value r, or number of successes, and p, the probability of a success, as follows:

$$p(N = n;r, p) = \frac{(n-1)!}{(r-1)!(n-r)!} \, p^r(1-p)^{n-r}, \qquad N \geqq r.$$

The basis for this rule can be shown as follows: Let $r = 2$, and let the probability of a success be p, with the probability of a failure $1 - p = q$. Now what is the probability of the second success occurring on the second trial, or $N = 2$? This corresponds to the sequence (S, S), with probability p^2. Note that this event can occur in only one way. For $N = 3$, however, the sequence of successes and failures might be (S, F, S) or it might be (F, S, S), each with a probability of $p^2 q$. Since there are two such sequences, the probability for $N = 3$ is $2p^2q$, or $\binom{2}{1}p^2q$. For $N = 4$, the possible sequences are (S, F, F, S), (F, S, F, S), and (F, F, S, S), each with probability $p^2 q^2$. Then the probability for $N = 4$ is $3p^4q^4$. Exactly the same sort of reasoning would lead to the probability for any other value of N, given r and p. Notice why the coefficient in the Pascal rule is $\binom{n-1}{r-1}$, rather than $\binom{N}{r}$ as in the binomial rule: The last term in any sequence of n trials must be a success since we are interested only in the event "the rth success occurs on trial n." However, in any sequence that qualifies for this event, there must be previous $n - 1$ trials that contain $r - 1$ successes, and hence there are $\binom{n-1}{r-1}$ possible sequences, rather than $\binom{n}{r}$ as you might otherwise expect.

For example, suppose that a fair coin is being tossed until the second head appears. What is the probability that the second head appears when $N = 2$? (Notice that the second head cannot appear when $N = 1$.) What is the probability that the second head appears when $N = 3$? $N = 4$? Here, $r = 2$, and N can take on any value 2, 3, 4, and so on. Then we find

$$p\left(N = 2;2, \frac{1}{2}\right) = \frac{(2-1)!}{(2-1)!(0)!}\left(\frac{1}{2}\right)^2\left(\frac{1}{2}\right)^0 = \frac{1}{4},$$

$$p\left(N = 3;2, \frac{1}{2}\right) = \frac{(3-1)!}{(2-1)!(3-2)!}\left(\frac{1}{2}\right)^2\left(\frac{1}{2}\right)^1 = \frac{1}{4},$$

$$p\left(N = 4;2, \frac{1}{2}\right) = \frac{(4-1)!}{(2-1)!(4-2)!}\left(\frac{1}{2}\right)^2\left(\frac{1}{2}\right)^2 = \frac{3}{16},$$

$$p\left(N = 5;2, \frac{1}{2}\right) = \frac{(5-1)!}{(2-1)!(5-2)!}\left(\frac{1}{2}\right)^2\left(\frac{1}{2}\right)^3 = \frac{4}{32},$$

and so on for any other possible value of N.

The plot of a Pascal distribution looks very different from the plot of a binomial distribution based on the same value of p. The probability mass function for a geometric distribution with $p = 1/2$ (Pascal distribution with $r = 1$ and $p = 1/2$), is shown in Figure 5.19.1. Notice the extremely long tail, or "skewness," to the right. Smaller values of p or larger values of r tend to reduce the skewness; larger values of p increase it. Notice also that the X axis, standing for

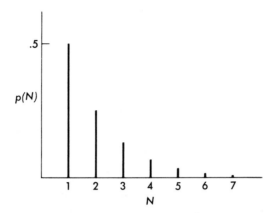

FIG. 5.19.1. Geometric probability mass function, for $p = .5$

values of N, is always open to the right. We never run out of possible values to which probabilities may be assigned, no matter how large the values or how small the associated probabilities.

Problems sometimes arise in which interest focuses not on the trial number on which the rth success occurs, but rather on the number of failures that occur before the rth success. However, when $Y = N - r$, the number of failures before the rth success, is the random variable, it is customary to refer to the distribution as the **negative binomial distribution,** and this formulation will sometimes be encountered in the statistical literature in place of the Pascal distribution. For all practical purposes these two kinds of distributions are the same, and problems can be solved in terms of either the number of failures before the rth success, $(N - r)$ or the trial number of the rth success, N.

5.20   Some uses of Pascal distributions

Like the binomial distribution, the Pascal distribution can be used as the basis for a variety of statistical techniques. Ordinarily, the kind of situation producing a random variable following a Pascal rule will be one involving **sequential sampling,** as opposed to simple random sampling. In sequential sampling no predetermined limit is set upon the number of observations. Rather, observations are made in sequence until sufficient data are accumulated to enable a decision to be made, according to a predetermined criterion. Since a Pascal variable is interpreted as the number of trials required in order to reach a given number of successes, this conception lends itself readily to the simplest kinds of sequential experiments.

For an example of how the idea of a Pascal variable can be used, consider the following: A psychologist is choosing items for inclusion in a test. He believes that one item is moderately easy, so that two out of three school children of a given age should pass the item. He interprets this to mean that the probability that any given child will pass is 2/3. However, he wants to test this belief against the possibility that the item may be harder. Time and expense are a factor, so he does not want to sample more children than absolutely necessary in order to arrive at a judgment. Thus he decides to administer the item to individual children, one at a time, selected at random. He also decides to test only until four children have successfully passed the item. Then, on the basis of the number of children tested in order to achieve four successes, he will reach a conclusion about his notion that the probability of a given child's passing is $2/3$. The more children it is necessary to test in order to achieve four successes, the more doubt will be cast upon $2/3$ as the value of p. For reasons that we will elaborate at length later on, he decides that he will reject the idea that $p = 2/3$ only if the number of children required to reach four successes is in some sense "excessive." In particular, he will reject $p = 2/3$ as the true probability only if n, the number of children actually tested, should be equalled or exceeded with probability of only .05 or less, given that $p = 2/3$ is the true probability of passing the test. Otherwise, he will retain $p = 2/3$ as a tenable hypothesis about the true value.

Now suppose that the psychologist begins to test children in the order of their random selection. Sure enough, he finds that the fourth child to pass the item is actually the eighth child tested. Thus, $n = 8$, and the fixed number of successes $r = 4$, while the value of p is believed to be $2/3$. The critical question then is the value of $p(N \geq 8; 4, 2/3)$, which is the same as $1 - F(7; 4, 2/3)$. Is this, or is it not, .05 or less?

The value of this probability is determined by the Pascal rule. Since

$$F(7; 4, 2/3) = p(4) + p(5) + p(6) + p(7),$$

it is necessary to find the probability for each of these values of the random variable. (Remember that one cannot have four successes in fewer than four trials.) Proceeding, we find

$$p(4) = \frac{(4-1)!}{(4-1)!(0)!} \, p^4 = (2/3)^4 = 16/81,$$

$$p(5) = \frac{(5-1)!}{(4-1)!(1)!} \, p^4(1-p) = 4(2/3)^4(1/3) = 64/243,$$

$$p(6) = \frac{(6-1)!}{(4-1)!(6-4)!} \, p^4(1-p)^2 = 10(2/3)^4(1/3)^2 = 160/729,$$

$$p(7) = \frac{(7-1)!}{(4-1)!(7-4)!} \, p^4(1-p)^3 = 20(2/3)^4(1/3)^3 = 320/2178.$$

Then

$$F(7; 4, 2/3) = 16/81 + 64/243 + 160/729 + 320/2178 = 1808/2178,$$

so that $F(7; 4, 2/3)$ is about .83. Thus, a number of trials equal to 8 or more can occur with probability of about .17. This is far in excess of the probability of .05 that the psychologist had decided upon. On this evidence he decides not to reject the hypothesis that $p = 2/3$.

On the other hand, had it been true that ten trials were required in

order to reach four successes, then $F(9;4, 2/3)$ would have been approximately
.96. This in turn implies that the probability of ten or more trials required to
reach four successes would have been .04, which is less than the criterion agreed
upon. In this case, the psychologist would have decided to reject the hypothesis
that $p = 2/3$. Either it is true that $p = 2/3$, and a rare event has occurred, or
$2/3$ is not the correct value of p. The psychologist has already decided to conclude
the latter, should such a rare event occur.

Notice what has been assumed here. First of all, we have assumed
that $p = 2/3$ for each and every one of the children tested, so that observation of
each child represents a trial in a stationary Bernoulli process. Furthermore, we
have assumed that the results for each child, or trial, are independent of those
for any other child. Finally, we have assumed that the order of selection of the
children was completely random. The failure of any of these assumptions to be
true could, of course, make a difference in the final conclusions.

5.21 THE POISSON DISTRIBUTION

Still another relative of the binomial family of distributions plays
a large role in theoretical and applied statistics. This is the Poisson distribution,
named after the nineteenth-century French mathematician S. Poisson. A random
variable following this rule is referred to as a Poisson variable, and the process
generating values of such a random variable is known as a Poisson process. The
probability function for a Poisson variable X follows the rule

$$p(x;m) = \begin{cases} \dfrac{e^{-m}m^x}{x!}, & x = 0, 1, 2, 3, \cdots ; m > 0; \\ 0, & \text{otherwise,} \end{cases}$$

where e is the mathematical constant, and m is a constant known as the "intensity" of the Poisson process.

Notice that like a Pascal variable, a Poisson variable X can take on
only integral or "whole" values—in this case from zero to an indefinitely large
value. This random variable thus can assume any of a countably infinite set of
values. It is, however, a discrete variable.

Although Poisson variables and the processes generating them can
be given a variety of useful interpretations, perhaps the simplest approach to the
study of the Poisson is to regard it as a special case of the binomial. A derivation
of the Poisson function from the binomial will now be sketched out. What we are
going to show is that if N is allowed to become extremely large while p is made
extremely small, and if Np remains constant, the binomial distribution approaches
the Poisson distribution. Take note, however, that this is just a sketch of the
general outline of a proof. A number of the mathematical qualifications really
necessary to such a proof will be omitted here.

Consider a binomial variable X, which takes on values with probabilities depending upon p, the probability of a success, and N, the number of
trials. Let us define $m = Np$. Then we can rewrite the binomial probability that
$X = x$ as

$$p(x;N, p) = \frac{N!}{x!(N-x)!} \, p^x(1-p)^{N-x},$$

$$p(x;N, m) = \frac{N!}{x!(N-x)!N^x} \, (m)^x \left(1 - \frac{m}{N}\right)^{N-x}$$

$$= \frac{(N)}{N} \frac{(N-1)}{N} \frac{(N-2)}{N} \cdots \frac{(N-x+1)}{N} \frac{m^x}{x!} \left(1 - \frac{m}{N}\right)^N \left(1 - \frac{m}{N}\right)^{-x}$$

Now we want to find the limiting value of this probability as N becomes indefinitely large and m remains constant. That is, we want to find

$$\underset{N \to \infty}{\text{limit}} \; p(x;N, m).$$

We can do so by examining what happens to the various terms in the expression above as N approaches the limit. First of all, as N is made indefinitely large, $1 - (1/N)$ approaches 1, $1 - (2/N)$ approaches 1, and so do all of the succeeding terms up through $1 - (x+1)/N$. The expression $[1 - (m/N)]^{-x}$ must also approach 1. The only other term involving N is $[1 - (m/N)]^N$. In the advanced calculus it is shown that

$$\underset{N \to \infty}{\text{limit}} \left(1 - \frac{m}{N}\right)^N = e^{-m}.$$

Hence, making these substitutions, we have

$$\underset{N \to \infty}{\text{limit}} \; p(x;N, m) = \frac{e^{-m}m^x}{x!}.$$

For fixed $m = Np$, and as N becomes indefinitely large, the limiting value of the binomial probability $p(x;N, m)$ is the Poisson probability $(e^{-m}m^x)/x!$.

Because of this connection between binomial and Poisson probabilities, Poisson sampling can be thought of as the act of making a vast number of trials from a stable and independent Bernoulli process for which the probability of a success is extremely small. The constant m is called the "intensity" of the Poisson process.

Not only does this connection with the binomial distribution permit one interpretation of the Poisson distribution; there are practical consequences as well. In cases in which N is large and p is relatively small, the binomial probabilities may be very laborious to calculate. In this instance, it is a much simpler matter to approximate the exact binomial probabilities through use of Poisson probabilities for the various values of x.

When a Poisson probability for some specific x value is desired, one simply calculates $m = Np$, and then applies the Poisson rule

$$p(x;m) = \frac{e^{-m}m^x}{x!}$$

for the value $X = x$ of interest. Table X in Appendix C gives selected values of e^{-m}, and Table VIII gives various values of factorials. For example, suppose that for $N = 3000$ and $p = .001$, we wished to find the probability for $X = 5$. Then, $m = (3000)(.001) = 3$, and we calculate the probability by taking

$$p(5\ ;3) = \frac{e^{-3}3^5}{5!}$$
$$= \frac{(.0498)(243)}{120}$$
$$= .1008.$$

This is an exact Poisson probability, $p(5;3)$. It is also an approximation of the binomial probability $p(5; 3000, .001)$. The Poisson probability will be approximately equal to the actual binomial probability only for very large N and very small p, of course. Nevertheless, the approximation is good enough to be useful even when N is only moderately large and p only relatively small. Table 5.21.1 contrasts binomial and Poisson probabilities for $N = 20$ and $p = .10$, and for $N = 30$, $p = .01$. In the first instance the Poisson probabilities approach but certainly do not equal the binomial values. In the second instance, the fit is somewhat better. The approximation would grow steadily better if we increased N and reduced p. Later we will have more examples of the use of the Poisson distribution to approximate the binomial distribution. However, first it will be of value to examine some other interpretations of a Poisson process.

A great many illustrations of Poisson processes occur in the physical and the biological sciences, as well as in everyday life. For example, the degeneration of a radioactive substance is regarded as a Poisson process. At any given instant the probability is very small that an alpha-particle will be emitted, while there are vast numbers of opportunities for such an event to occur. The distribution of bacteria on a Petri plate can be viewed as a Poisson process. Each tiny area on the plate can be viewed as a trial, and a bacterium may or may not occur on such area. The probability of such an occurrence on any given area is very

Table 5.21.1

A COMPARISON OF BINOMIAL PROBABILITIES
WITH POISSON PROBABILITIES WHERE $m = Np$

r	$p = .10, N = 20$		$p = .01, N = 30$	
	Binomial	*Poisson*	*Binomial*	*Poisson*
0	.1216	.1353	.7397	.7408
1	.2702	.2707	.2242	.2223
2	.2852	.2707	.0328	.0333
3	.1901	.1804	.0031	.0033
4	.0898	.0902	.0002	.0003
5	.0319	.0361	.0000	.0000
6	.0089	.0120		
7	.0020	.0034		
8	.0004	.0009		
9	.0001	.0002		
10 or more	.0000	.0000		

small indeed, but there are very many areas on such a plate. The distribution of misprints in a book can be studied as a Poisson process, as can the occurrence of accidents of a certain kind in a manufacturing plant.

The Poisson distribution is thus important in its own right, quite apart from its connection with the binomial distribution. A typical situation that is regarded as a Poisson process involves a continuum of time, which can be broken down into arbitrary small segments, or "instants." At any given instant an event or success may occur. The occurrence of a success at any instant is quite independent of the occurrence or nonoccurrence of a success on any prior or following instant. However, the successes occur at a given and constant rate, which is the intensity of the process. This rate is stated in terms of a time interval longer than an instant, although the interval may be of any size. Thus, the intensity may be stated in terms of the expected number of successes per minute, or the expected number of successes per hour, and so on. The random variable is then the actual number of successes in a minute, or the actual number of successes in an hour, or any other fixed time span of interest. This random variable, number of successes in a time interval, may take on any whole value from zero to an indefinitely large value.

As a simple example, imagine a checkout counter at a grocery store. The counter is open all the time, and a customer may or may not arrive at a given instant. Experience has shown that the customers tend to arrive at a rate of, say, 2 per minute. On the other hand, during any given minute, no customers may come, one may come, 500 may come, and so forth. Each possible number of customers that might arrive in a given minute has a probability, and this is given by the Poisson rule with $m = 2$, the intensity of the process.

Suppose that we wish to know the probability that 5 customers arrive at the counter in one minute. Then we must find

$$p(5;2) = \frac{e^{-2}2^5}{5!},$$

since $m = 2$, $X = 5$. with the aid of Table X of Appendix C we know that e^{-2} is about .135. The probability is then given by

$$p(5;2) = \frac{(.135)(32)}{5!}$$

$$= 4.32/120$$
$$= .036.$$

Only in about 36 out of a thousand minutes should we expect exactly 5 customers to arrive at the checkout counter.

If we wished to know the probability of *five or fewer* customers, we would find

$$F(5;2) = \frac{e^{-2}2^0}{0!} + \frac{e^{-2}2}{1!} + \frac{e^{-2}2^2}{2!} + \frac{e^{-2}2^3}{3!} + \frac{e^{-2}2^4}{4!} + \frac{e^{-2}2^5}{5!}$$

$$= e^{-2}\left(1 + 2 + \frac{4}{2} + \frac{8}{6} + \frac{16}{24} + \frac{32}{120}\right)$$

$$= (.135)(7.266)$$
$$= .983, \quad \text{approximately.}$$

It also follows that the probability of *six or more* customers is given by

$$1 - F(5;2) = 1 - .983 = .017, \quad \text{approximately.}$$

Poisson probabilities and cumulative probabilities can be determined relatively easily in the way just shown, given Table VIII for the values of factorials and Table X for values of e^{-m}. Many books on advanced statistics give extensive tables of Poisson probabilities, particularly when they are designed to be used in fields such as the physical sciences or industry. However, since our use of the Poisson will not be extensive, space will not be given to such tables here. Rather a method will be given for using another table to find cumulative probabilities that will be useful in solving problems involving Poisson variables.

Suppose that we wish to find the cumulative probability that a Poisson variable X takes on a value less than or equal to some specific value a. That is, we wish to find $F(a;m)$ for a Poisson variable, where the process has intensity m. We can proceed as follows: In Table IV, Appendix C, look in the column at the extreme left under the symbol "v." Find a value as close as possible to the value of $2(a + 1)$. Next, in the corresponding row, find in the body of the table the two values that fall to either side of $2m$. Then the cumulative Poisson probability lies between the two Q values given at the top of the table, corresponding to the two values to either side of $2m$. Thus, for example, suppose that we wished to know the value of $F(6;3)$. Here $a = 6$ and $m = 3$. In Table IV we look along the left-hand column until we locate $v = 2(6 + 1)$ or 14. Then in the row corresponding to 14 we look in the body of the table and find that the two values to either side of $2m = 6$ are 5.62872 and 6.57063. The first lies in a column headed $Q = .975$ and the second in the column headed $Q = .950$. Then we know that

$$.950 \not\angle F(6;3) \not\angle .975.$$

(As it happens, the actual value here is .97.) If a somewhat more accurate approximation is desired, the method of linear interpolation may be used. We shall employ this approximate method for finding cumulative Poisson probabilities in the next section.

5.22   SOLVING PROBLEMS THROUGH USE OF THE POISSON DISTRIBUTION

First we will consider an example in which the Poisson approximation to the binomial distribution is employed. Then, we will consider another example in which the random variable itself is presumed to follow a Poisson rule.

Suppose that in a study of the effect of training upon the ability of students to solve a difficult abstract problem, a psychologist used a problem that ordinarily can be solved only by about one in fifty students of the given age group.

He chose a random sample of 300 students and gave the prior training to each student. After the training, each student attempted to solve the problem. Twelve students solved it correctly. The psychologist wishes to know if results this far from what one ordinarily expects without training (about 6 students correct) cast any doubt on the hypothesis that the training has no effect. In particular, he decides to adopt the following rule for deciding: If the probability is .05 or less of getting 12 or more successes, given that p is actually .02, he will reject the hypothesis that $p = .02$; otherwise, he will not reject the hypothesis. In other words, if 12 or more successes given $p = .02$ is a rather rare event, then the psychologist will feel that sufficient doubts are cast by his results on the hypothesis that $p = .02$ to lead him to reject that hypothesis for his group of trained students.

As you can readily see, this has all of the elements of a binomial problem, since there is a fixed number of trials, $N = 300$, a fixed probability of a success, $p = .02$ under the hypothesis, and an obtained number of actual successes, $X = 12$. However, since N is relatively large, and p is relatively small, we can apply the Poisson approximation to the binomial. What we want to find is the probability of 12 or more successes. This is the same as finding 1.00 minus the probability of 11 or fewer successes, or

$$p(X \geq 12; 300, .02) = 1 - F(11; 300, .02).$$

In Poisson terms, letting $m = Np$, or $m = 300(.02) = 6$, we need to find

$$p(X \geq 12; m = 6) = 1 - F(11; 6).$$

Using Table IV of Appendix C as described above, we locate the entry $2(x + 1)$, or 24 in the column labeled v, and then in that row locate the two values closest to $2m$, or 12. These are 10.8564 and 12.4011. The Q value for the first is .990 and that for the second is .975. Thus, we can say that $F(11; 6)$ lies between .99 and .975. This means that $p(X \geq 12; 6)$ lies between .01 and .025. Since this probability is less than .05, our decision rule says that we reject the hypothesis that the probability of a success $p = .02$. The rule for deciding *how* to decide indicates that a result as rare as that actually obtained, given $p = .02$, casts sufficient doubt on the hypothesis to lead to its rejection. The psychologist would assert he had obtained significant evidence that the training improved performance on the problem.

Consider now this example of a rather different type. In an industrial plant, long experience has shown that the rate of accidents per standard working month is 8. However, it was decided that a new safety program should be instituted, with each worker receiving intensive safety training; the program would be evaluated in the month following its completion. The question was whether or not the number of accidents in that month was sufficiently low to permit the conclusion that the safety program was having an effect.

During the month after the program was completed, the number of accidents turned out to be 4. Thus the question was: "Is the occurrence of 4 or fewer accidents sufficient evidence to permit the conclusion that the rate is no longer 8 per month?" The decision was made to reject the hypothesis that $m = 8$ if the cumulative probability $F(4; 8)$ was .05 or less.

Again the probability will be evaluated by use of Table IV. The left-hand column gives us the entry $2(4 + 1) = 10$, and then in that row we find the two values closest to $2m$ or 16. These values are 15.9871 and 18.3070, corresponding to Q values of .10 and .05. Thus we can see that the probability of 4 or fewer successes (i.e., accidents), given that $m = 8$, is between .05 and .10. In fact, the value under $Q = .10$ is almost exactly equal to 16, so that in this instance we are safe in saying that $F(4;8) = .10$. This is not a small enough probability to permit us to reject the hypothesis that $m = 8$, and so we would conclude that the training program has not been effective enough to permit our saying that the accident rate has changed.

5.23 THE MULTINOMIAL DISTRIBUTION

The basic rationale underlying the binomial distribution can be generalized to situations with more than two event classes. This generalization is known as the "multinomial distribution," having the following rule:

Consider K classes, mutually exclusive and exhaustive, and with probabilities $p_1, p_2, \cdots, p_K$. If N observations are made independently and at random, then the probability that exactly n_1 will be of kind 1, n_2 of kind 2, $\cdots$, and n_K of kind K, where $n_1 + n_2 + \cdots + n_K = N$, is given by

$$\frac{N!}{n_1!\, n_2! \cdots n_K!} (p_1)^{n_1}(p_2)^{n_2} \cdots (p_K)^{n_K}.$$

Think once again of colored marbles mixed together in a box, where the following probability distribution holds:

Color	p
Black	.40
Red	.30
White	.20
Blue	.10
	1.00

Now suppose that 10 balls were drawn at random and with replacement. The sample shows 2 black, 3 red, 5 white, and 0 blue. What is the probability of a sample distribution such as this? On substituting into the multinomial rule, we have

$$\frac{10!}{(2!)(3!)(5!)(0!)} (.4)^2(.3)^3(.2)^5(.10)^0$$

$$= \frac{1 \cdot 2 \cdot 3 \cdot 4 \cdot 5 \cdot 6 \cdot 7 \cdot 8 \cdot 9 \cdot 10}{(1 \cdot 2)(1 \cdot 2 \cdot 3)(1 \cdot 2 \cdot 3 \cdot 4 \cdot 5)(1)} (.4)^2(.3)^3(.2)^5$$

since $0!$ and $(.1)^0$ are both equal to 1. Working out this number, we find that .087

is the probability of the *sample* distribution

Black	2
Red	3
White	5
Blue	0
	10

if the probability distribution given above is the true one. Using this multinomial rule, one could work out a probability for each *possible sample distribution*.

Although the multinomial rule is relatively easy to state, a tabulation or graph of this distribution is very complicated: here a sample result is not a single number, as for the binomial, but rather *an entire frequency distribution*.

This raises an interesting possibility. Given any discrete probability distribution, one might work out the probability of all possible sample distributions for N observations. Then, in terms of these probabilities, the disagreement of sample and theoretical distributions could be evaluated. Although this is possible in principle, the drawbacks should be obvious. For any but small numbers of possible events and small N, the sheer number of different sample distributions is fantastically large. Ordinarily we are not interested in the probability of our sample alone, but rather in that of samples *as deviant or more so* than ours. This makes for too much computation to be practical in most instances. Furthermore, we run into trouble in trying to formulate a similar scheme for *continuous* theoretical distributions. For these reasons, the multinomial distribution plays a rather small role in theoretical statistics. The probability of obtaining an entire sample distribution is not usually of interest; instead, we will be finding probabilities for various indices summarizing a sample distribution.

5.24 THE HYPERGEOMETRIC DISTRIBUTION

Another theoretical probability distribution deserves passing mention for the same reason. The multinomial rule (and the binomial, of course) assumes either that the sampling is done with replacement, or that the sample space is infinite, so that the basic probabilities do not change over the trials made. However, suppose that one were sampling from a finite space *without* replacement; then the probabilities would change for each observation made. By a series of arguments very similar to those used for finding probabilities of poker hands in Section 5.8, we could arrive at a new rule for finding the probabilities of sample results.

This rule describes the hypergeometric distribution, and can be stated as follows:

Given a sample space containing a finite number T of elements, suppose that the elements are divided into K mutually exclusive and exhaustive classes, with T_1 in class 1, T_2 in class 2, $\cdots$, T_K in class K. A sample of N observations is drawn at random without replacement, and is found to contain n_1 of class 1, n_2 of class 2, $\cdots$, n_K of class K. Then the probability of occurrence of such a sample is given by

$$\frac{\binom{T_1}{n_1}\binom{T_2}{n_2} \cdots \binom{T_K}{n_K}}{\binom{T}{N}}$$

where $n_1 + n_2 + \cdots + n_K = N$ and $T_1 + T_2 + \cdots + T_K = T$.

For an illustration of the use of the hypergeometric rule, let us return to the problem of drawing marbles at random from a box (Section 5.17), but this time *without* replacement. Suppose there had been 30 balls in the box originally, with the following distribution of colors:

Color	f
Black	12
Red	9
White	6
Blue	3
	30

Notice that the relative frequencies are the same as the probabilities in the previous example. Now ten balls are drawn at random without replacement. We want the probability that the *sample* distribution is

Color	f
Black	2
Red	3
White	5
Blue	0
	10

Using the hypergeometric rule, we get

$$\frac{\binom{12}{2}\binom{9}{3}\binom{6}{5}\binom{3}{0}}{\binom{30}{10}}$$

which works out to be about .0011. This is not, of course, the same probability as we found for this sample result using the multinomial rule, since the entire sampling scheme is assumed different in this second example. This illustrates that *the sampling scheme adopted makes a real difference in the probability of a given result.*

However, this is a practical consideration only when the basic sample space contains a finite and small number of elementary events. When there is a very large number of elementary events in the sample space, the selection and nonreplacement of a particular unit for observation has negligible effect on the probabilities of events for successive samplings. For this reason, the hypergeometric probabilities are very closely approximated by the binomial or multinomial probabilities when T, the total number of elements in the sample space, is extremely large. The distinction between these different distributions becomes practically important only when samples are taken from relatively small sets of potential units for observation.

Before we continue further into the problem of statistical inference, we must develop more vocabulary for discussing distributions. In particular, it is not always convenient or to the point to compare sample and population distribution directly, and we need to introduce certain features of distributions such as their "central tendency" and their measures of "spread" or variability. These indices will give us the ability to make more succinct statements about real or theoretical sets of data than does a display of the entire function. These topics will be the subject of the next chapter.

EXERCISES

1. In a fraternity house, three boys share a room with a single closet. Each boy can wear each of the other boys items of clothing, and they share freely. The closet contains 3 pairs of shoes, 7 shirts, 5 pairs of pants, 8 pairs of socks, and 4 coats. If each boy dresses in shoes, shirt, pants, socks, and coat, in how many combinations of clothing may the boys appear together? (Leave the answer in symbolic form, without bothering to work out the exact number.)

2. A college contains three departments, A, B, and C. Department A has 10 faculty members, B has 15 members, and C, 20 members. It is decided to form a college committee consisting of 1/5 of the members of each department. How many possible such committees could be formed? (Leave the answer in symbolic form.)

3. Suppose that N people work in the same office. What is the probability that 2 or more of these people were born in the same month? (Assume that the 12 possible months of birth are all equally probable.) Compute this probability for $N = 2$ and $N = 4$.

4. The legislature of a certain state decides that henceforth automobile license tags will consist of two letters followed by three digits. How many license plates of this form are possible? Suppose that the buyer of a set of license plates is equally likely to get any combination of letters and digits. (Count 0 through 9 as the possible digits, and include all 26 letters of the alphabet.) What is the probability that he will get a set of plates having:
   (a) The same two letters and the same three digits (for example BB-222)?
   (b) The same two letters and three different digits?
   (c) Two different letters and the same three digits?

5. Suppose that in the college mentioned in problem 2 above, it was decided to form a nine-member committee by selecting faculty members completely at random. What is the probability that such a committee would wind up containing:
   (a) Exactly 1/5 of the members of Department A?
   (b) Exactly 1/5 of Department B?

(c) Exactly 1/5 of Department C?

(d) Fewer than 1/5 of the members of Department A?

(e) *Only* members from Department C?

(Again, leave the answers in symbolic form.)

6. In a test of possible side effects of a new medication, a physician matched 22 pairs of persons on their physical characteristics. One member of each pair was given the medicine, and his mate was given a placebo. A "success" was recorded when the member receiving the medication showed more of the side effect than his mate, and a "failure" was recorded otherwise. There were 15 successes and 7 failures. If medication and placebo were equal in their tendency to produce the effect, how likely is a result this much or more deviant from what one should expect?

7. In a well-known game, three shells are used, one of which covers a pea. On each trial the shells are rearranged into a row and the subject's task is to locate the shell under which the pea lies. Suppose that the first, second, and third shells are equally likely to cover the pea on any trial. Suppose also that a subject is equally likely to guess any of the three shells on a trial. Under these conditions, assuming 10 independent trials, how probable is it that out of 10 trials that the subject guesses the correct shell two or fewer times? Eight or more times?

8. On a college football team, suppose that a passer has a probability of .60 of a completion on any attempt. If his passing performance reflects an independent and stable random process, what is the probability that in a game he gets his first completion on his fourth attempt? What is the probability of his first completion on his first attempt?

9. On a multiple choice examination, each item has exactly 5 options, of which the student must pick one. Only one of the options is correct for each item. If a student is merely guessing at the answer, each option should be equally likely to be chosen. Furthermore, the student's answer on a given item is believed to be independent of his answer on any other. Given that the test has 15 items, and that a student is just guessing on each item, find the probability of the following events:

   (a) 3 items correct.   (b) 7 items correct.

   (c) 4 items incorrect.   (d) more than 5 items correct.

   (e) fewer than 2 items correct.   (f) between 2 and 7 items correct (inclusive).

10. A machine that manufactures automobile parts is believed to produce defective parts with a frequency of about 5 in 100. We will assume that this process is stationary and independent. Given that 15 parts produced by this machine are sampled at random, find the probability that one or fewer parts will turn out to be defective.

11. Suppose that the machine in problem 10 above turned out 1000 parts, in a manner which represents a stable and independent process. Find the probability that the number of defective parts produced is 44 or fewer.

12. In a certain lottery, 40 percent of the tickets were purchased by men, and 60 percent by women. Each person purchased only one ticket. Ten tickets were drawn at random and with replacement. What is the probability that:

   (a) Four or more winners were women?

   (b) Two or fewer winners were women?

   (c) The winners were all of the same sex?

   (d) Exactly four men and six women were winners?

13. In the lottery situation of problem 12, what is the probability that the second win by a woman occurs on the fourth drawing? What is the probability that the first win by a man occurs on the sixth draw?

14. A Public Health officer in a certain area suspects that 2.5 percent of children in that

area are severely undernourished. He takes a sample of 200 children at random, and finds that 9 show severe malnutrition. What is the probability of 9 or more such children in the sample if the true proportion in the population is .025? What would the officer be inclined to conclude?

15. In a test of reaction time, husband and wife pairs were studied. Sixteen pairs chosen at random were given the same reaction-time test, and the individual time noted. In nine of the pairs, the wife showed the faster reaction time, and in seven pairs, the husband showed the shorter time. If husbands and wives tend to be about equal in reaction time, what is the probability of a result this much or more in favor of the wives?

16. For any stable and independent Bernoulli process, with probability p, show that the probability of at least one success out of N trials is $1 - (1 - p)^N$.

17. Imagine two independent Bernoulli processes, where each process is itself stable and independent. The first has probability p_1 for a success, and the second has probability p_2. Show that if N trials are made of each process, the resulting probability of at least one success is $1 - [(1 - p_1)(1 - p_2)]^N$.

18. A bank has discovered that about one out of every 30 checks it processes must be returned for insufficient funds. On a given day, what is the probability that a bank processes at least five checks before finding one which must be returned (leave the answer in symbolic form).

19. In an experiment two unmatched groups of ten individuals each were employed. Each individual in each group was selected independently and at random. The score of each individual on a certain perceptual test was determined. The results were as follows:

GROUP I	GROUP II
5	14
8	21
7	23
6	6
21	11
13	5
20	10
17	18
10	21
17	25

Before the experiment, the experimenter expected group II to have the larger scores. Use the median test to examine the hypothesis that the groups represented by these samples are actually equal in the scores one should expect.

20. Suppose that a certain door-to-door salesman believes that he has probability of .30 of making a sale on any given call. If his sales can be regarded as corresponding to events in a stable and independent Bernoulli process, what is the probability that he makes his fifth sale on his tenth call, given that $p = .30$? What is the probability that he makes his first sale on his fifth call?

21. A manufacturer of cloth has reason to believe that defects in his product occur at a rate of about 2 per bolt of cloth. If this is the case, and the occurrence of such defects represents a stable and independent random process, what is the probability that a bolt will contain no defects? Between and 1 and 3 defects?

22. Suppose that in problem 19 above Group II represented 12-year-old children and

Group I, 8 year olds. Imagine the children to be matched by rows, with each row representing one pair. Use the sign test to examine the hypothesis that these two age groups are identical in the ability reflected by these scores.

23. Suppose that a bag contains a large number of nickels, dimes, and quarters, all thoroughly mixed up. Ten coins are extracted at random and with replacement. If 40 percent of the coins are nickels, 30 percent dimes, and 30 percent quarters, what is the probability of drawing:

    (a) 6 nickels
        3 dimes
        1 quarter

    (b) all nickels or all dimes or all quarters

    (c) 3 nickels
        4 dimes
        3 quarters

24. Suppose that in problem 23 ten coins had been sampled at random *without* replacement. If the bag contained 50 coins, find the probability of drawing:

    (a) 5 nickels
        0 dimes
        5 quarters

    (b) all dimes or all quarters

6 Central Tendency and Variability

Any frequency distribution is a summarization of data, but for many purposes it is necessary to summarize still further. Rather than compare entire distributions of data with each other or with hypothetical distributions, it is generally more efficient to compare only certain characteristics of distributions. Two such general characteristics of any distribution, whether frequency or probability, obtained or theoretical, are its measures of central tendency and variability. Indices of central tendency are ways of describing the "typical" or the "average" value in the distribution. Indices of variability, on the other hand, describe the "spread" or the extent of difference among the observations making up the distribution.

The more basic of these two concerns is the description of central tendency. This will be treated first for obtained frequency distributions, and then the ideas will be extended to probability distributions as well. Next, the measurement of the dispersion or spread of a frequency distribution will be taken up, and a parallel treatment will once again be given probability distributions. Finally, an attempt will be made to show how two indices, the mean and standard deviation, form the cornerstone of most statistical inference.

6.1 THE SUMMATION NOTATION

In this chapter it will be necessary to employ the summation symbol Σ (capital Greek sigma). This is read as "the sum of," and tells us to take the sum of the values represented by the expressions following the symbol. Thus, for example, Σx stands for the sum over all the different values that the variable X can assume. Most simple statistical derivations involve various sums, and the use of this symbol introduces considerable economy of statement into these formulations.

There are a number of simple rules for the algebraic manipulation of the summation sign. These rules are given and illustrated in Appendix A. The student who has not already encountered summation notation in school algebra or elsewhere is urged to study these rules until he is thoroughly familiar with them. Actually, the rules themselves are easy, and a little practice at writing out the sums symbolized can familiarize one very quickly with the various ways sums can be manipulated. A little time spent in this way will greatly increase your ability to follow the simple mathematical arguments used in later sections.

6.2 Measures of central tendency

Imagine an obtained distribution of numerical scores. If you were asked to state *one value* that would best "capture" and communicate the distribution as a whole, which value should you choose? One way to answer this question is to find that score-value which is a good "bet" about any randomly selected case from this distribution. Such a score may not be exactly correct for any given case, but it should be a fairly good guess about the obtained score for that case. However, there are at least three different ways to specify what we mean by a "good bet" about any case: (1) the most frequent (most probable) measurement class, (2) the point exactly midway between the top and bottom halves of the distribution, and (3) the arithmetic average of the distribution. The first of these ways of defining the central tendency leads to the measure known as the **mode**; the second leads to the **median** of the distribution; the third is merely the familiar average, or **mean.**

6.3 The mode

The mode is certainly the most easily computed and the simplest to interpret of all the measures of central tendency. It is merely the midpoint or class name of the most frequent measurement class. If a case were drawn at random from the distribution, then that case is more likely to fall in the **modal class** than any other. So it is that in the graph of any distribution, the modal class shows the highest "peak" or "hump" in the graph. The mode may be used to describe any distribution, regardless of whether the events are categorized or numerical.

On the other hand, there are some disadvantages to the mode. One is that there may be more than one modal class: it is perfectly possible for two or more measurement classes to show frequencies that are equal to each other and higher than the frequency shown by any other class. In this case, there is ambiguity about which class gives *the* mode of the distribution, as two or more values are "most popular." Fortunately, this does not happen very often; even though there are two or more "humps" in the graph of a distribution, the class having the highest frequency or probability is still taken as *the* mode. Nevertheless, in such a distribution, where one measurement class is not clearly most popular, the mode loses its effectiveness as a characterization of the distribution as a whole.

Another disadvantage is that the mode is very sensitive to the size and number of class intervals employed when events are numerical; the value of the mode may be made to "jump around" considerably by changing the class intervals for a distribution.

Finally, the mode of a sample distribution forms a very undependable source of information about the mode of the basic probability distribution. For these reasons, our use of the mode will be restricted to situations where: (1) the data are truly nominal scale in nature; (2) only the simplest, most easily computed, measure of central tendency is needed.

6.4 THE MEDIAN

The median score in any set of observations or in a distribution is also, in a sense, a good bet about any case in the total set represented. Just as its name implies, the median is the score corresponding to the middle individual when all individual cases are arranged in order by scores, or to a score that divides the cases into two intervals having equal frequency. Thus, if you drew a case at random from any set of N observations, and guessed that this case showed the median score, you are just as likely to be guessing too high as too low.

The median for a set of observed data is ordinarily defined in slightly different ways depending upon whether N is odd or even, and upon whether raw data or a grouped frequency distribution is to be described. For a set of raw scores, **when N is odd the median corresponds to the score of individual number** $(N + 1)/2$, **when all individuals are arranged in order by scores; when N is even, the median is defined as the score-value midway between the scores for individual $(N/2)$ and individual $(N/2) + 1$ in order.** Then either for odd or even N it will be true that exactly as many cases fall above as fall below the median score.

On the other hand, this way of computing a median is not usually applicable if the data have been arranged into a grouped frequency distribution. **For any such grouped distribution, the median is defined as the point at or below which exactly 50 percent of the cases fall.** Consequently the first step in finding the median of a grouped distribution is to construct the *cumulative* frequency distribution. This is illustrated in the table shown on page 160.

The last column given shows the cumulative frequencies for the class intervals. Since, by definition, the median will be that point in the distribution at or below which 50 percent of the cases fall, the cumulative frequency *at* the median score should be .50N. Thus, the cumulative frequency is .50(200) = 100 at the median for this example. Where would such a score fall? It can be seen that it could not fall in any interval below the real limit of 43.5, since only 85 cases fall at or below that point. However, it does fall below the real limit 48.5, since 102 cases fall at or below that point. Hence, we have located the median as being in the class interval with real limits 43.5 and 48.5.

We must still ascertain the *score* that corresponds to the median, and this is where the process of interpolation comes in. The median score is somewhere in the interval 44–48. We assume that the 17 cases in that interval are evenly scattered over the interval width of 5 units, as in Figure 6.4.1.

The median score, which is greater than or equal to exactly 100 cases, must then exceed not only the 85 cases below 43.5, but also equal or exceed 15 cases above 43.5. In other words, the median is 15/17 of the way *up* the interval

Class	f	cf
74–78	10	200
69–73	18	190
64–68	16	172
59–63	16	156
54–58	11	140
49–53	27	129
44–48	17	102
39–43	49	85
34–38	22	36
29–33	6	14
24–28	8	8
	200	

from the lower limit. Next, what score is exactly 15/17 of the way between 43.5 and 48.5? Since the difference between these two limits is 5, the class interval size, we take (15/17)5, or 4.4 as the amount that must be added to the lower real limit to find the median score, so that the median must be 43.5 + 4.4 or 47.9.

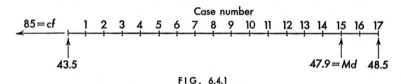

FIG. 6.4.1

A little computational formula that summarizes the steps just described is given by:

$$\text{median} = \text{lower real limit} + i\,\frac{(.50N - cf \text{ below lower limit})}{f \text{ in interval}} \qquad [6.4.1\dagger]$$

where the lower real limit used belongs to the interval containing the median, and the *cf* refers to the cumulative frequency *up to* the lower limit of the interval. For the example, we find

$$\text{median} = 43.5 + 5\,\frac{[.50(200) - 85]}{17} = 47.9.$$

If the median can fall only in some class interval with nonzero frequency, this method of interpolation gives a unique value. However, if the interval frequency happens to be zero, then actually any point in the interval serves equally well as the distribution median; here one usually takes the midpoint as the median.

Even when the raw data are available, it is often worthwhile to define and compute the median as for a grouped frequency distribution. In the first place, for a sizable N, ordering all of the cases by their score magnitudes can be a considerable chore, and it may be simpler to construct a grouped distribution. Secondly, a troublesome problem arises when two or more cases in the raw data are tied in order at the median position, and here it often makes sense to calculate the median by interpolation as for a grouped distribution.

In principle, a median may be found for any distribution in which the variable represents an interval or even an ordinal scale; it is not, in general, applied to a distribution in which the measurement classes are purely categorical, since such classes are unordered.

The median is considerably less sensitive to the distribution's grouping into class intervals than is the mode. Furthermore, when one is making inferences about a large "population" of potential observations from a sample, the median is generally more useful and informative than the mode, although the median itself is not ordinarily so useful as the mean, to be discussed next. We will have more to say about characteristics of the median in future sections. For the moment, we need to consider the third, and most used, of the indices of central tendency: the mean.

6.5 THE ARITHMETIC MEAN

By far the most used and familiar index of central tendency for a set of raw data or a distribution is the mean, or simple arithmetic average. Surely everyone knows that to take the average of a set of raw scores you simply add them all up and divide by the total number, N:

$$M = \frac{\sum_i x_i}{N}. \qquad [6.5.1*]$$

(Here, x_i stands for the score of the observation labeled i, and the sum is taken over all of the N different observations i.) Thus, equation **6.5.1** actually defines the mean for any set of raw data in the form of numerical scores.

Since expressions representing means will occur so frequently in all of the later sections, it is well to point out that we might also represent the arithmetic mean by

$$M = \sum_i \frac{x_i}{N},$$

standing for each value x_i first divided by N and then summed over the individual observations i. The value M represented by either of these expressions is precisely the same; this accords with rule 1 of Appendix A, that the sum of N observations each multiplied by a constant number is the same as the sum itself multiplied by that number. Therefore, in succeeding sections sometimes one and sometimes the other way of expressing the mean will be used, depending on the algebraic and typographical requirements of the particular discussion in which these expressions occur.

Incidentally, it should be mentioned that other texts in statistics often use other symbols for the sample mean. Frequently, the sample mean is shown as $\bar{X}$ or as $\bar{x}$, when X is the variable of interest. Since the bar notation has already been introduced here in another connection, we will continue to designate the sample mean by M.

The definition and computation of the arithmetic mean for raw data is simple enough, but the situation is slightly more complicated when one wishes to find the mean of a grouped distribution of scores. You will recall (Section 3.8) that when a distribution is grouped into class intervals the midpoint x of each class interval was taken to represent the score of each of the cases in the interval. Thus, in an interval 59–73 with midpoint 66 and frequency 16, the sum of the scores of the 16 cases falling into this interval is taken to be 66 summed 16 times, or $(66)(16) = xf$. Similarly, when all of the scores in any interval are assumed the same, their sum is the midpoint of that particular interval times the frequency for that interval, or xf. Then the sum of all of the scores in the distribution is taken to be the sum of the values of x times f over all of the respective intervals, and thus the mean is found from

$$M = \frac{\Sigma xf}{N}. \qquad [6.5.2^*]$$

(Note that here x is *any* midpoint, f is the frequency corresponding to that interval, and the sum is taken over *all intervals*.)

Table 6.5.1

Class	x	f	xf
74–78	76	10	760
69–73	71	18	1278
64–68	66	16	1056
59–63	61	16	976
54–58	56	11	616
49–53	51	27	1377
44–48	46	17	782
39–43	41	49	2009
34–38	36	22	792
29–33	31	6	186
24–28	26	8	208
		200	10,040 = Σxf

$$M = \frac{\Sigma xf}{N} = \frac{10040}{200} = 50.2$$

For example, consider once again the distribution shown in Section 6.5 (Table 6.5.1). The frequency of each class interval is multiplied by its midpoint, and these are then summed and divided by N to given the distribution mean.

The mean calculated from a distribution with grouped class intervals need not agree exactly with that calculated from raw scores. Information is lost and a certain amount of inaccuracy introduced when scores are grouped and treated as though each corresponded to the midpoint of some interval. The coarser the grouping, in general, the more likely is the distribution mean to differ from the raw-score mean. For most practical work the rule of ten to twenty class intervals gives relatively good agreement, however. Nevertheless, it is useful to think of the mean calculated from any given distribution as the mean of that *particular* distribution, a particular set of groupings with their associated frequencies.

6.6   THE MEAN AS THE "CENTER OF GRAVITY" OF A DISTRIBUTION

The mean of a distribution parallels the physical idea of a center of gravity, or balance point, of ideal objects arranged in a straight line. For example, imagine an ideal board having zero weight. Along this board are arranged stacks of objects at various positions. The objects have uniform weight and differ from each other only in their positions on the board. The board is marked off in equal units of some kind, and each object is assigned a number according to its position.

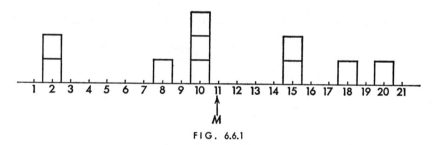

FIG. 6.6.1

This is shown in Figure 6.6.1. Now given this idealized situation, at what point would a fulcrum placed under the board create a state of balance? That is, what is the point at which the "push" of objects on one side of the board is exactly equal to the push exerted by objects on the other side? This is found from the mean of the positions of the various objects:

$$M = \frac{2 + 2 + 8 + 10 + 10 + 10 + 15 + 15 + 18 + 20}{10} = 11.$$

Here, the board would exactly balance if a fulcrum were placed at the position marked 11. Note that since there were piles of uniform objects at various positions on the board, this center of gravity was found in exactly the same way as for the mean of a distribution, since the position (midpoint of an interval) was in effect multiplied by the number of objects at that position (the class frequency), and then these values were summed and divided by the number of objects (the total frequency).

In short, the position of any object on the board is analogous to the score of a case, and each case is treated as having equal "weight" in our computations. The arithmetic mean is then like the center of gravity, or balance point. The mean is that score about which deviations in one direction exactly equal deviations in the other. The tendency for cases in a distribution to differ from the mean in one way is exactly balanced by the tendency to differ in the opposite way.

This property of the mean is bound up in the statement that **the sum of the deviations about the mean is zero in any distribution.** A deviation from the mean is simply the signed difference between the score for any case and the mean score:

$$d_i = (x_i - M). \qquad [6.6.1]$$

In order to show that the sum of these deviations must be zero, suppose that we had a frequency distribution consisting of the scores for N observed cases. For each observation's score x_i we find the deviation d_i from the mean and then find the sum over these N deviations:

$$\sum_i d_i = \sum_i (x_i - M).$$

By rule 4 in Appendix A this is the same as

$$\sum_i d_i = \sum_i x_i - \sum_i M,$$

or, since M is a constant number for each in every case, by rule 1, Appendix A,

$$\sum_i d_i = \sum_i x_i - NM.$$

However, since by definition $M = \dfrac{\sum_i x_i}{N}$,

then

$$NM = \sum_i x_i.$$

Thus,

$$\sum_i d_i = NM - NM = 0,$$

the sum of the signed deviations about the mean is always zero.

A simple consequence of this fact is that *the mean signed deviation from the mean is zero:*

$$\frac{\sum_i d_i}{N} = \frac{0}{N} = 0. \qquad [6.6.2*]$$

Suppose once again that you were told to *guess* the score of some case picked at random from a distribution. If you guessed the mean for each case you might be in error to some extent on each trial, since it need not be true that the mean is exactly the same as any obtained score. The *extent* of error for a given case is d, the departure of the true score from the mean. Over all possible cases that might be drawn from the distribution the average error would be $\sum_i d_i/N$, the mean deviation. But we have just seen that the mean deviation is always zero. Hence, the following statement is true: **if the mean is guessed as the score for any case drawn at random from the distribution, on the average the amount of signed error will be zero.** This is a most important interpretation of the mean: the mean is the best guess about the score of any case in the distribution, if one wishes his *average* signed error to be zero.

6.7 "BEST GUESS" INTERPRETATIONS OF THE MODE AND MEDIAN

We have just seen that for any distribution of cases the mean provides a best guess about any randomly selected case, provided that one is interested in making the signed error zero on the average. Furthermore, as we shall prove later in this chapter, **the mean has the unique property of making the average squared error as small as possible,** if one were to guess the mean in place of any score.

However, both the mode and the median can also be given the same kind of interpretation as best guesses if, instead of wishing to make average signed error zero, we want something else to be true about errors.

First of all, suppose there were some distribution where you had to guess the score of a case picked at random, and you wanted to be *absolutely right* with the highest possible probability. Then you should guess the mode rather than the mean; since it is the most frequent score, guessing the mode guarantees the greatest likelihood of hitting the score "on the nose" for a case drawn at random.

On the other hand, it might be that in guessing the score drawn at random you are not interested in being exactly right most often, nor in making signed error zero on the average, but rather wanted to make the smallest absolute error on the average. Here, the sign of the error is unimportant, but the size of the error is what matters. Then you should guess the median for any score. By doing so you would make the *smallest absolute error* on the average. The median is the typical score in this sense: it is closest on the average to all of the scores in the distribution.

There is really no way to say which is the best measure of central tendency in general terms. This depends very much on what one is trying to do and what one wants to communicate in summary form about the distribution. Each of the measures of central tendency is, in its way, a best guess about any score, but the sense of "best" differs with the way error is regarded. If both the size of the errors and their signs are considered, and we want zero error in the long run, then the mean serves as a best guess. If a miss is as good as a mile, and one wants to be exactly right as often as possible, then the mode is indicated. If one wants to come as close as possible on the average, irrespective of sign of error, then the median is a best guess.

From the point of view of purely descriptive statistics, as apart from inferential work, the median is a most servicable measure. Its property of representing the typical (most nearly like) score makes it fit the requirements of simple and effective communication better than the mean in many contexts.

On the other hand, the median is usually inferior to the mean when our purpose is to make inferences beyond the sample. The median has mathematical properties making it difficult to work with, whereas the mean is mathematically tractable. For this and other reasons, mathematical statistics has taken the mean as the focus of most of its inferential methods, and the median is relatively unimportant in inferential statistics. Nevertheless, as a description of a given set of data, the median is extremely useful in communicating the typical score.

6.8   CENTRAL TENDENCY IN DISCRETE PROBABILITY DISTRIBUTIONS

The ideas of mean, median, and mode apply to distributions of discrete random variables just as they do to frequency distributions. However, as we shall see, in probability distributions measures of central tendency such as the mean often play a much more important role: not only do such measures summarize the distribution, but also they may serve as parameters entering into the mathematical rule for the probability associated with a value or interval of values of the random variable.

For a discrete random variable, the mode is simply the *most probable* value. For example, in the discrete distribution shown in Table 6.8.1, the mode is 41, since this is the midpoint of the most probable class (recall that each case in any interval of a grouped distribution is ordinarily treated as though it had the value of the midpoint).

A median value for a discrete random variable need not be unique. Any value qualifies if the probability for X less than or equal to that value equals .50, $p(X \leq Md) = .5$. The median is any value that evenly divides the distribution. The median of a grouped probability distribution is found in exactly the same way as for a frequency distribution, except that probabilities take the place of frequencies in the computations:

$$Md = \text{lower real limit} + i \left[\frac{.50 - p(X \leq \text{lower real limit})}{p(\text{lower limit} \leq X \leq \text{upper limit})} \right] \quad [6.8.1]$$

where "lower limit" and "upper limit" refer to the interval containing the median. The example below shows the computation of the median in such a distribution.

Finally, the mean of a discrete distribution is found much as for a grouped frequency distribution: each distinct value of X, or midpoint of an interval, is multiplied by the *probability* that X takes on that value (or that X lies in the interval). Then the sum of these products is found:

$$\text{mean} = \Sigma x p(x) \qquad [6.8.2*]$$

summed over all possible values of X, or

$$\text{mean} = \Sigma x p(X \text{ in interval}) \qquad [6.8.3*]$$

where x is the midpoint of the interval, and the sum is over all intervals.

In general, **the mean of a discrete random variable is the sum of the products of the different values of X each times the probability that X takes on that value.**

All three central tendency indices are found in Table 6.8.1, an example of a probability distribution grouped into class intervals.

Table 6.8.1

Class interval	x	$p(X \text{ in interval})$	$xp(X \text{ in interval})$
74–78	76	.050	3.80
69–73	71	.090	6.39
64–68	66	.080	5.28
59–63	61	.080	4.88
54–58	56	.055	3.08
49–53	51	.135	6.89
44–48	46	.085	3.91
39–43	41	.245	10.05
34–38	36	.110	3.96
29–33	31	.030	.93
24–28	26	.040	1.04
			50.21

Mode $= 41 =$ (midpoint of interval with probability $= .245$)

Median $= 43.5 + \dfrac{5(.50 - .425)}{.085} = 47.91$

(where .425 is $p(X \leq 43.5)$)

Mean $= \Sigma x p(X \text{ in interval}) = 50.21$

6.9 THE MEAN OF A RANDOM VARIABLE AS THE EXPECTATION

A special term is often used to denote the mean of a probability distribution. This is the "expectation" or the "expected value" of a random variable X. The symbol $E(X)$ simply represents the mean of the probability distribution of X, and if X is discrete,

$$E(X) = \Sigma x p(x) = \text{mean of } X, \qquad\qquad [6.9.1^*]$$

the sum being taken over all values that X can assume.

The idea of expectation of a random variable is closely connected with the origin of statistics in games of chance. Gamblers were interested in how much they could "expect" to win in the long run in a game, and in how much they should wager in certain games if the game was to be "fair." Thus, expected value originally meant the expected long-run winnings (or losings) over repeated play; this term has been retained in mathematical statistics to mean the long-run average for any random variable over an indefinite number of samplings.

The use of expectation in a game of chance is easy to illustrate. For example, suppose that someone were setting up a lottery, selling 1000 tickets at \$1 per ticket. He is going to give a prize of \$750 to the winner of the first draw. Suppose now that you buy a ticket. How *good* is this ticket in the sense of *how much you should expect to gain?* Should you have bought it in the first place? You can think of your chances of winning and losing as represented in a probability distribution where the outcome of any drawing falls into one of two event categories:

Class	Prob.
win	1/1000
don't win	999/1000

Translated into the amount of money gained (the random variable X), and with a loss regarded as a negative gain, this distribution becomes

x	$p(x)$
\$749	1/1000
−\$ 1	999/1000

Since this is the distribution of a discrete random variable, the mean of the distribution can be found by the methods of Section 6.8; that is,

$$E(X) = \Sigma x p(x) = 749(1/1000) + (-1)(999/1000)$$
$$= .749 - .999 = -.25.$$

This amount, a *minus* 25 cents, is the amount which you can expect to gain by buying the ticket, meaning that if you played the game over and over indefinitely, in the long run you would be poorer by a quarter per play. Should you buy the lottery ticket? Probably not, if you are going to be strictly rational about it; the mean winnings (the expected value) is certainly not in your favor.

On the other hand, suppose that the prize offered were \$1000, so that the gain in winning is \$999. Now the expected value is

$$E(X) = 999(1/1000) + (-1)(999/1000) = .00.$$

Here the game is more worth your while, as there is at least no amount of money to be lost *or* gained in the long run. Such a game is often called "fair." Obviously, truly fair lotteries and other games of chance are hard to find, since their purpose is to make money for the proprietors, not to break even or lose money. If the prize were $2000, you would likely jump at the opportunity, as in this case the expected value would be exactly the *gain* of one dollar.

In figuring odds in gambling situations one uses the expected value to find what constitutes a fair bet; that is, a bet where the mean of the probability distribution of gains and losses is zero. For instance, it is known that in a particular game the odds are 4 to 1 *against* winning. This means that the probability of winning is 1/5 and that of losing is 4/5. It costs the player exactly $1 to play the game once. How much should the amount he gains by winning be in order to make this a fair game with expectation of zero? The expectation is

$$E(X) = \text{(gain value)}p(\text{win}) + \text{(loss value)}p(\text{lose})$$
$$= \text{(gain value)}(1/5) - \$1(4/5).$$

Setting $E(X)$ equal to zero and solving gives

$$\text{(gain value)}(1/5) - \$1(4/5) = 0$$

or

$$\text{gain value} = \$4.$$

In short, one should stand to gain $4 for $1 put up if the game is to be fair. In general, if odds are A to B against winning, the game or bet is fair when B dollars put up gains A dollars.

In betting situations, the random variable is, of course, gains or losses of amounts of money, or of other things having utility value for the person. Nevertheless, the same general idea applies to any random variable; the expectation is the long-run average value that one should observe.

In the discussion of preferences between lotteries in Chapter 2, it was assumed that such preferences were dictated by the value of any given lottery for the individual. Now it can be seen that the value of the lottery is simply the expected value. Choices among lotteries depend upon the relative magnitudes of their expected values as perceived by the individual.

6.10 Theoretical expectations: the means of the binomial, Poisson, and Pascal distributions

As an example of how the mean of a theoretical probability distribution may be deduced mathematically, we will consider the binomial distribution once again. Here we will see that the distribution rule alone dictates what the expectation of a binomial variable must be. What we will show is that if X is a binomial variable, then

$$E(X) = Np, \tag{6.10.1*}$$

the expectation is the number of observations times the probability of a success.

We start off with the definition of expectation for any discrete random variable

$$E(X) = \Sigma x p(x)$$

For the binomial distribution, the probability that the number of successes X takes on any value r is $\binom{N}{r} p^r q^{N-r}$ for $0 \leq X \leq N$. Thus, any value r multiplied by the probability $p(X = r)$ is

$$r \left[\frac{N!}{r! \, (N-r)!} \, p^r q^{N-r} \right]. \qquad [6.10.2]$$

Now notice that we could factor this expression somewhat, cancelling r in the numerator and denominator and bringing an N and a p outside the brackets:

$$Np \left[\frac{(N-1)!}{(r-1)! \, (N-r)!} \, p^{r-1} q^{N-r} \right]. \qquad [6.10.3]$$

For $r = 0$, expression **6.10.2** is equal to zero, and so there is no equivalent expression **6.10.3**. On substituting expression **6.10.3** into the expression for $E(X)$ we have

$$E(X) = \sum_{r=1}^{N} Np \left[\frac{(N-1)!}{(r-1)! \, (N-r)!} \, p^{r-1} q^{N-r} \right] \qquad [6.10.4]$$

with the sum going from $r = 1$ to $r = N$, since the term is zero for $r = 0$. By rule 1 in Appendix A, this is the same as

$$E(X) = Np \sum_{r=1}^{N} \frac{(N-1)!}{(r-1)! \, (N-r)!} \, p^{r-1} q^{N-r}. \qquad [6.10.5]$$

However, if we wrote out the various terms represented in expression **6.10.5** beyond the summation sign, we would find that each is a binomial probability for a distribution with parameters $N - 1$ and p, and thus their sum must be 1.00. Thus

$$E(X) = Np.$$

We have just proved that the mean of a binomial distribution depends only on the two parameters, N and p. If $N = 10$ and p is $1/2$, then the expectation or mean of the distribution is $(10)(1/2) = 5$. If $N = 25$ and $p = .3$, the mean is $(25)(.3) = 7.5$, and so on. Notice that the mean *can* be some value that X cannot take on, as in this last example. Nevertheless, the mean or expectation is a perfectly good statement about the "best guess" for any set of N observations, provided that we want our long-run error to be zero in guessing.

In an almost identical way, we can show that for a Poisson variable X, with intensity m,

$$E(X) = m.$$

Recall that if X is generated by a Poisson process, then the probability that $X = x$ is given by

$$p(x;m) = \frac{e^{-m}m^x}{x!}$$

Thus, in order to find the expectation, we must multiply each possible value of X by the probability of that value and sum these products:

$$E(X) = \sum_{x=0} x\frac{e^{-m}m^x}{x!}, \qquad x = 0, 1, 2, \ldots .$$

However, for $x = 0$ this product must be zero, and for all the other products, x cancels out the x factor in $x!$, leaving $(x - 1)!$ Hence we have

$$E(X) = \sum_{x=1} \frac{e^{-m}m^x}{(x - 1)!}.$$

If we divide and multiply each term by the constant m, and let the random variable $y = x - 1$, we have the following:

$$E(X) = m \sum_{x=1} \frac{e^{-m}m^{x-1}}{(x - 1)!}$$

$$= m \sum_{y=0} \frac{e^{-m}m^y}{y!}, \qquad y = 0, 1, 2, \cdots .$$

The last expression following the summation sign is simply a Poisson probability. Since the sum of Poisson probabilities over all values of the random variable must be 1.00, the sum of $e^{-m}m^y/y!$ must be equal to 1.00, for $y = 0, 1, 2, \cdots$. What we have left is m times 1.00, or

$$E(X) = m, \qquad \text{the intensity of the process.}$$

Furthermore, in a very closely related way one can show that the expectation of a Pascal variable N, with probability given by

$$p(N = n;r, p) = \binom{n - 1}{r - 1} p^r q^{n-r},$$

is

$$E(N) = \frac{r}{p}.$$

That is, the expected number of trials required in order to achieve the rth success from a Bernoulli process is r/p. Hence if $p = 1/2$ and $r = 5$, we expect that 10 trials will be required to reach the fifth success, and so on for any other r and p.

These are the best illustrations we have had of the use of the function rule for a probability distribution to derive characteristics of the distribution. In particular this is our first use of the expectation idea to derive a *formal* result in statistics. We will have many occasions in subsequent discussion to use the ideas of mathematical expectations in derivations, and we will find it a principal tool for deducing consequences about random variables and their distributions.

6.11   THE ALGEBRA OF EXPECTATIONS, APPENDIX B

Since the idea of expectation is so pervasive in theoretical statistics, it is very convenient to have available some list of the formal rules for dealing with expectations mathematically. These rules are summarized in Appendix B and should clarify the ways that expectations will be treated algebraically in other sections. The student is advised to become familiar with these rules, as with the rules of summation, which they greatly resemble. If he does so he should have very little trouble in following simple derivations in mathematical statistics such as this book contains.

6.12   THE EXPECTATION OF A CONTINUOUS RANDOM VARIABLE

The mean or expectation of a continuous random variable is defined in a way very similar to that for a discrete variable. However, since X may assume any of an infinite set of particular values, and because, for the reasons outlined in Section 3.18, it is necessary to discuss the *probability density* associated with any particular value of X, the actual definition of the mean is somewhat different. If we let $f(x)$ symbolize the probability density associated with any particular value of X, then

$$E(X) = \int_{-\infty}^{\infty} xf(x) \ dx. \qquad [6.12.1*]$$

Here, the integral sign indicates the infinite sum of x times a factor $f(x) \ dx$ for all real number values between the ultimate limits of $-\infty$ and $+\infty$. Notice that, much as in the discrete case, the expectation of the continuous random variable X is actually a sum of products. In the former the products were of values of X each times the probability of that value, but for the continuous case each value of X is weighted by a factor depending on the probability density at that value.

As simple examples of how one actually finds the expectation of continuous random variables, let us consider two very simple density functions. The first is the uniform distribution

$$f(x) = \begin{cases} 1/(b - a), & a \leq x \leq b; a < b; \\ 0, & \text{otherwise.} \end{cases}$$

For this distribution the expectation is found by taking

$$E(X) = \int_{a}^{b} \frac{x}{b - a} \ dx.$$

In the integral calculus it is shown that

$$\int_a^b kx \, dx = \frac{kx^2}{2} + C \bigg]_a^b = \frac{kb^2}{2} - \frac{ka^2}{2}$$

for any b and a. Hence we have

$$\int_a^b \frac{x}{b-a} \, dx = \frac{b^2 - a^2}{2(b-a)} = \frac{a+b}{2}$$

and the expectation of the uniform distribution is given by

$$E(X) = \frac{a+b}{2}.$$

Another very simple continuous random variable follows the rule

$$f(x) = \begin{cases} 2x & 0 \leq x \leq 1, \\ 0 & \text{otherwise.} \end{cases}$$

The expectation of this random variable is then found from

$$\int_0^1 x(2x) \, dx = \int_0^1 2x^2 \, dx.$$

From the calculus we find that

$$\int_0^1 2x^2 \, dx = \frac{2x^3}{3} + C \bigg]_0^1 = \frac{2}{3}.$$

Hence $E(X) = 2/3$ for this random variable.

The rules for expectations and their manipulations given in Appendix B are valid either for continuous or for discrete random variables, with a few minor exceptions that need not bother us in an elementary discussion. This is extremely convenient, since it makes it possible to demonstrate certain general features of statistics without having to qualify the result as pertaining to discrete or continuous variables.

6.13 THE MEAN AS A PARAMETER OF A PROBABILITY DISTRIBUTION

In many important instances of probability distributions, the mean or expectation is a parameter. That is to say, the mean enters into the function rule assigning a probability or probability-density to each possible value of X. A simple example is the binomial distribution of sample proportions (or, more properly, the distribution of sample proportions P, which can be found from the binomial distribution). Here, it is easy to show that $E(P) = p$, one of the two parameters figuring in the function rule for the binomial distribution.

In the preceding section we showed that, for a binomial distribution with parameters p and N,

$$E(X) = Np.$$

Now suppose the random variable were

$$P = \frac{X}{N}.$$

By rule 1 for expectations (Appendix B),

$$E(P) = E\left(\frac{X}{N}\right) = \frac{E(X)}{N} = \frac{Np}{N} = p. \qquad [6.13.1*]$$

For the binomial distribution of proportions P, the expectation is the parameter p.

     We have also just seen that the single parameter of a Poisson distribution, the intensity m, is also the expectation or mean of the distribution: $E(X) = m$.

     It will sometimes be convenient to use still another symbol when we are dealing with the mean of a random variable, especially in its role as a parameter. The small Greek letter mu, μ, will stand for the mean of a probability distribution of a random variable X. That is

$$E(X) = \mu = \text{mean of the distribution of } X. \qquad [6.13.2*]$$

     In much of what follows, small Greek letters will be used to indicate parameters of probability distributions, while Roman letters will stand for sample values. The word "parameter" will always indicate a characteristic of a probability distribution, and the word "statistic" will denote a summary value calculated from a sample. Thus, given some sample space and the random variable X, the mean of the distribution of X is a parameter, μ, and the mean of any given sample, M, is a statistic.

6.14   THE MODE AND MEDIAN OF A CONTINUOUS DISTRIBUTION

     When the random variable X is continuous, the mode and median can be defined much as in the discrete case. The mode of a continuous variable, provided that there is a single mode, is the value of X for which the probability density is greatest. In the smooth curve representing the distribution, the mode is the X value corresponding to the very top of the largest "hump" in the curve. There may be several values of X each associated with the same density, which is greater than any density elsewhere in the range of values, and in this instance there are two or more values that qualify as "the" mode.

     The median is that value a such that the probability that X is less than or equal to a is exactly .50. That is, a is the median value if and only if

$$F(a) = \text{prob.}(X \le a) = \int_{-\infty}^{a} f(x)\,dx = .5.$$

6.15   RELATIONS BETWEEN CENTRAL TENDENCY MEASURES AND THE "SHAPES" OF DISTRIBUTIONS

     In discussions either of obtained frequency or of theoretical probability distributions, it is often expedient to describe the general "shape" of the distribution curve. Although the terms used here can be applied to any distribution, it will be convenient to illustrate them by referring to graphs of continuous distributions.

First of all, a distribution may be described by the number of relative maximum points it exhibits, its "modality." This usually refers to the number of "humps" apparent in the graph of the distribution. Strictly speaking, if the density (or probability or frequency) is greatest at one point, then that value is *the* mode, regardless of whether other relative maxima occur in the distribution or not. Nevertheless, it is common to find a distribution described as **bi-modal** or **multimodal** whenever there are two or more pronounced humps in the curve, ever though there is only one distinct mode. Thus, a distribution may have no modes (Fig. 6.15.1), may be unimodal (Fig. 6.15.2), or may be multimodal (Fig. 6.15.3).

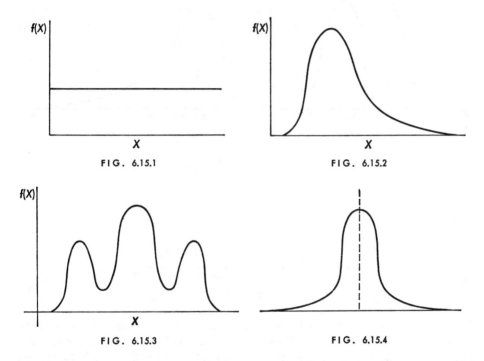

FIG. 6.15.1   FIG. 6.15.2

FIG. 6.15.3   FIG. 6.15.4

Once again, notice that the possibility of multimodal distributions lowers the effectiveness of the mode as a description of central tendency.

Another characteristic of a distribution is its symmetry, or conversely, its skewness. A distribution is **symmetric** only if it is possible to divide its graph into two "mirror-image halves," as illustrated in Figure 6.15.4. Note that in the graph of Figure 6.15.2 there is no point at which the distribution may be divided into two similar parts, as in the first example. When a distribution is symmetric, it will be true that the mean and the median are equal in value. It is not necessarily true, however, that the mode(s) will equal either the mean or the median; witness the example of Figure 6.15.5. On the other hand, a nonsymmetric distribution is sometimes described as **skewed,** which means that the length of one of the **tails** of the distribution, relative to the central section, is disproportionate to the other. For example, the distribution in Figure 6.15.6 is skewed to the right, or **skewed positively**. In a positively skewed distribution, the bulk of the cases fall into the lower part of the range of scores, and relatively few show extremely high values. This is reflected by the relation

$$\text{Mean} > \text{Median}$$

usually found in a positively skewed distribution.

On the other hand, it is possible to find distributions skewed to the left, or **negatively skewed**. In such a distribution, the long tail of the distribution occurs among the low values of the variable. That is, the bulk of the distribution shows relatively high scores, although there are a few quite low scores (Fig. 6.15.7).

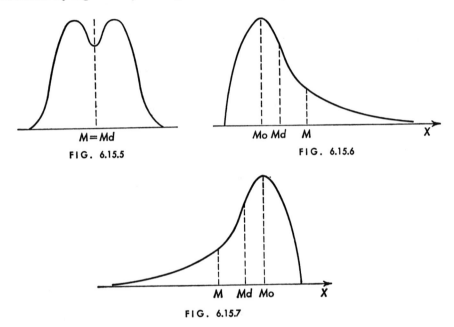

FIG. 6.15.5   FIG. 6.15.6

FIG. 6.15.7

Generally in a negatively skewed distribution, Median > Mean. Thus, a rough and ready way to describe the skewness of a distribution is to find the mean and median; if Mean > Median, then you can conclude that the distribution is skewed to the right (positively). If Median > Mean, then you should conclude that negative skewness exists. If a more accurate determination is needed, other indices reflecting skewness are available; however, this is seldom an important problem in psychological research. A word of warning: if a distribution is symmetric, then Mean = Median, but the fact that Mean = Median does not necessarily imply that the distribution is symmetric.

Describing the skewness of a distribution in terms of measures of central tendency again points up the contrast between the mean and the median as measures of central tendency. The mean is much more affected by the extreme cases in the distribution than is the median. Any alteration of the scores of cases at the extreme ends of a distribution will have no effect at all on the median so long as the rank order of the scores is roughly preserved; only when scores near the center of the distribution are altered is there likelihood of altering the median. This is not true of the mean, which is very sensitive to score changes at the extremes of the distribution. The alteration of the score for a single extreme case in a distribution may have a profound effect on the mean. It is evident that the mean follows the skewed tail in the distribution, while the median does so to less extent. *The occurrence of even a few very high or very low cases can seriously distort the impression of the distribution given by the mean, provided that one mistakenly interprets the mean as the typical value.* If you are dealing with a nonsymmetric distribution, and you want to communicate the typical value, you must report the median. On the other hand, in spite of the distribution's shape, the mean always communicates the same thing: the point about which the sum of deviations is zero.

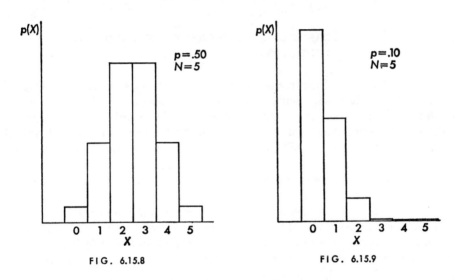

FIG. 6.15.8   FIG. 6.15.9

The choice of an index must depend on what the user is trying to get across about the distribution. (As an exercise, try to imagine a distribution of annual incomes in the United States. What measure of central tendency would you use if you were trying to show that people are well off? What measure would you use to show that people are poorly off?)

Symmetry and skewness in theoretical distributions may be illustrated by the two binominal distributions shown in Section 5.11. The graphs of these two distributions are shown in Figures 6.15.8 and 6.15.9.

In Figure 6.15.8, a binomial distribution with $p = .5$, the distribution is symmetric since the two halves of the histogram are exact mirror images when the division of the graph is made at the mean, $Np = 2.5$. The median of this distribution is also 2.5. Therefore, in the example, the mean equals the median, just as it must for any symmetric distribution.

On the other hand, in the second example with $p = .10$ and $N = 5$, the mode of the distribution occurs at $X = 0$, and the long tail of the distribution extends toward the high values of X. The mean, Np, is .5, and the median is .36

(by interpolation). The fact that the mean is larger than the median reflects the positive skewness of the distribution.

Differences in shape are only rough, qualitative ways of distinguishing among distributions. The only adequate description of a theoretical distribution is its **function rule.** Distributions that look similar in their graphic form may be very different functions. Conversely, distributions that appear quite different when graphed actually may belong to the same family; witness the two binomial distributions in the example. A description in terms of modality and skewness is sometimes useful for giving a general impression of what a distribution is like, but it does not communicate its essential character. Similarly, for obtained frequency distributions, an actual statement of the distribution contains far more information than is ever given by any report of central tendency or shape alone.

6.16   Measures of dispersion in frequency distributions

An index of central tendency summarizes only one special aspect of a distribution, be it mode, median, or mean. Any distribution has at least one more feature that must be summarized in some way. Distributions exhibit **spread** or **dispersion,** the tendency for observations to depart from central tendency. If central tendency measures are thought of as good bets about observations in a distribution, then measures of spread represent the other side of the question: dispersion reflects the "poorness" of central tendency as a description of a randomly selected case, the tendency of observations *not* to be like the average.

The mean is a good bet about the score of any observation sampled at random from a distribution, but *no observed case need be exactly like the mean.* A **deviation** from the mean expresses how "off" the mean is as a bet about a particular case, or how much *in error* is the mean as a description of this case:

$$d_i = (x_i - M)$$

where X_i is the value of a particular observation.

In the same way, we could talk about a deviation from the median, if we wished,

$$d_i' = (x_i - Md)$$

or even perhaps from the mode. It is quite obvious that the larger such deviations are, on the whole, from a measure of central tendency, the more do cases differ from each other and the more spread does the distribution show. What we need is an index (or set of indices) to reflect this spread or variability.

First of all, why not simply take the average of the deviations about the mean as our measure of variability:

$$\frac{\sum_i (x_i - M)}{N}$$

or

$$\frac{\Sigma(x - M) \text{ freq.}(x)}{N} \text{ ?}$$

This will not work, however, because in Section 6.6 it was shown that in any frequency distribution the mean deviation from the mean must be zero.

The device used to get around this difficulty is to take the *square* of each deviation from the mean, and then to find the average of these squared deviations:

$$S^2 = \frac{\sum_i (x_i - M)^2}{N} = \frac{\sum_i d_i^2}{N}. \qquad [6.16.1^*]$$

For any distribution, the index S^2, equal to the average of the squared deviations from the mean, is called the **variance** of the distribution. *The variance reflects the degree of spread, since S^2 will be zero if and only if each and every case in the distribution shows exactly the same score, the mean. The more that the cases tend to differ from each other and the mean, the larger will the variance be.*

The variance is defined in equation **6.16.1** *as though* the raw score of each of N cases were known, and can be computed by this formula when the raw data have not been grouped into a distribution. However, when data are in a grouped distribution, an equivalent definition of the variance is given by

$$S^2 = \frac{\Sigma(x - M)^2 \text{ freq.}(x)}{N} \qquad [6.16.2^*]$$

where x is, as usual, the midpoint of an interval. Here, for each interval, the deviation of the midpoint from the mean is squared, and multiplied by the frequency for that interval. When this has been done for each interval, the average of these products is the variance, S^2. Just as with the mean, the value of S^2 calculated for a grouped distribution need not agree exactly with that based on the raw scores; nevertheless, if a relatively large number of class intervals is used these two values should agree very closely.

6.17 THE STANDARD DEVIATION

Although the variance is an adequate way of describing the degree of variability in a distribution, it does have one important drawback. The variance is a quantity in squared units of measurement. For instance, if measurements of height are made in inches, then the mean is some number of inches, and a deviation from the mean is a difference in inches. However, the square of a deviation is in *square-inch units,* and thus the variance, being a mean squared deviation, must also be in square inches. Naturally, this is not an insurmountable problem: taking the positive square root of the variance gives an index of variability in the original units. **The square root of the variance for a distribution is called the standard deviation, and is an index of variability in the original measurement units.**

A Roman letter S will be used to denote the standard deviation for a frequency distribution:

$$S = \sqrt{S^2} = \sqrt{\frac{\sum_i (x_i - M)^2}{N}} \qquad [6.17.1^*]$$

or

$$S = \sqrt{\frac{\Sigma(x - M)^2 \text{ freq.}(x)}{N}}. \qquad [6.17.2^*]$$

6.18   THE COMPUTATION OF THE VARIANCE AND STANDARD DEVIATION

The variance and standard deviation can be computed from a list of raw scores by formulas **6.16.1** and **6.17.1**. This entails finding the mean, subtracting it successively from each score, squaring each result, adding the squared deviations together, and dividing by N to find the variance. Obviously, this is relatively laborious for a sizable number of cases. However, it is possible to simplify the computations by a few algebraic manipulations of the original formulas. This may be shown as follows: For any single deviation, expanding the square gives

$$(x_i - M)^2 = x_i^2 - 2x_iM + M^2,$$

so that, on averaging these squares, we have

$$\sum_i \frac{(x - M)^2}{N} = \sum_i \frac{(x_i^2 - 2x_iM + M^2)}{N}.$$

By rule 4 in Appendix A, the summation may be distributed, so that

$$\sum_i \frac{(x_i - M)^2}{N} = \sum_i \frac{x_i^2}{N} - 2\sum_i \frac{x_iM}{N} + \sum_i \frac{M^2}{N}.$$

However, wherever M appears it is a constant over the sum, and so, by rules 1 and 2 in Appendix A,

$$\sum_i \frac{(x_i - M)^2}{N} = \sum_i \frac{x_i^2}{N} - 2M\sum_i \frac{x_i}{N} + \frac{NM^2}{N}$$

or

$$S^2 = \sum_i \frac{x_i^2}{N} - 2M^2 + M^2 = \frac{\sum_i x_i^2}{N} - M^2. \qquad [6.18.1\dagger]$$

Finally,

$$S = \sqrt{\frac{\sum_i x_i^2}{N} - M^2}. \qquad [6.18.2\dagger]$$

These last formulas, **6.18.1** and **6.18.2**, give a way to calculate the indices S^2 and S with some saving in steps. These formulas will be referred to as the "raw-score computing forms" for the variance and standard deviation. For an example of the use of these forms, study the example shown in Table 6.18.1, based on seven cases. These two methods must always agree exactly, as in the example.

Table 6.18.1

Scores x_i	Deviation method		Raw-score method x_i^2
	$d_i = (x_i - M)$	d_i^2	
11	6	36	121
10	5	25	100
9	4	16	81
8	3	9	64
6	1	1	36
−4	−9	81	16
−5	−10	100	25
$35 = \sum_i x_i$	0	$268 = \sum_i d_i^2$	$443 = \sum_i x_i^2$

$$M = 5 \qquad S^2 = \frac{\sum_i d_i^2}{N} = \frac{268}{7} = 38.28 \qquad S^2 = \frac{\sum_i x_i^2}{N} - M^2$$

$$= \frac{443}{7} - 25 = 38.28$$

$$S = \sqrt{38.28} = 6.19$$

It is also possible to find similar computing forms for grouped distributions. Starting with the definition of the variance of a grouped distribution,

$$S^2 = \frac{\Sigma(x - M)^2 \, \text{freq.}(x)}{N}$$

we could develop the following computing forms:

$$S^2 = \frac{\Sigma x^2 \, \text{freq.}(x)}{N} - M^2 \qquad\qquad [6.18.3\dagger]$$

$$= \frac{\Sigma x^2 \, \text{freq.}(x)}{N} - \left[\frac{\Sigma x \, \text{freq.}(x)}{N}\right]^2. \qquad\qquad [6.18.4\dagger]$$

Then, as usual,

$$S = \sqrt{S^2} \qquad\qquad [6.18.5\dagger]$$

This little derivation is left to you as an exercise.

Table 6.18.2

Class	x	f	xf	x^2	x^2f
46–50	48	6	288	2304	13824
41–45	43	8	344	1849	14792
36–40	38	10	380	1444	14440
31–35	33	5	165	1089	5445
26–30	28	3	84	784	2352
21–25	23	1	23	529	529
		33	$1284 = \Sigma xf$		$51382 = \Sigma x^2f$

$$M = \frac{\Sigma xf}{N} = \frac{1284}{33} = 38.91$$

$$S^2 = \frac{51382}{33} - 1513.99 = 43.04 \qquad S = \sqrt{43.04} = 6.56$$
$$= 58.64$$

An example of the computation of the mean and standard deviation from a frequency distribution is given in Table 6.18.2.

6.19   SOME MEANINGS OF THE VARIANCE AND STANDARD DEVIATION

Since the variance and standard deviation will figure very largely in our subsequent work, it is well to gain some intuition about what they represent. One interpretation is provided by the fact that **the variance is directly proportional to the average squared difference between all pairs of observations:**

$$\sum_{(i,j)} \frac{(X_i - X_j)^2}{\binom{N}{2}} = \frac{4N}{N-1} S^2. \qquad [6.19.1]$$

(Here (i,j) indicates summation over all possible *pairs* of scores.) The variance summarizes *how different the various cases are from each other*, just as it reflects how different each case is from the mean. Given some N cases, the more that pairs of cases tend to be unlike in their scores, the larger the variance and standard deviation.

Still another way to think of the variance and standard deviation is by a physical analogy. A deviation from the mean can be identified with a certain amount of *force* exerted by a variety of factors making this case different from others in its group. Think of any score as composed of the mean, plus a deviation from the mean,

$$x_i = M + d_i,$$

where the deviation is the resultant force of these "influences." Picturing a deviation from the mean as we would show a physical force away from a point, we have:

$$M \xrightarrow{\hspace{3cm}} x_i.$$
$$\underbrace{\hspace{3cm}}_{d_i}$$

Now suppose that we think of two cases from the larger group, each of which has a deviation from the mean, or d_1 and d_2 respectively. These two cases are independent, so that we can represent their deviations as forces acting at right angles (Fig. 6.19.1). How would you find the *net force* away from the mean for these two cases? The rule of the parallelogram of forces shows the resultant force to be the *diagonal length* in Figure 6.19.1. This diagonal has length

$$\sqrt{\Sigma d^2} = \sqrt{(x_1 - M)^2 + (x_2 - M)^2}$$

by the Pythagorean theorem. If we divide by N before taking the square root, this looks much like the standard deviation.

Suppose that there were three independent cases. When their deviations are interpreted as forces away from the mean, the diagram shown in Figure 6.19.2 holds, and the resultant force is $\sqrt{d_1^2 + d_2^2 + d_3^2}$. The resultant force away from the mean per observation would again be calculated much like a standard deviation. In short, a physical analogy to the standard deviation is a resultant

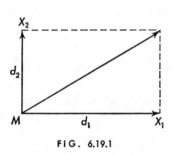

FIG. 6.19.1

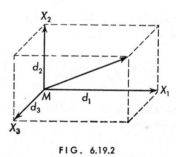

FIG. 6.19.2

force away from the mean per unit observation. A large standard deviation is analogous to a large "push" away from the mean per observation, due to all the factors making observations heterogeneous. In statistics, "error" is often viewed as such a resultant force away from homogeneity, and in terms of this physical analogy the standard deviation should reflect the net effect of such forces per observation.

The reader acquainted with elementary physics may recognize not only that the mean is the center of gravity of a physical distribution of objects, but also that the variance is the **moment of inertia** of a distribution of mass. Furthermore, the standard deviation corresponds to the radius of gyration of a mass distribution; this is the real basis for regarding the standard deviation as analogous to a resultant force away from the mean. These physical conceptions and their associated mathematical formulations have influenced the course of theoretical statistics very strongly, and have helped to shape the form of statistical inference as we will encounter it.

6.20   THE MEAN AS THE "ORIGIN" FOR THE VARIANCE

This question may already have occurred to the reader: "Why is the variance, and hence the standard deviation, always calculated in terms of deviations from the *mean?* Why couldn't one of the other measures of central tendency be used as well?" The answer lies in the fact that **the average squared deviation (i.e., the variance) is smallest when calculated from the mean.** That is, if the disagreement of any score with the mean is indicated by the square of its difference from the mean, then on the average the mean agrees better with the scores than any other single value one might choose.

This may be demonstrated as follows: Suppose that we chose some arbitrary real number C, and calculated a "pseudo-variance" S_C^2 by subtracting C from each score, squaring, and averaging:

$$S_C^2 = \sum_i \frac{(x_i - C)^2}{N}.$$

Adding and subtracting M for each score would not change the value of s_C^2 at all. However, if we did so, we could expand each squared deviation as follows:

$$(x_i - C)^2 = (x_i - M + M - C)^2$$
$$= (x_i - M)^2 + 2(x_i - M)(M - C) + (M - C)^2.$$

Substituting into the expression for s_C^2, and distributing the summation by rule 4, Appendix A, we have

$$S_C^2 = \sum_i \frac{(x_i - M)^2}{N} + 2 \sum_i \frac{(x_i - M)(M - C)}{N} + \sum_i \frac{(M - C)^2}{N}.$$

Notice that $(M - C)$ is the same for every score summed, so that by rules 1 and 2 in Appendix A

$$S_C^2 = \sum_i \frac{(x_i - M)^2}{N} + 2(M - C) \sum_i \frac{(x_i - M)}{N} + (M - C)^2.$$

However, the first term on the right above is simply s^2, the variance about the mean, and the second term is zero, since the average deviation from M is zero. On making these substitutions, we find

$$S_C^2 = S^2 + (M - C)^2.$$

Since $(M - C)^2$ is a squared real number it can be only positive or zero, and so S_C^2 must be greater than or equal to S^2. The value of S_C^2 can be equal to S^2 only when M and C are the same. In short, we have shown that the variance calculated about the mean will always be smaller than about any other point.

If we are going to use the mean to express the central tendency of a distribution, and if we let the standard deviation indicate the extent of error we stand to make in guessing the mean for any score, then this error-quantity is at its minimum when we guess the mean in place of any other single value for all scores. However, it is an important fact that if we appraise error by taking the absolute difference (disregarding sign) between a score and a measure of central tendency, then **the average absolute deviation is smallest when the median is used.** This is one of the reasons that squared deviations rather than absolute deviations figure in the indices of variability when the mean is used to express central tendency. When the median is used to express central tendency it is often accompanied by the average absolute deviation from the median to indicate dispersion, rather than by the standard deviation which is more appropriate to the use of the mean. The average absolute deviation is simply

$$A.D. = \sum_i \frac{|x_i - Md|}{N} \qquad [6.20.1]$$

where the vertical bars indicate a disregarding of sign. Analogically speaking this measure is to the median as the standard deviation is to the mean.

6.21   THE VARIANCE OF A DISCRETE RANDOM VARIABLE

The small Greek letter sigma, σ, is generally used to denote the standard deviation of a random variable, and σ^2 its variance. When the random variable is discrete, the variance is defined by

$$\sigma^2 = \Sigma[x - E(X)]^2 p(x) \qquad [6.21.1*]$$

which is equivalent to

$$\sigma^2 = E(X^2) - [E(X)]^2, \qquad [6.21.2*]$$

the expectation of the square of X minus the square of the expectation of X. Then the standard deviation σ of the random variable X is just the square root of the variance, exactly as in a frequency distribution.

The variance and standard deviation of a probability distribution have exactly the same interpretations as do the corresponding indices for a frequency distribution: each is a measure of the variability or spread, the former in squared units and the latter in the original units of the random variable. However, like the mean of a probability distribution, the variance (or standard deviation) often figures as a parameter entering into the function rule for the distribution.

6.22   AN EXAMPLE OF THE VARIANCE OF A DISCRETE RANDOM VARIABLE: THE BINOMIAL DISTRIBUTION ONCE AGAIN

As an example of the computation of the variance for a discrete random variable, let us take the binomial distribution with $p = 2/3$, and with $N = 6$:

r	$\binom{N}{r} p^r q^{N-r} = p(X = r)$
6	64/729
5	192/729
4	240/729
3	160/729
2	60/729
1	12/729
0	1/729
	729/729

(Notice that the distribution is negatively skewed, since the long tail of the distribution extends toward low values of X.)

In calculating the variance of this distribution, we will make use of the fact gained in Section 6.10, that the expectation of a binomial variable of number of successes is Np; thus,

$$E(X) = Np = (6)(2/3) = 4.$$

To find the variance, we take first the square of each value of X multiplied by its probability, and sum:

$$\Sigma x^2 p(x) = (6)^2(64/729) + (5)^2(192/729) + (4)^2(240/729) + (3)^2(160/729)$$
$$+ (2)^2(60/729) + (1)^2(12/729) + 0$$
$$= \frac{12636}{729} = 17.33.$$

Finally,

$$\sigma^2 = \Sigma x^2 p(x) - [E(X)]^2 = 17.33 - 16 = 1.33$$

and the standard deviation is

$$\sigma = \sqrt{1.33}.$$

This same result could have been found from a general expression for the variance of a binomial distribution. By algebraic manipulation of binomial probabilities substituted into the formula defining the variance of a discrete random variable (6.21.1), much as we did in Section 6.10, it can be shown that for any binomial distribution,

$$\sigma^2 = Npq \qquad\qquad [6.22.1*]$$

and that

$$\sigma = \sqrt{Npq}. \qquad\qquad [6.22.2*]$$

It is rather easy to demonstrate that the variance of a binomial distribution is equal to Npq, and we will proceed to do so. We begin by taking:

$$\sigma^2 = E(X^2) - [E(X)]^2$$

$$= \sum_{x=0}^{N} x^2 \binom{N}{x} p^x q^{N-x} - N^2 p^2.$$

When $x = 0$, the first term of the sum must be zero, and so we can show the sum as actually beginning with $x = 1$. Furthermore, the x in each of the summed terms cancels out the x factor in $x!$, leaving $(x - 1)!$ Finally, each term in the sum can be divided and multiplied by Np. Making these changes results in

$$\sigma^2 = Np \sum_{x=1}^{N} x \frac{(N-1)!}{(x-1)!(N-x)!} p^{x-1} q^{N-x} - N^2 p^2$$

Now let us define $y = x - 1$ and substitute y for x in the expression above. Note that when $x = 1, y = 0$, and when $x = N, y = N - 1$. We then have

$$\sigma^2 = Np \sum_{y=0}^{N-1} (y + 1) \frac{(N-1)!}{(y)!(N-1-y)!} p^y q^{N-1-y} - N^2 p^2$$

$$= Np \sum_{y=0}^{N-1} y \frac{(N-1)!}{(y)!(N-1-y)!} p^y q^{N-1-y} + Np \sum_{y=0}^{N-1} \binom{N-1}{y} p^y q^{N-1-y} - N^2 p^2.$$

The first sum corresponds to the expectation of a binomial variable where $N - 1$ is the number of trials and p the probability of a success. The second sum must be equal to 1.00. Hence we have

$$\sigma^2 = Np(N - 1)p + Np - N^2 p^2$$
$$= N^2 p^2 - Np^2 + Np - N^2 p^2$$
$$= Np(1 - p)$$
$$= Npq.$$

Therefore, for the example the variance must be

$$\sigma^2 = (6)(2/3)(1/3)$$
$$= 1.33.$$

Observe that, just as for the mean of a binomial distribution, the two parameters p and N determine the value of σ^2.

When the random variable is P, the proportion of successes in the sample of N observations, then

$$\sigma_P^2 = \frac{pq}{N}. \qquad [6.22.3*]$$

This can be shown rather easily: From **6.21.2**,

$$\sigma_P^2 = E(P^2) - [E(P)]^2$$
$$= \frac{E(X^2)}{N^2} - \frac{[E(X)]^2}{N^2},$$

by the definition of P and by rule 1, Appendix B.

Then,

$$\frac{E(X^2) - [E(X)]^2}{N^2} = \sigma_P^2$$

so that from **6.22.1**,

$$\sigma_P^2 = \frac{Npq}{N^2} = \frac{pq}{N}.$$

By an argument almost identical to that for the binomial one can show that the variance of a Poisson variable is m, the intensity of the process:

$$\sigma^2 = \sum_{x=0}^{\infty} x^2 \frac{m^x e^{-m}}{x!} - m^2$$

$$= m \sum_{y=0}^{\infty} (y+1) \frac{m^y e^{-m}}{y!} - m^2$$

$$= m^2 + m - m^2 = m.$$

Another, similar, argument shows that the variance of a Pascal distribution where r is the number of successes and p is the probability of a success is given by

$$\sigma^2 = \frac{r(1-p)}{p^2}$$

$$= rqp^{-2}.$$

6.23 THE VARIANCE OF A CONTINUOUS RANDOM VARIABLE

For a distribution of a continuous random variable the variance is defined in the same general way. As always for a continuous variable, the place of probability is taken by the probability density for each value that X can assume. The simplest definition of the variance for such a variable in simply **6.21.2** once again:

$$\sigma^2 = E(X^2) - [E(X)]^2,$$

the expectation of the square of the variable, minus the square of the expectation. This means that for a random variable X with probability density $f(x)$ at any point $X = x$, the variance is

$$\sigma^2 = \int_{-\infty}^{\infty} x^2 f(x) \, dx - [E(X)]^2.$$

As a simple example of how a variance is calculated for a continuous random variable, consider once again the uniform distribution introduced in Chapter 3. Recall that here the random variable is associated with the density function $f(x) = 1/(b-a)$ for $a \leq x \leq b$. We have already seen that the expectation for this random variable is $E(X) = (a+b)/2$. Now we find the variance of the random variable by taking

$$\sigma^2 = \int_a^b \frac{x^2}{b-a}\, dx - \frac{(a+b)^2}{4}$$
$$= \left. \frac{x^3}{3(b-a)} + C \right]_a^b - \frac{(a+b)^2}{4}$$
$$= \frac{b^3 - a^3}{3(b-a)} - \frac{(a+b)^2}{4}$$
$$= \frac{(b-a)^2}{12}.$$

Setting up the equation $\sigma^2 = E(X^2) - [E(X)]^2$ and applying the rules of the integral calculus gives the expression for the variance, which is $(b-a)^2/12$.

As another very simple example, consider once again the random variable with density function $f(x) = 2x$, $0 \leq x \leq 1$. We saw in a preceding section that in this instance $E(X) = 2/3$. Then we can find the variance of this random variable by taking

$$\sigma^2 = \int_0^1 (x^2)(2x)\, dx - (2/3)^2$$
$$= \left. \frac{2x^4}{4} + C \right]_0^1 - 4/9$$
$$= 1/2 - 4/9 = 1/18.$$

Although the mathematical details, such as the evaluation of the definite integral, are usually much more complicated for the sorts of random variables of interest in statistics, the same general principles used here apply in the calculation of any variance.

6.24 MOMENTS OF DISTRIBUTION

A truly mathematical treatment of distributions would introduce not only the mean and the variance, but also a number of other summary characteristics. These are the so-called "moments" of a distribution, which are simply **the expectations of different powers of the random variable.** Thus, the first moment about the origin of a random variable X is

$$E(X) = \text{the mean.}$$

The second moment about the origin is

$$E(X^2),$$

the third

$$E(X^3),$$

and so on. When the mean is subtracted from X before the power is taken, then the moment is said to be **about the mean;** the variance

$$E[X - E(X)]^2$$

is the second moment about the mean;

$$E[X - E(X)]^3$$

is the third moment about the mean, and so on.

Just as the mean describes the "location" of the distribution on the X axis, and the variance describes its dispersion, so do the higher moments reflect other features of the distribution. For example, the third moment about the mean is used in certain measures of degree of **skewness**: the third moment will be zero for a symmetric distribution, negative for skewness to the left, positive for skewness to the right. The fourth moment indicates the degree of "peakedness" or **kurtosis** of the distribution, and so on. These higher moments have relatively little use in elementary applications of statistics, but they are important for mathematical statisticians in the study of the properties of distributions and in arriving at theoretical distributions fitting observed data. The entire set of moments for a distribution will ordinarily determine the distribution exactly, and distributions are sometimes specified in this way when their general function rules are unknown or difficult to state.

6.25   Percentiles and percentile ranks

So far in this chapter we have been concerned with measures of central tendency and of variability. Such indices are summary measures of an entire distribution. However, another type of question is frequently asked. This concerns the relative location of a particular score value within the distribution. Next we will discuss percentiles and percentile ranks, which provide one way to answer this question of location of an individual or a score within any given distribution. After this topic has been explored, another way of approaching the problem of location will be introduced.

In any frequency distribution of numerical scores, the percentile rank of any specific value x is the percent of cases out of the total that fall at or below x in value. Thus, imagine a particular frequency distribution summarizing 345 cases. For the score value $x = 38$, some 115 cases fall at or below x in score value. Then the percentile rank of the value 38 is $(100)(115/345) = 100(.33) = 33$. If, in another distribution, 42 out of a total of 90 cases fall at or below $x = 6$, then $x = 6$ has a percentile rank of $100(42/90) = 46.7$. The values in any frequency distribution of numerical scores can be transformed in this way into a set of percentile ranks. If there is one distinct value that is the minimum score obtained, then its percentile rank must be $100 (1/N)$. The maximum score in the distribution must have percentile rank 100, since all scores in the distribution must be less than or equal to this maximum score.

In a probability distribution, the percentile rank of a given value x is simply $100F(x)$, or 100 times the cumulative probability associated with $X = x$. Note especially that the percentile rank associated with the median in any distribution must be 50, since 50 percent of all observed values in a frequency distribution must lie at or below the median, and $F(X = Md) = .50$ in a probability distribution.

One use of the idea of percentile ranks is in finding the relative locations of values within given distributions. The value need not actually have occurred in a frequency distribution in order for a percentile rank to be determined. One asks where that value would have fallen relative to the other cases if it actually had occurred. For example, suppose that in a distribution of, say, 36 cases we wish to find the percentile rank of the value $x = 20$. If that value actually occurs in the distribution, there is no problem, since we would simply see the number of cases falling at or below 20, divide by N, or 36, and multiply by 100 in order to find the percentile rank. However, suppose that $x = 20$ actually did not occur. Then we first find the value immediately below 20 that did occur. Let us say that this value was 15, and had percentile rank 52.7. Then we find the value closest above 20 that occurred. Suppose this was 23 and its percentile rank was 55.5. Then we can find the percentile rank for $x = 20$ by linear interpolation:

$$\frac{(x - \text{value below } x)}{(\text{value above } x - \text{value below } x)} (\%\text{ile rank above } x - \%\text{ile rank below } x)$$
$$+ \%\text{ile rank below} = \%\text{ile rank of } x,$$

where "%ile" is simply an abbreviation for "percentile." Accordingly, for the example we take

$$\%\text{ile rank of } x = \frac{(20 - 15)}{(23 - 15)} (55.5 - 52.7) + 52.7$$
$$= .625(2.8) + 52.7$$
$$= 54.45.$$

Thus, had $x = 20$ actually occurred among the values in this distribution, about 54.45 percent of the cases in the distribution would have been at or below this value. Such a use of percentile ranks is probably already familiar to most students, since this typically is the way that the test performance of an individual is compared to the performance of some reference group, or norm group.

The percent at or below a given value of x is the percentile rank of that value. Often, however, we are interested in the reverse problem: What score x cuts off the bottom G percent of the distribution? That score is then referred to the G percentile of the distribution. Thus, if a given value of x cuts off the bottom 25 of cases in some distribution, that value is the 25th percentile in the distribution. In order to find the value corresponding to the Gth percentile, we first look to see if there is any exact value that occurred in the distribution such that the number of cases falling at or below that value is $GN/100$. If so, the value is the Gth percentile. However, it may happen that no value can be located that corresponds exactly to the Gth percentile, among those values of X that actually occurred. In that case, linear interpolation is employed once again:

$$\frac{[(GN/100) - cf \text{ below})(\text{value above} - \text{value below})]}{(cf \text{ above} - cf \text{ below})} + \text{value below}$$
$$= G\text{th percentile,}$$

where "cf below" symbolizes the cumulative frequency nearest below the number $GN/100$ among the values of X that occurred, and "value below" is the actual

X value associated with that cumulative frequency. Similarly, "*cf* above" stands for the cumulative frequency just above $GN/100$ for one of the values of X that occurred, and "value above" is the actual value associated with that cumulative frequency. For example, suppose that in a distribution we desired to find the 30th percentile. There are 200 cases in the distribution, and so we want the value at or below which $(30)(200)/100$ or 60 cases would fall. Looking at the distribution, we find that up to and including $X = 22.5$ there are 58 cases. Then 58 is the cumulative frequency below, and 22.5 is the value below. On the other hand, up to and including 23.9 there are 62 cases. What is the value that would have cut off 60 cases—that is, the value of the 30th percentile? We take

$$\frac{(60 - 58)(23.9 - 22.5)}{(62 - 58)} + 22.5 = \frac{2}{4}(1.4) + 22.5$$
$$= 23.2,$$

so that the 60th percentile is 23.2 in this distribution.

A similar set of problems exist when one is trying to find percentile ranks and percentiles in a grouped frequency distribution. Suppose that N is the total number in the distribution, i is the class interval size, and we want to find the percentile rank for some value of X. First of all we locate the class interval in which $X = x$ must fall. That done, we take

$$\text{percentile rank} = \frac{(x - \text{lower limit})}{i}\frac{100(\text{frequency in interval})}{N}$$
$$+ \frac{100(\text{cumulative frequency below interval})}{N},$$

using the lower limit of the interval and i, the class interval size, or (upper real limit − lower real limit).

In order to find the Gth percentile value, we find the interval in which this percentile value must fall by taking $NG/100$, and seeing that the interval chosen has a cumulative frequency greater than or equal to $NG/100$, with the interval just below having a cumulative frequency less than $NG/100$. (Remember that the cumulative frequency of an interval is the number of cases in the total distribution falling at or below the *upper* real limit of the interval.) Then we take

$$\frac{(NG/100 - cf \text{ for interval below})}{\text{frequency in interval}}(i) + \text{lower real limit of interval}$$
$$= G\text{th percentile.}$$

Since we are using linear interpolation to find percentile ranks and percentiles in such a grouped distribution, it follows that we are assuming that within any given class interval, the scores falling into the interval are evenly spaced in value across the extent of the interval. This need not be true, of course, but this is the price we pay for the convenience of being able to work with class intervals, rather than with the original scores.

Occasionally one runs into reports of deciles, quartiles, or other "fractiles" of a distribution. Deciles are simply the values corresponding to the

10th, 20th, 30th, · · · percentiles: those values which divide the distribution into tenths. Similarly, quartiles divide the distribution into fourths, and are the values of the 25th, 50th, 75th, and 100th percentiles. Any other arbitrary division of a distribution might be worked out by percentiles as the occasion demanded.

Finally, it should be apparent that the percentile rank of any value $X = x$ in a distribution is merely the cumulative frequency expressed as a percentage of the total, and the value equivalent to a given percentile is that value for which the cumulative frequency as a percentage of N is a particular number. It follows that given a graph of the cumulative frequency distribution of X, we can find the percentile rank for any $X = x$ by locating the cumulative frequency corresponding to that point, and multiplying by a factor of $100/N$. In a similar way, we can find the value of any percentile G by taking $GN/100$ and finding the corresponding point on the X axis. If there are a great many percentile ranks or percentiles to be found for a given distribution, then the most efficient way of proceeding is through the construction of a relative-frequency polygon, and the location of the required values on the graph in the way just described. In a probability distribution, the percentile rank for $X = x$ is simply $100F(x)$, and the Gth percentile is simply that x for which $G/100 = F(x)$. In this situation, graphic methods are particularly useful.

Next, a different approach to the problem of location of a score value will be presented. Here, rather than percentile ranks, the score value is transformed into a standardized score, relative to the mean and the standard deviation.

6.26 THE RELATIVE LOCATION OF A SCORE IN A FREQUENCY DISTRIBUTION: STANDARDIZED SCORES

A major use of the mean and standard deviation is in transforming a raw score into a **standardized score,** showing *the relative status* of that score in a distribution. If you are given the information, "John Doe has a score of 60," you really know very little about what this score means. Is this score high, low, middling, or what? However, if you know something about the distribution of scores for the group including John Doe, you can judge the location of the score in the distribution. The score of 60 gives quite a different picture when the mean is 30 and the standard deviation 10, than when the mean is 65 and the standard deviation is 20.

Each value in any distribution can be converted into a standardized score, or z **score,** expressing the deviation from the mean in standard deviation units:

$$z = \frac{x - M}{S}. \qquad \qquad [6.26.1*]$$

The z score tells how many standard deviations away from the mean is x. The two distributions mentioned above give quite different z scores to the score of 60:

$$z_1 = \frac{60 - 30}{10} = 3$$

$$z_2 = \frac{60 - 65}{20} = -.25.$$

The conversion of raw scores to z scores is handy when one wishes to emphasize the *location* or *status* of a score in the distribution, and in future sections we will deal with standardized scores when this aspect of any score is to be discussed.

Changing each of the scores in a distribution to a standardized score creates a distribution having a "standard" mean and standard deviation: **the mean of a distribution of standardized scores is always 0, and the standard deviation is always 1.** This is easily shown as follows:

$$\text{mean of } z \text{ scores} = M_z = \sum_i \frac{z_i}{N} = \sum_i \frac{(x_i - M)}{NS}. \qquad [6.26.2]$$

Since N and S are constant over the summation, this becomes

$$M_z = \frac{1}{NS} \sum_i (x_i - M) = 0 \qquad [6.26.3^*]$$

since the sum of deviations about the mean is always zero.

In a similar fashion,

$$\text{variance of } z \text{ scores} = S_z^2 = \sum_i \frac{z_i^2}{N} = \sum_i \frac{(x_i - M)^2}{NS^2}$$

$$= \frac{S^2}{S^2} = 1, \qquad [6.26.4^*]$$

$$\text{standard deviation of } z \text{ scores} = S_z = 1.$$

Although it is true that the mean and standard deviation of a distribution of z scores will always be 0 and 1 respectively, *changing the scores in any distribution to z scores does not alter the shape (or mathematical form) for the distribution. The frequency of any given z score is exactly that of the X score corresponding to it in the distribution.*

6.27 STANDARDIZED SCORES IN PROBABILITY DISTRIBUTIONS

Standardized scores have exactly the same use in probability distributions as in frequency distributions. If X is a value of a random variable, then the corresponding standardized score, relative to the probability distribution, is

$$z = \frac{x - E(X)}{\sigma} = \frac{x - \mu}{\sigma}. \qquad [6.27.1^*]$$

The standardized score is a deviation from expectation, relative to the standard deviation. The standardized score corresponding to any value of X is found in precisely the same way whether the distribution is discrete or continuous.

Furthermore, the mean of the standardized scores is zero for any probability distribution:

$$E(z) = E\left[\frac{X - E(X)}{\sigma}\right] = \frac{E(X) - E(X)}{\sigma} = 0 \qquad [6.27.2^*]$$

since $E(X)$ and σ are constants over the various possible values of X. The standard deviation of standardized scores is always 1 in a probability distribution:

$$\sigma_z^2 = E(z^2) - [E(z)]^2 = E\left[\frac{X - E(X)}{\sigma}\right]^2 = E\frac{[X - E(X)]^2}{\sigma^2}$$

$$= \frac{\sigma^2}{\sigma^2} = 1 \qquad [6.27.3^*]$$

and $\qquad \sigma_z = 1.$

The form of the probability distribution is not changed at all by the transformation to z scores, in the sense that the probability (or probability density) of any value of z is simply the probability (or probability density) of the corresponding value of X.

6.28 TCHEBYCHEFF'S INEQUALITY

There is a very close connection between the size of deviations from the mean and probability, holding for distributions having finite expectation and variance. The following relation is called Tchebycheff's inequality, after the Russian mathematician who first proved this very general principle:

$$\text{prob.}(|X - \mu| \geqq b) \leqq \frac{\sigma^2}{b^2}, \qquad [6.28.1^*]$$

the probability that a random variable X will differ absolutely from expectation by b or more units $(b > 0)$ is *always* less than or equal to the ratio of σ^2 to b^2. Any deviation from expectation of b or more units can be *no more probable* than σ^2/b^2.

This relation can be clarified somewhat by dealing with the deviation in σ units, making the random variable a standardized score. If we let $b = k\sigma$, then the following version of the Tchebycheff inequality is true:

$$\text{prob.}\left(\frac{|X - \mu|}{\sigma} \geqq k\right) \leqq \frac{1}{k^2}, \qquad [6.28.2^*]$$

the probability that a standardized score drawn at random from the distribution has *absolute* magnitude greater than or equal to some positive number k is *always* less than or equal to $1/k^2$. Thus, given a distribution with some mean and variance, the probability of drawing a case having a standardized score of 2 or more (disregarding sign) must be *at most* 1/4. The probability of a standardized score of 3 or more must be no more than 1/9, the probability of 10 or more can be no more than 1/100, and so on.

This last form of the Tchebycheff inequality (**6.28.2**) is quite simple to prove for a discrete random variable. However, one very important principle must be kept in mind during the course of this proof. If we have some event A with probability $p(A)$, and then we somehow create a 1-to-1 function $g(A)$, so that the occurrence of the event $g(a)$ depends only on the occurrence of the event $A = a$, then the probability of the event $g(a)$ is the same as the probability for the event $A = a$, or $p(a)$. Thus, for example, suppose that the toss of the coin can produce either the event "Heads" or the event "Tails," each with some probability $p(H)$ or $p(T)$. Now we define the 1-to-1 function $g(H) = 3$, $g(T) = 17$. Since "3" can occur if and only if "Heads" occurs, then the probability of "3" must be $p(H)$. Similarly, since "17" can occur if and only if "Tails" occurs, then the probability of "17" must be equal to $p(T)$. The probabilities remain the same because the actual events have remained the same; only the names of the events have been changed—in this case converted from nouns to numbers.

Similarly, for any 1-to-1 function $g(x)$ of a random variable X, the probability (or probability density) of $g(x)$ must be the same as the probability (or probability density) for $X = x$. Hence, if we have the random variable defined as follows:

x	$p(x)$
4	1/3
3	1/6
2	3/8
1	1/8
elsewhere	0

and then define the function of X as $g(x) = (4x - 19)/3$, the distribution of values of $g(x)$ will be

$g(x)$	$p[g(x)]$
-1	1/3
$-7/3$	1/6
$-11/3$	3/8
-5	1/8
elsewhere	0

Here, the event $g(x) = -1$ is actually the same as the original event $X = 4$, and hence has probability 1/3. In the same way, $g(x) = -7/3$ can occur if and only if $X = 3$ occurs, $g(x) = -11/3$ can occur if and only if $X = 2$ occurs, and so on. **The original probability or density function of X defines the probability or density function of any 1-to-1 function of X.**

Now, after this digression, we are ready to resume our proof. Let

$$z = \frac{x - \mu}{\sigma}.$$

Then $p(z) = p(x)$, $E(z) = 0$, and $\sigma_z^2 = 1$. Now the probability of a z value equaling or exceeding any arbitrary positive number in absolute value is, for a discrete variable,

$$\text{prob.}(|z| \geqq k) = \sum_{(z \geqq k)} p(z) + \sum_{(z \leqq -k)} p(z)$$

so that, multiplying both sides of this equation by k^2 we have

$$k^2 \text{ prob.}(|z| \geqq k) = \sum_{(z \geqq k)} k^2 p(z) + \sum_{(z \leqq -k)} k^2 p(z).$$

However, each and every value of z represented in the sum on the right of the equation above is greater than or equal to k in absolute value. Thus,

$$\sum_{(z \geqq k)} k^2 p(z) + \sum_{(z \leqq -k)} k^2 p(z) \leqq \sum_{(z \geqq k)} z^2 p(z) + \sum_{(z \leqq -k)} z^2 p(z).$$

Furthermore,

$$\sum_{(z \geqq k)} z^2 p(z) + \sum_{(z \leqq -k)} z^2 p(z) \leqq \sigma_z^2,$$

by the definition of σ_z^2 for a discrete random variable **(6.27.3)**. Then

$$k^2 \text{ prob.}(|z| \geqq k) \leqq \sigma_z^2$$

and

$$\text{prob.}(|z| \geqq k) \leqq \frac{1}{k^2}.$$

Since we have just proved expression **6.28.2** to be true, it is easy enough to work backward and prove **6.28.1**, using the principle outlined above. Note that $X - \mu = z\sigma$ is a 1-to-1 function of z. Hence, the probability for any value of $X - \mu$ is the same as the probability of $z\sigma$, or of z. Furthermore, the event $|z| \geqq k$ is the same as the event $|X - \mu| \geqq k\sigma$. Then, if we let $k\sigma = b$, so that $k = b/\sigma$, we have

$$\text{prob.}(|X - \mu| \geqq k\sigma) \leqq \frac{1}{k^2},$$

$$\text{prob.}(|X - \mu| \geqq b) \leqq \frac{1}{(b/\sigma)^2},$$

$$\text{prob.}(|X - \mu| \geqq b) \leqq \frac{\sigma^2}{b^2},$$

which is the Tchebycheff inequality expressed as in **6.28.1**.

Once the Tchebycheff inequality has been proved for a discrete variable, it is very easy to show that Bernoulli's theorem must be true. It has just been shown that the mean of the distribution of values of P is p, the probability of a "success" on a single trial. Furthermore, in Section 6.22 the variance of the distribution of the random variable P was found to be $\frac{pq}{N}$. Now substituting these values into the Tchebycheff inequality, we find

$$\text{prob.}(|P - p| \geq k) \leq \frac{pq}{Nk^2}.$$

Now suppose that k is some arbitrarily small positive number, and that N is allowed to become indefinitely large. The larger the value of N, the smaller the ratio to the right of the expression becomes; as N approaches an infinite value, then the probability represented on the left of the inequality must grow closer and closer to zero. This is equivalent to the statement that

$$\text{prob.}(|P - p| < k) \to 1.00, \text{ as } N \to \infty,$$

which is actually Bernoulli's theorem.

This way of proving Bernoulli's theorem rests ultimately on the fact that as N becomes larger, the value of the variance of the distribution of P must become smaller, other things being equal. For this reason, no matter how small k may be, as N increases the probability that the absolute difference between P and p is greater than k must approach zero.

When the value of k^2 is less than or equal to 1.00, the Tchebycheff inequality itself is quite trivial. Here, the value of 1 divided by k^2 must be a number no less than 1.00, and we have a statement that we know *must* be true for any probability:

$$\text{prob.}(|z| \geq k) \leq v, \text{ for any } v \geq 1.00.$$

On the other hand, for any value of k greater than 1.00, the Tchebycheff inequality is not trivial, and it does set broad limits to the probability associated with extreme intervals of z values, or of deviations of X values from the mean in terms of standard deviation units.

As a fairly concrete example of how this principle might be applied, imagine a personality test that has been constructed to have a mean of 100 and a standard deviation of 25 in the population of American adults. Unfortunately, we know nothing more of the score distribution over this set of people. We give the test to a person sampled at random, and find that he has a score of 175. What can we say about the probability of observing a score this much or more deviant (in either direction) from the mean? The standardized score relative to mean of 100 and σ of 25 is

$$z = \frac{175 - 100}{25} = 3.$$

Now by setting $k = 3$ in the Tchebycheff inequality, we find

$$\text{prob.}(|z| \geq 3) \leq \frac{1}{9}.$$

The probability of observing a score 3 or more standard deviations from the mean is *no more* than 1/9, regardless of the distribution of scores.

Within how many standard deviations from the mean must *at least one half* the population of American adults fall on this test? That is, we want the absolute value k of z for which

$$p(|z| \geqq k) \leq \frac{1}{2}.$$

By **6.28.2** the value is $\sqrt{2}$, or about 1.4. Regardless of how the scores are actually distributed, we can state that one half or more of the population must lie within the approximate limits $100 - (1.4)(25)$ and $(100) + (1.4)(25)$, or 65 to 135.

Although this principle is very important theoretically, it is not extremely powerful as a tool in applied problems. The Tchebycheff inequality can be strengthened somewhat if we are willing to make assumptions about the general form of the distribution, however. For example, if we assume that the distribution of the random variable is both *symmetric* and *unimodal*, then the relation becomes

$$p(|z| \geqq k) \leq \frac{4}{9}\left(\frac{1}{k^2}\right). \qquad [6.28.3*]$$

For such distributions, we can make somewhat "tighter" statements about how large the probability of a given amount of deviation may be. For the example of the personality test, if the distribution were unimodal-symmetric a score differing by three or more standard deviations from the mean should be observed with probability *no greater than* $(4/9)(1/9)$, or about .05. Similarly, we could find that at least $1 - 4/9$ or $5/9$ of American adults fall within one standard deviation to either side of the mean. Furthermore, $1 - (4/9)(1/4)$ or $8/9$ must fall within two standard deviations, and so on.

Thus, by adding assumptions about the form of the distribution to the general principle relating standardized values to probabilities we are able to make stronger and stronger probability statements about a sample result's departures from the mean. In order to make very precise statements one has to make even stronger assumptions about the distribution of the random variable, unless he is dealing with very large samples of cases, as we shall see. This chapter has concluded with the introduction of the Tchebycheff inequality to suggest that the mean and the standard deviation play a key role in the theory of statistical inference. The mean and standard deviation are, of course, useful devices for summarizing data if that is our purpose, although there are situations where other measures of central tendency and variability may do equally well or better. The really paramount importance of mean and standard deviation does not emerge until one is interested in making *inferences*, involving the estimation of parameters, or assigning probabilities to sample results. Here, we shall find that most "classical" statistical theory is erected around these two indices, together with their combination in standardized scores.

EXERCISES

1. A nutritionist was studying the eating patterns of early adolescent boys and girls. She used two matched groups, the first consisting of 50 thirteen-year-old boys, and the second of 50 thirteen-year-old girls. The data follows in the form of two frequency distributions, showing the number of calories (in hundreds) consumed daily by the members of each group. Compute the median and the modal number of calories consumed by the boys and by the girls: leave the answer in hundreds of calories.

CLASS INTERVAL	BOYS	GIRLS
50.00–54.99	1	1
45.00–49.99	2	0
40.00–44.99	4	2
35.00–39.99	16	5
30.00–34.99	12	10
25.00–29.99	7	20
20.00–24.99	5	8
15.00–19.99	2	2
10.00–14.99	1	2
	50	50

2. For the distributions of problem 1, calculate the mean number of calories consumed by each group. What do these means reveal about the two groups?

3. An experiment produced the following set of data. Calculate the mean and the median:

20	24	20	27	18	22	22	35	26	25	30	18	25	26	30	18
19	25	27	24	26	20	22	28	24	21	24	32	26	34	28	16

4. Calculate the mean and the median for the data shown in the frequency distribution of table 3.9.2 in the text. Is this "shape" of that distribution reflected in the relationship of the mean and the median? If so, how?

5. Compute the mean and the median for the data shown in table 3.7.1 in the text. Compare these values with the mean and the median computed from table 3.5.1. Did any difference in these values appear to result from a difference in class interval size? Which set of values is likely to be the more accurate?

6. Using the data of problem 3, calculate the sum of the deviations (i.e. disregarding sign) of these values about their median. Then calculate the average absolute deviations of the scores about the mean. Demonstrate that the average absolute deviation about the median is smaller than that about the mean.

7. Take the average squared deviation of the scores in problem 3 about their median. Show that this gives a result larger than the average squared deviation about the mean.

8. Take the probability distribution of heights in section 3.14 and determine the expected value of this distribution.

9. Using the data of problem 1, find (separately for boys and for girls) the following percentile values:
   (a) the 25th percentile

(b) the 75th percentile
(c) the 10th percentile
(d) the 90th percentile

10. For the data of problem 1, what are the percentile ranks of the following numbers of calories consumed:

   (a) 2750 for boys   (b) 2750 for girls
   (c) 3200 for boys   (d) 3200 for girls
   (e) 1800 for boys   (f) 1800 for girls
   (g) 4800 for boys   (h) 4800 for girls

11. A discrete random variable X can take on any of the integer values 1, 2, 3, . . . , N. The rule for the distribution of the random variable is

$$p(x) = \begin{cases} \dbinom{N}{x} \Big/ (2^N - 1) & x = 1, 2, \ldots, N \\ 0 & \text{elsewhere} \end{cases}$$

What is the expected value of this random variable if $N = 5$? If $N = 7$?

12. See if you can show that for the random variable defined in problem 11 the expected value must be equal to $\dfrac{2^{N-1}}{2^N - 1} N$. (Hint: notice that $N!/[(x - 1)!(N - x)]$ $= [N(N - 1)!]/[(x - 1)!(N - 1 - x + 1)!]$ and that $x - 1$ can equal any of the values 0, 1, 2, 3, 4.)

13. Compute the variance and standard deviation for the data of problem 3.

14. Compute the variance and standard deviation for the grouped distributions shown in tables 3.7.1 and 3.5.1. Do these values differ for the distributions? Why?

15. Compute the standard deviation for the two distributions shown in problem 1. What would you conclude about the variability of caloric consumption among the boys and among the girls?

16. For the data of problem 3, choose some arbitrary value and compute the average squared deviation for these scores from that value. How does this compare with the variance found in problem 13? Why must this be so?

17. Given the theoretical random variable of problem 11, compute the variance for $N = 5$ and for $N = 7$.

18. Two fair dice are tossed independently and at random, yielding the number of "spots" on each trial ranging from 2 to 12. Find the variance and standard deviation of the random variable "number of spots coming up on the dice."

19. Find the mean and standard deviations of the following theoretical distributions:

   (a) Binomial, with $N = 17$, $p = .35$
   (b) Binomial, with $N = 75$ and $p = .80$
   (c) Poisson, with $m = 1.25$
   (d) Pascal, with $r = 3$ and $p = .40$
   (e) Geometric, with $p = .19$
   (f) Binomial, with $N = 2000$ and $p = .01$
   (g) Poisson, with $m = 20$

20. Recall that a random variable which follows the rule for the rectangular or "uniform" distribution has the rule

$$f(x) = \begin{cases} k & a \leq x \leq b \\ 0 & \text{elsewhere.} \end{cases}$$

Within the range $a \leq x \leq b$ this is a continuous random variable. By use of the following mathematical results, show that the mean of such a random variable must be $(a + b)/2$ and the variance must be $(b - a)^2/12$:

$$\int_a^b k\, dx = k(b - a); \int_a^b kx\, dx = k(b^2 - a^2)/2; \int_a^b x^2 k\, dx = \frac{k(b^3 - a^3)}{3}$$

21. Convert the two distributions of problem 1 into distributions of standardized or z-score values. Determine the following:
    (a) What is the percentile rank corresponding to a z-value of -1 among the boys? Among the girls?
    (b) An individual consumes 3300 calories per day. Where does this place him among the boys, in standardized score terms? Among the girls?
    (c) Between what two standardized score values do the middle 90 percent of the girls lie? Between what two z-values do the middle 90 percent of the boys lie?
    (d) Is a caloric intake of 4200 relatively more deviant for a boy or for a girl?

22. A certain probability distribution has a mean of 68 and a standard deviation of 11. What is the maximum probability that a case drawn at random from this distribution will show a value greater than or equal to 115.5 or less than or equal to 20.5? In this distribution, what is the maximum probability of observing a case which is more than 1.7 standard deviations from the mean? What is the smallest proportion of cases that we should expect to fall in the interval 53.7 to 82.3 in this distribution?

23. A person bought two tickets to a very large lottery. In that lottery the first prize carried $100, the second prize $25, and there were no other prizes. The probability for the first prize was .01, that for the second prize was .09. The lottery was so large that the drawings could be regarded as independent and with replacement. The individual could win as many as two first prizes, two second prizes, or a combination of first and second prizes. On the other hand, any given ticket might be worth nothing at all. About how much should the individual have paid for the two tickets, if he expected to break even? (Hint: first calculate the possible combinations of prizes and worthless tickets and their monetary values, and proceed from there.)

7 Sampling Distributions and Point Estimation

7.1 Populations, parameters, and statistics

Formerly we have called the entire set of elementary events the *sample space*, since this term is useful and current in probability theory. However, in psychology and other fields using statistics it is more common to find the word **population** used to mean the totality of potential units for observation; these potential units for observation are very often real or hypothetical sets of people or animals, and *population* provides a very appropriate alternative to *sample space* in such instances. Nevertheless, whenever the term "population" is used in the following, we shall mean only the sample space of elementary events from which samples are drawn.

Given a population of potential observations, we shall think of the particular numerical score assigned to any particular unit observation as a value of a random variable; the distribution of this random variable is the **population distribution.** This distribution will have some mathematical form, with a mean μ, a variance σ^2, and all the other characteristic features of any distribution. If you like, you can usually think of the population distribution as a frequency distribution based upon some large but finite number of cases. However, population distributions are almost always discussed as though they were theoretical probability distributions; the process of random sampling of single units with replacement insures that the long-run relative frequency of any value of the random variable is the same as the probability of that value. Later we shall have occasion to idealize the population distribution and treat it as though the random variable were continuous. This is impossible for real-world observations, but we shall assume that it is "true enough" as an approximation to the population state of affairs.

Population values such as μ and σ^2 will be called **parameters of the**

261

population (or sometimes, **true values**). Strictly speaking, a parameter is a value entering as an arbitrary constant in the particular function rule for a probability distribution, although the term is used more loosely to mean any value summarizing the population distribution. Just as parameters are characteristics of populations, so are **statistics** associated with samples.

Some amplification of this idea of a statistic is called for, however. We have already seen that the same sample of data may be used as the basis for a wide variety of statistics. Thus, even in samples from a Bernoulli process, we have examined both the number of successes out of N trials and the proportion of successes, as well as the trial number on which the rth success occurred (as in Pascal sampling) and the number of runs in a given sequence. Each of these ways of regarding the same basic set of data provides a useful statistic. However, these are not the only statistics that might have been formed. We might, for example, have taken the product of the proportion of successes and the proportion of failures, or perhaps the logarithm of the proportion of successes. In some contexts we might have been interested in the reciprocal of the proportion of successes. In still another context we might have wished to form a statistic by letting $x = 1$ for a success and $x = -1$ for a failure, and then taking

$$y = \sum_{i=1}^{N} ix,$$

where i stands for the trial numbers, $i = 1, 2, \cdots, N$. Here values of y would range between $-(N)(N + 1)/2$ and $(N)(N + 1)/2$.

The point is that there is no limit to the number of ways in which statistics can be constructed and associated with samples, even for samples as simple as binomial sequences. Not all of these statistics would have been very useful perhaps, but we are perfectly free to define them. **A statistic is simply a function on samples, such that any sample is paired with a value of that statistic.** For samples of numerical data we ordinarily construct and use familiar statistics such as means, variances, medians, percentile ranks and the like because they happen to be simple and useful. However, in some situations we would want to use still other statistics, such as the sum of the logarithms of all of the values, or perhaps the sum of the reciprocals of the values observed, or the difference between the highest and lowest values, and so on for any other way of combining or transforming the values in the sample. Each possible function relating each sample to some new "summarizing" value is legitimate.

Moreover, a statistic need not use all of the information in a sample. Certainly the median, along with the other percentiles, appears to be based on less of the information in a sample than is the mean or the variance. If we wished, we could use even less of the information in a sample in defining a statistic; thus, we might define our statistic as merely the value of the fifth observation made, for example, and ignore the rest of the sample values. Indeed, we might even let the value of a statistic be a constant, having no relationship to the sample values themselves, so that none of the information in the sample is used in the formation of the statistic.

This point has been stressed in order to bring out a related point: Since there are no "natural" statistics among the wide variety possible, and if an unlimited variety of statistics might be associated with any given sample, we need criteria for choosing among such statistics. In the last chapter some descriptive statistics were compared on the basis of what they tell, and what we wish to communicate, about a given sample distribution of values. Now, however, we will examine sample statistics against the criteria of what they tell about population distributions. While it is often true that the best way to gain information about a population parameter is through the use of the analogous sample statistic (that is, population mean and sample mean, population variance and sample variance), this is not always or necessarily true. It may be that some other statistic with a different form contains more, or more useful, information about the population from which the sample came than does a familiar statistic. The main business of this chapter is, then, to examine how population distributions determine (or induce) distributions of sample statistics, and then to outline some of the criteria that have been developed for the choice of statistics to be used in inferences about populations.

7.2 SAMPLING DISTRIBUTIONS

In actual practice, random samples seldom consist of single observations. Almost always some N observations are drawn from the same population. Furthermore, the value of some statistic is associated with the sample. Our interest then lies in the distribution of values of this statistic across all possible samples of N observations from this population. This means that we must distinguish still another kind of theoretical distribution, called a sampling distribution.

A sampling distribution is a theoretical probability distribution that shows the functional relation between the possible values of a given statistic based on a sample of N cases, and the probability (density) associated with each value, for all possible samples of size N drawn from a particular population.

In general, the sampling distribution of values for a particular sample statistic will not be the same as the distribution of the random variable for the population. We shall see, however, that the sampling distribution always depends in some specifiable way upon the population distribution, provided the probability structure underlying the occurrence of samples is known.

Notice that we have not restricted this definition to simple random samples, even though in most simple applications it will be assumed that samples are drawn at random from the population. Nevertheless, some probability structure linking the occurrence of the possible samples with the population must exist and be known if the population distribution is to be related to the sampling distribution of any statistic. For our elementary purposes this probability structure will be that of simple random sampling, in which each possible sample of size N has exactly the same probability of occurrence as any other. However, in more advanced work assumptions other than simple random sampling are sometimes made.

We have already used sampling distributions in the preceding chapters. For example, a binomial distribution is a sampling distribution. Recall that a binomial distribution is based on a two-category population distribution, or Bernoulli process. A sample of N independent cases is drawn at random from such a distribution, and the number (or proportion) of successes is calculated for each sample. Then the binomial distribution is the sampling distribution showing the relation between each possible sample result and the theoretical probability of occurrence. The binomial distribution is *not* the same as the Bernoulli process unless N is 1; however, given the Bernoulli process and the size of the sample, N, the binomial distribution may be worked out. In a similar way, the Pascal and the Poisson distributions can be regarded as sampling distributions.

Still another example of a sampling distribution already introduced is the multinomial distribution. Here, the population distribution has several event categories. A sample of size N is drawn at random with replacement and summarized according to how many of each measurement class occurred. The multinomial distribution gives the probability of each possible sample distribution.

Other examples of sampling distributions will now be given. A most important distribution we shall employ is the sampling distribution of the mean. Here, samples of N cases are drawn independently and at random from some population and each observation is measured numerically. For each sample drawn the sample mean M is calculated. **The theoretical distribution that relates the possible values of the sample mean to the probability (density) of each over all possible samples of size N is called the sampling distribution of the mean.** Furthermore, for each sample of size N drawn, the sample variance S^2 may be found. The theoretical distribution of sample variances in relation to the probability of each is the sampling distribution of the variance. By the same token, *the sampling distribution of any summary characteristic* (mode, median, range, etc.) *of samples of N cases may be found, given the population distribution and the sample size N.*

Any function on samples has a sampling distribution, which is induced by the population distribution from which the samples are drawn. Some of the functions on samples, such as the sample mean, or the sample variance, or the proportion of successes, are familiar and frequently used. The sampling distributions of these statistics all depend on the distribution of the particular populations from which the samples are drawn. And, in the same way, many other, less familiar statistics may be associated with samples. Each of these statistics also will conform to its own sampling distribution, where that sampling distribution depends upon the particular population distribution from which the samples are generated.

Perhaps a simple example will illustrate this point. We are already familiar with the number of successes out of N Bernoulli trials as a statistic, and with the binomial distribution as the sampling distribution of this statistic. The *particular* one of the family of binomial distributions that applies depends directly on the Bernoulli process (population distribution) underlying the samples. Now let us construct a new statistic for a sample of N independent Bernoulli trials and show that this statistic has a sampling distribution that is not a binomial distribution, but that does depend on the underlying Bernoulli process.

Assume a Bernoulli process with two events, success and failure, and imagine a series of N independent trials. Let us denote the trial number by $i = 1$, $2, 3, \cdots, N$. Then on any trial i, let $X_i = 1$ if the trial is a success, and let $X_i = -1$ if that trial results in a failure. Then let the statistic be

$$U = \sum_{i=1}^{N} iX_i.$$

As a concrete example, let $N = 5$. Then if a sequence turns out as, say, $S\,S\,F\,S\,F$, we find the value of U by taking $1(1) + 2(1) + 3(-1) + 4(1) + 5(-1)$, since there was a success on trial 1, a success on trial 2, a failure on trial 3, and so on. Here the value of the statistic U would be $1 + 2 - 3 + 4 - 5$ or -1. In order to find the sampling distribution of U, we take all possible sequences and find their U values and their associated probabilities, given the independent Bernoulli process with p and q. Thus, for $N = 5$ we have the values tabulated below.

Sequence	U value	Probability	Sequence	U value	Probability
$S\ S\ S\ S\ S$	15	p^5	$F\ F\ F\ F\ F$	-15	q^5
$F\ S\ S\ S\ S$	13	p^4q	$S\ F\ F\ F\ F$	-13	pq^4
$S\ F\ S\ S\ S$	11	p^4q	$F\ S\ F\ F\ F$	-11	pq^4
$S\ S\ F\ S\ S$	9	p^4q	$F\ F\ S\ F\ F$	-9	pq^4
$S\ S\ S\ F\ S$	7	p^4q	$F\ F\ F\ S\ F$	-7	pq^4
$S\ S\ S\ S\ F$	5	p^4q	$F\ F\ F\ F\ S$	-5	pq^4
$F\ F\ S\ S\ S$	9	p^3q^2	$S\ S\ F\ F\ F$	-9	p^2q^3
$F\ S\ F\ S\ S$	7	p^3q^2	$S\ F\ S\ F\ F$	-7	p^2q^3
$F\ S\ S\ F\ S$	5	p^3q^2	$S\ F\ F\ S\ F$	-5	p^2q^3
$F\ S\ S\ S\ F$	3	p^3q^2	$S\ F\ F\ F\ S$	-3	p^2q^3
$S\ F\ F\ S\ S$	5	p^3q^2	$F\ S\ S\ F\ F$	-5	p^2q^3
$S\ F\ S\ F\ S$	3	p^3q^2	$F\ S\ F\ S\ F$	-3	p^2q^3
$S\ F\ S\ S\ F$	1	p^3q^2	$F\ S\ F\ F\ S$	-1	p^2q^3
$S\ S\ F\ F\ S$	1	p^3q^2	$F\ F\ S\ S\ F$	-1	p^2q^3
$S\ S\ F\ S\ F$	-1	p^3q^2	$F\ F\ S\ F\ S$	1	p^2q^3
$S\ S\ S\ F\ F$	-3	p^3q^2	$F\ F\ F\ S\ S$	3	p^2q^3

The sampling distribution of the statistic U is then:

U value	Probability	U value	Probability
15	p^5	-1	$p^3q^2 + 2p^2q^3$
13	p^4q	-3	$p^3q^2 + 2p^2q^3$
11	p^4q	-5	$2p^2q^3 + pq^4$
9	$p^4q + p^3q^2$	-7	$p^2q^3 + pq^4$
7	$p^4q + p^3q^2$	-9	$p^2q^3 + pq^4$
5	$p^4q + 2p^3q^2$	-11	pq^4
3	$2p^3q^2 + p^2q^3$	-13	pq^4
1	$2p^3q^2 + p^2q^3$	-15	q^5

It is obvious that the sampling distribution of the statistic U is very different from the binomial distribution. Yet, this sampling distribution was induced by the same basic Bernoulli process that induces the binomial sampling distribution for number of successes. Just as the binomial distribution depends upon N, p, and q, so does the distribution of U. Furthermore, U is only one of a wide variety of statistics that might be defined for sequences of N trials from a Bernoulli process. The sampling distribution of each such statistic would depend upon the Bernoulli process itself.

In the same way, for samples drawn from other basic population distributions or processes, any of a variety of statistics can be defined. Then over all possible results for samples of size N, the various possible values of the statistic will have probabilities or probability densities. These depend upon the distribution of the original random variable in the population, together with the probability structure according to which samples were drawn. This fact is of crucial importance in the theory of inferential statistics, since the theory of sampling distributions permits one to judge the probability that a given value of some statistic arose by chance from some particular population distribution. The bulk of the material in the remainder of this chapter will deal with various characteristics and uses of statistics and their sampling distributions.

7.3 CHARACTERISTICS OF SINGLE-VARIATE SAMPLING DISTRIBUTIONS

A sampling distribution is a theoretical probability distribution, and like any such distribution, is a statement of the functional relation between the values or intervals of values of some random variable and probabilities. Sampling distributions differ from population distributions in that *the random variable is always the value of some statistic based on a sample of N cases*, such as the sample mean, or the sample variance, the sample median, etc. Thus, a plot of a sampling distribution, such as that in Figure 7.3.1, always has for the abscissa (or horizontal axis) the different sample statistic values that might occur. Figure 7.3.1, for example, shows a theoretical distribution for sample variances for all possible samples of size 7 drawn from a particular population. Any point on the horizontal axis is a possible value of a sample variance, and the height of the curve on the vertical axis gives the probability density $f(S^2)$, for that particular value.

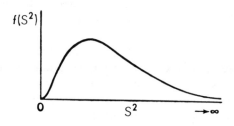

FIG. 7.3.1. A theoretical sampling distribution of S^2 for $N = 7$.

Like population distributions, sampling distributions may be either continuous or discrete. The binomial distribution is discrete, although in applied problems it is sometimes treated as though it were continuous. Most of the commonly encountered sampling distributions based on a continuous population distribution will be continuous.

It should be clear that for any given population sampled there will be any number of different sampling distributions, depending upon the particular statistic associated with each sample. Thus there will be a theoretical sampling distribution of means, a different distribution that is the theoretical sampling distribution of variances, still another distribution that is the sampling distribution of medians, and so on, for as many ways as a statistic can be computed for a sample of N cases. Obviously, we should seldom choose to describe a sample in *some* ways, such as describing the entire sample by the square root of the score of the third case observed; even so, bear in mind that the theoretical distributions of even such esoteric statistics can be worked out.

It is a good idea to reiterate at this point: so far three distinct kinds of distributions have been introduced. The first was *the sample distribution*, which is a frequency distribution summarizing a given set of data, based on a randomly selected subset of a population. The second kind of distribution with which we dealt was the *population distribution*. This is a theoretical distribution, which describes the relative frequency associated with each of the various measurement classes, or intervals of values of a numerical variable, into which an entire set of possible observations may be mapped. Finally there was the *sampling distribution*. This is a theoretical probability distribution which relates various values or intervals of values of some sample statistic to their probabilities of occurrence over all possible samples of size N given a specific population distribution, and some probability structure underlying the selection of samples. In order to know any sampling distribution exactly, we must specify the population distribution, but the population distribution will, in general, be unlike the sampling distribution. Only when the sample size is 1, and the statistic is simply the score-value, will the sampling distribution be exactly like the population distribution. The sampling distribution, in turn, must be known before the probability of the occurrence of a particular sample statistic value or interval of values can be calculated. It is very important to bear these distinctions in mind, since all three kinds of distributions will be considered in the sections to follow.

7.4   The mean and variance of a sampling distribution

Except for a few statistical curiosities that need not concern us here, any random variable will have a determinable mean and variance. Since a sample statistic is a random variable, the mean and variance of any sampling distribution are defined in the usual way. That is, let G be any sample statistic; then if the sampling distribution of G is discrete, its expectation or mean is

$$E(G) = \mu_G = \Sigma g p(g) \qquad\qquad [7.4.1*]$$

(see Section 6.8). If the variable G is continuous, then

$$E(G) = \mu_G = \int_{-\infty}^{\infty} g f(g)\,dg, \qquad\qquad [7.4.2*]$$

just as for any other continuous random variable (Section 6.12).

In the same way, the variance of a sampling distribution for some statistic G can be defined:

$$\sigma_G^2 = E(G - \mu_G)^2 \qquad\qquad [7.4.3*]$$

or

$$\sigma_G^2 = E(G^2) - [E(G)]^2. \qquad\qquad [7.4.4*]$$

The variance of the statistic G gives a measure of the dispersion of particular sample values about the average value of G over all possible samples of size N. The standard deviation σ_G of the sampling distribution reflects the extent to which sample G values tend to be *unlike* the expectation, or are *in error*. To aid in distinguishing the standard deviation of a sampling distribution from the standard deviation of a population distribution, a standard deviation such as σ_G is usually called the "standard error" of the statistic G. When one speaks of the standard error of the mean, he is referring to the standard deviation of the distribution of possible sample means, for all possible samples of size N drawn from a specified population. Similar meanings hold for the standard error of the median, the standard error of the standard deviation, the standard error of the range, and so on.

7.5   Sample statistics as estimators

Some population parameters have obvious parallels in sample statistics. The population mean μ has its sample counterpart in M, the variance σ^2 in the sample variance S^2, the population proportion p in the sample proportion P, and so on. On the other hand, the relationship between population parameters and sample statistics is not necessarily 1 to 1. It is entirely possible for a population distribution to depend upon parameters that are not directly paralleled in one of the common sample statistics, and for sample statistics to be used that are not direct parallels of population parameters.

It is true, however, that a sample of cases drawn from a population contains information about the population distribution and its parameters. Furthermore, a statistic computed from the data in the sample contains some of that information. Some statistics contain more information than others, and some statistics may contain more information about certain parameters than about others.

A central problem of inferential statistics is point estimation, the use of the value of some statistic to infer the value of a population parameter. The value of some statistic (or point in the "space" of all possible values) is taken as the "best estimate" of the value of some parameter of the population distribution.

How does one go from a sample statistic to an inference about the population parameter? In particular, which sample statistic does one use, if it is to give an estimate that is in some sense "best"?

The fact that the sample represents only a small subset of observations drawn from a much larger set of potential observations makes it nearly impossible to say that any estimate is exactly like the population value. As a matter of fact they very probably will not be the same, as all sorts of different factors of which we are in ignorance may make the sample a poor representation of the population. Such factors we lump together under the general rubrics *chance* or *random effects*. In the long run such samples should reflect the population characteristics. However, practical action can seldom wait for "in the long run"; things must be decided here and now in the face of limited evidence. We need to know how to use the available evidence in the best possible ways to infer the characteristics of the population.

In the following, it will be well to keep in mind the difference between an "estimator" of some population parameter, and an "estimate" of the value of that parameter. An estimator is a formula for combining the values occurring in the data. Hence an estimator is a random variable, which takes on values dependent upon the sample data. On the other hand, the particular value that results from the application of that formula is an estimate of the population parameter in question. Viewed as a random variable, with a value arrived at in a certain way from the sample data, the sample mean is an estimator. On the other hand, a particular value of the sample mean, based on a particular sample, is an estimate. Our main interest will lie in the properties of statistics as estimators— that is, as random variables with certain characteristics.

In order to meet the need for criteria for choosing among the statistics that might be used as estimators, statisticians rely upon a number of well-established principles. One of these is the very important **principle of maximum likelihood,** which will be discussed next. Then, we will discuss some of the properties that a statistic should have in order to be a good estimator. We shall see that:

1. The estimate should be *unbiased.*
2. The estimate should be *consistent.*

3. The method of estimation should be relatively *efficient*.
4. The set of estimators should be *sufficient*.

Although few statistics satisfy all of these desirable properties as estimators, each criterion is an important property for an estimator to possess if possible.

7.6 THE PRINCIPLE OF MAXIMUM LIKELIHOOD

Even though a formal presentation of the maximum-likelihood principle is far beyond our scope, some intuitive feel for this idea may help you understand the methods that follow.

Basically, the problem of the person using statistics is, "Given several possible population situations, any one of which might be true, how shall I bet on the basis of the evidence so as to be as confident as possible of being right?" Everyone who uses statistical inference is faced with this question. He knows that any number of things might be true of the population. Fortunately, he can "snoop" on the population to a certain extent by taking a sample, but he knows that the evidence he gains will be faulty. Since this evidence is all there is, the person must use it nevertheless to form an opinion or make up his mind. How does one decide on the basis of evidence that is probably erroneous? The principle of maximum likelihood gives a general strategy for such decisions, and may be paraphrased as follows.

Suppose that a random variable X has a distribution that depends only upon some population parameter θ (symbolized by Greek "theta"). The form of the density function will be assumed known, but not the value of θ. A sample of N independent observations is drawn, producing the set of values $(x_1, x_2, \cdots, x_N)$. Let

$$L(x_1, \cdots, x_N; \theta)$$

represent the likelihood, or probability (density), of this particular sample result *given* θ. For each possible value of θ, the likelihood of the sample result will be different, perhaps. Then, **the principle of maximum likelihood requires us to choose as our estimate that possible value of θ making $L(x_1, \cdots, x_N; \theta)$ take on its largest value.**

In effect this principle says that when faced with several parameter values, any of which might be the true one for a population, the best "bet" is that parameter value which would have made the sample actually obtained have the highest prior probability. When in doubt, place your bet on that parameter value which would have made the obtained result most likely.

This principle may be illustrated very simply for the binomial distribution. Suppose that a sample is drawn at random from a population of college graduates. Each graduate is classified either as a "literary-college major" or as a graduate from some other university college. Now three possibilities or hypotheses are entertained about the proportion of college graduates from literary colleges. Hypothesis 1 states that .5 of all graduates are from such colleges, hypothesis 2 states that .4 of the graduates are from such schools, and hypothesis 3 states that .6 are literary-college graduates.

Now suppose that some 15 college graduates are drawn at random and classified according to "literary-college degree" versus "nonliterary degree." The result is that 9 out of the 15 hold a literary-college degree (we are supposing that no one in the population holds more than one degree). What decision about the three available hypotheses should we reach? This is a simple binomial sampling problem, so that the probability of each sample result may be calculated for each of the three hypothetical values of the parameter p: $p = .5$, $p = .4$, or $p = .6$.

If $p = .5$ the prior probability of a sample result such as that obtained would be

$$\binom{15}{9} (.5)^9 (.5)^6 = .153.$$

For $p = .4$ the prior probability becomes

$$\binom{15}{9} (.4)^9 (.6)^6 = .061.$$

Finally, if p were .6 the prior probability of 9 successes out of 15 would be

$$\binom{15}{9} (.6)^9 (.4)^6 = .207.$$

The use of the principle of maximum likelihood to decide among these three possibilities leads to the choice of hypothesis 3, that .6 is the population proportion, since this is the parameter value among the possibilities considered that would have made the obtained sample result most likely, a priori.

This principle has a great deal of use in theoretical statistics, since general methods exist for finding the value of θ that maximizes the likelihood of a sample result. Statistics chosen as estimators because their values substituted for the parameter maximize the likelihood of the sample result are called **maximum-likelihood estimators.** For example, if there were no prior information at all about the number of literary-college graduates, so that any number between 0 and 1.00 might be entertained as a hypothesis about the value of p, the maximum likelihood estimate of p would be the sample P, or 9/15, since among all possible values of p this value makes the occurrence of the actual result have greatest a priori likelihood (the student may check a few values of p for himself to see that this is true). Later on, we will show that for populations having a normal distribution, the sample mean M is a maximum-likelihood estimator of μ, since if $M = \mu$, the obtained sample is made more likely than for any other possible value of μ, other things being equal. The principle of maximum likelihood gains further importance from the fact that if a *sufficient* estimator of a parameter exists (as defined in the section to follow) the maximum-likelihood estimate is based on this sufficient statistic. Thus, the principle of maximum likelihood provides a fairly routine way of finding estimators having "good" properties.

However, the principle of maximum likelihood is introduced here not only because of its importance in estimation, but also because of the general point of view it represents about inference. This point of view is that *true popula-*

tion situations should be those making our empirical results likely; if a theoretical situation makes our obtained data have very low prior likelihood of occurrence, then doubt is cast on the truth of the theoretical situation. Theoretical propositions are believable to the extent that they accord with actual observation. If a particular result should be very unlikely given a certain theoretical state of affairs, and we do get this result nevertheless, then we are led back to examine our theory. Good theoretical statements accord with observation by giving predictions having high probability of being observed.

Naturally, the results of a single experiment, or even of any number of experiments, cannot prove or disprove a theory. Replications, variants, different ways of measuring the phenomena, must all be brought into play. Even then, proof or disproof is never absolute. Nevertheless, the principle of maximum likelihood is in the spirit of empirical science, and it runs throughout the methods of statistical inference.

7.7   UNBIASED ESTIMATORS

Suppose that one is interested in estimating the value of population parameter θ, and he is considering the use of some sample statistic G as an estimate of the value of θ. Then an estimate of the parameter θ made from the sample statistic G is said to be an unbiased estimate if

$$E(G) = \theta. \qquad [7.7.1*]$$

That is, the sample quantity G is unbiased as an estimate of θ if the expectation of G is θ; in the long run, G averaged over all possible random samples is exactly θ.

For example, consider the mean of a sample as an estimator of the mean of the population. Here

$$G = M = \frac{\sum_i x_i}{N}$$

and

$$\theta = \mu.$$

Is the sample mean an unbiased estimate of the population mean μ? *Yes*, since it is quite easy to show that

$$E(M) = \mu. \qquad [7.7.2*]$$

This may be shown as follows:

$$E(M) = E\frac{(x_1 + x_2 + \cdots + x_N)}{N}.$$

Using rules 5 and 2 for expectations, we have

$$E(M) = \frac{E(X_1) + E(X_2) + \cdots + E(X_N)}{N}.$$

But any $E(X)$ is μ by definition, for observations taken at random from the same population. Hence,

$$E(M) = \frac{NE(X)}{N} = \mu.$$

The mean of a random sample is an unbiased estimate of the population mean μ.

In exactly the same way, for independent samples from a Bernoulli process where each observation is put into one of two categories, it can be shown that the **sample proportion P of cases in a given category is an unbiased estimate of the population proportion p:**

$$E(P) = p. \qquad\qquad [7.7.3]$$

(As a matter of fact, this was shown in Section 6.13.)

It is important to observe that equation 7.7.3 is really just a special case of the equation $E(M) = \mu$. Thus, in a Bernoulli process, let us define the random variable X, such that

$$\begin{cases} X = 1 & \text{for the event "success,"} \\ X = 0 & \text{for the event "failure."} \end{cases}$$

Then in the Bernoulli process itself, the mean μ of this random variable is, by definition,

$$\mu = 1p(X = 1) + 0p(X = 0) = p(\text{Success}) = p.$$

Furthermore, for any sample of N trials from the Bernoulli process, we find the mean of the random variable:

$$M = \sum_{i=1}^{N} \frac{x_i}{N},$$

where X_i stands for the value of X on the trial number i. Then

$$M = (\text{number of successes})/N = P.$$

Hence, for this particular random variable, defined on a Bernoulli process, $E(P) = p$ is the same as $E(M) = \mu$.

What would an example of a *biased* estimate be like? An example is provided by a sample variance. The sample variance S^2 is a biased estimate of the population variance σ^2, since

$$E(S^2) \neq \sigma^2. \qquad\qquad [7.7.4^*]$$

This can be demonstrated as follows: the expectation of a sample variance is

$$E(s^2) = E\left(\frac{\sum_i x_i^2}{N} - M^2 \right) = E\left(\frac{\sum_i x_i^2}{N} \right) - E(M^2).$$

Let us consider the two terms on the extreme right separately. By rules 5 and 2 for expectations,

$$E\left(\frac{\sum\limits_i x_i^2}{N}\right) = \frac{\sum\limits_i E(x_i^2)}{N}.$$

From the definition of the variance of the population,

$$\sigma^2 = E(x^2) - \mu^2$$

so that

$$E(x^2) = \sigma^2 + \mu^2$$

for any observation i. Thus

$$E\left(\frac{\sum\limits_i x_i^2}{N}\right) = \frac{\sum\limits_i (\sigma^2 + \mu^2)}{N} = \sigma^2 + \mu^2. \qquad [7.7.5]$$

Now the variance of the sampling distribution of means is, from **7.4.4,**

$$\sigma_M^2 = E(M^2) - \mu^2$$

so that

$$E(M^2) = \sigma_M^2 + \mu^2. \qquad [7.7.6]$$

Putting these two results (**7.7.5** and **7.7.6**) together, we have

$$E(S^2) = \sigma^2 - \sigma_M^2, \qquad [7.7.7^*]$$

the expectation of the sample variance is the *difference* between the population variance σ^2 and the variance of the sampling distribution of means σ_M^2. In general this difference will not be the same as σ^2, since ordinarily the variance σ_M^2 will not be zero, and so the sample variance is biased as an estimator of the population variance. In particular, the sample variance is, on the average, smaller than the population variance σ^2. A way of correcting the sample variance for this bias will be given in Section 7.14.

7.8   CONSISTENCY

For an estimate to be a good one, **the sample estimate should have a higher probability of being close to the population value** θ **the larger the sample size** N. Statistics that have this property are called consistent estimators. More formally: the statistic G is a consistent estimator of θ if for any arbitrary positive number ϵ,

$$\text{prob.}(|G - \theta| \angle \epsilon) \rightarrow 1, \text{ as } N \rightarrow \infty. \qquad [7.8.1^*]$$

The probability that G is within a certain distance ϵ of the parameter θ approaches 1 as the sample size N increases without bound, however small the size of the positive number ϵ.

The sample mean, the sample variance, and many other statistics are consistent estimators, as they tend in likelihood to get closer to the true population value the larger the sample. It is, however, possible to create sample statistics that are not consistent estimators. For example, suppose that one wished to estimate the population mean, and he decided to take the score of the second observation made in any sample as the estimate. Would this be a consistent estimate? No, since the second observation's score would not tend in likelihood to get any closer to the mean value of the population the larger the sample. According to the criterion of consistency, this would not be a good way to estimate the mean. (Would this be an unbiased estimate of the mean, however?)

7.9 RELATIVE EFFICIENCY

A third criterion for evaluating a statistic G as an estimator of a parameter θ is that G be efficient relative to other statistics that might be used to estimate θ. You will recall that the standard deviation (or standard error as it is called when the distribution it describes is a sampling distribution) represents the extent of the difference that chance factors tend to create between a sample estimate and a true parameter value. Good estimators should have sampling distributions with small standard errors, given the N of the sample.

Suppose that there were two different sample statistics G and H calculated from the same data, and that these two statistics were each unbiased estimators of the same population parameter θ. The theoretical sampling distribution of G figured from samples of size N has a standard error σ_G. There is also a theoretical sampling distribution of H, having a standard error σ_H. Each of these two standard errors reflects the tendency of the sample statistic to deviate by chance from the population value. Then, for any N, the efficiency of the statistic G relative to the statistic H, both as estimators of θ, is given by

$$\frac{\sigma_H^2}{\sigma_G^2} = \text{efficiency of } G \text{ relative to } H. \qquad [7.9.1^*]$$

The more efficient estimator has the smaller standard error, so that the ratio is greater than 1.00 when the variance of the more efficient estimator appears in the denominator. Other things, such as sample size, being equal, relatively inefficient statistics have relatively larger standard errors than other estimators of the same parameter value.

For example, in the family of unimodal symmetric "normal" distributions the population mean and the population median both have the same value, $Md = \mu$. We wish to estimate this value μ by drawing a random sample of N cases. We could use either the sample mean (statistic G, let us say) or we could use the sample median (statistic H). In any given sample, these values will very likely not be the same. Which should we use to estimate μ? The answer involves efficiency. If the standard error of the mean is represented by σ_M and the standard error of the distribution of sample medians is represented by σ_{Md}, then it is true that for normal population distributions σ_{Md} is greater than σ_M for a given sample size, $N > 2$. For such normal populations, and for a reasonably large N, the magnitude of error in any given estimate of the population mean from the sample mean is likely to be less than in an estimate of μ from the sample median. This says that for such situations

$$\frac{\sigma^2_{Md}}{\sigma^2_M} > 1.00,$$

the relative efficiency of the mean will be greater than 1.00, indicating that the mean is a more efficient estimator than the median for normal population distributions. This is one reason that so much of statistical inference deals with the mean rather than the median; the median is less efficient than the mean as an estimator of μ for a normal population distribution, and such distributions are historically very important in statistical work. On the other hand, there are other families of distributions in which just the opposite is true, and the median turns out to be a more efficient estimator of the population parameter μ than is the mean. The point is that the relative efficiency of one statistic as against another depends very much upon the family of population distributions under consideration.

7.10   Sufficiency

A concept of major importance throughout the modern theory of statistics is that of sufficiency. Here we will restrict ourselves to the concept as it applies to estimation. Consider a random variable which depends only on a parameter θ. Then a statistic G is said to be a sufficient estimator of the parameter θ if G contains *all* the information available in the data about the value of θ. A sufficient statistic G is a "best" estimator of θ in the sense that G cannot be improved by considering any other aspects of the sample data not already included in the statistic G itself.

A slightly more formal definition of sufficiency may be given as follows: Consider the estimator G of θ once again, and then let H be every other sample statistic. We can consider the *conditional* distribution of H given the value of G. Then if the conditional distribution of every H given G does not depend in any way on the value of θ, G is a sufficient statistic. Furthermore, if G is a set of estimators of the set of parameters $\theta_1, \theta_2, \cdots, \theta_t$, and the conditional distribution of every other statistic H given the set G does not depend in any way upon the parameters $\theta_1, \theta_2, \cdots, \theta_t$, then the set of statistics G is said to be sufficient.

An example of a sufficient estimator is the sample proportion P, when samples are drawn independently and at random from a Bernoulli process. The sampling distribution of P can be found from the binomial rule in this situation, of course. If we are estimating the population value p, then P is a sufficient statistic, since there is no information we can add to P, given a fixed sample size N, that would make it a better estimator of p. In this situation, the Bernoulli process actually has only one parameter, p; other population distributions may have two or more parameters. In such other population distributions, where specification of two or more parameters is required in order to distinguish the particular member of the family of distributions that applies, two or more statistics are required for sufficiency. Thus, in the family of normal distributions, the particular distribution depends on the specification of two parameters, μ and σ^2. For random samples from such a normal distribution, M and S^2 are sufficient for μ and σ^2. In such instances we must talk about the set of sufficient statistics, rather than a single sufficient estimator.

A sufficient estimator, or set of sufficient estimators, need not be unique. Thus, in the binomial situation, it is true that P is a sufficient estimator, but then so is P^2, or $P^{1/2}$, or $-\log P$, or $k(1 + P)$, and a host of other functions of P. The existence of a sufficient statistic or a set of sufficient statistics does not, in and of itself, tell which form of the statistic to use. Other criteria are still called for.

Sufficient statistics do not always even exist, and situations can be constructed in which no sufficient set of estimators can be found for a set of parameters. Nevertheless, sets of sufficient estimators, when they do exist, are important, since if one can find a set of sufficient estimators, then it is ordinarily possible to find unbiased and efficient estimators based upon that sufficient set. In particular, as already indicated, when a set of sufficient statistics exist, then the maximum likelihood estimators will be based upon that set.

In most of the work to follow, the estimators will be the sample mean and the corrected sample variance, both of which will fulfill these criteria for good estimators in the particular situations where they will be used. This does not imply, however, that other estimators failing to meet one or more of these criteria are useless. In particular, special situations exist where other methods of estimating central tendency may be better than either mean or median, and variance estimates other than S^2 are called for. Other statistics are useful on occasion, but we will focus most attention on the mean and the corrected variance, since they occupy a central place in the "classical" statistical methods we will be treating.

7.11   THE SAMPLE MEAN AS AN ESTIMATOR

We have just seen that the sample mean comes off very well by the criteria for a good estimator. In general, the sample mean is unbiased, is consistent, and in an important set of circumstances is both efficient relative to other statistics and, taken with S^2, sufficient.

For our immediate purposes, the most important of these properties is that of being unbiased: the expectation of the sample mean is the population mean,

$$E(M) = E(X) = \mu.$$

For random sampling from any population, you can be sure that the long-run average of sample means will be identical to the population mean. In fact, this principle generally applies even when samples are drawn on some probability basis other than simple random sampling.

This same statement can be interpreted in another way: **the mean of the sampling distribution of means is the same as the population mean.** However, it is not true that the sampling distribution of the mean will be the same as the population distribution; in fact, these distributions will ordinarily be quite different, depending particularly on sample size. In the first place, the variance of the sampling distribution of means will not be the same as σ^2, but instead will be smaller than σ^2 for samples of size 2 or larger. This will be shown in the next section.

7.12   THE VARIANCE AND STANDARD ERROR OF THE MEAN

Intuitively it seems quite reasonable that the larger the sample size the more confident we may be that the sample mean is a close estimate of μ. When it was asserted that the mean is a consistent estimator (Section 7.7) this was another way of saying that the mean is a better estimator for large than for small samples. Now we can put that intuition on a firm basis by looking into the effect of sample size on the variance and standard deviation of the distribution of sample means.

By definition, the variance of the distribution of means, from samples drawn at random from a population with mean μ and variance σ^2, is

$$\sigma_M^2 = E(M^2) - [E(M)]^2$$
$$= E(M^2) - \mu^2. \qquad [7.12.1*]$$

Bear in mind that we are assuming that the sample mean is based on N *independent* observations. Let us call any pair of these observations i and j, with scores X_i and X_j. The square of the sample mean is

$$M^2 = \frac{(x_1 + x_2 + \cdots + x_N)^2}{N^2}$$

$$= \frac{1}{N^2}(x_1^2 + \cdots + x_N^2 + 2\sum_{i<j} x_i x_j), \qquad [7.12.2]$$

the sum of the squared scores, plus twice the sum of the products of all pairs of scores, all divided by N^2. For any single observation i,

$$E(X_i^2) = \sigma^2 + \mu^2 \qquad [7.12.3]$$

since, from Section 6.21,

$$\sigma^2 = E(X_i^2) - \mu^2.$$

For a pair of *independent* observations, i and j, rule 6 in Appendix B gives

$$E(X_i X_j) = E(X_i)E(X_j) = \mu^2. \qquad [7.12.4]$$

Thus, putting these two facts (**7.12.3** and **7.12.4**) together, we have

$$E(M^2) = \frac{1}{N^2}[E(X_1^2) + E(X_2^2) + \cdots + 2\sum_{i<j} E(X_iX_j)]$$

$$= \frac{N\sigma^2 + N\mu^2 + (N)(N-1)\mu^2}{N^2}$$

$$= \frac{\sigma^2}{N} + \mu^2. \qquad [7.12.5]$$

Making this substitution in expression **7.12.1** we find

$$\sigma_M^2 = E(M^2) - \mu^2 = \frac{\sigma^2}{N}. \qquad [7.12.6^*]$$

The variance of the sampling distribution of means for independent samples of size N is always the population variance divided by the sample size, σ^2/N.
This is a most important fact, and gives direct support to our feeling that large samples produce better estimators of the population mean than do small. When the sample size is only 1, then the variance of the sampling distribution is exactly the same as the population variance. If however, the sample mean is based on two cases, $N = 2$, then the sampling variance is only 1/2 as large as σ^2. Ten cases give a sampling distribution with variance only 1/10 of σ^2, $N = 500$ gives a sampling distribution with variance 1/500 of σ^2, and so on. If the sample size approaches infinity, then σ^2 approaches zero. **If the sample is large enough to embrace the entire population, there is no difference between the sample mean and μ.**
In general, the larger the sample size, the more probable it is that the sample mean comes arbitrarily close to the population mean. This fact is often called **the law of large numbers,** and is closely allied both to Bernoulli's theorem and the Tchebycheff inequality.
Given the Tchebycheff inequality and the variance of the mean, it is easy to see that the law of large numbers must be true. From the Tchebycheff inequality (**6.28.1**), it is true that for any random variable,

$$\text{prob.}(|X - \mu| < k) \geqq 1 - \frac{\sigma^2}{k^2}, \qquad k > 0. \qquad [7.12.7]$$

Let the random variable be M, and the variance be σ_M^2. Then

$$\text{prob.}(|M - \mu| < k) \geqq 1 - \frac{\sigma_M^2}{k^2}. \qquad [7.12.8^*]$$

When N becomes very large, σ_M^2 approaches zero. Thus, regardless of how small k is, the probability approaches 1 that the value of M will be within k units of μ when sample size grows large.

7.13   Standardized scores corresponding to sample means

The standard error of the mean is

$$\sigma_M = \sqrt{\sigma_M^2} = \frac{\sigma}{\sqrt{N}}. \qquad [7.13.1*]$$

so that when a sample mean is put into standardized form, we have

$$z_M = \frac{M - \mu}{\sigma_M} = \frac{M - \mu}{\sigma/\sqrt{N}}. \qquad [7.13.2*]$$

The larger the standard score of a mean, relative to μ and the standard error, the *less likely* is one to observe a mean this much or more deviant from μ. Any given degree of departure of M from μ corresponds to a larger absolute standard score value, and hence a less probable class of result, the larger the sample size N.

For example, suppose that in sampling from some population a sample mean was found differing by 10 points from the population mean. Suppose that σ were 5 and the sample size were 2. Then the standard score z_M would be

$$z_M = \frac{10}{5/\sqrt{2}} = 2.8,$$

disregarding sign. The Tchebycheff inequality tells us that means this deviant or more so from expectation can occur with probability no greater than about $1/(2.8)^2$ or .13.

Suppose, however, that the sample size had been 200. Here

$$z_M = \frac{10}{5/\sqrt{200}} = 28.28$$

so that a sample M deviating this much or more from expectation could occur only with probability no greater than about $1/(28.28)^2$ or .0013. An extent of deviation of M from the true mean that could occur relatively often for small samples is rare when the sample size is large. Notice that knowing the standard error of the mean is essential if we are going to judge the agreement between a sample and a population mean in terms of the probability of occurrence for a given extent of deviation. For the moment we are assuming that this information is simply given to us.

7.14   Sample size and estimation

Suppose than an experimenter is interested in estimating the mean of some population by use of the sample mean. If he happens to know the exact form of the sampling distribution of the mean induced by the population distribution, then he can choose a sample size that will be sufficient to give him any degree of accuracy he may desire. That is, since the larger the sample size the smaller the error he should expect in estimating μ from M, he can choose N in such a way as to make the measure of that error, or σ_M, arbitrarily small. Even so, some value of N should be a *sufficient* sample size, which is just large enough to meet the experimenter's requirement of accuracy of estimation, but not so large as to exceed it.

Unfortunately, the experimenter may not know the exact form of the sampling distribution. He may, perhaps, have some partial information about the form of the sampling distribution, or he may have no information about it at all. Under these limitations, is there any way at all for the experimenter to make a judgment about the sample size that he should use to achieve a given degree of accuracy? The answer is yes, by use of the law of large numbers. For any given sampling distribution (and induced sampling distribution) there will be some sufficient sample size that will guarantee a given degree of accuracy. For some distributions this sample size will be one thing; for other distributions a different sample size will be called for. What the experimenter can find out, in the absence of any other information, is the *largest* of these sufficient sample sizes, over all of the possible distributions that might apply. He can determine the "maximum-minimum" or the maximum sufficient sample size by use of the law of large numbers, as embodied in the Tchebycheff inequality. This is best illustrated concretely.

Suppose that in some experiment it is desired to estimate the population mean to within one-tenth of a population standard deviation. Naturally, the experimenter can never be absolutely sure that his estimate will come this close to the population value, but he would like the probability to be quite high that his estimate actually is within $.1\sigma$ of the value of μ. He knows nothing at all about the sampling distribution. What is a sufficient number of cases for him to observe if he wants the probability to be $.95$ or less that his estimate comes within $.1\sigma$ of the population value μ? He knows full well that if he had more information about the sampling distribution, a smaller number of observations would very likely be sufficient, but in the absence of such information he will use the maximum-minimum, or the maximum-sufficient, number of cases.

The answer can be found from the Tchebycheff inequality, as given by **7.12.8**, with the value of k set equal to $.1\sigma$: we want

$$\text{prob.}(|M - \mu| < .1\sigma) \geqq 1 - \frac{\sigma^2}{N(.1\sigma)^2}$$

when the right hand side of the inequality is replaced by $.95$. Actually setting the right hand of the inequality equal to $.95$, and solving for N, we have

$$1 - \frac{1}{N(.01)} = .95$$

$$\frac{1}{N} = .0005$$

$$N = 2000.$$

A sufficient number of cases that could be observed to meet these qualifications is 2000, and this is the largest number that would have to be observed. If the experimenter actually observes 2000 cases, then he can be sure that the probability is at least .95 that his estimate comes within .1σ of μ, even in the absence of any information about the sampling distribution. If he had such additional information, then fewer observations would likely be required.

Note well that the number arrived at, or 2000, is actually a *sufficient* number of observations for the experimenter to take, if he wishes the probability to be .95 that M is within .1σ of μ, and given no information about the sampling distribution. If he chooses to take more than 2000 observations, then he can be even more confident that his requirements are met. On the other hand, the number 2000 is also the *largest* sufficient number of cases, as compared with the numbers required when something more is known about the sampling distribution of the mean. Given such information, he might be able to achieve the same degree of confidence in the accuracy of the estimate through the use of fewer observations. For instance, suppose that he has reason to believe that the sampling distribution of M is unimodal and symmetric, as it would tend to be if the population itself had a unimodal and symmetric distribution. As one consequence of the Tchebycheff inequality in Section 6.28, it can be shown that for a unimodal symmetric distribution

$$\text{prob.}(|X - \mu| < k) \geqq 1 - \frac{4\sigma^2}{9k^2}.$$

Applying this information, and assuming that a sampling distribution for means is unimodal-symmetric, the experimenter would find that no more than 889 cases are required in the sample to make the probability at least .95 that the value of M comes within .1σ of μ. As usual in matters of statistical inference, the more one can assume true about the population or about the sampling distribution of a particular statistic, the "tighter" are the statements one can deduce from general statistical principles.

These values for N arrived at from the Tchebycheff inequality are useful only in setting *maximum* sufficient sample size requirements. In most instances the sample sizes given by the Tchebycheff inequality are much too large, and they are employed only when the experimenter cannot commit himself at all about the basic form of the sampling distribution.

Now we can look ahead to a principle that makes it considerably less important whether or not the form of the population distribution is known. In the next chapter the **central limit theorem** will be introduced. This is a very general principle that lets us know the approximate form of the distribution of the sample mean from a wide variety of population distributions, provided that the sample size is relatively large. That is, we will find that for most population distributions of interest, as sample size is made relatively large, the sampling distribution of the mean approaches the unimodal, symmetric form known as the **normal distribution**. Although the sampling distribution of means from a given population need never be exactly like the normal distribution, nevertheless with a sufficiently large N per sample, the normal distribution provides an excellent approximation to the exact sampling distribution of the mean. This makes it possible to solve problems such as those of sufficient sample size much more accurately and efficiently than does the Tchebycheff inequality. For example, by use of the normal distribution it can be shown that for large samples

$$\text{prob.}\left(\left|\frac{M - \mu}{\sigma_M}\right| < 1.96\right) = .95.$$

Now if we wish the probability to be .95 that the difference between M and μ falls within $.1\sigma$, and if we know by the central limit theorem that the sampling distribution of M is approximately normal, we can solve for the required N as follows:

$$\frac{M - \mu}{\sigma_M} = 1.96$$

$$\frac{.1\sigma}{\sigma/\sqrt{N}} = 1.96$$

$$N = (19.6)^2 = 384.$$

If the experimenter takes a sample of 384 cases, then he can state that the probability is approximately .95 that M falls within $.1\sigma$ of μ. This statement is only approximate, because the sampling distribution of M is only approximately like the normal distribution. Nevertheless, if he takes as many as 384 cases the approximation is extremely good, and the experimenter can feel confident that M estimates μ within the desired limits. Contrast this estimated sample size, a required 384 cases, with the values 2000 and 889 cases set by use of the Tchebycheff inequality! This is but one instance of the enormous practical utility of the central limit theorem in statistics. We will have a great deal more to say about this principle, but this must wait until Chapter 8.

7.15 CORRECTING THE BIAS IN THE SAMPLE VARIANCE AS AN ESTIMATOR

As shown in Section 7.7, the sample variance is a biased estimator of the population variance σ^2. At first this may seem a serious drawback, since our only knowledge of the population variance, and hence of the standard error of the mean, must come from samples. However, now that the variance of the sampling distribution of the mean has been found, we can work out a simple way to correct the sample variance as an estimator of σ^2.

It will be shown first that the expectation of the variance from samples of size N is

$$E(S^2) = \frac{N - 1}{N} \sigma^2. \qquad [7.15.1*]$$

We found in Section 7.6 that the expectation of the sample variance is actually a difference between two variances:

$$E(S^2) = \sigma^2 - \sigma^2_M.$$

However, we have just shown that

$$\sigma^2_M = \frac{\sigma^2}{N}.$$

On making this substitution, we have

$$E(S^2) = \sigma^2 - \frac{\sigma^2}{N} = \left(\frac{N-1}{N}\right)\sigma^2,$$

so that on the average the sample variance is *too small* by a factor of $\dfrac{N-1}{N}$.

Since this is true, a way emerges for correcting the variance of a sample to make it an unbiased estimator. **The unbiased estimate of the variance based on any sample of N independent cases is**

$$s^2 = \frac{N}{N-1}\, S^2. \qquad\qquad [7.15.2*]$$

(In all that follows, we will reserve the symbol S^2 to stand for the uncorrected variance of a sample, as originally defined, and use the lower-case letter s^2 to indicate the corrected variance. Similarly, S will stand for the standard deviation based upon S^2, and s for the standard deviation based upon s^2.)

It is simple to show that s^2 is indeed unbiased:

$$E(s^2) = \frac{N}{N-1}\, E(S^2) = \frac{N}{(N-1)}\,\frac{(N-1)}{N}\,\sigma^2 = \sigma^2. \qquad [7.15.3*]$$

Quite often it is convenient to calculate the unbiased variance estimate s^2 directly, without the intermediate step of calculating S^2. This is done either by the formula

$$s^2 = \frac{\sum_i x_i^2 - NM^2}{N-1} \qquad\qquad [7.15.4\dagger]$$

$$s^2 = \frac{\sum_i x_i^2}{N-1} - \frac{\left(\sum_i x_i\right)^2}{N(N-1)} \qquad\qquad [7.15.5\dagger]$$

Even though we will be using the square root of s^2, or s, to estimate the population σ, it should be noted that s is not *itself* an unbiased estimate of σ, and that

$$E(s) \neq \sigma$$

in general. The correction factor used to make s an unbiased estimate of σ depends upon the form of the population distribution; thus, for the unimodal symmetric distribution known as the "normal" distribution (Chapter 8) an unbiased estimate for large N is provided by

$$\text{unbiased estimate of } \sigma = \left[1 + \frac{1}{4(N-1)} \right] s. \qquad [7.15.6]$$

Furthermore, special tables exist for correcting the estimate of σ for relatively small samples from such populations (Dixon and Massey, 1957). However, the problem of estimating σ from s is bypassed, in part, by the methods we will use in Chapter 11 and elsewhere, and, if the sample size is reasonably large, the amount of bias in s as an estimator of σ ordinarily is rather small. For these reasons, we will not trouble to correct for the bias in s found from the *unbiased* estimate s^2 of σ^2.

Some modern statistics texts completely abandon the idea of the sample variance S^2 as used here, and introduce only the unbiased estimate s^2 as *the* variance of a sample. However, this is apt to be confusing in some work, and so we will follow the older practice of distinguishing between the sample variance S^2 as a descriptive statistic, and s^2 as the unbiased estimate of σ^2.

7.16   ESTIMATING THE STANDARD DEVIATION OF THE POPULATION AND THE STANDARD ERROR OF THE MEAN

In most elementary work the standard deviation of the population, σ, is estimated simply by taking the square root of the unbiased estimate s^2,

$$\text{estimated } \sigma = \sqrt{s^2}. \qquad [7.16.1]$$

In the same way, one estimates the standard error of the mean by using the unbiased estimate s^2:

$$\text{estimated } \sigma_M = \frac{\text{estimated } \sigma}{\sqrt{N}} = \sqrt{\frac{s^2}{N}}. \qquad [7.16.2\dagger]$$

On the other hand, if the sample variance (not the unbiased estimate) is used, then the estimate may be arrived at as follows:

$$\text{estimated } \sigma_M = \sqrt{\frac{s^2}{N}} = \sqrt{\frac{N}{(N-1)}\frac{S^2}{N}} = \frac{S}{\sqrt{N-1}}. \qquad [7.16.3\dagger]$$

7.17   PARAMETER ESTIMATES BASED ON POOLED SAMPLES

Sometimes it happens that one has several independent samples, and that each provides an estimate of the same parameter or set of parameters. The most usual situation occurs when one is estimating μ or σ^2, or both. When this happens there is a real advantage in pooling the sample values to get an unbiased estimate. These pooled estimates are actually weighted averages of the estimates from the different samples. The big advantage lies in the fact that the sampling error will tend to be smaller for the pooled estimate than that for any single sample's value taken alone.

Suppose that there were two independent samples, based on N_1 and N_2 observations respectively. From each sample a mean M is calculated, each of which estimates the same value μ. Then the pooled estimate of the mean is

$$\text{pooled } M = \text{est. } \mu = \frac{N_1 M_1 + N_2 M_2}{N_1 + N_2}. \qquad [7.17.1\dagger]$$

It is easy to see that this must give an unbiased estimate of μ:

$$E\left[\frac{N_1 M_1 + N_2 M_2}{N_1 + N_2}\right] = \frac{N_1 E(M_1) + N_2 E(M_2)}{N_1 + N_2}$$
$$= \frac{\mu(N_1 + N_2)}{N_1 + N_2} = \mu.$$

For some J different and independent samples, where any given sample j gives an estimate M_j of μ, the pooled estimate is

$$\text{est. } \mu = \frac{\sum\limits_{j} N_j M_j}{\sum\limits_{j} N_j}. \qquad [7.17.2\dagger]$$

The fact that the pooled estimate is likely to be better than the single sample estimates taken alone is shown by the standard error of the pooled mean. For two independent samples, each composed of independent observations drawn from the same population,

$$\sigma_M = \frac{\sigma}{\sqrt{N_1 + N_2}}, \qquad [7.17.3]$$

which *must* be smaller than the standard error either of M_1 or of M_2. Similarly, for any number J of independent samples, with $N_1, N_2, \cdots, N_J$ as the respective sample sizes,

$$\sigma_M = \frac{\sigma}{\sqrt{\sum\limits_{j} N_j}}. \qquad [7.17.4*]$$

Naturally, these standard errors refer to the situation where the respective samples all come from populations with the same variance σ^2.

In the same general way, it is possible to find pooled estimates of σ^2, for populations with the same variance σ^2, even though the population means may be different. Thus for two independent samples

$$\text{est. } \sigma^2 = \frac{(N_1 - 1)s_1^2 + (N_2 - 1)s_2^2}{(N_1 - 1) + (N_2 - 1)},$$ [7.17.5*]

which is an unbiased estimate of the variance σ^2 of each of the populations.

For any number J of independent samples, drawn from populations with the same variance σ^2 (though not necessarily with the same mean), the corresponding estimate of σ^2, is given by

$$\text{est. } \sigma^2 = \frac{\sum_j (N_j - 1)s_j^2}{\sum_j (N_j - 1)} = \frac{\sum_j (N_j - 1)s_j^2}{N - J},$$ [7.17.6*]

where N is the total number of observations. We are going to make a great deal of use of these pooling principles in Chapters 10 and 12.

7.18   SAMPLING FROM FINITE POPULATIONS

It has been mentioned repeatedly that most sampling problems deal with populations so large that the fact that samples are taken without replacement of single cases can safely be ignored. However, it may happen that the population under study is not only finite, but relatively small, so that the process of sampling without replacement has a real effect on the sampling distribution.

Even in this situation the sample mean is still an unbiased estimate regardless of the size of the population sampled. Hence no change in procedure for mean estimation is needed.

However, for finite populations the sample variance is biased as an estimator of σ^2 in a way somewhat different from the former, infinite-population, situation. When samples are drawn without replacement of individuals, the unbiased estimate of σ^2 is

$$\text{est. } \sigma^2 = \frac{N(T - 1)}{(N - 1)T} S^2$$

$$= \frac{(T - 1)}{T} s^2$$ [7.18.1]

where T is the *total number* of elements in the population.

Another difference from the infinite-population situation comes with the variance of the sampling distribution of means. For a population with T cases in all, from which samples of size N are drawn, the sampling variance of the mean is

$$\sigma_M^2 = \left(\frac{T - N}{T - 1}\right)\frac{\sigma^2}{N}.$$ [7.18.2†]

The variance of the mean tends to be *somewhat smaller* for a fixed value of N when sampling is from a finite population than when it is from an infinite population.

Note that here the size of σ_M^2 depends both upon T, the total number in the population, and upon N, the sample size. An unbiased estimate of the variance of the mean is thus given by

$$\text{est. } \sigma_M^2 = \left(\frac{T - N}{T - 1}\right)\frac{s^2}{N}. \qquad [7.18.3\dagger]$$

The square root of this value gives the estimate of σ_M, for sampling from a finite population.

Sometimes the number $C = N/T$ is called the "sampling fraction," and when T is at least moderately large, the variance of the mean, σ_M^2, is approximately

$$\frac{\sigma^2}{N}(1 - C).$$

When T is so large that C is virtually zero, the sampling variance is the same as for the infinite-population situation.

7.19   THE SAMPLE PROPORTION AS AN UNBIASED ESTIMATOR

When the observations made in an experiment fall into qualitative classes, then the sample proportion P in each class is the usual summary statement of the sample result. This is an unbiased estimate of the population proportion p in that class:

$$E(P) = p. \qquad [7.19.1]$$

This is true for any number of classes.

When there are exactly two classes, the sampling distribution of the proportion P is found from the binomial, with variance

$$\sigma_P^2 = \frac{pq}{N} \qquad [7.19.2]$$

and standard error

$$\sigma_P = \sqrt{\frac{pq}{N}}. \qquad [7.19.3*]$$

Notice that unlike the situation for the mean, once the population p is specified, and given N, the standard error of P is known completely. When p is given there is ordinarily no reason to estimate the standard error separately from sample data, as one must when using the mean.

On the other hand, suppose that p is not given. In this case, the usual estimate of p is given by P, the sample proportion. Indeed, the sample proportion is simply the mean of the sample, just as p is the mean of the population generating the sample. Thus, if we let the random variable $X = 1$ for a success, and $X = 0$ for a failure, in the usual counting way, we have

$$M = \frac{\sum_i x_i}{N} = P$$

and

$$S^2 = \frac{\sum_i x^2}{N} - M^2$$
$$= PQ.$$

Hence, it seems reasonable to estimate σ_P^2 by taking the sample value PQ/N. However, is the estimate of the variance of a proportion biased in this case? The answer is yes, since

$$E\left(\frac{PQ}{N}\right) = \frac{1}{N}\,[E(P) - E(P^2)]$$
$$= \frac{1}{N}\,[p - (\sigma_P^2 + p^2)]$$
$$= \frac{1}{N}\left[p - \frac{pq}{N} - p^2\right]$$
$$= \frac{1}{N}\left[\frac{(N-1)pq}{N}\right]$$
$$= \frac{N-1}{N}\left(\frac{pq}{N}\right).$$

Hence, when P is used to estimate the unknown p, an unbiased estimate of σ_P^2 is given by $PQ/(N-1)$, so that σ_P is estimated from $\sqrt{PQ/(N-1)}$.

7.20   THE SAMPLING DISTRIBUTION OF THE SAMPLE VARIANCE AND OF OTHER STATISTICS

We have reiterated at length in this chapter that the ideas of sampling distributions and of the use of statistics for estimation apply to any of the limitless variety of statistics that can be associated with samples. Because of the simplicity of the mathematics involved, the mean has been used as the principal example, but most of the considerations discussed apply to other statistics as well. Some of these other sampling distributions will be discussed in other chapters. In particular, although we have seen that the mean of the distribution of S^2 is $(N-1)\sigma^2/N$ and that of s^2 is σ^2, we have not yet really explored the sampling distribution of the variance. Since some special problems arise in discussing the variance, we will defer a further consideration of its sampling distribution until Chapter 11. For the nonce, we will turn to a consideration of some general features of the use of sampling distributions in statistical inference.

7.21   THE USES OF SAMPLING DISTRIBUTIONS

Almost all of the remainder of this text will be devoted to problems in using sampling distributions for statistical inference. Parameter estimation is but the starting place for inference. The really essential features of sampling distributions, the standard errors and the probabilities associated with intervals of sample values, have not yet been brought into systematic use. It has been suggested in the preceding chapter that a sampling distribution, such as the binomial, may be used to assess the probability of a certain kind of sample result, given some hypothetical population situation. However, a set of formal procedures still must be given that will permit a similar kind of inference to be made about *any* population or population characteristic. This set of procedures using specific sampling distributions is known as **hypothesis-testing** and will be discussed in Chapter 9.

Specific sampling distributions also give a way to carry on the second important part of estimation, the so-called **interval estimation.** It is not enough to estimate population values from sample statistics; the statistician must provide some estimate of the sampling error involved. Moreover, he uses his knowledge of the sampling distribution to set up a *range of values* which, over all possible samples, has a high probability of covering the true population value—these are called **confidence intervals** for the population value. A much stronger inference may be made using confidence intervals than using the point estimate alone, since a very positive statement can be made about a *range* of values very likely to contain the true value. This method of interval estimation will also be considered in Chapter 9.

7.22 OTHER KINDS OF SAMPLING

The model of simple random sampling is the basis of almost all the discussion to follow. The classical procedures of statistical inference rest upon the sampling scheme in which each and every population element sampled is independent of every other, and is equally likely to be included in any sample.

However, statistical inference is not at all limited to such equally probable samples. Samples may be drawn according to any other probability structure, just so long as the probability of occurrence of any particular sample is known or can be calculated. Samples are drawn by some method consistent with these probabilities, and the probabilities are then taken into account in the treatment of any information gained from the sample. The generic term for the process of drawing samples according to some known probability structure is, simply, **probability sampling.** Simple random sampling is then a special case of probability sampling; the probability structure in simple random sampling dictates equal probability for each possible sample. Many methods exist for treating samples drawn according to other probability structures, and often these methods differ from those we will discuss.

One common scheme is called **stratified sampling.** Here, the population is divided into a number of parts, called "strata." A sample is drawn independently and at random in each part. Given the sizes of the various strata, one can make inferences about the total population represented. Such a scheme is very good for insuring a representative sample, and it may reduce the error in estimation, although it does require special methods over and above those given here.

7.23   To what populations do our inferences refer?

Most psychologists who use inferential statistics in research rely on the model of simple random sampling. Yet how does one go about getting such a "truly" random sample? It is not easy to do, unless, as in all probability sampling, each and every potential member of the population may somehow be listed. Then, by means of a device such as random number tables, individuals may be assigned to the sample with approximately equal probabilities.

However, in behavioral sciences such as psychology, interest often lies in experimental effects that, presumably, should apply to very large populations of men or other things. Such a listing procedure is simply not possible. Still other experiments may refer to all possible measurements that *might* be made of some phenomenon under various experimental conditions, where estimated true values may be sought from the experimental observation of a few instances. Here, the population is not only infinite, it is hypothetical, since it includes all *future* or *potential* observations of that phenomenon under the different conditions. In sampling from such experimental populations, where there is no possibility of listing the elements for random assignment to the sample, the only recourse of the experimenter is to draw his basic experimental units in some more or less random, "haphazard," way, and then *make sure that in his experiment only random factors determine which unit gets which experimental treatment.* In other words, there are two ways in which randomness is important in an experiment: the first is in the selection of the sample as a whole, and the second is in the allotment of individuals to experimental treatments. Each kind of randomness is important for the "generalizability" of the experimental results, so that when one does an experiment he usually takes pains to see that both kinds of randomness are present. However, even given that individual cases are assigned to experimental manipulations at random, the possible inferences are still limited by the fundamental population from which the total sample is drawn.

How does one know the population to which statistical inferences drawn from a sample apply? If random sampling is to be assumed, then the population is defined by the manner in which the sample is drawn. The only population to which the inferences strictly apply is that for which the *units* sampled have equal (and nonzero) likelihood of appearing in any sample. In some contexts a unit may be one thing, and in other contexts quite another: thus, for example, if the units sampled are families inhabiting the same dwelling, it may be that each such family-unit has equal probability of appearing in a sample of N such units. However, it would not then be true that individual persons have equal probability for appearing in any sample. On the other hand, if individual persons are sampled at random, it will not be true that family-units have equal probability of appearing, since families contain different numbers of people. It is most important to have a clear definition of the unit that is to be sampled, since in random sampling the totality of such possible units with equal probability of appearing in any given sample must make up the population.

Unfortunately, "population" *is* an awkward term in many respects, since it seems to imply that a sampling unit will somehow be an individual. While this is often the case, it is not always true, and units may be sampled for observation that are not individuals in the usual sense. Thus, although "population" is (and will continue to be) used to mean the totality of units available for sampling, it is best to forget the common-sense associations that this term carries along with it, and to think of "population" only as the basic and total set of units available for observation, either by random sampling or by the use of some other probability structure.

It should be obvious that simple random samples from one population may not be random samples of another population. For example, suppose that some one wishes to sample American college students. He obtains a directory of college students from a Midwestern university and, using a random number table, takes a sample of these students. He is not, however, justified in calling this a random sample of the population of American college students, although he may be justified in calling this a random sample of students at *that* university. *The population is defined not by what he said, but rather by what he did to get the sample.* For any sample, one should always ask the question, "What is the set of potential units that could have appeared in my sample with equal probability?" If there is some well-defined set of units that fits this qualification, then inferences may be made to that population. However, if there is some population whose members could not have been represented in the sample with equal probability, then inferences do not *necessarily* apply to that population when methods based on simple random sampling are used. Any generalization beyond the population actually sampled at random must rest on extrastatistical, scientific, considerations.

From a mathematical-statistical point of view the assumption of random sampling makes it relatively simple to determine the sampling distribution of a particular statistic given some particular population distribution. As we noted above, it is possible to use other probability structures for the selection of samples, and in some situations there are advantages in doing so. The point is that *some* probability structure must be known or assumed to underlie the occurrence of samples if statistical inference is to proceed. This point is belabored only because it is so often overlooked, and statistical inferences are so often made with only the most casual attention to the process by which the sample was generated. The assumption of some probability structure underlying the sampling is a little "price-tag" attached to a statistical inference. It is a sad fact that if one knows nothing about the probability of occurrence for particular samples of units for observation, he can use very little of the machinery we are describing here. This is why our assumption of random sampling is not to be taken lightly. All the techniques and theory that we will discuss apply to random samples, and do not necessarily hold for any data collected in any way. In practical situations, the experimenter may be hard put to show that a given sample is "truly" random, but he must realize that he is acting *as if* this were a random sample from some well-defined population when he applies methods of statistical inference. Unless this assumption is at least reasonable, the results of inferential methods mean very little, and these methods might as well be omitted. Data that are not the product of random sampling may yield extremely valuable conclusions in their own right, but there is usually little to be gained from the application of inferential methods to such data. Certainly, the application of some statistical method does not somehow magically make a sample random, and the conclusions therefore valid. Inferential methods apply to probability samples, drawn either by simple random sampling or in accordance with some other probability structure. There is no guarantee of their validity in other circumstances.

EXERCISES

1. In a study of the occurrence of accidents in a plant, the investigators had reason to believe that the incidence of accidents within a fixed time period corresponded to a stable and independent Poisson process, with some parameter value m. Within one sample time period 7 accidents occurred. Suppose that m were actually one of the values 4, 6, 8, 10. According to the principle of maximum likelihood, and in the light of the data, which of these values is the best estimate of m?

2. A banker is confronted with a very large bag of coins. He knows that the bag contains one of four mixtures of quarters and half dollars:

> 50% quarters, 50% half-dollars
> 30% quarters, 70% half-dollars
> 70% quarters, 30% half-dollars
> 90% quarters, 10% half-dollars

He draws 15 coins from the bag at random with replacement. Of the fifteen coins, 12 turn out to be quarters and the rest half dollars. Which mixture of coins should the banker guess that the bag contains?

3. Consider a discrete random variable having the following distribution:

x	$P(x)$
3	$1/2$
2	$1/3$
1	$1/6$

An experiment is conducted in such a way that three independent observations are made at random, and as a result of each observation some value of X occurs. Furthermore, the mean of each group of three observations is calculated. Find the theoretical sampling distribution of these means.

4. Given the theoretical sampling distribution of means found in problem 3, calculate its mean and variance. Check the values obtained against those which one would expect to obtain given the original distribution of X.

5. Suppose that the same population is sampled three times, on the first occasion 10 observations were made independently and at random, on the second 20 observations were made independently and at random, and on the third occasion 15 observations were similarly made. The results were

SAMPLE 1	SAMPLE 2	SAMPLE 3
$M_1 = 96$	$M_2 = 105$	$M_3 = 103$
$S_1^2 = 22$	$S_2^2 = 29$	$S_3^2 = 31$

Estimate the mean and the variance of the population. What is the estimated standard error of this estimate of the population mean?

6. A population is known to have a variance of 18.5. Two samples of 90 observations each are made independently and at random from this population. One sample produces a mean of 65 and a variance of 16 and the second independent sample a mean of 70, and a variance of 20. Estimate the population mean, and find the standard error of this estimate.

7. In a certain sampling distribution of means based on 44 cases each, a sample mean of 438 is known to correspond to a z-value of 1.75 in the sampling distribution. If the standard error of the mean is 9, what is the population mean? What is the population variance?

8. In a certain sampling distribution of the mean based on samples of size N_1, a sample mean of 100 corresponds to a z value of 1.5. However, when samples of size N_2 are taken, a sample mean of 100 corresponds to a z-value of 2.00. How large is N_2 relative to N_1?

9. Two independent samples were drawn from the same population with a standard deviation of 9. The first sample contained 10 cases and the second sample contained 20 cases. The mean of the first sample corresponded to a z-score of 1.8 in its sampling distribution of means, while the second sample mean corresponded to a z-score of .5. How much larger was the first mean than the second? What z-score would correspond to the mean of the pooled samples?

10. A random variable X is generated by a stable and independent Poisson process with intensity 10. Five independent observations of values of X are made, and a mean calculated. Find the mean and standard deviation of the sampling distribution of these means.

11. Suppose that a random variable X can take on only the values 1 and 0, such that the probability that $X = 1$ is p and that $X = 0$ is q. Some N independent observations are made of this random variable, and the mean calculated. Describe the sampling distribution of these means.

12. Suppose that in a certain experiment, it was decided that only the third, fourth, and fifth observation out of any N would be recorded to form the basis for estimating the mean and standard deviation of the population. Comment on the characteristics of such an estimator of the mean (i.e. is it unbiased, consistency, and the like).

13. In a certain experiment, 55 subjects drawn at random from a given population were used. Each produced a score value independently and the mean and sample variance were computed. These turned out to be $M = 78.33$ and $S^2 = 27.91$. If the mean of the population were actually 60, what do you estimate the z-score corresponding to this sample mean is to be in the sampling distribution? Are z-scores this much or more deviant from 0 relatively likely?

14. In a survey of ethical principles among college students, some 1000 students responded to a confidential questionnaire. Among the respondents, 635 admitted that they had been guilty of plagiarism in a term paper on at least one occasion. A similar survey carried out 25 years before had shown about 55 percent of students making the same

admission. Given that the population of today's students is really like that of a generation ago, what z-score does this result correspond to in the sampling distribution of proportions? Given no difference in the populations, does the z-value suggest a relatively unlikely sample result?

15. A discrete random variable can take on only the values 1, 2, 3, and 4, each with equal probability. Suppose that observations were made independently and at random, each yielding a value of this random variable. Furthermore, suppose that means of these observed values were taken for each set of observations. Construct the sampling distribution of means for samples of size 2, 3, and 4. Compare each of these sampling distributions with the original distribution of X (in this case, the population distribution). How do the sampling distributions tend to change with steadily increasing sample size? (It will be useful to keep in mind that for $N = 2$, $p(x_1 + x_2) = p(x_1)p(x_2)$, for $N = 3$, $p(x_1 + x_2 + x_3) = p(x_1 + x_2)p(x_3)$, and so on.)

16. Repeat problem 15, but use the random variable following the rule

$$p(x) = \begin{cases} x/10 & x = 1, 2, 3, 4 \\ 0 & \text{elsewhere} \end{cases}$$

Under these conditions, how do the sampling distributions of the mean tend to change with increasing sample size?

17. Show that for each of the sampling distributions found in problem 15 above, the mean of the sampling distribution equals the mean of the parent distribution, and the variance of the sampling distribution is the original distribution's variance divided by N.

8 Normal Population and Sampling Distributions

8.1 THE NORMAL DISTRIBUTION

Heretofore, the discrete binomial distribution has been used as our prime example of a sampling distribution. Now we consider a type of distribution having a domain which is *all* of the real numbers. This is the so-called "normal" or "Gaussian" distribution. The normal distribution is but one of a vast number of mathematical functions one might invent for a distribution; it is purely theoretical. At the very outset let it be clear that, like binomial or other probability functions, the normal distribution is not a fact of nature that one actually observes to be exactly true. Rather, the normal distribution is a theory about what might be true of the relation between intervals of values and probabilities for some variable. There is nothing magical about the normal distribution; it happens to be only one of a number of theoretical distributions that have been studied and found useful as an idealized mathematical concept. Normal distributions do not really exist, however, and in applied situations the closest we can come to finding a normal distribution will never quite correspond to the requirements of the mathematical rule. Many concepts in mathematics and science that are never quite true give good practical results nevertheless, and so does the normal distribution.

Like most theoretical functions, a normal distribution is completely specified only by its mathematical rule. Quite often, the distribution is symbolized by a graph of the functional relation generated by that rule, and the general picture that a normal distribution presents is the familiar bell-shaped curve of Figure 8.1.1. The horizontal axis represents all the different values of X, and the vertical axis $f(x)$ their densities. The normal distribution is continuous for all values of X between $-\infty$ and ∞, so that each conceivable nonzero *interval* of real numbers has a probability other than zero. For this reason, the curve is shown as never quite touching the horizontal axis; the tails of the curve show decreasing probability densities as values grow extreme in either direction from the mode, but any interval representing *any* degree of deviation from central tendency is possible in this theoretical distribution. The distribution is absolutely symmetric and unimodal, and mean, median, and mode all have the same value of X. Bear in mind that since the normal distribution is continuous, the height of the curve shows the probability *density* for each X value. However, as for any continuous distribution, **the area cut off beneath the curve by any interval is a probability,** and the entire area under the curve is 1.00.

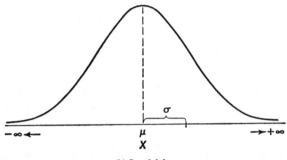

FIG. 8.1.1

The student is warned not to let the graph of any distribution lead him into jumping to conclusions about the kind of distribution represented. Although the normal distribution graphs as a bell-shaped curve, just any bell-shaped curve is not necessarily a normal distribution. The kind of distribution, the family to which it belongs, depends absolutely on the function rule, and on nothing else. Only if each possible value of X is paired with a density in the way provided by the normal function rule can one say that the distribution is normal. Other rules may give similar pictures or probabilities that are close to their normal counterparts but, by definition, these distributions are not exactly normal.

The mathematical rule for a normal density function is as follows:

$$f(x; \mu, \sigma^2) = \frac{1}{\sqrt{2\pi\sigma^2}} e^{-(x-\mu)^2/2\sigma^2}. \qquad [8.1.1*]$$

This rule pairs a probability density $f(x)$ with each and every possible value x. This rule looks somewhat forbidding to the mathematically uninitiated, but actually it is not. The π, the e, and the 2, of course, are simply positive numbers acting as mathematical constants. The "working" part of the rule is the exponent

$$-\frac{(x-\mu)^2}{2\sigma^2} \qquad [8.1.2]$$

where the particular value of the variable X appears, along with the two parameters μ and σ^2.

The more that the value of X differs from μ, the larger will the quantity in the numerator of this exponent be. However, this deviation of x from μ enters as a squared quantity, so that *two* different x values showing the same absolute deviation from μ have the same probability density according to this rule. This dictates the symmetry of the normal distribution. Furthermore, the exponent as a whole has a negative sign, meaning that the larger the absolute deviation of x from μ, the smaller will the density assigned to x be by this rule. This dictates that either tail of the distribution shows decreasing density, since the wider the departure of x values from μ the lower will be the height of this function's curve. However, a little thought will convince you that no real and finite value of x can possibly make the density itself negative or exactly zero: the normal function curve never touches the X axis, indicating that any interval of numbers will have a nonzero probability. On the other hand, any number, such as e, raised to the zero power is 1; when X exactly equals μ, the density is

$$\frac{1}{\sqrt{2\pi\sigma^2}} \qquad\qquad [8.1.3]$$

which is the *largest* density value any X may have. This fact implies that the function curve must be unimodal, with maximum density at the mean, μ. Note that, unlike a probability, a density value such as expression **8.1.3** can be greater than 1.00.

It is very important to notice that the precise density value assigned to any X by this rule cannot be found unless the two parameters μ and σ are specified. The parameter μ can be any finite number, and σ can be any finite *positive* number. Thus, like the binomial, the normal distribution rule actually specifies a family of distributions. Although each distribution in the family has a density value paired with each x by this same general rule, the *particular* density that is paired with a given X value differs with different assignments of μ and σ. Thus, normal distributions may differ in their means (Fig. 8.1.2); in their standard deviations (Fig. 8.1.3); or in both means and standard deviations (Fig. 8.1.4). Never-

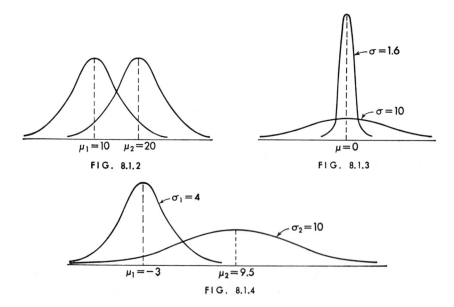

FIG. 8.1.2

FIG. 8.1.3

FIG. 8.1.4

theless, given the mean and standard deviation of the distribution, the rule for finding the probability density of any value of the variable is the same.

Since the normal distribution is really a family of distributions, statisticians constructing tables of probabilities found by the normal rule find it convenient to think of the variable in terms of standardized or z scores. That is to say, if the random variable is a standardized score, z, so that $\mu = 0$ and $\sigma = 1.00$, then the rule becomes simpler:

$$f(z) = \frac{1}{\sqrt{2\pi}} e^{-z^2/2}. \qquad [8.1.4*]$$

For standardized normal variables, the density depends only on the *absolute* value of z; since both z and $-z$ give the same value z^2, they both have the same density. The higher the z in absolute value, the less the associated density. The standardized-score form of the distribution makes it possible to use one table of densities for any normal distribution, regardless of its particular parameters. Thus, if we want to know the density of a score x in a normal distribution, with mean equal to 80 and standard deviation equal to 5, we can look up the density associated with a standardized score $(x - 80)/5$, and this gives the desired number.

8.2 Cumulative probabilities and areas for the normal distribution

In Section 3.19 it was pointed out that when a distribution is continuous the probability that X takes on some exact value x is, in effect, zero. For this reason, *all probability statements we will make using a normal distribution will be in terms either of cumulative probabilities or the probabilities of intervals.*

The cumulative probability

$$F(a) = p(X \leq a)$$

can be thought of as the area under the normal curve in the interval bounded by $-\infty$ and the value a (Fig. 8.2.1).

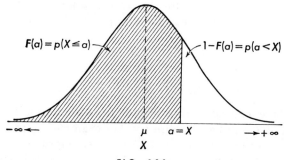

FIG. 8.2.1

In terms of the density function, this area is given by

$$F(a) = \int_{-\infty}^{a} f(x)\, dx$$

where $f(x)$ is the normal density of any value in the interval from $-\infty$ to a.

These cumulative probabilities can be used to find the probability of any interval. For example, suppose that in some normal distribution we want to find the probability that X lies between 5 and 10, given some exact values for μ and σ. This is the probability represented by the area shown in Figure 8.2.2. The

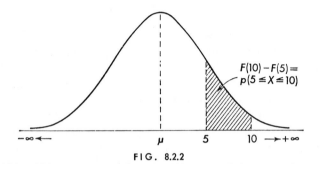

FIG. 8.2.2

cumulative probability *up to and including* 10, or $F(10)$, minus the cumulative probability up to and including 5, or $F(5)$, gives the probability in the interval:

$$p(5 \leq X \leq 10) = F(10) - F(5).$$

In the same way, the probability of any other interval with limits a and b can be found from cumulative probabilities:

$$P(a \leq X \leq b) = F(b) - F(a).$$

8.3   THE USE OF TABLES OF THE NORMAL DISTRIBUTION

Most tables of the normal distribution give the cumulative probabilities for various *standardized* values. That is, for a given z score the table provides the cumulative probability *up to and including that standardized score* in a normal distribution: for example:

$$F(2) = p(z \leq 2)$$
$$F(-1) = p(z \leq -1)$$

and so on.

If the cumulative probability is to be found for a positive z, then Table I in Appendix C can be used directly. For example, suppose that a normal distribution is known to have a mean of 50 and a standard deviation of 5. What is the cumulative probability of a score of 57.5? The corresponding standardized score is

$$z = \frac{57.5 - 50}{5} = 1.5.$$

A look at Table I shows that for 1.5 in the first column, the corresponding cumulative probability in the column labeled $F(z)$ is .933, approximately. This is the probability of observing a score *less than or equal to* 57.5 in this particular distribution. The cumulative probability for any other positive z score can be found in the same way.

If the z score is negative, this procedure is changed somewhat. Since the normal distribution is symmetric, the density associated with any z is the same as for the corresponding value with a negative sign, or $-z$. However, the cumulative probability for a negative standardized score is 1 minus the cumulative probability for the z value with a positive sign:

$$F(-z) = 1 - F(z)$$

where z is the positive standardized score of the same magnitude. Figure 8.3.1 will

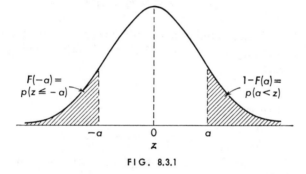

FIG. 8.3.1

clarify this point. The shaded area on the left is the cumulative probability for a value $z = -a$. The table gives only the area falling below the point $z = a$, the positive number of the same absolute value as $-a$. However, the shaded area on the *right*, which is $1 - F(a)$, is the same as the shaded area on the *left*, $F(-a)$.

For example, suppose that in a distribution with mean equal to 107 and standard deviation of 70, we want the cumulative probability of the score 100. In standardized form,

$$z = \frac{100 - 107}{70} = -.1.$$

We look in the table for z equal to *positive* .1, which has a cumulative probability of approximately .5389. The cumulative probability of a z equal to $-.1$ must be approximately

$$F(-.1) = 1 - .5389 = .4611.$$

We may also answer questions about the probabilities of various intervals from the table. For example, what proportion of cases in a normal distribution must lie within one standard deviation of the mean? This is the same as the probability of the interval $(-1 \leq z \leq 1)$. The table shows that $F(1) = .8413$, and we know that $F(-1)$ must be equal to $1 - .8413$ or $.1587$.

Thus,

$$p(-1 \leq z \leq 1) = F(1) - F(-1) = .8413 - .1587 = .6826.$$

About 68 percent of all cases in a normal distribution must lie within one standard deviation of the mean.

To find the proportion lying between 1 and 2 standard deviations *above* the mean, we take

$$p(1 \leq z \leq 2) = F(2) - F(1).$$

From the table, these numbers are $.9772$ and $.8413$, so that

$$p(1 \leq z \leq 2) = .9772 - .8413 = .1359.$$

About 13.6 percent of cases in a normal distribution lie in the interval between 1σ and 2σ above the mean. By the symmetry of the distribution we know immediately that

$$p(-2 \leq z \leq -1) = .1359$$

as well.

Beyond what number must only 5 percent of all standardized scores fall? That is, we want a number b such that the following statement is true for a normal distribution:

$$p(b < z) = .05.$$

This is equivalent to saying that

$$p(z \leq b) = .95.$$

A look at the table shows that roughly $.95$ of all observations must have z scores at or below 1.65. Thus,

$$p(1.65 < z) = .05, \text{ approximately.}$$

If intervals are mutually exclusive, their probabilities can be added by the "or" rule to find the probability that X falls into either interval. For example, we want to know

$$p(z < -2.58 \text{ or } 2.58 < z),$$

which is the same as

$$p(2.58 < |z|).$$

For the first of these intervals, we find

$$F(2.58) = .995$$

so that

$$F(-2.58) = 1 - .995 = .005.$$

For the other interval, the probability is also

$$p(2.58 < z) = 1 - F(2.58) = .005.$$

The two intervals mutually define

$$p(z < -2.58 \text{ or } 2.58 < z) = .005 + .005 = .01.$$

Put in the other way,

$$p(2.58 < |z|) = .01,$$

the probability that z exceeds 2.58 in absolute value is about one in one hundred. This probability is given by the extreme tails of the distribution in the graph of Figure 8.3.2.

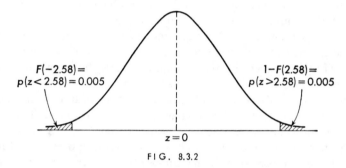

FIG. 8.3.2

It is interesting to compare the probability just found with that given by the Tchebycheff inequality for *any* symmetric, unimodal distribution (Section 6.27):

$$p(2.58 \leq |z|) \leq \frac{4}{9}\left[\frac{1}{(2.58)^2}\right].$$

On working out the number on the right, we find about .067. That is, for *any* unimodal symmetric distribution the probability of a score deviating 2.58 or more standard deviations from expectation is less than about seven in one hundred. By specifying that the distribution is *normal*, we pin this probability down to the exact value .01. The more one assumes or knows about the population distribution, the stronger is the statement he can make about the probability of any degree of deviation from expectation.

8.4 THE IMPORTANCE OF THE NORMAL DISTRIBUTION

The normal is by far the most used distribution in inferential statistics. There are at least four very good reasons why this is true.

A population may be assumed to follow a normal law because of what is known or presumed to be true of the measurements themselves. There are two rather broad instances in which the random variable is a measurement of some kind, and the distribution is conceived as normal. The first is when one is considering a hypothetical distribution of "errors," such as errors in reading a dial, in discriminating between stimuli, or in test performance. Any observation can be assumed to represent a "true" component plus an error. Each error has a magnitude, and this number is thought of as a reflection of "pure chance," the resultant of a vast constellation of circumstances operative at the moment. Any factor influencing performance at the moment contributes a tiny amount to the size of the error and to its direction. Furthermore, such errors are appropriately considered to push the observed measurement up or down with equal likelihood, to be independent of each other over samplings, and to "cancel out" in the long run. Thus, in theory, it makes sense that errors of measurement or errors of discrimination follow something like the normal rule.

In other instances, the distribution of *true magnitudes* of some trait may be thought of as normal. For example, human heights form approximately a normal distribution, and so, we believe, does human intelligence. Indeed, there are many examples, especially in biology, where the distribution of measurements of a natural trait seems to follow something closely resembling a normal rule. No obtained distribution is ever exactly normal, of course, but some distributions of measurements have relative frequencies very close to normal probabilities. However, the view that the distribution of almost any trait of living things should be normal, although prominent in the nineteenth century, has been discredited. The normal distribution is not, in any sense, "nature's rule," and all sorts of distributions having little resemblance to the normal distribution occur in all fields.

It may be convenient, on mathematical grounds alone, to assume a normally distributed population. Mathematical statisticians do not devote so much attention to normal distributions just because they think bell-shaped curves are pretty! The truth of the matter is that the normal function has important mathematical properties shared by no other theoretical distribution. Assuming a normal distribution gives the statistician an extremely rich set of mathematical consequences that he can use in developing methods. Very many problems in mathematical statistics either are solved, or can be solved, *only* in terms of a normal population distribution. We will find this true especially when we come to methods for making inferences about a population variance. The normal distribution is the "parent" of several other important theoretical distributions that figure in statistics. In some practical applications the methods developed using normal theory work quite well even when this assumption is not met, despite the fact that the problem can be given a *formal* solution only when a normal population distribution is assumed. In other instances, there just is not any simple way to solve the problem when the normal rule does not hold for the population, at least approximately.

The normal distribution may serve as a good approximation to a number of other theoretical distributions, having probabilities that are either laborious or impossible to work out exactly. For example, in the following section it will be shown that the normal distribution can be used to approximate binomial probabilities under some circumstances, even though the normal distribution is continuous, and the binomial discrete. For very large N, binomial probabilities are troublesome to work out exactly; on the other hand, intervals based on the normal function give probabilities that can be used as though they were binomial probabilities.

There is a very intimate connection between the size of sample N and the extent to which a sampling distribution approaches the normal form. Many sampling distributions based on large N can be approximated by the normal distribution even though the population distribution itself is definitely not normal. This is the extremely important principle that we will call the **central limit theorem.** The normal distribution is the *limiting form* for large N for a very large variety of sampling distributions. This is one of the most remarkable and useful principles to come out of theoretical statistics.

Thus, you see, it is no accident that the normal distribution is the workhorse of inferential statistics. The assumption of normal population distributions or the use of the normal distribution as an approximation device is not as arbitrary as it sometimes appears; this distribution is part of the very fabric of inferential statistics. These features of the normal distribution will be illustrated in the succeeding sections.

8.5 THE NORMAL APPROXIMATION TO THE BINOMIAL

One of the interpretations given to the normal distribution is that it is the limiting form of a binomial distribution as $N \rightarrow \infty$ for a fixed p. If one drew samples of N from a Bernoulli distribution with the probabilities of the two categories being p and q, respectively, he should expect a binomial distribution of number of successes. Suppose that first he does this for all possible samples of size 10. Next he draws all possible samples with $N = 10,000$, then repeats the sampling with an N of 10,000,000, and so on. For each sample size, there will be a different binomial distribution he should observe when *all possible* samples of that size have been drawn.

How does the binomial distribution change with increasing sample size? In the first place, the actual range of the possible number of successes grows larger, since the whole numbers 0 through N form a larger set with each increase in N. Second, the expectation, Np, is larger for each increase in N for a fixed p, and the standard deviation $\sqrt{Npq}$ also increases with N. Finally, the probability associated with any given exact value of X tends to *decrease* with *increase* in N.

However, suppose that for any value of N the number of successes X were put into standardized form:

$$z = \frac{x - Np}{\sqrt{Npq}} = \frac{x - E(X)}{\sigma}.$$

Then regardless of the size of N, the mean of the z scores for a binomial distribution will be 0, and the standard deviation 1. The probability of any z score is the same as for the corresponding x. Any z score interval can be given a probability in the particular binomial distribution, and the same interval can also be given a probability for a normal distribution. We can compare these two probabilities for any interval; if the two probabilities are quite close over all intervals, the normal distribution gives a good approximation to the binomial probabilities. On making this comparison of binomial and normal interval probabilities, we should find that the larger the sample size N the better is the fit between the two kinds of probabilities. **As N grows infinitely large, the normal and binomial probabilities become identical for any interval.**

For example, imagine that $p = .5$, and $N = 5$. Here the expectation is $(5)(.5)$, or 2.5, and the standard deviation is $\sqrt{5(.5)(.5)} = \sqrt{1.25}$ or about 1.12. The binomial distribution for these values of p and N is given below:

x	$p(x)$
5	.0312
4	.1563
3	.3125
2	.3125
1	.1563
0	.0312

For the moment, let us pretend that this is actually a continuous distribution, and that the possible numbers of successes are midpoints of class intervals. Furthermore, let us turn each X into its standardized value z, by taking

$$z = \frac{x - E(X)}{\sigma} = \frac{x - 2.5}{1.12}.$$

Then the distribution is

z real limits	z midpoint	$p(x)$	Normal
(1.79 to 2.68)	2.23	.0312	.0367
(.89 to 1.79)	1.34	.1563	.1500
(.00 to .89)	.45	.3125	.3132
(−.89 to .00)	−.45	.3125	.3132
(−1.79 to −.89)	−1.34	.1563	.1500
(−2.68 to −1.79)	−2.23	.0312	.0367

Here, the z intervals were found by converting each real limit for the distribution of successes (regarded as continuous) into its equivalent z score: thus, $1.79 = (4.5 - 2.5)/1.12$, and so on. Note that the probabilities are the same for these intervals as for the original X values.

The last column gives the probabilities that these same intervals would have in a normal distribution, found by using Table I in Appendix C. Thus, the interval .89 to 1.79 has probability

$$F(1.79) - F(.89) = .96327 - .81327 = .1500$$

and the other probabilities are found in the same way. The top and bottom intervals are given the probabilities

$$1 - F(1.79) = .0367$$

and

$$F(-1.79) = .0367$$

in order to make the probabilities over all intervals sum to 1.00.

Observe how closely the normal probabilities approximate their binomial counterparts: each normal probability is correct to two decimal places as an approximation to a binomial probability. The difference in the two probabilities is smallest for the middle intervals, and is larger for the extremes. Thus, even when N is only 5, the normal distribution gives a respectable approximation to binomial probabilities for $p = .5$.

Now suppose that for this same example, N had been 15. Here the expectation is 7.5, and the standard deviation is $\sqrt{(15)(.25)} = 1.94$. The distribution, with z score intervals and both binomial and normal probabilities, is shown in Table 8.5.1.

Table 8.5.1

x	z intervals	$p(x)$	Normal probabilities
15	3.608 to 4.124	.00003	.0002
14	3.092 to 3.608	.0005	.0012
13	2.577 to 3.092	.0032	.0036
12	2.061 to 2.577	.0139	.0147
11	1.546 to 2.061	.0416	.0409
10	1.030 to 1.546	.0916	.0909
9	.515 to 1.030	.1527	.1501
8	.000 to .515	.1964	.1984
7	− .515 to .000	.1964	.1984
6	−1.030 to − .515	.1527	.1501
5	−1.546 to −1.030	.0916	.0909
4	−2.061 to −1.546	.0416	.0409
3	−2.577 to −2.061	.0139	.0147
2	−3.092 to −2.577	.0032	.0036
1	−3.608 to −3.092	.0005	.0012
0	−4.124 to −3.608	.00003	.0002

For this distribution the normal approximation gives an even better fit to the exact binomial probabilities; the average absolute difference in probability over the intervals is about .001, whereas when N was only 5, the average absolute difference was about .004. In general, as N is made larger, the fit between normal and binomial probabilities grows increasingly good. In the limit, when N approaches infinite size, the binomial probabilities are exactly the same as the normal probabilities for any interval, and thus one can say that the normal distribution is the limit to the binomial. This is true regardless of the value of p.

On the other hand, for any finite N, the more p departs from .5, the less well does the normal distribution approximate the binomial. When p is not exactly equal to .5, the binomial distribution is somewhat skewed for any finite sample size N, and for this reason the normal probabilities will tend to fit less well than for $p = .5$, which always gives a symmetric distribution. For example, notice that for $N = 5$ and $p = .3$, illustrated below, the correspondence is not so good as before:

x	$p(x)$	Normal probabilities
5	.00243	.0017
4	.02835	.0239
3	.13230	.1379
2	.30870	.3365
1	.36015	.3365
0	.16807	.1635

Here the normal probabilities are still reasonably close to the binomial, but the fit is not as good as for $p = .5$.

Given a sufficiently large N, normal probabilities may always be used to approximate binomial probabilities irrespective of the value of p, the true probability of a success. The more that p departs in either direction from .5, the less accurate is this approximation for any given N. In practical situations where the sampling distribution is binomial, the normal approximation may ordinarily be used safely if the *smaller* of Np, the number of successes expected, or Nq, the expected number of failures, is 10 or more. Otherwise, binomial probabilities can either be calculated directly or found from tables that are readily available. Obviously, this rule requires a larger sample size the smaller the value of either p or q, and if one must use the normal approximation to make an inference about the value of p, he is wise to plan on a relatively large sample.

In a more advanced study of the normal distribution we would show that as N becomes infinite with fixed p the mathematical rule for a binomial distribution becomes identical with the rule for a normal density function, with $\mu = Np$ and $\sigma = \sqrt{Npq}$. Unfortunately, this sort of demonstration is far beyond our managing at this point.

In any practical use of the normal approximation to binomial probabilities, it is important to remember that here we regard the binomial distribution as though it were continuous, and actually find the normal probability associated with an interval with real lower limit of $(X - .5)$ and real upper limit of $(X + .5)$, where X is any given number of successes. Thus, in order to find the normal approximation to the binomial probability of X, we take

$$\text{prob.} \left(\frac{x - Np - .5}{\sqrt{Npq}} \le z \le \frac{x - Np + .5}{\sqrt{Npq}} \right)$$

by use of the standardized normal tables. Similarly, in terms of sample $P = X/N$, we have

$$\text{prob.} \left(\frac{P - p - .5/N}{\sqrt{pq/N}} \le z \le \frac{P - p + .5/N}{\sqrt{pq/N}} \right).$$

This adjustment, by which the probability of an interval of values is taken in place of the exact value of X or P, gives rise to the so-called **correction for continuity**. That is, in later sections when the normal distribution is to be used to approximate a binomial probability, we will deal with the z value

$$z = \frac{x - Np - .5}{Npq}$$

when X is larger than Np, and with

$$z = \frac{x - Np + .5}{Npq}$$

when X is less than Np. This will allow for the fact that the normal distribution is being used in order to approximate the probability in a discrete distribution. Only when the sample size N is relatively large does this correction become unimportant enough to ignore.

As noted above, the normal distribution need not be particularly good as an approximation to the binomial if either p or q is quite small, and if either p or q is extremely small the normal approximation may not be satisfactory even when N is quite large. In these circumstances another theoretical distribution, the Poisson, provides a better approximation to binomial probabilities. This approximation of binomial probabilities by use of Poisson probabilities was outlined in Chapter 5.

8.6 THE THEORY OF THE NORMAL DISTRIBUTION OF ERROR

The fact that the limiting form of the binomial is the normal distribution actually provides a rationale for thinking of random error as distributed in this normal way. Consider an object measured over and over again independently and in exactly the same way. Imagine that the value Y obtained on any occasion is a sum of two independent parts:

$$Y = T + e.$$

That is, the obtained score Y is a sum of a constant *true* part T plus a random and independent *error* component. However, the error portion can also be thought of as a sum:

$$e = g[e_1 + e_2 + e_3 + \cdots + e_N]$$

Here, e_1 is a random variable that can take on only two values

$$e_1 = 1, \text{ when factor 1 is operating}$$
$$e_1 = -1, \text{ when factor 1 is not operating.}$$

The g is merely a constant, reflecting the "weight" of the error in Y. Similarly, the other random errors are attributable to different factors, and take on only the values 1 and -1. Now imagine a vast number of influences at work at the moment of any measurement. Each of these factors operates independently of each of the others, and whether any factor exerts an influence at any given moment is purely a chance matter. If you want to be a little anthropomorphic in your thinking about this, visualize old Dame Fortune tossing a vast number of coins on any occasion, and from the result of each deciding on the pattern of factors that will operate. When coin one comes up heads, e_1 gets value 1, and if tails, value -1, and the same principle determines the value for every other error portion of the observed score.

Now under this conception, the number of factors operating at the moment one observes Y is a number of successes in N independent trials of a Bernoulli experiment. If this number of successes is X, then

$$e = gX - g[N - X] = g[2X - N].$$

The value of e is exactly determined by X, the number of "influences" in operation at the moment, and the probability associated with any value of e must be the same as for the corresponding value of X. **If N is very large, then the distribution of e must approach a normal distribution.** Furthermore, if any factor is equally likely to operate or not operate at a given moment, so that $p = 1/2$, then

$$E(e) = g[2E(X) - N],$$

so that, since $E(X) = Np = N/2$,

$$E(e) = 0.$$

In the long run, over all possible measurement occasions the errors all "cancel out." This makes it true that

$$E(Y) = E(T) + E(e) = T,$$

the long-run expectation of a measurement Y is the true value T, provided that error really behaves in this random way as an additive component of any score.

This is a highly simplified version of the argument for the normal distribution of errors in measurement. Much more sophisticated rationales can be invented, but they all partake of this general idea. Moreover, the same kind of reasoning is sometimes used to explain why distributions of natural traits, such as height, weight, and size of head, follow a more or less normal rule. Here, the mean of some population is thought of as the "true" value, or the "norm." However, associated with each individual is some departure from the norm, or error, representing the culmination of all the billions of chance factors that operate on him, quite independently of other individuals. Then by regarding these factors as generating a binomial distribution, we can deduce that the whole population should take on a form like the hypothetical normal distribution. However, this is only a theory about how errors might operate and there is no reason at all why errors must behave in the simple additive way assumed here. If they do not, then the distribution need not be normal in form at all.

8.7   STATISTICAL PROPERTIES OF NORMAL POPULATION DISTRIBUTIONS

As suggested earlier in this chapter, the normal distribution has mathematical properties that are most important for theoretical statistics. For the moment, we are going to discuss only two of these general properties, both of which will be useful to know in later sections. The first has to do with the independence of the sample mean and variance, and the second concerns the distribution of combinations of random variables.

8.8   INDEPENDENCE OF SAMPLE MEAN AND VARIANCE

Any sample consisting of N independent observations of the same random variable provides both a sample mean M and a sample variance. The sample mean estimates μ and the unbiased sample variance, s^2, estimates σ^2. These two values (M, s^2) obtained from any sample can be thought of as a joint event. But are these two estimates independent? That is, is the conditional distribution of s^2 given M the same as the marginal distribution of s^2, and is the distribution of M given s^2 like the marginal distribution of M?

The answer to these questions is provided by the following important principle:

Given random and independent observations, the sample mean M and the sample variance (either S^2 or s^2) are independent if and only if the population distribution is normal.

The information contained in the sample mean in no way dictates the value of the sample variance, and vice-versa, when a normal population is sampled. Furthermore, **unless the population actually is normal, these two sample statistics are not independent across samples.**

This is a most important principle, since a great many problems concerned with a population mean can be solved only if one knows something about the value of the population variance. At least an estimate of the population variance is required before particular sorts of inferences about the value of μ can be made. Unless the estimate of the population variance is statistically independent of the estimate of μ made from the sample, no simple way to make these inferences may exist. This question will be considered in more detail in Chapters 10 and 11. For the moment, suffice it to say that this principle is one of the main reasons for statisticians' assuming normal population distributions.

8.9   DISTRIBUTIONS OF LINEAR COMBINATIONS OF SCORES

So far we have emphasized the sample mean as a description of a particular set of data, and as an estimator of the corresponding population mean.

However, in many situations the sample mean and the mean of the population do not really tell the experimenter what he wishes to find out from the data. It well may be that other ways of weighting and summing the scores or means obtained in one or more samples will answer particular questions about the phenomena under study. Therefore, we need to know something about the sampling distributions of weighted combinations of sample data.

Such weighted sums of sample scores are thought of as values of **linear combinations of random variables.** As a very simple case of a linear combination, imagine two distinct random variables, labeled X_1 and X_2 to show that their values need not be the same. We draw samples of two observations at a time, obtaining one value of X_1 and one of X_2. Then we combine these two numbers into a new value Y in some way such as

$$Y = 3X_1 - 2X_2.$$

We continue to sample X_1, X_2 pairs of values, and each time we combine the results in the same way, turning each pair of values into a single combined value. This gives a new random variable Y. The range of possible values of Y depends, as you can see, on the ranges of X_1 and X_2. Furthermore, over all possible such samples the probabilities of the various Y values must depend on the *joint* probabilities of (X_1, X_2) pairs. Over all possible samples of (X_1, X_2) pairs drawn at random there is some probability distribution of Y.

In general, given any n random variables, $(X_1, \cdots, X_n)$, a *linear combination* of values of those variables is a weighted sum,

$$Y = c_1X_1 + c_2X_2 + \cdots + c_nX_n, \qquad [8.9.1*]$$

where $(c_1, \cdots, c_n)$ is any set of n real numbers (not all zero) used as weights.

In short, each sample produces a value for each of n random variables. These are weighted and summed in the same way for each sample to produce a new random variable Y, the linear combination.

The weights in a linear combination can be any set of n real numbers, provided that at least one weight is not zero. For example, for two random variables, a linear combination might be

$$Y = X_1 + X_2$$

where $c_1 = 1$ and $c_2 = 1$; or

$$Y = 7X_1 - 10X_2$$

where $c_1 = 7$ and $c_2 = -10$; or

$$Y = -4/5X_1 + 7/8X_2,$$

with $c_1 = -4/5$, $c_2 = 7/8$, and so on.

Perhaps values for six random variables are represented in a sample, and the linear combination chosen is

$$Y = 5X_1 - 3X_2 + 7X_3 + 3/4X_4 - .8X_5 + 0X_6.$$

Here, $(c_1, \cdots, c_6) = (5, -3, 7, 3/4, -.8, 0)$.

One very familiar linear combination of random variables is the mean of a random sample: given that X_1 is the score of the first case, X_2 that of the second, and so on, then

$$M = \frac{1}{N} X_1 + \frac{1}{N} X_2 + \cdots + \frac{1}{N} X_N,$$

where each value obtained is weighted by $1/N$ and the sum taken. This linear combination is also a random variable, and its distribution is the sampling distribution of the mean.

Now for the important principle involving the normal distribution:

Given some n independent random variables $X_1, \cdots, X_n$, each normally distributed, then any linear combination of these variables is also a normally distributed random variable.

In other words, if one takes samples, obtaining n independent values in each, and if the random variable represented by each value has a normal distribution, then any weighted sum of those values gives another normally distributed random variable. Thus, if X_1 and X_2 are independent and have normal distributions, then so does the variable Y, whose values are

$$Y = 3X_1 - 2X_2,$$

or any other variable obtained by weighting and summing X_1 and X_2 values in some constant way over all samples.

An immediate and useful consequence of this principle is this:

Given random samples of N independent observations each drawn from a normal population, then the distribution of the sample means is normal, irrespective of the size of N.

We have just seen that the sample mean of N cases is a linear combination, where the score of case 1, representing random variable 1, is given the weight $1/N$, and so on for all other scores. Hence, if the scores of the N individuals are independent of each other and are drawn from a normal population, the sampling distribution of the mean is exactly normal, regardless of how large or small N may be. As we saw in the last chapter, the standard error of this distribution, σ_M, will be smaller the larger the N, and so the particular probabilities of various sample values will differ with N. On the other hand, the distribution *rule*, given μ and σ_M, will be that for the normal function.

We are going to make a great deal of use of this principle. When we can assume that a population with known μ and σ has a normal distribution a major problem in statistical inference is solved, since we can then give a probability to a sample mean's falling into any interval also in terms of a normal distribution. This will be illustrated both at the end of this chapter, and in the next.

8.10 MEANS AND VARIANCES OF LINEAR COMBINATIONS

While on the subject of linear combinations, we should look into the question of the mean and variance of the sampling distribution of a linear combination of random variables. These two principles do not depend upon the normal distribution, however. **Given n random variables, and some linear combination,**

$$Y = c_1 X_1 + \cdots + c_n X_n,$$

the expected value of Y over all random samples is

$$E(Y) = c_1 E(X_1) + c_2 E(X_2) + \cdots + c_n E(X_n). \qquad [8.10.1*]$$

The expectation of any linear combination is the same linear combination of the expectations.

This really follows quite simply from rules 5 and 2 in Appendix B, since the expectations of a sum is always a sum of expectations, and the expectation of a constant times a variable is the constant times the expectation.

A special case of this principle gives us the following:

Given n sample values from the same distribution with mean μ, then the expectation of any linear combination of those sample values is

$$E(Y) = \mu(c_1 + c_2 + \cdots + c_n). \qquad [8.10.2*]$$

This principle accords with our earlier result that the expectation of the sample mean is the population mean:

$$E(M) = E\left(\frac{1}{N} X_1 + \cdots + \frac{1}{N} X_N\right) = \mu\left(\frac{1}{N} + \frac{1}{N} + \cdots + \frac{1}{N}\right) = \mu.$$

However, suppose that samples of two cases were drawn, and the linear combination were

$$Y = X_1 - X_2.$$

In this instance it would not be true that the expectation of this combination equals μ; instead we would have $E(Y) = \mu(1 - 1) = 0$.

As a concrete example, suppose that samples of two cases were observed, one case being a male drawn from the population of all males, with $\mu = 25$, and the other a female drawn from the female population, with $\mu = 27$. The score of the male is represented by X_1 and that of the female by X_2. For each sample, the difference between the male's and the female's scores is taken:

$$Y = X_1 - X_2.$$

Then, over all such samples, the expectation is

$$E(Y) = E(X_1) - E(X_2) = \mu_1 - \mu_2 = -2.$$

Furthermore, if the scores of the two cases are independent, and if both populations have a normal distribution, the sample difference Y is normally distributed as well.

The variance of the sampling distribution of any linear combination can also be found by the following rule:

Given n independent random variables, with variances $\sigma_1^2, \sigma_2^2, \cdots, \sigma_n^2$ respectively, and the linear combination

$$Y = c_1 X_1 + c_2 X_2 + \cdots + c_n X_n,$$

then the distribution of Y has variance given by

$$\sigma_Y^2 = c_1^2 \sigma_1^2 + c_2^2 \sigma_2^2 + \cdots + c_n^2 \sigma_n^2. \qquad [8.10.3*]$$

The variance of a linear combination of independent random variables is a weighted sum of their separate variances, each weight being the *square* of the original weight given the variable.

Thus, in the preceding example, if the population of males has a variance of 10, and that of females a variance of 12, the variance of the difference, $X_1 - X_2$, is

$$\sigma_Y^2 = (1)^2 \sigma_1^2 + (-1)^2 \sigma_2^2 = 10 + 12 = 22.$$

As a more complicated example, suppose that three independent observations at a time are made, drawn respectively from populations represented by the three random variables

$$X_1, \text{ with } \mu_1 = 20, \sigma_1^2 = 5$$
$$X_2, \text{ with } \mu_2 = 16, \sigma_2^2 = 9$$
$$X_3, \text{ with } \mu_3 = 25, \sigma_3^2 = 7.$$

If all three random variables are normally distributed, what will the distribution of the linear combination

$$Y = \frac{X_1}{2} + \frac{X_2}{2} - X_3$$

be like? In the first place, we know from the principle of Section 8.9 that Y will be normally distributed. Second, the mean of the distribution of Y will be

$$E(Y) = \frac{\mu_1}{2} + \frac{\mu_2}{2} - \mu_3 = 10 + 8 - 25 = -7.$$

The variance of Y must be

$$\sigma_Y^2 = \left(\frac{1}{2}\right)^2 \sigma_1^2 + \left(\frac{1}{2}\right)^2 \sigma_2^2 + (-1)^2 \sigma_3^2$$

$$= \frac{5 + 9}{4} + 7 = 10.5.$$

Under these conditions, the three principles of Sections 8.9 and 8.10 allow us to specify the distribution of Y completely, and we could proceed to evaluate the probability that an obtained value of Y falls into any given interval.

The variance of the sampling distribution of the mean, found in Section 7.12 to be σ^2/N, can also be found by using principle **8.10.3**. Here, samples of N independent observations are made from the same distribution, with some true mean μ and with some true variance σ^2. The sample mean itself is, as we have seen, a linear combination of the scores, each weighted by $1/N$. Then, by the application of the principle,

$$\sigma_M^2 = \left(\frac{1}{N}\right)^2 \sigma^2 + \left(\frac{1}{N}\right)^2 \sigma^2 + \cdots + \left(\frac{1}{N}\right)^2 \sigma^2$$
$$= \frac{N}{N^2}\sigma^2 = \frac{\sigma^2}{N}.$$

You are probably wondering why anyone would be interested in linear combinations other than the sample mean. Actually, as we shall see in Chapter 14, forming linear combinations of the data is one way of answering particular questions about what goes on in an experiment based on two or more samples. Each item of information about different experimental treatments that the experimenter gains from his data usually corresponds to one way of weighting and summing means of the various samples. The linear combination applied to the set of sample means gives an estimate of the same combination applied to the corresponding set of population means. Thus, the study of the sampling distribution of linear combinations forms an important part of the theory of sampling.

The most important point that should emerge from this discussion of linear combinations is that, given normal population distributions, the sampling distribution of the mean or of any other linear combination of independent values or of independent sample means must be normal. Thus, the assumption of a normal distribution for the population settles the question of the exact form of the sampling distribution for any linear combination of independent values in the sample data, including the mean. This is true quite irrespective of the sample size. Do not lose sight of the fact that the size of the sample does, however, determine how small the variance of the sampling distribution will be.

8.11   THE CENTRAL LIMIT THEOREM

It is quite common for the psychologist to be concerned with populations where the distribution should *definitely not be normal*. We may know this either from empirical evidence about the distribution, or because some theoretical issue makes it impossible for scores to be this kind of random variable. Illustrations are distributions of intelligence scores among graduate engineers, and of measures of socially nonconforming behavior among normal adults. In principle both these distributions should be extremely skewed, though for quite different reasons: in the first, there is a selection in the education process making it rather unlikely for anyone in the middle or low intelligence ranges to qualify as an engineer; in the second, social nonconformity is something that most people must show in very small degree. The assumption of a normal distribution does not make sense in such instances.

Nonetheless, very often an inference must be made about the mean of such a population. To do this effectively the experimenter needs to know the sampling distribution of the mean, and to know this exactly, he has to be able to specify the particular form of the population distribution. However, if we had enough evidence to permit this, we would likely have an extremely good estimate of the population mean in the first place, and would not need any other statistical methods!

The way out of this apparent impasse is provided by the **central limit theorem,** which can be given an approximate statement as follows:

If a population has a finite variance σ^2 and mean μ, then the distribution of sample means from samples of N independent observations approaches a normal distribution with variance σ^2/N and mean μ as the sample size N increases. When N is very large, the sampling distribution of M is approximately normal.

Absolutely nothing is said in this theorem about the form of the population distribution. Regardless of the population distribution, if sample size N is large enough, the normal distribution is a good approximation to the sampling distribution of the mean. This is the heart of the theorem, since, as we have already seen, the sampling distribution will have a variance σ^2/N and mean μ for any N.

The sense of the central limit theorem is illustrated by Figures 8.11.1–4. The solid curve in Figure 8.11.1 is a very skewed population distribution in z-score form, and the dotted curve is a standardized normal distribution. The

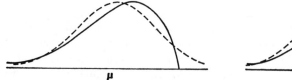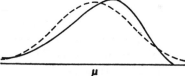

A negatively skewed population distribution (Fig. 8.11.1, left) and a sampling distribution of M for N = 2 (Fig. 8.11.2, right), with comparable normal distributions

other figures show the standardized form of the sampling distribution of means for samples of size 2, 4, and 10 respectively from this population, together with the corresponding standardized normal distributions. Notice that it is *not* the number of samples drawn, but rather the *size* of each sample, that is the effective part of the central limit theorem. A vast number of samples of any size N may be drawn; the larger N is for each sample, the more nearly is the distribution of means normal.

Even with relatively small sample sizes it is obvious that each increase in sample size gives a sampling distribution more nearly symmetric and tending more toward the normal distribution with the same mean and variance.

This symmetry increases with increasing sample size until, in the limit, the normal distribution is reached.

Sampling distributions of M for N = 4 (Fig. 8.11.3, left) and N = 10 (Fig. 8.11.4, right), with comparable normal distributions

It must be emphasized that in most instances the tendency for the sampling distribution of M to be like the normal distribution is very strong, even for samples of moderate size. Naturally, the more similar to a normal distribution the original population distribution, the more nearly will the sampling distribution of M be like the normal distribution for any given sample size. However, even extremely skewed or other nonnormal distributions may yield sampling distributions of M that can be approximated quite well by the normal distribution, for samples of at least moderate size. In the examples shown in Figures 8.11.1–4 the correspondence between the exact probabilities of intervals in the sampling distribution and intervals in the normal distribution is fairly good even for samples of only 10 observations; for a rough approximation, the normal distribution probabilities might be useful even here in some statistical work. In a great many instances in psychological research, a sample size of 30 or more is considered large enough to permit a satisfactory use of normal probabilities to approximate the unknown exact probabilities associated with the sampling distribution of M. Thus, even though the central limit theorem is actually a statement about what happens *in the limit* as N approaches an infinite value, the principle at work is so strong that in many instances the theorem is practically useful even for moderately large samples.

The mathematical proof of this theorem is extremely advanced, and many eminent mathematicians over the centuries contributed to its development before it was finally proved in full generality. However, some intuitive feel for why it should be true can be gained from the following example. Here we will actually work out the sampling distribution of the mean for a special and very simple population distribution.

Imagine a random variable with the following distribution:

x	$p(x)$
5	1/3
4	1/3
3	1/3

The μ of this little distribution is 4. However, instead of dealing directly with this random variable, let us consider the deviation d from μ, or

$$d = (X - \mu).$$

The distribution of d has a form identical to the distribution of X itself, although the mean d is zero. Since X can assume only three values, each equally probable, there are three possible deviation values of d, also equally probable.

Suppose that sample observations were taken two at a time, independently and at random. The deviation value of the first observation made is d_1 and that of the second d_2. Corresponding to each joint event (d_1, d_2) there is a mean deviation, $(d_1 + d_2)/2$, which is equal to $M - \mu$. The following table shows the mean deviation value that is produced by each possible joint event:

d_2			
1	0	.5	1
0	$-.5$	0	5
-1	-1	$-.5$	0
	-1	0	1 d_1

Since X_1 and X_2 are independent, then d_1 and d_2 are also independent, according to Section 4.9. The probability associated with each cell in the table is thus $p(d_1)p(d_2)$ or $1/9$.

Now let us find the probabilities of the various values of mean d. The value -1 can occur in only one way, and so its probability is $1/9$. On the other hand, $-.5$ can occur in two mutually exclusive ways, giving a probability of $2/9$. In this same way we can find the other probabilities, and form the distribution of the mean d and the mean X values, as follows:

M	Mean d	$p(M) = p(mean\ d)$
5.0	1.0	1/9
4.5	0.5	2/9
4.0	0	3/9
3.5	-0.5	2/9
3.0	-1.0	1/9

where M = mean $d + \mu$. Now notice that whereas in the original distribution there was no distinct mode, in this sampling distribution there *is* a distinct mode at 4, with exact symmetry about this point. What causes the sampling distribution to differ from the population distribution in this way? Look at the table of mean deviations corresponding to joint events once again: there are simply more possible ways for a sample to show a small than an extreme deviation from μ. There are more ways for a sample joint event to occur where the deviations tend to "cancel out" than where deviations tend to cumulate in the same direction.

Exactly the same idea could be illustrated for any discrete population distribution, whatever its form. The small and "middling" deviations of M

from μ always have a numerical advantage over the more extreme deviations. This superiority in number of possibilities for a small mean deviation increases as N is made larger. Regardless of how skewed or otherwise irregular a population distribution is, by increasing sample size one can make the advantage given to small deviations so big that it will overcome any initial advantage given to extreme deviations by the original population distribution. The numerical advantage given to relatively small deviations from μ effectively swamps any initial advantage given to other deviations by the form of the population distribution, as N becomes large.

Perhaps this simple example gives you some feel for why it is that the sampling distribution of the mean approaches a unimodal, symmetric form. Of course, this is a far cry from showing that the sampling distribution must approach a *normal* distribution, which is the heart of the central limit theorem. Nevertheless, the basic operation of chance embodied in this theorem is of this general nature: for large N it is relatively much easier to get a small deviation of M from μ than a large one.

8.12   THE CENTRAL LIMIT THEOREM AND LINEAR COMBINATIONS OF MEANS

As suggested above, it is very common for an experimenter to be interested in the means of several samples put into some kind of weighted sum, or linear combination. The central limit theorem applies to such weighted combinations as well:

Given K independent samples, containing N_1, N_2, $\cdots$, N_K independent observations respectively, then the sampling distribution of any linear combination of means of those samples approaches a normal distribution as the size of each sample grows large.

It is easy to see that the central limit theorem applies to the sampling distribution of each mean separately as the sample size increases. In the limit, each mean has a normal sampling distribution. Then by the principle described in Section 8.9, any linear combination of normal random variables is normal so that the sampling distribution of the linear combination of means approaches a normal distribution as the sample sizes all grow large.

This is but one of the most elementary extensions of the central limit theorem. It can be shown that a great many of the sampling distributions of all sorts of sample characteristics also approach a normal distribution with increasing sample size. However, in elementary work we will have most occasions to use the principle as it applies to means.

8.13   USING THE NORMAL DISTRIBUTION IN INFERENCES ABOUT MEANS

We have seen that there are two circumstances in which the sampling distribution of the mean can be considered normal: when the population distribution itself is normal, and when the population is not normally distributed but the sample size is large. In either case, the normal tables can be used to find the probability that the sample mean will fall into any interval, given μ and σ_M.

For an example of the first situation, suppose that samples of size 25 were drawn independently and at random from a normal population, so that the sampling distribution of the mean is *exactly* normal. Given that $\mu = 138$ and $\sigma = 20$ for the population, what is the probability of observing a sample mean in the interval 136–140? To find this probability, one needs, first of all, σ_M, given by

$$\sigma_M = \frac{\sigma}{\sqrt{N}} = \frac{20}{5} = 4.$$

The standardized score corresponding to the *lower limit* of the interval is

$$z = \frac{136 - 138}{4} = -.5$$

and for the *upper limit*,

$$z = \frac{140 - 138}{4} = .5.$$

The probability of the interval is, from Table I,

$$p(136 \leq M \leq 140) = F(.5) - F(-.5) = .6915 - .3085$$
$$= .3830.$$

Only about thirty-eight times in one hundred random samples should one expect to observe a sample mean in this particular interval if $\mu = 138$, $\sigma = 20$, and $N = 25$.

On the other hand, we might have the situation where the population distribution is not normal, but the sample size is relatively quite large, say 900. Here we could be sure that the normal approximation to the exact probabilities is very good. In this instance, suppose that we have a hunch or hypothesis that the value of μ is 35.7 with $\sigma = 1.3$. If these parameter values are true, what is the probability that an observed sample mean should exceed 35.8? Since N is so large, σ_M is quite small,

$$\sigma_M = \frac{1.3}{30} = .043.$$

The value 35.8 thus corresponds to

$$z = \frac{35.8 - 35.7}{.043} = \frac{.1}{.043} = 2.33.$$

What we want is the area on the upper tail of the normal curve cut off by a z of 2.33. This area or probability is

$$p(2.33 < z) = 1 - F(2.33) = 1 - .9901 = .0099$$

or about .01. Thus, only about one time in one hundred samples with $N = 900$ should we expect to observe a sample M value this deviant from expectation, if the expected value is 35.7 and $\sigma = 1.3$. Granted that our hypothetical expectation and standard error are both correct and we do, nevertheless, observe the value exceeding 35.8, then we have encountered a value qualifying for a "rare" interval of events. On the other hand, if μ were, say, 35.78, a sample mean exceeding 35.80 could hardly be classed as unusual. In this instance,

$$z = \frac{35.80 - 35.78}{.043} = .465, \text{ or about } .47.$$

From the table we find that

$$F(.47) = .6808$$

and so

$$p(35.80 \leq M) = 1 - F(.47) = .32, \text{ approximately.}$$

Deviations of M from expectation this big or bigger should occur in about 32 percent of all such samples.

At this point it should be reiterated that a large sample, in the sense of the central limit theorem, does not necessarily mean a *vast* sample. The theorem deals with the limit when N is infinite, but the normal approximation is usually good long before that point is reached. When the population distribution is unimodal and symmetric, then even quite small samples (10 cases or so) afford a fairly good fit between the exact sampling distribution and the normal approximation. Obviously, the more skewed or otherwise irregular the population distribution, the poorer the fit will be, and the relatively larger the sample size must be to permit use of the normal approximation. On the other hand, unless the population distribution is quite skewed, a sample size of some thirty or more cases is almost always adequate to permit the use of normal probabilities in the work we will be doing. Be sure to notice that even the normal probabilities cannot be found unless one specifies both μ and σ for the population.

8.14   THE GAMMA AND BETA FAMILIES
OF DISTRIBUTIONS

Although, as we have seen, the normal distribution is of extreme importance in both theoretical and applied statistics, many other families of continuous distributions also are of use. These may be quite different from the normal distribution in terms of the range of the random variable, or in terms of the form of its distribution. Here we will mention only two of these additional families of continuous distributions. These are known as the **gamma** and the **beta** distributions.

In order to introduce the gamma distribution, let us consider a random variable X ranging over the positive real numbers and zero, so that X can take on any value $x \geq 0$. Then the random variable is said to have a gamma distribution if the probability density at any value $X = x$ is given by the rule

$$f(x;m, r) = \begin{cases} \dfrac{e^{-mx}(mx)^{r-1}m}{(r-1)!} & x \geq 0, m > 0, r > 0, \\ 0 \cdot & \text{elsewhere.} \end{cases}$$ [8.14.1]

Here the parameters r and m are *positive* constants. When r is a whole number, then $(r-1)!$ is defined in the usual way: $(r-1)! = (1)(2)(3) \cdots (r-1)$. However, when r is not a whole number, $(r-1)!$ is evaluated by the so-called "gamma function," studied in the advanced calculus:

$$\Gamma(r) = (r-1)! = \int_{z=0}^{\infty} e^{-z} z^{r-1}.$$

This function, which has been extensively studied and tabled, gives its name to our present probability density function. When r is an integer, then $\Gamma(r)$ is simply $(r-1)!$. The value of $\Gamma(r)$ is evaluated by the definite integral when r is not an integer.

The mean and the variance of a gamma variable both depend upon the parameters r and m:

$$E(X) = \frac{r}{m}, \qquad \text{Var}(X) = \frac{r}{m^2}.$$

In form, a gamma density function is unimodal, and skewed to the right, with the "short" tail of the distribution in the direction of zero, and a long tail of steadily decreasing density values trailing away indefinitely toward the higher reaches of X. For $r = 1$ the mode is at 0, but for higher values of r the mode moves steadily away from zero; for very large values of r, the gamma distribution begins to approach a symmetric form. In fact, for large r a reasonably good approximation to the cumulative probability for any value $X = x$, given r and m, may be found by taking the statistic

$$u = \sqrt[3]{r}\left(\sqrt{\frac{mx}{r}} - \frac{9r-1}{9r}\right)$$

and then treating this statistic exactly as though it were a standardized z value in a normal distribution. A method for finding cumulative probabilities for gamma variables with smaller values of r will be given a little later.

If you recall the Poisson distribution discussed in Chapter 5, you should already have been struck by the similarity between the rule for that discrete distribution and the continuous gamma distribution function. There is a most important difference between these two function rules, however. In the Poisson case the intensity m is the only parameter, and the number of successes r is the random variable. Here, in the gamma case, r is a parameter and is fixed, and the place formerly occupied by the parameter m is now taken by the random variable mx. Even so, the relationship between the two types of distributions is very close, and this will be exemplified by the following.

Suppose that we are observing a stable and independent Poisson process that produces successes at the average rate of m per unit of time (or other continuous dimension). Rather than taking a fixed amount of time and noting how many successes occur, as formerly, we now set a fixed number of successes, r, and record how much time elapses until r successes have occurred. This extent of elapsed time (or other dimension) is the random variable X. Since we assume that it takes no time at all for an event to occur, the amount of elapsed time x may be as small as 0, or x may be indefinitely large. The parameter m is simply the intensity of the process for a single unit of time, and X is the total number of such units elapsed until r successes have occurred.

Strong analogies exist between the gamma distribution and the Pascal and negative binomial distributions. Recall that when samples are drawn from a stable and independent Bernoulli process, the number of trials required for the rth success to occur follows the Pascal distribution rule. Similarly, the number of failures preceding the rth success follows the negative binomial rule, which is closely related to the Pascal rule. If one thinks of the underlying process as Poisson, where, over time, successes occur with a given intensity m, any instant without a success is like a failure. Then the total elapsed time before the rth success is analogous to the total number of failures to the rth success, and also analogous to the total number of trials before the rth success. On the other hand, variables such as total number of trials or number of failures before the rth success are discrete, even though the set of their possible values is countably infinite. A variable such as elapsed time before the rth success must necessarily be continuous, and extend over all the nonnegative real numbers.

As an example, the staff of the obstetrics unit in a certain hospital believes that triplets tend to occur with an intensity of about one set every four years, or .25 sets per year. If we begin keeping records for this hospital at some given point in time, what is the distribution of time that might pass before the first set of triplets is born? If we can assume that triplet events occur at this hospital in accordance with a Poisson process, then the distributions of time to the first set of triplets should follow a gamma law. If X is the random variable, "Time elapsed up to the first set of triplets," then

$$f(x;\ .25,\ 1)\ =\ \frac{e^{-.25x}(.25x)^0(.25)}{0!},$$

since here $r = 1$ and $m = .25$. This distribution will have a mean of $(r/m) = 1/.25$ or 4 years, and a variance of r/m^2 or $1/.0625$. If we are interested in the elapsed time until the third set of triplets is born, then

$$f(x;\ .25,\ 3)\ =\ \frac{e^{-.25x}(.25x)^{3-1}(.25)}{(3-1)!},$$

and so on for any other r value of interest.

As another example, suppose that we are concerned with testing the wearing power of automobile tires of a particular kind under rough running conditions. Each tire is run separately under standard and constant conditions until a blowout occurs. The experiment is such that we may assume that blowouts occur according to a stable and independent Poisson process. Here, over the continuous miles run (equivalent in this instance to continuous time run), blowouts can be expected to occur with some given intensity m. Now suppose that we take four tires, and run each tire separately until it blows out. Then we will let the random variable X be the total miles elapsed, summed over all four tires, until all four tires have blown out. We assume that the *same* stable and independent Poisson process underlies the occurrence of blowouts for each tire. For our argument's sake, let us say that for tires of this type and these running conditions, blowouts are believed to occur on the average at 10,000 miles, so that the intensity of the occurrence of blowouts on a per mile basis is 1/10000 or .0001. Then, for this experiment, $r = 4$ and $m = .0001$, and the probability density for any number of miles elapsed x is given by

$$f(x;.0001, 4) = \frac{e^{-(.0001)x}(.0001x)^{4-1}(.0001)}{(4 - 1)!}.$$

Remember that x is the total number of miles run by the four tires separately, r is the number of successes (blowouts), and m is the intensity of occurrence of blowouts on a per-mile basis. We also know that the expected number of miles before four blowouts is

$$E(X) = \frac{r}{m} = \frac{4}{.0001} = 40,000$$

and the variance is

$$\text{Var}(X) = \frac{r}{m^2} = \frac{4}{(.0001)^2} = (20,000)^2.$$

In many examples we are interested in the time or other dimension elapsed only up to the point of the first success. This would have been the case in the example above if we had been testing only one tire. When the parameter r in a gamma distribution is set at 1, so that the amount of time, distance, etc. up to the first success is the random variable, the distribution is often known as the **exponential distribution**. The rule now becomes

$$f(x; m, 1) = \begin{cases} e^{-mx}m, & x \geqq 0, m > 0, \\ 0, & \text{elsewhere.} \end{cases} \qquad \text{[8.14.2]}$$

One property of the gamma family of distributions is especially important in theoretical statistics: If X is distributed according to the gamma rule with parameters r and m, and Y is also distributed according to the gamma rule with parameters t and m, then if X and Y are independent, the sum $Z = X + Y$ is distributed according to the gamma rule with parameters $r + t$ and m. That is, for independent X and Y, with parameters r and m, and t and m respectively:

$$f(z = x + y; m, t + r) = \frac{e^{-mz}(mz)^{r+t-1}m}{(r + t - 1)!}. \qquad \text{[8.14.3]}$$

In other words, independent gamma variables are **additive** in the sense that the distribution of their sum also follows a gamma rule with parameters $r + t$ and m (provided that the parameter m is the same for both of the original variables). This principle also extends to three or more mutually independent gamma variables as well. In fact, the justification for the example of the tires just given lies in this principle. The time to blowout for each tire follows a gamma distribution with parameters m and 1; the distribution of the sum of the elapsed times over all four tires follows a gamma distribution with parameters $r = 4$ and m.

Although tables of the cumulative gamma function are available, we will not use this function enough to make their inclusion here worthwhile. Nevertheless, we can solve problems involving gamma distributions by use of a table in Appendix C. Since the gamma distribution is so closely related to the Poisson, it should not be especially surprising that this table is the same as that employed for cumulative Poisson probabilities: the upper percentage points of the χ^2 (chi-square) distribution. In order to find the value of a gamma variable cutting off the upper Q proportion of its distribution, given r and m, turn to Table IV of Appendix C and proceed as follows: Let $\nu = 2r$ and let $\chi^2 = 2mx$. Then look in the table under the appropriate Q value and find the corresponding chi-square value for the row labeled ν. Then simply substitute in the expression $\chi^2 = 2mx$ and solve for x. Thus, in our previous example involving the tires, we had $m = .0001$ and $r = 4$. Suppose that we wish to find the value of x cutting off the upper 5 percent of its distribution. To do this we take $\nu = 2r = 8$ and look in the table under $Q = .05$. This gives $\chi^2 = 15.5073$. Next, solving $2(.0001)x = 15.5073$ gives us $x = 77536.5$ miles. This means that if the intensity of occurrence of blowouts is such that on the average one occurs every 10000 miles, or $m = .0001$, then the probability is no more than .05 that a total of 77536.5 miles will elapse until all four tires have blown out. The value 77536.5 cuts off the upper 5 percent of the gamma distribution with $r = 4$ and $m = .0001$.

In this same way we can find the value of X cutting off, say, the lower 1 percent of the same distribution. Again letting $\nu = 2r$, we find the χ^2 value under the column $Q = .99$, which turns out to be 1.646482. Then $2mx = 1.646482$ yields the value $x = 8232$, approximately. This means that in a gamma distribution with $r = 4$ and $m = .0001$, 1 percent of x values lie at or below 8232. The same general method lends itself, of course, to other percentage points and other gamma distributions.

Much as the gamma distribution bears close affinities to the Poisson, the **beta** family of distributions follow a rule very similar in form to that for the binomial family. However, unlike the discrete binomial variable, the random variable X in a beta distribution is continuous, and can take on any value in the interval $0 \leq x \leq 1$. The standardized beta density function is defined by

$$f(x; N, r) = \begin{cases} \dfrac{(N - 1)!}{(r - 1)!(N - r - 1)!}\, x^{r-1}(1 - x)^{N-r-1}, & 0 \leq x \leq 1, 0 < r < N, \\ 0, & \text{elsewhere.} \end{cases}$$

$$[8.14.4]$$

Although the expression in factorials in this definition may appear at first glance to restrict r and N to integral values, it actually can be generalized to cover nonintegral values of r and N as well. In this case, one employs the so-called **beta function** studied in the advanced calculus and defined by

$$B(r, N - r) = \frac{(r - 1)!(N - r - 1)!}{(N - 1)!} = \int_{z=0}^{1} z^{r-1}(1 - z)^{N-r-1} \, dz.$$

This function, which has been extensively studied and tabled, gives the name to the present density function. The beta function equals the expression in factorials when r and N are integers. Otherwise, the value is given by the integral expression. For our immediate purposes the beta function can be thought of as the ratio of factorials given above, so long as we bear in mind that it is also defined for any r and N, where $0 < r < N$.

This beta density function involves two parameters, r and N, and the mean and variance of the random variable X depend upon the values of these parameters as follows:

$$E(X) = \frac{r}{N}, \qquad \mathrm{Var}(X) = \frac{r(N - r)}{N^2(N + 1)}.$$

There is, of course, a very striking similarity between the form of the rule for a beta density function and that for a binomial probability distribution. On the other hand, the random variable in a binomial distribution is r, the number of successes, which is discrete. In a beta distribution, r is a fixed parameter, and the random variable X plays the role held by p in a binomial distribution. The value of X can be any number in the interval between 0 and 1 inclusive, and thus X is continuous. In one sense, a beta density function corresponds to a binomial probability mass function when the number of successes is fixed and the probability of a success is allowed to vary. This feature of the beta distribution is important in the Bayesian approach to certain statistical problems, and we will use it in this way in Chapter 19.

On the other hand, beta distributions are important in theoretical and applied statistics quite apart from the analogies between the beta and the binomial rules; one such use will be illustrated a little later on.

The graphs of beta density functions may be strikingly different from each other, depending upon the values taken for the parameters r and N. Figure 8.14.1 illustrates the graphs produced by different members of the family

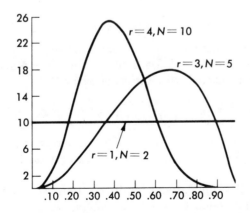

FIG. 8.14.1. Beta distributions with various values of r and N.

of beta distributions. The distribution is symmetric when $r = N/2$ and N is greater than 2. It becomes a uniform distribution when $r = 1$ and $N = 2$. If $r < N/2$, the distribution is positively skewed, and negatively skewed when $r > N/2$. The distribution can even become U-shaped for some values of $r < 1$ and $N < 2$. Truly the beta is the chameleon among theoretical density functions, since its form is so flexible, depending upon r and N. We will discuss some of these possibilities in more detail when we come to actual applications of beta distributions in later sections.

Although the beta family of distributions finds many different interpretations and uses in statistics, elementary examples of its use are not too easy to find. One use of the beta distribution depends on the following principle: If W is distributed as a gamma variable with parameters r and m_1, and Y is independently distributed as a gamma variable with parameters $N - r$ and m_2, then the random variable X defined by

$$x = \frac{w/y}{(w/y) + (m_2/m_1)}$$

follows a beta density function rule with parameters r and N. Given two independent gamma variables, the beta distribution may be used to make inferences about the distribution of their ratio.

For example, a psychologist has a notion that in a continuous tracking task, in which a pointer must be held so as to remain in a groove on a moving tape, a subject under a certain condition will tend to make errors at twice the rate of another (control) subject. That is, if for subject 1 m_1 is the intensity of error production over time, and if m_2 is the intensity of error production for subject 2, the experimenter tends to believe that $m_1 = 2m_2$. He thinks of the process generating errors over time for each subject as a stationary Poisson process, so that the elapsed time up until the rth error for subject 1 is a gamma variable, and the same is true for subject 2. Within either subject, errors made are assumed to be independent of each other. Then, if the experiment is continued for each subject until five errors are made, and the elapsed time noted for each, the variable x is

$$x = \frac{w/y}{(w/y) + 1/2}$$

where w is the time elapsed to the fifth error for subject 1, and y is the time elapsed to the fifth error for subject 2. Then, since $r = 5$ and $N - r = 5$, the probability density associated with any value x is

$$f(x) = \frac{(10 - 1)!}{(5 - 1)!(5 - 1)!} (x)^{5-1}(1 - x)^{10-5-1}.$$

The expected value of x is r/N or $1/2$. Values far larger than $1/2$ would tend to suggest that $m_1 < 2m_2$, whereas obtained values of x far less than $1/2$ suggest that $m_1 > 2m_2$.

Given the cumulative beta distribution, we could find $F(x)$ for any x, and translate that value back into a cumulative probability for w/y, the obtained ratio.

Extensive tables of the cumulative beta distribution are available, most notably in the Biometrika Tables for Statisticians, edited by E. S. Pearson and H. O. Hartley. We will not have enough use for such tables to warrant their inclusion here. However, it is possible to estimate certain percentage points in a beta distribution by use of Table V given in Appendix C. Following the discussion of the F distribution in Chapter 11, mention will be made of how that table may be used to estimate cumulative beta probabilities.

If a table of binomial probabilities is available, then the cumulative probability for a beta variable can also be found as follows:

$$F_\beta(x;r, N - r) = 1 - F_{\text{bin}}(r - 1;p = x, N - 1), \qquad x \leq 1/2,$$
$$F_\beta(x;r, N - r) = F_{\text{bin}}(N - r - 1;q = 1 - x, N - 1), \qquad x \geq 1/2,$$

where F_β stands for the cumulative probability for the beta variable at value x, and F_{bin} stands for the cumulative probability of a binomial variable at value $r - 1$, when $p = x$ and the number of trials is $N - 1$, or when the value of the binomial variable is $N - r - 1$, $q = 1 - x$, and the number of trials is $N - 1$. For example, suppose that in the experiment described above, the value of x turned out to be .39. What is the probability of an x value this low or lower when $m_1 = 2m_2$, and $r = N - r = 5$? The p value closest to $p = x = .39$ in the binomial table is .40, so this will be used. We then take the number of trials as $N - 1 = 9$ and look up the cumulative probability for $r - 1 = 4$. This cumulative probability $F_{\text{bin}}(4;.40,9)$ is about .73. Hence, we calculate the cumulative probability for the beta value to be about $1 - .73$ or .27. (The exact value turns out .25, when a table of the beta distribution is used.) Hence, when N and r are small, binomial probabilities do give a way to approximate beta probabilities, according to the rule above.

EXERCISES

1. A teacher believes that the class scores on a final examination should be approximately normally distributed, provided the class is large enough. If his belief is correct, to what percentile ranks should the following z-values correspond:
   (a) -1.2 (b) .96
   (c) 1.88   (d) -1.78
   (e) $-.43$ (f) 2.15

2. Another teacher wishes to curve her class grades in order to make them correspond to values in a normal distribution with a mean of 100 and a standard deviation of 10. In order to do this, she first converts each score into a percentile rank, converts these percentile ranks into the corresponding z-values in a normal distribution, and proceeds from there. Carry out this process with the following set of scores:
   18, 21, 23, 25, 30, 31, 36, 38, 40, 42, 45, 46, 48, 50, 52, 57, 60, 75

3. A mean from a sample of 36 cases has a value of 100. How probable is a sample mean of 100 or more when the population being sampled is normal with the following parameters:
   (a) mean 103, standard deviation 10
   (b) mean 99, standard deviation 4
   (c) mean 80, standard deviation 50

(d) mean 98, standard deviation 24

(e) mean 110, standard deviation 80

4. Under each of the conditions outlined for a population in problem 3, how probable is it that a sample mean based upon 36 independent observations will fall in the interval of values 95 − 105?

5. Using the method illustrated in section 8.5 in the text, compare the binomial probabilities for $N = 8$, $p = .5$ with the corresponding normal probabilities. Then compare $N = 8$, $p = .4$ with normal probabilities. Is the latter appreciably the poorer?

6. Prior to a presidential primary in a certain state, a candidate predicted that he would get approximately 45 percent of the votes of members of his party. A newspaper took a random sample of 500 registered voters in that party and found that only 37 percent indicated they would vote for the candidate. If the true proportion of voters favoring him is .45, how probable is a sample result showing 37 percent or fewer in favor? Would you say that there is good reason to doubt the candidate's assertion?

7. A researcher was interested in the age at first marriage of American women. (Only women married in the calendar year just past were included.) He was fairly confident that this distribution of ages is not symmetric, but rather tends to be skewed toward the higher ages. However, he did want to see if this age tends to be about the same as it was twenty years ago, when the average age was 23.6. A sample of 200 women chosen at random showed a mean age of 24.1, with a standard deviation S of 5.6. How likely is he to get a sample result this much or more deviant from 23.6 (in either direction) if the mean age at marriage is truly 23.6? How can you justify the calculation of this probability, given the nature of this population?

8. A known population has an average height of 68.2 inches. The population distribution is normal, and it is known that the middle fifty percent of the population have heights between 66.5 and 69.9 inches. If a sample of 25 individuals were drawn independently and at random from this population, how likely is it that their mean height would exceed 72 inches? How likely is it that their mean height would be less than 60 inches? Beyond what mean height should only about 5 percent of sample means fall?

9. Four observations were made independently and at random from a normal distribution with mean of 100 and standard deviation of 10. Let X_1 symbolize the value of the first observation, X_2 the second, and so on through X_4. A linear combination of these observed values creates a new variable Y such that

$$Y = 5X_1 + 2X_2 - 2X_3 - 5X_4$$

Describe the sampling distribution of Y, along with its mean and variance. Within what two possible values of Y should the middle 99 percent of all sample values fall? The middle 50 percent?

10. Two populations are being sampled independently. Population I has a mean of 151 and a standard deviation of 15, while population II has a mean of 155 and a standard deviation of 12. Fifty observations are made independently and at random from each population. Interest lies in the difference in the means of the two samples. What is the probability of a difference $M_1 - M_2$ of 2 points or less? What is the probability that the sample difference will fall into the interval −1 to +1? What is the probability of a sample difference falling outside the interval −5 to −3?

11. In a study of children's reactions to a particular form of medication, a researcher believed that two different populations of children should show about the same proportion of individuals who react unfavorably to this medication. Therefore, two samples, each of 200 children, were selected independently and at random from the population, and were given the medication. In Population I, 34 percent of the children had unfavorable reactions, and in Population II, 32 percent showed this reaction. Is a dif-

ference between the proportions of this absolute magnitude or greater something that is likely to have arisen by chance, even though the populations are actually the same in terms of the true proportion? (Hint: since by hypothesis the populations are identical, estimate the true proportion in either population by pooling the estimates of the samples.)

12. Using the results of problems 3 and 15 from Chapter 7, compare the probabilities for the various values of the mean given $N = 3$ with corresponding normal probabilities. What does this show about applications of the central limit theorem when sample sizes are relatively small?

13. An airport has an information counter at which people tend to arrive at the rate of four per minute. The process is stable throughout the day, and the arrivals and departures of people at the counter are independent. A young lady takes her post at the counter at a given moment. What is the probability that two or more minutes will pass before she has her first request for information? What is the probability that no more than two minutes elapse before she has her tenth request?

14. Suppose that the airport described in problem 13 has several information counters, and that at each counter people tend to arrive at a stable rate of 4 per minute. Let Y be the sum of the times elapsed at all five counters from the time each opens until each has had its first request for information. Then the mean elapsed time is $Y/5$. What is the probability that the mean elapsed time is less than 30 seconds? What is the probability that the mean elapsed time is 2 minutes or more?

15. Notice that if the number of counters in problems 13 and 14 had been made very large, one would expect the mean elapsed time to conform to the central limit theorem, just as most other means based on independent observations made at random. Does this suggest a connection between the normal and the gamma distributions? What is this connection?

16. In a study of cognitive processes, a teacher gave each member of a group of ten children a puzzle to solve. The puzzle required a certain amount of insight to solve; some children could do it right away, and others took more time. The teacher recorded the time elapsed until all ten children solved the puzzle. Then another independent group of ten children was given the same puzzle and the time recorded. For Group I the time required was 150 seconds. For Group II the time was 122 seconds. Let us assume that the process of solving the problem by insight is like a stable Poisson process, and the children were working independently. If the children in the two groups were actually solving the problem at the same true rate, what is the probability that Group I should perform this slowly or more so, relative to Group II?

17. Consider a continuous random variable X. Some T independent observations are made, yielding the values $x_1, x_2, \cdots, x_T$, where x_1 is the smallest value observed in the sample, and x_T is the largest value observed. Now we define a new random variable Y, such that

$$y = F(x_T) - F(x_1).$$

Here $F(x_T)$ is the cumulative probability for x_T, and $F(x_1)$ that for x_1. Thus the value of Y associated with any sample is the area cut off under the density function for X by the largest and the smallest values in the sample. It is the proportion of the population lying between the two extreme limits observed in the sample. The value of Y must range between 0 and 1.00 inclusive. Now it can be shown that Y is distributed as a beta variable with parameters $r = T - 1$ and $N = T + 1$. Given this fact, what is the probability that the largest and the smallest values in a random sample of ten independent observations will cover at least ninety percent of the population? Ten percent or less?

9 Hypothesis Testing and Interval Estimation

9.1 STATISTICAL TESTS

Let us assume that the general motivation for inferential statistics is clear to you by now, and you are ready for some concrete procedures. In each of the preceding chapters the author has referred repeatedly to the theme underlying our whole discussion: how do you say something about the population given only the sample evidence? We have seen that certain sample statistics are capable of giving good estimates of particular population parameters; ordinarily, these point estimates are the first and foremost statistical inferences one makes from his data. On the other hand, there is almost always a dark side to the picture: virtually any estimate will be in error to some extent due to fluctuations of sampling, and so we discussed the concept of a sampling distribution of a particular statistic based on a sample of size N. In particular we focused on the mean of the sampling distribution as the expected value of the statistic, and the standard deviation of the sampling distribution as the standard error. Finally, in the last chapter we found that sampling distributions of statistics such as M can often be regarded as normal, particularly when N is large. Given the form of the population distribution together with the appropriate parameter values such as μ and σ, then we can find the probability that the sample value for a statistic such as M will fall into any given interval of values in the sampling distribution. In short, the point has now been reached where most of the basic pieces that fit into the theory of significance tests and interval estimation have been introduced.

The title of this chapter might well be "How to decide how to decide," or "How much evidence is enough?" A psychologist or anyone else who samples from a population is trying to decide, or at least form an opinion on, something about the population. Quite often he wants to decide if some hypothetical population situation appears reasonable in the light of the sample evidence. Sometimes the problem is to judge which of several possible population situations is best supported by the evidence at hand. In either instance, one is trying to make up his mind from evidence. However, as we shall see, there are many possible ways of making this kind of decision, depending on what information in the sample is actually used, the various possibly true population situations being compared, and especially the risk one is willing to take of being wrong in his decision.

The first problem to be faced in making a decision from data is that of choosing the relevant and appropriate statistic for the particular purpose. Obviously, different combinations of the sample data give different kinds and amounts of information about the population. Reaching a conclusion about some population characteristic requires effective use of. the right information in the sample, and various statistics differ in their relevance to different questions about the population. Sometimes the best statistic to permit us to form a judgment about the population also happens to be the best estimate of the particular population parameter under study, but in other instances statistics may be used which are not estimates of any population parameter at all. Such "test statistics" will be introduced in Chapter 10. In this chapter we will use only the simple statistics M or P, since they happen to be best for illustrative purposes, but elsewhere quite different statistics will provide the information we require from the sample.

Granting for the moment that we know the information we need from the sample and how to extract it, how do we judge the tenability of a particular hypothesis about the population? It seems reasonable that a tenable or "good" hypothesis about the population should provide us with a "good" expectation about the sample result in terms of the relevant statistic. That is, the hypothesis gives a good fit to the data if, as a consequence of that hypothesis' being true, we can deduce an expected value of the statistic that agrees quite well with the value actually observed; if the hypothesis is true, there is some value that we should expect the appropriate sample statistic to show. Naturally, in any given sample the obtained value of the statistic will not exactly equal the expectation dictated by the hypothesis, and there is a theoretical sampling distribution of values for the statistic that can be worked out once the hypothetical population situation is specified. Nevertheless, if the hypothesis is a good one, our obtained sample result should fall into a region of values relatively close to the expected value the truth of the hypothesis dictates for our statistic.

A deviation between the obtained value of the statistic and its expected value under the hypothesis implies one of two things: either the hypothesis is right, and the difference in value between the statistic and its expectation is the product of chance, or the hypothesis is wrong and we were not led to expect the right thing of our sample statistic. In particular, if the sample value falls so far from expectation that it lies in an interval of values very improbable given the hypothesis, but this sort of sample result would be made probable if some alternative hypothesis were true, then doubt is cast on the original hypothesis and we reject it in favor of the alternative. Obviously, however, these notions of "disagree with expectation," "improbable," and "probable" are relative matters; when do we say that a sample result disagrees "enough" with expectation under some hypothesis to warrant rejection of the hypothesis itself?

One of the stickiest problems in comparing sample results with expectation given by some hypothesis is deciding what we mean by a sufficient kind and degree of disagreement. How and by how much must the sample statistic's value disagree with expectation before we decide that one hypothesis should be abandoned in favor of another? What we need is a decision-rule, a guide giving the conclusion we will reach depending on how the data turn out, and which can be formulated even before the data themselves are seen. However, there is literally no end to the number of decision-rules that one might formulate for a particular problem. Some of these ways of deciding might be very good in a particular circumstance, but not so good in others, and so we need to decide how to decide on the basis of what we wish to accomplish in a specific situation. The branch of mathematics known as "decision theory" treats of this problem of choosing a decision-rule, and we are going to apply some of its elementary principles here. We will find that there is no rigid formula supplying the right way to decide in all situations, and, indeed, we often lack the information necessary for choosing an optimal decision-rule. For this reason psychologists and others using statistical inference most often fall back on accepted conventions for evaluating evidence, and these conventions may have very little to do with the principles of decision theory. Nevertheless, the application of principles from decision theory to the problem of statistical inference does shed some light on the theory underlying tests of statistical hypotheses.

The process of comparing two hypotheses in the light of sample evidence is usually called a "statistical test" or a "significance test." The competing hypotheses are stated in terms of the form, or of one or more parameters, of the distribution for the population. Then the statistic containing the relevant information is chosen, having a sampling distribution dictated by the population distribution specified in the particular hypotheses. The hypothesis that dictates the particular sampling distribution against which the obtained sample value is compared is said to be "tested." The significance test is based on the sampling distribution of the statistic given that the particular hypothesis is true. If the observed value of the statistic falls into an interval representing a kind and degree of deviation from expectation which is improbable given the hypothesis, but which would be relatively probable given the other, alternative, hypothesis, then the first hypothesis is said to be rejected. The actual decision to entertain or to reject any hypothesis is thus based on whether or not the sample statistic falls into a particular region of values in the sampling distribution dictated by that hypothesis.

A few words of warning before we go further: In the first place, although we will speak of testing some single hypothesis, in practice the experimenter always acts as though he is deciding between *two* hypotheses. A great many issues connected with the use of significance tests can be understood only if this decision-making task is presupposed. Second, even though the form of significance testing is based on decision making, this does not mean that the experimenter must actually make such a decision, particularly in a situation where he himself is not able to apply principles for actually choosing the best decision-rule. When, in following sections, we speak of the necessity for the experimenter to decide between two hypotheses, we will mean only that the particular conclusions follow if we act as if this were the experimenter's concern; the standard practices can be justified only if we suppose this to be the experimenter's task. Thus, in the following sections significance testing will be treated in this decision-making setting, since the problem is most easily discussed in these terms, and most psychologists and others using inferential statistics have come to think in this way. In addition, presenting hypothesis testing in a decision context will also make it possible to point out some of the inadequacies and misuses of this approach from the psychologist's point of view.

Now we are ready to look into this problem of the choice of a decision-rule, using the mean M and the proportion P as examples of statistics on which decisions are based. However, before we can consider these ideas more closely, some standard terminology must be given.

9.2 STATISTICAL HYPOTHESES

A statistical hypothesis is usually a statement about one or more population distributions, and specifically about one or more parameters of such population distributions. It is always a statement about the population, not about the sample. The statement is called a hypothesis because it refers to a situation that *might* be true. Statistical hypotheses are almost never equivalent to the hypotheses of science, which are usually statements about phenomena or their underlying bases. Quite commonly, statistical hypotheses grow out of or are implied by scientific hypotheses, but the two are seldom identical. The statistical hypothesis is usually a concrete description of one or more summary aspects of one or more populations; there is no implication of *why* populations have these characteristics.

We shall use the following scheme to indicate a statistical hypothesis: a letter H followed by a statement about parameters, the form of the distribution, or both, for one or more specific populations. For example, one hypothesis about a population could be written

H: the population in question is normally distributed with $\mu = 48$ and $\sigma = 13$,

and another

H: the population in question has a Bernoulli (two-class) distribution with $p = .5$.

Hypotheses that completely specify a population distribution are known as **simple hypotheses**. In general, the sampling distribution of any statistic also is completely specified given a simple hypothesis and N, the sample size.

We will also encounter hypotheses such as:

H: the population is normal with $\mu = 48$.

Here, the exact population distribution is not specified, since no requirement was put on σ, the population standard deviation. When the population distribution is not determined completely, the hypothesis is known as **composite.**

Hypotheses may also be classified by whether they specify *exact* parameter values, or merely *a range* or *interval* of such values. For example, the hypothesis

$$H : \mu = 100$$

would be an exact hypothesis, although

$$H : \mu \gneqq 100$$

would not be exact.

9.3   ASSUMPTIONS IN HYPOTHESIS TESTING

In applied situations very seldom will a hypothesis specify the sampling distribution completely. Simple hypotheses are just not of interest or are not available in most practical situations. On the other hand, it *is* necessary that the sampling distribution be completely specified before any precise probability statement can be made about the sample results. For this reason assumptions usually are made, which, taken together with the hypothesis itself, determine the relevant statistic and its sampling distribution and justify a test of the hypothesis. These assumptions differ from hypotheses in that they are rarely or never tested against the sample data. They are assertions that are simply assumed true, or the evidence is collected in such a way that they must be true.

First and foremost among such assumptions is that the sample is random. In all discussion of significance tests and confidence methods to follow, we shall assume simple random sampling unless it is specifically indicated otherwise. Independence of two or more random variables is another assumption frequently made. At other points we shall assume that the population distribution is normal, that the variances of different populations are equal, and so on. Wherever such assumptions underlying a test are important they will be pointed out.

These assumptions are necessary for the formal justification of many of the methods to be discussed. It is not true, however, that these assumptions are always or even ever realized in practice. The results of any data analysis leading to statistical inference can always be prefaced by "If such and such assumptions are true, then . . . ." The effects of the violation of these assumptions on the conclusions reached can be very serious in some circumstances, and only minor in others. In later sections the importance of the most common assumptions will be discussed. Nevertheless, these assumptions should always be kept in mind, along with the conditional character of any result subject to the assumptions being true.

9.4   TESTING A HYPOTHESIS IN THE LIGHT OF SAMPLE EVIDENCE

Every test of a hypothesis involves the following features:

1. The hypothesis to be tested is stated, together with an alternative hypothesis.
2. Additional assumptions are made, permitting one to specify the sample sta-

tistic that is most relevant and appropriate to the test, as well as the sampling distribution of that statistic.
3. Given the sampling distribution of the test statistic when the hypothesis to be tested is true, a **region of rejection** is decided upon. This is an interval of possible values deviant from expectation if the hypothesis were true, but which more or less accord with expectation if the alternative were true. This region of rejection contains values relatively improbable of occurrence if the first hypothesis were true, but relatively probable given the alternative. The risk one is willing to take in rejecting the tested hypothesis *falsely* determines the size of the region of rejection, and the alternative hypothesis determines its location.
4. The sample itself is obtained. If the computed value of the test statistic falls into the region of rejection, then doubt is cast on the hypothesis, and it is said to be rejected in favor of the alternative. If the result falls out of the region of rejection, then the hypothesis is not rejected, and the experi·menter may choose either to accept the hypothesis or to suspend judgment, depending on the circumstances. A sample result falling into the region of rejection is said to be **statistically significant, or to depart significantly from expectation** under the hypothesis.

As an example, suppose that an experimenter entertains the hypothesis that the mean of some population is 75:

$$H_0: \mu = 75.$$

This is actually a composite hypothesis, since nothing whatever is said about the form of the population distribution nor about other parameters such as the standard deviation. The experimenter is, however, prepared to assume that whatever is true about the mean the population has a normal distribution, and that the standard deviation for the population is 10; for the moment, let us imagine that both of these assumptions are reasonable in the light of what the experimenter already knows about his problem. In effect, the addition of these two assumptions to the original hypothesis makes this hypothesis simple:

$$H_0: \mu = 75, \sigma = 10, \text{ normal distribution.}$$

Now, what of the alternative hypothesis? It might be that the alternate hypothesis is

$$H_1: \mu \neq 75.$$

Note that this alternative happens to be inexact, and really asserts that some (unspecified) value of μ other than 75 is true. However, even if H_1 is true and H_0 false, the experimenter will still assume that the distribution is normal with a standard deviation of 10.

It has already been decided that 25 independent observations will be made. Given this sample size N, the information in the hypothesis H_0, and the assumptions, the sampling distribution of the mean is known completely: this must be a normal distribution with a mean of 75, and a standard error of $10/\sqrt{25}$ or 2, if the hypothesis H_0 is true.

Now in accordance with H_1, the experimenter decides that he will reject this hypothesis H_0 only if his sample mean falls among the extremely deviant ones in either direction from the value 75; in fact, he will reject H_0 only if the probability is .05 or less for the occurrence a sample mean as deviant or more so from 75. Thus, the region of rejection for this hypothesis contains only 5 percent of all possible sample results when the hypothesis H_0 is actually true. Such sample

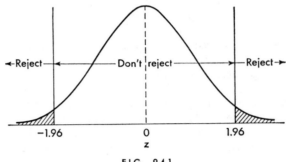

FIG. 9.4.1

values depart widely from expectation and are relatively improbable given that H_0 is true; on the other hand, samples falling into these regions are relatively more probable if a situation covered by H_1 is true. These regions of rejection in the sampling distribution are shown in Figure 9.4.1.

From tables of the normal distribution, he finds that a z score of 1.96 shows a cumulative probability $F(1.96)$ of about .975. Hence only .025 of all sample means should lie *above* 1.96 standard errors from μ. Similarly, it must be true that about .025 of all sample means will lie below -1.96 standard errors from μ. In short, the region of rejection includes all sample means such that, in this sampling distribution,

$$1.96 \leq z_M$$

or

$$z_M \leq -1.96$$

giving a total probability of .025 + .025 or .05 for the combined intervals. A value of the sample mean that lies exactly on the boundary of a region of rejection is called a **critical value** of M; the critical z_M values here are -1.96 and 1.96, so that the critical values of M are

$$-1.96\sigma_M + \mu = -1.96(2) + 75 = 71.08$$

and

$$1.96\sigma_M + \mu = 1.96(2) + 75 = 78.92.$$

Now the sample is drawn and turns out to have a mean of 79. The corresponding z_M score is

$$z_M = \frac{79 - \mu}{\sigma_M} = \frac{79 - 75}{2} = 2.$$

Notice that this z_M score *does* fall into the region of rejection, since it is greater than a critical value of 1.96. Then the experimenter says that the sample result is **significant beyond the 5 percent level.** Less than 5 percent of all samples should show results this deviant (or more so) from expectation under H_0, if H_0 is actually true. If the experimenter has decided in advance that samples falling into this region show sufficient departure from expectation to be called *improbable* results, then doubt is cast on the truth of the hypothesis, and H_0 is said to be rejected.

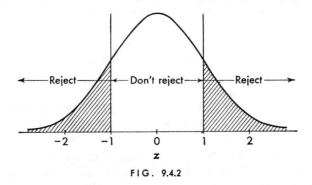

FIG. 9.4.2

In its essentials this simple problem exemplifies all significance tests. Naturally, the details vary with the particular problem. Hypotheses may be about variances, or any other characteristic feature of one or more populations. Other statistics may be appropriate to a test of the particular hypothesis. Other assumptions may be made to permit the specification of the sampling distribution of the relevant statistic. Other regions of rejection may be chosen in the sampling distribution. *But once these specifications are made, the experimenter knows, even before he sees the sample, how he will make up his mind, if he must, in the light of what the sample shows.*

Here the region of rejection was simply specified without any particular explanation. In this step, however, lies a key problem of significance tests. In the choice of the region of rejection the experimenter is deciding how to decide from the data. It is perfectly obvious that there are very many ways that the experimenter could have chosen a rejection region. He could have decided to reject the hypothesis if the sample result fell beyond one standard error of the mean specified by the hypothesis, as in Figure 9.4.2, or if the sample result fell between 3 and 4 standard errors below the hypothetical mean, as in Figure 9.4.3, and so on

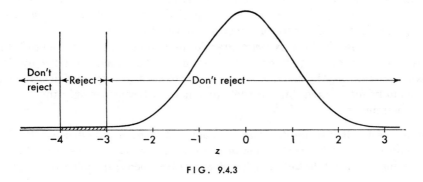

FIG. 9.4.3

ad infinitum, for any interval or set of intervals. There is no law against using any decision-rule for rejection, although on intuitive grounds alone the one actually used here seems somehow more sensible than the others suggested.

In the next section we are going to address ourselves to this problem of choosing a rejection region. This can be discussed most simply if we consider a choice between *two exact hypotheses*. Admittedly, this is an extremely artificial situation, and almost never does one actually have this kind of choice to make. Nevertheless, the terminology and procedures of hypothesis testing can best be understood by acting as if this were the experimenter's problem.

9.5   CHOOSING A WAY TO DECIDE BETWEEN TWO EXACT HYPOTHESES

An experimenter has two alternative theories about some behavior. According to Theory I, exactly .80 of all human subjects should exhibit this behavior in his experimental situation, whereas in Theory II, .40 should be the proportion doing this. The task of the experimenter is to form a judgment about the two theories. Imagine, if you can, that the two theories exhaust the logical possibilities accounting for what he observes: either one or the other must be true.

These two theories suggest the two competing hypotheses:

$$H_0: p = .80$$
$$H_1: p = .40.$$

The subscripts 0 and 1 have no particular meaning here; they are merely indices to let us tell the hypotheses apart.

Some N subjects are to be drawn at random and independently for introduction into the experimental situation. The occurrence of the behavior for any subject is a "success," and for either hypothesis there is a binomial sampling distribution giving probabilities for various possible values of P, the sample *proportion* of successes. Of course, the probabilities of different sample P values differ under the two hypotheses, since $p = .8$ in one and only .4 in the other.

For the sake of simplicity, let us suppose that there are to be only ten cases in the sample. Then the sampling distributions of P found from the binomial distribution under the two hypotheses are as shown in Table 9.5.1.

The experimenter knows the probability of each possible sample P result given the possible truth of either of the two hypotheses.

How shall the experimenter decide upon seeing the evidence? What he needs is a decision-rule, some way that lets him judge which of the two hypotheses is actually favored by the sample result. There are any number of decision-rules that could be formulated, but for the moment we will consider only the following three possibilities:

Decision-rule 1:   If P is greater than or equal to .8, choose H_0; otherwise H_1.

Decision-rule 2:   If P is greater than or equal to .6, choose H_0; otherwise H_1.

Decision-rule 3:   If P falls between .2 and .8 inclusive, choose H_0; otherwise H_1.

Table 9.5.1

	$p(P)$	
P	*if* $p = .8$	*if* $p = .4$
1.0	.107	.0001 $(+)$
.9	.268	.002
.8	.302	.011
.7	.201	.042
.6	.088	.111
.5	.026	.200
.4	.006	.251
.3	.001	.215
.2	.000 $(+)$	.121
.1	.0000 $(+)$	.040
.0	.0000 $(+)$	.006

Regardless of which decision rule is selected there are two ways that the experimenter *could* be right, and two ways of his making an *error*. This is diagrammed below:

TRUE SITUATION
H_0 H_1

H_0 correct   error

DECISION

H_1 error   correct

If he decides H_0, and H_1 is actually true, an error is made. Furthermore, he can make an error if he decides on H_1 and H_0 is actually true. The other possibilities lead, of course, to correct decisions.

Given the sampling distributions under each of the two hypotheses, and any decision-rule, we can find the *probabilities* of these two kinds of error. The experimenter has no idea which hypothesis is true, and all he can go by is the occurrence of a P value leading him to reject one or the other of the hypotheses. Suppose that decision rule 1 were adopted. This rule requires that the occurrence of any P of .8 or more lead automatically to a decision that H_0 is true. What is the probability of observing such a sample result *if H_1 is true*? Notice that whenever this happens, there will be a wrong decision and so this probability is that of *one kind* of error. The binomial distribution for $p = .4$ shows a probability of about .013 for P greater than or equal to .8, and so the probability is .013 of *wrongly* deciding that H_0 is true.

In the same way, we can find the probability of the error made in choosing H_1 when H_0 is true. By decision-rule 1, H_1 is chosen whenever P is less than or equal to .7; in the distribution under H_0, this interval of values has a probability of about .323. Thus, the probabilities of error and correct decisions are, under decision-rule 1,

TRUE SITUATION

H_0 H_1

	H_0	H_1
H_0	.677	*.013*
H_1	*.323*	.987

DECISION

The probability of a correct decision is $1 - p$(error) for either of the possibly true situations (the two error probabilities appear in italics). Notice that the experimenter is far more likely to make an erroneous judgment by using this rule when H_0 is true than for a true H_1.

Exactly the same procedure gives the erroneous and correct decision probabilities under rule 2:

	H_0	H_1
H_0	.966	*.166*
H_1	*.034*	.834

By this second rule, the probability of error is relatively smaller when H_0 is true than it is for rule 1. However, look what happens to the probability of the other error! This illustrates a general principle in the choice of decision-rules: *any change in a decision-rule that makes the probability of one kind of error smaller will ordinarily make the other error probability larger* (other things, such as sample size, being equal).

Before trying to choose between rules 1 and 2, let us write down the probabilities for decision-rule 3.

	H_0	H_1
H_0	.625	*.952*
H_1	*.375*	.048

Even on the face of it this decision-rule does not look sensible. The probability of error is large when H_0 is true, and the experimenter is almost sure to make an error if H_1 is the true situation! This illustrates that not all decision-rules are reasonable if the experimenter has any concern at all with making an error; here regardless of what is true the experimenter has a larger chance of making an error using this rule than in using either rules 1 or 2. Rules such as this are called **inadmissible** by decision theorists. We need confine our attention only to the two relatively "good" rules 1 and 2.

There is a real problem in deciding between the rules 1 and 2; rule 1 is good for making error probability small when H_1 is true, but risky when H_0 represents the true state of affairs. On the other hand, rule 2 makes for small chance of error when H_0 is true, but gives relatively large probability of error when H_1 holds. Obviously, any choice between these two rules must have something to do with the relative importance of the two kinds of errors. It might be that making an error is a minor matter when H_0 is true, but very serious given H_1. In this case rule 1 is preferable. On the other hand, were error very serious given H_0, rule 1 could be disastrous. In short, any rational way of choosing among rules must involve some notion of the loss involved in making an error.

9.6 ERRORS AND LOSSES

Decision theory, which grew out of problems of economic decision-making, is concerned with the possible outcomes of an action. Any decision-maker has several courses of action open to him, and once he has chosen a course of action some event or chain of events is going to occur. However, the decision-maker does not necessarily know what this outcome will be. Instead, there are various "states of the world" that might be true, and the outcome of any decision to take action depends both on what is decided and what is really true.

As a simple example, imagine a man interested in buying a sweater. His interest at the moment is in finding a sweater that will wash without shrinking. He locates a sweater that he likes, but he does not know if it is washable. For this particular sweater, two courses of action are open to him: "buy" or "don't buy." However, there are also two possibilities for the "real" nature of the sweater: either it won't shrink, or it will. The courses of action, the states of the world, and the different action-outcome possibilities are shown in the following table:

		STATE OF THE WORLD	
		sweater won't shrink	*sweater will shrink*
POSSIBLE ACTIONS	*buy*	(buy, good sweater)	(buy, sweater shrinks)
	don't buy	(don't buy, miss a good sweater)	(don't buy, miss a shrunken sweater)

In two of these action-outcome possibilities there is possibly a real economic gain to the buyer; the man does well if either the combination (buy, good sweater) or the combination (don't buy, miss a shrunken sweater) occurs. On the other hand, two possibilities for error exist as well: (buy, sweater shrinks), and (don't buy, miss a good sweater). In principle, each of these action-outcome possibilities has some value or utility for the buyer, and his choice to buy or not to buy should depend upon these relative values.

With only this to go on the buyer still has the problem of making up his mind. However, just as the experimenter gathers data in order to "snoop" on the possible states of the world, the buyer decides that he will do a little snooping on his own, and he looks for the cleaning instructions tag on the sweater. He happens to know that only once in about one hundred times does a sweater factory label a garment "washable" when it is not. On the other hand, he knows that among washable sweaters, only nine in ten bear instructions that definitely say so. Thus, if he decides solely on the basis of what the tag says, he has the following probabilities of error and right decisions (assuming that this is a random sample from among one of the two sorts of sweaters):

	will not shrink	*will shrink*
tag says washable (buy)	.90	.01
tag does not say washable (don't buy)	.10	.99

Here, the chances are much smaller of making the mistake of buying a sweater that will shrink, than of failing to buy a sweater that will not shrink. Is this a good way to decide? That depends upon the value to him of the various outcomes. If the sweater is very expensive and buying a sweater that shrinks means a big loss, then he probably should use this rule. On the other hand, if the man needs a sweater very badly and the price is low, so that much would be lost if this opportunity were mistakenly passed up, the rule is not so good. Here the various outcomes of the actions have a very real, and perhaps even a dollar-and-cents, value to the buyer. Not only the probabilities, but also the possible losses attached to the different outcomes, should figure in the buyer's choice of a way to decide.

In business situations, it is frequently true that a given action must be followed by one of a fairly small and specifiable set of outcomes. The decision-maker does not know what the particular outcome of his action will be, but at least he is able to state the possible outcomes. The one that occurs may depend upon economic conditions in general, quality of product, competition, and so on. Furthermore, each of these possible outcomes to any action can be assigned a profit or loss value, at least in principle.

On the other hand, the possible outcomes of actions taken by the scientist are far more difficult to specify. It is clear that some definite courses of action are usually open to the experimenter: he may decide to publish or withhold his findings, to collect or not to collect more data, to pursue or abandon the line of investigation, to ask for funds for further research or to forget the whole matter. By stretching a point, we might say that finding out what the true situation is constitutes an outcome following any action. However, this outcome may not apply to the experimenter at all, but rather may serve as an outcome for the science he represents. It is important to recognize that the scientist, human being that he is, may be capable of foreseeing only the short-term monetary, prestige or other outcome possibilities of his actions, which are insignificant compared to the long-range impact of his action on knowledge and human welfare.

At any rate, let us assume that all actions available to the scientist are contingent upon a few basic actions he may take: he may decide that any one of the set of hypotheses entertained is true, or he may suspend judgment pending further evidence. Parenthetically, we note that this last possibility distinguishes much of the scientist's decision-making from that of the businessman: few people could stay in business by suspending judgment almost indefinitely, and yet this is almost an ideal in science. Unfortunately, even the modern scientist cannot suspend judgment forever, and sooner or later he must take a course of action equivalent to the decision that "such and such is true."

Whatever the set of actions open to the scientist, each is accompanied by some set of potential outcomes and it is reasonable that some outcomes are good and others bad from his point of view. Each possible outcome has some "weight" for him and his science. Experimentation costs time and money, not to speak of the mental exertion involved. A person living in the twentieth century does not need to be reminded that scientific actions can lead directly to miraculous advances in social welfare, as well as genuine catastrophes. Fortunately, these are the exceptions; but even pursuing dead ends on the basis of faulty evidence or failing to reap the benefits of potential knowledge are costly matters for the scientist.

9.7   EXPECTED LOSS AS A CRITERION FOR CHOOSING A DECISION-RULE

Let us imagine each action-outcome combination as having some numerical value for the experimenter. The exact nature of the value need not concern us here: it may reflect dollars, time, prestige, pleasure, or any of a variety of other things. Let us assume only that some function exists assigning one number to each action-outcome combination for a particular problem. Thus, the combination "decide H_0 and H_1 turns out to be true" is accompanied by the numerical value

$$u(H_0;H_1).$$

For our purposes we need consider these values only as *losses*. Each loss value is some positive number or zero given to an action-outcome combination. If the combination represents the best available action under the true circumstances, then the decision is *right* and the loss-value is zero. On the other hand, if the action is less desirable than the best available action, then this is an *error* and the loss-value is positive. (A more technical term for such a value is *opportunity loss*.)

Now let us return to the poor experimenter who is faced with two hypotheses, and who was left in the lurch in Section 9.5. Suppose that the potential losses connected with the two erroneous decision-possibilities were

<table>
<tr><td></td><td></td><td colspan="2">TRUE SITUATION</td></tr>
<tr><td></td><td></td><td>H_0</td><td>H_1</td></tr>
<tr><td></td><td>H_0</td><td>0</td><td>5</td></tr>
<tr><td>POSSIBLE ACTIONS</td><td>decide</td><td></td><td></td></tr>
<tr><td></td><td>H_1</td><td>10</td><td>0</td></tr>
</table>

so that

$$u(H_0;H_1) = 5$$

$$u(H_1;H_0) = 10.$$

Each possible decision rule D can be given an *expected loss value*

$$E(u;D,H_0) = \text{(expected loss given } D \text{ and true } H_0\text{)}$$

for any specified true situation such as H_0. Here

$$E(u;D,H_0) = u(H_1;H_0)p(\text{decide } H_1 \text{ by rule } D \text{ given } H_0 \text{ actually true})$$
$$+ u(H_0;H_0)p(\text{decide } H_0 \text{ true given } H_0 \text{ true}).$$

Furthermore,

$$E(u;D,H_1) = u(H_0;H_1)p(\text{decide } H_0 \text{ true given } H_1 \text{ actually true})$$
$$+ u(H_1;H_1)p(\text{decide } H_1 \text{ true given } H_1 \text{ true}).$$

For example, consider our decision-rule 1 once again. Here

$$E(u;D_1,H_0) = 10(.323) = 3.23$$

which is just the loss associated with an error multiplied by its probability under D_1. When H_1 is true, then

$$E(u;D_1,H_1) = 5(.013) = .065.$$

The two expected losses for D_1 can be written together as the first row in Table 9.7.1.

Table 9.7.1

	TRUE	
	H_0	H_1
D_1	3.23	.065
DECISION-RULE D_2	.34	.83
D_3	3.75	4.76

Notice that the expected loss using this rule is rather high when H_0 is true, but low when H_1 is true. The other rows of the table can be filled out using decision-rules 2 and 3:

$$E(u;D_2,H_0) = 10(.034) = .34$$
$$E(u;D_2,H_1) = 5(.166) = .83$$

and

$$E(u;D_3,H_0) = 10(.375) = 3.75$$
$$E(u;D_3,H_1) = 5(.952) = 4.76.$$

Suppose that we adopt the following point of view: *A good decision-rule is one giving low expected loss regardless of what the true situation turns out to be.* Let us apply this criterion.

An inspection of this table shows, once again, the absurdity of rule 3. Given either true situation (a column of Table 9.7.1) the expected loss for rule 3 is greater than that for either rule 1 or rule 2, and hence this rule is far less desirable than either of the others. In general, any rule that gives a higher expected loss than another *regardless* of the true situation is disqualified (or is inadmissable).

Can we choose between rules 1 and 2, however, on the basis of expected loss? A device within decision theory for choosing between rules is known as the "minimax" principle. Applying this principle leads us to choose that rule showing the *minimum maximum-expected-loss* over all possible true situations. For expected-loss tables such as 9.7.1, a minimax decision-rule is one having a largest entry in its row that is smaller than the largest entry in any other row of the table. Notice that in this situation, rule 2 has .83 as the largest value in its row; this is smaller than the largest value for rule 1, or 3.23. Thus, applying the minimax principle, we find that rule 2 is the one to use. Using rule 2, the largest that we expect long-run loss to be is smaller than for either rule 1 or rule 3.

This minimax criterion for choosing among decision-rules is historically important in the theory of games and in decision theory, and it does give a way to compare decision-rules on the basis of their expected losses. However, this criterion is no longer taken very seriously by statisticians. Among other things, it is unduly conservative, and this way of choosing a decision-rule often produces results that go against our intuition. The minimax criterion seems especially to be out of the spirit of scientific research, since it focusses only on the extreme potential loss in making errors, and not on the very large gains one can make through increased knowledge *despite* the risk of being wrong on occasion. An alternative to the minimax criterion is given below.

9.8 SUBJECTIVE EXPECTED LOSS AS A CRITERION FOR CHOOSING AMONG DECISION-RULES

One of the considerations left completely out of our discussion so far is the fact that the experimenter usually knows something (or at least believes something) about the hypotheses before the experiment takes place. The scientist does not operate in a vacuum; for any experimental problem many sources of prior information are open to him. There are related experiments (or even repetitions of the same experiment) done by others, there are theories that differ in their logical tightness and plausability, there are the scientist's own informal observations of the world about him, and so on. The net result is that the experimenter has some initial ideas about how credible each of the hypotheses is in the light of what he already knows. He is not certain about which one of the available hypotheses is true or otherwise he would not trouble to do the experiment. However, if asked to bet about which is true, he may be able to give odds he considers fair for this bet. In short, there are prior beliefs, or opinions, or objective knowledge that make some of the possibilities for a true situation a better bet than others for the experimenter.

There is a school of thought among statisticians holding that these prior considerations must be brought into the choice of a decision-rule. It is possible to speak of the "personal probability" of the experimenter, reflecting the degree of prior confidence he has that any hypothesis is true *before* seeing the evidence. That is, imagine that we could assign numbers between 0 and 1 to hypotheses such as H_0 and H_1, standing for the experimenter's personal probability for each. If you will, these numbers index his degree of belief in the respective hypotheses. For the moment, let us distinguish these personal probabilities from "ordinary" probabilities interpreted as relative frequencies by use of the symbols

$$\mathcal{P}(H_0) \text{ and } \mathcal{P}(H_1)$$

to stand for these numbers, where $\mathcal{P}(H_0) + \mathcal{P}(H_1) = 1.00$.

Suppose that the experimenter is totally in the dark about the true situation, knowing only that either H_0 or H_1 must be true. If he is willing to bet with 50–50 odds that either hypothesis will turn out true, then

$$\mathcal{P}(H_0) = \mathcal{P}(H_1) = 1/2.$$

On the other hand, suppose that you bet with the experimenter about which will turn out eventually to be true, and he were willing to put up nine dollars to your one dollar on the bet that H_1 is true. In other words, he thinks that the correct odds are 9 to 1 against H_0 being true. In terms of personal probabilities, this is like saying

$$E(\$) = \$1\mathcal{P}(H_1) - \$9\mathcal{P}(H_0) = 0$$

so that

$$\mathcal{P}(H_1) = 9/10 \text{ and } \mathcal{P}(H_0) = 1/10,$$

if from his point of view this bet is fair.

Given the personal probabilities, it is possible to compare decision-rules in still another way. We can form the "subjective expected loss" for each decision-rule:

$$SE(u;D) = E(u;D,H_0)\mathcal{P}(H_0) + E(u;D,H_1)\mathcal{P}(H_1)$$

which is just a weighted sum of the expected losses for a given rule, the weights being the personal probabilities. For example, for decision-rule 1, Section 9.5,

$$SE(u;D_1) = 3.23(1/10) + (.065)(9/10)$$
$$= .382,$$

and for rule 2,

$$SE(u;D_2) = .34(1/10) + .83(9/10)$$
$$= .781.$$

Decision-rules may be compared by their subjective expected losses; the lower the subjective expected loss, the better the rule. In this case, given the loss-values, and given the personal probabilities, rule 1 is clearly superior to rule 2. Even though error has a fairly high probability by this rule when H_0 is true, the experimenter's own weighting of what he expects to be true of the situation tends to discount the likelihood of such errors.

In case you have forgotten again, we left our experimenter long ago, lost in thought and trying to decide how to decide between his two hypotheses. Now we have come up with an answer. Or have we? *Provided* that the experimenter can furnish us with the loss-values of the possible errors, and *provided* that we can find out his personal probabilities, then the subjective expected loss gives an unequivocal way for him to pick a decision-rule. Unfortunately, these two provisions introduce a problem as big as before.

9.9   THE FAILURE OF THE DECISION-THEORY APPROACH IN PSYCHOLOGICAL RESEARCH

This excursion into some of the rudiments of decision theory has been introduced to point up some of the shortcomings of hypothesis-testing in psychological research. In the sections to follow procedures will be given that are in common use in such research. Most such procedures use a rule for deciding when a result is significant that *might* be justified by a decision-making argument very similar to the one just given. The experimenter wants to avoid errors in inference, presumably because such errors lead to losses, at least in time and effort. Decision-rules differ in the expected losses they give, and it behooves the experimenter to choose a decision-rule that minimizes expected loss. The act of deciding on a region of rejection is the choice of a decision-rule, and should be subject to the same considerations.

Unfortunately, the cost of erroneous decisions is almost never considered in psychological research. Indeed, in most instances in psychological research it seems very unlikely that a numerical value could ever be assigned to the loss incurred in an erroneous decision. Just how bad is it to be wrong in a scientific inference? This problem is not so difficult in many other fields using statistical inference, especially in research on applied problems. In business decisions it is often possible to give a value in dollars to the outcomes that might result from various decisions, and statistical decision theory then provides guides for choosing an effective decision-rule. Even in some applied areas in psychology, losses involved in errors might be reckoned in the same general way. Thus, in studying methods for selecting people for various jobs the cost of an error may be a calculable thing, and the decision-rule may be chosen on that basis. Similarly, this possibility may exist for studies of diagnostic methods, of training effectiveness, and so on. By and large, however, most research psychologists would not know how to assign loss-values to decision errors. Even in a relatively clear-cut situation such as diagnosis or selection, the costs to the people involved or even to the community may be extremely important. For example, consider a diagnostic method for potential suicides. Errors of the two different types (false positives and false negatives) have enormously different consequences for the person and for society. Such costs are most difficult to assess, and the effort required is seldom expended. Even so, the decision rules obtained by omitting such incalculable costs could be disastrous. Such an omission is sometimes called the "accountant's error," although that term fails to convey the importance of the problem.

In the same way, even if we admit the existence of personal probabilities or $\mathcal{P}$ values for the experimenter, typically he is not in a position to assess objectively what these values should be, even though this is possible in principle. Perhaps in the future a standard part of statistical analysis will be the statement of prior personal probabilities before the collection of the data. In principle, personal probabilities can be found by the odds the experimenter is willing to accept in betting on the truth of a hypothesis, and it might be possible to make this a standard part of statistical practice. This "Bayesian" approach to inference will be examined in more detail in Chapter 19.

In short, psychology uses much of the terminology of statistical decision theory without its main feature, the choice of a decision-rule having optimal properties for a given purpose. Instead, the psychologist uses conventional decision-rules, completely ignoring questions of the loss involved in errors and the degree of prior-certainty of the experimenter. These conventional rules can be justified by decision theory in some contexts, but they are surely not appropriate to every situation.

The author deplores this lack as much as anybody else. However, he sees very little chance for change in psychologists' behavior in the near future. Research psychologists apparently understand these conventions, and the majority seem to be able to work within them. The problem of error-loss is something that is different for different kinds of research questions, and any solution requires real effort that no one is yet willing to apply. Finally, the inclusion of personal probability in statistical decision making is by no means universally endorsed among statisticians. This question currently arouses considerable interest and study in mathematical statistics, but the matter is not resolved. Once again, an exploration of this matter in more depth awaits Chapter 19.

9.10 THE OPTION OF SUSPENDING JUDGMENT

Must the psychologist actually make a decision about what is true from his data? Naturally, choosing to suspend judgment and wait for more evidence is a decision to adopt a course of action. However, why cannot the psychologist adopt this time-honored strategy of the scientist more often than he apparently does? As we have seen, he is usually cut off from the possibility of choosing an appropriate decision-rule, using either the minimax principle or minimum subjective-expected-loss. He has no recourse except to fall back on some conventional rule. No convention can possibly be ideal or even sensible for all the problems to which it may be applied, as we shall illustrate below. If the psychologist cannot make up his mind how to decide in some rational way, then why *must* he decide that something is true or false?

Suppose that we give our experimenter the option of suspending judgment. That is, he does not have to decide that either H_0 or H_1 is true, but rather can suspend judgment until more evidence is collected. One rule he might adopt is

Decision-rule 4:   If P is .8 or more, decide H_0; if P is .4 or less decide H_1; if P is between .5 and .7 inclusive, then suspend judgment.

From Table 9.5.1, we can find that the probabilities of the various action-outcome possibilities are, by this rule,

		TRUE	
		H_0	H_1
DECIDE	H_0	.677	.013
	H_1	.007	.633
SUSPEND JUDGMENT		.316	.354

The action "suspend judgment" is technically an error if either H_0 or H_1 is true, and assuredly there is going to be some loss connected with making such an error. However, it well could be that this loss is much less than that incurred by falsely concluding that either H_0 or H_1 is true. Suppose that the loss values were

		TRUE	
		H_0	H_1
DECIDE	H_0	0	5
	H_1	10	0
SUSPEND JUDGMENT		1	1

Now in this case,

$$E(u;D_4,H_0) = (0)(.677) + (10)(.007) + (1)(.316) = .323$$
$$E(u;D_4,H_1) = (5)(.013) + (0)(.633) + (1)(.354) = .419.$$

Compare these expected losses with those in Table 9.7.1. When either H_0 or H_1 is true, the expected loss in using rule 4 is less than that for rule 2, and when H_0 is true, the expected loss is much less than that for rule 1 as well, although rule 1 has a smaller expected loss when H_1 is true. By the minimax-loss criterion, rule 4 is clearly superior to both rules 1 and 2 in this situation. In short, a rule involving suspension of judgment is relatively good here, and if the minimax criterion is adopted, is the best of those considered. It is entirely possible that the course of action represented by "no decision, suspend judgment" will be best in a given situation, especially if the penalty involved in waiting to decide is small compared to the loss involved in an incorrect decision.

On the other hand, in many applied situations the opportunity to suspend judgment is just not available to the experimenter. He must decide here and now on the basis of what little evidence he has. Fortunately, however, scientists often do have the privilege of waiting for more evidence.

The occurrence of a significant result in a significance test is not a command to decide something. All that the significance test per se can give is a probability statement about obtaining such and such result if the given hypothesis is true. If one is using the rule represented by the rejection region, then this probability statement is also a statement of the probability of one kind of error. But here the direct contribution of statistics stops; the actual decision should depend on other factors, such as potential losses and personal probabilities, that are not a part of the formal mechanism of the statistical test. It does not make much sense to go about making decisions among hypotheses when one has no idea about the goodness of the decision-rule he uses. If the psychologist has enough information about potential losses to be able to see that the conventional rule is not appropriate, then he should be able to figure out a better rule. On the other hand, if the situation in which he finds himself is so vague that he has no conception of the risk of loss through error, then how does he know that deciding among the hypotheses is better than suspending judgment regardless of the evidence? At least, suspending judgment is usually a conservative course of action.

Why, then, do psychologists bother with significance tests at all? *Regardless of what one is going to do with the information—change his opinion, adopt a course of action, or what not—he needs to know relatively how probable is a result like that obtained, given a hypothetical true situation. Basically, a significance test gives this information, and that is all.* The conventions about significance level and regions of rejection can be regarded as ways of defining "improbable." The occurrence of a significant result in terms of these conventions is really a signal: "Here is a direction and degree of deviation which falls among those relatively unlikely to occur given that the tested hypothesis is true, but which is relatively more likely given the truth of some other hypothesis." If one should decide to reject the original hypothesis on the basis of a significant result, then at least he knows the probability of error in doing so. This does not mean that he must decide against the hypothesis simply because some conventional level of significance was met. Other options, such as suspending judgment, may actually be better actions under the circumstances regardless of the result of the conventional significance test. Even more emphatically, the occurrence of a nonsignificant result does not mean that you must accept the hypothesis as true. As we shall see, one often has not the foggiest idea of the error probability in saying that the tested hypothesis is true; here, making a decision to accept the tested hypothesis is absurd in the light of the unknown, and perhaps very large, probability of such an error. On occasion, however, the probability of such an error can be assessed, and here one can feel safe in asserting that the tested hypothesis is true when this error-probability is small. In other circumstances where this probability of error cannot be found, and the experimenter is under no duress to make an immediate choice between hypotheses, then suspending judgment is a relatively safe way out. In still other circumstances the loss-values of errors may be so small that it really does not make much difference what the experimenter decides.

The thoughtless application of the decision-theory terminology to conventional significance-testing can introduce more problems than it solves. Thus, the student is advised to think of hypothesis testing in the way that it is actually used by most scientists: a conventional signalling device saying, for a significant result: "Here is something relatively unlikely given the situation initially postulated, but which is rendered relatively much more likely under an alternative situation. The probability is remote of this being a chance occurrence under the initial hypothesis, and something interesting may be here."

It is perfectly all right to discuss hypothesis testing in the decision-theoretic language, but the user is speaking as though the conventional rule were a good way to decide, which it may not be. *There is no God-given rule about when and how to make up your mind in general!*

9.11 DECIDING BETWEEN TWO HYPOTHESES ABOUT A MEAN

Before we leave the problem of deciding between two exact hypotheses, one more example will be given, this time involving the mean of a population. Furthermore, in this example, something about losses will be known, so that the applicability of a conventional rule to the problem can be studied.

A psychologist working in industry is assigned to study the possibility of using women in a certain job. Heretofore, only men have been used, and it is known that for the population of men on this job, the average performance score is 138, with a standard deviation of 20. If women perform exactly like men then the population of women should have the same mean and standard deviation as the men. On the other hand, if women have mean performance score of at least 142, a considerable profit will accrue to the management by employing them. From what the experimenter already knows about women's performance in other situations, he believes that one of these two possibilities will turn out to be true.

Accordingly, the experimenter frames two hypotheses, both of which deal with a population of women on this job:

$$H_0: \mu = 138$$
$$H_1: \mu = 142.$$

It is assumed that the population distribution in either situation will have a standard deviation of 20.

Now a sample of 100 women is to be drawn at random and put into this job situation on an experimental basis. The psychologist feels that this sample is large enough to permit using the normal approximation to the sampling distribution of the mean.

The following conventional rule is used:

If the sample result falls among the highest 5 percent of means in a normal distribution given H_0, then reject H_0; otherwise, reject H_1.

This region of rejection for H_0 is shown in Figure 9.11.1.

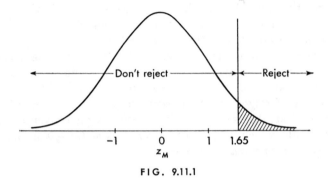

FIG. 9.11.1

In this instance, there are once again two kinds of errors that can be made. The probabilities for each kind are labeled α and β respectively:

		TRUE	
		H_0	H_1
DECIDE	H_0	$1 - \alpha$	β
	H_1	α	$1 - \beta$

In this situation with two *exact* hypotheses, it does make sense to talk about accepting or deciding in favor of H_0, since both probabilities can be known. In essence, the conventional rule says to fix α at .05, and the region of rejection chosen does just that.

However, notice that assigning the value of .05 to the α probability of error automatically fixes β as well. Given that H_0 is true, then the region of rejection must be bounded by a z_M score such that

$$F(z_M) = .95, \text{ or } 1 - F(z_M) = .05.$$

The normal tables show that this z_M score is 1.65. In terms of a sample mean

$$z_M = \frac{M - 138}{\sigma_M} = \frac{M - 138}{2}.$$

Thus, the critical value of M is

$$M = 138 + 1.65\sigma_M = 138 + 3.30 = 141.30.$$

However, what would the z_M score for this critical mean be if H_1 were true?

$$z_M = \frac{141.3 - 142}{2} = -.35.$$

In a normal distribution, $F(-.35) = .36$, approximately, and so we can see that $\beta = .36$. Thus, the two error probabilities are

$$\alpha = .05$$
$$\beta = .36.$$

In short, the β probability of making an error is much greater than α. Our experimenter is rather unlikely to decide on H_1 when H_0 is true, but has considerable chance of concluding H_0 when H_1 is true. Nevertheless, he can draw his sample at random, compute the mean, and decide by this rule: if the sample has a mean that is more than 1.65 standard errors away from 138 then he concludes H_1 is true; otherwise H_0.

Suppose, however, that the possible losses to the management could be figured in this way: a decision that H_1 is true leads the management to hire women for the job, whereas women are not hired if the decision favors H_0. If the test indicates H_1 even though H_0 is true, the management is out the cost of the experiment, $10,000, but no other loss is incurred, since the women hired should perform like the men. On the other hand, if the second kind of error is made, and women are not hired even though they are really better than the men, the management is out not only the cost of the experiment but also the $80,000 profit that might have been made in a year by hiring the women, resulting in a total loss of $90,000. Clearly, the two errors do not have the same consequences for the management. And yet, this rule for decision almost guarantees that the less important error will not be made, while permitting large risk of making the expensive error.

This difficulty can be shown most clearly in terms of subjective expected loss. If the personal probabilities of the experimenter actually were

$$\mathcal{P}(H_0) = \mathcal{P}(H_1) = 1/2, \text{ then}$$
$$SE(u;D) = (10,000)(.05)(.5) + (90,000)(.36)(.5)$$
$$= \$16,250,$$

the subjective expected loss would be over sixteen thousand dollars.

On the other hand, suppose that another rule were adopted, where β is set equal to .05, making α equal to .36. In this circumstance,

$$SE(u;D) = (10,000)(.36)(.5) + (90,000)(.05)(.5)$$
$$= \$4050.$$

Obviously, this second rule is more desirable than the first if the expected loss is to be low. Still better rules exist if the probability β is made even smaller. In this situation, where losses can be considered, the adoption of an arbitrary decision-rule is not very wise.

On the other hand, if a sample does show a mean which is more than 1.65 standard errors away from 138, the experimenter can say that such samples are "improbable" given $\mu = 138$, regardless of whether or not this statement has any connection with a decision-rule he is actually going to use. If he *must* use the arbitrary rule, then the probability α is given by the probability of finding a sample mean in the rejection region for H_0. Nevertheless, the probability β may be so big that this rule is unreasonable for the actual decision.

9.12 CONVENTIONAL DECISION-RULES

In psychology, given some hypothesis H_0 to be tested, the region of rejection is usually found as follows:

CONVENTION: Set α, the probability of falsely rejecting H_0, equal to some small value. Then, in accordance with the alternative H_1, choose a region of rejection such that the probability of observing a sample value in that region is equal to α when H_0 is true. The obtained result is significant beyond the α level if the sample statistic falls within that region.

Ordinarily, the values of α used are .05 or .01, although, on occasion, larger or smaller values are employed. Even though only one hypothesis, H_0, may be exactly specified, and this determines the sampling distribution employed in the test, in choosing the region of rejection one acts as though there were two hypotheses, H_0 and H_1. The alternate hypothesis H_1 dictates which portion or **tail** of the sampling distribution contains the rejection region for H_0. In some problems, the region of rejection is contained in only one tail of the distribution, so that only extreme deviations in a given direction from expectation lead to rejection of H_0. In other problems, big deviations of either sign are candidates for the rejection region, so that the region of rejection lies in both tails of the sampling distribution.

In all that follows, these conventional rules will be used for finding a rejection region. Some hypothesis will be designated as H_0, and an arbitrary α value will be chosen, which, taken together with H_1, determines the sample values that lead to a possible rejection of H_0.

A major problem remains, however. Visualize the graph of a continuous random variable. The line representing the possible values of the random variable is to be divided in such a way that in one region the probability is exactly α, and outside that region the probability is exactly $1 - \alpha$. In how many ways may the X axis be divided into regions satisfying these requirements? A little thought should convince you that an infinite number of such regions might be chosen. What is needed is still another criterion for the selection of the exact interval or intervals to serve as the region of rejection. This criterion is provided by the concept of the power of a test, to be discussed in the next section. We will see that a region of rejection may be selected in such a way as to maximize the power of the statistical test.

9.13 THE POWER OF A STATISTICAL TEST

Given only the one exact hypothesis H_0, and the required value of α, one may still think of an inference involving two hypotheses:

H_0, the hypothesis actually being tested
H_1, the hypothesis, whatever it may be, that is true.

Once any true H_1 is specified, then we can determine β, the probability of an error in decision given the true H_1. Similarly, one may compute the probability $1 - \beta$, which is the probability of being right in rejecting H_0 given that H_1 is true. This probability, $1 - \beta$, is often called the **power** of the statistical test. It is literally the probability of finding out that H_0 is wrong, given the decision-rule and the true value under H_1.

For example, consider the hypotheses in the preceding problem. When α was set equal to .05, then the hypothesis actually being tested was H_0: $\mu = 138$. Suppose, however, that the true situation is represented by H_1: $\mu = 142$. By the rule used, H_0 is rejected when a sample exceeds 141.3. This gives a probability of error β shown by the shaded region in Figure 9.13.1, if H_1 is true.

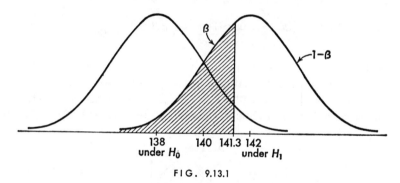

FIG. 9.13.1

The unshaded area under the curve is $1 - \beta$, or the power of the test, given H_1. In this instance the power is about .64, so that the chances of correctly rejecting H_0 are about six in ten. Notice, however that *the power of a test of H_0 cannot be found until some true situation H_1 is specified.*

9.14 POWER OF TESTS AGAINST VARIOUS TRUE ALTERNATIVES

The power of a test of H_0 is not unlike the power of a microscope. It reflects the ability of a decision-rule to detect from evidence that the true situation differs from a hypothetical one. Just as a high-powered microscope lets us distinguish gaps in an apparently solid material that we would miss with low power or the naked eye, so does a high-powered test of H_0 almost insure us of detecting when H_0 is false. Pursuing the analogy further, any microscope will reveal "gaps" with more clarity the larger these gaps are; the larger the departure of H_0 from the true situation H_1, the more powerful is the test of H_0, other things being equal.

For example, suppose that the hypothesis to be tested is

$$H_0: \mu = \mu_0$$

but that the true hypothesis is

$$H_1: \mu = \mu_1.$$

Here, μ_0 and μ_1 symbolize two different possible numerical values for μ. It will be assumed that under either hypothesis the sampling distribution of the mean is normal with σ known. Suppose now that the α probability of error is set at .05, and the rejection region is on the *right tail* of the distribution. The decision-rule is such that the critical value of M (the smallest M leading to the rejection of H_0) is $\mu_0 + 1.65\sigma_M$. The power of the test can then be figured for any possible true H_1.

First of all, suppose that μ_1 were actually $\mu_0 + \sigma_M$. Then in the distribution under H_1, the critical value of M has a z score given by

$$z_M = \frac{M - \mu_1}{\sigma_M}$$

$$= \frac{(\mu_0 + 1.65\sigma_M) - (\mu_0 + 1\sigma_M)}{\sigma_M} = .65.$$

The probability of a sample's falling into the region of rejection for H_0 is found from Table I to be .26. This probability is $1 - \beta$, the power of the test.

This may be clarified by Figure 9.14.1. The region of rejection for H_0 is the segment of the horizontal line to the right of $M = \mu_0 + 1.65\sigma_M$. The shaded

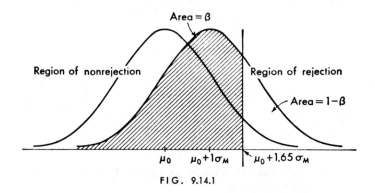

FIG. 9.14.1

portion under the curve for $\mu = \mu_0 + 1\sigma_M$ denotes β, the probability of *failing* to reject H_0, and the power is the unshaded part of the curve.

If another alternative hypothesis were true, say,

$$H_1: \mu_1 = \mu_0 + 3\sigma_M$$

then the power would be much larger. Here, relative to the sampling distribution based on the *true* mean, or $\mu_0 + 3\sigma_M$, the critical value of M would correspond to a z_M score of

$$z_M = \frac{(\mu_0 + 1.65\sigma_M) - (\mu_0 + 3\sigma_M)}{\sigma_M} = -1.35.$$

Table I shows us that above a z score of -1.35 lie .91 of sample means in a normal sampling distribution, so that the power is .91.

The power of the test for any true value of μ_1 can be found in the same way. Often, to show the relation of power to the true value of μ_1, so called **power-functions** or **power curves** are plotted. One such curve is given in Figure 9.14.2, where the horizontal axis gives the possible values of true μ_1 in terms of μ_0 and σ_M, and the vertical axis the value of $1 - \beta$, the power for that alternative. Notice that for this particular decision-rule, the power curve rises for increasing values of μ_1, and approaches 1.00 for very large values. On the other hand, for this decision-rule, when true μ_1 is less than μ_0 the power is very small, and approaches 0 for decreasing μ_1 values. In any statistical test where the region of rejection is in the direction of the true value covered by H_1, the greater the discrepancy between the tested hypothesis and the true situation, the greater the power.

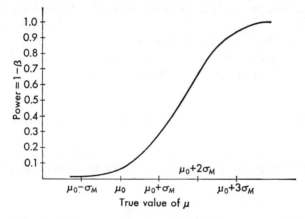

FIG. 9.14.2. Power curve for a one-sided test of a mean, $\alpha = .05$

9.15 POWER AND THE SIZE OF α

Since β will ordinarily be small for large α, as we have seen in the discussion above, it follows that setting α larger makes for relatively more powerful tests of H_0. For example, the two power curves given below for the same decision-rule show that if α is set at .10 rather than at .05, the test with $\alpha = .10$ will be more powerful than that for $\alpha = .05$ over all possibly true values under alternative H_1 (note Figure 9.15.1). Making the probability of error in rejecting H_0 larger has the effect of making the test more powerful, other things being equal.

In principle, if it is very costly to make the mistake of overlooking a true departure from H_0, but not very costly to reject H_0 falsely, one could (and

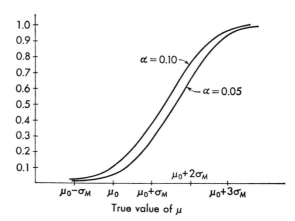

FIG. 9.15.1. Power curves for two tests with equal sample sizes but with different values of α

perhaps should) make the test more powerful by setting the value of α at .10, .20, or more. This is not ordinarily done in psychological research, however. There are at least two reasons why α is seldom taken to be greater than .05: In the first place, as observed in Section 9.9, the problem of relative losses incurred by making errors is seldom faced in psychological research; hence conventions about the size of α are adopted. The other important reason is that given some fixed α the power of the test can be increased either by increasing sample size or by reducing the standard error of the test statistic in some other way.

9.16   THE EFFECT OF SAMPLE SIZE ON POWER

Given a population with true standard deviation σ, the standard error of the mean depends inversely upon the square root of sample size N. That is,

$$\sigma_M = \frac{\sigma}{\sqrt{N}}.$$

When N is large, then the standard error is smaller than when N is small. Provided that $1 - \beta > \alpha$ to begin with, *increasing the sample size increases the power of a given test of H_0 against a true alternative H_1.*

For example, suppose that

$$H_0: \mu = 50$$
$$H_1: \mu = 60$$

where H_1 is true. Let us assume that true $\sigma = 20$. If samples of size 25 are taken, then

$$\sigma_M = \frac{20}{5} = 4.$$

Now the α probability for this test is fixed at .01, making the critical value of M be

$$M = \mu_0 + 2.33\sigma_M = 50 + (2.33)(4) = 59.32.$$

In the true sampling distribution (under H_1) this amounts to a z score of

$$z_M = \frac{59.32 - 60}{4} = \frac{-.68}{4} = -.17$$

so that .57 of all sample means should fall into the rejection region for H_0. The power here is thus .57.

Now let the sample size be increased to 100. This changes the standard error of the mean to

$$\sigma_M = \frac{20}{10} = 2$$

and the critical value of M to

$$M = 50 + (2.33)(2) = 54.66.$$

The corresponding z score when $\mu = 60$ is

$$z_M = \frac{54.66 - 60}{2} = -2.67$$

making the power now in excess of .99. With this sample size we would be almost certain to detect correctly that the H_0 is false when this particular H_1 is true. With only 25 cases, we are quite likely not to do so.

The disadvantages of an arbitrary setting of α can thus be offset, in part, by the choice of a large sample size. Other things being equal and regardless of the size chosen for α, the test may be made powerful against any given

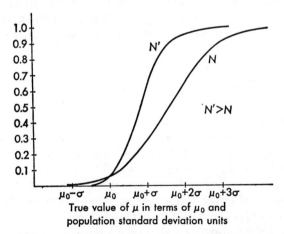

FIG. 9.16.1. Power curves for two tests with $\alpha = .05$ but with different sample sizes, $N' > N$

alternative H_1 in the direction of the rejection region, provided that sample N can be made very large. (See Figure 9.16.1.)

Once again, however, it is not always feasible to obtain very large samples. In most psychological research, samples of substantial size are costly for the experimenter, if not in money, then in effort. Our ability to attain power through large samples only partly offsets the failure to choose a decision-rule according to cost of error in such research; large samples may not really be necessary in some research using them, especially if the error thereby made improbable is actually not very important. This matter of sample size and power will be considered once again in Chapter 10.

9.17 POWER AND "ERROR VARIANCE"

Even given a fixed sample size, the experimenter has one more device for attaining power in tests of hypotheses. *Anything that makes σ, the population standard deviation, small will increase power, other things being equal.* This is one of the reasons for the careful control of conditions in good experimentation. By making conditions constant, the experimenter rules out many of the factors that contribute to variation in his observations. Statistically this amounts to a relative reduction in the size of σ for some experimental population. When the experimenter rules out some of the error variance from his observations he is *decreasing* the standard error of the mean, and thus *increasing* the power of the test against whatever hypothesis H_1 is true. Experiments in which the variability attributable to experimental or sampling error is small are said to be **precise;** the result of such precision is that the experimenter is quite likely to be able to detect when something of interest is happening. The application of experimental controls is like restricting inferences to populations with smaller values of σ^2 than otherwise, and thus control over error variation through careful experimentation implies powerful statistical tests. It follows that controlled experiments in which there is little "natural" variation in the materials observed can attain statistical power with relatively few observations, while those involving extremely variable material may require many observations to attain the same degree of power.

9.18 TESTING INEXACT HYPOTHESES

The primary notions of errors in inference and the power of a statistical test have just been illustrated for an extremely artificial situation, in which a decision must be made between two exact alternatives. Such situations are almost nonexistent in psychological research. Instead, the experimenter is far more likely to be called on to evaluate inexact hypotheses, each of which encompasses a whole range of possibly true values. What relevance, then, does the discussion of the exact two-alternative case have to what psychologists actually do? The answer is that the researcher makes his inferences as though he were deciding between two exact alternatives, even though his interest lies in judging between inexact hypotheses. Thus, the mechanism we have been using for decisions between exact hypotheses is exactly the same as for any other set of alternatives.

An example will probably clarify this point. As a fairly plausible situation, imagine a psychologist who has constructed a new test of adult intelligence. This test has been very carefully standardized on a population of American adults. The standardization has been so carried out that the mean score is 100 and the standard deviation of the test is 15. Furthermore, the distribution of scores in the standardization population is approximately normal.

The psychologist is now interested in the application of the test to residents of England. From his knowledge of the abilities required by the test he has some reason to suspect that English adults may, on the average, score somewhat higher than American adults. So, in order to arrive at some idea if this may be the case, and especially to see if the test may require some scoring modification for administration in England, he decides to give the test to a sample of English adults. Basically, the question to be answered is "Do English adults tend to score higher on this test than Americans?" We will suppose that there is no reason to question the standard deviation of scores among the English, nor the general form of the distribution of scores.

The answer to his question is tantamount to a decision between two *inexact* hypotheses:

$$H_0: \mu \leq 100$$
$$H_1: \mu > 100$$

i.e., the English population has either a mean score less than or equal to the American mean, or a mean greater than the American mean.

In choosing the decision-rule he will use, the experimenter decides to set α equal to .01. A result greater than 100 tends to favor H_1, and so he takes as his region of rejection all z scores greater than or equal to 2.33, since this value cuts off the highest 1 percent of sample means in a normal sampling distribution.

What, however, is the hypothesis actually being tested? As written, H_0 is inexact, since it states a whole region of possible values for μ. One exact value is specified, however: this is $\mu = 100$. Actually, then, the hypothesis tested is $\mu = 100$ against some unspecified alternative greater than 100. In effect, the decision-rule can be put in the following form:

TRUE SITUATION

$\mu = 100 \quad \mu > 100$

	$\mu = 100$	$\mu > 100$
$\mu = 100$	.99	(β)?
DECIDE		
$\mu > 100$	.01	$(1 - \beta)$?

The α probability of error can be specified in advance as .01, but β and the power are unknown, depending as they do upon the true situation. The experimenter has no real interest in the hypothesis that $\mu = 100$; he may even feel extremely confident that the true mean is not precisely 100. Nevertheless, this is a useful dummy hypothesis, in the sense that *if he can reject $\mu = 100$ with $\alpha = .01$, then he can reject any other hypothesis that $\mu < 100$ with $\alpha < .01$.* In other words, by this decision-rule, if he can be confident that he is not making an error in rejecting the hypothesis actually tested, then he can be even more confident in rejecting any other hypothesis covered by H_0.

But what does this do to β? Given some true mean μ_1 covered by H_1, *the power of the test of $\mu = 100$ is less than the power for any other hypothesis covered by H_0 with fixed α.* If he had chosen to test any other exact hypothesis embodied in H_0, such as $\mu = 90$, then neither α nor β would exceed those for $\mu = 100$. Testing the exact hypothesis with given α and β probabilities can be regarded as testing *all* hypotheses covered by H_0, with *at most* α and β probabilities of error (the β depending on the *true* mean, of course). This is illustrated in Figure 9.18.1.

Suppose now that a sample of 200 cases yields a mean of 103.5. The standard error of the mean is

$$\sigma_M = \frac{15}{\sqrt{200}} = 1.06.$$

Thus, the z score corresponding to the sample mean is

$$z_M = \frac{103.5 - 100}{1.06} = 3.30.$$

Since this z score exceeds the critical value of 2.33, the hypothesis that $\mu = 100$ can be rejected with $\alpha < .01$, and any other hypothesis covered by H_0 can be rejected with α less than .01. If the experimenter decides that the true mean is greater than 100 he could be wrong in this decision, but the probability of such an error is less than .01. Notice that we did not say that the experimenter must decide in this way, but only that the α probability of error is slight should he make this decision.

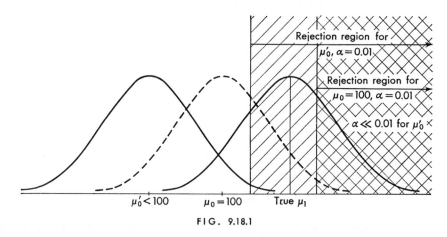

FIG. 9.18.1

On the other hand, suppose that the sample mean had been only 102. Here the z_M score would have been only 1.89, which is not large enough to place the sample in the region of rejection for H_0 if the probability α is to be .01. Under the decision-rule using $\alpha = .05$ this would lead to a rejection of the hypothesis, but not by the rule originally decided upon.

The same problem can be put on a much more realistic basis, particularly with regard to sample size, if the experimenter has an idea of some minimal value of μ he would like to be very sure to detect as a significant result if true. That is, suppose the experimenter in this problem knows that it is absolutely essential that he restandardize the test if the mean for the population of English adults is *at least* 103. This is important enough to him that he wants the power of the test to be at least .95 when μ is at least 103. Does his choice of a sample size and of a rejection region meet this qualification?

For $N = 200$, the standard error of the mean is 1.06. Consequently, the critical value of the mean is

$$\text{critical } M = 2.33(1.06) + 100$$
$$= 102.47.$$

If the value of μ were actually 103, then in this sampling distribution under the true value of μ the critical value of M would correspond to

$$z_M = \frac{102.47 - 103}{1.06} = -.5.$$

In a normal distribution, some 69 percent of cases must fall at or above $-.5$ standard errors from the mean. Hence, the power of this test against the true alternative $\mu = 103$ is only .69. If the experimenter wants this power to be .95 for $\mu = 103$, then he must take a larger sample. In fact, the sample size required can be found from

$$-1.65 = \frac{M - 103}{\sigma/\sqrt{N}}$$

and

$$2.33 = \frac{M - 100}{\sigma/\sqrt{N}}.$$

Solving for N, we find that the required sample size is 396. Then for this sample size we can find the following probabilities of error and correct decisions:

		TRUE SITUATION		
		$\mu \leq 100$	$100 < \mu < 103$	$103 \leq \mu$
DECIDE	H_0	$1 - \alpha \geq .99$	$.99 > \beta > .05$	$.05 \geq \beta$
	H_1	$\alpha \leq .01$	$.01 < 1 - \beta < .95$	$.95 \leq 1 - \beta$

In circumstances where there *is* such an important range of possible values of μ, representing a situation we want to be very sure to detect if true, then we can adjust the sample size or α value or both so as to make the power against such alternatives as great as necessary.

Obtaining a nonsignificant result (one not leading to a rejection of H_0 in favor of H_1) often puts the experimenter in a quandary. He cannot make the decision to reject H_0 and still stay within the bounds on α set by his decision-rule. Does this mean, however, that he must decide that H_0 is true? Most emphatically not! In the first place the experimenter does not have to decide anything; he may be content merely to report "H_0 not rejected," or "nonsignificant result," and let it go at that. On the other hand, he might want to adopt some more definite course of action, and here the decision to accept H_0 as true really makes sense only if the probability of error in such a decision is known. In the examples above, where the choice lay between two exact hypotheses, the probability for each kind of error could be calculated, and so the risk in accepting H_0 as true could be known. In examples such as the present one, the β probability of error is unknown unless we are willing to postulate some exact value of μ as true. When one is operating with completely unknown error-probabilities, and the cost of an error may be great, then the decision to accept H_0 may be very unwise.

In the case of nonsignificant results a wise decision may be to suspend judgment and wait for more evidence. This is true particularly when the error of falsely accepting H_0 could be very costly. When the sample result is in the direction indicated by H_1, but is not sufficiently deviant from H_0 to warrant rejection, a mere increase in sample size might very well permit one to reject H_0. Thus, in many experimental contexts, failing to reject H_0 might appropriately be interpreted as suspension of judgment, since the experimenter has considerable stake in avoiding *either* kind of error, and wants to be sure to adopt a conservative course of action when H_0 cannot be rejected.

On the other hand, occasions exist when accepting H_0 makes quite good sense, even when the hypotheses are both inexact. In these circumstances the experimenter is able to calculate the probability of a sample such as the one he obtained, given that one of the values covered by H_1 were true. Here, he asks the probability of a sample result differing this much or more from expectation *in the direction of H_0*, given that a value covered by H_1 is true. Perhaps an example will make this point clear. Consider the developer of the intelligence test once again. Suppose that the sample mean turns out to be 98 in the test of $H_0: \mu \leq 100$, against $H_1: \mu > 100$. When it is true that $\mu = 100$ this sample result represents a standardized value of

$$z_M = \frac{98 - 100}{1.06} = -1.89.$$

This result falls quite far from the region of rejection, and so one does not accept H_1 as true; in fact, our best estimate of the value of μ is not 100, but rather 98, a value not even included in the hypothesis H_1. Each and every possible value of μ covered by H_1 is greater than 100, and so the z_M value for any μ in H_1 must be *less* than -1.89.

Now suppose that the experimenter decides to *accept* $H_0: \mu \leq 100$ as true in the light of evidence such as this. How much risk is he taking? In a normal distribution, $F(-1.89)$ is approximately .03, so that the probability of error is *less* than .03 in saying that H_0 is true on evidence such as this when a value covered by H_1 is true. In other words, the experimenter who decides to accept H_0 on evidence this much or more in disagreement with H_1 is running very little risk of error. When the alternative hypothesis is "directional," as in this example, the experimenter may be quite justified in saying that H_0 is true, provided that he can assess the error probability involved in making this statement on this kind of evidence.

In summary, if there is no way at all to gain some idea of the risk run in falsely *accepting* H_0, and such errors are at all important, then suspension of judgment may be the best policy following a nonsignificant result. On the other hand, if the obtained result does agree with expectation under values covered by H_0, but would fall among the deviant and unlikely results given any value covered by H_1, then H_0 may very well be accepted as true. Either decision is possible, and how one interprets the occurrence of a nonsignificant result depends very much on what can be learned about the error-probabilities involved. There is really no rule about this; nevertheless, the experimenter is wise to make no actual decision between H_0 and H_1 unless he has some definite notion of the error probabilities involved. The risk run in accepting H_1 (rejecting H_0) is, of course, known to be α, but the risk run in accepting H_0 (the probability β) is not, in general, known. For this reason, the decision to accept H_0 as true is reserved for situations where the experimenter has some idea of the risk he is taking.

9.19   TYPE I AND TYPE II ERRORS

The time has come to dignify with names the two kinds of errors we have been discussing. **Type I error is that made when H_0 (the tested hypothesis) is falsely rejected.** One makes a Type I error whenever the sample result falls into the rejection region even though H_0 is true. Thus the probability α gives the risk run of making a Type I error.

The errors of Type II are those made by not rejecting H_0 when it is false. When a sample result does not fall into the rejection region, even though some H_1 is true, we are led to make a Type II error. Thus for a given true alternative H_1, β is the probability of Type II error. It follows that

$$\text{power} = 1 - \text{probability of Type II error.}$$

9.20   THE BASIS FOR THE CONVENTIONS ABOUT TYPE I ERRORS

The conventions about the permissible size of the α probability of Type I error actually grew out of a particular sort of experimental setting. Here it is known in advance that one kind of error is extremely important and is to be avoided. In this kind of experiment these conventional procedures do make sense when viewed from the decision-making point of view. Furthermore, designation of the hypothesis H_0 as the "null" hypothesis and the arbitrary setting of the level of α can best be understood within this context.

As an example of an experimental setting where Type I error is clearly to be avoided, imagine that one is testing a new medicine, with the goal of deciding if the medicine is safe for the normal adult population. By "safe" we will mean that the medicine fails to produce a particular set of undesirable reactions on all but a very few normal adults. Now in this instance, deciding that the medicine is safe when actually it tends to produce reactions in a relatively large proportion of adults is certainly an error to be avoided. Such an error might be called "abhorrent" to the experimenter and the interests he represents. Therefore, the hypothesis "medicine unsafe" or its statistical equivalent is cast in the role of the null hypothesis, H_0, and the value of α chosen to be extremely small, so that the abhorrent Type I error is very unlikely to be committed. A great deal of evidence against the null hypothesis is required before H_0 is to be rejected. The experimenter has complete control over Type I error, and regardless of any other feature his study of the medicine may have, he can be confident of taking very little risk of asserting that H_1, or "medicine safe" is true when actually H_0, or "medicine unsafe" is true.

In other words, the conventional practice of arbitrarily setting α at some very small level is based on the notion that one kind of error is extremely important and must be avoided if possible. This is quite reasonable in some contexts, such as the study of the safety of a new medicine.

On the other hand, the experimenter always has some control over the values of β and the power of the test over the various possibly true alternatives. By the choice of an appropriate sample size and by exercising control over the size of σ^2 in the population considered, the value of β for any possibly true alternative to H_0 may be made as small (or conversely, the power as large) as desired. If a number of different ways to test H_0 are available in terms of different sample statistics, some of these ways may give more powerful tests than others regardless of what the true situation actually is, and the experimenter chooses the best procedure among those available. However, the essential points are that the experimenter is absolutely free to set α at any level he chooses, and that the conventional levels are dictated by the notion that Type I errors are bad and must be avoided. The experimenter does not have this same freedom with respect to Type II errors; power must be bought in terms of sample size and other features of the test procedure. Power can always be made large against any given alternative, but only at some cost to the experimenter. An inappropriately small value of α makes it more difficult than otherwise to achieve a powerful test.

Within contexts such as the test of a new medication, where Type I error is abhorrent, setting α extremely small is manifestly appropriate. Here, considerations of Type II error are actually secondary. In some instances in psychology a similar situation exists, where one kind of error is clearly to be avoided, and from the outset the experimenter wants to be sure that this kind of error is very improbable. The designation of one hypothesis as H_0 should rest on which kind of error is to have the small probability α. In the example of Section 9.11 the experimenter using the conventional rule should have designated $\mu = 142$ as H_0 for this very reason. On the other hand, in some psychological research it is very hard to see exactly why the particular hypothesis tested, or H_0, should be the one we are loath to abandon, and why Type I errors necessarily have this drastic character. Granting that scientific discretion is commendable, the mistaken conclusion that "something really happened" is not *necessarily* worse than overlooking a real experimental phenomenon. In some situations, perhaps, we should be far more attentive to Type II errors and the power of our tests, and less attentive to setting α at one of the conventional levels. Furthermore, if the conventional α levels are to be used, a little more thought might be given to deciding exactly what *is* the null hypothesis we want to be so careful not to reject falsely.

Incidentally, there is an impression in some quarters that the term "null hypothesis" refers to the fact that in experimental work the parameter value specified in H_0 is very often zero. Thus, in many experiments the hypothetical situation "no experimental effect" is represented by a statement that some mean or difference between means is exactly zero. However, as we have seen, the tested hypothesis can specify any of the possible values for one or more parameters, and this use of the word "null" is only incidental. It is far better for the student to think of the null hypothesis H_0 as simply designating that hypothesis actually being tested, the one which, if true, determines the sampling distribution actually referred to in the test, and which specifies the kind of error called Type I.

9.21 ONE-TAILED REJECTION REGIONS

In the example of Section 9.18 two inexact hypotheses were compared, having the form

$$H_0: \mu \leq \mu_0$$
$$H_1: \mu > \mu_0.$$

Here the entire range of possible values for the parameter under study (in this case, the population mean) was divided into two parts, that above and that below (or equal to) an exact value μ_0. The interest of the experimenter lay in placing the true value of μ either above or below μ_0.

In this instance the appropriate region of rejection for H_0 consists of values of M relatively much larger than μ_0. Such values have a rather small probability of representing true means covered by the hypothesis H_0, but are more likely to represent true means in the range of H_1. Such a rejection region consisting of sample values in a particular direction from the expectation given by the exact value included in H_0 are called **directional** or **one-tailed** rejection regions. For the particular hypotheses compared in Section 9.18, the rejection region was one-tailed, since only the right or "high-value" tail of the sampling distribution under H_0 was considered in deciding between the hypotheses.

For some questions, the two inexact hypotheses are of the form

$$H_0: \mu \geq \mu_0$$
$$H_1: \mu < \mu_0.$$

Once again the region of rejection is one-tailed, but this time the lower or left tail of the sampling distribution contains the region of rejection for H_0. The choice of the particular rejection region thus depends both on α and the alternative hypothesis H_1.

Tests of hypotheses using one-tailed rejection regions are also called **directional**. The direction or sign of the value of the statistic (such as z_M) is important in directional tests since the sample result must show not only an extreme departure from expectation under H_0 but also a departure in the right direction to be considered strong evidence against H_0 and for H_1.

Directional hypotheses are implied when the basic question involves terms like "more than," "better than," "increased," "declined." The essential question to be answered by the data has a clear implication of a difference or change in a specific direction. For example in the problem of Section 9.18 the experimenter wanted specifically to know if Englishmen make higher scores than Americans, indicating a directional hypothesis.

Many times, however, the experimenter goes into a problem without a clearly defined notion of the direction of difference to expect if H_0 is false. He asks "Did something happen?" "Is there a difference?" or "Was there a change?" without any specification of expected direction. Next we will examine techniques for nondirectional hypothesis-testing.

9.22 Two-tailed tests of hypotheses

Imagine a study carried out on the "optical dominance" of human subjects. There is interest in whether or not the dominant eye and the dominant hand of a subject tend to be on the same or different sides. Subjects are to be tested for both kinds of dominance, and then classified as "same side" or "different side" in this respect. We will use the letters "S" and "D" to denote these two classes of subjects.

The experimenter knows that in 70 percent of subjects in this population the right hand is dominant. He also knows that the right eye is dominant for 70 percent of subjects. From this knowledge he already has about the relative frequency of each kind of dominance the experimenter reasons that if there actually is no tendency for eye dominance to be associated with hand dominance, then in a particular population of subjects he should expect 58 percent S and 42 percent D. (Why?) However, he knows little about what to expect if there is some connection between the two kinds of dominance. To try to answer this question, our experimenter draws a random sample of 100 subjects, each with a full set of eyes and hands, and classifies them.

The question at hand may be put into the form of two hypotheses, one exact and one inexact:

$$H_0: p = .58$$
$$H_1: p \neq .58.$$

The first hypothesis represents the possibility of no connection between the two kinds of dominance, and the inexact alternative is simply a statement that H_0 is not true, since H_1 does not specify an exact value of population p.

The α chosen is .05. The experimenter then is faced with the choice of a rejection region for the hypothesis H_0. Either a very high percentage of S subjects in his sample or a very low percentage would tend to discount the credibility of the hypothesis that $p = .58$, and would lend support to H_1. Thus he wants to arrange his rejection region so that the H_0 will be rejected when extreme departures from expectation of *either* sign occur. This calls for a rejection region on *both* tails of the sampling distribution of sample P when $p = .58$. Since the sample size is relatively large the binomial distribution of sample P may be

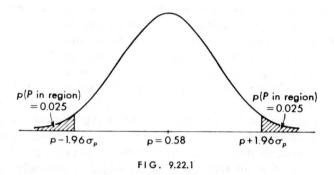

p(P in region) = 0.025   p(P in region) = 0.025

$p - 1.96\sigma_p$ $p = 0.58$ $p + 1.96\sigma_p$

FIG. 9.22.1

approximated by a normal distribution, and the two rejection regions may be diagrammed as shown in Figure 9.22.1.

Since the total probability of Type I error has been set at .05, the rejection region on the upper tail of the distribution will contain *the highest 2.5 percent,* and the lower tail *the lowest 2.5 percent of sample proportions,* given that $p = .05$. In short, each region should contain exactly $(1/2)\alpha$ or $\alpha/2$ proportion of all samples under H_0. Consequently, either a very large or very small sample P will lead to a rejection of H_0, and the total probability of Type I error is .05. The z score cutting off the upper rejection region is 1.96, and that for the lower is -1.96. Any sample giving a z beyond these two limits will lead to a rejection of H_0.

Since the basic sampling distribution is binomial, the standard error of P is

$$\sigma_P = \sqrt{\frac{pq}{N}} = \sqrt{\frac{(.58)(.42)}{100}}$$
$$= .049$$

when H_0 is true (Section 6.22). Using the normal approximation to the binomial we find

$$z = \frac{P - p}{\sigma_P}.$$

Now suppose that the proportion of S subjects is .69. Then

$$z = \frac{.69 - .58}{.049} = 2.24.$$

This value exceeds the critical z score of 1.96, and so the result is said to be significant beyond the 5 percent level. If the experimenter concludes that eye and hand dominance do tend to be related then he runs a risk of less than .05 of being wrong.

Suppose, however, that the sample result had come out to be only .53. Then

$$z = \frac{.53 - .58}{.049} = -1.02$$

making the result nonsignificant. Here, the experimenter should most likely suspend judgment pending further evidence. He should not be willing to assert that

H_0 is true: the risk in that assertion is unknown. Indeed, his best guess about the true value of p is .53, not .58, and given enough sample observations, a P value of .53 might be enough to warrant rejection. Thus, for a nonsignificant result such as this the experimenter cannot wisely accept H_0 as true. His best available choice may be to suspend judgment, and look for more evidence.

One additional refinement of this test must be pointed out here. The actual sampling distribution of concern here is the binomial, which is, of course, discrete. On the other hand, we are approximating this discrete distribution by a continuous normal distribution. In order that the fit between binomial probabilities and the probabilities for intervals under a normal curve be as good as possible, the "correction for continuity" (Section 8.5) is generally made. Such a correction consists in regarding a given P value as simply the midpoint in an interval of values with limits $P - .5/N$ and $P + .5/N$. Then the z value is computed, not from the difference between P and p, but for the difference between the limiting value of the interval and p. That is, we take

$$z = \frac{P - p - (.5/N)}{\sigma_P}$$

if P happens to be greater in value than p, or we take

$$z = \frac{P - p + (.5/N)}{\sigma_P}$$

If P happens to be smaller in value than p. For the example above, the correction for continuity amounts to .5/100 or .005. This has the result of making the z value in the test 2.14 rather than the 2.24 found previously.

Obviously, the correction for continuity can make a difference in the conclusions reached from a test, although it happened in this example not to do so. In general, the correction really should be used, especially when the sample size N is relatively small.

Although the example just concluded dealt with a hypothesis about a proportion, much the same procedure applies to two-tailed tests of means. An exact and an inexact hypothesis are opposed:

$$H_0: \mu = \mu_0$$
$$H_1: \mu \neq \mu_0.$$

The exact hypothesis is tested by forming two regions of rejection, each region containing exactly $\alpha/2$ proportion of sample results when H_0 is true, and lying in the higher and lower tails of the distribution. A sample result falling into either of these regions of rejection is said to be significant at the α level. If the result is not significant, then judgment is suspended.

9.23   RELATIVE MERITS OF ONE- AND TWO-TAILED TESTS

In deciding whether a hypothesis should be tested with a one- or two-tailed rejection region, the primary concern of the experimenter must be his original question. Is he looking for a directional difference between populations, or a difference only in kind or degree? By and large, most significance tests done

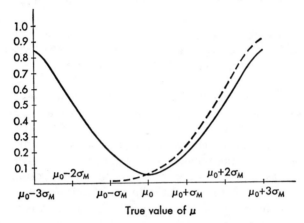

FIG. 9.23.1. Power curves for a two-tailed and a one-tailed test of a mean, $\alpha = .05$

in psychological research are nondirectional, simply because research questions tend to be framed this way. However, there are situations where one-tailed tests are clearly indicated by the question posed.

The powers of one- and two-tailed tests of the same hypothesis will be different, given the same α level and the same true alternative. If a one-tailed test is used, and the true alternative is in the direction of the rejection region, then the one-tailed test is more powerful than the two-tailed over all such possibly true values of μ. In a way, we get a little statistical credit in the one-tailed test for asking a more searching question. On the other hand, if the true alternative happens to be on the tail opposite the rejection region in a one-tailed test, the power is very low; in fact, the power will always be less than α in the test of a mean. If you will, we are penalized for framing a stupid question. Power curves for one- and two-tailed tests for means are compared in Figure 9.23.1.

In many circumstances calling for one-tailed tests the form of the question or of the sampling distribution makes it clear that the only alternative of logical or practical consequence must lie in a certain direction. For example, when we turn to the problem of testing many means for equality simultaneously (Chapter 12) it will turn out that the only rejection region making sense lies on one tail of the particular sampling distribution employed. In other circumstances the question involves considerations of "which is better" or "which is more," where the discovery of a difference from H_0 in one direction may have real consequences for practical action, although a mean in another direction from H_0 may indicate nothing. For example, consider a treatment for some disease. The cure rate for the disease is known, and we want to see if the treatment *improves* the cure rate. We have not the slightest practical interest in a possible decrease in cures; this, like no change, leads to nonadoption of the treatment. The thing we want to be sure to detect is whether or not the new treatment is really better than what we have. Thus, we can safely ignore the low power for detecting a poorer treatment, so long as we have high power for detecting a really better treatment. In many such problems where different practical actions depend on the sign of the deviation from expectation, a one-tailed test is clearly called for. Otherwise, the two-tailed test is safest for the experimenter asking only "What happens?"

9.24   SOME REMARKS ON THE GENERAL THEORY UNDERLYING TESTS OF HYPOTHESES

For purposes of exposition the kinds of hypotheses considered in this chapter have been very elementary ones concerning the mean μ or the population proportion p. The hypotheses about μ have actually been composite as stated, but these have, in effect, been turned into simple hypotheses by our assumption that both σ^2 and the form of the population distribution are known. Working with these elementary examples has permitted us to use the familiar normal and binomial sampling distributions, and to illustrate decision-rules, error probabilities, and power in the simplest possible terms. Finally, and most important, we pretended that we knew that the best way to test a particular hypothesis was in terms either of the statistic M or the statistic P; that is, the question of the relevant and appropriate statistic was not allowed to come up.

As it happens, M and P actually are the appropriate statistics for the particular hypotheses tested; however, as we shall see in succeeding chapters, in testing other hypotheses (particularly composite hypotheses) various other statistics that are not necessarily estimators for any population parameters may provide us the best ways of carrying out the tests. Such statistics, whose primary role is that of providing a test of some hypothesis, are usually called **test statistics,** and the hypothesis test itself is carried out in terms of the sampling distribution of the appropriate test statistic, rather than that of some estimator such as M, $\hat{s}^2$, or P.

In mathematical statistics a general theory of hypothesis testing exists that extends the elementary ideas discussed here to much more complex situations. Among other things this theory specifies desirable characteristics that a test statistic should have and indicates methods for finding such test statistics, all within the general framework outlined here: some particular H_0 and H_1 are formulated and a value of α is chosen arbitrarily by the experimenter; some appropriate test statistic is selected and the actual decision-rule is formulated in terms of rejection regions in the sampling distribution of this statistic. Much of the theory of hypothesis testing deals with the choice of the *right* statistic for the purpose.

Among the desirable properties of a test based on some statistic is that of **unbiasedness**: a test statistic (and decision-rule) is said to afford an unbiased test of the hypothesis H_0 if the use of this statistic (and rule) makes the probability of rejecting H_0 be at its smallest when H_0 is actually true. An unbiased test will have *minimum* power when H_0 is true.

Another property of a good test is **consistency**: in somewhat imprecise terms, a consistent test is one for which the probability of rejecting a *false* H_0 approaches 1.00 as the sample size approaches ∞. That is, a consistent test always gains power against any true alternative as the sample size is increased.

Still another criterion of a good test is that it be **most powerful for the particular true alternative** to the hypothesis H_0. That is, among all of the different ways that one could devise to test some particular H_0 against some particular true alternative value covered by H_1, the most powerful test affords the smallest probability of Type II error, other things such as sample size being equal. **Uniformly most powerful tests** are those which give a smaller probability of Type II error than any other test regardless of the value which happens to be true among those covered by H_1, other things being equal.

The general theory of hypothesis testing first took form under the hand of Sir Ronald Fisher in the nineteen-twenties, but it was carried to a high state of development in the work of J. Neyman and E. S. Pearson, beginning about 1928. Neyman and Pearson introduced a device for finding "good" tests of hypotheses, a device known as the "likelihood ratio" test. This is a general procedure for finding the test statistic that will have optimal properties for testing any of a broad class of hypotheses, particularly the statistic that will give a most powerful test for a specific hypothesis. The theory underlying the likelihood ratio is very closely allied to the maximum-likelihood principle mentioned in Section 7.20, and involves the maximum likelihood of the particular sample result *given* the hypothesis H_0 relative to the maximum likelihood of the sample result over all possible values of the relevant parameters. Unfortunately, space and the mathematical level assumed here do not permit us to go further into this topic; the student with some command of the elementary calculus will find the likelihood ratio explained clearly in the text by Mood and Graybill (1963). For the moment, suffice it to say that the "classical" tests of hypotheses to be discussed in Chapter 10 and thereafter are all equivalent to likelihood ratio tests. These are "best" available tests for the hypotheses considered, meaning that when the assumptions underlying these tests are true, no other available procedure would answer our question about H_0 and H_1 better than the one we will use.

9.25 CONFIDENCE INTERVALS

Hypothesis testing is but one form of statistical inference commonly used in psychological research. In many circumstances the primary purpose of data collection is not to test a hypothesis, but rather to obtain an estimate of some parameter. It was shown in Chapter 7 that the sample mean is an unbiased estimate of the population mean, and in the maximum-likelihood sense, the sample mean is our best estimate of its normal population counterpart.

Nevertheless, it is clear that sample mean ordinarily will not be equal to the true population value because of sampling error. It is necessary to qualify our estimate in some way to indicate the general magnitude of this error. Usually this is done by showing a confidence interval, which is **an estimated range of values with a given high probability of covering the true population value.** When there is a large degree of sampling error the confidence interval calculated from any sample will be large; the range of values likely to cover the population mean is wide. On the other hand, if sampling error is small the true value is likely to be covered by a small range of values; in this case one can feel confident that he has "trapped" the true value within a small range of values calculated from the sample.

In any normal distribution of sample means, with population mean μ and standard deviation σ_M, the following statement is true: **Over all samples of N, the probability is approximately .95 that**

$$-1.96\sigma_M \leq \mu - M \leq +1.96\sigma_M. \qquad [9.25.1^*]$$

That is, very nearly 95 percent of all possible sample means from the population in question must lie within 1.96 standard errors to either side of the true mean. You can check this for yourself from Table I.

However, the statement above is still true if we alter the inequality by adding M to each term: **Over all sample values of M, the probability is approximately .95 that**

$$M - 1.96\sigma_M \leq \mu \leq M + 1.96\sigma_M. \qquad [9.25.2^*]$$

That is, over all possible samples, the probability is about .95 that the range between $M - 1.96\sigma_M$ and $M + 1.96\sigma_M$ will include the true mean, μ.

This range of values between $M \pm 1.96\sigma_M$ is called the **95 percent confidence interval for** μ. The two boundaries of the interval, or $M + 1.96\sigma_M$ and $M - 1.96\sigma_M$, are called **the 95 percent confidence limits.**

Although the best estimate of μ is M, of course, the confidence interval based on any sample can be thought of as a range of "good" estimates. One can think of the confidence interval as containing all the hypotheses about μ that could not be rejected at the .05 level (two-tailed) in the light of the evidence M.

Be sure to notice that μ is not a random variable and the probability statement is really not about μ, but about *samples*. The population either does or does not have a μ equal to a given number, and if we use the ordinary relative-frequency definition of probability it does not make sense to say "the probability is such and such that the true mean takes on value x." Thus, one must be very careful about how he interprets the confidence interval. Before any sample mean is seen, one may decide to compute the 95 percent confidence interval for each and every sample. The actual range of numbers obtained for a sample will depend on the value of M. In short, over all possible samples, there will be many possible 95 percent confidence intervals. Some of these confidence intervals will represent the event "covers the true mean," and others will not. If one such confidence interval were sampled at random, then the probability is .95 that it covers the true mean.

The 99 percent confidence interval for the mean is given by

$$M - 2.58\sigma_M \leq \mu \leq M + 2.58\sigma_M. \qquad [9.25.3^*]$$

Notice that the 99 percent confidence interval is larger than the 95 percent interval. More possible values are included, and thus one can assert with more confidence that the true mean is covered. A 100 percent confidence interval would be

$$M - \infty\sigma_M < \mu < M + \infty\sigma_M.$$

Here, one can be completely confident that the true value is covered, since the range includes all possible values of μ. On the other hand, the 0 percent interval is simply M; for a continuous sampling distribution of M, the probability is zero that M exactly equals μ.

In the particular situation where the population σ is known, then an analogy to the idea of finding confidence intervals for the mean is tossing rings of a certain size at a post. The size of the confidence interval is like the diameter of a ring, and random sampling is like random tosses at the post. The predetermined size of the ring determines the chances that the ring will cover the post on a given try, just as the size of the confidence interval governs the probability that the true value μ will be covered by the range of values in the estimated interval. Reducing the size of σ_M by increasing N is like improving our aim, since we are more likely to cover the true value with an interval of a given size in terms of σ_M. Later when we deal with situations where the population σ is unknown, the analogy becomes that of tossing rings with a certain *distribution* of sizes at a post; here, the probability that the ring covers the post on a given try depends on the distribution of ring-sizes used, just as the probability that the confidence interval covers the true mean depends on the distribution of estimated confidence intervals over random samples. This idea will be elaborated upon in the next chapter.

It is interesting to note that, in principle, the 95 percent, or 99 percent, or any other confidence interval for a mean might be defined in other ways. For example, for a normal sampling distribution of M it is true that the probability is approximately .95 that

$$-1.75\sigma_M \leq \mu - M \leq 2.33\sigma_M.$$

This implies that a 95 percent confidence interval might possibly be found with limits

$$M - 1.75\sigma_M \text{ and } M + 2.33\sigma_M.$$

In a similar way, still another 95 percent confidence interval might be found having limits

$$M - 2.06\sigma_M \text{ and } M + 1.88\sigma_M,$$

and so on. In principle, 95 percent confidence limits for μ might be found in any number of ways, since any number of ways exist to find areas in a normal distribution equivalent to probabilities of .95.

If so many different ways of defining the 95 percent confidence interval are possible, what is the advantage of defining this confidence interval to have limits

$$M - 1.96\sigma_M \text{ and } M + 1.96\sigma_M?$$

The answer is that, in the particular case of the mean, this way of defining the confidence interval gives the *shortest* possible range of values such that the probability statement holds. Naturally, there is an advantage in pinning the population parameter within the narrowest possible range with a given probability, and this dictates the form of the confidence interval we use. A secondary consideration is that this way of defining the confidence interval corresponds to the standard two-tailed test of the mean, so that a parameter value falling outside the 95 percent confidence interval for the mean for a particular sample would be rejected as a hypothesis (at the .05 level, two-tailed) in the light of this sample. Later, in Chapter 11, we will encounter standard confidence intervals for other parameters that do not necessarily yield the shortest possible such intervals.

In many ways the presentation of a confidence interval in a data report is much more informative than reporting a hypothesis test or only a significance level. The reader can test any hypothesis he may entertain simply by looking at the confidence interval; only if the μ value in question lies outside the $100(1 - \alpha)$ percent confidence interval can the hypothesis be rejected beyond the α level in a two-tailed test. The confidence interval for μ not only specifies the unbiased estimate of the mean, but also includes the error variation qualifying that estimate.

As an example of a confidence interval for the mean when σ_M is known, consider a sample of 50 cases giving a mean of 143. The population standard deviation, σ, is 35. The 95 percent confidence interval will be found. First of all,

$$\sigma_M = \frac{35}{\sqrt{50}} = 4.9.$$

Then the desired interval is

$$143 - (1.96)(4.9) \leq \mu \leq 143 + (1.96)(4.9)$$
$$133.39 \leq \mu \leq 152.60.$$

In succeeding sections methods for finding confidence intervals will be given not only for means, but also for proportions, for differences between means, and for variances. It is extremely good practice to compute and report confidence intervals, and the student should get in the habit of reporting them routinely for his data. Ordinarily, far more of interest is communicated to the reader of a research report by a confidence interval or by several simultaneous confidence intervals than by any other statistical summary of results.

9.26   Approximate Confidence Intervals for Proportions

Forming an exact confidence interval for a proportion is a considerably more complicated business than it is for a single mean. The trouble is that although the standard error of the mean does not depend upon the true mean's value, the standard error of a proportion P does depend upon true p. From a sample, however, we have only P, an *estimate* of p, and so the true standard error is unknown unless p is known.

However, when samples are large, so that the sampling distribution of the proportion is approximately normal, the $100(1 - \alpha)$ percent confidence limits for a proportion are given approximately by

$$\frac{N}{N + z^2}\left[P + \frac{z^2}{2N} \pm z \sqrt{\frac{PQ}{N} + \frac{z^2}{4N^2}}\right] \qquad [9.26.1\dagger]$$

where z is the standard score in a normal distribution cutting off the *upper* $\alpha/2$ proportion of cases. These limits define the $100(1 - \alpha)$ percent confidence interval for p.

This rather mysterious-looking form for finding the confidence limits for a proportion can be justified as follows: Since the sampling distribution of P for large samples can be assumed approximately normal, the appropriate confidence limits should be approximately

$$P - z \sqrt{\frac{(pq)}{N}} \text{ and } P + z \sqrt{\frac{(pq)}{N}}.$$

The possible values of p corresponding to the limits thus satisfy the relation

$$\frac{N(P - p)^2}{(p - p^2)} = z^2.$$

Working out this quadratic equation in terms of p we have

$$(N + z^2)p^2 - (2NP + z^2)p + NP^2 = 0.$$

Solving for p by the rule for a quadratic equation,

$$\frac{-b \pm \sqrt{b^2 - 4ac}}{2a}$$

where $a = N + z^2$
$ b = -(2NP + z^2)$
$ c = NP^2$

and arranging terms gives expression **9.26.1**. Since the roots of such a quadratic equation may be either real or imaginary, we must assure ourselves that 9.26.1 will involve only real numbers. This is simple enough to do. Considering only the expression under the radical, we have:

$$\begin{aligned}
\sqrt{b^2 - 4ac} &= \sqrt{(2NP + z^2)^2 - 4(N + z^2)(NP^2)} \\
&= \sqrt{4N^2P^2 + 4NPz^2 + z^4 - 4N^2P^2 - 4z^2NP^2} \\
&= \sqrt{4z^2NP(1 - P) + z^4} \\
&= z\sqrt{4NPQ + z^2}.
\end{aligned}$$

Since $4NPQ$ is positive, and z^2 must also be positive, the entire expression under the radical must be positive. This guarantees that the roots of the original quadratic equation must be real, given $0 \leq P \leq 1$ and some value for z.

Then, substituting into the expression for the roots of a quadratic equation, and rearranging terms, we obtain expression 9.26.1. If N is very large, this confidence interval may be replaced by the simpler approximation

$$P - z \sqrt{\frac{PQ}{N - 1}} \leq p \leq P + z \sqrt{\frac{PQ}{N - 1}} \qquad [9.26.2\dagger]$$

Finally, a conservative confidence interval may always be found by setting σ_P at its largest possible value, or $.5/\sqrt{N}$. Then the probability is no less than .95 that p is covered by the interval with limits

$$P \pm 1.96(.5/\sqrt{N}),$$

and so forth for other values of $(1 - \alpha)$. Such intervals are conservative in the sense that the exact interval, given $p \neq .5$, would be shorter than the interval actually calculated.

9.27   SAMPLE SIZE AND THE ACCURACY OF ESTIMATION OF THE MEAN

The width of any confidence interval for the mean μ depends upon σ_M, the standard error, and anything that makes σ_M proportionately smaller reduces the width of the interval. Thus, any increase in sample size, operating to reduce σ_M, makes the confidence interval shorter. For example, the σ value for some normal population is 45. When samples of size 9 are taken the 95 percent confidence interval based on any sample mean has limits

$$M - 1.96(15)$$

and

$$M + 1.96(15).$$

However, if samples of size 81 are taken, these limits come closer together,

$$M - 1.96(5)$$
$$M + 1.96(5),$$

so that the interval itself is smaller for any sample. Samples of 900 cases give a confidence interval which is relatively quite narrow: the limits are

$$M - 1.96(1.5)$$
$$M + 1.96(1.5).$$

Continuing in this way, if the sample size becomes extremely large, the width of the confidence interval approaches zero.

A practical result of this relation between the standard error of the mean and sample size is that *the population mean may be estimated within any desired degree of precision, given a large enough sample size.* This principle has already been introduced in terms of the Tchebycheff inequality (Section 7.12), but it can be stated even more strongly when the sampling distribution has some known form, such as the normal distribution. For instance, how many cases should one sample if he wants the probability to be .99 that his sample mean lies within $.1\sigma$ of the true mean? That is, the experimenter wants

$$\text{prob.}(|M - \mu| \leq .1\sigma) = .99.$$

Assuming that the sampling distribution is nearly normal, which it should be if sample size is large enough, this is equivalent to requiring that the 99 percent confidence interval should have limits

$$M - .1\sigma$$

and
$$M + .1\sigma.$$

However, this is the same as saying that

$$.1\sigma = 2.58\sigma_M$$

so that

$$.1\sigma = 2.58 \frac{\sigma}{\sqrt{N}}.$$

Solving for N, we find

$$\sqrt{N} = 25.8$$
$$N = 665.64.$$

In short, if the experimenter makes 666 independent observations, he can be sure that the probability of his estimate's being wrong by more than $.1\sigma$ is only one in one hundred. Notice that we do not have to say exactly what σ is in order to specify the desired accuracy in σ units, and to find the required sample size.

If the experimenter is willing to take a somewhat larger chance of an absolute error exceeding $.1\sigma$, then the required N is smaller. If the probability of an error exceeding $.1\sigma$ in absolute magnitude is to be .05,

$$.1\sigma = 1.96\sigma_M$$

$$.1\sigma = 1.96 \frac{\sigma}{\sqrt{N}}$$

so that $N = 384$.

This idea may be turned around to find the accuracy *almost* insured by any given sample size. Given that N is, say, 30, within how many σ units of the true mean will the sample estimate fall with probability .95? Here, letting k stand for the required number of σ units,

$$k\sigma = 1.96 \frac{\sigma}{\sqrt{N}}$$

or
$$k = \frac{1.96}{\sqrt{30}} = \frac{1.96}{5.48} = .358.$$

For $N = 30$, the chances are about 95 in 100 that our estimate of μ falls within about .36 population standard deviations of the true value, provided that the sampling distribution is approximately normal.

It is obvious that the accuracy of estimation in a psychological experiment could be judged in this way. If the sample size is known, then one can judge how "off" the estimate of the mean is likely to be in terms of the population standard deviation. Furthermore, if one has some idea of the desirable accuracy of estimation, then he can set his sample size so as to attain that degree of accuracy with high probability. This is a very sensible and easy way to determine necessary sample size. Unfortunately, it does require the experimenter to think to some extent about the population he is sampling and what he is going to do with the result. For the problem at hand, just how much error in estimation can he tolerate?

9.28   EVIDENCE AND CHANGE IN PERSONAL PROBABILITY

In some research settings, a very good case can be made that the purpose of data collection is neither to permit an immediate decision to be made nor to obtain point-estimates of parameters. Instead, the purpose of data collection is to modify the experimenter's *degrees of belief* in the various situations that may exist. The experimenter starts out with various hypotheses about the true situation. He does not begin his experimentation in a state of total ignorance, however. For various theoretical and empirical reasons he has reason to believe in some of these hypotheses more strongly than others, *before* the data are seen. Then as a result of the data he adjusts his beliefs, so that some may be strengthened, some weakened, and still others unchanged. In short, some statisticians and psychologists would argue that scientific investigation does not usually deal with decision making, but rather with alteration in personal probabilities. As evidence accumulates some beliefs become so strong that these propositions are regarded as proved, and become part of the body of empirical science. Other beliefs fall away as evidence fails to support them.

The idea of personal probability has already been introduced in the discussion of decision-rules. Let us suppose once again that it is possible to assign numerical values to various hypotheses, each standing for the experimenter's personal probability that a given proposition is true.

For example, in the situation described in Section 9.5, the experimenter knows that the true population proportion is either

$$H_0: p = .80$$
or
$$H_1: p = .40,$$

so that we may regard the personal probability for any other H to be zero. Suppose that in the light of what the experimenter already knows he has good reason to believe much more strongly in H_0 than in H_1. Imagine the personal probabilities assigned to these hypotheses are

$$\mathcal{P}(H_0) = .75$$
$$\mathcal{P}(H_1) = .25.$$

The experimenter is willing to give odds of three to one that H_0 is true. These personal probabilities of the experimenter are *prior*, since they reflect his degree of belief in the two alternatives before the evidence. After the data are observed, something should happen to these personal probabilities; the degree of belief in each of the hypotheses should change somewhat.

The personal probability for H_0, given the evidence y, will be designated by $\mathcal{P}(H_0|y)$, and that for H_1 by $\mathcal{P}(H_1|y)$. These are *posterior* values, since they *do* depend upon what is observed.

How should evidence affect the degree of belief in a given hypothesis? The key is given by Bayes' theorem, discussed in Section 4.5. If personal probabilities operate like ordinary, relative-frequency, probabilities, then by Bayes' theorem,

$$\mathcal{P}(H_0|y) = \frac{p(y|H_0)\mathcal{P}(H_0)}{p(y|H_0)\mathcal{P}(H_0) + p(y|H_1)\mathcal{P}(H_1)}.$$

The $p(y|H_0)$ and $p(y|H_1)$ are the usual probabilities dealt with in statistical inference: the conditional probability of evidence given that the particular hypothesis is true. Notice that, other things being equal, the more likely is y given H_0, and the less likely is y given H_1, the larger is $\mathcal{P}(H_0|y)$. Under this conception, evidence made likely by one hypothesis and not by the other always strengthens belief in the first, and weakens belief in the second.

However, the effective *strength* of the evidence also depends upon the *prior personal probabilities*. The amount that evidence changes degree of belief in H_0 and H_1 depends on the *prior* $\mathcal{P}(H_0)$ and $\mathcal{P}(H_1)$ values. For example, suppose that the experimenter observes a sample proportion P of .6 among 10 cases. From the sampling distributions given in Section 9.5, we can see that

$$p(.6|H_0) = .088$$
$$p(.6|H_1) = .111.$$

Applying Bayes' theorem, we find

$$\mathcal{P}(H_0|.6) = \frac{p(.6|H_0)\mathcal{P}(H_0)}{p(.6|H_0)\mathcal{P}(H_0) + p(.6|H_1)\mathcal{P}(H_1)}$$

$$= \frac{.066}{.066 + .028}$$

$$= .70$$

$$\mathcal{P}(H_1|.6) = .30.$$

This evidence lowers the personal probability for H_0 somewhat, since it is less probable under H_0 than H_1. However, the posterior personal probability for H_0 is *still* much greater than for H_1, reflecting the original discrepancy in the degree of belief the experimenter holds for each.

On the other hand, suppose that the sample evidence is a P of .3. Here,

$$p(.3|H_0) = .001; \, p(.3|H_1) = .215.$$

Notice that this value of P is *very* improbable if H_0 were true. Now,

$$\mathcal{P}(H_0|.3) = \frac{(.001)(.75)}{(.001)(.75) + (.215)(.25)}$$

$$= \frac{.00075}{.00075 + .05375} = .01$$

$$\mathcal{P}(H_1|.3) = 1 - (H_0|.3) = .99.$$

In spite of the original weighting given the evidence by the prior personal probabilities, this time the P obtained is so unlikely given H_0 that the experimenter is almost certain that H_1 is true in the light of the evidence.

This way of evaluating evidence actually uses the raw materials of ordinary tests of hypotheses; as we have seen, the net result of any significance test is a statement of the likelihood of the sample result given that H_0 is true. However, a new ingredient is added: the prior personal probabilities of the experimenter. Intuitively, this approach has much to recommend it. In many ways it seems a far better description of what the scientist actually does with the statistical result than is the usual discussion of a significance test, burdened as it is with irrelevant decision terminology. In this conception, the scientist is viewed as changing his mind in the light of evidence, but he is not thought of as making some phony decision.

On the other hand, mathematical statisticians are not always enthusiastic about this approach. The general idea is very old, and at one time was carried to extreme and unreasonable lengths. Even metaphysical questions were treated by this general method. Much of modern theoretical statistics grew out of attempts to get away from the idea of prior probabilities, and it is not surprising that many statisticians look at such uses of Bayes' theorem with a jaundiced eye.

This is not an ideal place to go further into this approach to inferential statistics. However, in Chapter 19 this approach will be discussed in considerably more detail. There, both the strengths and the weaknesses of the Bayesian approach will be examined.

9.29   Significance tests and common sense

Stripped of the language of decision theory and of concern with personal probabilities, all that a significant result implies is that one has observed something relatively unlikely given the hypothetical situation, but relatively more likely given some alternative situation. Everything else is a matter of what one does with this information. Statistical significance is a statement about the likelihood of the observed result, nothing else. It does not guarantee that something important, or even meaningful, has been found.

By shifting our point of view slightly, we can regard the statistical significance or nonsignificance of a result as a measure of the "surprisal value" of that result. That is, the surprisal value of a result can be thought of as the reciprocal of the probability (or likelihood) of the result given H_0. If a result is relatively likely given H_0, the surprisal value of this result is quite small. On the other hand, if the result falls among unlikely events given H_0, then it has great surprisal value. An impossible result should lead to infinite surprise. Then, since we tend to favor those hypotheses that lead to unsurprising results, H_0 is rejected only when the surprisal value of our result, given H_0, is quite high.

Furthermore, the simple matter of parsimony should lead one to be cautious in rejecting H_0. If trivial deviations from expectation are to lead us to rejection of a given hypothesis, our ideas about what is true might flit all over the place. The conservativism of the conventional rules for hypothesis testing act as a kind of brake on our tendency to follow trivial or ephemeral tendencies in the data.

These are very good arguments indeed for the traditional procedures of hypothesis-testing. However, these arguments to not apply with equal force to all situations, nor do they engage the full conceptual machinery of decision-making.

If a psychologist is seriously concerned with making a decision from his data, then he should pay attention to the possible losses the various outcomes represent for him. There is no guarantee that the use of any of the conventionalized rules for deciding significance is appropriate for his particular purpose. On the other hand, so long as the real costs of decision errors in scientific research remain as obscure as they are at present, then these conventions may not be a bad thing. In a young science, still mapping out its area of study, the error of pursuing a "phantom" result *is* costly, perhaps even more so than failing to recognize a real experimental effect when we see one. If these conventions are actually used to determine which lines of research are pursued and which are abandoned, then at least we tend to follow up the big departures from expectation that our data show.

However, conventions about significant results should not be turned into canons of good scientific practice. Even more emphatically, a convention must not be made a superstition. It is interesting to speculate how many of the early discoveries in physical science would have been statistically significant in the experiments where they were first observed. Even in the crude and poorly controlled experiment, some departures from expectation stand out simply because they are interesting and suggest things to us that we might not be able to explain. These are matters that warrant looking into further regardless of what the conventional rule says to decide. Statistics cannot do the scientist's basic job—looking and wondering and looking again.

It is a grave error to evaluate the "goodness" of an experiment only in terms of the significance level of its results. Regardless of the scientist's convictions about personal probability, even a little bit of evidence coming out of a careful experiment is far more persuasive than a great deal from a sloppy one. To what population of subjects do the conclusions refer? Has a significant result been "bought" by restricting the population so much that the result fails to have any generality at all? In an effort to achieve a large sample N has the experimenter sacrificed all claim that his sample is random? Finally, how potent is the finding in a *predictive* sense: how much do these results permit us to reduce our uncertainty about the status of a given individual in this situation? This is by no means the same as statistical significance! As we shall see in later chapters, it is entirely possible for a highly significant result to contribute nothing to our ability to predict behavior, and for a nonsignificant result to mask an important gain in predictive ability.

It is very easy for research psychologists, particularly young psychologists, to become overconcerned with statistical method. Sometimes the problem itself seems almost secondary to some elegant method of data-analysis. Significant results are often confused with good results. But overemphasizing the role of statistical significance in research is like confusing the paintbrush with the painting. This form of statistical inference is a valuable tool in research, but it is never the arbiter of good research.

The remainder of this book is about statistical inference, and largely about significance tests. Since a text in statistics must necessarily deal mostly with statistics, there just is not room to discuss all the *if's*, *and's*, and *but's* that accompany the use of this tool in research. The skills of doing good research are acquired slowly, and often painfully, although a good share both of native curiosity and of common sense help matters along. Statistical texts can display the tools, but they cannot give a guide to their use in every conceivable situation. If the user has some idea of how the tool works perhaps he can see its uses and limitations in a given situation. Expert help is available and should always be sought. But if there is ever a conflict between the use of a statistical technique and common sense, then common sense comes first. Careful observation is the main business of empirical science, and statistical methods are useful only so long as they help, not hinder, the systematic exploration of data and the cumulation and coordination of results.

EXERCISES

1. A certain random variable is normally distributed and has a standard deviation of 4.2. Twenty-six observations were made independently and at random, and yielded a sample mean value of 31 for this random variable. Test the hypothesis that the mean of the random variable is 28.6 against the alternative hypothesis that the mean has some other value. (Use the .05 level for significance.)

2. A psychological test was standardized for the population of tenth-grade students in such a way that the mean must be 500 and the standard deviation 100. A sample of 90 twelfth-grade students was selected independently and at random, and each given the test. The sample mean turned out to be 506.7. On this basis, can one say that the population distribution for twelfth-grade students would differ from that for tenth-graders?

3. For a particular task given to subjects in an experiment, the researcher theorized that about 25 percent of this population of subjects should be able to complete the task within the allotted time. In order to test out this hunch, he took twenty subjects chosen at random, and gave each the task independently. Of this group, 45 percent actually did finish within the time allotted. Test the hypothesis that the true proportion is .25 or less against the alternative hypothesis that the proportion is greater than .25.

4. Suppose that in problem 3 above the experimenter had taken 200 subjects in order to test the hypothesis, and that the sample proportion had come out to be .28. Test the hypothesis that the true proportion is .25 against the alternative that the true proportion is not .25.

5. In a certain part of the country, medical records over a long period suggest that only about one in every 500 otherwise normal children have a particular birth defect. However, among the 1000 otherwise normal births during the past year in this area, six such defects have been noted. If the occurrences of the defect can be considered outcomes of a stable and independent random process, would you conclude that the rate of such occurrences has increased, as compared to former times? (Hint: does the

normal approximation to the binomial seem to be appropriate in this instance?)

6. Some eighty rats selected at random were taught to run a maze. All of them finally succeeded in learning the maze, and the average number of trials to perfect performance was 15.91. However, long experience with a population of rats trained to run a similar maze shows that the average number of trials to success is 15, with a standard deviation of 2. Would you say that the new maze appears to be harder for rats to learn than the older, more extensively used maze?

7. A teacher wished to study the change in student attitude toward the Federal Government which he had produced in his Political Science course. At the beginning of the class, therefore, he gave a specially constructed attitude test, and then repeated the test at the end of the course. The student's score was the difference between his scores on the second and the first test. A total of 368 students took the tests on both occasions. The mean change score was 2.3 points, and the sample standard deviation S was 10.5. Assuming that the students in the course constitute a random sample, would you say that there was a significant change in attitude? (Hint: since the sample is large, use the unbiased estimate of the population variance as taken from the sample.)

8. Given a normal distribution with a standard deviation of 10, suppose that exactly one of the following two hypotheses must be true:

$$H_0: \mu = 100, \quad H_1: \mu = 105$$

If the probability of Type I error is to be .10, and the probability of Type II is also to be .10, what sample size should be used, and what is the critical value of the sample mean leading one to choose H_0 or H_1 as true?

9. Given the conditions of problem 8, suppose that Type I error probability is to be fixed at .01, and Type II error probability at .10. Now what is the sample size that should be used, and what is the critical value of M?

10. In problem 1, what is the power of the test against the alternative hypothesis that the mean is 32? Against the hypothesis that the mean is 25? (Use .05 as the probability of Type I error.)

11. In problem 2, what is the power of the test against the hypothesis that the mean is 510? Against the hypothesis that the mean is 480? Against the hypothesis that the mean is 520? (Use .05 as the probability of Type I error.)

12. In problem 6, what is the power of the test against the hypothesis that the mean is 16? That the mean is 20?

13. In problem 6, if the experimenter had wished to have a power of .95 against the alternative hypothesis that the true mean is 16, what number of subjects would have been sufficient? (Use .05 as the probability of Type I error.)

14. Sketch the power function for a test of hypothesis about a mean which is two-tailed for $\alpha = .10$ and for $\alpha = .05$. (Assume a normal population distribution and, for simplicity, let $\sigma/\sqrt{N} = 1.00$. Divide the X axis of the plot into units of .5 or less.) What does this plot show about the relationship of power to the probability of Type I error?

15. In problem 5 above see if you can calculate the approximate power of the test against the alternative hypothesis that the rate of birth defect is 5 in 1000 children.

16. For example of problem 2 above, find the 95 percent confidence interval for the mean of twelfth-grade students. Assume that the standard deviation is 100.

17. In problem 3 above, determine the 99 percent confidence limits for the true proportion.

18. In problem 7 above, calculate the 95 percent confidence limits for the mean change in attitude.

19. Suppose that there were two normal populations. Population I has mean μ_I and

Population II has mean μ_{II}. Let us assume that each population has the same standard deviation, σ. Now, separately and independently we sample each population, and find a sample mean M_I and a sample mean M_{II}. Each sample has the same number of cases N. How would you find the 95 percent confidence interval for the difference between the population means, $\mu_I - \mu_{II}$? (Remember that a difference between two sample means is a linear combination of those means.)

20. Suppose that an experimenter were trying to decide among three hypotheses:

$$H_0: \mu = 200, \quad H_1: \mu = 210, \quad H_2: \mu = 200$$

One and only one of these hypotheses must be true. Drawing a sample of data, he finds that if H_0 is true, such a result has a likelihood of .15; if H_1 is true the sample result has a likelihood of .23; and if H_2 is true the sample result has a likelihood of .20. Prior to the experiment, the experimenter believes that the probability that H_0 is true is 1/2, and that H_1 and H_2 each has probability of only .25. Describe the experimenter's probabilities for these three hypotheses *after* the sample results are in.

21. Suppose that an experimenter is entertaining three hypotheses, one and only one of which must be true:

$$H_0: \mu = 200 \quad H_1: \mu = 210 \quad H_2: \mu = 220$$

He knows that the population is normal, and that the population standard deviation is 20. A sample of 25 cases is to be drawn and the sample mean computed. He formuates Decision Rule A: If $M \leq 205$, accept H_0; if $205 < M < 215$, accept H_1; if $215 \leq M$ accept H_2. Find the probabilities of his being correct and of his being in error under this rule, and display these probabilities in a table similar to that in section 9.5.

22. Suppose that the experimenter of problem 21 above formulates a new rule, Decision Rule B: If $M \leq 207$ accept H_0; if $207 < M < 214$ accept H_1; if $M \geq 214$ accept H_2. Calculate the probabilities of correct decisions and of errors under Decision Rule B.

23. The experimenter in problem 22 learns that losses are connected with erroneous decision in the following way:

		TRUE STATE		
		H_0	H_1	H_2
	ACCEPT H_0	0	5	20
DECISION	ACCEPT H_1	10	0	10
	ACCEPT H_2	20	5	0

Under the criterion of making the expected loss as small as possible, is Decision **Rule A** or **Rule B** the better rule? Why?

10 Inferences about Population Means

In this chapter we are going to discuss ways for making inferences about means, first for a single population and then for a difference between the means of two populations. The procedures and conventions of significance testing were emphasized in the previous chapter, and now we are going to apply these procedures to tests of hypotheses about means. Remember that a significant result will always be one falling among those that are extremely deviant from expectation and that are improbable if the null hypothesis were true, but that agree relatively well with expectation and have relatively higher probability if some situation covered by the alternative hypothesis were true. Before the test is carried out, some α level for significance is chosen as a specification of "improbable event, given H_0," for this situation. The conventional rules of the game determine which of the values of α are chosen.

10.1 LARGE-SAMPLE PROBLEMS WITH UNKNOWN POPULATION σ^2

In most of the examples of hypothesis testing up to this point we have actually "fudged" a bit on the usual situation: we have assumed that σ^2 is somehow known, so that the standard error of the mean is also known exactly. In these examples the author did not explain how σ^2 became known, largely because he could not think up a good reason. Now we must face the cold facts of the matter: for inferences about the population mean, σ^2 is seldom known. Instead, we must use the only substitute available for σ^2, which is our unbiased estimate s^2, calculated from the sample.

Notice that this problem does not exist for hypotheses about a population proportion p, since the existence of an exact hypothesis about p specifies what the value of the standard error of P, the sample proportion, must be. Therefore, the special techniques of this chapter apply only to inferences about means, and not to inferences about proportions.

From what we have already seen of the relation between sample size and accuracy of estimation, it makes sense that for large samples s^2 should be a very good estimate of σ^2. *In general, for very large samples, there is rather little risk of a sizable error when one uses s in place of σ in estimating the standard error of the mean.*

Hence, when the sample size is quite large, tests of hypotheses about a single mean are carried out in the same way as when σ is known, except that the standard error of the mean is estimated from the sample:

$$\text{est. } \sigma_M = \frac{s}{\sqrt{N}} = \frac{S}{\sqrt{N-1}}.$$

The standardized score corresponding to the sample mean is then referred to the normal distribution. This step is justified by the central limit theorem when N is large, regardless of the population distribution's form.

For example, consider the following problem. A small rodent characteristically shows hoarding behavior for certain kinds of foodstuffs when the environmental temperature drops to a certain point. Numerous previous experiments have shown that in a fixed period of time, and given a fixed food supply, the mean amount of food hoarded by an animal is 9 grams. The experimenter is currently interested in possible effects that early food deprivation may have upon such hoarding behavior in the animal as an adult. So, the experimenter takes a random sample of 175 infant animals and keeps them on survival rations for a fixed period while they are at a certain age, and on regular rations thereafter. When the animals are adults he puts each one in an experimental situation where the lowered temperature condition is introduced. The amount of food each hoards is recorded, and a score is assigned to each animal.

What is the null hypothesis implied here? The basic experimental question is "Does the experimental treatment (deprivation) tend to affect the amount of food hoarded?" The experimenter has no special reason to expect either an increase or a decrease in amount, but is interested only in finding out if a difference from normal behavior occurs. This question may be put into the form of a null and an alternative hypothesis:

$$H_0\colon \mu_0 = 9 \text{ grams}$$
$$H_1\colon \mu \neq 9 \text{ grams}.$$

Suppose that the conventional level chosen for α is .01, so that the experimenter will say that the result is significant only if the sample mean falls among either the upper .005 or the lower .005 of all possible results, given H_0. Reference to Table I shows that .005 is the probability of z score in a normal distribution falling at or below -2.58, and the probability is likewise .005 for a z equal to or exceeding $+2.58$. Accordingly, the sample result will be significant only if

$$z_M = \frac{M - E(M)}{\text{est. } \sigma_M}$$

equals or exceeds 2.58 in absolute magnitude (disregarding sign). When the null hypothesis is true, $E(M) = 9$, and for a sample this large the value of the standard error of the mean should be reasonably close to $\frac{s}{\sqrt{N}}$ or $\frac{S}{\sqrt{N-1}}$, the value of the sample estimate.

Everything is now set for a significance test except for the sample results. The sample shows a mean of 8.8 grams of food hoarded, with a standard deviation, S, of 2.3. The estimated standard error of the mean is thus

$$\text{est. } \sigma_M = \frac{2.3}{\sqrt{175 - 1}} = \frac{2.3}{13.23} = .1738$$

The standardized score of the mean is found to be

$$z_M = \frac{8.8 - 9}{.174} = \frac{-.2}{.174} = -1.149.$$

This result does not qualify for the region of rejection for $\alpha = .01$. Since the experimenter feels that he can afford to reject H_0 only if the α probability of error is no more than .01, then he cannot do so on the basis of this sample. On the other hand, the risk run in accepting H_0 is unknown, so he might well suspend judgment, pending more evidence.

10.2 CONFIDENCE INTERVALS FOR LARGE SAMPLES WITH UNKNOWN σ^2

Confidence intervals may also be found by the methods of Chapter 9. However, either when σ^2 is unknown, or when the population distribution has unknown form, a normal sampling distribution is assumed only for large samples. Just as in significance tests, the estimated standard error of the mean can be used in place of σ_M in finding confidence limits when the sample is relatively large.

For example, the experimenter studying hoarding behavior computes the approximate 99 percent confidence limits in the following way:

$$M - 2.58 \ (\text{est. } \sigma_M)$$
and
$$M + 2.58 \ (\text{est. } \sigma_M)$$

so that for this problem, the numerical confidence limits are

$$8.8 - 2.58(.174) \text{ or } 8.35$$
and
$$8.8 + 2.58(.174) \text{ or } 9.25.$$

The experimenter can say that the probability is approximately .99 that the true value of μ is covered by an interval such as that between 8.35 and 9.25.

Notice that the value $\mu_0 = 9$ falls between these limits, reflecting the fact that the hypothesis H_0 cannot be rejected if α is set at .01 (two-tailed).

10.3 THE PROBLEM OF UNKNOWN σ^2 WHEN SAMPLE SIZE IS SMALL

Just as for any statistic used to estimate a parameter value, the estimated standard error of the mean will very likely not be exactly equal to σ_M. This is not a particular problem when sample size is large, since we can at least be sure that est. σ_M is very likely to be close to the true σ_M in value.

On the other hand, we simply cannot have this confidence in our estimate of the standard error when sample size is small. Our estimate is almost bound to be in error to some extent, and if the sample size is very small, we can expect the size of this error to be substantial in any given sample. This necessitates a different approach to the problem of testing hypotheses and establishing confidence intervals for the population mean for small samples.

In inferences about μ, the ratio we would like to evaluate and refer to a normal sampling distribution is the standardized score

$$z_M = \frac{M - E(M)}{\sigma_M}. \qquad [10.3.1^*]$$

However, when we have only an *estimate* of σ_M, then the ratio we really compute and use is not a normal standardized score at all, although it has much the same form. The ratio actually used is

$$t = \frac{M - E(M)}{\text{est. } \sigma_M}. \qquad [10.3.2^*]$$

There is an extremely important difference between the two ratios, z_M and t. For z_M, the numerator $(M - E(M))$ is a random variable, the value of which depends upon the particular sample drawn from a given population situation; on the other hand, the denominator is a constant, σ_M, which is the same regardless of the particular sample of size N we observe. Now contrast this ratio with the ratio t: just as before, the numerator of t is a random variable, but the denominator is also a random variable, since the particular value of s— and hence the estimate of σ_M— is a sample quantity. Over several different samples, the same value of M must give us precisely the same value of z_M; however, over different samples, the same value of M will give us different t values. Similar intervals of t and z_M values should have different probabilities of occurrence. For this reason it is risky to use the ratio t as though it were z_M unless the sample size is very large.

10.4 THE DISTRIBUTION OF t

The solution to the problem of the nonequivalence of t and z_M rests on the study of t itself as a random variable. That is, suppose that the t ratio were computed for each conceivable sample of N independent observations drawn from some normal population distribution with true mean μ. Each sample would have some t value,

$$t = \frac{M - E(M)}{\text{est. } \sigma_M} = \frac{M - \mu}{s/\sqrt{N-1}}. \qquad [10.4.1^*]$$

Over the different samples the value of t would vary, of course, and the different possible values would each have some probability-density. A random variable such as t is an example of a *test*-statistic, so called to distinguish it from an ordinary descriptive statistic or estimator, such as M or s^2. The t value depends on other sample statistics, but is not itself an estimate of a population value. Nevertheless, such test-statistics have sampling distributions just as ordinary sample statistics do, and these sampling distributions have been studied extensively.

   In order to find the exact distribution of t, one must assume that the basic population distribution is normal. The main reason for the necessity of this assumption is that only for a normal distribution will the basic random variables in numerator and denominator, sample M and s, be statistically independent; this is a use of the important fact mentioned in Section 8.8. Unless M and s are independent, the sampling distribution of t is extremely difficult to specify exactly. On the other hand, for the special case of normal populations, the distribution of the ratio t is quite well known. In order to learn what this distribution is like, let us take a look at the rule for the density function associated with this random variable.

   The density function for t is given by the rule:

$$f(t; \nu) = G(\nu) \left[1 + \frac{t^2}{\nu} \right]^{-(\nu+1)/2} \quad \begin{array}{l} -\infty < t < \infty \\ 0 < \nu \end{array} \qquad [10.4.2^*]$$

Here, $G(\nu)$ stands for a constant number which depends *only* on the parameter ν (Greek nu), and how this number is found need not really concern us. Let us focus our attention on only the "working part" of the rule, which involves only ν and the value of t. This looks very different from the normal distribution function rule in Section 8.1. As with the normal function rule, however, a quick look at this mathematical expression tells us much about the distribution of t (for $\nu > 1$).

   First of all, notice that the particular value of t enters this rule only as a squared quantity, showing that the distribution of sample t values must be symmetric, since a positive and a negative value having the same absolute size must be assigned the same probability-density by this rule. Second, since all the constants in the function rule are positive numbers, and the entire term involving t is raised to a negative power, the largest possible density value is assigned to $t = 0$. Thus $t = 0$ is the distribution mode. Furthermore, although it is not quite so apparent from an examination of the function rule, the distribution is unimodal and "bell-shaped." If we inferred from the symmetry and unimodality of this distribution that the mean of t is also 0, we should be quite correct. In short, the t distribution is a unimodal, symmetric, bell-shaped distribution having a graphic form much like a normal distribution, even though the two function rules are quite dissimilar. Loosely speaking, the curve for a t distribution differs from the standardized normal in being "plumper" in extreme regions and "flatter" in the central region, as Figure 10.4.1 shows. (Note that both t and the standardized normal distribution have a mean of zero, $\nu > 1$.)

The most important feature of the t distribution will appear if we return for a look at the function rule. Notice that the only unspecified constants in the rule are those represented in **10.4.2** by ν and $G(\nu)$, which depends only on ν. This is a **one-parameter distribution**: the single parameter is ν, called *the degrees of freedom. Ordinarily, in most applications of the t distribution to problems involving a single sample, ν is equal to $N - 1$, one less than the number of independent observations in the sample.* For samples of N independent observations from any normal population distribution, the exact distribution of sample t values depends only on the degrees of freedom, $N - 1$. Remember, however, that the value of $E(M)$ or μ must be specified when a t ratio is computed, although the true value of σ need not be known.

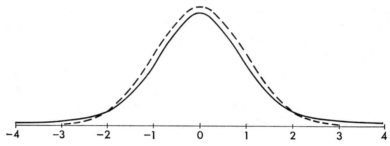

FIG. 10.4.1. Distribution of t with $\nu = 4$, and standardized normal distribution

In principle, the value of ν can be any positive number, and it just happens that $\nu = N - 1$ is the value for the degrees of freedom for the particular t distributions we will use first. Later we will encounter problems calling for t distributions with other numbers of degrees of freedom. Like most theoretical distributions, the t distribution is actually a family of distributions, with general form determined by the function rule, but with particular probabilities dictated by the parameter ν. For any value of $\nu > 1$, the mean of the distribution of t is 0. For $\nu > 2$ the variance of the t distribution is $\nu/(\nu - 2)$, so that the smaller the value of ν the larger the variance. As ν becomes large the variance of the t distribution approaches 1.00, which is the variance of the standardized normal distribution.

Incidentally, the random variable t is often called "Student's t," and the distribution of t, "Student's distribution." This name comes from the statistician W. S. Gosset, who was the first to use this distribution in an important problem, and who first published his results in 1908 under the pen-name "Student." Distributions of the general "Student" form have a number of important applications in statistics. One such application will occur in Chapter 19. It should also be noted that Student distributions are closely related to the beta family of distributions discussed in Chapter 8. This connection will be developed more fully in the next chapter.

10.5 THE t AND THE STANDARDIZED NORMAL DISTRIBUTION

As we have seen, the "shape" of the t distribution is not unlike that of the normal distribution. Just as for the standardized normal, the mean of the distribution of t is 0 for $\nu > 1$ although the variance of t is greater than 1.00 for finite $\nu > 2$. Given any extreme interval of fixed size on either tail of the t distribution, the probability associated with this interval in the t distribution is larger than that for the corresponding normal distribution of z_M. The smaller the value of ν, the larger is this discrepancy between t and normal probabilities at the extreme ends of each distribution. This reflects and partly explains the danger of using a t ratio as though it were a z ratio: extreme values of t are relatively more likely than comparable values of z_M. A small sample size corresponds to a small value of ν, or $N - 1$, and thus there is serious danger of underestimating the probability of an extreme deviation from expectation when sample size is small. This is apparent in the illustration (Figure 10.4.1) showing the distribution of t together with the standardized normal function.

Suppose that a sample of 5 observations is drawn, and from this sample we compute a ratio, t, using the estimate of σ_M from the sample. Furthermore, suppose that

$$t = \frac{M - E(M)}{\text{est. } \sigma_M} \geq 2.13.$$

That is, we obtain a value for t greater than or equal to 2.13. In the t distribution for samples of size 5 ($\nu = 4$), this interval has probability of .05. That is, when sample size is 5, so that degrees of freedom are 4, the probability of obtaining a ratio in this interval of values is $1/20$. However, if the ratio is interpreted as a z_M variable, then the normal probability for this interval is .0166. Incorrectly considering a t ratio as a standardized normal variable leads one to underestimate the probability of values in extreme intervals, which are really the only intervals of interest in significance tests.

On the other hand, notice what should happen to the distribution of t as ν becomes large (sample size grows large), as suggested both by Figure 10.4.1 and by the variance of a t distribution. *As sample size N grows large, the distribution of t approaches the standardized normal distribution. For large numbers of degrees of freedom, the exact probabilities of intervals in the t distribution can be approximated closely by normal probabilities.*

The practical result of this convergence of the t and the normal probabilities is that the t ratio *can* be treated as a z_M ratio, provided that the sample size is substantial. The normal probabilities are quite close to—though not identical with—the exact t probabilities for large ν. On the other hand, when sample size is small the normal probabilities cannot safely be used, and instead one uses a special table based on the t distribution.

How large is "large enough" to permit use of the normal tables? If the population distribution is truly normal, even forty or so cases permit a fairly accurate use of the normal tables in confidence intervals or tests for a mean. If really good accuracy is desired in determining interval probabilities, the t distribution should be used even when the sample size is around 100 cases. Beyond this number of cases, the normal probabilities are extremely close to the exact t probabilities. For example, in the "hoarding" experiment just discussed, use of t rather than z values would have given confidence limits of $M \pm 2.6(\text{est } \sigma_M)$ instead of $M \pm 2.58(\text{est } \sigma_M)$, a very slight difference.

Recall that the stipulation is made that the population distribution be *normal* when the t distribution is used, even when the normal approximations are substituted for the exact t-distribution probabilities. As we have already seen, for a normal population the distribution of sample means must be normal anyway; the difficulty with the use of a normal sampling distribution for small N comes solely from the fact that our estimate of the standard error is a random variable rather than a constant over samples, and this is the reason we must use the t distribution. Thus the t distribution is related to the normal distribution in two distinct ways: the parent distribution must be normal if t probabilities are to be found exactly, and for sufficiently large N, the distribution of t approaches the normal sampling distribution in form.

10.6   THE APPLICATION OF THE t DISTRIBUTION WHEN THE POPULATION IS NOT NORMAL

It is apparent that the requirement that the population be normal limits the usefulness of the t distribution, since this is an assumption that we can seldom really justify in practical situations. Fortunately, when sample size is fairly large, and provided that the parent distribution is roughly unimodal and symmetric, the t distribution apparently still gives an adequate approximation to the exact (and often unknown) probabilities of intervals for t ratios under these circumstances. However, one should insist on a relatively *larger* sample size the *less* confident that he is that the normal rule holds for the population, if he plans to use the t distribution. In effect, if the sample size is large enough so that the normal probabilities are good approximations to the t probabilities anyway, then the form of the parent distribution is more or less irrelevant. However, often the sample size is so small that the t distribution must be used and here it *is* somewhat risky to make inferences from t ratios unless the population is more or less normally distributed. This is an especially serious problem when one-tailed tests of hypotheses are made, since a very skewed population distribution can make the t probabilities for one-tailed tests considerably in error. Once again, it is wise to plan on somewhat larger samples when one is considering a one-tailed test using the t distribution and the population is not assumed normal.

10.7   TABLES OF THE t DISTRIBUTION

Unlike the table of the standardized normal function, which suffices for all possible normal distributions, tables of the t distribution must actually include many distributions each depending on the value of ν, the degrees of freedom. Consequently, tables of t are usually given only in abbreviated form; otherwise, a whole volume would be required to show all the different t distributions one might need.

Table III in Appendix C shows selected percentage points of the distribution of t, in terms of the value of ν. Different ν values appear along the left-hand margin of the table. The top margin gives values of Q, which is $1 - p(t \leq a)$, one minus the cumulative probability that t is less than or equal to a specific value a, for a distribution within the given value for ν. A cell of the table then shows the value of t cutting off the upper Q proportion of cases in a distribution for ν degrees of freedom.

This sounds rather complicated, but an example will clarify matters considerably: suppose that $N = 10$, and we want to know the value *beyond which* only 10 percent of all sample t values should lie. That is, for the distribution of t shown in Figure 10.7.1, we want the t value that cuts off the shaded area in the curve, the upper 10 percent:

First of all, since $N = 10$, $\nu = N - 1 = 9$. So, we enter the table for the row marked 9. Now since we want the upper 10 percent, we find the column for which $Q = .1$. The corresponding cell in the table is the value of t we are looking for, $t = 1.383$. We can say that in a t distribution with 9 degrees of freedom,

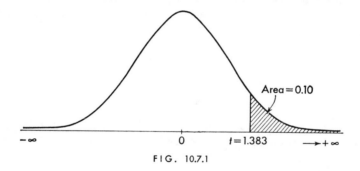

FIG. 10.7.1

the probability is .10 that a t value *equals or exceeds* 1.383. Since the distribution is symmetric, we also know that the probability is .10 that a t value *equals or falls below* -1.383. If we wanted to know the probability that t equals or exceeds 1.383 in absolute value, then this must be $.10 + .10 = .20$, or $2Q$.

Suppose that in a sample of 21 cases, we get a t value of 1.98. We want to see if this value falls into the upper .05 of all values in the distribution. We enter row $\nu = 21 - 1 = 20$, and column $Q = .05$; the t value in the cell is 1.725. Our obtained value of t is larger than this, and so the obtained value does fall among the top 5 percent of all such values. On the other hand, suppose that the obtained t had been -3. Does this fall either in the top .001 or the bottom .001 of all such sample values? Again with $\nu = 20$, but this time with $Q = .001$, we find a t value of 3.552. This means that at or above 3.552 lie .001 of all sample values, and also at or below -3.552 lie .001 of all sample values. Hence our sample value does not fall into either of these intervals; we can say that the sample value does not fall into the rejection region for $\alpha = .002$.

The very last row, marked ∞, shows the z scores that cut off various areas in a normal distribution curve. If you trace down any given column, you find that as ν gets larger the t value bounding the area specified by the column comes closer and closer to this normal deviate value, until finally, for an infinite sample size, the required value of t is the same as that for z.

For one-tailed tests of hypotheses, the column Q values are used to find the t value which exactly bounds the rejection region. If the region of rejection is on the upper tail of the distribution, then Q is the probability of a sample value's falling into the region greater than or equal to the tabled value of t. If the region is on the lower tail, the t value in the table is given a negative sign, and Q is the probability of a sample's falling at or below the negative t value. If a two-tailed region is to be used, then the total α probability of error is $2Q$, and the number in the table shows the absolute value of t that bounds the rejection region on *either* tail.

10.8 THE CONCEPT OF DEGREES OF FREEDOM

Before we proceed to the uses of the t distribution, it is well to examine the notion of degrees of freedom. The degrees of freedom parameter reflects the fact that a t ratio involves a sample standard deviation as the basis for estimating σ_M. Recall the basic definitions of the sample variance and the sample standard deviation:

$$S^2 = \frac{\Sigma(x - M)^2}{N}$$

and

$$S = \sqrt{\frac{\Sigma(x - M)^2}{N}}.$$

The sample variance and standard deviation are both based upon a sum of squared deviations from the sample mean. However, recall another fact of importance about deviations from a mean: in Section 6.6 it was shown that

$$\sum_i (x_i - M) = 0,$$

the sum of deviations about the mean must be zero.

These two facts have an important consequence: Suppose that you are told that $N = 4$ in some sample, and that you are to guess the four deviations from the mean M. For the first deviation you can guess any number, and suppose you say

$$d_1 = 6.$$

Similarly, quite at will, you could assign values to two more deviations, say

$$d_2 = -9$$
$$d_3 = -7.$$

However, when you come to the fourth deviation value, you are *no longer free* to guess any number you please. The value of d_4 *must* be

$$d_4 = 0 - d_1 - d_2 - d_3$$

or
$$d_4 = 0 - 6 + 9 + 7 = 10.$$

In short, given the values of any $N - 1$ deviations from the mean, which could be any set of numbers, the value of the last deviation is completely determined. Thus we say that there are $N - 1$ degrees of freedom for a sample variance, reflecting the fact that only $N - 1$ deviations are "free" to be any number, but that given these free values, the last deviation is completely determined. It is not the sample size per se that dictates the distribution of t, but rather the number of degrees of freedom in the variance (and standard deviation) estimate. We will consider the degrees of freedom again in the next chapter, where the variance will be studied in more detail, and also in Chapter 14.

10.9 SIGNIFICANCE TESTS FOR SINGLE MEANS USING THE t DISTRIBUTION

For the moment you can relax; there is really nothing new to learn! When the null hypothesis concerns a single mean, then the test is carried out just as before, except that the table of t (Appendix C, Table III) is used instead of the normal table (Appendix C, Table I). The α level is chosen, and the value (or values) of t corresponding to the region of rejection can be determined from the t table. The number of degrees of freedom used is simply $\nu = N - 1$. Then, the ratio

$$t = \frac{M - E(M)}{\text{est. } \sigma_M} \qquad [10.9.1\dagger]$$

obtained from the sample is compared with values in the rejection region specified by Table III. If the obtained t ratio falls into the rejection region chosen, the sample result is said to be significant beyond the α level.

If the sample size is large, then the only difference in procedure is in the use of the normal tables to establish the region of rejection. Naturally, all the considerations hitherto discussed, especially the assumed normal distribution of the population, should be faced before the sample size and rejection region are decided upon. If large samples are available then the assumption of a normal population is relatively unimportant; on the other hand, this matter should be given some serious thought if you are limited to a very small sample size.

10.10 CONFIDENCE LIMITS FOR THE MEAN USING t DISTRIBUTIONS

The t distribution may also be used to establish confidence limits for the mean. For some fixed percentage representing the confidence level, $100(1 - \alpha)$ percent, the sample confidence limits depend upon three things: the sample value of M, the estimated standard error, or est. σ_M, and the number of degrees of freedom, ν. For some specified value of ν, then the $100(1 - \alpha)$ percent confidence limits are found from

$$M - t_{(\alpha/2;\, \nu)} \ (\text{est. } \sigma_M)$$
$$M + t_{(\alpha/2;\, \nu)} \ (\text{est. } \sigma_M).$$

[10.10.1 †]

Here, $t_{(\alpha/2;\, \nu)}$ represents the value of t that bounds the upper $\alpha/2$ proportion of cases in a t distribution with ν degrees of freedom. In Table II this is the value listed for $Q = \alpha/2$ and ν. Thus, if one wants the 99 percent confidence limits, the value of $\alpha = .01$, and one looks in the table for $Q = .005$.

For example, imagine a study using 8 independent observations drawn from a normal population. The sample mean is 49 and the estimated standard error of the mean is 3.7. Now we want to find the 95 percent confidence limits. First of all, $\alpha = .05$, so that $Q = .025$. The value of ν is $N - 1$, or 7. The table shows a t value of 2.365 for $Q = .025$ and $\nu = 7$, so that $t_{(\alpha/2;\, \nu)} = 2.365$. The confidence limits are

$$49 - (2.365)(3.7) = 40.25$$
and $$49 + (2.365)(3.7) = 57.75.$$

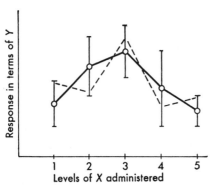

FIG. 10.10.1. Confidence intervals for means based on five independent samples

Over all random samples, the probability is .95 that the true value of μ is covered by an interval such as that between 40.25 and 57.75, the confidence interval calculated for this sample.

In summary, confidence limits are calculated in much the same way, and have the same general interpretation, when based on the t distribution as for the normal distribution. The essential difference is that values of t corresponding to $\alpha/2$ and ν must be used instead of normal z values.

One important application of confidence intervals in psychology occurs when there is a set of some J independent means, each based on a different sample given exactly one of a set of J experimental treatments. In particular, the experimental treatments may represent some quantitative experimental variable (represented here by X), and the experimenter may be trying to infer the general form of relationship between the amount of treatment applied and the average or expected response of a subject in terms of variable Y. Here, he may choose to construct a confidence interval around *each* of the sample means on variable Y. Figure 10.10.1 represents a set of such means, with the 95 percent confidence interval shown for each. In this figure, the horizontal axis represents the different levels or quantities of the treatment administered, and the open circles the corresponding means of the samples on the dependent variable, Y. The vertical bars extending to either side of a mean point symbolize the 95 percent confidence interval based on that sample's mean. The experimenter's best guess about the general form of the function relating the experimental variable to the dependent variable is symbolized by the heavy line in the figure: this is simply a plot joining the sample means, since his best guess about the population mean under any given treatment is the sample mean. Nevertheless, it may well be true that the form of relationship in the population is something like that shown by the broken line. The experimenter has no basis at all for discounting the possibility of some such true relation on the basis of the obtained relation alone.

How sure can the experimenter be that the set of J confidence intervals based on independent means all *simultaneously* cover the population values? In other words, how confident can he be that he has narrowed the possible relationships between the experimental and dependent variables to those symbolized by graphs joining points *within* the various intervals of Figure 10.10.1? A little thought should convince you that the probability is *not* .95 that all J of the confidence intervals simultaneously cover the true means; we can, however, work out an approximation to the value of this probability.

Suppose that both the means and the estimated σ_M values, and thus the obtained values for the confidence limits themselves, are independent across samples. We can consider the event "confidence interval covers true μ" as though it were a "success" in a binomial experiment for any of the samples. The probability of any such success is .95 for a 95 percent confidence interval. Then the probability that all J independent confidence intervals simultaneously cover the true means is simply the probability of exactly J out of J possible successes in a binomial experiment:

prob.(all J of the 95 percent confidence intervals cover true values

$$\text{simultaneously}) = \binom{J}{J} (.95)^J (.05)^0$$
$$= (.95)^J.$$

For the example in Figure 10.10.1, $J = 5$, so that the probability that the true means all are covered by the indicated intervals is

$$(.95)^5 = .77.$$

The experimenter can have considerably less "confidence" in the statement that *all* of the confidence intervals simultaneously cover the true means than in the statement that *any one* confidence interval covers the true mean.

The probability of .77 calculated for this example was based on the assumption that each of the confidence limits obtained for a sample is independent of the corresponding limits obtained for the other samples. However, this is not a reasonable assumption in a great many instances, because the same estimated value of σ or of σ_M may be used for determining each of the confidence intervals. Nevertheless, even when the confidence limits for the various samples are not independent, the probability that all J of the $100(1 - \alpha)$ percent confidence intervals simultaneously cover the true values must lie between $1 - \alpha$ and $1 - J\alpha$. Conversely, given any J such confidence intervals, the probability that *at least one* fails to cover the true value is between α and $J\alpha$. For J independent confidence intervals, the probability that at least one of the set fails to cover the true value is exactly $1 - (1 - \alpha)^J$. The practical implication is clear: given enough confidence intervals calculated from a set of data the probability can be quite high that *at least one* fails to cover the true parameter value. For similar reasons, given enough significance tests carried out on a set of data, each with some conventional value for α, the probability can be much greater than α that *at least one* of these tests results in a Type I error. This point is an important one, and will recur in Chapters 12 and 14.

10.11   QUESTIONS ABOUT DIFFERENCES BETWEEN POPULATION MEANS

Examples of hypotheses about single means often sound rather "phony" in their experimental contexts, and the reason for this is not hard to find. In most experimental work it is not true that the experimenter knows about one particular population in advance and then draws a single sample for the purpose of comparing some experimental population to the known population. Rather, it is far more common to draw two samples, to only one of which the experimental treatment is applied; the other sample is given no treatment, and stands as a control group for comparison with the treated group. In other situations, two different treatments may be compared. The advantages of this method over the single sample procedure are obvious; the experimenter can exercise the same experimental controls on both samples, making sure that insofar as possible they are treated in exactly the same way, with the only systematic experimental difference being in the fact that something was done to representatives of one sample which was not done to members of the other. Then, if a very large difference appears between the two samples he can rest assured that the difference is a product of the experimental treatments and not just a peculiarity introduced by the way in which his data were gathered.

Each treatment group is a sample from a potential population of observations made under that treatment. A difference between the treatment populations should exist if the treatment is having an effect; but what can the experimenter infer from a sample difference? *His best estimate (based on these data alone) is that the population means are different to the same extent as the sample means. Regardless of the significance level given by any test he may apply, the actual difference obtained is always the best estimate he can make of the true difference between the population means.*

As always, this estimate is in error to some unknown extent, and although the obtained difference between the sample means is the best guess the experimenter can make, there is absolutely no guarantee that this estimate is exactly correct. It could well be true that the difference the experimenter observes has no real connection with the treatment administered, and is purely a chance result.

What is needed is a way of applying statistical inference to differences between means of samples representing two populations. First, large sample distributions of *differences* between sample means will be studied. Then, the application of the t distribution to small sample differences will be introduced.

10.12   THE SAMPLING DISTRIBUTION OF DIFFERENCES BETWEEN MEANS

Suppose that we wished to test a hypothesis that two populations have means which differ by some specified amount, say 20 points. This is tested against the hypothesis that the population means do not differ by that amount. In our more formal notation:

$$H_0: \mu_1 - \mu_2 = 20$$
$$H_1: \mu_1 - \mu_2 \neq 20.$$

We draw a sample of size N_1 from population 1, and an *independent* sample of size N_2 from population 2, and consider the difference between their means, $M_1 - M_2$. Now suppose that we kept on drawing pairs of independent samples of these sizes from these populations. For each pair of samples drawn, the difference $M_1 - M_2$ is recorded. What is the distribution of such sample *differences* that we should expect in the long run? In other words, what is the sampling distribution of the difference between two means?

You may already have anticipated the form of the sampling distribution of the difference between two means, since all the groundwork for this distribution has been laid in Section 8.9. The difference between sample means drawn from independent samples is actually a linear combination:

$$(1)M_1 + (-1)M_2.$$

Let us apply the results of Sections 8.9 and 8.10 to this problem. In the first place,

$$E(M_1 - M_2) = E(M_1) - E(M_2) = \mu_1 - \mu_2, \qquad [10.12.1*]$$

which accords with principle **8.10.1** for any linear combination. Second, what is the standard error of the difference between two independent sample means? By principle **8.10.3**,

$$\text{var.}(M_1 - M_2) = (1)^2\sigma^2_{M_1} + (-1)^2\sigma^2_{M_2}$$
$$= \sigma^2_{M_1} + \sigma^2_{M_2}. \qquad [10.12.2*]$$

Hence, the standard error of the difference, $\sigma_{\text{diff.}}$, is

$$\sigma_{\text{diff.}} = \sqrt{\sigma^2_{M_1} + \sigma^2_{M_2}} = \sqrt{\frac{\sigma^2_1}{N_1} + \frac{\sigma^2_2}{N_2}} \qquad [10.12.3*]$$

provided that samples 1 and 2 are completely independent.

Actually, we could have found this last result quite easily without invoking principle **8.10.2**. It may be instructive to do so.

By definition:

$$\text{var.}(M_1 - M_2) = E[(M_1 - M_2) - (\mu_1 - \mu_2)]^2$$
$$= E[(M_1 - \mu_1) - (M_2 - \mu_2)]^2.$$

For any given pair of samples, expanding the square gives

$$[(M_1 - \mu_1) - (M_2 - \mu_2)]^2 = (M_1 - \mu_1)^2 + (M_2 - \mu_2)^2 - 2(M_1 - \mu_1)(M_2 - \mu_2).$$

Now let us take the expectation of each of these terms separately:

$$E(M_1 - \mu_1)^2 = \sigma_{M_1}^2$$

and
$$E(M_2 - \mu_2)^2 = \sigma_{M_2}^2$$

by the definition of the variance of a sampling distribution of the mean. Furthermore,

$$E[(M_1 - \mu_1)(M_2 - \mu_2)] = 0$$

by rule 6, Appendix B, since M_1 and M_2 are independent. Thus, combining these results, we find that

$$\text{var.}(M_1 - M_2) = \sigma_{M_1}^2 + \sigma_{M_2}^2$$

or
$$\sigma_{\text{diff.}} = \sqrt{\frac{\sigma_1^2}{N_1} + \frac{\sigma_2^2}{N_2}}.$$

Notice that there is no requirement at all that the samples be of equal size. Regardless of the sample sizes, the expectation of the difference between two means is always the difference between their expectations, and the variance of the difference between two *independent* means is the *sum* of the separate sampling variances.

Furthermore, these statements about the mean and the standard error of a difference between means are true regardless of the form of the parent distributions. However, the form of the sampling distribution can also be specified under either of two conditions:

If the distribution for each of two populations is normal then the distribution of differences between sample means is normal.

This follows quite simply from principle 8.9 for linear combinations. When we can assume both populations normal, the form of the sampling distribution is known to be *exactly* normal.

On the other hand, one or both of the original distributions may not be normal; in this case the central limit theorem comes to our aid:

As both N_1 and N_2 grow infinitely large, the sampling distribution of the difference between means approaches a normal distribution, regardless of the form of the original distributions.

In short, when we are dealing with two very large samples, then the question of the form of the original distributions becomes irrelevant, and we can approximate the sampling distribution of the difference between means by a normal distribution.

10.13   AN EXAMPLE OF A LARGE-SAMPLE SIGNIFICANCE TEST FOR A DIFFERENCE BETWEEN MEANS

An experimenter working in the area of motivational factors in perception was interested in the effects of deprivation upon the perceived size of objects. Among the studies carried out was one done with orphans, who were compared with nonorphaned children on the basis of the judged size of parental figures viewed at a distance. Each child was seated at a viewing apparatus in which cut-out figures appeared. Each figure was actually of the same size and at the same distance from the viewer, although he was not told that the figures had the same size. A device was provided on which the child could actually judge the apparent sizes of the different figures in numerical terms. Several of the figures in the set viewed were obviously parents, whereas others were more or less neutral, such as milkmen, postmen, nurses, and so on. Each child was given a score, which was itself a difference in average judged size of parental and nonparental figures.

Now two independent randomly selected groups were used. Sample 1 was a group of orphaned children without foster parents. Sample 2 was a group of children having a normal family with both parents. Both populations of children sampled showed the same age level, sex distribution, educational level, and so forth.

The question asked by the experimenter was, "Do deprived children tend to judge the parental figures relatively larger than do the nondeprived?" In terms of a null and alternative hypothesis,

$$H_0: \mu_1 - \mu_2 \leq 0$$
$$H_1: \mu_1 - \mu_2 > 0.$$

The α level for significance decided upon was .05. The actual results were

Sample 1	Sample 2
$M_1 = 1.8$	$M_2 = 1.6$
$s_1 = .7$	$s_2 = .9$
$N_1 = 125$	$N_2 = 150$

These sample sizes are rather large, and the experimenter felt safe in using the normal approximation to the sampling distribution, even though he had no idea about the distribution form for the two populations sampled. The t ratio used was

$$t = \frac{(M_1 - M_2) - E(M_1 - M_2)}{\text{est. } \sigma_{\text{diff.}}} \qquad [10.13.1*]$$

In this problem, $E(M_1 - M_2) = 0$, under the hypothesis tested. It was obviously necessary for the experimenter to estimate the standard error of the difference, since both σ_1 and σ_2 were unknown to him. This estimate was found by first estimating $\sigma_{M_1}^2$ and $\sigma_{M_2}^2$:

$$\text{est. } \sigma_{M_1}^2 = \frac{s_1^2}{N_1} = \frac{.49}{125} = .004 \qquad [10.13.2\dagger]$$

$$\text{est. } \sigma_{M_2}^2 = \frac{s_2^2}{N_2} = \frac{.81}{150} = .005. \qquad [10.13.3\dagger]$$

Then,

$$\text{est. } \sigma_{\text{diff.}} = \sqrt{\text{est. } \sigma_{M_1}^2 + \text{est. } \sigma_{M_2}^2} = \sqrt{.004 + .005} = .095. \quad [10.13.4\dagger]$$

On making these substitutions, the experimenter found

$$t = \frac{1.8 - 1.6}{.095} = 2.11.$$

The rejection region implied by the alternative hypothesis is on the *upper* tail of the sampling distribution. For a normal distribution the upper 5 percent is bounded by $z = 1.65$. Thus, the result is significant; deviations this far from zero have a probability of less than .05 of occurring by chance alone when the true difference is zero.

The experimenter may conclude that a difference exists between these two populations, *if* an α value less than .05 is a small enough probability of error to warrant this decision. However, the experimenter does not necessarily conclude that parental deprivation causes an increase in perceived size. The statistical conclusion suggests that it *might* be safe to assert that a particular direction of numerical difference exists between the mean scores of the two populations of children, but the statistical result is absolutely noncommittal about the reason for this difference, if such exists. The experimenter takes the step of advancing a reason at his own peril. The statistical test as a mathematical tool is absolutely neutral about what these numbers measure, the level of measurement, what was or was not represented in the experiment, and, most of all the cause of the experimenter's particular finding. As always, the test takes the numerical values as given, and cranks out a conclusion about the conditional probability of such numbers, given certain statistical conditions.

The general procedure for hypotheses about two means when sample size is quite large is represented by this example. The test statistic is

$$t = \frac{(M_1 - M_2) - E(M_1 - M_2)}{\text{est. } \sigma_{\text{diff.}}}.$$

This t value may be referred to a normal distribution. The expected difference depends upon the hypothesis tested, and the estimated $\sigma_{\text{diff.}}$ is found directly from the estimate σ_M^2 for each sample by **10.13.4.**

The exact hypothesis actually tested is of the form

$$H_0: \mu_1 - \mu_2 = k,$$

where k is any difference of interest. Quite often, as in the example, the experimenter is interested only in $k = 0$, but it is entirely possible to test any other meaningful difference value. The alternative hypothesis may be directional,

$$H_1: \mu_1 - \mu_2 > k$$

or

$$H_1: \mu_1 - \mu_2 < k,$$

or nondirectional,

$$H_1: \mu_1 - \mu_2 \neq k,$$

depending on the form of the original question.

As an illustration of a situation where some value other than zero figures in the null hypothesis, and also as an illustration of a one-tailed test, take the following example: a manufacturer is considering introducing a change in training procedure for his new employees. However, it is more expensive than the old, and he feels that he cannot afford it unless the average output of a man trained in the new way is more than 50 units per hour better than that of a man trained under the old procedure. The null hypothesis is

$$H_0: \mu_1 - \mu_2 \leq 50,$$

since the exact value that the null hypothesis requires is given by 50 units per hour. The alternative hypothesis states

$$H_1: \mu_1 - \mu_2 > 50.$$

Notice how the null and the alternative hypotheses are framed so as to correspond to the alternative practical decisions that our manufacturer may make: if the null hypothesis is true, he will not adopt the new training procedure since it does not meet the requirement he set up. He has no interest in the training procedure if it is less than 50 units better than the old. If, however, the alternative hypothesis is true, then he will adopt the new procedure. In this instance, where clear-cut courses of action depend on the evidence, the one-tailed test of a nonzero hypothesis makes sense. Subjects are assigned at random to two groups, one getting the new and the other the old training. Given large samples, the t ratio is computed just as in the previous example, except that here $E(M_1 - M_2) = 50$. A significant result gives the manufacturer considerable assurance in saying that one procedure is on the average more than 50 units better than the other.

10.14 LARGE-SAMPLE CONFIDENCE LIMITS FOR A DIFFERENCE

When both samples are large, as in the example in Section 10.13, confidence limits are found exactly as for a single mean, except that $(M_1 - M_2)$ and est. $\sigma_{\text{diff.}}$ are substituted for M and est. σ_M respectively. Thus, 95 percent confidence limits for a difference with large samples are

$$M_1 - M_2 - 1.96 \ (\text{est. } \sigma_{\text{diff.}}) \qquad \text{[10.14.1*]}$$
$$M_1 - M_2 + 1.96 \ (\text{est. } \sigma_{\text{diff.}}).$$

For the example in the preceding section, the 95 percent limits are

$$.2 - 1.96(.095)$$
$$.2 + 1.96(.095)$$

or .014 and .386. Notice that since the value $\mu_1 - \mu_2 = 0$ does not fall within these values this value can be rejected as a hypothesis beyond the .05 level (two-tailed).

10.15 USING THE t DISTRIBUTION TO TEST HYPOTHESES ABOUT DIFFERENCES

Given the assumption that both populations sampled have normal distributions, any hypothesis about a difference can be tested using the t distribution, regardless of sample size. However, one additional assumption becomes necessary: *in order to use the t distribution for tests based on two (or more) samples, one must assume that the standard deviations of both (or all) populations are equal.* The basis for this assumption will be discussed in the next chapter.

Given these assumptions, then the distribution of t for a difference has the same form as for a single mean, except that the degrees of freedom are

$$\nu = N_1 - 1 + N_2 - 1 = N_1 + N_2 - 2.$$

When samples are drawn from populations with equal variance, then the estimated standard error of a difference takes a somewhat different form. First of all, when $\sigma_1 = \sigma_2 = \sigma$,

$$\sigma_{\text{diff.}} = \sqrt{\frac{\sigma^2}{N_1} + \frac{\sigma^2}{N_2}} = \sqrt{\sigma^2\left(\frac{1}{N_1} + \frac{1}{N_2}\right)}. \qquad [10.15.1*]$$

Now, as we showed in Section **7.17**, when one has two or more estimates of the same parameter σ^2, the *pooled* estimate is actually better than either one taken separately. From **7.17.5** it follows that

$$\text{est. } \sigma^2 = \frac{(N_1 - 1)s_1^2 + (N_2 - 1)s_2^2}{N_1 + N_2 - 2}$$

is our best estimate of σ^2 based on the two samples. Hence

$$\text{est. } \sigma_{\text{diff.}} = \sqrt{\text{est. } \sigma^2\left(\frac{1}{N_1} + \frac{1}{N_2}\right)}$$

$$= \sqrt{\left(\frac{(N_1 - 1)s_1^2 + (N_2 - 1)s_2^2}{N_1 + N_2 - 2}\right)\left(\frac{N_1 + N_2}{N_1 N_2}\right)} \qquad [10.15.2\dagger]$$

This estimate of the standard error of the difference ordinarily forms the denominator of the t ratio when the t distribution is used for hypotheses about a difference.

10.16   An example of inferences about a difference for small samples

Two random samples of subjects are being compared on the basis of their scores on a motor learning task. The subjects are allotted to two experimental groups, with five subjects in the first and seven in the second. In the first group a subject is rewarded for each correct move made, and in the second each incorrect move is punished. The score is the number of trials to reach a specific criterion of performance. The experimenter wishes to find evidence for the question, "Does the kind of motivation employed, reward or punishment, affect the performance?" This question implies the null and alternative hypotheses:

$$H_0: \mu_1 - \mu_2 = 0$$
$$H_1: \mu_1 - \mu_2 \neq 0.$$

The experimenter is willing to assume that the population distributions of scores are normal, and that the population variances are equal. The probability of Type I error decided upon is .01. Since this is a two-tailed test, a glance at Table II shows that for $N_1 + N_2 - 2$ or $5 + 7 - 2 = 10$ degrees of freedom, and for $2Q = .01$, the required t value is 3.169. Thus an obtained t ratio equaling or exceeding 3.169 in absolute value is grounds for rejecting the hypothesis of no difference between population means.

The sample results are

$$M_1 = 18 \qquad M_2 = 20$$
$$s_1^2 = 6.00 \qquad s_2^2 = 5.83$$

The estimated standard error of the difference is found by the pooling procedure given in the last section:

$$\text{est. } \sigma_{\text{diff.}} = \sqrt{\text{est. } \sigma^2 \left(\frac{1}{N_1} + \frac{1}{N_2} \right)}$$
$$= \sqrt{\frac{(4)(6) + (6)(5.83)}{10} \left(\frac{12}{35} \right)}$$
$$= \sqrt{2.02}$$
$$= 1.42.$$

Thus, the t ratio is

$$t = \frac{(M_1 - M_2) - E(M_1 - M_2)}{\text{est. } \sigma_{\text{diff.}}} = \frac{-2}{1.42} = -1.41.$$

This value comes nowhere close to that required for rejection, and thus if α must be no more than .01 the experimenter does not reject the null hypothesis. His best choice may be to suspend judgement, pending more evidence.

Confidence intervals are found just as for a single small sample mean: the limits are

$$(M_1 - M_2) - t_{(\alpha/2; \, \nu)} \, (\text{est. } \sigma_{\text{diff.}}) \qquad\qquad \text{[10.16.1*]}$$
$$(M_1 - M_2) + t_{(\alpha/2; \, \nu)} \, (\text{est. } \sigma_{\text{diff.}}).$$

For this example, the 99 percent limits are

$$-2 - (3.169)(1.42)$$

and

$$-2 + (3.169)(1.42)$$

or approximately -6.5 and 2.5. The probability is .99 that the true *difference*, $\mu_1 - \mu_2$, is covered by an interval such as this. Once again, notice that this interval *does* contain the value 0, indicating the hypothesis entertained above is not rejected.

10.17   THE IMPORTANCE OF THE ASSUMPTIONS IN A *t* TEST OF A DIFFERENCE

In order to justify the use of the t distribution in problems involving a difference between means, one must make two assumptions: the populations sampled are normal, and the population variances are homogeneous, σ^2 having the same value for each population. Formally, these two assumptions are essential if the t probabilities given by the table are to be exact. On the other hand, in practical situations these assumptions are sometimes violated with rather small effect on the conclusions.

       The first assumption, that of a normal distribution in the populations, is apparently the less important of the two. So long as the sample size is even moderate for each group quite severe departures from normality seem to make little practical difference in the conclusions reached. Naturally, the results are more accurate the more nearly unimodal and symmetric the population distributions are, and thus if one suspects radical departures from a generally normal form then he should plan on larger samples. Furthermore, the departure from normality can make more difference in a one-tailed than in a two-tailed result, and once again some special thought should be given to sample size when one-tailed tests are contemplated for such populations. By and large, however, this assumption may be violated almost with impunity provided that sample size is not extremely small.

       On the other hand, the assumption of homogeneity of variance is more important. In older work it was often suggested that a separate test for homogenity of variance be carried out before the t test itself, in order to see if this assumption were at all reasonable. However, the most modern authorities suggest that this is not really worth the trouble involved. In circumstances where they are needed most (small samples), the tests for homogeneity are poorest. Furthermore, for samples of equal size relatively big differences in the population variances seem to have relatively small consequences for the conclusions derived from a t test. On the other hand, when the variances are quite unequal the use of different sample sizes can have serious effects on the conclusions. The moral should be plain: given the usual freedom about sample size in experimental work, *when in doubt use samples of the same size.*

       However, sometimes it is not possible to obtain an equal number in each group. Then one way out of this problem is by the use of a correction in the value for degrees of freedom. This is useful when one cannot assume equal population variances and samples are of different size. In this situation, however, the t ratio is calculated as in Section 10.13, where the separate standard errors are computed from each sample and the pooled estimate is not made. Then the corrected number of degrees of freedom is found from

$$\nu = \frac{(\text{est. } \sigma_{M_1}^2 + \text{est. } \sigma_{M_2}^2)^2}{(\text{est. } \sigma_{M_1}^2)^2/(N_1 + 1) + (\text{est. } \sigma_{M_2}^2)^2/(N_2 + 1)} - 2. \qquad [10.17.1\dagger]$$

This need not result in a whole value for ν, in which case the use of the nearest whole value for ν is sufficiently accurate for most purposes. When somewhat greater accuracy is desired, the approximate formula for critical values of t given in Section 14.17 is useful. When both samples are quite large, then both the assumptions of normality and of homogeneous variances become relatively unimportant, and the method of Section 10.13 can be used.

10.18 THE POWER OF *t* TESTS

The idea of the power of a statistical test was discussed in the preceding section only in terms of the normal distribution. Nevertheless, the same general considerations apply to the power of tests based on the *t* distribution. Thus, the power of a *t* test increases with sample size, increases with the discrepancy between the null hypothesis value and the true value of a mean or a difference, increases with any reduction in the true value of σ, and increases with any increase in the size of α, given a true value covered by H_1.

Unfortunately, the actual determination of the power for a *t* test against any given true alternative is more complicated than for the normal distribution. The reason is that when the null hypothesis is false, each *t* ratio computed involves $E(M)$ or $E(M_1 - M_2)$, which is the exact value given by the null (and false) hypothesis. If the true value of the expectation could be calculated into each *t* ratio, then the distribution would follow the *t* function tabled in the appendix. However, when H_0 is false, each *t* value involves a false expectation; this results in a somewhat different distribution, called the **noncentral *t* distribution**. The probabilities of the various *t*'s cannot be known unless one more parameter, δ, is specified beside ν. This is the so-called noncentrality parameter, defined by

$$\delta^2 = \left(\frac{\mu - \mu_0}{\sigma_M}\right)^2. \qquad [10.18.1]$$

The parameter δ^2 expresses the squared difference between the true expectation μ and that given by the null hypothesis, or μ_0, in terms of σ_M. For a hypothesis about a difference and for samples of equal size,

$$\delta^2 = \left[\frac{(\mu_1 - \mu_2) - (\mu_{0_1} - \mu_{0_2})}{\sigma_{\text{diff}}}\right]^2. \qquad [10.18.2]$$

The value of the parameter δ is then the positive square root of δ^2.

The matter is made even more complex by the fact that a noncentral *t* distribution not only has an additional parameter that must be specified; the form of a noncentral t' distribution differs from that of a central *t* distribution. Hence, rather detailed tables become necessary for each pair of parameter values ν and δ if exact determinations of power are to be made. Such tables are provided in some advanced texts on statistics.

Fortunately, when great accuracy is not required, an approximation based upon the normal distribution can be used. This approximation, given by Scheffé (1959), provides the cumulative probability that the variable t' is less than or equal to some value x, given the noncentral distribution with parameters ν and δ. This is found by use of the expression

$$\Pr(t'_{(\nu,\delta)} \leqq x) = \Pr\left\{z \leqq (x - \delta)\left(1 + \frac{x^2}{2\nu}\right)^{-1/2}\right\},$$

where z is a value in a normal distribution with mean 0 and variance 1.00.

The use of this approximation can be demonstrated in terms of the preceding problem (Section 10.16). There, the null hypothesis was that of no difference between the two population means. Let us determine the power of this test against the alternative that $\mu_1 - \mu_2 = 4$. That is, given that $\mu_1 - \mu_2 = 4$, what is the probability that the obtained t' value would fall outside the interval with limits -3.169 to 3.169?

We start off by calculating the value of the noncentrality parameter δ. We really need to know the true value of the standard error of the difference, $\sigma_{\text{diff.}}$, but in the absence of this information we will use the estimate from the samples. This was found to be 1.42. Then the value of δ corresponding to a difference of 4 is given by

$$\delta = \left| \frac{4}{\sigma_{\text{diff.}}} \right| = \frac{4}{1.42} = 2.816.$$

Then

$$\Pr(t'_{(\nu,\delta)} \le 3.169) = \Pr \left\{ z \le (3.169 - 2.816) \left[1 + \frac{(3.169)^2}{2(10)} \right]^{-1/2} \right\}$$

$$= \Pr \left(z \le \frac{.353}{\sqrt{1.5}} \right)$$

$$= \Pr(z \le .288)$$
$$= .614, \quad \text{approximately.}$$

Thus, we have found that if the true difference between the means is 4, the probability of an obtained t' value less than 3.169 is approximately the same as the probability of a normal z value less than .288. This probability is about .614.

Since this is a two-tailed test, we must also consider the possibility of an obtained t value that is less than -3.169. Then the probability of a Type II error will be the probability that $t \le 3.169$ minus the probability that $t \le -3.169$ (i.e., the probability that t falls in the region of nonrejection for H_0, even though the true difference is 4). Hence we take

$$\Pr(t'_{(\nu,\delta)} \le -3.169) = \Pr \left\{ z \le \frac{(-3.169 - 2.816)}{\sqrt{1.5}} \right\}$$

$$= \Pr(z \le -4.88).$$

This probability is virtually zero in a normal distribution. We then take the probability that $-3.169 \le t' \le 3.169$ to be approximately .614, and this is the probability of a Type II error when $\mu_1 - \mu_2 = 4$. The power of the t test against this alternative is then approximately $1 - .614$ or .386. If we desired, we could keep applying this method and construct the entire power function of the test for the various alternatives to the null hypothesis.

It should be kept in mind that this method depends upon the usual assumptions underlying the use of a t distribution being satisfied. That is, one still assumes that the parent distributions underlying the data are normal, that the observations are made independently and at random, and that, if two distributions are involved, each has the same variance. The noncentral variable t' differs from the central t variable only in that its distribution depends upon the new parameter δ. All of the other requirements for the use of a t distribution must be met.

10.19 TESTMANSHIP, OR HOW BIG IS A DIFFERENCE?

When an experimenter assigns subjects at random to two experimental groups, giving a different treatment to subjects in each group, he is usually looking for evidence of a statistical relation. Here, the independent variable represents the various experimental treatments and the dependent variable is the score of any subject within a group. Each treatment group is a random sample of all potential subjects given that treatment. The sample space is conceived as the set of all possible treatment-subject combinations, and the statistical relation itself is defined in terms of this sample space.

As we saw in Chapter 4, the complete absence of a statistical relation, or no association, occurs only when the conditional distribution of the dependent variable is the same regardless of which treatment is administered. Thus if the independent variable is not associated at all with the dependent variable the population distributions must be identical over the treatments. If, on the other hand, the means of the different treatment populations *are* different, the conditional distributions themselves must be different and the independent and dependent variables must be associated. The rejection of the hypothesis of no difference between population means is tantamount to the assertion that the treatment given does have some statistical association with the dependent variable score.

However, the occurrence of a significant result says nothing at all about the strength of the association between treatment and score. A significant result leads to the inference that some association exists, but in no sense does this mean that an important degree of association necessarily exists. Conversely, evidence of a strong statistical association can occur in data even when the results are not significant. The game of inferring the true degree of statistical association has a joker: this is the sample size. The time has come to define the notion of the strength of a statistical association more sharply, and to link this idea with that of the true difference between population means.

Just as in our discussion of relations in Chapters 1 and 4, let us call the experimental variable (or the independent variable) X once again. Here, X may symbolize a number standing for a quantity of some treatment, or it may simply represent any one of a set of qualitatively different treatments. In either circumstance, X stands for the status of the individual observation on the experimental factor, the condition manipulated by the experimenter. The dependent variable is Y, which here stands for a numerical score. If we conceive the sample space as comprising the outcomes of our observing all of a population of individuals under each of the possible set of treatments X, then each possible observation in the experiment is some (x, y) event. Furthermore, if individuals from the population of potential subjects are sampled at random, and assigned at random to the various possible treatments X in the experiment, then the occurrence of any individual in the treatment x has a probability $p(x)$. For our purposes, it will be convenient to assume that

$$p(x) = \frac{\text{number of individuals observed under treatment } x}{\text{total number of individuals observed}}.$$

When does it seem appropriate to say that a strong association exists between the experimental factor X and the dependent variable Y? Over all of the different possibilities for X there is a probability distribution of Y values, which is the **marginal** distribution of Y over (x,y) events. The existence of this distribution implies that we do not know exactly what the Y value for any observation will be; we are always uncertain about Y to some extent. However, given any particular X, there is also a **conditional** distribution of Y, and it may be that in this conditional distribution the highly probable values of Y tend to "shrink" within a much narrower range than in the marginal distribution. If so, we can say that the information about X tends to *reduce uncertainty* about Y. *In general we will say that the strength of a statistical relation is reflected by the extent to which knowing X reduces uncertainty about Y.*

One of the best indicators of our uncertainty about the value of a variable is σ^2, the variance of its distribution. The marginal distribution of Y has variance σ_Y^2, and given any X, the conditional distribution has variance $\sigma_{Y|X}^2$. For the time being, let us assume that $\sigma_{Y|X}^2$ is the same regardless of which X we specify. This is exactly the assumption of equal variances made in the t test, since each population distribution is actually a conditional distribution, given some treatment specification. The reduction in uncertainty provided by X is then proportional to

$$\sigma_Y^2 - \sigma_{Y|X}^2, \qquad [10.19.1^*]$$

the difference between the marginal and the conditional variance of Y.

It is convenient to turn this reduction in uncertainty into a **relative reduction** by dividing by σ_Y^2, giving

$$\omega^2 = \frac{\sigma_Y^2 - \sigma_{Y|X}^2}{\sigma_Y^2}. \qquad [10.19.2^*]$$

The relative reduction in uncertainty about Y given by X is shown by the index ω^2 (Greek omega, squared). Sometimes the value ω^2 is called **the proportion of variance in Y accounted for by X.** Viewed either as a relative reduction in uncertainty, or as a proportion of variance accounted for, the index ω^2 represents the strength of association between independent and dependent variables. (The index ω^2 is almost identical to two other indices to be introduced later, *the intraclass correlation* and the *correlation ratio*, usually represented by the symbols ρ_I and η^2 respectively. However, since these indices were developed for and are used in somewhat different contexts, it seems better to use the relatively neutral symbol ω^2 here, to avoid later confusion.)

This index reflects the predictive power afforded by a relationship: when ω^2 is zero, then X does not aid us at all in predicting the value of Y. On the other hand, when ω^2 is 1.00, this tells us that X lets us know Y exactly. All intermediate values of the index represent different degrees of predictive ability. Notice that for any functional relation, $\omega^2 = 1.00$, since there can be only one Y for each possible X. A value less than unity tells us that precise prediction is not possible, although X nevertheless gives *some* information about Y unless $\omega^2 = 0$.

About now you should be wondering what the index ω^2 has to do with the difference between population means. It can be shown, by methods we shall use in Chapter 12, that when $p(x_1) = p(x_2) = 1/2$,

$$\sigma_Y^2 = \sigma_{Y|X}^2 + \frac{(\mu_1 - \mu_2)^2}{4} \qquad [10.19.3*]$$

where μ_1 is the mean of population 1, μ_2 that of population 2, and

$$\frac{\mu_1 + \mu_2}{2} = \mu,$$

the mean of the marginal distribution.

On substituting into **10.19.2,** we find

$$\omega^2 = \frac{(\mu_1 - \mu_2)^2}{4\sigma_Y^2}.$$

For two treatment-populations with equal variances the strength of the statistical association between treatment and dependent variable varies directly with the squared difference between the population means, relative to the unconditional, marginal, variance of Y.

When the difference $\mu_1 - \mu_2$ is zero, then ω^2 must be zero. In the usual t test for a difference, the hypothesis of no difference between means is equivalent to the hypothesis that $\omega^2 = 0$. On the other hand, when there is any difference at all between population means, the value of ω^2 must be greater than 0. In short, a true difference is "big" in the sense of predictive power only if the square of that difference is large relative to σ_Y^2. However, in significance tests such as t, we compare the difference we get with an estimate of $\sigma_{\text{diff.}}$. The standard error of the difference can be made almost as small as we choose if we are given a free choice of sample size. Unless sample size is specified, there is no *necessary* connection between significance and the true strength of association.

This points up the fallacy of evaluating the "goodness" of a result in terms of statistical significance alone, without allowing for the sample size used. All significant results do not imply the same degree of true association between independent and dependent variables.

It is sad but true that researchers have been known to capitalize on this fact. There is a certain amount of "testmanship" involved in using inferential statistics. *Virtually any study can be made to show significant results if one uses enough subjects, regardless of how nonsensical the content may be.* There is surely nothing on earth that is completely independent of anything else. The strength of an association may approach zero, but it should seldom or never be exactly zero. If one applies a large enough sample of the study of any relation, trivial or meaningless as it may be, sooner or later he is almost certain to achieve a significant result. Such a result may be a valid finding, but only in the sense that one can say with assurance that some association is not exactly zero. The degree to which such a finding enhances our knowledge is debatable. If the criterion of strength of association is applied to such a result, it becomes obvious that little or nothing is actually contributed to our ability to predict one thing from another.

For example, suppose that two methods of teaching first grade children to read are being compared. A random sample of 1000 children are taught to read by method I, another sample of 1000 children by method II. The results of the instruction are evaluated by a test that provides a score, in whole units, for each child. Suppose that the results turned out as follows:

Method I	Method II
$M_1 = 147.21$	$M_2 = 147.64$
$s_1^2 = 10$	$s_2^2 = 11$
$N_1 = 1000$	$N_2 = 1000$

Then, the estimated standard error of the difference is about .145, and the z value is

$$z = \frac{147.21 - 147.64}{.145} = -2.96.$$

This certainly permits rejection of the null hypothesis of no difference between the groups. However, does it really tell us very much about what to expect of an individual child's score on the test, given the information that he was taught by method I or method II? If we look at the group of children taught by method II, and assume that the distribution of their scores is approximately normal, we find that about 45 percent of these children fall *below* the mean score for children in group I. Similarly, about 45 percent of children in group I fall above the mean score for group II. Although the difference between the two groups is significant, the two groups actually overlap a great deal in terms of their performances on the test. In this sense, the two groups are really not very different at all, even though the difference between the means is quite significant in a purely statistical sense.

Putting the matter in a slightly different way, we note that the grand mean of the two groups is 147.425. Thus, our best bet about the score of any child, not knowing the method of his training, is 147.425. If we guessed that any child drawn at random from the combined group should have a score above 147.425, we should be wrong about half the time. However, among the original groups, according to method I and method II, the proportions falling above and below this grand mean are approximately as follows:

	Below 147.425	Above 147.425
Method I	.51	.49
Method II	.49	.51

This implies that if we know a child is from group I, and we guess that his score is below the grand mean, then we will be wrong about 49 percent of the time. Similarly, if a child is from group II, and we guess his score to be above the grand mean, we will be wrong about 49 percent of the time. If we are not given the group to which the child belongs, and we guess either above or below the grand mean, we will be wrong about 50 percent of the time. Knowing the group does reduce the probability of error in such a guess, but it does not reduce it very much. The method by which the child was trained simply doesn't tell us a great deal about what the child's score will be, even though the difference in mean scores is significant in the statistical sense.

This kind of testmanship flourishes best when people pay too much attention to the significance test and too little to the degree of statistical association the finding represents. This clutters up the literature with findings that are often not worth pursuing, and which serve only to obscure the really important predictive relations that occasionally appear. The serious scientist owes it to himself and his readers to ask not only, "Is there any association between X and Y?" but also, "How much does my finding suggest about the power to predict Y from X?" Much too much emphasis is paid to the former, at the expense of the latter, question.

10.20   Estimating the Strength of a Statistical Association from Data

It is quite possible to estimate the amount of statistical association implied by any obtained difference between means. The ingredients for this kind of estimation are essentially those used in a t test. The problems connected with the sampling distribution of this estimate will be deferred until Chapter 12, and for the moment we shall consider only how this estimate is made and used.

(A number of ways have been proposed for estimating the strength of a statistical association from obtained differences between means. For reasons to be elaborated later, none of these methods is entirely satisfactory. The method to be introduced here is thus only one of the ways that may be encountered in the statistical literature, but it seems to have as much to recommend it as any other.)

For samples from two populations, each of which has the same true variance, $\sigma_{Y|X}^2$, a rough estimate of ω^2 is provided by

$$\text{est. } \omega^2 = \frac{t^2 - 1}{t^2 + N_1 + N_2 - 1}. \qquad [10.20.1]$$

(A more general form for estimating ω^2 will be given in Chapter 12.) Notice that if t^2 is less than 1.00, then this estimate is negative, although ω^2 cannot assume negative values. In this situation the estimate of ω^2 is set equal to zero.

Let us consider an example using this estimate. Imagine a study involving two groups of 30 cases each. Subjects are assigned at random to these two groups, and each set of subjects is given a different treatment. The results are

Group 1	Group 2
$M_1 = 65.5$	$M_2 = 69$
$s_1^2 = 20.69$	$s_2^2 = 28.96$
$N_1 = 30$	$N_2 = 30$

First of all the t ratio is computed in the usual way (Section 10.16):

$$\text{est. } \sigma^2 = \frac{(29)(20.69 + 28.96)}{58} = 24.83$$

and

$$\text{est. } \sigma_{\text{diff.}} = \sqrt{\frac{24.83(2)}{30}} = 1.29.$$

Thus,

$$t = \frac{65.5 - 69}{1.29} = -2.71.$$

For a two-tailed test with 58 degrees of freedom, this value is significant beyond the .01 level. Thus, we are fairly safe in concluding that some association exists.

What do we estimate the true degree of association to be? Substituting into 10.20.1, we find

$$\text{est. } \omega^2 = \frac{(2.71)^2 - 1}{(2.71)^2 + (60 - 1)} = .096.$$

Our rough estimate is that X (the treatment administered) accounts for about 10 percent of the variance of Y (the obtained score).

Suppose, however, that the groups had contained only 10 cases each, and that the results had been:

Group 1	Group 2
$M_1 = 65.5$	$M_2 = 69$
$s_1^2 = 5.55$	$s_2^2 = 7.78$
$N_1 = 10$	$N_2 = 10$

Here

$$\text{est. } \sigma^2 = \frac{9(5.55 + 7.78)}{18} = 6.67$$

$$\text{est. } \sigma_{\text{diff.}} = \sqrt{6.67 \left(\frac{2}{10}\right)} = 1.15$$

so that

$$t = \frac{-3.5}{1.15} = -3.04.$$

For 18 degrees of freedom, this value is also significant beyond the .01 level (two-tailed), and once again we can assert with confidence that some association exists.

Again, we estimate the degree of association represented by this finding:

$$\text{est. } \omega^2 = \frac{(3.04)^2 - 1}{(3.04)^2 + 19} = .29.$$

Here, our rough estimate is that X accounts for about 29 percent of the variance in Y. Even though the difference between the sample means is the same in these two examples, and both results are significant beyond the .01 level, the second experiment gives a much higher estimate of the true association than the first.

The point of this discussion should be evident by now: statistical significance is not the only, or even the best, evidence for a strong statistical association. A significant result implies that it is safe to say some association exists, but the estimate of ω^2 tells how strong that association appears to be. It seems far more reasonable to decide to follow up a finding that is *both* significant *and* indicates a strong degree of association than to tie this course of action to significance level alone. Conversely, when a result fails to attain significance and there is no ready way to estimate the β probability, the experimenter really has at least two courses of action: he can suspend judgment temporarily and actually collect more data, or he can suspend judgment permanently by forgetting the whole business. If the estimated strength of association is relatively small it may not be worthwhile to spend more time and effort in this direction. Regardless of the courses of action open to the experimenter, on the whole it is reasonable that a better decision can be made in terms of both significance level and estimated strength of relation than by either taken alone. In most experimental problems we want to find and refine relationships that "pay off," that actually increase our ability to predict behavior. When the results of an experiment suggest that the strength of an association is very low, then perhaps the experimenter should ask himself whether this matter is worth pursuing after all, regardless of the statistical significance he may attain by increased sample size or other refinements of the experiment.

10.21 STRENGTH OF ASSOCIATION AND SAMPLE SIZE

Of all the questions that psychologists carry to statisticians, surely the most frequently heard is, "How many subjects do I need in this experiment?" The response of the statistician is very likely to be unsatisfactory, the gist being "How big is a difference that you consider important?" This is a question that can be answered only by the psychologist, and then only if he has given the matter some serious thought. If the experimenter cannot answer this question the statistician really cannot help him. Perhaps framing the essential point in terms of strength of a relation rather than the size of a difference will make it a little easier to grasp.

Basically, the question of sample size depends upon the strength of association the experimenter *wants* to detect as significant. Actually, this matter is properly discussed in terms of the t distribution, but for the kinds of rough determinations most psychologists need to make, the normal approximation will suffice.

Recall the basic definition of ω^2 in terms of two population means:

$$\omega^2 = \frac{(\mu_1 - \mu_2)^2}{4\sigma_Y^2}.$$

From this definition, we can derive the fact that

$$\frac{|\mu_1 - \mu_2|}{\sigma_{Y|X}} = 2\sqrt{\frac{\omega^2}{1 - \omega^2}} = \Delta. \qquad [10.21.1*]$$

Given any value of ω^2, we can find the ratio of the absolute difference between population means to the standard deviation of either population. The symbol Δ

(capital Greek delta) will stand for this absolute difference between means in standard deviation units.

If we want to discuss the difference between means in units of the standard error of the difference, then for samples of the same size, n,

$$\frac{|\mu_1 - \mu_2|}{\sigma_{\text{diff.}}} = \Delta \sqrt{\frac{n}{2}}.$$

Now suppose that an experiment is being planned which involves two groups, each of size n. The experimenter wants to be very sure that he will detect a significant difference if the true degree of association ω^2 is k or more in value. How large should n be in each sample?

Several things must be specified: the value of $k = \omega^2$, the α probability, and the probability $1 - \beta$, which is the power of the test when the true degree of association is equal to k. Given these three specifications, then one can approximate the required size of n by taking

$$\sqrt{\frac{n}{2}} = \frac{[z_{(1-\alpha/2)} - z_{(\beta)}]}{\Delta} \qquad [10.21.2\dagger]$$

or

$$n = \frac{2[z_{(1-\alpha/2)} - z_{(\beta)}]^2}{\Delta^2} \qquad [10.21.3\dagger]$$

where $z_{(1-\alpha/2)}$ is the value of a standardized score in a normal distribution cutting off the lower $(1 - \alpha/2)$ proportion of cases, and $z_{(\beta)}$ is the standardized score cutting off the lower β proportion of cases. The value of Δ is found from the required value of ω^2 by substitution into **10.21.1**.

For example, suppose that the experimenter wants to be very sure to detect a true association when X actually accounts for 25 percent or more of the variance of Y, so that ω^2 is .25 or more. He wants the test to have a power of .99 when $\omega^2 = .25$, and he has already decided that α must be .01. How many cases should he include in each sample?

First of all, solving from **10.21.1** for Δ, we find

$$\Delta = 2 \sqrt{\frac{.25}{1 - .25}}$$

or

$$\Delta = 1.15.$$

The value of $z_{(1-\alpha/2)}$ is 2.58, and that for $z_{(\beta)} = -2.33$. Thus,

$$n = \frac{2(2.58 + 2.33)^2}{(1.15)^2} = 36.5.$$

In order to have $\alpha = .01$, and to have a test with power of about .99 for a significant result when $\omega^2 = .25$, the experimenter should plan on about 37 subjects in each sample, a total of 74 subjects in all.

This may be more subjects than the experimenter can manage to obtain. He can reduce his estimate of the required number either by *lowering* his requirements for the power of the test, making the power, say, .95 for $\omega^2 = .25$, or by *raising* the probability α to, say, .05. Suppose that he adopts the latter course, making $\alpha = .05$. In this instance, his revised estimate of sample size is given by

$$n = \frac{2(1.96 + 2.33)^2}{(1.15)^2} = 27.9$$

showing that these requirements are approximately satisfied if he takes around 28 subjects in each group, for a total of about 56 cases in all.

These estimates of required sample size are only approximate, since we use the normal rather than the t distribution. They are to be regarded only as rough guides to the general sample sizes required. Unless the sample size estimates turn out rather large as in the example, and if it is important that the experimenter fulfill the requirements he has set himself about ω^2, α, and $1 - \beta$, he is very wise to take samples somewhat larger than his estimate suggests.

It is remarkable how few studies reported in psychology seem to be based on sample sizes chosen in any systematic way. Certainly there are situations where real limits exist about how large a sample can be, and here the experimenter merely does the best he can. However, there is usually some freedom of choice within fairly broad limits. One does not have to look very far to find the reason this question is often ignored: all too seldom is the experimenter prepared to state the strength of association that he feels he *must* be sure to detect as a significant result. To decide this requires a great deal of thought about the potential applications, or the experimental follow-up that should be implied by a significant finding. On the other hand, unless this thought is expended in planning a study there is simply no way to determine required sample size. If psychologists are going to use conventions for deciding significance of results then perhaps a few conventions are called for about the strength of association it is desirable to *detect* as significant.

Regardless of the sample size actually chosen, the experimenter can form a rough estimate of the sensitivity of the experiment for detecting statistical association. For two relatively large samples of equal size, the expression

$$\Delta^2 = \frac{2[z_{(1-\alpha/2)}]^2}{n} \qquad [10.21.4*]$$

can be solved for Δ^2. Then by the relation

$$\omega^2 = \frac{\Delta^2}{\Delta^2 + 4} \qquad [10.21.5*]$$

one finds the strength of association for which the power is approximately .50. One can be reasonably sure that if the true degree of association is greater than the value of ω^2 found by this procedure, then he has better than a fifty-fifty chance of detecting this fact as a significant result. Conversely, if the true association is less than the value of ω^2 found the chances are about .50 or better that he will not detect this as a significant result.

For example, suppose that 25 subjects are used in each of two experimental groups. The α level chosen is .01, two-tailed. Then

$$\Delta^2 = \frac{2(2.58)^2}{25} = .53$$

so that

$$\omega^2 = \frac{.53}{.53 + 4} = .12.$$

The experimenter can say that if the true degree of association is about .12 or more he has at least a fifty-fifty chance of detecting this as a significant result.

Had the experimenter used 100 subjects per group, then the value of ω^2 in **10.21.4** would have been about .03. For this relatively large sample size the experimenter has about a fifty-fifty chance of detecting a significant difference when the experimental variable X accounts for no more than about 3 percent of the variance of Y. The larger the sample size, the smaller the proportion of variance accounted for that we can safely expect to be detected as a significant result.

The discussion of sample size in this section has been conducted in terms of two-tailed tests, since these are most comon in psychological research. However, the same idea applies in approximating the required sample size for a one-tailed test as well. Instead of using $z_{(1-\alpha/2)}$ in the computations, one simply substitutes $z_{(1-\alpha)}$ where α is the chosen error probability in the one-tailed region. For one-tailed tests the statements made about degrees of association and their detection are valid only for differences in the direction of the region of rejection, of course.

10.22   CAN A SAMPLE SIZE BE TOO LARGE?

In one sense, even posing this question sounds like heresy! Psychologists are often trained to think that large samples are *good things*, and we have seen that the most elegant features within theoretical statistics actually are the limit theorems, each implying a connection between sample size and the goodness of inferences.

Nevertheless, it seems reasonable that sample size can never really be discussed apart from what the experimenter is trying to do, and the stakes that he has in the experiment. *So long as the experimenter's primary interest is in precise estimation, then the larger the sample the better.* When he wants to come as close as he possibly can to the true parameter values, he can always do better by increasing sample size.

This is not, however, the main purpose of some experiments. These studies are, in the strict sense, exploratory. The experimenter is trying to map out the main relationships in some area. His study serves as a guide for directions that he will pursue in further, more refined, studies. He wants to find those statistical associations that are relatively large and that give considerable promise that a more or less precise relationship is there to be discovered and refined. He does not want to waste his time and effort by concluding an association exists when the degree of prediction actually afforded by that association is negligible. In short, the experimenter would like a significant result to represent not only a nonzero association, but an association of considerable size.

When this is the situation it is advisable to look into the effects of sample size on the probability of finding a significant result given a *weak* association. For example, an experimenter has decided to use 30 subjects in each of two experimental groups. However, he does not want to waste his time with a significant result when the true degree of association is .01 or less. He decides that he wants the probability of a significant result to be .05 or less when the true ω^2 is .01 or less. He has already decided that α must be .01, and he knows that this must be the probability of a significant result when $\omega^2 = 0$ (the true difference is zero). For a two-tailed test, he cannot make the power of the test less than α. However, the experimenter's requirement is that his test have power of only .05 or less when ω^2 is less than or equal to .01.

Using **10.21.1** from the previous section,

$$\Delta = 2\sqrt{\frac{.01}{1 - .01}}$$
$$= 2(.10) = .2, \text{ approximately.}$$

In this problem, $1 - \beta = .05$, and so, by **10.21.3**,

$$n = \frac{2(2.58 - 1.65)^2}{(.2)^2}$$
$$= 43.$$

If he uses *no more* than about 43 subjects in each group then the experimenter can be quite sure that a significant result is not likely to occur when the true degree of association is .01 or less. However, if he uses more subjects, he cannot be this confident of *not* detecting a very small association.

What does setting maximum sample size at 43 dictate about the ω^2 values he *will* detect as significant? How large a ω^2 will he detect as a significant result 95 percent of the time? This is found from

$$\Delta^2 = \frac{2(2.58 + 1.65)^2}{43}$$
$$= .83$$

so that

$$\omega_1^2 = \frac{(.83)^2}{(.83)^2 + 4}$$
$$= .15.$$

Even with the sample size of 43 cases per group, the experimenter knows that there is a probability of about .95 of finding a significant result when the true proportion of variance accounted for is as small as .15. This is not necessarily a negligible degree of association, and in some contexts it may be very important to account for as much as .15 proportion of variance. Nevertheless, in this instance the experimenter is interested in large degrees of association, and is content to rule out of consideration proportions of variance accounted for as small or smaller than .15.

Trivial associations may well show up as significant results when the sample size is very large. If the experimenter wants significance to be very likely to reflect a sizable association in his data, and also wants to be sure that he will not be led by a significant result into some blind alley, then he should pay attention to both aspects of sample size. Is the sample size *large* enough to give confidence that the big associations will indeed show up, while being *small* enough so that trivial associations will be excluded from significance?

10.23   PAIRED OBSERVATIONS

Sometimes it happens that subjects are actually sampled in pairs. Even though each subject is experimentally different in one respect (nominally, the independent variable) from his pair-mate and each has some distinct dependent variable score, the scores of the members of a pair are not necessarily independent. For instance, one may be comparing scores of husbands and wives; a husband is "naturally" matched with his wife, and it makes sense that knowing the husband's score gives us some information about his wife's, and vice versa. Or individuals may be matched on some basis by the experimenter, and within each matched pair the members are assigned at random to experimental treatments. This matching of pairs is one form of experimental control, since each member of each experimental group must be identical (or nearly so) to his pair-mate in the other group with respect to the matching factor or factors, and thus the factor or factors used to match pairs is less likely to be responsible for any observed difference in the groups than if two unmatched groups are used.

Given two groups matched in this pairwise way, either by the experimenter or otherwise, it is still true that the difference between the means is an unbiased estimate of the population difference (in two matched populations):

$$E(M_1 - M_2) = \mu_1 - \mu_2.$$

However, the matching, and the consequent *dependence* within the pairs, changes the standard error of the difference. This can be shown quite simply: By definition, the variance of the difference between two sample means is

$$\sigma^2_{\text{diff.}} = E(M_1 - M_2 - \mu_1 + \mu_2)^2$$

which is the same as

$$E[(M_1 - \mu_1) - (M_2 - \mu_2)]^2.$$

Expanding the square, we have

$$E(M_1 - \mu_1)^2 + E(M_2 - \mu_2)^2 - 2E(M_1 - \mu_1)(M_2 - \mu_2).$$

The first of these terms is just $\sigma^2_{M_1}$, and the second is $\sigma^2_{M_2}$. However, what of the third term? From rule 6 Appendix B we find that the expectation of this product must be zero when the variables are independent. On the other hand, when variables are dependent the expectation is *not* ordinarily zero. Let us denote this last term above as cov.(M_1, M_2), the **covariance** of the means. Then, for matched groups,

$$\sigma^2_{\text{diff.}} = \sigma^2_{M_1} + \sigma^2_{M_2} - 2 \text{ cov.}(M_1, M_2).$$

In general, for groups matched by pairs, this covariance is a positive number, and thus the variance and standard error of a difference between means will usually be *less* for matched than for unmatched groups. This fact accords with the experimenter's purpose in matching in the first place: to remove one or more sources of variability, and thus to lower the sampling error.

On the other hand, some caution must be exercised in this matching process. In the first place, it can be true that the factor on which subject-pairs are matched is such that the means are *negatively* related. Thus, for example, suppose that one had an effective measure of the dominance of personality of an individual. It just might be that highly dominant women tend to marry men with low dominance, and vice versa, so that among husband-wife pairs, dominance scores are negatively related. Then, if our interest is basically that of comparing men and women generally on such scores, it would be a mistake to match, since the negative relationship would lead to a larger, rather than a smaller, standard error of the difference than would a comparison of unmatched groups.

Furthermore, such matching may be less efficient than the comparison of unmatched random groups, unless the factor used in matching introduces a relatively strong positive relationship between the means. While a positive relationship, reflected in a positive covariance term, does reduce the standard error of the difference, this procedure also *halves* the number of degrees of freedom. Dealing with a sample of N pairs gives only half the number of degrees of freedom available when we deal with two independent groups of N cases each. Thus, if the factor entering into the matching is only slightly relevant to the differences between the groups, or is even irrelevant to such differences, matching is not a desirable procedure. The experimenter should have quite good reasons for matching before he adopts this procedure in preference to the simple comparison of two randomly selected groups.

Let us leave these considerations for a moment and return to the actual procedure for matched groups. The unknown value of cov. (M_1, M_2) could be something of a problem, but actually it is quite easy to bypass this difficulty altogether. Instead of regarding this as two samples, we simply think of the data coming from one sample of *pairs*. Associated with each pair i is a difference

$$D_i = (y_{i1} - y_{i2}),$$

where Y_{i1} is the score of the member of pair i who is in group 1, and Y_{i2} is the score of the member of pair i who is in group 2. Then an ordinary t test for a *single* mean is carried out using the scores D_i. That is,

$$M_D = \frac{\sum_i D_i}{N}$$

and
$$s_D^2 = \frac{\sum_i D_i^2}{N-1} - \frac{N(M_D)^2}{N-1}.$$

Then
$$\text{est. } \sigma_{M_D} = \frac{s_D}{\sqrt{N}}$$

and t is found from

$$t = \frac{M_D - E(M_D)}{\text{est. } \sigma_{M_D}}$$

with $N - 1$ degrees of freedom. *Be sure to notice that here N stands for the number of differences, which is the number of pairs.*

Naturally, the hypothesis is about the true value of $E(M_D)$, which is always $\mu_1 - \mu_2$. Thus any hypothesis about a difference can be tested in this way, provided that the groups used are matched *pairwise*. Similarly, confidence limits are found just as for a single mean, using M_D and σ_{M_D} in place of M and σ_M.

An example will now be given of this method of computation for a test of the difference in means of two matched groups. Not only will this example illustrate the method; the data have also been "rigged" to illustrate the point made above, that in some situations it may be less efficient to match than to simply take two randomly selected groups for comparison. This example will involve matched groups where a negative relationship exists among the pairs.

Consider once again the question of scores on a test of dominance. The basic question has to do with the mean score for men as opposed to the mean score for women. In carrying out the experiment, the investigator decided to sample eight husband-wife pairs at random. The members of each pair were given the test of dominance separately, and the data turned out as follows:

Pair	Husband	Wife	D	D²
1	26	30	−4	16
2	28	29	−1	1
3	28	28	0	0
4	29	27	2	4
5	30	26	4	16
6	31	25	6	36
7	34	24	10	100
8	37	23	14	196
			31	369

Then

$$M_D = \frac{31}{8} = 3.87, \qquad s_D^2 = \frac{369}{7} - \frac{(31)^2}{7(8)} = 35.59,$$

$$\text{est.} \sigma_{M_D} = \sqrt{\frac{35.59}{8}} = \sqrt{4.45} = 2.109.$$

The t test is thus given by

$$t = \frac{3.87 - 0}{2.109} = 1.835.$$

For 7 degrees of freedom, this result is not significant (two-tailed test) for $\alpha = .05$ or less.

Now let us change our frame of reference slightly. Suppose that it had been true that these data came from two independent groups, one of men and one of women, each drawn at random. In this case, the men's group (formerly the husbands) would show a mean of 30.37, and an unbiased estimate of the variance s^2 of 13.00. The women's group would have a mean of 26.50, with an s^2 value of 6.02. Then the standard error of the difference would be estimated from

$$\text{est.}\sigma_{\text{diff}} = \sqrt{\frac{7(13 + 6.02)}{16 - 2}\left(\frac{2}{16}\right)}$$

$$= \sqrt{1.189} = 1.09, \quad \text{approximately.}$$

Then, in this instance the t would be given by

$$t = \frac{(30.37 - 26.50) - 0}{1.09} = 3.55,$$

which, for 14 degrees of freedom, is significant well beyond the .01 level, two-tailed. Why this very different result from the same set of numbers? The answer is given by the relationship of the scores when they are regarded as paired. Notice that high scores for husbands are paired with low scores for wives, and vice versa. This implies a negative relationship of such scores, leading to a negative covariance term in the estimate of the standard error. This, in turn, actually *increases* the size of the standard error relative to that for unmatched groups. When such a situation actually exists, there is a distinct disadvantage to matching.

Now do not get the idea that the option open to the author here is open to an experimenter. The sample is either drawn from a population of pairs or from two independent populations, and one does not have the right to change his mind about the nature of the sample after the fact. This procedure was strictly for illustrative purposes! Furthermore, when such matching or pairing is used, prior evidence or sheer common sense should suggest whether or not a positive relationship among the scores should exist or not. If such a positive relationship should exist, then the matching procedure may well reduce the sampling variance sufficiently to offset the loss in degrees of freedom, and a matching procedure may thus be desirable. The point is that this is not an automatic consequence of such matching. As with any question of experimental design, no routine procedure is advantageous for all situations, and the experimenter must bring his judgment and knowledge to bear on such decisions.

10.24 SIGNIFICANCE TESTING IN MORE COMPLICATED EXPERIMENTS

Only in the very simplest experimental problems does the experimenter confine himself to two treatment groups. It is far more usual to find experiments that involve a number of qualitatively or quantitatively different treatments. However, the basic conception of what the experimenter is doing remains the same: he is looking for evidence of a statistical relation between experimental and dependent variables. When there are several groups it is no longer possible to make a simple and direct connection between the degree of statistical association and the difference between any pair of means; here there are any number of pairs of means that may be different and thus imply association, and the mechanism of the simple t test breaks down. Thus, we will introduce this problem once again in somewhat different terms in Chapter 12. However, before we can discuss methods general enough to handle multi-group data, we need to study two more theoretical sampling distributions, both of which grow out of problems of inference about population variances. The next chapter is devoted to the study of these two distributions.

Exercises

1. A random sample of 300 American women were asked to record their body temperatures twice a day for a full month. From their records an average value was found for each woman. The mean of these values was 98.7 with a standard deviation S of .95. Test the hypothesis that the mean body temperature of such American women is 98.6 against the alternative that the mean is some other value.
2. Find the 99 percent confidence interval for the mean in problem 1.
3. In a study of truth in advertising, a government agency opened 500 boxes selected at random of a well-known brand of raisin bran. For each box the actual number of raisins was counted. The mean number of raisins was 32.4, with a standard deviation $S = 4.1$. Evaluate the company's claim that each box contains 34 raisins on the average, against the alternative of fewer raisins than claimed.
4. Find the 95 percent confidence interval for the mean in problem 3.
5. Suppose that the body weight at birth of normal children (single births) within the United States is approximately normally distributed and has a mean of 115.2 ounces. A pediatrician believes that the birth weights of normal children born of mothers who are habitual smokers may be lower on the average than for the population as a whole. In order to test this hypothesis, he secures records of the birth weights of a random sample of 20 children from mothers who are heavy smokers. The mean of this sample is 114.0 with $S = 4.3$. Evaluate the pediatrician's hunch.
6. Reevaluate the data of problem 5 on the assumption that a sample of 80 children had been used.
7. For the results of problem 5, find the 99 percent confidence interval for the mean birth weight of normal children from smoking mothers.
8. Suppose that in a certain large community the number of hours that a TV set is turned on in a given home during a given week is approximately normally distributed. A sample of 26 homes was selected, and careful logs were kept of how many hours per week the TV set was on. The mean number of hours per week in the sample turned out to be 36.1 with a standard deviation S of 3.3 hours. Find the 95 percent confidence interval for the mean number of hours that TV sets are played in the homes of this community.
9. For the data of problem 8, test the hypothesis that the true mean number of hours is 35. Test the hypothesis that the mean number of hours is 30.

10. Four random samples are taken independently from a population. For each random sample, the ninety percent confidence interval for the mean is found. What is the probability that at least one of those confidence intervals fails to cover the population mean? What is the probability that two or more confidence intervals fail to cover the population value?

11. The same government agency referred to in problem 3 has decided to compare two well-known brands of raisin bran with respect to the numbers of raisins each contain on the average. Some 100 boxes of Brand A were taken at random, and the same number of boxes of Brand B were randomly selected. On the average the Brand A boxes contained 38.7 raisins, with $S = 3.9$, and Brand B contained an average of 36 raisins with $S = 4$. Test the hypothesis that the two brands are actually identical in the average number of raisins that their boxes contain. Let H_1 be "not H_0."

12. For problem 11, find the 99 percent confidence interval for the difference in average number of raisins for Brands A and B.

13. The editor of a journal in Psychology tends to believe that the contributors to that journal now use shorter sentences on the average than they did a few years ago. In order to test this hunch, he takes a random sample of 150 sentences from journal articles written ten years ago and a random sample of 150 sentences from articles published within the last two years. The first sample showed a mean length of 127 type spaces per sentence, whereas the second sample showed a mean length of 113 type spaces. The first standard deviation $S = 41$, and the second standard deviation $S = 45$. Should he conclude that the recent articles do tend to have shorter sentences?

14. Find the 95 percent confidence interval for difference in sentence length from problem 13.

15. In an experiment, subjects were assigned at random between two conditions, five to each. Their scores turned out as follows:

CONDITION A	CONDITION B
128	123
115	115
120	130
110	135
103	113

Can one say that there is a significant difference between these two conditions? What must one assume in carrying out this test?

16. Find the 99 percent confidence interval for the difference between Conditions A and B in problem 8. On the evidence of this confidence interval, could one reject the hypothesis that the true mean of Condition B is five points higher than that of condition A?

17. In an experiment, the null hypothesis is that two means will be equal. The variance of each population is believed to be equal to 16. If $\alpha = .05$, two-tailed, and the test is to have a power of .90 against the alternative that $M_1 - M_2 = 3$, about how many cases should one take in each experimental group?

18. Suppose that two brands of gasoline were being compared for mileage. Samples of each brand were taken and used in identical cars under identical conditions. Nine tests were made of Brand I and six tests of Brand II. The following miles per gallon were found.

BRAND I	BRAND II
16	13
18	15
15	11
23	17
17	12
14	13
19	
21	
16	

Are the two brands significantly different? What must be assumed here in order to carry out the test?

19. An experimenter was interested in dieting and weight losses among men and among women. He believed that in the first two weeks of a standard dieting program, women would tend to lose more weight than men. As a check on this notion, a random sample of 15 husband-wife pairs were put on the same strenuous diet. Their weight losses after two weeks showed the following:

PAIR	HUSBANDS	WIVES
1	5.0 lbs	2.7 lbs
2	3.3	4.4
3	4.3	3.5
4	6.1	3.7
5	2.5	5.6
6	1.9	5.1
7	3.2	3.8
8	4.1	3.5
9	4.5	5.6
10	2.7	4.2
11	7.0	6.3
12	1.5	4.4
13	3.7	3.9
14	5.2	5.1
15	1.9	3.4

Did wives lose significantly more than husbands? What are we assuming here?

11 The Chi-Square and the *F* Distributions

The essential ideas of inferential statistics are most easily discussed in terms of inferences about means or proportions, and so attention has been focused almost exclusively on these matters in the preceding chapters. However, population distributions can be compared in terms of variability as well as central tendency, and it is important to have inferential methods for the variance at our disposal.

The three basic sampling distributions used so far (the binomial, the normal, and the t distribution) no longer apply directly when the variance of a population is under study. Rather, we must turn to two new theoretical distributions. The first of these is called the **chi-square distribution,** or the distribution of the random variable χ^2 (small Greek chi, squared). We will use this distribution first in making inferences about a single population variance, although it has many other applications. The second distribution we will consider is usually called the *F* **distribution,** or the distribution of the random variable F (after Sir Ronald Fisher, who developed the main applications of this distribution). The study of this theoretical distribution grows out of the problem of comparing two population variances. The uses of both of these distributions extend far beyond the problems for which they were originally developed, since, like the normal distribution, they provide good approximations to a large class of other sampling distributions that are not easy to determine exactly. These five theoretical distributions, the three already studied plus the two to be introduced in this chapter, make up the arsenal of theoretical functions from which the statistician draws most heavily; almost all the elementary methods of statistical inference rest on one or more of these theoretical distributions. Furthermore, these five theoretical functions have very close connections with each other, and after we conclude our discussion of the chi-square and *F* distributions, some of these relationships will be pointed out.

11.1   THE CHI-SQUARE DISTRIBUTION

Suppose that there exists a population having a normal distribution of scores Y. The mean of this distribution is $E(Y) = \mu$, and the variance is

$$E(Y - \mu)^2 = \sigma^2.$$

Now cases are sampled from this distribution *one* at a time, $N = 1$. For each case sampled the **squared standardized score**

$$z^2 = \frac{(y - \mu)^2}{\sigma^2} \qquad [11.1.1]$$

is computed. Let us call this squared standardized score $\chi^2_{(1)}$ so that

$$\chi^2_{(1)} = z^2. \qquad [11.1.2^*]$$

Now we will look into the sampling distribution of this variable $\chi^2_{(1)}$.

First of all, what is the range of values that χ^2 might take on? The original normal variable Y ranges over all real numbers, and this is also the range of the standardized variable z. However, $\chi^2_{(1)}$ is always a squared quantity, and so its range must be all the *nonnegative* real numbers, from zero to a positive infinity. We can also infer something about the form of this distribution of $\chi^2_{(1)}$; the bulk of the cases (about 68 percent) in a normal distribution of standardized scores must lie between -1 and 1. Given a z between -1 and 1, the corresponding $\chi^2_{(1)}$ value lies between 0 and 1, so that the bulk of this sampling distribution will fall in the interval between 0 and 1. This implies that the form of the distribution of $\chi^2_{(1)}$ will be very skewed, with a high probability for a value in the interval from 0 to 1, and relatively low probability in the interval with lower bound 1 and approaching positive ∞ as its upper bound. The graph of the distribution of $\chi^2_{(1)}$ is represented in Figure 11.1.1.

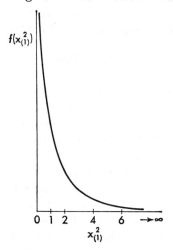

FIG. 11.1.1. The distribution of χ^2 for $\nu = 1$

Figure 11.1.1 pictures **the chi-square distribution with one degree of freedom. The distribution of the random variable $\chi^2_{(1)}$, where**

$$\chi^2_{(1)} = \frac{(Y - \mu)^2}{\sigma^2},$$

and Y is normally distributed with mean μ and variance σ^2 is a chi-square distribution with 1 degree of freedom.

Now let us go a little further. Suppose that samples of *two* cases are drawn independently and at random from a normal distribution. We find the squared standardized score corresponding to each observation:

$$z_1^2 = \frac{(y_1 - \mu)^2}{\sigma^2}$$

$$z_2^2 = \frac{(y_2 - \mu)^2}{\sigma^2}.$$

If the *sum* of these two squared standardized scores is found over repeated independent samplings, the resulting random variable is designated as $\chi_{(2)}^2$:

$$\chi_{(2)}^2 = \frac{(Y_1 - \mu)^2}{\sigma^2} + \frac{(Y_2 - \mu)^2}{\sigma^2} = z_1^2 + z_2^2. \qquad [11.1.3*]$$

If we look into the distribution of the random variable $\chi_{(2)}^2$ we find that the range of possible values extends over all nonnegative real numbers. However, since the random variable is based on *two* independent observations, the distribution is somewhat less skewed than for $\chi_{(1)}^2$; the probability is not so high that the sum of two squared standardized scores should fall between 0 and 1. This is illustrated in Figure 11.1.2. This illustrates a **chi-square distribution with two degrees of freedom.**

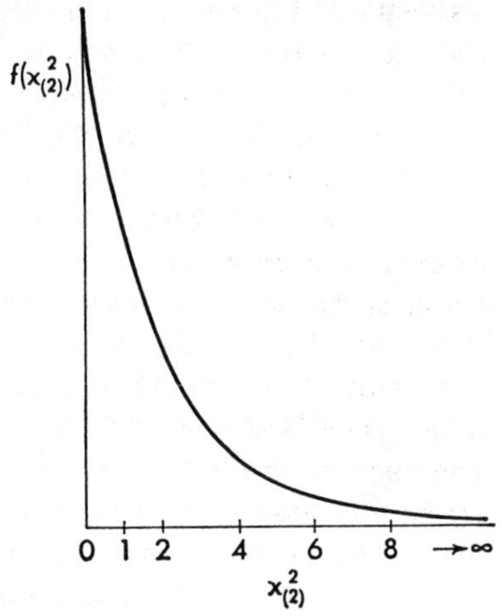

Finally, suppose that we took N independent observations at random from a normal distribution with mean μ and variance σ^2 and defined the random variable

$$\chi_{(N)}^2 = \frac{\sum_{i}^{N} (Y_i - \mu)^2}{\sigma^2} = \sum_{i} z_i^2. \qquad [11.1.4*]$$

The distribution of this random variable will have a form that depends upon the number of independent observations taken at one time. **In general, for N independent observations from a normal population, the sum of the squared standardized scores for the observations has a chi-square distribution with N degrees of freedom.** Notice that the standardized scores must be relative to the *population mean* and the *population standard deviation*.

Another way of expressing the definition of a chi-square variable is as follows: **a random variable has a chi-square distribution with N degrees of freedom if it has the same distribution as the sum of the squares of N independent random variables, each normally distributed, and each having expectation 0 and variance 1.** Chi-square distributions for larger numbers of degrees of freedom have the form illustrated in Figure 11.1.3.

FIG. 11.1.2. The distribution of χ^2 for $\nu = 2$

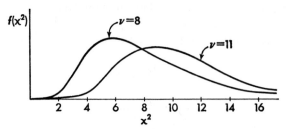

FIG. 11.1.3. The general form of the distribution of χ^2 for larger numbers of degrees of freedom

11.2   THE FUNCTION RULE AND THE MEAN AND VARIANCE FOR A CHI-SQUARE DISTRIBUTION

The function rule assigning a probability density to each possible value of χ^2 is given by

$$f(\chi^2;\nu) = h(\nu)e^{-\chi^2/2}(\chi^2)^{(\nu/2)-1}, \qquad \begin{array}{l} \text{for } \chi^2 \geqq 0 \\ \nu > 0. \end{array} \qquad [11.2.1^*]$$

As shown in Figures 11.1.1 through 11.1.3, the plot of this density function always presents the picture of a very positively skewed distribution, at least for relatively small values of ν, with a distinct mode at the point $\nu - 2$ for $\nu > 2$. However, as ν is increased, the form of the distribution appears to be less skewed to the right. In this function rule the value $h(\nu)$ is a constant depending only on the parameter ν. Unlike the normal and the t distributions, it is not so easy to infer the characteristics of the distribution of χ^2 from this function rule. However, the thing to notice is that there is only one value other than χ^2 that must be specified in order to find the density: this is the parameter ν. Like the t distribution, the distribution of χ^2 depends only on the degrees of freedom, the parameter ν. The family of chi-square distributions all follow this general rule, but the exact form of the distribution depends on the number of degrees of freedom, ν. In principle, the value of ν can be any positive number, but in the applications we will make of this distribution in this chapter, ν will depend only on the sample size, N.

From the original definition of χ^2, it is quite easy to infer what the mean of this distribution must be. For N independent observations we defined the random variable

$$\chi^2_{(N)} = \frac{\sum_i^N (Y_i - \mu)^2}{\sigma^2} = \sum_i^N z_i^2$$

where the random variable Y is normal. If we take the expectation of $\chi^2_{(N)}$ over all possible samples of N independent observations, we find

$$E[\chi^2_{(N)}] = E\left(\sum_i^N z_i^2\right)$$

$$= \sum_i^N E(z_i^2). \qquad [11.2.2]$$

However, for any i, $E(z_i^2) = 1$, by the definition of a standardized score. Thus

$$E[\chi^2_{(N)}] = N. \qquad [11.2.3]$$

If a chi-square variable has a distribution with ν degrees of freedom, then the expectation of the distribution is simply ν. In our derivation, $\nu = N$, the number of independent observations, and so the mean of the distribution is the number of degrees of freedom, or N.

The variance of a chi-square distribution with ν degrees of freedom is always

$$\text{var.}[\chi^2_{(\nu)}] = 2\nu. \qquad [11.2.4]$$

Giving the degrees of freedom ν actually gives all the information one needs to specify the particular chi-square distribution completely.

11.3   TABLES OF THE CHI-SQUARE DISTRIBUTION

Like the t distribution, the particular distribution of χ^2 depends on the parameter ν, and it is difficult to give tables of the distribution for all values of ν that one might need. Thus, Table IV in Appendix C, like Table III, is a condensed table, showing values of χ^2 that correspond to percentage points in various

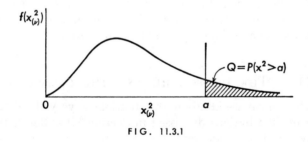

FIG. 11.3.1

distributions specified by ν. The rows of Table IV list various degrees of freedom ν and the column headings are probabilities Q, just as in the t table. The numbers in the body of the table give the values of χ^2 such that in a distribution with ν degrees of freedom, *the probability of a sample chi-square value this large or larger is Q* (Fig. 11.3.1).

For example, suppose that we are dealing with a chi-square distribution with 5 degrees of freedom. Find the row ν equal to 5, and the column headed .05. The value in this row and column is approximately 11.071, showing that for 5 degrees of freedom, random samples showing a chi-square value of 11.071 *or more* should in the long run occur about five times in a hundred. As another example, look at the row labeled $\nu = 2$ and the column headed .10. The entry in this combination of row and column indicates that in a distribution with 2 degrees of freedom, chi-square values of 4.605 or more should occur with probability of about 1 in 10, under random sampling. Finally, look at the row for $\nu = 24$, and the column for .001; the cell entry indicates that chi-square values of 51.179 or more occur with probability of only about .001 in a distribution with 24 degrees of freedom.

11.4   THE ADDITION OF CHI-SQUARE VARIABLES

When several independent random variables each have a chi-square distribution, then the distribution of the *sum* of the variables is also known. This is a most important and useful property of variables of this kind, and can be stated more precisely as follows:

If a random variable $\chi^2_{(\nu_1)}$ has a chi-square distribution with ν_1 degrees of freedom, and an independent random variable $\chi^2_{(\nu_2)}$ has a chi-square distribution with ν_2 degrees of freedom, then the new random variable formed from the sum of these variables

$$\chi^2_{(\nu_1 + \nu_2)} = \chi^2_{(\nu_1)} + \chi^2_{(\nu_2)} \qquad [11.4.1*]$$

has a chi-square distribution with $\nu_1 + \nu_2$ degrees of freedom.

In short, the new random variable formed by taking the sum of two independent chi-square variables is itself distributed as χ^2, with degrees of freedom equal to the sum of those for the original distributions. A little thought will convince you that this must be so: if $\chi^2_{(\nu_1)}$ is the sum of squares of ν_1 independent variables, each normally distributed with mean 0 and variance 1, and $\chi^2_{(\nu_2)}$ is based on the sum of ν_2 independent such squares, then $\chi^2_{(\nu_1 + \nu_2)}$ must also be a sum of $\nu_1 + \nu_2$ squared values of independent normal variables each with mean 0 and variance 1. Therefore $\chi^2_{(\nu_1 + \nu_2)}$ qualifies as a chi-square variable by definition.

11.5   THE CHI-SQUARE AND THE GAMMA DISTRIBUTIONS

The chi-square family of distributions is very closely related to the gamma family of distributions discussed in Section 8.14. Recall that a gamma distribution involves two parameters, m and r, each of which are positive values. Now if we define

$$m = \tfrac{1}{2}, \qquad r = \tfrac{1}{2}\nu,$$

the gamma function becomes

$$f(x; \tfrac{1}{2}\nu, \tfrac{1}{2}) = \frac{\tfrac{1}{2}e^{-x/2}(\tfrac{1}{2}x)^{(1/2)r-1}}{(\tfrac{1}{2}\nu - 1)!}, \qquad \begin{matrix} x \geq 0, \\ \nu > 0 \end{matrix}$$

which, letting $x = \chi^2$, is simply the chi-square density function. In a sense, then, the chi-square distribution is a member of the gamma family of distributions.

Use of this fact was made in Section 8.14, when the tables of chi-square were used to find percentage points for the gamma distribution. Recall also the close connection of the gamma and the Poisson distributions. This, in turn, permitted us to use these same chi-square tables to find percentage points for a given Poisson distribution.

One important property that the chi-square distribution shares with its relatives in the gamma family is that of additivity. We have just seen that the sum of two independent chi-square variables with ν_1 and ν_2 degrees of freedom follows a chi-square function rule with $\nu_1 + \nu_2$ degrees of freedom. This accords with the fact, also mentioned in Section 8.14, that the sum of two independent gamma variables, with parameters r and m and s and m, respectively, is a gamma variable with parameters $r + s$ and m.

Finally, the mean and variance of a chi-square distribution can be found in terms of the mean and variance of a gamma distribution. A gamma distribution has a mean $E(X) = r/m$, and a variance $\text{Var}(X) = r/m^2$. Substituting $r = \frac{1}{2}\nu$ and $m = \frac{1}{2}$, we have for a chi-square variable

$$E(\chi^2) = \frac{\frac{1}{2}\nu}{\frac{1}{2}} = \nu, \qquad \text{Var}(\chi^2) = \frac{\frac{1}{2}\nu}{(\frac{1}{2})^2} = 2\nu.$$

11.6 THE DISTRIBUTION OF THE SAMPLE VARIANCE FROM A NORMAL POPULATION

In this section, the sampling distribution of the sample variance will be studied. It will be assumed that the population actually sampled is normal, and the results in this section apply, strictly speaking, only when this assumption is true.

Recall that in Section 6.16 the sample variance was defined by

$$S^2 = \frac{\sum_i^N (y_i - M)^2}{N}.$$

Suppose, however, that we actually knew the value of μ. For any single deviation from μ, $(y_i - \mu)$, it is true that

$$(y_i - \mu) = (y_i - M) + (M - \mu) \qquad [11.6.1]$$

since adding and subtracting the same number to any given number does not change the value. Now, for any squared deviation from μ,

$$
\begin{aligned}
(y_i - \mu)^2 &= [(y_i - M) + (M - \mu)]^2 \\
&= (y_i - M)^2 + (M - \mu)^2 + 2(y_i - M)(M - \mu). \qquad [11.6.2]
\end{aligned}
$$

Summing over all of the squared deviations from μ in the sample, we have

$$\sum_i^N (y_i - \mu)^2 = \sum_i (y_i - M)^2 + \sum_i (M - \mu)^2 + 2 \sum_i (y_i - M)(M - \mu). \qquad [11.6.3]$$

The value of $(M - \mu)^2$ is constant over all individuals i, so that

$$\sum_i^N (M - \mu)^2 = N(M - \mu)^2. \qquad [11.6.4]$$

Furthermore, since $(M - \mu)$ is the same for all i, then

$$\sum_i^N (y_i - M)(M - \mu) = (M - \mu) \sum_i (y_i - M)$$
$$= 0, \qquad [11.6.5]$$

because the sum of the deviations about a sample mean must be 0. Hence,

$$\sum_i (y_i - \mu)^2 = \sum_i (y_i - M)^2 + N(M - \mu)^2. \qquad [11.6.6*]$$

If we divide the entire expression by σ^2, this becomes

$$\frac{\sum_i (y_i - \mu)^2}{\sigma^2} = \frac{\sum_i (y_i - M)^2}{\sigma^2} + \frac{N(M - \mu)^2}{\sigma^2}. \qquad [11.6.7]$$

Now look at the term on the left in this expression: given samples of N independent cases from a normal population distribution,

$$\frac{\sum_i (Y_i - \mu)^2}{\sigma^2} = \chi^2_{(N)}, \qquad [11.6.8]$$

so that this sum of squared deviations from μ, divided by σ^2, qualifies as a chi-square variable with N degrees of freedom.

Furthermore, if the population distribution is normal, the sampling distribution of M must be normal, with mean μ and variance σ^2/N. Thus,

$$\frac{N(M - \mu)^2}{\sigma^2} = \frac{(M - \mu)^2}{(\sigma^2/N)} = \chi^2_{(1)}. \qquad [11.6.9*]$$

It follows that the last term in **11.6.7** qualifies as a chi-square variable with 1 degree of freedom, since this is just z_M^2 from a normal distribution. Putting these two facts together, we have

$$\chi^2_{(N)} = \frac{\sum_i (Y_i - M)^2}{\sigma^2} + \chi^2_{(1)} \qquad [11.6.10]$$

or

$$\chi^2_{(N)} = \frac{NS^2}{\sigma^2} + \chi^2_{(1)}. \qquad [11.6.11*]$$

The term on the left is a chi-square variable with N degrees of freedom, and the term on the extreme right is a chi-square variable with 1 degree of freedom. Finally, as S^2 is independent of M, it can be shown that

$$\frac{NS^2}{\sigma^2} = \chi^2_{(N-1)}, \qquad [11.6.12*]$$

the ratio NS^2/σ^2 is a chi-square variable with $N-1$ degrees of freedom.
If, instead of S^2, we use the unbiased estimate s^2, we find that

$$\frac{(N-1)s^2}{\sigma^2} = \chi^2_{(N-1)},$$ [11.6.13*]

meaning that the ratio of $(N-1)s^2$ to σ^2 is a random variable distributed as
chi-square with $N-1$ degrees of freedom. This fact is the basis for inferences
about the variance of a single normally distributed population. Note that, in
every instance of a chi-square variable discussed on this page, the numerator is
a sum of squares. This fact will be of considerable importance to us in the next
chapter.

This same idea applies to the sampling distribution of the estimate
of a variance based upon the pooling of independent estimates, provided that
the basic population sampled is normal. Thus, it is also true that when the pooled
estimate of σ^2 is made from two independent samples of N_1 and N_2 cases,

$$\frac{(N_1 + N_2 - 2) \text{ est. } \sigma^2}{\sigma^2} = \chi^2_{(N_1+N_2-2)}.$$ [11.6.14*]

The pooled sample estimate of σ^2, multiplied by the degrees of freedom and divided
by the true value of σ^2, is a chi-square variable with $\nu = N_1 + N_2 - 2$.

The sampling distribution of the unbiased estimate s^2 is actually the
distribution of the variable

$$s^2 = \frac{\chi^2_{(\nu)}\sigma^2}{\nu}, \text{ or } \frac{\text{est. } \sigma^2}{\sigma^2} = \frac{\chi^2_{(\nu)}}{\nu}$$ [11.6.15*]

a chi-square variable multiplied by σ^2 and divided by degrees of freedom. However, inferences are usually made in terms of the distribution of $\dfrac{(N-1)s^2}{\sigma^2}$ since
this is just a chi-square variable; thus one can get by with only one chi-square
table for this as well as for other uses of this distribution.

11.7   TESTING EXACT HYPOTHESES ABOUT A SINGLE VARIANCE

Just as for the mean, it is possible to test exact hypotheses about a
single population variance (and, of course, a standard deviation). The exact
hypothesis tested is

$$H_0 : \sigma^2 = \sigma_0^2$$

where σ_0^2 is some specific positive number. The alternative hypothesis may be
either directional or nondirectional, depending, as always, on the original question.

As usual, some value of α is decided upon, and a region of rejection adopted depending both on α and the alternative H_1. The test statistic itself is

$$\chi^2_{(N-1)} = \frac{(N-1)s^2}{\sigma_0^2}. \qquad [11.7.1\dagger]$$

The value of this test statistic, computed from the s^2 actually obtained and the σ_0^2 dictated by the null hypothesis, is referred to the distribution of χ^2 for $N-1$ degrees of freedom.

For example, there is some evidence that women tend to be a less variable, more homogenous group than do men. One might ask this question about height: "It is well known that men and women in the United States differ in terms of their mean height; is it true, however, that women show less variability in height than do men?" Now let us assume that from the records of the Selective Service System we actually *know* the mean and standard deviation of height for American men between the ages of twenty and twenty-five years. However, such complete evidence is lacking for women. Assume that the standard deviation of height for the population of men twenty to twenty-five years is 2.5 inches. For this same age range, we want to ask if the population of women shows this same standard deviation, or if women are *less* variable, with their distribution having a smaller σ. The null and alternative hypotheses can be framed as

$$H_0: \sigma^2 \geq 6.25 \qquad (\text{or } [2.5]^2)$$
$$H_1: \sigma^2 < 6.25.$$

Imagine that we plan to draw a sample of 30 women at random, each between the ages of twenty and twenty-five years, and measure the height of each. The test statistic will be

$$\chi^2_{(29)} = \frac{(29)s^2}{6.25}.$$

What, however is the region of rejection? Here, *small* values of χ^2 tend to favor H_1, that women actually are less variable than men. Hence, we want to use a region of rejection on the left (or small-value) tail of the chi-square distribution with 29 degrees of freedom. If $\alpha = .01$, then from Table IV in Appendix C, the value of χ^2 leading to rejection of H_0 should be less than the value given by the row with $\nu = 29$, and the column for $Q = .99$. This value is 14.257.

Now the actual value of s^2 obtained turns out to be 4.55, so that

$$\chi^2_{(29)} = \frac{(29)(4.55)}{6.25} = 21.11.$$

This value is larger than the critical value decided upon, and we cannot reject H_0 if α is to be .01.

This example illustrates that, as with the t and the normal distribution, either or both tails of the chi-square distribution can be used in testing a hypothesis about a variance. Had the alternative hypothesis in this problem been

$$H_1: \sigma^2 \neq 6.25,$$

then the rejection region would lie in both tails of the chi-square distribution. The rejection region on the lower tail of the distribution would be bounded by a chi-square value corresponding to $\nu = 29$ and $Q = .995$, which is 13.121; the rejection region on the upper tail would be bounded by the value for $\nu = 29$ and $Q = .005$, which is 52.336. Any obtained χ^2 value falling *below* 13.121 or *above* 52.336 would let one reject H_0 beyond the .01 level.

11.8 CONFIDENCE INTERVALS FOR THE VARIANCE AND STANDARD DEVIATION

Finding confidence intervals for the variance is quite simple, provided that the normal distribution rule holds for the population.

Suppose that we had a sample of some N independent observations, and we wanted the 95 percent confidence limits for σ^2. For samples from a normal distribution, it must be true that

$$\text{prob.} \left[\chi^2_{(N-1;.975)} \leq \frac{(N-1)s^2}{\sigma^2} \leq \chi^2_{(N-1;.025)} \right] = .95 \qquad [11.8.1^*]$$

where $\chi^2_{(N-1;.975)}$ is the value in a chi-square distribution with $N-1$ degrees of freedom cutting off the upper $.975$ of sample values, and $\chi^2_{(N-1;.025)}$ the value cutting off the upper $.025$ of sample values. This inequality can be manipulated to show that it is also true that

$$\text{prob.} \left[\frac{(N-1)s^2}{\chi^2_{(N-1;.025)}} \leq \sigma^2 \leq \frac{(N-1)s^2}{\chi^2_{(N-1;.975)}} \right] = .95. \qquad [11.8.2^*]$$

That is, the probability is $.95$ that the true value of σ^2 will be covered by an interval with limits found by

$$\frac{(N-1)s^2}{\chi^2_{(N-1;.025)}}$$

and

$$\frac{(N-1)s^2}{\chi^2_{(N-1;.975)}}.$$

Suppose, for example, that a sample of 15 cases is drawn from a normal distribution. We want the 95 percent confidence limits for σ^2. The value of s^2 is 10; this is our best single estimate of σ^2, of course. For $\nu = 14$ and for $Q = .025$, $\chi^2_{(14;.025)}$ is found from Table IV to be 26.12, and the value of $\chi^2_{(14;.975)}$ is 5.63. Thus the confidence limits for σ^2 are

$$\frac{(14)(10)}{26.12} \text{ or } 5.36$$

and

$$\frac{(14)(10)}{5.63} \text{ or } 24.87.$$

We can say that the probability is $.95$ that an interval such as 5.36 to 24.87 covers the true value of σ^2.

In general, to find the $100(1 - \alpha)$ percent confidence limits, we take

$$\frac{(N-1)s^2}{\chi^2_{(N-1;\alpha/2)}}$$

and
$$\frac{(N-1)s^2}{\chi^2_{(N-1;1-\alpha/2)}}.$$

These limits can always be turned into confidence limits for σ by replacing each limiting value by its square root. This is due to the fact that the event of the squared limits' including σ^2 is the same as the event of the square roots of these limits' including σ.

On the other hand, this procedure for finding confidence limits for σ^2 contains a serious difficulty. When a confidence interval for μ is found by the methods of Section 9.25 or of Section 10.10, the confidence limits are taken to be equally distant from the obtained value of M. Since the normal sampling distribution of M is symmetric, and the distribution of t is also symmetric, these methods of constructing confidence intervals for M give *shortest* intervals. As mentioned in Section 9.25, it is perfectly possible to construct 95 percent, 99 percent, or any other confidence intervals for M in a variety of ways, but making the confidence limits equidistant from M gives the shortest of all possible such intervals. In other words, by the methods of 9.25 and 10.10, we "pin down" the value of μ within the smallest possible estimated range of values.

Unfortunately, this is not true when we find a confidence interval for σ^2 by the method just introduced. Here, the confidence limits are determined by values of χ^2 cutting off highest and lowest intervals having the same probability, $\alpha/2$, in the asymmetric chi-square distribution. This does make it true that the *failure* of the confidence interval to cover any hypothetical value of σ_0^2 is equivalent to the rejection of σ_0^2 as a hypothesis at the α level, two-tailed. It does *not*, however, make it true that we have the shortest possible such interval of possible values of σ^2 for a given value of $(1-\alpha)$, nor that we have the shortest possible interval of "good bets" about the value of σ^2. For this reason, the confidence interval for σ^2, although corresponding to procedures for testing hypotheses about σ^2, is not particularly useful when our interest lies in obtaining the best possible estimates of σ^2 from the data. More advanced methods do exist for finding this shortest possible confidence interval for σ^2, and useful tables for finding these intervals are given in Tate and Klett (1959).

Regardless of whether confidence intervals are found by the method given here, or by a method giving shortest intervals, it is true that the larger the sample size, the narrower is the interval with a given probability of covering the true value of σ^2. For this reason, when we are estimating σ^2 for any purpose the larger the sample size the more confident we can be of coming close to the value of σ^2 with our sample estimate s^2.

11.9   THE IMPORTANCE OF THE NORMALITY
ASSUMPTION IN INFERENCES ABOUT σ^2

In the preceding chapter it was pointed out that although the rationale for the t test demands the assumption of a normal population distribution of scores, in practice the t test may be applied when the parent distribution is not normal, provided that sample N is at least moderately large. *This is not the case for inferences about the variance, however.* One runs a considerable risk of error in using the chi-square distribution either to test a hypothesis about a variance or to find confidence limits unless the population distribution is normal, or approximately so. The effect of the violation of the normality assumption is usually minor for large N, but can be quite serious when inferences are made about variances, even for moderate N. Indeed, for some population forms, the effect of using the chi-square statistics may actually grow more serious for larger samples. Thus, the assumption of a normal distribution is important in inferences about the variance. This principle is an important one and will be emphasized several times in succeeding sections.

11.10 THE NORMAL APPROXIMATION TO THE CHI-SQUARE DISTRIBUTION

As the number of degrees of freedom grows infinitely large, the distribution of χ^2 approaches the normal distribution. You should hardly find this principle surprising by now! Actually, the same mechanism is at work here as in the central limit theorem: for any ν greater than 1, the chi-square variable is equivalent to a *sum* of ν independent random variables. Thus, like the mean, M, given enough summed terms the sampling distribution approaches the normal form in spite of the fact that each component of the sum does not have a normal distribution when sampled alone.

This fact is of more than theoretical interest when very large samples are used. For very large ν, the probability for any interval of values of $\chi^2_{(\nu)}$ can be found from the normal standardized scores

$$z = \frac{\chi^2_{(\nu)} - \nu}{\sqrt{2\nu}} \qquad [11.10.1*]$$

since the mean and variance of a chi-square distribution are ν and 2ν respectively. However, this approximation is not good unless ν is extremely large. A somewhat better approximation procedure is to find the value of $\chi^2_{(\nu)}$ that cuts off the *upper Q* proportion of cases by taking

$$\chi^2_{(\nu;Q)} = \frac{1}{2} \{z_Q + \sqrt{2\nu - 1}\}^2. \qquad [11.10.2*]$$

This can be used to find chi-square values for ν greater than 100, which are not given in Table IV. The necessary z_Q values are listed at the bottom of Table IV, and can also be found from Table I.

11.11 THE *F* DISTRIBUTION

It is rather rare to find a problem in psychology that centers on the value of a single population variance. Somewhat more common, however, is the situation where the variances of two populations are compared for equality. That is, we ask if the variability of population 1 is precisely equal to that of population 2, so that $\sigma_1^2 = \sigma_2^2$. The F distribution is used to make inferences of this sort, although we will also use this same distribution for quite a different kind of problem in the next chapter.

Imagine two distinct populations, each showing a normal distribution of the variable Y. The means of the two populations may be different, but each population shows the same variance, σ^2. We draw two independent random samples: the first sample, from population 1, contains N_1 cases, and that from population 2 consists of N_2 cases. From each sample, an unbiased estimate of σ^2 is made, based on s_1^2 and s_2^2 for each pair of samples. From the results of Section 11.6 we know that

$$\frac{(N_1 - 1)s_1^2}{\sigma^2} = \chi^2_{(N-1)}$$

so that

$$s_1^2 = \frac{\sigma^2 \chi^2_{(\nu_1)}}{\nu_1}.$$

In the same way, we know that

$$s_2^2 = \frac{\sigma^2 \chi^2_{(\nu_2)}}{\nu_2}.$$

For each possible pair of samples, one from population 1, and one from population 2, we take the *ratio* of s_1^2 to s_2^2 and call this ratio the random variable F:

$$F = \frac{s_1^2}{s_2^2} = \frac{\text{est. } \sigma_1^2}{\text{est. } \sigma_2^2}. \tag{11.11.1*}$$

What sort of distribution should the random variable F actually have when the hypothesis that $\sigma_1^2 = \sigma_2^2$ is true? First of all notice that when we put the two variance estimates in ratio, this is actually the ratio of two independent chi-square variables, each divided by its degrees of freedom:

$$F = \frac{[\chi^2_{(\nu_1)}/\nu_1]}{[\chi^2_{(\nu_2)}/\nu_2]} \tag{11.11.2*}$$

since

$$\frac{s_1^2}{s_2^2} = \frac{(s_1^2/\sigma_1^2)}{(s_2^2/\sigma_2^2)}$$

provided that $\sigma_1^2 = \sigma_2^2$ (that is to say, the hypothesis of the equality of the population variances is true).

Showing F as a ratio of two independent chi-square variables each divided by its ν is actually a way of defining the F variable:

A random variable formed from the ratio of two independent chi-square variables, each divided by its degrees of freedom, is said to be an F ratio, and to follow the rule for the F distribution.

When this definition is satisfied, meaning that both parent populations are normal, have the same variance, and that the samples drawn are independent, then the theoretical distribution of F values can be found.

Formally, the density function for the random variable F is defined by

$$
f(F;\nu_1,\nu_2) = \begin{cases} \left[\dfrac{(\frac{1}{2}\nu_1 + \frac{1}{2}\nu_2 - 1)!}{(\frac{1}{2}\nu_1 - 1)!(\frac{1}{2}\nu_2 - 1)!} \right] \dfrac{F^{(1/2)\nu_1-1}(\nu_2/\nu_1)^{(1/2)\nu_2}}{(F + \nu_2/\nu_1)^{(1/2)\nu_1+(1/2)\nu_2}} \\ \qquad \text{for } 0 < \nu_1,\, 0 < \nu_2,\, 0 \leq F, \\[4pt] 0 \qquad \text{otherwise.} \end{cases}
$$

(Note that here, as in sections 10.4 and 11.2, we are violating our conventions about symbolizing random variables). Obviously, this function is much too complicated for us to gain an impression of the form of a given distribution from the function rule alone. However, it is worth noting that for the values of ν_1 and ν_2 that will usually concern us (that is, integral values, with $\nu_1 < \nu_2$) the F distribution is skewed positively, and is unimodal for $\nu_1 > 2$. On the other hand, the distribution can take on other forms when other conditions are set for ν_1 and ν_2.

The mean and the variance of an F distribution also depend on the values of ν_1 and ν_2. Thus,

$$
E(F) = \frac{\nu_2}{\nu_2 - 2} \qquad\qquad \text{for } \nu_2 > 2
$$

and

$$
\mathrm{Var}(F) = \frac{2\nu_2^2(\nu_1 + \nu_2 - 2)}{\nu_1(\nu_2 - 2)^2(\nu_2 - 4)} \qquad \text{for } \nu_2 > 4.
$$

Note that as the value of ν_2 is made increasingly large, the mean of the F distribution approaches 1.00.

Mathematically, the F distribution *is* rather complicated. However, for our purposes it will suffice to remember that the density function for F depends only upon two parameters, ν_1 and ν_2, which can be thought of as *the degrees of freedom associated with the numerator and the denominator of the F ratio*. The range for F is *nonnegative real numbers*. The expectation of F is $\nu_2/(\nu_2 - 2)$ for $\nu_2 > 2$. In general form, the distribution for any fixed ν_1 and ν_2 is nonsymmetric, although the particular "shape" of the function curve varies considerably with changes in ν_1 and ν_2.

Before we can apply this theoretical distribution to an actual problem of comparing two variances, the use of F tables must be discussed.

11.12 The use of F tables

Since the distribution of F depends upon two parameters ν_1 and ν_2, it is even more difficult to present tables of F distributions than those of χ^2 or t. Tables of F are usually encountered only in drastically condensed form.

Such tables give only those values of F that cut off the *upper* proportion Q in an F distribution with ν_1 and ν_2 degrees of freedom. The only values of Q given are those that are commonly used as α in a test of significance (that is, the

.05, the .025, and the .01 values). Table V in Appendix C shows F values required for significance at the $\alpha = Q$ level, given ν_1 and ν_2. The columns of each table give values of ν_1, the degrees of freedom for the numerator, and the rows give values of ν_2, the degrees of freedom for the denominator. Each separate table represents one value of Q. The entries in the body of the table are the values of F required for significance at this level.

The use of Table V can be illustrated in the following way: Suppose that two independent samples are drawn, containing $N_1 = 10$ and $N_2 = 6$ cases, respectively. The degrees of freedom associated with the two variances are $\nu_1 = 10 - 1 = 9$ and $\nu_2 = 6 - 1 = 5$. For the ratio

$$F = \frac{s_1^2}{s_2^2}$$

the degrees of freedom must be 9 for the numerator, and 5 for the denominator. Now suppose that the obtained $F = 7.00$. Does this fall into the upper .05 of values in an F distribution, with 9 and 5 degrees of freedom? We turn to Table V and find the page for .05. Then we look at the column for $\nu_1 = 9$ and the row for $\nu_2 = 5$. The tabled value is 4.77. Our obtained F value of 7.00 exceeds 4.77, and thus our sample result falls among the upper 5 percent in an F distribution.

As another example, suppose that $\nu_1 = 1$, and $\nu_2 = 45$. For the problem at hand, the rejection region contains the *upper* .01 proportion in an F distribution. What value must our sample result equal or exceed in order to be significant? Table V shows that for the .01 level, the required F for this number of degrees of freedom is between 7.31 (the F for $\nu_2 = 40$) and 7.08 (the F for $\nu_2 = 60$). It is difficult to find the exact value of F required, but we are reasonably sure that the required F is somewhere around 7.3. If a little more accuracy is desired, linear interpolation may be used. That is, let F' be the desired value. Then in this example, calculate

$$F' = F_{(1,40)} - \frac{(45 - 40)}{(60 - 40)} \left[F_{(1,40)} - F_{(1,60)} \right]$$

This gives an F' value of about 7.25. This is still not the exact value required, of course, but it may be sufficient for most practical purposes.

In both of these illustrations we have dealt only with the upper tail of the F distribution. This would be appropriate if we were doing a one-tailed test, where $H_1: \sigma_1^2 > \sigma_2^2$. However, it is possible, although somewhat more difficult, to find F values corresponding to two-tailed rejection regions.

Let $F_{(\nu_1; \nu_2)}$ stand for an F ratio with ν_1 and ν_2 degrees of freedom, and let $F_{(\nu_2; \nu_1)}$ be an F ratio with ν_2 degrees of freedom in the *numerator*, and ν_1 degrees of freedom in the *denominator*. The numbers of degrees of freedom in numerator and denominator are simply reversed for these two F ratios. Then it is true that, for any positive number C,

$$p[C \leq F_{(\nu_1; \nu_2)}] = p \left[F_{(\nu_2; \nu_1)} \leq \frac{1}{C} \right] \qquad [11.12.1]$$

the probability that $F_{(\nu_1;\nu_2)}$ is greater than or equal to some number C is the same as the probability that the reciprocal of $F_{(\nu_1;\nu_2)}$ is less than or equal to the *reciprocal of C*. Practically, this means that the *value required for F on the lower tail of some particular distribution can always be found by finding the corresponding value required on the upper tail of a distribution with numerator and denominator degrees of freedom reversed, and then taking the reciprocal*.

For instance, we want the F value that cuts off the *bottom* .05 of sample values in a distribution with 7 and 10 degrees of freedom. We find this by first locating the value which cuts off the *top* .05 in a distribution with 10 and 7 degrees of freedom: this is 3.64 from Table **V**. Then the F value we are looking for on the *lower* tail when 7 and 10 are the degrees of freedom is the reciprocal:

$$F_{(.95;\,7,10)} = \frac{1}{F_{(.05;\,10,7)}} = \frac{1}{3.64} = .27.$$

Any obtained F value that has 7 and 10 degrees of freedom, and that has a value of .27 *or less* represents a result occurring no more than .05 of the time in this distribution of possible results.

Since the tables of F given in the appendix are for the .05, .025, and the .01 significance levels, *one-tailed*, the only *two-tailed* F values that can be found from these tables are the .10, the .05, and the .02 significance levels. Suppose that $\alpha = .05$ in a problem where $\nu_1 = 20$, and $\nu_2 = 30$. A two-tailed test is desired; what are the limits to the region of rejection? First of all, from Table **V** for $Q = .025$, we find that $F_{(20;\,30)} = 2.20$, which is the boundary for the *upper* rejection region. Then we look up the value in the same table for $F_{(30;\,20)}$, which is 2.35. The boundary value for the lower rejection region is found to be $1/2.35$, or .43. Thus, if our sample F ratio is larger than 2.20, or smaller than .43, the sample value falls in a region of rejection.

11.13   THE F AND THE BETA FAMILIES OF DISTRIBUTIONS

Just as the chi-square distribution is closely related to the gamma family of distributions, the F distribution is related to the beta family. You may already have guessed this from our use of F tables in the discussion of probabilities in a beta distribution in Section 8.14.

Recall from Section 8.14 that a beta distribution has two parameters, r and N, such that $0 < r < N$. The density function is defined by

$$f(x; N, r) = \begin{cases} \dfrac{(N-1)!}{(r-1)!(N-r-1)!}\, x^{r-1}(1-x)^{N-r-1}, & 0 \le x \le 1,\, 0 < r < N, \\ 0, & \text{elsewhere.} \end{cases}$$

In order to go from such a beta distribution function to that for the F variable, we let $r = \frac{1}{2}\nu_1$ and $N = \frac{1}{2}(\nu_1 + \nu_2)$, and define the random variable x to be $x = F/(F + \nu_2/\nu_1)$. After these substitutions, the resulting function must be adjusted in order to make the total area under its curve equal to 1.00 (a so called "integrand transformation" required by the fact that we wish to regard F rather than x as the random variable). This is accomplished by multiplying the function by $(v_2/v_1)/(F + v_2/v_1)^2$, which then gives the F function

$$f(F;v_1,\ v_2) = \left[\frac{(\frac{1}{2}v_1 + \frac{1}{2}v_2 - 1)!}{(\frac{1}{2}v_1 - 1)!(\frac{1}{2}v_2 - 1)!} \right] \frac{F^{(1/2)\,v_1-1}(v_2/v_1)^{(1/2)\,v_2}}{(F + v_2/v_1)^{(1/2)\,v_1+(1/2)\,v_2}}.$$

Thus, in this sense the F family of distributions may be regarded as members of the more general beta family. This was the basis of our use of F tables to find beta probabilities in Section 8.14. Sometimes the subfamily of beta distributions to which F distributions belong are called "inverted beta distributions." Speaking rather loosely, in the inverted beta distributions the random variable can take on any positive value and zero, as distinguished from the beta variables originally discussed, which can range only in the interval between zero and one. Even so, these inverted beta distributions may be transformed into "standardized beta" distributions relatively simply.

One practical consequence of the connection between beta and F variables is that tables of percentage points of the F distribution, such as those in Table V, Appendix C, can also be used to find percentage points for a beta distribution. That is, if a certain value of F in a distribution with v_1 and v_2 degrees of freedom cuts off the upper G proportion of its distribution, the value of the beta variable

$$x = \frac{v_1 F}{v_1 F + v_2}$$

must cut off G proportion of its distribution, where $r = \frac{1}{2}v_1$ and $N - r = \frac{1}{2}v_2$.

Now suppose we wish to find what value of a beta variable X in a distribution with $r = 4$ and $N - r = 15$ cuts off the upper 5 percent of its distribution. First we look in the F table (Table V) and locate the value cutting off the upper 5 percent when $v_1 = 2r = 8$ and $v_2 = 2(N - r) = 30$. This value turns out to be 2.27. Then the corresponding value of the beta variable is found to be

$$x = \frac{(8)(2.27)}{8(2.27) + 30} = .377.$$

In a beta distribution $r = 4$ and $N - r = 15$, about 5 percent of the distribution lies above the value .377. The same method can be used to find any other percentage point in a beta distribution for which the corresponding F table is available.

You may also recall that in section 8.14 a way was given for using tables of the binomial distribution in order to approximate cumulative probabilities in a beta distribution. The close connection between the F distribution and the beta distribution also makes it possible to use cumulative binomial probabilities to approximate cumulative F probabilities. It may happen that the experimenter is interested in F percentage points that are not among those ordinarily tabled. Most frequently this occurs when questions of the power of an F test are at issue. Then, one can approximate the percentage point represented by a given F, in a distribution with v_1 and v_2 degrees of freedom, as follows: Let

$$x = \frac{\nu_1 F}{\nu_1 F + \nu_2}.$$

Then X is a beta variable, with $r = \frac{1}{2}\nu_1$ and $N - r = \frac{1}{2}\nu_2$. Now suppose that we are interested in the cumulative probability that X takes on some specific value x, where $x \leq \frac{1}{2}$. Then this cumulative probability is one minus the cumulative *binomial* probability for $r - 1$ in a distribution with $p = x$ and $N - 1$, where $N = \frac{1}{2}(\nu_1 + \nu_2)$. This is also the cumulative probability for the F value in question, corresponding to the value of x.

If $x \geq \frac{1}{2}$, then the cumulative beta probability for x is simply the cumulative binomial probability for the value $N - r - 1$, where $q = 1 - x$, and $N - 1$ as before. This cumulative beta probability is thus also the cumulative probability for the associated F value.

For example, suppose that we wished to find the proportion of an F distribution lying below .50 when $\nu_1 = 2$ and $\nu_2 = 18$. Then $r = \frac{1}{2}(2) = 1$, $N - r = \frac{1}{2}(18) = 9, N = 1 + 9 = 10$. We take $x = .50/(.50 + 9) = .053$. Then, we look in the binomial table (Table II) for $r - 1 = 0$ and $N - 1 = 9$, and for a p value as close as possible to .053. This turns out to be $p = .05$. The cumulative probability at zero is .6302 in the binomial table. However, since $x < \frac{1}{2}$, we take 1 minus this probability, giving $1 - .63$ or .37 as the approximate cumulative probability at the value .50 in the F distribution.

We have also just seen that about 5 percent of an F distribution with 8 and 30 degrees of freedom lies above a value of 2.27. This means that the cumulative probability for $F = 2.27$ and these degrees of freedom must be about .95. Let us check this by use of the binomial tables (Table II). First we take

$$x = \frac{(8)(2.27)}{8(2.27) + 30}$$
$$= .377$$

and $r = \frac{1}{2}\nu_1 = 4$ and $N - r = \frac{1}{2}\nu_2 = 15$. Then, we wish to find the cumulative binomial probability for $r - 1$ or 3, given $p = .377$ and $N - 1 = 4 + 15 - 1 = 18$. We look in Table II for 3 successes out of 18 trials, with a p value as close as possible to .377. This turns out to be $p = .40$. Summing the tabled probabilities from zero successes up to and including 3 successes, we obtain a cumulative probability of .0328. Subtracting this from 1.00, we find that the approximate cumulative probability for $F = 2.27$ is .9623, as compared with the actual value of .95. If we had taken the table showing $p = .35$, we would have found an approximate value of .9217. The true value of .95 does fall between these two estimates. For a great many practical purposes, such approximations are quite adequate, particularly since very extensive tables of F may be hard to find.

11.14 USING THE F DISTRIBUTION TO TEST HYPOTHESES ABOUT TWO VARIANCES

In an investigation of the effect of stress on children's performance of a reasoning test, it is felt that competition with peers represents one form of stressful situation to a child. However, the experimenter suspects that competition has different effects on different children. He conjectures that bright children might be stimulated to do even better than otherwise by the competitive atmosphere, but that the relatively dull child will appear even more at a disadvantage. One implication of this notion is that if groups of children are sampled from a population having a normal distribution of ability, but the groups are given different amounts of "competitive stress," the group subjected to the greater stress should show a relatively *greater variance* among the scores.

Letting population 1 stand for the potential group of children tested under stress, and 2 for the control, or nonstressed, population, we frame the following hypotheses:

$$H_0: \sigma_1^2 \leq \sigma_2^2$$
$$H_1: \sigma_1^2 > \sigma_2^2.$$

The rejection region decided upon is one-tailed, reflecting the experimenter's prior "hunch" about effects of stress. The α level to be used is .05.

Thirty-two children are selected at random, and assigned at random to two experimental groups of sixteen cases each. Group 1 is given the stress treatment, and group 2 is not. The results are

$$s_1^2 = 5.8, \; s_2^2 = 1.7.$$

If the exact null hypothesis is true, so that both of these values are estimates of the same population value σ^2, the ratio

$$F = \frac{s_1^2}{s_2^2}$$

should be distributed as the F distribution with $N_1 - 1 = 15$ and $N_2 - 1 = 15$ degrees of freedom. The value required for significance at the .05 level, one-tailed, is found from Table IV to be 2.40. However, the obtained value of F is

$$F = \frac{5.8}{1.7} = 3.41,$$

which exceeds the value required. On this evidence we can reject the null hypothesis of equal variances at the .05 level. We are fairly safe in saying that the experimental increase in stress seems to increase the variability of scores.

Had the alternative hypothesis been two-tailed, then we would have had to consider the required value of F on both tails of the distribution, of course. In this instance, the procedure given in Section 11.12 would be used.

11.15 THE ASSUMPTION OF THE NORMAL DISTRIBUTION IN TESTING HYPOTHESES ABOUT TWO VARIANCES

Despite the risk of needless repetition, it is well to emphasize the importance of the normal-distribution assumption in inferences about population variances. Neither the chi-square nor the F distributions can safely be used for *variance* hypotheses unless the population distribution is normal or the sample sizes are quite large. This warning does not necessarily extend to all uses of these two distributions, but it is valid when the primary question to be answered is about one or more variances.

Occasionally you will find the F test for equality of variance used along with a t test, the experimenter's idea being to test the assumption of equal variances before doing the t test itself. However, this is not an especially good idea, considering the sensitivity of the variance test to nonnormality. The variance-ratio test is least effective for this purpose when it is needed most: with small samples. What one really wants to find out is something about means, and this inference may be quite valid in *spite* of the evidence for unequal variances given by the F test. This consideration applies with even greater force to inferences about several variances; methods exist for comparing several variances simultaneously, and sometimes one is advised to carry out such a test as a preliminary to the simultaneous comparison of several means (by use of the method we will study in the next chapter). Once again, however, the test of equality of variances is quite sensitive to nonnormality, although this apparently makes little difference in tests concerning means. Therefore, one can easily do himself a disservice if he interprets a significant result from a test of variances as a prohibition against the use of a test of means. Although assumptions about equal variances are made in the use of most methods for testing means, there does not seem to be much profit in trying routinely to test the validity of this assumption, at least by the methods generally available.

11.16 Relationships among the Theoretical Distributions

Now that the major sampling distributions have been introduced, some of the connections among these theoretical distributions can be examined in more detail. (The theoretical connections between chi-square and gamma distributions, and between F and beta distributions, have already been mentioned and will not be repeated here.) Over and over again, the binomial, normal, t, chi-square, and F distributions have proved their utility in the solution of problems in statistical inference. Remember, however, that none of these distributions is empirical in the sense that someone has taken a vast number of samples and found that the sample values actually do occur with exactly the relative frequencies given by the function rule. Rather, it follows mathematically (read "logically") that if we are drawing random samples from certain kinds of populations, various sample statistics *must* have distributions given by the several function rules. Like any other theory, the theory of sampling distributions deals with "if-then" statements. This is why the assumptions we have introduced are important; if we wish to apply the theory of statistics to making inferences from samples, then we cannot expect the theory necessarily to provide us with correct results unless the conditions specified in the theory hold true. As we have seen, from a practical standpoint these assumptions may be violated to some extent in our use of these theoretical distributions as approximations, especially for large samples. However, these assumptions are quite important for the *mathematical* justification of our methods, in spite of the possible applicability of the methods to situations where the assumptions are not met.

Apart from the general requirement of simple random sampling of independent observations, the most usual assumption made in deriving sampling distributions is that the population distribution is normal. The chi-square, the _t_, and the _F_ distributions all rest on this assumption. The normal distribution is, in a real sense, the "parent" distribution to these others. This, as mentioned in Chapter 8, is one of the main reasons for the importance of the normal distribution: the normal function rule not only provides probabilities that are often excellent approximations to other probability (density) functions, but it also has highly convenient mathematical properties for deriving other distribution functions based on normal populations.

The chi-square distribution rests directly upon the assumption that the population is normal. As you will recall from Section 11.1, the chi-square variable is basically a sum of squares of independent *normal* variables, each with mean 0 and with variance 1. At the elementary level, the problem of the distribution of the sample variance can be solved explicitly only for normal populations; this sampling distribution depends on the distribution of χ^2, which in turn rests on the assumption of a normal distribution of single observations Furthermore, in the limit the distribution of χ^2 approaches a normal form.

There are close connections in theory between the _F_ distribution and both the normal and chi-square distributions. Basically, the _F_ variable is a ratio of two independent chi-square variables, each divided by its degrees of freedom. Since a chi-square variable is itself defined in terms of the normal distribution, then the _F_ distribution also rests on the assumption of two (or more) normal populations.

The _t_ distribution has links with the _F_, chi-square, and normal distributions. The _t_ ratio for a single mean can be written as

$$t = \frac{(M - \mu)/\sigma}{\sqrt{(s^2/N\sigma^2)}} = \frac{(M - \mu)/\sigma_M}{\sqrt{(s^2/\sigma^2)}}.$$ [11.16.1]

The numerator of the _t_ ratio is obviously merely z_M, a standardized score in the normal sampling distributions of means. However, consider the term in the denominator. Since, from Section 11.6 we know that

$$\frac{(N - 1)s^2}{\sigma^2} = \chi^2_{(N-1)}$$

$$\frac{s^2}{\sigma^2} = \frac{\chi^2_{(N-1)}}{N - 1},$$

it follows that

$$t = \frac{z_M}{\sqrt{\chi^2_{(N-1)}/(N - 1)}}.$$ [11.16.2*]

In general, a t variable is a standardized normal variable z_M in ratio to the square root of a chi-square variable divided by ν. Let us look at t^2 for a single mean in the light of this definition:

$$t^2 = \frac{z_M^2}{\chi^2/(N-1)}. \qquad [11.16.3]$$

The numerator o̶ ̶ ̶ ̶ ̶ ̶ definition, a chi-square variable with 1 degree of freedom, and the de̶ ̶ ̶ ̶ ̶ chi-square variable divided by its degrees of freedom, $\nu = N -$ ̶ ̶ ̶ two chi-square variables are independent, by
the principle ̶ ̶ ̶ opulation, s^2 is independent of M (Section 8.8). Th̶ ̶ ̶ ̶ ̶alifies as an F ratio, with 1 and $N-1$
degrees of f

$$_{1,\nu_2)} \qquad \nu = \nu_2; \qquad [11.16.4^*]$$

the squa̶ ̶ ̶ ̶edom is an F variable with 1 and ν degrees of
freedor

ν_2 de̶ for yourself by examining the column for 1 and
can̶ ̶ ̶ ̶ble. In the table of F values required for signifi-
en̶ ̶ ̶ e entries in this column is simply the square of the
er̶ ̶ ̶ ̶ and $t_{.025}$. Similarly, in the table for $\alpha = .01$, each
i̶ ̶ ̶ ν_2 degrees of freedom is the square of the correspond-
̶ ̶ ribution of t.
̶ ̶ ̶between t^2 and F lets us illustrate the importance of the
assumption th̶ ̶ populations have the same true σ^2 when a difference
between means is tes̶ ̶ ̶ For n cases in each of two independent groups, the value
of z^2 represented by the numerator of t^2 must be

$$z_{\text{diff.}}^2 = \frac{(M_1 - M_2)^2}{\sigma_{\text{diff.}}^2}. \qquad [11.16.5]$$

When $\mu_1 = \mu_2$, this is a chi-square variable with one degree of freedom. Furthermore, the square of the denominator term in the t ratio must correspond to

$$\frac{\text{est. } \sigma_{\text{diff.}}^2}{\sigma_{\text{diff.}}^2} \qquad [11.16.6]$$

for this to be a chi-square variable divided by its degrees of freedom. When each
population has the same true variance, σ^2, the denominator term we compute is
equivalent to such a variable, and the value t^2 is then the square of the ratio we
actually find:

$$t^2 = \frac{(M_1 - M_2)^2}{\text{est. } \sigma_{\text{diff.}}^2} = \frac{(\chi_{(1)}^2/1)}{(\chi_{(\nu_2)}^2/\nu_2)} = F. \qquad [11.16.7]$$

The ratio one calculates, not knowing the true $\sigma_{\text{diff.}}^2$, actually is distributed exactly
as t (or its square distributed as F) when the true variances are equal. However, if
the values of the two variances are not the same, the ratio of the estimated $\sigma_{\text{diff.}}^2$ to

the true $\sigma^2_{\text{diff.}}$ is equivalent to

$$\frac{\text{est. } \sigma^2_{\text{diff.}}}{\text{true } \sigma^2_{\text{diff.}}} = \frac{\chi_a^2 \sigma_1^2 + \chi_b^2 \sigma_2^2}{(n-1)(\sigma_1^2 + \sigma_2^2)} \qquad [11.16.8]$$

where χ_a^2 and χ_b^2 symbolize two possibly different values of a chi-square variable with $n-1$ degrees of freedom. This ratio is not necessarily distributed as a chi-square variable divided by its degrees of freedom. Thus, when variances are unequal for the two populations, the ratio we compute is not really distributed *exactly* like the random variable t, since the square of the ratio we compute cannot be equivalent to an F ratio. For this reason, a correction procedure, such as that in Section 10.17, is required when the variances are unequal and sample size is small.

The important relationships among these theoretical distributions are summarized in Table 11.16.1, showing how the distribution represented in the column depends for its derivation upon the distribution represented in the row.

Table 11.16.1

Distribution	Chi-square	F	t
Normal	Parent, and limiting form as $\nu \to \infty$. Defined as sum of normal and independent z^2 values.	Parent, making values in numerator and denominator independent χ^2/ν values.	Parent, and limiting form as $\nu \to \infty$. Numerator is normal z.
Chi-square		Variables in numerator and denominator are independent χ^2/ν.	Denominator is $\sqrt{\chi^2/\nu}$.
F			$t^2_{(\nu)} = F_{(1,\nu)}$

All four of these theoretical distributions will figure in the next three chapters as we work through the rationale for the analysis of variance and comparisons among means. Having some idea of the interrelations of these distributions will be of help in understanding how the F distribution can be used to test a hypothesis about *several* means.

EXERCISES

1. A sample of seven observations drawn independently and at random from a normal distribution gave the following results:

$$22, 2, 0, 30, 28, 26, 32$$

   Test the hypothesis that the population standard deviation is 10 against the alternative that it has some other value.
2. For the data of problem 1, find the 95 percent confidence limits for the population standard deviation.

3. Suppose that a sample of 20 independent observations is drawn at random from a normal population with standard deviation of 30. What is the probability that the sample value S will turn out to be less than or equal to 25?

4. A certain achievement test is standardized in such a way that a score value of 80 lies one standard deviation below the mean score value of 100 in the normally distributed population. A sample of 30 scores is drawn at random and independently from this population. What is the approximate probability that the sample standard deviation S will be at least 1.3 times larger than the population standard deviation?

5. In problem 4 above, what is the expected value of the ratio of the sample variance S^2 to the population variance σ^2? What is the *most likely* value of that ratio?

6. An experimenter drew random samples from two normal distributions, with different means, but with the same variance. The first sample contained 18 independent observations, and the second contained 11. If the two sample variances turned out to be $S_1^2 = 92$ for the first sample, and $S_2^2 = 86$ for the second, what are the 99 percent confidence limits for the variance of either population?

7. Six observations are drawn independently and at random from a normal population having a known mean of 100. The values obtained are

$$106, 98, 97, 103, 101, 99.$$

Find the 95 percent confidence interval for the population variance.

8. In a study of the size of the vocabularies of six-year-old normal children, the experimenter entertained the notion that size of vocabulary should increase from ages five to six, but that the variability of these two age groups should be about the same in this respect. Previous studies had convinced him that the standard deviation of five-year-olds was about 30 words. He also felt sure that size of vocabularly is approximately normally distributed among six-year-olds. A sample of 250 children showed that $S^2 = 945$. Does the standard deviation of the population of six-year-olds seem to be equal to about 30?

9. How is the fact that independent chi-square variables are additive related to the fact that with increasing numbers of degrees of freedom, the chi-square distribution approaches a normal distribution?

10. Suppose that a test is being carried out on the variance of a sample of observations drawn independently and at random from a normal population. The null hypothesis states that $\sigma^2 = 50$ against the alternative hypothesis that $\sigma^2 > 50$. How would one calculate the power of this test against the alternative that $\sigma^2 = 100$, given that $\alpha = .05$? See if you can find an approximate value for the power against this alternative.

11. Two normally distributed populations are being compared in order to see if they have the same values for their variances. The alternative hypothesis is that Population I has the larger variance. A sample of 13 independent observations are drawn from Population I and a separate random sample of 9 independent observations are made of Population II. Sample I shows that $S_I^2 = 141$, while Sample II shows $S_{II}^2 = 123$. What can one conclude about the variances of the two populations?

12. Is it possible to test a hypothesis of the form $H_0: \sigma_I^2 = k\sigma_{II}^2$. Explain why it should or should not be possible, and, if possible, how one would go about doing so.

13. Suppose that independent random samples each consisting of 16 cases were drawn at random from two normal populations. The first sample produced a sample standard deviation $S = 38.2$ and the second sample a standard deviation $S = 59.2$. The null hypothesis was $H_0: \sigma_I^2 = \sigma_{II}^2$ and the alternative was "not H_0." Test the null hypothesis.

14. An experimenter faced the following problem: A series of 4 independent random

samples each of size 5 had been drawn from a certain normally distributed population, and for each sample, the mean had been computed. Now the experimenter has drawn an independent sample of 15 observations, which he believes to be from the same population. The mean of the sample means tends to agree with the mean of his new sample. However, the variance S^2 of the 5 sample means about their grand mean is 84, while the variance S^2 of his sample of 15 observations is 380. Is this significant evidence that the previous set of means did not come from the same population as did the new sample?

15. See if you can describe the formal relationships among the t, F, chi-square, and the beta, and gamma distributions.

12 The Analysis of Variance: Model I; Fixed Effects

In the examples of Chapter 10, the presence of a statistical relation was inferred from the difference between only two sample means. Of course, such data come only from the simplest kinds of experiments, involving an experimental group and a control group, or perhaps two different experimental treatments. Most psychological experiments are far more complicated than this, employing several groups given different treatments, or groups given different combinations of treatments. However, the experimenter's problem is essentially the same: here he needs a method for the simultaneous comparison of many means in order to decide if some statistical relation exists between the experimental and the dependent variables. The most commonly used procedure for this purpose is the analysis of variance, which is the topic of this chapter.

Bear in mind that the independent or experimental variable X may stand for a qualitatively different treatment administered to each group, or for a numerical amount of some treatment. Both in Chapter 10 and in this chapter we deal with the first instance, in which the different groups represent *qualitatively* distinct classes. The study of experiments involving a quantitative independent variable will be deferred until Chapter 15, although the basic ideas underlying the analysis of variance also will apply in that instance.

The methods and examples discussed in this chapter are a very small subset of the potential applications of the analysis of variance. Only two very simple kinds of experiments will be considered here: first to be considered are experiments having *one qualitative experimental variable or factor* and *one quantitative dependent variable*. In this situation, we will apply the *one-way* analysis of variance (that is to say, the problem is to compare sample groups differing systematically in only one way). Then we will consider the situation where each distinct group of n cases is given a different combination of *two different qualitative* treatments. Here the *two-way* analysis of variance will apply (experimental groups

457

may differ systematically in two ways). There are several good reasons for limiting this discussion to these simple experimental situations. In the first place, a beginning student in psychology is likely to be doing rather simple experiments, or at least his experiments *ought* to be simple if they are to be done with care. In the second place, treating the more complex applications of the analysis of variance must take us fairly far into experimental design, a topic quite capable of occupying a large book of its own. Third, the basic reasoning underlying our discussion of the simple one-way and two-way situations extends quite easily to more complicated situations, so that the student who can follow this discussion should have relatively little trouble with most elementary texts on experimental design and the applications of analysis of variance.

It is important to distinguish between two different sampling situations to which the analysis of variance applies. These differ both in the way that experimental treatments are selected and in the kinds of inferences one makes from the analysis. The formal statistical models applying to these two sampling situations have become known as "Model I" or the "fixed effects model," and "Model II," the "random effects model." The situations calling for these two different models will be compared briefly in the next section.

12.1 MODEL I: FIXED EFFECTS

Imagine a situation where several experimental treatments are to be administered. Suppose that there are J different such treatments, and that each treatment is to be administered to one and only one experimental group. Each of the J groups consists of individuals chosen at random and independently and assigned at random to the groups. For example, four different tranquilizing drugs are to be compared for the effect each has on driving skill. Some N subjects are chosen at random and independently and allotted at random to four non-overlapping groups. The individuals in the first group are given drug 1, those in the second drug 2, and so on. Then the groups are to be compared on the dependent variable Y, a score on driving skill. Or, perhaps, six different methods of teaching second-grade arithmetic are known. School children from a specific population are sampled and allotted at random to six different groups, each group representing one of the six instructional methods. The groups are then compared on their average achievement after a year's instruction. In both examples, members of some small set of treatments are to be compared, and each treatment of interest is actually used in the experiment.

Experiments to which Model I applies are distinguished by the fact that inferences are to be made only about differences among the J treatments actually administered, and about no other treatments that might have been included. In advance of the actual experiment, the experimenter decides that he wants to see if differences in effect exist among some fairly small set of treatments or treatment combinations. He is interested in these treatments or combinations, and *no* others. Each treatment of immediate interest to the experimenter is actually included in the experiment, and the set of treatments or treatment combinations applied exhausts the set of treatments about which the experimenter wants to make inferences. The effect of any treatment is "fixed," in the sense that it must appear in any complete repetition of the experiment on new subjects.

Two examples of experiments in which the fixed effects model does not apply may help to clarify this distinction. Consider the question of the effect of the individual experimenter himself on the outcome of an experiment. Perhaps the experiment is that of persuasion of a group of subjects to change their attitudes on some controversial subject. Not only will the experience given to an individual subject have some effect upon his attitude, but also the personality and mannerisms of the experimenter may be important in how the subject tends to change. Our interest is not in any particular set of experimenters, but rather in *experimenters*, broadly defined. Thus, from the set of all possible such experimenters, a random sample of individuals is selected, and then each of these experimenters carries out the same procedure. In this way it is possible to decide if variation in experimenter personality accounts for some of the variation in the performance of subjects. Here, the levels of the factor "experimenter" are randomly sampled, not fixed.

As another example to which the fixed effects model does not apply, consider an experiment involving a projective test. This test consists of ten different stimulus cards given in turn to the subject, who gives verbal responses to each. Among the things recorded about the behavior of a subject is the mean time between the presentation of a card and his first response. The experiment is designed to study the effect of the order of presentation of the cards to a subject upon his average first response time. However, there are 10! or 3,628,800 different possible orders of presentation of the cards. The experimenter takes a random sample of, say, twenty such orders, and tests a different group of randomly selected subjects under each. Here, the experimenter is not really interested in the twenty orders actually administered so much as he is in the possible effects of order *in general*. Thus, the fixed effects model is inappropriate for this problem.

The random effects model (Model II) applies when the experiment involves only a random sample of the set of treatments about which the experimenter wants to make inferences. The various treatments actually applied do not exhaust the set of all treatments of interest. Here, the effect of a treatment is not regarded as fixed, since any particular treatment itself need not be included each time the experiment is carried out: on each repetition of the entire experiment a new sample of treatments is to be taken. The experimenter may not actually plan to repeat the experiment, but conceptually each repetition involves a fresh sample of treatments. This second model will be considered in more detail in the next chapter. For the moment, we will confine our attention to Model I, for the fixed effects situation.

12.2 THE IDEA OF AN EFFECT

It is time to be more specific about the meaning of "effect" as used here. Common sense suggests that if various experimental treatments are having different systematic influences on groups, the means of the groups should tend to be different. In particular, if there really exists a functional relationship between treatment X and the value of variable Y when all possible other factors are controlled, and if the influence of any uncontrolled factors can be regarded as random error, "canceling out" in the long run, a functional relationship should exist between X and *mean Y* over the various treatment populations. The presence of different means for different experimental populations is, as we have seen, an indicator of a statistical relation between the experimental and dependent variables. As the term is used in analysis of variance, **an effect is a reflection of a difference among population means;** this idea will now be given a more formal statement.

If we think of a treatment population j as the hypothetical set of all possible unit observations that might be made under treatment j, then the experimental group to which treatment j is actually applied represents a sample from this hypothetical population. Let M_j be the sample mean of the group to which treatment j was actually applied, and let the mean of the potential treatment population be μ_j. Furthermore, consider the grand population, formed by the pooling of the J treatment populations. Let the proportionate representation of any population j in the grand population be represented by $p(x_j)$, so that the probability that any observation drawn at random from the grand population belongs to population j is $p(x_j)$. If the J treatment populations are equally represented in the grand population, then for every j the value of $p(x_j) = 1/J$.

Now under these conditions, the mean μ of the grand population would be the weighted average of the several population means:

$$\mu = \sum_j p(x_j)\mu_j.$$

In the special case where $p(x_j) = 1/J$ for all j, the grand mean would be

$$\mu = \frac{\sum_j \mu_j}{J}.$$

The effect of treatment j is defined as the deviation of the mean of population j, μ_j, from the grand population mean, μ:

$$\text{effect of treatment } j = (\mu_j - \mu)$$
or
$$\alpha_j = (\mu_j - \mu). \qquad [12.2.1*]$$

This symbol α_j (*not* to be confused with the alpha standing for the probability of Type I error) will stand for the effect of any single treatment j.

Since the grand population mean μ is also the weighted sum of all of the treatment population means, it follows that the weighted sum of all of the effects must be zero:

$$\sum_j p(x_j)\alpha_j = \sum_j p(x_j)(\mu_j - \mu) = \mu - \mu = 0.$$

In the special case where $p(x_j) = 1/J$ for all treatment populations j, then the sum of the effects is zero:

$$\frac{\sum_j \alpha_j}{J} = 0 \quad \text{implies that} \quad \sum_j \alpha_j = 0.$$

Now suppose that there is absolutely no effect associated with any treatment. This means that

$$\alpha_j = 0 \qquad\qquad\qquad\qquad [12.2.3]$$

for each and every treatment population j. This is equivalent to the statement that

$$\mu_1 = \mu_2 = \cdots = \mu_J = \mu,$$

where the index numbers 1, 2, $\cdots$, J designate the various treatments. *The complete absence of effects is equivalent to the absolute equality of all of the population means.*

Notice that when there are no treatment effects

$$\sum_j p(x_j)\alpha_j^2 = 0, \qquad\qquad\qquad [12.2.4^*]$$

the weighted sum of the α_j^2 must be zero, since each and every α_j is zero when no treatment effects exist.

12.3   THE LINEAR MODEL IN THE FIXED EFFECTS SITUATION

Now we are ready to state a model, or "theory," about the composition of any observed score Y_{ij}, corresponding to the ith individual observed in condition j. This is called a *linear* model, since it states that the score of any individual in any treatment group is composed from a simple sum:

$$Y_{ij} = \mu + \alpha_j + e_{ij}. \qquad\qquad\qquad [12.3.1^*]$$

This model asserts that the score for observation i in group j is based on the sum of three components: the grand mean of all of the J different treatment populations, the effect associated with the particular treatment j, and another part which is strictly peculiar to the ith observation made under treatment j. This last term, e_{ij}, is the random error associated with that particular treatment-subject combination.

Notice that any individual i observed under treatment j is a random sample of one from the corresponding treatment population j, so that the expectation over all individuals given treatment j is

$$E(Y_{ij}) = \mu_j.$$

By the linear model, the expected score value over all individuals i in population j is

$$E(Y_{ij}) = \mu + \alpha_j + E(e_{ij}),$$

or

$$E(Y_{ij}) = \mu_j + E(e_{ij}).$$

Thus, for any population j, the expectation of e_{ij} over all individuals is zero:

$$E(e_{ij}) = 0.$$

Furthermore, let M_j be the mean of the sample given the particular treatment j, and let M stand for the mean over all individuals in all treatments for the experiment. Then over all possible samples of n_j individuals from population j,

$$E(M_j) = \mu + \alpha_j + E(M_{\bullet j}),$$

where

$$M_{\bullet j} = \frac{\displaystyle\sum_i e_{ij}}{n_j}.$$

That is, the expected value of M_j depends on μ, on the effect α_j, and on the expected mean error over the n_j observations in group j. Since

$$E(M_j) = \mu_j,$$

it must be true that

$$E(M_{\bullet j}) = 0$$

for any group j.

Now before going further, let us once again distinguish two cases that may apply. In the first case, which we will treat next, each sample is of the same size n, or $n_j = N/J$. After consideration of this case, we will discuss the more general situation where the sample size n_j differs among the treatment groups.

Consider J treatment groups, each of size n. Then the grand mean M over all of the J groups is

$$M = \frac{\displaystyle\sum_j M_j}{J} = \frac{\displaystyle\sum_j \sum_i y_{ij}}{nJ}.$$

Since

$$E(M) = E\left[\frac{\displaystyle\sum_j \sum_i (\mu + \alpha_j + e_{ij})}{nJ}\right]$$

$$= \mu + \frac{\displaystyle\sum_j \alpha_j}{J} + E\left(\frac{\displaystyle\sum_j M_{\bullet j}}{J}\right)$$

and since

$$E(M) = \mu, \qquad \frac{\sum_j \alpha_j}{J} = 0,$$

it must be true that

$$E(M_e) = 0,$$

where

$$M_e = \frac{\sum_j \sum_i e_{ij}}{nJ} = \frac{\sum_j M_{e_j}}{J}.$$

In other words, the expectation of the average error over all observations in all treatments is zero.

In the instance in which the sample size n_j may be different from treatment group to treatment group, the grand mean of the samples is

$$M = \frac{\sum_j n_j M_j}{N} = \frac{\sum_j \sum_i y_{ij}}{N},$$

so that

$$E(M) = E\left[\frac{\sum_j \sum_i (\mu + \alpha_j + e_{ij})}{N}\right]$$

$$= \mu + \frac{\sum_j n_j \alpha_j}{N} + E\left(\frac{\sum_j n_j M_{e_j}}{N}\right).$$

Here, if we take $n_j/N = p(x_j)$ for any treatment group j, then

$$\frac{\sum_j n_j \alpha_j}{N} = \sum_j p(x_j)\alpha_j = 0$$

and

$$E(M) = \mu,$$

so that

$$E(M_e) = E\left(\frac{\sum_j n_j M_{e_j}}{N}\right) = 0.$$

That is, if we assume that the proportional representation of the various treatment groups is the same as the proportional representation of the various treatment populations in the total population, then the mean error in the total sample must have an expectation of zero. We will deal with this assumption at more length in later sections. For the nonce, either we deal either with groups of equal size, or we treat the proportional makeup of the sample as representative of the proportional makeup of the entire population under study.

Since both individual errors and mean errors have expectations of zero in this model, an unbiased estimate of the effect of any treatment j can be found by taking

$$\text{est. } \alpha_j = (M_j - M) \tag{12.3.2*}$$

so that

$$E(\text{est. } \alpha_j) = E(M_j - M) = \alpha_j. \tag{12.3.3*}$$

To gain some feel for this linear model, imagine three samples consisting of three observations each. Suppose that these three samples represent identical population distributions, and that there is *no* variability (that is, no error) within any of the populations. If the mean of each of the populations were $\mu = 40$, then our sample results should look like this:

Sample 1	Sample 2	Sample 3
40	40	40
40	40	40
40	40	40

There should be no differences either between or within samples if this were the true situation. When this is true the linear model becomes simply

$$Y_{ij} = \mu,$$

since $\alpha_j = 0$, $e_{ij} = 0$ for all i and j.

Now suppose that the three samples are given different treatments, and that treatment effects exist, but that there is once again no variability within a treatment population (again, no error). Our results might look like this:

Sample 1	Sample 2	Sample 3
$40 - 2 = 38$	$40 + 6 = 46$	$40 - 4 = 36$
$40 - 2 = 38$	$40 + 6 = 46$	$40 - 4 = 36$
$40 - 2 = 38$	$40 + 6 = 46$	$40 - 4 = 36$

Here there are differences between observations in different treatments, but there are no differences within a treatment sample. The linear model here is

$$Y_{ij} = \mu + \alpha_j,$$

since $\alpha_j \neq 0$ while $e_{ij} = 0$, for any i and j.

In actuality there is always variability in a population, so that there is sampling error. The actual data we might obtain would undoubtedly look something like this:

Sample 1	*Sample 2*	*Sample 3*
$40 - 2 + 5 = 43$	$40 + 6 - 5 = 41$	$40 - 4 + 3 = 39$
$40 - 2 + 2 = 40$	$40 + 6 + 1 = 47$	$40 - 4 - 2 = 34$
$40 - 2 - 3 = 35$	$40 + 6 + 8 = 54$	$40 - 4 + 1 = 37$
$M_1 = 39.3$	$M_2 = 47.3$	$M_3 = 36.7 \quad M = 41.1$

Here, a random error component has been added to the value of μ and the value of α_j in the formation of each score. The linear model in this situation is

$$Y_{ij} = \mu + \alpha_j + e_{ij}.$$

Notice that not only do differences exist between observations in different treatments, but also between observations in the same treatment.

If we estimate the effect of treatment 1 by taking

$$\text{est. } \alpha_1 = M_1 - M = 39.3 - 41.1 = -1.8$$

it happens in this example that we are almost right, since the data were fabricated so that $\alpha_1 = -2$. Likewise, our estimate of α_2 is in error by .2 and our estimate of α_3 in error by $-.4$. Although these errors may seem rather slight in this example, there is no guarantee in any given experiment that they will not be very large. Thus we need to evaluate how much of the apparent effect of any experimental treatment is, in fact, due to error before we can decide that something systematic is actually occurring.

This example should suggest that evidence for experimental effects has something to do with the differences *between* the different groups relative to the differences that exist *within* each group. Next, we will turn to the problem of separating the variability among observations into two parts: the part that should reflect both experimental effects and sampling error, and that part that should reflect sampling error alone.

12.4 THE PARTITION OF THE SUM OF SQUARES FOR ANY SET OF J DISTINCT SAMPLES

In this section we are going to leave the study of population effects for a while, and show how the variability in any set of J experimentally different samples may be partitioned into two distinct parts. Actually, we will do this in terms of the sum of squared deviations about the grand mean for the samples, rather than the sample variance itself.

Any score y_{ij} in sample j exhibits some deviation from the grand sample mean of all scores, M. The extent of deviation is merely

$$(y_{ij} - M).$$

This deviation can be thought of as composed of two parts,

$$(y_{ij} - M) = (y_{ij} - M_j) + (M_j - M),\qquad [12.4.1]$$

the first part being the deviation of y_{ij} from the mean of group j, and the second being the deviation of the group mean from the grand mean. Notice that if the groups in question are actually entire populations, equation 12.4.1 is equivalent to the statement that

$$(y_{ij} - \mu) = e_{ij} + \alpha_j.$$

Now suppose that we square the deviation from M for each score in the entire sample (12.4.1), and sum these squared deviations across all individuals i in all sample groups j:

$$\sum_j \sum_i (y_{ij} - M)^2 = \sum_j \sum_i [(y_{ij} - M_j) + (M_j - M)]^2$$

$$= \sum_j \sum_i (y_{ij} - M_j)^2 + \sum_j \sum_i (M_j - M)^2$$

$$+ 2 \sum_j \sum_i (y_{ij} - M_j)(M_j - M). \quad [12.4.2]$$

Now look at the last term on the right in equation 12.4.2 above:

$$2 \sum_j \sum_i (y_{ij} - M_j)(M_j - M) = 2 \sum_j (M_j - M) \sum_i (y_{ij} - M_j)$$

$$= 0$$

since the value represented by the term $(M_j - M)$ is the same for all i in group j, and the sum of $(y_{ij} - M_j)$ must be zero when taken over all i in any group j.

Furthermore,

$$\sum_j \sum_i (M_j - M)^2 = \sum_j n_j (M_j - M)^2$$

since, once again, $(M_j - M)$ is a constant for each individual i figuring in the sum. Putting these results together, we have

$$\sum_j \sum_i (y_{ij} - M)^2 = \sum_j \sum_i (y_{ij} - M_j)^2 + \sum_j n_j (M_j - M)^2. \quad [12.4.3*]$$

This identity is usually called the "partition of the sum of squares," and is true for any set of J distinct samples. Verbally, this fact can be stated as follows: the total sum of squared deviations from the grand mean can always be separated into two parts, the sum of squared deviations within groups, and the weighted sum of squared deviations of group means from the grand mean. It is convenient to call these two parts

$$\text{SS within} = \sum_j \sum_i (y_{ij} - M_j)^2 \qquad [12.4.4]$$

for **sum of squares within groups,** and

$$\text{SS between} = \sum_{j} n_j (M_j - M)^2 \qquad\qquad [12.4.5]$$

for **sum of squares between groups.** Thus, it is a true statement that

$$\text{SS total} = \text{SS within} + \text{SS between}. \qquad\qquad [12.4.6\dagger]$$

The meaning of this partition of the sum of squares into two parts can easily be put into common-sense terms: Individual observations in any sample will differ from each other, or show variability. These obtained differences among individuals can be due to two things. Some pairs of individuals are in different treatment groups, and their differences are due either to the different treatments, or to chance variation, or to both. The sum of squares between groups reflects the contribution of different treatments, as well as chance, to intergroup differences. On the other hand, individuals in the *same* treatment groups can differ only because of chance variation, since each individual within the group received exactly the same treatment. The sum of squares within groups reflects these intragroup differences due only to chance variation. Thus, in any sample two kinds of variability can be isolated: the sum of squares between groups, reflecting variability due to treatments *and* chance, and the sum of squares within groups, reflecting chance variation alone.

12.5   ASSUMPTIONS UNDERLYING INFERENCES ABOUT TREATMENT EFFECTS

The partition of the sum of squares is possible for any set of J distinct samples, and no special assumptions about populations or sampling are necessary in its derivation. However, before we can use sample data to make inferences about the existence of population effects, several assumptions must be made. These are as follows.

1. For each treatment population j, the distribution of e_{ij} is assumed normal.
2. For each population j, the distribution of e_{ij} has a variance σ_e^2, which is assumed to be the same for each treatment population.
3. The errors associated with any pair of observations are assumed to be independent. A consequence of this assumption is that if h and i stand for any pair of observations, and j and k for any pair of treatments, then

$$E(e_{ij}e_{hj}) = 0$$

and

$$E(e_{ij}e_{hk}) = 0.$$

In short, we are going to regard our observations as independently drawn from normal treatment populations each having the same variance, and with error components independent across all pairs of observations.

12.6   THE MEAN SQUARE BETWEEN GROUPS

The next question is how to use the partition of the sum of squares in making inferences about the existence of treatment effects. First of all, we will examine the expectation of the sum of squares between groups.

For any group j, a simple substitution from **12.3.1** above shows that

$$M_j = \frac{\sum_i y_{ij}}{n_j} = \mu + \alpha_j + M_{ej}$$

and

$$M = \frac{\sum_j \sum_i y_{ij}}{N} = \mu + M_e.$$

since we are assuming that

$$\sum_j \frac{n_j \alpha_j}{N} = \sum_j p(x_j)\alpha_j = 0.$$

Thus, the deviation of any sample group mean from the grand sample mean is actually

$$(M_j - M) = \alpha_j + (M_{ej} - M_e).$$

From this it follows that

$$\text{SS between} = \sum_j n_j(M_j - M)^2 = \sum_j n_j[\alpha_j + (M_{ej} - M_e)]^2. \qquad [12.6.1]$$

On taking the expectation of the SS between, we find

$$E(\text{SS between}) = E \sum_j n_j[\alpha_j + (M_{ej} - M_e)]^2$$

$$= \sum_j n_j \alpha_j^2 + E \sum_j n_j(M_{ej} - M_e)^2, \qquad [12.6.2]$$

since

$$E \sum_j 2n_j\alpha_j(M_{ej} - M_e) = \sum_j 2n_j\alpha_j E(M_{ej} - M_e) = 0.$$

Bear in mind that each α_j is conceived as fixed over samples, and the fact that $E(e_{ij}) = 0$.

Now turn your attention for a moment to the last term on the right in expression **12.6.2**. First of all, on squaring and distributing the summation we find that

$$\sum_j n_j(M_{ej} - M_e)^2 = \sum_j n_j M_{ej}^2 - 2M_e \sum_j n_j M_{ej} + \sum_j n_j M_e^2 = \sum_j n_j M_{ej}^2 - N M_e^2$$

since

$$M_e = \frac{\sum_j n_j M_{ej}}{N}$$

Hence
$$E \sum_j n_j (M_{ej} - M_e)^2 = \sum_j n_j E(M_{ej}^2) - N E(M_e^2).$$

Because $E(M_{ej}) = E(M_e) = 0$, then for any j,

$$E(M_{ej}^2) = \sigma_{M_{ej}}^2 = \frac{\sigma_e^2}{n_j}, \qquad [12.6.3]$$

the variance of the sampling distribution of *mean errors* for samples of size n_j.

Furthermore,
$$E(M_e^2) = \sigma_{M_e}^2 = \frac{\sigma_e^2}{N}, \qquad [12.6.4]$$

the variance of the sampling distribution of mean errors for samples of size N. Thus, combining the results of **12.6.3** and **12.6.4** we have

$$E \sum_j n_j (M_{ej} - M_e)^2 = \sum_j n_j \frac{\sigma_e^2}{n_j} - \frac{N \sigma_e^2}{N}$$
$$= (J - 1)\sigma_e^2. \qquad [12.6.5]$$

On making this substitution (into **12.6.2**) we finally arrive at the result we were seeking:

$$E(\text{SS between}) = \sum_j n_j \alpha_j^2 + (J - 1)\sigma_e^2. \qquad [12.6.6^*]$$

Ordinarily, we deal with the *mean square between*,

$$\text{MS between} = \frac{\text{SS between}}{J - 1}. \qquad [12.6.7\dagger]$$

Then

$$E(\text{MS between}) = \sigma_e^2 + \frac{\sum_j n_j \alpha_j^2}{J - 1}. \qquad [12.6.8^*]$$

The mean square between groups is an unbiased estimate of σ_e^2, the error variance, plus a term that can be zero only when there are no treatment effects at all. When the hypothesis of no treatment effects is absolutely true, then,

$$E(\text{MS between}) = \sigma_e^2. \qquad [12.6.9^*]$$

If any true treatment effects at all exist, then,

$$E(\text{MS between}) > \sigma_e^2. \qquad [12.6.10^*]$$

Accordingly, we can see that the mean square between groups gives one piece of the evidence needed to adjudge the existence of treatment effects. The sample value of MS between should be an unbiased estimate of error variance alone when no treatment effects exist. On the other hand, the value of MS between must be an estimate of σ_e^2 *plus* a positive quantity when any treatment effects exist.

Naturally, MS between is always a sample quantity, and thus it must have a sampling distribution. However, it is easy to see what this sampling distribution must be: when there are no treatment effects, MS between is an unbiased estimate of σ_e^2. For this estimate, as for many estimates of σ^2 for normal populations,

$$\frac{(\text{est. } \sigma_e^2)}{\sigma_e^2} = \frac{\chi_{(\nu)}^2}{\nu}.$$

The ratio of MS between to σ_e^2 must be a chi-square variable divided by degrees of freedom, *when* there are no treatment effects *and* the parent populations are normal (assumption 1, Section 12.5).

What is the number of degrees of freedom for MS between? There are really only J different sample values that go into the computation of MS between: these are the J values of M_j. Thus, *there are $J - 1$ degrees of freedom for MS between.*

As yet we have no idea of the value of σ_e^2, so that the sampling distribution of MS between cannot be used directly to provide a test of the hypothesis of no treatment effects. Now, however, let us investigate the sampling distribution of MS within.

12.7 THE MEAN SQUARE WITHIN GROUPS

What population value is estimated by the mean square within groups? Under the fixed effects model it is obvious that the treatments administered cannot be responsible for differences that occur among observations within any given group. This kind of within-groups variation should be a reflection of random error alone. Keeping this in mind, let us find the expectation of the sum of squares within groups.

$$E(\text{SS within}) = E\left[\sum_j \sum_i (Y_{ij} - M_j)^2\right].$$

For any given sample j,

$$E\frac{\left[\sum_i (Y_{ij} - M_j)^2\right]}{n_j - 1} = \sigma_e^2 \qquad [12.7.1]$$

since for any sample j this value is an unbiased estimate of the population error variance, σ_e^2. Thus,

$$
\begin{aligned}
E(\text{SS within}) &= \sum_j E \sum_i (Y_{ij} - M_j)^2 \\
&= \sum_j (n_j - 1)\sigma_e^2 \\
&= (N - J)\sigma_e^2. \qquad [12.7.2*]
\end{aligned}
$$

If we define

$$\text{MS within} = \frac{\text{SS within}}{N - J},$$ [12.7.3†]

then

$$E(\text{MS within}) = \sigma_e^2.$$ [12.7.4*]

The expectation of MS within is σ_e^2. The mean square within groups is thus an unbiased estimate of the error variance within each treatment population. This is true regardless of the possible existence of treatment effects.

12.8 THE SAMPLING DISTRIBUTION OF MS WITHIN

We have just shown that the mean square within groups is a pooled estimate of the value of the variance σ_e^2, which was assumed to be the same for each population. Thus, once again for normal populations, it must be true that

$$\frac{\text{MS within}}{\sigma_e^2} = \frac{\chi_{(\nu)}^2}{\nu}.$$

Here, however, the value of the degrees of freedom ν is quite different than that for MS between. This chi-square variable is actually a *sum* of independent chi-square variables, each of which has some $n_j - 1$ degrees of freedom. The addition property mentioned in Section 11.4 must apply, and so **the degrees of freedom for MS within is**

$$\sum_j (n_j - 1) = N - J.$$ [12.8.1†]

Surely you can anticipate the turn the argument takes now! We have MS between, which estimates σ_e^2 when there are no treatment effects, but a value greater than σ_e^2 when effects exist. Moreover, we have another estimate of σ_e^2 given by MS within, which does *not* depend on the presence or absence of effects. Two variance estimates which *ought* to be the same under the null hypothesis suggest the F distribution, and this is what we use to test the hypothesis.

12.9 THE F TEST IN THE ANALYSIS OF VARIANCE

The usual hypothesis tested using the analysis of variance is

$$H_0: \mu_1 = \cdots = \mu_j = \cdots = \mu_J,$$

the hypothesis that all treatment population means are equal. The alternative is just

$$H_1: \text{not } H_0,$$

implying that some of the population means are different from others. As we have seen, these two hypotheses are equivalent to the hypothesis of no-effects and its contrary:

$$H_0: \alpha_j = 0, \text{ for all } j$$
$$H_1: \alpha_j \neq 0, \text{ for some } j.$$

The argument in Sections 12.6 through 12.8 has shown that *when H_0 is true,*

$$E(\text{MS between}) = \sigma_e^2$$

and

$$E(\text{MS within}) = \sigma_e^2,$$

both the mean square between and the mean square within are unbiased estimates of the same value, σ_e^2. On the other hand, *when the null hypothesis is false,* then

$$E(\text{MS within}) < E(\text{MS between}).$$

Since both of these mean squares divided by σ_e^2 are distributed as chi-square variables divided by their respective degrees of freedom when H_0 is true, it follows that their ratio should be distributed as F, provided that MS between and MS within are *independent* estimates of σ_e^2. From the principle of Section 8.8 the following can be proved: **For J samples of independent observations, each drawn from a normal population distribution, MS between and MS within are statistically independent.** For each sample, the mean M_j is independent of the variance estimate s_j^2, provided that the population distribution is normal. By an extension of the principle given in Section 8.8, MS between, based on the J values M_j, must be independent of MS within, based on the several s_j^2 values; each piece of information making up MS between is independent of the information making up MS within, given normal parent distributions.

Finally, we have all the justification needed in order to say that the ratio

$$\frac{(\text{MS between}/\sigma_e^2)}{(\text{MS within}/\sigma_e^2)} = \frac{\text{MS between}}{\text{MS within}} \qquad [12.9.1\dagger]$$

is distributed as F with $J - 1$ and $N - J$ degrees of freedom, *when the null hypothesis is true.* This statistic is the ratio of two independent chi-square variables, each divided by its degrees of freedom, and thus is exactly distributed as F when H_0 is true.

The F ratio used in the analysis of variance always provides a *one-tailed* test of H_0 in terms of the sampling distribution of F. Evidence for H_1 must show up as an F ratio greater than 1.00, and an F ratio less than 1.00 can signify nothing except sampling error (or perhaps nonrandomness of the samples or failure of the assumptions). Therefore, for the analysis of variance, the F ratio obtained can be compared directly with the one-tailed values given in Table V.

An α level is chosen in advance, and this value determines the section of the table one uses to determine the significance of the obtained F ratio.

12.10 COMPUTATIONAL FORMS FOR THE SIMPLE ANALYSIS OF VARIANCE

Although the argument given above dealt with sums of squares defined as follows:

$$\text{SS total} = \sum_j \sum_i (y_{ij} - M)^2$$

$$\text{SS within} = \sum_j \sum_i (y_{ij} - M_j)^2$$

$$\text{SS between} = \sum_j n_j (M_j - M)^2$$

most users of the analysis of variance find it more convenient to work with equivalent, but computationally simpler, versions of these sample values. These computational forms will be given below.

First of all, the total sum of squares can be shown to be equal to

$$\text{SS total} = \sum_j \sum_i y_{ij}^2 - \frac{\left(\sum_j \sum_i y_{ij} \right)^2}{N}. \qquad [12.10.1\dagger]$$

It is easy to show that this is true:

$$\text{SS total} = \sum_j \sum_i (y_{ij} - M)^2 = \sum_j \sum_i (y_{ij}^2 - 2y_{ij}M + M^2)$$

$$= \sum_j \sum_i y_{ij}^2 - 2M \sum_j \sum_i y_{ij} + \sum_j \sum_i M^2,$$

by rules 3 and 4 in Appendix A. This last expression reduces further to

$$\sum_j \sum_i y_{ij}^2 - 2M(NM) + NM^2$$

or

$$\sum_j \sum_i y_{ij}^2 - NM^2,$$

by the definition of the sample grand mean, $M = \sum_j \sum_i y_{ij}/N$. Making one last substitution for M gives the computing formula, **12.10.1**.

The computing formula for the sum of squares between groups can be worked out in a similar way:

$$\text{SS between} = \sum_j n_j (M_j - M)^2 = \sum_j n_j (M_j^2 - 2MM_j + M^2)$$

$$= \sum_j n_j M_j^2 - 2M \sum_j n_j M_j + M^2 \sum_j n_j$$

$$= \sum_j \frac{\left(\sum_i y_{ij}\right)^2}{n_j} - 2NM^2 + NM^2$$

or

$$\text{SS between} = \sum_j \frac{\left(\sum_i y_{ij}\right)^2}{n_j} - \frac{\left(\sum_j \sum_i y_{ij}\right)^2}{N}. \qquad [12.10.2\dagger]$$

Finally, the computing formula for the sum of squares within groups is found by

$$\text{SS within} = \text{SS total} - \text{SS between}$$

$$= \sum_j \sum_i y_{ij}^2 - \frac{\left(\sum_j \sum_i y_{ij}\right)^2}{N} - \sum_j \frac{\left(\sum_i y_{ij}\right)^2}{n_j} + \frac{\left(\sum_j \sum_i y_{ij}\right)^2}{N}$$

$$= \sum_j \sum_i y_{ij}^2 - \sum_j \frac{\left(\sum_i y_{ij}\right)^2}{n_j}. \qquad [12.10.3\dagger]$$

Ordinarily, the simplest computational procedure is to calculate both the sum of squares total and the sum of squares between directly, and then to subtract SS between from SS total in order to find the SS within.

12.11 A COMPUTATIONAL OUTLINE FOR THE ONE-WAY ANALYSIS OF VARIANCE

It is natural for the beginner in statistics to be a little staggered by all of the arithmetic that the analysis of variance involves. However, take heart! With a bit of organization and with the aid of a desk calculator simple analyses can be done quite quickly. The important thing is to form a clear mental picture of the different sample quantities you will need to compute, and how they combine. Below is an outline of the steps to follow:

1. Start with a listing of the raw scores separated by columns into the treatment groups to which they belong.
2. Square each score (y_{ij}^2) and then add these squared scores over all individuals in all groups. The result is $\sum_j \sum_i y_{ij}^2$. Call this quantity A.

3. Now sum the *raw* scores over all individuals in all groups to find $\sum_j \sum_i y_{ij}$.

   Call the resulting value B (on some desk calculators it is possible to find A and B simultaneously).

4. Now for a single group, say group j, sum all of the raw scores in that group and square the sum, to find $\left(\sum_i y_{ij}\right)^2$. Divide by the number in that group: $\left(\sum_i y_{ij}\right)^2 / n_j$.

5. Repeat step 4 for each group, and then sum the results across the several groups to find $\sum_j \dfrac{\left(\sum_i y_{ij}\right)^2}{n_j}$. Call this quantity C.

6. The **sum of squares total** is found from $A - B^2/N$.
7. The **sum of squares between** is $C - B^2/N$.
8. The **sum of squares within** is

$$\text{SS total} - \text{SS between} = A - C.$$

9. Divide SS between by $J - 1$ to give **MS between.**
10. Divide SS within by $N - J$ to give **MS within.**
11. Divide MS between by MS within to find the F **ratio.**
12. Carry out the test by referring the F ratio to a table of the F distribution with $J - 1$ and $N - J$ degrees of freedom.

12.12   THE ANALYSIS OF VARIANCE SUMMARY TABLE

The results of an analysis of variance are often (though not invariably) displayed in a table similar to Table 12.12.1.

Table 12.12.1

Source	SS	df	MS	F
Treatments (between groups)	$\sum\limits_{j} \dfrac{\left(\sum_i y_{ij}\right)^2}{n_j} - \dfrac{\left(\sum_j \sum_i y_{ij}\right)^2}{N}$	$J - 1$	$\dfrac{\text{SS between}}{J - 1}$	$\dfrac{\text{MS between}}{\text{MS within}}$
Error (within groups)	$\sum\limits_{j} \sum\limits_{i} y_{ij}^2 - \sum\limits_{j} \dfrac{\left(\sum_i y_{ij}\right)^2}{n_j}$	$N - J$	$\dfrac{\text{SS within}}{N - J}$	
Totals	$\sum\limits_{j} \sum\limits_{i} y_{ij}^2 - \dfrac{\left(\sum_j \sum_i y_{ij}\right)^2}{N}$	$N - 1$		

In practice, the column labeled SS in Table 12.12.1 contains the actual values of the sums of squares computed from the data. In the d.f. column appear the numbers of degrees of freedom associated with each sum of squares; these numbers of degrees of freedom must sum to $N - 1$. The MS column contains the values of the mean squares, each formed by dividing the sum of squares by its degrees of freedom. Finally, the F statistic is formed from the ratio of the mean square between groups to the mean square within groups.

The student does well to form the habit of arranging the results of an analysis of variance in this way. Not only is it a good way to display the results for maximum clarity, but it also forms a convenient device for organizing and remembering the computational steps.

12.13   An example

An experiment was carried out to study the effect of a small lesion introduced into a particular structure in a rat's brain on his ability to perform in a discrimination problem. The particular structure studied is bilaterally symmetric, so that the lesion could be introduced into the structure on the right side of the brain, the left side, both sides, or neither side (a control group). Four groups of randomly selected rats were formed, and given the various treatments. Originally the control group contained 7 rats and each of the experimental groups 14 rats, but due either to death or postoperative incapacity only the following numbers were actually observed in the discrimination situation. The experimenter assumed that the death or incapacity of a rat operated randomly over the treatments. The basic design of the experiment was then as follows:

Group

I	II	III	IV
no lesion	left	right	both left and right
(7 rats)	(11 rats)	(13 rats)	(9 rats)

After a period of postoperative recovery, each rat was given the same series of discrimination problems. The dependent variable score was the average number of trials it took each rat to learn the task to some criterion level.

The null hypothesis was that the four treatment populations of rats are identical in their average ability to learn this task:

$$H_0: \mu_1 = \mu_2 = \mu_3 = \mu_4,$$

as against the hypothesis that treatment differences exist:

$$H_1: \text{not } H_0.$$

The alpha level chosen for the experiment was .05. It was assumed that scores for each population of rats was normally distributed with the same variance. The data are shown in Table 12.13.1.

Table 12.13.1

	Group		
I	II	III	IV
20	24	20	27
18	22	22	35
26	25	30	18
19	25	27	24
26	20	22	28
24	21	24	32
26	34	28	16
—	18	21	18
159	32	23	25
	23	25	—
	22	18	223
	—	30	
	266	32	
		—	
		322	

Now the simple analysis of variance will be illustrated for these data. Given this listing of the raw scores according to treatment groups, we first square and sum over all individuals in all groups (step 2 above):

$$A = \sum_j \sum_i y_{ij}^2 = 20^2 + 18^2 + \cdots + 18^2 + 25^2 = 24424.$$

Next, the raw scores over all observations are summed, and the result squared (step 3):

$$B = \left(\sum_j \sum_i y_{ij} \right) = (20 + 18 + \cdots + 18 + 25) = 970$$

$$\frac{B^2}{N} = \frac{(970)^2}{40} = 23522.5.$$

Using steps 4 and 5 above we find

$$C = \sum_j \frac{\left(\sum_i y_{ij} \right)^2}{n_j} = \frac{(159)^2}{7} + \frac{(266)^2}{11} + \frac{(322)^2}{13} + \frac{(223)^2}{9} = 23545.1.$$

Then (steps 6, 7, and 8),

$$\text{SS total} = A - \frac{B^2}{N} = 24424 - 23522.5 = 901.5$$

$$\text{SS between} = C - \frac{B^2}{N} = 23545.1 - 23522.5 = 22.6$$

$$\text{SS within} = 901.5 - 22.6 = 878.9.$$

Steps 9, 10, and 11 are represented in the following summary table:

Source	SS	df	MS	F
Treatments (between groups)	22.6	3	7.5	$\dfrac{7.5}{24.4}$
Error (within groups)	878.9	36	24.4	
Totals	901.5	39		

Ordinarily, at this point, the obtained F ratio would be compared with the value shown in Table V for 3 and 36 degrees of freedom and the specified α level. For $\alpha = .05$, the required F is 2.84, with 40 degrees of freedom used as the value nearest to 36. However, this step is really not necessary for this example, since the obtained F value is less than one, and the null hypothesis cannot be rejected. There is not enough evidence to warrant the conclusion that mean differences, or effects, truly exist among these treatment populations.

12.14 The F test and the t test

When only two independent groups are being compared in the experiment, and a nondirectional alternative hypothesis is being considered, it makes no difference whether the analysis of variance or the t test shown in Section 10.15 is used. As noted in Section 11.16, the square of a variable distributed as t with $N - 2$ degrees of freedom will be distributed as F with 1 and $N - 2$ degrees of freedom. A simple analysis of variance for two groups always yields an F ratio that is the same as the *square* of the t ratio calculated as in 10.15 for the same data. If the obtained F value is significant for any α, the corresponding value of $t = \sqrt{F}$ will be significant at the same α level in a two-tailed test. However, if the alternative hypothesis is directional, the sign of the difference between the two means must be considered; in this situation if F is significant at the α level, the one-tailed t test will show significance at the $\alpha/2$ level, provided that the sign of the obtained difference is appropriate to the alternative hypothesis.

This direct parallel between the F test in the analysis of variance and the t test for a difference in means holds only for the case of two groups, with an important exception that will be discussed in Chapter 14. One is never really justified in carrying out all the $\binom{J}{2}$ different t tests for differences among J groups, and then regarding this as some kind of substitute for the analysis of variance. Such t tests carried out on all pairs of means must necessarily extract redundant, overlapping, information from the data, and as a result a complicated pattern of dependency must exist among the tests. Furthermore, the apparent levels of significance found from a set of such tests have neither a simple interpretation nor a simple connection with the hypothesis tested by the F test in the analysis of variance. Thus, for example, one might suppose that the hypothesis of equality among the J population means could be rejected if at least one of the $\binom{J}{2}$ t tests between means turned out significant at the α level. However, even if each one of these tests were independent of each of the others, it would still be true that the probability of at least one significant result by chance alone would be $1 - (1 - \alpha)^G$, where $G = \binom{J}{2}$; this probability is closer to $G\alpha$ than to α when α is small. Furthermore, the tests, as pointed out above, are not independent, and thus the exact probability of one or more significant such results is not easy to determine. All in all, there is very little to recommend such a multiple t test procedure.

Nevertheless, situations do arise when it is important for the experimenter to be able to compare certain means in the data, in order to answer particular questions about the experiment. Chapter 14 will be devoted to ways for making these individual comparisons among sets of means, either as a substitute for the analysis of variance, or as a supplement to a significant F test.

12.15   ANOTHER EXAMPLE OF A SIMPLE, ONE-WAY, ANALYSIS OF VARIANCE

An experimenter is interested in "level of aspiration" as the dependent variable in an experiment. He has developed an experimental task consisting of a difficult game apparently involving motor skill, yielding a numerical score that can be attached to a person's performance. But this appearance is deceptive: unknown to the subject, the game is actually under the control of the experimenter, so that each subject is made to obtain exactly the same score. After a fixed number of trials, during which the subject unknowingly receives the pre-assigned score, he is asked to predict what his score will be on the next group of trials. However, before he predicts, the subject is given "information" about how his score compares with some fictitious norm group. In one experimental condition the subject is told that his performance is *above average* for the norm group, in the second that he is *average*, and in the third that he is *below average*. There are thus three possible experimental treatments or "standings" that might be given to any subject.

The actual dependent variable score y is based on the report the subject makes about his anticipated performance in the next group of trials. Since each subject has obtained the same score, this anticipated score on the next set of trials is treated as equivalent to a level of aspiration that the subject has set for himself. Each subject is tested privately, and no communication is allowed between subjects until the entire experiment is completed.

A random sample of sixty men in college is drawn, and the subjects are assigned individually and at random to the three experimental groups. It is decided to test the hypothesis of no treatment effects at the .05 level.

Table 12.15.1

INDIVIDUAL ANTICIPATED SCORES BY GROUPS, ACCORDING
TO INFORMATION GIVEN ABOUT STANDING

	Above average	Average	Below average
	52	28	15
	48	35	14
	43	34	23
	50	32	21
	43	34	14
	44	27	20
	46	31	21
	46	27	16
	43	29	20
	49	25	14
	38	43	23
	42	34	25
	42	33	18
	35	42	26
	33	41	18
	38	37	26
	39	37	20
	34	40	19
	33	36	22
	34	35	17
Totals	832	680	392

The data are given in Table 12.15.1. Note that here, $J = 3$, and $n_j = 20$ for each group j. We calculate

$$\sum_j \sum_i y_{ij}^2 = 66872 = A$$

$$\sum_j \sum_i y_{ij} = 1904 = B$$

$$\sum_j \frac{\left(\sum_i y_{ij}\right)^2}{20} = \frac{(832)^2 + (680)^2 + (392)^2}{20} = 65414.4 = C.$$

$$\text{SS total} = A - \frac{B^2}{60} = 66872 - \frac{(1904)^2}{60} = 6451.7$$

$$\text{SS between} = C - \frac{B^2}{60} = 65414.4 - 60420.3 = 4994.1$$

$$\text{SS within} = A - C = 66872 - 65414.4 = 1457.6.$$

Finally, the analysis of variance is summarized in the following table:

Source	SS	df	MS	F
Between groups	4994.1	$3 - 1 = 2$	2497.1	97.5
Within groups	1457.6	$60 - 3 = 57$	25.6	
Totals	6451.7	$60 - 1 = 59$		

The hypothesis of no treatment effects can definitely be rejected, far beyond the .05 level chosen for α. For the .05 level, with 2 and 57 degrees of freedom, an F value of only about 3.2 is necessary, and the value for F found here is very much greater than this. One can feel very confident in asserting that the "standing" information given to the subject has some effect on aspiration level, as reflected in the score each subject predicts for himself.

12.16 THE IMPORTANCE OF THE ASSUMPTIONS IN THE FIXED EFFECTS MODEL

In the development of the fixed effects model for the analysis of variance a number of assumptions were made. These assumptions help to provide the theoretical justification for the analysis and the F test. On the other hand, it is sometimes necessary to analyze data when these assumptions clearly are not met; indeed, it seldom stands to reason that they are exactly true. In this section we will examine the consequences of the application of the analysis of variance and the F test when these assumptions are not met.

In the first place, note that the inferences made in this chapter are about *means*. The models to be described in the next chapter provide inferences about *variances*, and *the remarks made in this section apply only to the fixed effects model*.

The first assumption listed in Section 12.5 specifies a normal distribution of errors, e_{ij}, for any treatment population j. This is equivalent to the assumption that each population has a normal distribution of scores, Y_{ij}. What are the consequences for the conclusions reached from the analysis when this assumption is not true? It can be shown that, other things being equal, *inferences made about means that are valid in the case of normal populations are also valid even when the forms of the population distributions depart considerably from normal, provided that the n in each sample is relatively large*. This is really but another instance of the principle underlying the central limit theorem, discussed in Section 8.11. Consequently, we need not worry unduly about the normality assumption so long as we are dealing with relatively large samples. In circumstances where the assumption of normality appears more or less unreasonable, the experimenter might do well to take a somewhat larger number of observations than otherwise. The more severely the population distributions are thought to depart from normal form, the relatively larger should the n per sample be. In particular, when each population is supposed to have the same, nonnormal, form, the F test is relatively unaffected.

The second assumption listed in Section 12.5 states that the error variance, σ_e^2, must have the same value for all treatment populations. Ordinarily, other things being equal, *this assumption of homogeneous variances can be violated without serious risk, provided that the number of cases in each sample is the same*. On the other hand, *when different numbers of cases appear in the various samples, violation of the assumption of homogeneous variances can have very serious consequences for the validity of the final inference*. The moral is again plain: whenever possible, an experiment should be planned so that the number of cases in each experimental group is the same, unless the assumption of equal population variances is eminently reasonable in the experimental context.

The third assumption in Section 12.5 requires statistical independence among the error components, e_{ij}. The assumption of independent errors is most important for the justification of the F test in the analysis of variance, and, unfortunately, violations of this assumption have important consequences for the results of the analysis. *If this assumption is not met, very serious errors in inference can be made.* In general, great care should be taken to see that data treated by the fixed effects analysis of variance are based on independent observations, both within and across groups (that is, each observation in no way related to any of the other observations). This is most likely to present a problem in studies where repeated observations are made of the same experimental subjects, perhaps with each subject being observed under each of the experimental treatments. In some experiments of this sort there is good reason to believe that the performance of the subject on one occasion has a systematic effect on his subsequent performances under the same or another experimental condition. In the fixed effects model, such systematic connections or dependencies among observations amount to a lack of statistical independence among errors, in violation of the assumption. For this reason, some authors suggest that data based on repeated observations should never be treated under the fixed effects model for analysis of variance. However, this seems to be a point of experimental technique on which the statistician must tread very lightly. The circumstance that observations were repeated does not, ipso facto, imply that observations must be regarded as statistically dependent; a shrewd experimenter can sometimes get his subject into a stable state very early by pre-experimental warm-up techniques, and in such situations it may be quite reasonable to assume that repeated observations are statistically independent. In other kinds of experiments, still other grounds may exist for assuming independence of repeated observations. The point, as always, is that statistics is limited in its ability to legislate experimental practice. Statistical assumptions must not be turned into prohibitions against particular kinds of experiments, although these assumptions must be borne in mind by the experimenter exercising thought and care in matching the particular experimental situation with an appropriate form of analysis. More will be said about the problem of repeated observations in Chapter 13.

Some further comments are in order about the meaning of an obtained F ratio with a value less than 1.00. When the null hypothesis is true, we should expect the F ratio to have a value close to 1.00 (more precisely, a value given by $\nu_2/(\nu_2 - 2)$, where ν_2 symbolizes the degrees of freedom for the denominator). Nonetheless, by chance alone it is entirely possible to obtain an F ratio much less than 1.00, or even an F value of exactly zero, regardless of whether H_0 or H_1 is true. Although such values can occur by chance, the occurrence of a very small F ratio can also serve as a signal that the experimenter needs to think more carefully about the experimental situation itself. Very commonly, a close analysis of the experimental situation shows a systematic, but unanalyzed, factor to be in operation, resulting in a mean square within groups that reflects something other than error variation alone. The existence of such an uncontrolled but nonrandom factor is, of course, a failure of the assumptions, but before the experimenter appeals to chance or to some less obvious failure of assumptions, he might entertain the possibility that the experiment itself is open to suspicion.

Finally, a word about the general assumption embodied in the fixed effects model: each score is assumed to be a *sum*, consisting of a general mean, plus a treatment effect, plus an independent error component. The appealing simplicity of this model notwithstanding, many situations exist where we know that an additive model such as this is not realistic. For example, in some experimental problems it may be far more reasonable to suppose that random errors serve to multiply treatment effects, rather than add to them. In such instances, expert statistical advice should be sought. It is often possible to transform the original scores by, for example, taking their logarithms, so as to make the transformed scores correspond to the linear model. Such transformations should not be undertaken by the novice, however, without expert statistical guidance.

Other special problems arise when the observed scores are proportions, or rates, or time scores, or any of a wide variety of ratios. Once again, expert help should be sought in the choice of a transformation of the raw data making such scores correspond to the additive, fixed effects, model.

In summary, for ordinary numerical scores, regarded as a continuous random variable, the assumption of normality may safely be ignored when the n in each sample is fairly large, and the assumption of equal variances appears to be relatively unimportant when the number of observations in each sample is the same. The assumption of independence of errors cannot, however, be ignored without severe risk of false conclusions. Each of these principles applies only to inferences about means under Model I, and need not hold for the analysis of variance models to be discussed in the next chapter.

12.17 THE PROBLEM OF TESTING FOR HOMOGENEITY OF VARIANCE

Since the analysis of variance is based on the assumption of equal variances, it may seem quite sensible to carry out a test for homogeneous variances on the sample data and then use to the result of that test to decide if the analysis of variance is legitimate. Such tests for the homogeneity of several variances exist, and some statistics books advocate these procedures. However, the standard tests for equality of several variances are extremely sensitive to any departure from normality in the populations. The statistician says that these tests with outcomes that depend heavily on incidental assumptions are not "robust." It could easily turn out that one would refrain from carrying out the analysis of variance because variances were apparently unequal, when a test of equality of *means* would actually be quite justifiable. Consequently, a test for homogeneity of variance before the analysis of variance has rather limited practical utility, and modern opinion holds that the analysis of variance can and should be carried on without a preliminary test of variances, especially in situations where the number of cases in the various samples can be made equal (see Box, 1953, 1954).

On the other hand, sometimes there is direct interest in a comparison among a set of variances, rather than in a comparison among means. The experimenter may be trying to find out if different treatments produce different degrees of variability. In such instances, it is sometimes possible to carry out an analysis of variance procedure on the logarithms of the sample variances. The reader is referred to Scheffé (1959, p. 83) for the details of this test, although help should be sought in its application.

12.18   ESTIMATING THE STRENGTH OF A STATISTICAL RELATION FROM THE ONE-WAY ANALYSIS OF VARIANCE

The size of the effect α_j associated with any treatment j can always be estimated in the fixed effects model by taking

$$\text{est. } \alpha_j = M_j - M.$$

In the example of Section 12.15 the best available estimate of the effect of the "above average" information on the individual subject's score is

$$\text{est. } \alpha_1 = \frac{832}{20} - \frac{1904}{60} = 41.6 - 31.7 = 9.9.$$

In the same way, effects can be estimated for the other treatments.

However, it is useful to have some idea of the over-all strength of association between independent and dependent variables that a significant finding actually represents. The index ω^2 introduced in Section 10.20 can be extended to serve this purpose when there are several distinct treatment groups or "levels" of the independent variable.

Just as before, we would like the index ω^2 to represent the relative reduction in variance of Y afforded by the information that an observation belongs in independent variable category X. Over all possible (x, y) events, the variance of the marginal distribution of Y will be called, once again, σ_Y^2. The variance within any conditional distribution of Y for a fixed X category, $\sigma_{Y|X}^2$, will be assumed to equal σ_e^2, the variance within any treatment population.

Suppose that the probability of any observation unit's falling into the treatment category x_j is $p(x_j)$. Then

$$\sigma_Y^2 = E(Y_{ij} - \mu)^2 = E(Y_{ij} - \mu_j + \alpha_j)^2$$
$$= \sigma_e^2 + \sum_j \alpha_j^2 p(x_j). \qquad [12.18.1^*]$$

When each observation unit has equal probability of falling into each treatment category x_j, then $p(x_j) = 1/J$, and

$$\sigma_Y^2 = \sigma_e^2 + \sum_j \frac{\alpha_j^2}{J}. \qquad [12.18.2^*]$$

On the other hand, if each of the *relative* sample sizes in the various treatment groups is the same as the probability of observing a case in the corresponding population, so that

$$p(x_j) = \frac{n_j}{N},$$

then

$$\sigma_Y^2 = \sigma_e^2 + \sum_j \frac{n_j \alpha_j^2}{N}. \qquad [12.18.3^*]$$

In either circumstance, we can define

$$\omega^2 = \frac{\sigma_Y^2 - \sigma_e^2}{\sigma_Y^2}$$
$$= \frac{\displaystyle\sum_j \alpha_j^2 p(x_j)}{\sigma_e^2 + \displaystyle\sum_j \alpha_j^2 p(x_j)}. \qquad [12.18.4^*]$$

This population index should reflect how much knowing the particular X_j category represents in increased ability to predict the Y value for an observation. Notice that if α_j is zero for each j, then ω^2 must be zero.

In order to estimate the value of ω^2 from sample data, we recall from Sections 12.6 and 12.8 that

$$E(\text{MS between}) = \sigma_e^2 + \sum_j \frac{n_j \alpha_j^2}{J - 1}$$

and

$$E(\text{MS within}) = \sigma_e^2.$$

Therefore, when the probability $p(X_j) = n_j/N$, so that the proportional representation of cases in the J samples is the same as the proportions in the respective populations, a reasonable (though rough) estimate of ω^2 is given by

$$\text{est. } \omega^2 = \frac{\text{SS between} - (J - 1)\text{ MS within}}{\text{SS total} + \text{MS within}}. \qquad [12.18.5\dagger]$$

This estimate of ω^2 can also be written as

$$\text{est.}\omega^2 = \frac{F' - 1}{\dfrac{\nu_2 + 1}{\nu_1} + F'}.$$

Here, F' is the obtained ratio (MSB/MSW), which has a noncentral distribution when the null hypothesis is not true, and thus should be distinguished from the central F variable, which obtains when the null hypothesis is true.

A different, somewhat more sophisticated approach to the estimation of ω^2 is also possible. Bear in mind that when the null hypothesis is false, so that $0 < \omega^2$, the ratio F' follows the noncentral F distribution, with parameters ν_1, ν_2, and δ^2. Following Patnaik (1949), we may express the mean of such a noncentral F distribution by

$$E(F') = \left(\frac{\nu_2}{\nu_2 - 2}\right)\left(1 + \frac{\delta^2}{\nu_1}\right),$$

so that

$$\frac{\nu_1(\nu_2 - 2)}{\nu_2} E(F') - \nu_1 = \delta^2.$$

However, since, by our assumptions

$$\frac{\delta^2}{N} = \frac{\sum\limits_{j} \alpha^2 p(x_j)}{\sigma_e^2} = \frac{\sum\limits_{j} n_j \alpha^2}{N \sigma_e^2},$$

we may express the value of ω^2 as

$$\omega^2 = \frac{\delta^2}{N + \delta^2}.$$

It follows that if the expected value of F' is replaced by the sample value of (MSB/MSW) actually obtained, an estimate of ω^2 is given by

$$\text{est.}\omega^2 = \frac{\dfrac{(\nu_2 - 2)F'}{\nu_2} - 1}{\dfrac{\nu_2 + 1}{\nu_1} + \dfrac{(\nu_2 - 2)F'}{\nu_2}}.$$

For reasonably large numbers of degrees of freedom within groups, or ν_2, this estimate is virtually identical to that given by **12.18.5.**

It must be said that there is no universally approved way either of defining an index such as ω^2 for a population or of estimating such an index. A number of such indices and estimators are to be found in the literature (one, historically important index, η^2, will be discussed in Chapter 16). In terms of the usual standards for statistical estimators, very little is known about such statistics; it may well be that none of the available methods is very satisfactory. One of the reasons for this lack is the tremendous difficulty involved in dealing directly with noncentral F distributions, on which all such estimators must ultimately depend within Model I of the analysis of variance. Even so, there does seem to be a need for indices such as ω^2 and for ways of estimating such population indices from data. Perhaps in time a more solidly based set of procedures will become available. In the interim, we will continue to make do with these rough-and-ready versions.

It should be pointed out that the ω^2 introduced in Section 10.20 is actually identical with that of **12.18.5**. Here, $t^2 = F = $ MS between/MS within, and $J = 2$.

Estimates of the strength of the statistical relation can be negative, even though the population quantity symbolized must be greater than or equal to zero. When a negative estimate occurs, the population value is simply taken to be equal to zero. On the other hand, a significant F guarantees a nonnegative value for the estimate of ω^2.

For example, let us apply the estimate of ω^2 to the data of Section 12.15. Here, the result was very significant; how much does knowing the particular treatment group to which an individual belongs let us reduce our uncertainty about his Y score? For this example,

$$\text{est. } \omega^2 = \frac{4994.1 - (2)(25.6)}{6451.7 + 25.6}$$
$$= .76.$$

The estimate implies that this statistical association is *very* strong. The independent variable X (the norm information given) is estimated to account for 76 percent of the variance in the Y scores. Naturally, this example is fabricated, and it is rather rare to find evidence for statistical association this strong in psychological data. Nevertheless, the example does show how the estimate of ω^2 can reinforce the meaning of a significant finding: not only is there evidence for *some* association between independent and dependent variables; our rough estimate suggests that this association is very strong.

By way of contrast, suppose that another study employed 6 groups of 30 observations each, and that the analysis of variance turned out as follows:

Source	SS	df	MS	F
Between groups	1500	5	300	4.5
Within groups	11605.8	174	66.7	
Totals	13105.8	179		

For 5 and 174 degrees of freedom, this F value is significant just beyond the .01 level. Now we will see how much statistical association is apparently represented by this finding.

$$\text{est. } \omega^2 = \frac{1500 - 333.5}{13105.8 + 66.7} = .089.$$

In this instance, only between 8 and 9 percent of the variance in the dependent variable seems to be accounted for by the independent variable. This still may be enough to make the statistical association an important one from the experimenter's point of view, but the example shows, nevertheless, that a significant result need not correspond to a *very* strong association.

The comments in Section 10.21 apply equally well to the analysis of variance situation. A result that is both significant *and* that gives an estimate of relatively strong association is usually far more informative than a significant result taken alone. Furthermore, even though a result is not significant, estimating whether or not a fairly high degree of association may in fact be present gives a clue to the experimenter about the advisability of a possible increase in sample size, or a refinement in the experimental procedure, when and if this experiment is repeated.

12.19   Strength of association and the power of the F test

The rationale for taking the ratio of MS between to MS within as the test statistic makes sense only if each of these sample quantities has the *same* expectation. That is, both MS between and MS within are unbiased estimates of σ_e^2 *when* the null hypothesis is true, and under these circumstances, their ratio *is* distributed exactly as an F variable. On the other hand, when the null hypothesis is *not* true, the MS between has a different expectation, as we have seen, and sample ratios of MS between to MS within are *not* distributed as F.

Instead, when the null hypothesis is not true, the sample ratio actually computed has a distribution known as **noncentral F**. That is, let us designate the ratio $F' = (\text{MSB/MSW})$, under conditions in which the null hypothesis is false, as a noncentral F variable. The distribution of F' depends upon three parameters, ν_1, ν_2, and δ^2, the so-called noncentrality parameter. This additional parameter is defined as

$$\delta^2 = \sum_j \frac{n_j \alpha_j^2}{\sigma_e^2}$$

(given normal population distributions, each with variance σ_e^2, and J independent samples of size n_j, respectively).

The distribution of F', the noncentral F variable, thus depends upon the value of δ^2, which in turn depends on the sizes of the squared effects relative to the error variance. Unless one has some idea of the value of δ^2, he has no way to calculate the power of the F test against the alternative which that value of δ^2 represents. The particular member of the family of noncentral F distributions that must be used to find the power against a given alternative depends upon the specification of the value of δ^2. In this light it is not too surprising that most psychological reports involving the analysis of variance have little or nothing to say about the power of the test, since it is rare to find much consideration given to the sizes of the squared effects relative to σ_e^2.

Fairly extensive tables of noncentral F exist, and there are useful charts for the power of the F tests in texts such as that of Scheffé (1959). A major use of these charts is in finding the sample size, n per group, necessary to ensure a certain amount of power against a particular alternative, expressed in terms of δ^2. If the experimenter has decided upon a value of δ^2 that represents a situation he wants or needs to detect as significant, these charts are relatively easy to use. As we shall see later in this section, the power of the F test may also be approximated in other ways.

There is a very close connection between the strength of association in a population, as defined by our measure ω^2, and the noncentrality parameter δ^2 in a noncentral F distribution. That is, we can write ω^2 in terms of δ^2 as follows:

$$\omega^2 = \frac{\delta^2}{N + \delta^2}.$$

Similarly,

$$\delta^2 = \frac{N\omega^2}{1 - \omega^2}.$$

Now when the null hypothesis is not true, so that $\omega^2 > 0$ and $\delta^2 > 0$, the F ratio, MS between/MS within, actually follows a noncentral F distribution with parameters ν_1, ν_2, and δ^2. However, because of the connection between δ^2 and ω^2, we can just as well think of the noncentral distribution of F as depending upon the true degree of association that exists in the population. In some instances it may be much easier to think about the state of affairs in the population in terms of the strength of association than in terms of a weighted sum of squared effects.

Incidentally, somewhat different versions of δ^2 and ω^2 as sometimes defined may be useful in discussions of the power of the F test. For example, some tables of the power of the F test deal with a parameter φ rather than the parameter δ^2, where φ is something like the standard deviation of the weighted effects:

$$\varphi = \sqrt{\frac{\delta^2}{J}} = \sqrt{\frac{\Sigma n_j \alpha_j^2}{J\sigma_e^2}}.$$

Charts of the power of the F test developed by E. S. Pearson and H. O. Hartley, and reproduced in Scheffé (1959), are given in terms of ν_1, ν_2, and φ.

Charts for the power of the F test require a considerable amount of space, and since we will not be making extensive use of power calculations in the following chapters, these charts will not be included here. As noted above, such charts are available in the volume by Scheffé and elsewhere. However, a somewhat rough-and-ready estimate of the power of an F test against a particular value of ω^2 is relatively easy to obtain. This method will be outlined briefly below.

Consider an F test to be undertaken with ν_1 degrees of freedom in the numerator and ν_2 degrees of freedom in the denominator. For the α level of significance, let the critical value of F required be designated by F_α. This value is, of course, found from Table V in Appendix C. It is desired to estimate the power of this test against a specific value of ω^2. We proceed as follows: Let

$$A = \nu_1 + \frac{(\delta^2)^2}{\nu_1 + 2\delta^2}, \qquad \text{where } \delta^2 = \frac{N\omega^2}{1 - \omega^2},$$

$$B = \frac{F_\alpha}{F_\alpha + \frac{(\nu_2/\nu_1)(\nu_1 + 2\delta^2)}{\nu_1 + \delta^2}}.$$

Then we take the cumulative binomial probability, found from the binomial probabilities of Table II, and calculate

$$\text{power} = \text{cum.pr.}_{\text{bin}}[\tfrac{1}{2}A - 1; p = B, N' = \tfrac{1}{2}(A + \nu_2) - 1] \qquad \text{if } B \leq \tfrac{1}{2},$$

and

$$\text{power} = 1 - \text{cum.pr.}_{\text{bin}}[\tfrac{1}{2}\nu_2 - 1; q = 1 - B, N' = \tfrac{1}{2}(A + \nu_2) - 1] \qquad \text{if } B > \tfrac{1}{2}.$$

When the value of $N' = \tfrac{1}{2}(A + \nu_2) - 1$ is less than or equal to 20, Table II may be used to calculate these power values quite readily.

      When cumulative binomial probabilities are difficult to obtain, a somewhat less satisfactory, but still adequate approximation may be had by calculating A and B as above, and also calculating $C = A + \nu_2 - 2$ and $D = BC$. Then if $B \leq \tfrac{1}{2}$, look up the value of D in the chi-square table (Table IV, Appendix C). Look under a degrees of freedom equal to a whole number *just below* the value of $A + 2$. The value of D should fall between two columns, each headed by a Q value. The power of the test is estimated to lie between these two Q values.

      If $B > \tfrac{1}{2}$, then take $D = (1 - B)C$ and proceed as before, using ν_2 as the number of degrees of freedom in the chi-square table. After finding the two Q values, take $1 - Q$ for each. The power of the test is estimated to fall between the two $1 - Q$ values.

      This method of approximating the power of an F test is basically due to Patanik (1949), who showed that a noncentral F distribution (which obtains when the null hypothesis is not true) can be approximated by the usual central F distribution under certain conditions. The use of the binomial distribution and of the chi-square distribution depends upon the close connections between F, beta, the binomial, and the gamma distributions.

      As an example of the approximation of the power of an F test through use of binomial probabilities, consider the following. An F test is to be carried out where $\nu_1 = 6$ and $\nu_2 = 21$. The level chosen for significance is .01. This means that an obtained F value of 3.81 or more is required in order that the null hypothesis may be rejected. The experimenter is interested in the power of this test in the situation where the true value of ω^2 is .55. Since $N = 28$, this means that

$$\delta^2 = \frac{N\omega^2}{1 - \omega^2} = \frac{28(.55)}{.45} = 34.22.$$

We take

$$A = 6 + \frac{(34.22)^2}{6 + 2(34.22)} = 21.77,$$

$$B = \frac{3.81}{3.81 + \dfrac{21(6 + 68.44)}{6(6 + 34.22)}} = .3704.$$

Since B is less than $1/2$, we find the power from the following cumulative binomial probability:

$$\text{power} = \text{cum.pr.}_{\text{bin}}[\tfrac{1}{2}(21.77) - 1; p = .3704, N' = \tfrac{1}{2}(21.77 + 21) - 1].$$

That is, we look in Table II for $\frac{1}{2}(21.77) - 1$ or about 9 successes when p is as close as possible to .37 and N' is about 20. Adding up the probabilities up to and including 9, we find that the cumulative probability is .8782. This is the power of the test, as approximated by this method. (Actual power charts for the F test show that the power in this case is about .88.)

As a second example, suppose that $\nu_1 = 3$ and $\nu_2 = 4$, and that the α level in the F test is to be .05. Then $F_\alpha = 6.59$. We wish to approximate the power of the test when the true value of $\omega^2 = .10$. This means that the value of δ^2 is about .9. We compute

$$A = 3 + (.9)^2/(3 + 1.8) = 3.17,$$

$$B = 6.59 \bigg/ \left[6.59 + \frac{4(3 + 1.8)}{3(3 + .9)} \right] = .80.$$

Since B is greater than $\frac{1}{2}$, we must find the cumulative binomial probability most closely corresponding to

$$\text{cum.pr.}_{\text{bin}}[\tfrac{1}{2}(4) - 1; q = 1 - .80, N' = \tfrac{1}{2}(3.17 + 4) - 1]$$

In Table II, the binomial distribution for $q = .20$ and $N' = 3$ seems to be that corresponding most closely to the required distribution. Then after finding the cumulative probability for 1 or fewer successes given 3 trials, with probability .20, we subtract this cumulative probability from 1.00. This gives

$$\text{power} = 1 - .8960 = .104.$$

(Power tables for F show that for $\delta^2 = .9$, and for these numbers of degrees of freedom, the power is about .10.)

Although in very refined work the standard power charts for F should probably still be used, the method just demonstrated can be quite useful, particularly in decisions about the sample size required in order to reach a particular level of power against a particular value of ω^2. If ω^2 is specified, along with a fixed number of groups J, so that $N = nJ$, then it is a fairly simple matter to explore the effects upon power of different choices of n. Notice that regardless of the method used to determine power, the experimenter must still decide something about ω^2, or about the relative magnitudes of squared effects that he *wants* to detect as significant if such actually do exist. As always, a considerable amount of prior knowledge and thought must go into the design of an experiment if one is going to take advantage of the full resources of statistical analysis.

12.20   THE TWO-WAY ANALYSIS OF VARIANCE WITH REPLICATION

We now have enough background to discuss how the simple one-way analysis of variance can be extended to cover a more complicated experimental set-up in which there are two different sets of treatments. Suppose, once again, that an experimenter is interested in level of aspiration as the dependent variable, just as in Section 12.15. Just as before, the experimental game is under the control of the experimenter, so that each subject actually obtains the same score. After a fixed number of trials, during which the subject gets the preassigned score, he is asked to predict what his score will be on the next group of trials. Before he predicts, the subject is given "information" about how his score compares with some norm group. In one experimental condition he is told that his performance is above average for the norm group, in the second condition he is told that his score is average, and in the third condition he is told that his score is below average for the norm group. Once again, there are three experimental treatments in terms of "standings": "above average," "average," and "below average." However, unlike the previous example, this time *two different norm groups* are used in the information given subjects. One half of the subjects are told that they are being compared with college men, and the other half are told that they are being compared with professional athletes. Hence, there are two additional experimental treatments: "college norms," and "professional athlete norms."

In this example interest is focused on two distinct experimental factors: the information about standing that the subject is given, and the norm group used for comparison. Either or both of these experimental factors might possibly influence the value that a subject anticipates as his next score (his level of aspiration). A random sample of sixty male college students is selected and assigned at random to each of the six possible treatment *combinations*. This can be diagrammed as follows:

Table 12.20.1

ASSIGNMENT OF SUBJECTS TO TREATMENT COMBINATIONS,
BY NUMBERS IN EACH COMBINATION

NORM INFORMATION	Athlete	10	10	10
	College	10	10	10
		Above	Average	Below
			STANDING	

This experiment represents an instance where two different sets of experimental treatments are crossed, or given in every combination. Here there are six distinct sample groups, each group being given a particular combination of two kinds of treatments. *Three* questions are of interest:

1. Are there systematic effects due to experimental set alone (averaged over the norm group)?
2. Are there systematic effects due to norm information alone (averaged over the experimental set)?
3. Are there systematic effects due neither to norm information alone, nor to experimental set alone, but attributable only to the *combination* of a particular norm group with a particular experimental set?

Notice that this study could be viewed as two separate experiments carried out on the same set of subjects: There are three groups of twenty subjects each, differing only in experimental set or "standing"; exactly the same set of "norm-group" conditions are represented in each experimental "standing" group. On the other hand, looking at Table 12.20.1 by rows rather than by columns, we see that there are also two samples of thirty subjects each, differing systematically by "norm-group." Each "norm-group" sample has exactly the same representation of the other experimental conditions within it.

Question 3 above cannot, however, be answered by the comparison of "norm-groups" alone or by the comparison of experimental "set" groups alone. This is a question of "interaction," *the unique effects of combinations of treatments.* This is the important new feature of the two-way (or higher) analysis of variance: we will be able to examine *main effects* of the separate experimental variables or factors just as in the one-way analysis, as well as *interaction effects*, differences apparently due only to the unique combinations of treatments.

12.21 Crossing and nesting in experiments

Before we go further into the topic of experiments where two experimental factors are to be studied, two convenient terms should be introduced. Consider the experiment described in the preceding section: two experimental factors are under study, the first being the "standing information" given the subject, and the second being the "norm information." Each subject is to get a combination of a category of "standing information" and a category of "norm information." In experiments such as this, where two experimental factors are present and each category or level of one factor occurs with each level of the other, the two factors are said to be **completely crossed.** All possible combinations of levels of two (or more) experimental factors occur in a completely crossed experiment. Furthermore, if all possible combinations of factor levels occur an equal number of times, the experiment is said to be **balanced.** Note that the experiment described above is thus completely crossed and balanced.

On the other hand, in some experiments categories or levels of one factor occur only *within* levels of another factor. Thus, we might be comparing three different teaching methods, and another factor of interest might be the particular classrooms in which the methods were tried. It is impracticable to cross the factors of methods and classrooms, and so it is decided to apply the same method to each of two different classrooms, for a total of six classrooms, two per method. Here, the factor of "classroom" is said to be **nested** within the factor of "teaching method." A particular classroom occurs in the experiment only in association with one particular method. In a nested experiment such as this, any comparison of methods that we make is also a comparison of sets of classrooms; the only meaningful evidence for the possible effects of classrooms, over and above the effects of the methods, must come from comparison of classrooms *within* the particular methods. Many experiments calling for a simple one-way analysis of variance can be thought of as nested designs, where a factor corresponding to "individual subjects" is nested within the main treatments factor of the experiment. For example, in the experiment of Section 12.13 a factor of "particular rats" can be thought of as nested within the factor corresponding to "type of lesion."

The remainder of this chapter will be devoted to experiments where there are two completely crossed and balanced factors, and then a different random sample of subjects is assigned to each particular *combination* of the other two factors: that is, each combination of treatments is administered to a separate and distinct set of randomly selected subjects. However, this is but one of the simplest kinds of experimental designs, and it is entirely possible to design and analyze a wide variety of experiments where different factors are either crossed or nested.

12.22   THE FIXED EFFECTS MODEL IN THE TWO-WAY ANALYSIS

Just as in the fixed effects model for the one-way analysis of variance, in the corresponding model for a two-way analysis it is assumed that each observed dependent variable score is a sum of systematic effects associated with experimental treatments, plus random error:

$$Y_{ijk} = \mu + \alpha_j + \beta_k + \gamma_{jk} + e_{ijk} \qquad [12.22.1*]$$

where α_j is the effect of treatment j

$$\alpha_j = \mu_j - \mu, \qquad [12.22.2*]$$

and β_k is the effect of treatment k

$$\beta_k = \mu_k - \mu. \qquad [12.22.3*]$$

Here, μ_j is the mean of the population given treatment j and pooled over all of the K different treatments k, and μ_k is the mean of the population given treatment k and pooled over all of the J different treatments j. The grand mean μ is the mean of the population formed by pooling all of the different populations given the possible treatment combinations j and k.

The new feature of equation 12.22.1 is the inclusion of a term representing the interaction effect, γ_{jk} (small Greek gamma). The interaction effect is the experimental effect created by the combination of treatments j and k over and above any effects associated with treatments j and k considered separately:

$$\begin{aligned} \gamma_{jk} &= \mu_{jk} - \mu - \alpha_j - \beta_k \\ &= \mu_{jk} - \mu_j - \mu_k + \mu. \end{aligned} \qquad [12.22.4*]$$

The interaction effect γ_{jk} is thus equal to the mean of the population given both of the treatments j and k, minus the mean of the treatment population j, minus the mean of the population given treatment k, plus the grand mean.

Some intuition about the meaning of interaction effects may be gained by examining another set of artificial data. In the experiment outlined in Section 12.20, suppose that the only effect (that is nonzero effect) is associated with experimental "standing." Furthermore, suppose that the value of each error e_{ijk} is zero. Then the *means* of each of the six groups given different treatment combinations should look something like this:

Table 12.22.1

	College	28	33	35
NORM GROUP				
	Athlete	28	33	35

	Above	Average	Below	$M = 32$
		STANDING		

Here the effect associated with treatment "above" is $28 - 32 = -4$, the effect of the treatment "average" is $33 - 32 = 1$, and the effect of the treatment "below" is $35 - 32 = 3$. Note that the columns of the table differ from each other, but that the rows within each column show identical values. Each score is simply $Y_{ijk} = 32 + \alpha_j$.

On the other hand, it might turn out that effects exist only for the norm groups, so that no effects are associated with the column treatments or with the row by column interaction. Once again, if the data were errorless, we should observe something like:

Table 12.22.2

	College	34	34	34
NORM GROUP				
	Athlete	30	30	30

	Above	Average	Below
		STANDING	

Here the rows differ from each other, but the values for the columns within each row are identical. The effect of the first row is $34 - 32 = 2$, and that of the second row is $30 - 32 = -2$. Each score fits the rule

$$y_{ijk} = 32 + \beta_k.$$

Now suppose that there were *both* column and row effects, but that no interaction or error effects exist. In this case

$$y_{ijk} = 32 + \alpha_j + \beta_k,$$

so that the means for the groups each represent a sum of effects like this

Table 12.22.3

	College	$32 + 2 - 4$ $= 30$	$32 + 2 + 1$ $= 35$	$32 + 2 + 3$ $= 37$
NORM GROUP				
	Athlete	$32 - 2 - 4$ $= 26$	$32 - 2 + 1$ $= 31$	$32 - 2 + 3$ $= 33$

	Above	Average	Below
		STANDING	

In this instance the six treatment combinations yield means differing across the different **cells** of the table. However, the effect of a combination, $\mu_{jk} - \mu$, associated with cell jk, is exactly equal to the effect associated its row, β_k, plus the effect associated with its column α_j, so that $y_{ijk} = \mu + \alpha_j + \beta_k$. **When there is no interaction, effects are said to be additive, since the effect of a combination is the sum of the effects of the treatments involved.** Notice that the difference between a particular pair of columns is the same over the rows, and that any row difference is constant over columns, within Table 12.22.3 representing the condition of no interaction.

Finally, in the table below, interaction effects have also been added to the data of Table 12.22.3. Once again, assume that these data are errorless:

Table 12.22.4

		Above	Average	Below
NORM GROUP	College	$32 + 2 - 4 - 2$ $= 28$	$32 + 2 + 1 + 6$ $= 41$	$32 + 2 + 3 - 4$ $= 33$
	Athlete	$32 - 2 - 4 + 2$ $= 28$	$32 - 2 + 1 - 6$ $= 25$	$32 - 2 + 3 + 4$ $= 37$

STANDING

The effect associated with a combination of treatments is now no longer the simple sum of the effects of its row and its column, an indication that interaction effects are present. **Notice that columns are different in different ways within rows, and vice-versa, when interaction is present.**

Naturally, for any real data there will be random error as well. This implies that the problem is now threefold: we must find out (1) if there are effects of the treatments represented by the columns, (2) if there are effects of treatments represented by the rows, and (3) if there are effects which are attributable neither to rows (irrespective of columns) nor columns (irrespective of rows) but rather to interaction. (As an exercise, show that if the grand mean is subtracted from each cell, then the resulting row mean from its row, and the resulting column mean from its column, so that row, column, and total means are zero, the entries in the cells consist of interaction effects + error.)

12.23 THE IMPORTANCE OF INTERACTION EFFECTS

The presence or absence of interaction effects, as inferred from the F test for interaction, can have a very important bearing on how one interprets and uses the results of an experiment. When the presence of column effects is inferred, this implies that the populations represented by the columns have means that differ; the amount and direction of difference between sample means for any pair of columns provides an estimate of the corresponding difference between population means. When interaction effects are absent, differences among the means representing different column-treatment populations have the same size and sign, even though the populations are conceived as receiving still another treatment represented by one of the rows. This suggests that the difference between a pair of column means in the data is our best bet about the difference to be expected between a pair of individuals given different column treatments, quite irrespective of the particular row treatment that might have been administered. On the other hand, when interaction effects exist, *varying differences* exist between the means of populations representing different column treatments, depending on the particular row treatment that is applied. It is still true that differences between column means in the data provide an estimate of the difference we should expect between individuals given the particular treatments, but only on the average, over all of the different row treatments that might have been applied. When a particular row treatment is specified, it may be that quite another size and direction of difference should be expected between individuals given different column treatments. In short, interaction effects lead to a qualification on the estimate one makes of the differences attributable to different treatments; when interaction effects exist, the best estimate one can make of a difference attributable to one factor depends on the particular level of the other factor.

For example, suppose that an experimenter is comparing two methods of instruction in golf. Let us represent these two methods as the column treatments in the data table. The other factor considered is the sex of the student; the study employs a group of 50 boys and a group of 50 girls, with 25 subjects in each group taught by Method I and the remainder by Method II. After a fixed period of instruction by one or the other method, each member of the sample is given a proficiency test. Suppose that the sample means for the four subgroups turn out as follows:

	METHOD		
	I	II	
Girls	55	65	60
Boys	75	45	60
	65	55	

For a small enough estimated error variance, such data would lead to the conclusion that no difference exists between boys and girls in terms of performance on the proficiency test, but that *both column effects and interaction effects do exist.* Now suppose that the experimenter wants to decide which method to use for the instruction of an individual student. If *low* scores indicate good performance, but he doesn't know or doesn't wish to specify the sex of the student, then Method II clearly is called for, since the experimenter's best estimate is that Method I gives a higher mean than Method II over both sexes. However, suppose that he knows that the individual to be instructed is a *girl:* in this case, the experimenter does *much* better to choose Method I, since he has evidence that *within the population of girls, mean II is higher than mean I.*

Significant interaction effects usually reflect a situation very like this: over-all estimates of differences due to one factor are fine as predictors of average differences over *all possible levels of the other factor*, but it will not necessarily be true that these are good estimates of the differences to be expected when information about the category on the other factor is given. Significant interaction serves as a warning: treatment differences *do* exist, but to specify exactly *how* the treatments differ, and especially to make good individual predictions, one must look *within* levels of the *other* factor. The presence of interaction effects is a signal that in any predictive use of the experimental results, effects attributed to particular treatments representing one factor are best qualified by specifying the level of the *other* factor. This is extremely important if one is going to try to use estimated effects in forecasting the result of applying a treatment to an individual; when interaction effects are present, the best forecast can be made only if the individual's status on *both* factors is known.

In the language of Chapter 1, tests for interaction permit one to decide if variable Z is relevant to the relationship between X and Y. If there is no interaction, then the relationship between X and Y does not depend upon the specification of Z, and that between Z and Y does not depend upon X. The presence of interaction effects suggests the contrary: a functional relation between X and Y will not appear unless Z is controlled. The presence only of interaction suggests that the only nontrivial functional relationship between X or Z and Y involves (x, z) *pairs* in relation to Y, or else control over variable Z. In a very real sense, the study of interaction sheds light on the relevancy of different variables in the search for predictive functional relationships.

For these reasons, the presence of interaction effects can be most important to the interpretation of the experiment. The estimated effects of any given treatment are not "best bets" about any randomly selected individual when interaction effects are present; the best prediction entails knowing the other treatment or treatments administered.

Although it is necessary to consider possible interaction effects even in fairly simple experiments, the subject of interaction and of the interpretation that should be given to significant tests for interaction is neither elementary nor fully explored. To a very large extent, the presence or absence of interactions in an experiment is governed by the scale of measurement used for the dependent variable. Thus, in terms of the original scale of measurement interaction may be present, but if, for example, the values are transformed into their respective logarithms, interaction effects may vanish. It is clear that in many circumstances evidence for interaction reflects not so much a state of nature as our own inability to find the proper measurement scales for the phenomena we study. Since simple additive models are so much more tractable theoretically and practically than models including all the qualifications introduced by interaction, it is often desirable to transform the original data to eliminate interactions. Such considerations are, however, far beyond our limited scope.

Interaction effects can be studied separately only in a two-way (or higher) analysis of variance with crossed factors, where the experiment is carried out **with replication.** Furthermore, the procedures we will develop in the remainder of this chapter apply only to the situation where the experimental design is **orthogonal.** We need to have a look at what these terms "with replication" and "orthogonal" imply about the way the experiment is designed.

12.24 THE IDEA OF REPLICATION

The discussion of the two-way analysis of variance will be limited here to replicated experiments. For our purposes this means that **within each treatment combination there are at least two independent observations made under identical experimental circumstances.** The requirement that the experiment be replicated is introduced here so that an error sum of squares will be available, permitting the study of tests both for treatment effects and for interaction. If there were only one observation for each treatment combination, we would not be able to test separately for interaction effects, since in this situation there is no direct way to estimate error variance apart from interaction effects. Occasionally, experiments are carried out where only one observation is made per treatment combination; under the fixed effects model this makes it necessary to know or to assume that no interaction effects exist if a test for main treatment effects is to be carried out. This assumption is often very questionable, and most circumstances requiring a nonreplicated experiment will fit into one of the models to be discussed in Chapter 13. For this reason, our discussion for the fixed effect model will be confined to replicated experiments.

12.25 ORTHOGONAL DESIGNS FOR EXPERIMENTS

An orthogonal design for an experiment can be defined as a way of collecting observations that will permit one to estimate and test for the various treatment effects and for interaction effects separately. The potential information in the experiment can be "pulled apart" for study in an orthogonal design. Any experimental layout can be regarded as an orthogonal design provided that: (1) the observations within a given treatment *combination* are sampled at random and independently from a normal population, and (2) the number of observations in *each possible combination* of treatments is the same. Thus, the usual procedure in setting up an experiment to be analyzed by the two-way (or higher) analysis of variance is to assign subjects at random and independently to each combination of treatments so as to have an equal number in each combination. This means that in a table representing the experimental groups, such as Table 12.20.1, the cells of the table all contain the same number of observations. Let us call the number of "row" treatments R, and the number of "column" treatments C. For experiments of this sort where each cell in an $R \times C$ data-table contains the same number n of cases, each *row* will contain Cn cases, and each *column* Rn cases. If at all possible, experiments should be set up in this way, not only to insure orthogonality, but also to minimize the effect of nonhomogeneous population variances should they exist (see Section 12.17).

It is also possible to design two-factor experiments that will be orthogonal even though the numbers of cases in the various cells differ, provided that **proportionality** holds within the cells of any given row or column. This means that for any treatment combination jk,

$$n_{jk} = \frac{n_j n_k}{N},$$

where n_{jk} is the number of observations in the combination j and k, n_j is the total number of observations in treatment j, n_k is the total number of observations in treatment k, and N is the total number of observations in all. However, in this circumstance, the ordinary computational procedures for the analysis of variance must be altered somewhat. For the sake of simplicity, the discussion to follow is restricted to the case of *equal numbers of observations* in the cells of a $R \times C$ table. The beginning student is advised to seek expert help when his data do not conform to this pattern, especially when unequal cell numbers appear because of "lost" or otherwise missing data.

12.26   Sums of Effects in the Fixed Model

Just as in the simple one-way analysis of variance under the fixed effects model, in the two-way situation it is assumed that the experimental treatments and treatment combinations are fixed, and that the only inferences to be made are about those treatments and treatment combinations actually represented in the experiment.

The following equalities are defined to be true of the effects:

$$\sum_j \alpha_j = 0$$

$$\sum_k \beta_k = 0$$

$$\sum_j \gamma_{jk} = 0$$

$$\sum_k \gamma_{jk} = 0.$$

The effects, being deviations from a grand mean μ, sum to zero over all the different "levels" of a given kind of treatment. However

$$\sum_j \alpha_j^2 > 0$$

unless $\alpha_j = 0$ for each j;

$$\sum_k \beta_k^2 > 0$$

unless $\beta_k = 0$ for each k;

$$\sum_j \sum_k \gamma_{jk}^2 > 0$$

unless $\gamma_{jk} = 0$ for each and every combination j,k.

12.27 THE PARTITION OF THE SUMS OF SQUARES FOR THE TWO-WAY ANALYSIS OF VARIANCE

Once more we start off by looking at how the total sum of squares can be partitioned for a set of data. For any individual i in any treatment combination jk, the deviation of the score y_{ijk} from the sample grand mean M can be written as

$$y_{ijk} - M = (y_{ijk} - M_{jk}) + (M_j - M) + (M_k - M) \\ + (M_{jk} - M_j - M_k + M). \quad [12.27.1]$$

If the deviation for each score is squared, and these squares are summed over all individuals in all combinations j and k, we have

$$\sum_j \sum_k \sum_i (y_{ijk} - M)^2 = \sum_j \sum_k \sum_i (y_{ijk} - M_{jk})^2 \\ + \sum_j n_j(M_j - M)^2 + \sum_k n_k(M_k - M)^2 \\ + \sum_j \sum_k n_{jk}(M_{jk} - M_j - M_k + M)^2. \quad [12.27.2]$$

(The algebraic argument for this statement is almost exactly the same as in Section 12.4 and will not be repeated here; the student, however, may find it profitable to try to derive this for himself.)

Now let us examine the various individual terms on the right of expression 12.27.2. We call

$$\sum_j \sum_k \sum_i (y_{ijk} - M_{jk})^2 = \text{SS error} \quad [12.27.3]$$

the **sum of squares for error,** since it is based on deviations from a cell mean for individuals treated in exactly the same way; the only possible contribution to this sum of squares should be error variation.

Next, consider

$$\sum_j n_j(M_j - M)^2 = \text{SS columns,} \quad [12.27.4]$$

which is the **sum of squares between columns.** Here the deviations of column treatment means from the grand mean make up this sum of squares. This sum of squares reflects two things: the treatment effects of the columns *and* error. Notice that this sum of squares is identical to the sum of squares between groups found in the one-way analysis, if the different experimental groups are regarded as columns in the table.

The third term is

$$\sum_k n_k (M_k - M)^2 = \text{SS rows},$$ [12.27.5]

which is the **sum of squares between rows.** It is based upon deviations of the row means from the grand mean, and thus reflects both row-treatment effects and error. This is the same as the sum of squares between groups if data were regarded as coming only from experimental groups corresponding to the rows.

Finally, the fourth term

$$\sum_j \sum_k n_{jk} (M_{jk} - M_j - M_k + M)^2 = \text{SS interaction}$$ [12.27.6]

the **sum of squares for interaction.** This sum of squares involves only *interaction effects and error.*

The partition of the sum of squares for a two-way analysis can be written in the following schematic form:

$$\text{SS total} = \text{SS error} + \text{SS columns} + \text{SS rows} + \text{SS interaction}. \quad [12.27.7\dagger]$$

Whereas in the one-way analysis the total sum of squares can be broken into only two parts, a sum of squares between groups and a sum of squares within groups (error), in the two-way analysis with replication the total sum of squares can be broken into *four* distinct parts. The principle generalizes to experimental layouts with any number of treatment and treatment combinations, but we shall stop with the two-way situation.

12.28   Assumptions in the two-way fixed effects model

Before we turn to an examination of the sampling distribution of the various mean squares, the assumptions we must make to determine these sampling distributions will be stated:

1. The errors e_{ijk} are normally distributed with expectation of zero for each treatment-combination population jk.
2. The errors e_{ijk} have exactly the same variance σ_e^2 for each treatment-combination population.
3. The errors e_{ijk} are independent, both within each treatment combination and across treatment combinations.

You will note that these are essentially the same assumptions made for the one-way model, except that now we deal with treatment-combination populations, the entire set of potential observations to be made under any combination of treatments.

12.29   THE MEAN SQUARES AND THEIR EXPECTATIONS

We begin by finding the expectation of the *error* sum of squares: substituting from **12.27.3,** we have

$$E \sum_j \sum_k \sum_i (Y_{ijk} - M_{jk})^2 = E \sum_j \sum_k \sum_i (\mu + \alpha_j + \beta_k + \gamma_{jk}$$
$$+ e_{ijk} - \mu - \alpha_j - \beta_k - \gamma_{jk} - M_{ejk})^2$$
$$= E \sum_j \sum_k \sum_i (e_{ijk} - M_{ejk})^2. \qquad [12.29.1]$$

By rule 5 of Appendix B, and by assumption 2, Section 12.28,

$$E \sum_j \sum_k \sum_i (Y_{ijk} - M_{jk})^2 = \sum_j \sum_k E \sum_i (e_{ijk} - M_{ejk})^2$$
$$= \sum_j \sum_k (n_{jk} - 1)\sigma_e^2$$
$$= RC(n - 1)\sigma_e^2. \qquad [12.29.2]$$

(Since the number n_{jk} in each cell is assumed to be the same, hereafter n will be written to signify this number. Remember that the R represents the number of rows and the C the number of columns; the number of observations in any row is Cn, and in any column is Rn.) Let

$$\text{MS error} = \frac{\text{SS error}}{RC(n - 1)}. \qquad [12.29.3\dagger]$$

Then, by **12.29.2** above,

$$E(\text{MS error}) = \frac{E(\text{SS error})}{RC(n - 1)} = \frac{RC(n - 1)\sigma_e^2}{RC(n - 1)} = \sigma_e^2. \qquad [12.29.4*]$$

The expected value of the mean square error is simply the error variance σ_e^2.

Now look at the mean square between columns. Since there are C columns, the mean square between columns is found from

$$\text{MS columns} = \frac{\text{SS columns}}{C - 1} \qquad [12.29.5\dagger]$$

in exactly the same way as for the MS between in a one-way analysis of variance. Then

$$E(\text{MS columns}) = E\left[\sum_j \frac{n_j(M_j - M)^2}{C - 1} \right]. \qquad [12.29.6]$$

Since $\sum_k \beta_k = 0$ and $\sum_k \gamma_{jk} = 0$, it follows that

$$E(\text{MS columns}) = E \sum_j Rn \frac{(\alpha_j + M_{ej} - M_e)^2}{C - 1} = \sigma_e^2 + \frac{Rn \sum_j \alpha_j^2}{C - 1}. \qquad [12.29.7*]$$

When the hypothesis of no column effects is true,

$$E(\text{MS columns}) = \sigma_e^2,$$ [12.29.8*]

but when the hypothesis is false,

$$E(\text{MS columns}) > \sigma_e^2.$$ [12.29.9*]

The mean square between columns and the mean square error are independent and unbiased estimates of the same variance σ_e^2 when the null hypothesis of no column effects is true. This hypothesis can be tested by the F ratio,

$$F = \frac{\text{MS columns}}{\text{MS error}}$$ [12.19.10†]

with $C - 1$ and $RC(n - 1)$ degrees of freedom. The rationale is precisely the same as that given in Sections 12.6 through 12.9.

In the same way we examine the expectation of the mean square for rows:

$$\text{MS rows} = \sum_k \frac{Cn(M_k - M)^2}{R - 1}.$$ [12.29.11†]

Since $\sum_j \alpha_j = 0$ and $\sum_j \gamma_{jk} = 0$, this expectation is

$$E(\text{MS rows}) = E \sum_k Cn \frac{(\beta_k + M_{ek} - M_e)^2}{R - 1} = \sigma_e^2 + \frac{Cn \sum_k \beta_k^2}{R - 1}.$$ [12.29.12*]

The expectation of the mean square rows can be exactly σ_e^2 only when the hypothesis of no row effects is true; otherwise,

$$E(\text{MS rows}) > \sigma_e^2.$$ [12.29.13*]

The mean square between rows is an unbiased estimate of σ_e^2 when the null hypothesis of no row effects is true, and it is independent of the mean square error. The hypothesis of no row effects is then tested by the ratio

$$F = \frac{\text{MS rows}}{\text{MS error}}$$ [12.29.14†]

with $R - 1$ and $RC(n - 1)$ degrees of freedom.

Finally, the expectation of the sum of squares for interaction may be found, although this requires a little more work:

$$E \sum_j \sum_k n(M_{jk} - M_j - M_k + M)^2$$

$$= E\left[\sum_j \sum_k n(\gamma_{jk} + M_{ejk} - M_{ej} - M_{ek} + M_e)^2 \right]$$

$$= \sum_{j} \sum_{k} n\gamma_{jk}^2 + E \sum_{j} \sum_{k} n(M_{ejk} - M_{ej} - M_{ek} + M_e)^2 \quad [12.29.15]$$

since the expectation of any of the error terms is always zero. Now consider the last term on the right in **12.29.15**:

$$E \sum_{j} \sum_{k} n(M_{ejk} - M_{ej} - M_{ek} + M_e)^2$$

$$= E \sum_{j} \sum_{k} nM_{ejk}^2 - E \sum_{j} RnM_{ej}^2 - E \sum_{k} CnM_{ek}^2 + E(RCnM_e^2)$$

$$= \frac{RCn\sigma_e^2}{n} - \frac{RCn\sigma_e^2}{Rn} - \frac{RCn\sigma_e^2}{Cn} + \frac{RCn\sigma_e^2}{RCn}$$

$$= \sigma_e^2(RC - C - R + 1) = \sigma_e^2(R - 1)(C - 1). \quad [12.29.16*]$$

If we define

$$\text{MS interaction} = \frac{\text{SS interaction}}{(R - 1)(C - 1)} \quad [12.29.17\dagger]$$

then

$$E(\text{MS interaction}) = \frac{E(\text{SS interaction})}{(R - 1)(C - 1)}$$

$$= \sigma_e^2 + \frac{\sum_{j} \sum_{k} n\gamma_{jk}^2}{(R - 1)(C - 1)}. \quad [12.29.18*]$$

When there are no interaction effects *at all*, then

$$E(\text{MS interaction}) = \sigma_e^2 \quad [12.29.19*]$$

the mean square for interaction is also an unbiased estimate of the error variance σ_e^2. Otherwise,

$$E(\text{MS interaction}) > \sigma_e^2. \quad [12.29.20*]$$

The mean square for interaction is independent of the mean square for error, and so the hypothesis of no interaction effects may be tested by

$$F = \frac{\text{MS interaction}}{\text{MS error}} \quad [12.29.21\dagger]$$

with $(R - 1)(C - 1)$ and $RC(n - 1)$ degrees of freedom.

Thus we see that it is possible to make separate tests of the hypothesis of no row effects, the hypothesis of no column effects, and the hypothesis of no interaction effects, all from the same data. Furthermore, under the fixed effects model, and given an orthogonal experimental design, estimates of the three different kinds of effects are independent of each other.

12.30 COMPUTING FORMS FOR THE TWO-WAY ANALYSIS WITH REPLICATIONS

In carrying out an analysis of variance the following computing forms are generally used. These sums of squares are algebraically equivalent to those given in Section 12.27.

$$\text{SS total} = \sum_j \sum_k \sum_i y_{ijk}^2 - \frac{\left(\sum_j \sum_k \sum_i y_{ijk} \right)^2}{N} \qquad [12.30.1\dagger]$$

$$\text{SS rows} = \frac{\sum_k \left(\sum_j \sum_i y_{ijk} \right)^2}{Cn} - \frac{\left(\sum_j \sum_k \sum_i y_{ijk} \right)^2}{N} \qquad [12.30.2\dagger]$$

$$\text{SS columns} = \frac{\sum_j \left(\sum_k \sum_i y_{ijk} \right)^2}{Rn} - \frac{\left(\sum_j \sum_k \sum_i y_{ijk} \right)^2}{N} \qquad [12.30.3\dagger]$$

$$\text{SS error} = \sum_j \sum_k \sum_i y_{ijk}^2 - \frac{\sum_j \sum_k \left(\sum_i y_{ijk} \right)^2}{n} \qquad [12.30.4\dagger]$$

$$\text{SS interaction} = \frac{\sum_j \sum_k \left(\sum_i y_{ijk} \right)^2}{n} - \frac{\sum_k \left(\sum_j \sum_i y_{ijk} \right)^2}{Cn}$$

$$- \frac{\sum_j \left(\sum_k \sum_i y_{ijk} \right)^2}{Rn} + \frac{\left(\sum_j \sum_k \sum_i y_{ijk} \right)^2}{N}$$

$$= \text{SS total} - \text{SS rows} - \text{SS columns} - \text{SS error}. \quad [12.30.5\dagger]$$

Notice that the sum of squares for columns is calculated just as for a one-way analysis of data arranged into columns. Furthermore, the sum of squares for rows is identical to the sum of squares between groups when the data are arranged into a table where the experimental groups are designated by rows. The sum of squares total is also calculated in exactly the same way as for a one-way analysis. The only new features here are the computations for error and for interaction. Generally, the error term is calculated directly, and then the interaction term is found by subtracting the sum of squares, rows, columns, and error all from the sum of squares total.

12.31 A COMPUTATIONAL OUTLINE FOR THE TWO-WAY ANALYSIS UNDER THE FIXED EFFECTS MODEL

1. Arrange the data into an $R \times C$ table, in which the R rows represent the R different treatments of one kind, and the C columns the C different treatments of the other kind. Each cell in the table should contain the

same number n of observations. There are $N = RCn$ distinct observations in all.

2. Square each raw score and sum over all individuals in all cells to find $\sum_j \sum_k \sum_i y_{ijk}^2$. Call this quantity A.

3. Sum the raw scores in a given *cell jk* to find $\sum_i y_{ijk}$. Do this for *each cell*, and reserve these values for use in later steps.

4. Now sum the resulting values (step 3) over *all cells* to find $\sum_j \sum_k \sum_i y_{ijk}$. Call this quantity B. Find the **sum of squares total** by $A - \dfrac{B^2}{N}$.

5. Next take the RC different values found in step 3 and sum the cell totals for *a given row across columns* to find $\sum_j \sum_i y_{ijk}$. The result for any row k will be designated by D_k.

6. Having carried out step 5 for *each row*, square each of the D_k, sum over all of the various rows to find $\sum_k D_k^2$. Divide this quantity by Cn, the number of observations per row. Then

$$\frac{\sum\limits_k D_k^2}{Cn} - \frac{B^2}{N}$$

is the **sum of squares for rows**.

7. Now return to the quantities found in step 3. This time sum the cell totals for *a given column across rows* to find $\sum_k \sum_i y_{ijk}$ and call this value for column j, G_j.

8. Having carried out step 7 for each column, square each of the G_j and sum across the various columns to find $\sum_j G_j^2$. Divide this quantity by Rn, the number of observations per column. Then

$$\frac{\sum\limits_j G_j^2}{Rn} - \frac{B^2}{N}$$

is the **sum of squares for columns**.

9. Once again return to the cell totals found in step 3. For a given cell jk call the total H_{jk}. Now square H_{jk} for each cell and sum across *all cells* to find $\sum_j \sum_k H_{jk}^2$. Divide this by n, the number of observations per cell. Then

$$A - \frac{\sum\limits_j \sum\limits_k H_{jk}^2}{n}$$

is the **sum of squares for error**.

10. Find the **sum of squares for interaction** by taking

$$\text{SS total} - \text{SS rows} - \text{SS columns} - \text{SS error}$$

or

$$\frac{\sum_j \sum_k H_{jk}^2}{n} - \frac{\sum_k D_k^2}{Cn} - \frac{\sum_j G_j^2}{Rn} + \frac{B^2}{N}.$$

11. Enter these sums of squares in the summary table.
12. Divide the SS rows by $R - 1$ to find MS rows.
13. Divide the SS columns by $C - 1$ to find MS columns.
14. Divide the SS interaction by $(R - 1)(C - 1)$ to find MS interaction.
15. Divide the SS error by $RC(n - 1)$ to find MS error.
16. The hypothesis of no row effects is tested by

$$F = \frac{\text{MS rows}}{\text{MS error}}$$

with $R - 1$ and $RC(n - 1)$ degrees of freedom.
17. The hypothesis of no column effects is tested by

$$F = \frac{\text{MS columns}}{\text{MS error}}$$

with $C - 1$ and $RC(n - 1)$ degrees of freedom.
18. The hypothesis of no interaction is tested by

$$F = \frac{\text{MS interaction}}{\text{MS error}}$$

with $(R - 1)(C - 1)$ and $RC(n - 1)$ degrees of freedom.

12.32 THE SUMMARY TABLE FOR A TWO-WAY FIXED EFFECTS ANALYSIS OF VARIANCE WITH REPLICATIONS

For the fixed effects model, the results of a two-way analysis of variance are displayed in a summary table (Table 12.32.1).

Naturally, when the summary table is used to report an analysis of actual data, the algebraic expressions and symbols are replaced by the corresponding values obtained.

12.33 AN EXAMPLE

Suppose that the experiment on level of aspiration, outlined in Section 12.20, had actually been carried out, and the data shown in Table 12.33.1 obtained. We wish to examine three null hypotheses: (1) there is no effect of the standing given the subject, corresponding to the hypothesis of no column effects;

Table 12.32.1

Source	SS	d.f.	MS	F
Rows	$$\dfrac{\sum_{k}\left(\sum_{j}\sum_{i} y_{ijk}\right)^{2}}{Cn} - \dfrac{\left(\sum_{j}\sum_{k}\sum_{i} y_{ijk}\right)^{2}}{N}$$	$R-1$	$\dfrac{\text{SS rows}}{R-1}$	$\dfrac{\text{MS rows}}{\text{MS error}}$
Columns	$$\dfrac{\sum_{j}\left(\sum_{k}\sum_{i} y_{ijk}\right)^{2}}{Rn} - \dfrac{\left(\sum_{j}\sum_{k}\sum_{i} y_{ijk}\right)^{2}}{N}$$	$C-1$	$\dfrac{\text{SS col.}}{C-1}$	$\dfrac{\text{MS col.}}{\text{MS error}}$
Inter-action	$$\dfrac{\sum_{j}\sum_{k}\left(\sum_{i} y_{ijk}\right)^{2}}{n} - \dfrac{\sum_{k}\left(\sum_{j}\sum_{i} y_{ijk}\right)^{2}}{Cn}$$ $$- \dfrac{\sum_{j}\left(\sum_{k}\sum_{i} y_{ijk}\right)^{2}}{Rn} + \dfrac{\left(\sum_{j}\sum_{k}\sum_{i} y_{ijk}\right)^{2}}{N}$$	$(R-1)(C-1)$	$\dfrac{\text{SS int.}}{(R-1)(C-1)}$	$\dfrac{\text{MS int.}}{\text{MS error}}$
Error (within cells)	$$\sum_{j}\sum_{k}\sum_{i} y_{ijk}^{2} - \dfrac{\sum_{j}\sum_{k}\left(\sum_{i} y_{ijk}\right)^{2}}{n}$$	$RC(n-1)$	$\dfrac{\text{SS error}}{RC(n-1)}$	—
Totals	$$\sum_{j}\sum_{k}\sum_{i} y_{ijk}^{2} - \dfrac{\left(\sum_{j}\sum_{k}\sum_{i} y_{ijk}\right)^{2}}{N}$$	$RCn-1$	—	—

(2) the actual norm group given the subjects has no effect, corresponding to the hypothesis of no row effects; and (3) the norm–group-standing combination has no unique effect, corresponding to the hypothesis of no row-column interaction. The α level chosen for each of these three tests will be .05.

Following the computational outline given in Section 12.31 we first find the square of each of the scores, and sum:

$$A = \sum_{j}\sum_{k}\sum_{i} y_{ijk}^{2} = (52)^{2} + (48)^{2} + \cdots + (22)^{2} + (17)^{2} = 66872.$$

The sum of the scores in each cell (step 3) is given in the table above. Taking the sum of the cell sums gives the total sum,

$$B = \sum_{j}\sum_{k}\sum_{i} y_{ijk} = 464 + 302 + \cdots + 214 = 1904.$$

Hence the total sum of squares is

$$A - \frac{B^{2}}{N} = 66872 - \frac{(1904)^{2}}{60} = 6451.7.$$

Now the cell totals are summed for each row:

$$D_{1} = \sum_{j}\sum_{i} y_{ij1} = 464 + 302 + 178 = 944$$

Table 12.33.1

| Norms | Standing | | |
	Above	Average	Below
College men	52	28	15
	48	35	14
	43	34	23
	50	32	21
	43	34	14
	44	27	20
	46	31	21
	46	27	16
	43	29	20
	49	25	14
	464	302	178
Professional athletes	38	43	23
	42	34	25
	42	33	18
	35	42	26
	33	41	18
	38	37	26
	39	37	20
	34	40	19
	33	36	22
	34	35	17
	368	378	214

$$D_2 = \sum_j \sum_i y_{ij2} = 368 + 378 + 214 = 960.$$

The sum of squares for rows is found from

$$\frac{\sum_k D_k^2}{Cn} - \frac{B^2}{N} = \frac{(944)^2 + (960)^2}{30} - \frac{(1904)^2}{60}$$
$$= 4.2.$$

In a similar way, we find the sum of squares for columns by first summing cell totals for each column

$$G_1 = \sum_k \sum_i y_{i1k} = 464 + 368 = 832$$

$$G_2 = \sum_k \sum_i y_{i2k} = 302 + 378 = 680$$

$$G_3 = \sum_k \sum_i y_{i3k} = 178 + 214 = 392.$$

The sum of squares for columns is found from

$$\frac{\sum_j G_j^2}{Rn} - \frac{B^2}{N} = \frac{(832)^2 + (680)^2 + (392)^2}{20} - \frac{(1904)^2}{60}$$
$$= 4994.1.$$

Next, the sum of squares for error will be calculated. We begin by squaring and summing the *cell totals:*

$$\sum_j \sum_k H_{jk}^2 = (464)^2 + (302)^2 + \cdots + (214)^2 = 662288.$$

The sum of squares for error is

$$A - \frac{\sum_j \sum_k H_{jk}^2}{n} = 66872 - \frac{662288}{10} = 643.2.$$

The only remaining value to be calculated is the sum of squares for interaction; this is done by subtraction, as follows:

$$\text{SS total} - \text{SS rows} - \text{SS cols.} - \text{SS error} = 6451.7 - 4.2 - 4994.1 - 643.2$$
$$\text{SS interaction} = 810.2.$$

Table 12.33.2 is the summary table for this analysis of variance.

Table 12.33.2

Source	SS	df	MS	F
Rows (norm groups)	4.2	1	4.2	.35
Columns (standings)	4994.1	2	2497.05	209.8
Interaction	810.2	2	405.1	34.0
Error (within cells)	643.2	54	11.9	
Totals	6451.7	59		

The hypothesis of no row effects cannot be rejected, since the F value is less than unity. For the hypothesis of no column effects, an F of approximately 3.15 is required for rejection at the 5 percent level; the obtained F of 209 far exceeds this, and so may conclude with considerable confidence that column effects exist. In the same way, the F for interaction effects greatly exceeds that required for rejecting the null hypothesis, and so there seems to be reliable evidence for such interaction effects.

Our conclusions from this analysis of variance make it reasonably safe to make the following assertions:

1. There is apparently little or no effect of norm group alone on level of aspiration.
2. The experimental standing does seem to affect level of aspiration when considered over the different norm groups.
3. There is apparently an interaction between norm group and standing, meaning that the magnitude and direction of the effects of standing differ for different norm groups.

In short, the standing one is told that he has makes a difference in his aspiration level, but the kind and extent of difference that it makes depends upon the norm group to which he is being compared.

The different column effects can be estimated from the column means and the over-all mean:

$$\text{est. } \alpha_1 = 41.6 - 31.7 = 9.9$$
$$\text{est. } \alpha_2 = 34.0 - 31.7 = 2.3$$
$$\text{est. } \alpha_3 = 19.6 - 31.7 = -12.1.$$

(Because of rounding error these do not quite total zero, as they should.) In a similar way, interaction effects may be estimated from the means of cells, the rows, and the columns:

$$\text{est. } \gamma_{11} = 46.4 - 31.4 - 41.6 + 31.7 = 6.1$$
$$\text{est. } \gamma_{12} = 30.2 - 31.4 - 34.0 + 31.7 = 3.1$$

and so on. The estimated total effect of the "above" treatment on a subject in the college norm group is thus

$$\text{est. } (\alpha_1 + \gamma_{11}) = 9.9 + 6.1 = 16.0.$$

Note that for an individual selected at random from group 1 ("above" standing) with unspecified norm group, the best guess we can make about the effect on *him* of this treatment is 9.9 units. However, if we are told that he belongs to the group given the "college men" norms, our best bet is 16.0 as the amount of effect. In the same way, the effect of any column treatment j within a row-treatment population k is estimated $\alpha_j + \gamma_{jk}$.

12.34   ESTIMATING STRENGTH OF ASSOCIATION FROM A TWO-WAY ANALYSIS OF VARIANCE

From the experimenter's point of view, it may be informative to assess the strength of association represented either by significant treatment or interaction effects. This may be done once again by rough estimates of the index ω^2, although the form of the estimation is somewhat different, since there are three different such indices for a two-way experimental design.

Imagine a sample space in which there are three kinds of events X, Y, and Z, and each elementary event belongs to some joint event class (x, z, y). The event X stands for the column treatment given, the event Z is the row treatment, and the event Y is the value of the random variable standing for the dependent variable score. Suppose that the probability of an event x_j, the probability of an observation's being made in column treatment j, is $p(x_j) = 1/C$. Let $p(z_k)$, the probability of an observation's being in row treatment k, be $1/R$. Furthermore, let the probability of an observation in the combination jk be $p(x_j, z_k) = 1/RC$.

Under these circumstances, for Model I the variance σ_Y^2 of the **marginal** distribution (Sections 4.7 and 4.10) of Y is

$$\sigma_Y^2 = \sigma_e^2 + \frac{\sum_j \alpha_j^2}{C} + \frac{\sum_k \beta_k^2}{R} + \frac{\sum_j \sum_k \gamma_{jk}^2}{RC}. \qquad [12.34.1^*]$$

The definition of $\omega_{Y|X}^2$, the proportion of variance accounted for by X alone in the population is

$$\omega_{Y|X}^2 = \frac{\left(\sum_j \alpha_j^2\right)/C}{\sigma_Y^2}. \qquad [12.34.2^*]$$

Similarly, we can define

$$\omega_{Y|Z}^2 = \frac{\left(\sum_k \beta_k^2\right)/R}{\sigma_Y^2}, \qquad [12.34.3^*]$$

and

$$\omega_{Y|XZ}^2 = \frac{\left(\sum_j \sum_k \gamma_{jk}^2\right)/RC}{\sigma_Y^2}. \qquad [12.34.4^*]$$

This last index is the proportion of variance accounted for uniquely by the combination of *both* X and Z.

Given these definitions, and our results about the expectations of mean squares for the two-way analysis of variance (Section 12.29), we can estimate these values of ω^2 by taking

$$\text{est. } \omega_{Y|X}^2 = \frac{\text{SS columns} - (C-1)\text{ MS error}}{\text{MS error} + \text{SS total}} \qquad [12.34.5\dagger]$$

$$\text{est. } \omega_{Y|Z}^2 = \frac{\text{SS rows} - (R-1)\text{ MS error}}{\text{MS error} + \text{SS total}} \qquad [12.34.6\dagger]$$

$$\text{est. } \omega_{Y|XZ}^2 = \frac{\text{SS interaction} - (R-1)(C-1)\text{ MS error}}{\text{MS error} + \text{SS total}}. \qquad [12.34.7\dagger]$$

For the example in Section 12.33, these estimated values are

$$\text{est. } \omega_{Y|X}^2 = \frac{4994.1 - (2)(11.9)}{11.9 + 6451.7} = .77$$

$$\text{est. } \omega_{Y|XZ}^2 = \frac{810.2 - (2)(11.9)}{11.9 + 6451.7} = .12.$$

Since the F ratio shows a value less than 1.00 in the test for row differences, the

514   Analysis of Variance: Model I; Fixed Effects

estimate of $\omega_{Y|Z}^2$ is set equal to zero. These estimates suggest that a very strong association exists between the treatments symbolized by X and the dependent variable Y. Knowing X alone tends to reduce our "uncertainty" about Y by about 77 percent. Notice that this is almost the same estimated value found in Section 12.18, which was actually based on these same data. However, since here we are dealing with a two-way design, we can also find out something more: there is apparently a further accounting for around 12 percent of the variance of Y if one knows *both* of the categories represented by X and Z, the treatment combination. In other words, we may safely conclude not only that association exists between independent and dependent variables but also that this association is quite sizable in a predictive sense, for any population situation corresponding to our experiment.

As always, we cannot be sure that any association at all exists: the validity of this statement depends upon the assumptions being correct, and on these data not representing a chance result representing some really quite empty situation. However, the significance level assures us that the probability of error in such a statement is rather small, and our estimates of the strength of association are the best guesses we are able to make about the association's magnitude. Estimates of the size of effects and of strength of association are aids to the experimenter in trying to figure out what went on in the experiment and the meaning of the results. The F test per se is capable of indicating merely that something systematic seems to have happened. Only a careful examination of the data can make the meaning of the experiment clear, and this is why estimation of effects or of association-strength forms an important and informative part of any experimental analysis. (In Chapter 14 methods will be given for further exploration of differences in means after the analysis of variance has shown overall significance.)

12.35 THE IMPORTANCE OF THE ASSUMPTIONS IN THE TWO-WAY (OR HIGHER) ANALYSIS OF VARIANCE WITH FIXED EFFECTS

The list of assumptions in Section 12.28 was an almost exact parallel to the list in Section 12.5. Similar assumptions are made for more complex experiments requiring a higher-order analysis, provided that the fixed effects model is appropriate.

As you may have anticipated, the same relaxation of assumptions is possible in the two- or multi-way analysis as in the one-way analysis. For experiments with a relatively large number of observations per cell, the requirement of a normal distribution of errors seems to be rather unimportant. In an experiment where it is suspected that the parent distributions of dependent variable values are very unlike a normal distribution, perhaps a correspondingly large number of observations per cell should be used.

When the data table represents an equal number of observations in each cell, the requirement of equal error variance in each treatment combination population may also be violated without serious risk, at least in terms of Type I error. However, in some circumstances the power of the F test may be affected. On the whole, there are two good reasons for planning experiments with equal n per cell: the experimental design will thus be orthogonal (Section 12.25) and the possible consequences of nonhomogeneous variances on the probability of Type I error will be minimized.

Regardless of the simplicity or complexity of the experiment, however, the error portions entering into the respective observations should be independent if the fixed effects model is to apply. This seems to be the one requirement that can be violated only with grave risk of erroneous conclusions. For this reason considerable caution must be exercised in the planning and analysis of experiments involving repeated observations of the same subjects or of subjects matched in certain ways if the methods of this chapter are to be used. Ordinarily, the methods to be discussed in the next chapter are preferable for the analysis of such data.

12.36 The Analysis of Variance as a Summarization of Data

It may appear that the main use of the analysis of variance, particularly for two-factor or multi-factor experiments, is in generating a number of F tests on the same set of data, and that the partition of the sum of squares is only a means to this end. However, this is really a very narrow view of the role of this form of analysis in experimentation. The really important feature of the analysis of variance is that it permits the separation of all of the potential information in the data into distinct and nonoverlapping portions, each reflecting only certain aspects of the experiment. For example, in the simple, one-way analysis of variance, the mean square between groups reflects both the systematic differences among observations that are attributable to the experimental manipulations, as well as the chance, unsystematic differences attributable to all of the other circumstances of the experiment. On the other hand, the mean square within the groups reflects only these latter, unsystematic, features. Under the fixed effects model, these two statistics are independent, completely nonoverlapping, ways of summarizing the data. The information contained in one is nonredundant with the information contained in the other. Estimates of the effects of the treatments are independent of estimates of error-variability. The mechanics of the analysis of variance allow the experimenter to arrange and summarize his data in these nonredundant ways, in order to decide if effects exist and to estimate how large or important those effects may be.

Similarly, for a two-factor experiment we arrive at a mean square for one treatment factor, and a separate mean square for the other. These two mean squares reflect quite nonredundant aspects of the experiment, even though they were each based on the same basic data: the first sum of squares reflects only the effects attributable to the first experimental factor (plus error), and the second those effects attributable to the other (plus error). Under the statistical assumptions we make these two mean squares are independent of each other. Furthermore, the mean squares for interaction and for error are independent of each other and of the treatment mean squares. The analysis of variance lets the experimenter "pull apart" the factors that contribute to variation in his experiment, and identify them exclusively with particular summary statistics. For experiments of the orthogonal, balanced type considered here, the analysis of variance is a routine method for finding the statistics that reflect particular, meaningful, aspects of the data.

In short, it is useful to think of the analysis of variance as a device for "sorting" the information in an experiment into nonoverlapping and meaningful portions. As mentioned in Section 12.12, multiple t tests carried out on the same data do not provide this feature; the various differences between means do overlap in the information they provide, and it is not easy to assess the evidence for over-all existence or importance of treatment effects from a complete set of such differences. On the other hand, the analysis of variance packages the information in the data into neat, distinct "bundles," permitting a relatively simple judgment to be made about the effects of the experimental treatments. The real importance of the analysis of variance lies in the fact that it routinely provides such succinct over-all "packaging" of the data.

In Chapter 14 the information that goes into the various mean squares will be studied in more detail. We will find that each distinct degree of freedom in the analysis corresponds to one nonredundant, independent, nonoverlapping question that one *might* ask of the data via an estimate and accompanying t test, and that the F test itself provides an over-all or "omnibus" conclusion about a set of such questions. Furthermore, a significant F may be interpreted as a signal to look further into the data, to try to answer particular questions about what went on in the experiment. We will go much further into these matters in Chapter 14.

For the moment, however, let us consider the several F tests obtained from a two- or multi-factor experiment. The sums of squares and mean squares for columns, for rows, for interaction, and for error are all, under the assumptions made, independent of each other. However, are the three or more F tests themselves independent? Does the level of significance shown by any one of the tests in any way predicate the level of significance shown by the others? Unfortunately, it can be shown that such F tests are *not* independent. Some connection exists among the various F values and significance levels. This is due to the fact that each of the F ratios involves the same mean square for error in the denominator; the presence of this same value in each of the ratios creates some statistical dependency among them. The practical result is much the same as for the multiple confidence intervals discussed in Section 10.10. If three F tests are carried out, and these tests actually are independent, then one should expect about $3(.05)$ or $.15$ of these tests to show significance at the $.05$ level by chance alone. Furthermore, the probability is $1 - (.95)^3$ or about $.14$ that *at least one* of the tests will show spurious significance. However, for the usual situation where the tests are not independent, one has no ready way to calculate the number that should be expected by chance, and one knows only that the probability that at least one is spuriously significant is somewhere between $.05$ and $.15$.

For really complicated analyses of variance, the problem becomes much more serious, since a fairly large number of F tests may be carried out, and the probability may be quite large that one or more tests gives spuriously significant results. The matter is further complicated by the fact that the F tests are not independent, and the number to be expected by chance is quite difficult to calculate exactly. For this reason, when large numbers of F tests are performed, the experimenter should not pay too much attention to isolated results that happen to be significant. Rather, the pattern and interpretability of results, as well as the strength of association represented by the findings, form a more reasonable basis for the over-all evaluation of the experiment. When the number of degrees of freedom for the mean square error is very large, then the various F tests may be regarded as approximately independent, and the number of significant results at the .05 level to be expected should be close to 5 percent. Even here, however, the importance of a particular result is very difficult to interpret on the basis of significance level alone. A great deal of thought must go into the interpretation of a complicated experiment, quite over and above the information provided by the significance tests.

12.37 ANALYZING EXPERIMENTS WITH MORE THAN TWO EXPERIMENTAL FACTORS

The essential features of any fixed effects analysis of variance have now been discussed. In experiments involving three or more different experimental factors, the total sum of squares is partitioned into even more parts, but the basic ideas of the partition, the mean squares, and the F tests are the same. In a three-factor experiment, not only are there mean squares representing the interactions of particular pairs of the experimental factors, but also a mean square representing the simultaneous interaction of all three of the factors. The higher the order of the experimental design, the larger becomes the number of possible interactions representing every combination of two or more factors. Each and every significant interaction represents a new qualification on the meaning of the results. Moreover, as suggested above, if many significance tests are carried out on the same data, the probability of at least one spuriously significant result may be very large, and this probability, as well as the number of Type I errors to be expected by chance, cannot be determined routinely. For these reasons, very complicated experiments with many factors are somewhat uneconomical to perform, since they require a large number of observations as a rule, and a complete analysis of the data yields so many statistical results that the experiment as a whole is often very difficult to interpret in a statistical light. In planning an experiment, it is a temptation to throw in many experimental treatments, especially if the data are inexpensive and the experimenter is adventuresome. However, this is *not* always good policy if the psychologist is interested in finding meaning in his results; other things being equal, the simpler the psychological experiment the better will be its execution, and the more likely will one be able to decide what actually happened and what the results actually mean.

The two-way experiments discussed in this chapter as illustrating Model I are examples of "factorial" designs, since they involve several experimental "factors" each represented at several "levels." In a complete factorial experiment the set of experimental factors is completely crossed, so that every possible combination of factor levels is observed. The fixed effects model applies when the different levels of each factor are chosen in advance of the experiment and when the only conclusions to be drawn concern *those particular* levels and combinations. By far the most commonly encountered experiments are of this type, and the student is referred to the references at the end of this book giving a further exposition of the planning and analysis of factorial experiments. The varieties of experimental design are almost limitless, however, and many designs in common use are quite different from the factorial arrangement.

Far and away, the fixed effects analysis of variance is the statistical technique encountered most in current experimental psychology. Its advantages are many: the general technique is extremely flexible and applies to a wide variety of experimental arrangements. Indeed, the availability of a statistical technique such as the analysis of variance has done much to stimulate inquiry into the logic and economics of the *planning* of experiments. Statistically, the F test in the fixed effects analysis of variance is relatively *robust;* as we have seen, the failure of at least two of the underlying statistical assumptions does not necessarily disqualify the application of this method in practical situations. Computationally the analysis is relatively simple and routine and provides a condensation of the main statistical results of an experiment into an easily understood form.

However, the application of the analysis of variance never transforms a sloppy experiment into a good one, no matter how elegant the experimental design appears on paper, nor how neat and informative the summary table appears to the reader. Furthermore, there are psychological experiments where the fixed effects model is manifestly inappropriate; as mentioned above, often these are experiments in which treatments are sampled, in which subjects are somehow matched within groups, or in which repeated observations are made on the same individuals. The techniques of analysis of variance and its connections with experimental design were first developed by Sir Ronald Fisher and others mainly for problems in the biological sciences, most notably in agriculture, where assumptions such as independence among errors are often simple to meet. The unique problems of psychological research are not always provided for either in analysis of variance assumptions or in standard experimental designs. Quite often, the psychologist is able to tailor his experimental problem to fit the methodology available to him *without losing the essence of the problem in so doing.* However, experiments should be planned so as to capture the phenomena under study in its clearest, most easily understood form, and this does not necessarily mean that one of the "textbook" experimental designs, nor a treatment by the analysis of variance will best clarify matters. The experimental *problem* must come first in planning, and not the requirements of some particular form of analysis, even though, ideally, both should be considered together from the outset. If it should come to a choice between preserving the essential character of the experimental problem, or using a relatively elegant technique such as the analysis of variance, then the problem should come first.

The point has been repeated several times, but it bears repetition: statistics should aid in the clarification of meaning in an experimental situation, but the production of a statistical summary in some impressive and elegant form should never be the primary goal of a research enterprise. If the analysis of variance does not fit the problem, do not use it. If *no* inferential statistical techniques are available to fit the problem, do not alter the problem in essential ways to make some pet or fashionable technique apply. Above all, do not "jam" the data willy-nilly into some wildly inappropriate statistical analysis simply to get a significance test; there is little or nothing to gain by doing this. Thoughtless application of statistical techniques makes the reader wonder about the care that went into the experiment itself. A really good experiment, carefully planned and controlled, often speaks for itself with little or no aid from inferential statistics.

EXERCISES

1. Following the pattern of section 12.3 construct the scores for an experiment with 4 samples of 4 observations each where $\mu = 100$, $\mu_1 = 95$, $\mu_2 = 104$, $\mu_3 = 98$, and $\mu_4 = 103$. Let the error terms be as follows:

I	II	III	IV
−5	−2	0	3
1	5	−4	−6
0	6	−1	4
2	−2	3	−7

From these "data," estimate μ and the separate effects, and compare them with the true values.

2. By use of the artificial data of problem 1, establish whether or not it is necessarily true that

$$SS \text{ between} = \sum_j n_j \alpha_j^2 + (M_{e_j} - M_e)^2$$

for any given set of samples.

3. Carry out an analysis of variance on the following data, corresponding to two independent experimental groups of ten cases each.

I		II	
19	16	18	21
20	20	19	19
24	20	19	23
20	18	24	17
22	24	18	18

4. Test the significance of the difference between the means of the two groups in problem 3 by using the t-test. How does this compare with the results of problem 3?
5. Carry out a one-way analysis of variance for the following data based on four inpendent groups:

A	B	C	D
29	19	31	33
41	13	37	47
27	21	23	33
17	23	21	25
33	17	31	37

6. For the data of problem 5, estimate the various treatment effects and the strength of association between the experimental and dependent variables.
7. The following experiment utilized three groups of differing numbers of cases. Test the hypothesis of the equality of all of the population means.

I	II	III
.5	.6	.8
.4	.6	.9
.6	.5	.8
1.0	1.0	1.1
.6	.9	1.2
.5	.4	1.1
1.1	.7	

8. Estimate the true effects, and the strength of association between experimental and dependent variables in problem 7.
9. Clergymen of the same, large denomination from six large, geographical areas of the United States were sampled at random. Each clergyman sampled was given an attitude test, providing a score on the "liberalism" of his attitudes toward modern life. The following data resulted:

S.E.	S.W.	N.E.	N.W.	MIDWEST	FAR WEST
27	29	34	44	32	45
43	49	43	36	28	50
40	27	30	30	54	30
30	46	44	28	50	33
42	26	32	42	46	35
29	48	42		36	47
30	28	41		41	
41	30	33			
28	47	31			
	50	40			

Test the hypothesis that the means of these populations are equal. What is the estimate of ω^2, the proportion of the variance in "liberalism" accounted for by geographical region?

10. Consider the following sets of data:

GROUP 1	GROUP 2	GROUP 3	GROUP 4
1.69	1.82	1.71	1.69
1.53	1.93	1.82	1.82
1.91	1.94	1.75	1.86
1.82	1.60	1.64	1.90
1.57	1.78	1.52	1.39
1.77	1.85	1.73	1.56
1.94	1.98	1.86	1.74
1.60	1.72	1.68	1.83
1.74	1.83	1.54	1.47
1.74	1.75	1.75	1.64

We wish to carry out an analysis of variance on these data, and test for equality of means ($\alpha = .05$). However, we will simplify our computations by subtracting 1.00 from each number and multiplying by 100. Complete the analysis and carry out the F-test. Should the transformation of the numbers ($X' = 100(X - 1)$) affect the results of our F test? Does it affect the values of M.S. between and M.S. within? How?

11. Estimate ω^2 for the data given in problem 10 above, as well as the effects associated with the four groups.

12. In the construction of a projective test, 40 more or less ambiguous pictures of two or more human figures were used. In each picture, the sex of at least one of the figures was only vaguely suggested. In a study of the influence of the introduction of extra cues into the pictures, one set of 40 was retouched so that the vague figure looked slightly more like a woman, in another set each was retouched to make the figure look slightly more like a man. A third set of the original pictures was used as a control. The forty pictures were administered to a group of 18 male college students and an independent group of 18 female college students. Six members of each group saw the pictures with female cues, six the pictures with male cues, and six the original pictures. Each subject was scored according to the number of pictures in which he interpreted the indistinct figure as a female. The results follow:

	FEMALE CUES		MALE CUES		NO CUES	
FEMALE SUBJECTS	29	36	14	5	22	25
	35	33	8	7	20	30
	28	38	10	16	23	32
MALE SUBJECTS	25	35	3	5	18	7
	31	32	8	9	15	11
	26	34	4	6	8	10

Complete the analysis of variance.

13. For the data of problem 16, estimate the ω^2 values for rows, columns, and interaction. Furthermore, estimate row, column, and interaction effects based on these data.

14. In a study of post-meningitic and post-encephalitic brain damage, each of 36 subjects was given a battery of tests, providing a composite score for each. Low scores on this composite measure presumably represented a considerable degree of residual brain damage. The subjects were divided into 3 groups according to type of initial infection, and into 3 crossed groups according to time since apparent physical recovery from the illness. The data follow:

	1–2 YEARS		3–5 YEARS		7–10 YEARS	
POST-ENCEPHALITIC	76	73	69	53	59	43
	75	62	72	55	41	57
POST-MENINGITIC	81	89	82	70	68	50
	83	75	91	74	75	47
POST-OPERATIVE CONTROL	75	84	85	79	98	100
	65	63	76	87	82	79

Do there seem to be significant differences in performance among the post-encephalitic, post-meningitic, and control groups? Among the groups according to time since recovery? Is there apparent interaction between type of illness and time since recovery? (Use $\alpha = .05$ here.) State, verbally, your conclusions from these data.

15. Estimate the power of the F-test of problem 5, if the true value of $\omega^2 = .40$, and $\alpha = .05$.

16. An experiment was carried out on the relation between the size and the wall color of a room used for a standardized interview, and the measured anxiety level of the respondent. The following results were obtained:

ROOM SIZE		RED	YELLOW	GREEN	BLUE
	SMALL	160	134	104	86
		155	139	175	71
		170	144	96	112
	MEDIUM	175	150	83	110
		152	156	89	87
		167	159	79	100
	LARGE	180	170	84	105
		154	133	86	93
		141	128	83	85

ROOM-COLOR

Complete the analysis of variance on these data.

17. Estimate the power of the F-test with $\alpha = .05$ in the situation in which 5 independent groups of 5 independent subjects each are used, and the true value of $\omega^2 = .20$. How

does that power change as the number in each group is increased to 10?

18. In a certain experiment, two sets of experimental factors were used. The first (columns) contained 3 levels, and the second (rows) 4 levels. Twelve independent subjects assigned at random to the treatment combinations provided the following data. Test for significant row and column effects. What must one assume in doing so?

	I	II	III
A	92	40	24
B	−13	98	16
C	12	−8	64
D	82	83	46

19. Within the analysis of variance, could one test a specific hypothesis about the total population mean, such as H_0: $\mu = k$? If so, what form would the F-test take for this hypothesis? Does this have anything to do with the fact that the total number of degrees of freedom is usually taken to be $N - 1$ instead of N?

20. At what point in the analysis of variance does it become critical that the population variances are homogeneous?

21. For a given set of sample data, is the following statement true

$$S^2_{total} = S^2_{between} + S^2_{within}$$

where S^2 is the average squared deviation from the mean?

13 The Analysis of Variance: Models II and III, Random Effects and Mixed Models

In the preceding chapter, the analysis of variance was discussed only for the fixed effects model. At the beginning of that chapter it was mentioned that the fixed effects model is appropriate when the experimental treatments actually administered are thought of as exhausting all treatments of interest. That is, given any experimental factor, any "level" or category of that factor figuring in the experimental question is observed. The only inferences to be drawn from the experiment concern the effects of those levels actually represented. Similarly, in an experiment with two or more crossed factors, each combination of factor levels ordinarily is represented in the experiment, and the only inferences drawn concern those observed levels and their combinations.

Now we turn to a model for experiments in which inferences are to be drawn about an entire set of distinct treatments or factor levels, including some not actually observed. For such experiments, many more categories or levels of a factor are possible than actually occur as observations in the experiment itself. The experimenter is interested in the *whole range* of possible levels, and what he observes as factor levels or experimental treatments is only a random sample of the potential set he might have observed. Before the experiment a sample is drawn from among all possible levels of a particular experimental factor, and then inferences are made about the effects of all such levels from the sample of factor levels.

For example, suppose that in a psychological experiment we suspect that the personality of the experimenter himself may be having an effect on the results. In principle there are a vast number of people, each presumably having a distinct personality, who might possibly serve as the person who conducts the experiment. Obviously, it is reasonable to narrow this population of potential experimenters down to people having the requisite technical skills to conduct the experiment (for example, English-speaking psychologists), but even this leaves a great many people from which to choose. To put it mildly, trying out each such person in our experiment would complicate matters! So, instead, we draw a random sample from among English-speaking psychologists. Suppose that five different persons are chosen; each of the five persons conducts the experiment on a different sample of n subjects assigned at random to him. Each experimenter constitutes an experimental treatment given to one group of subjects; in all other respects subjects are treated exactly the same. Since the experimental treatments employed are themselves a random sample, this experiment fits the random effects, rather than the fixed effects, model: an inference is to be made about experimenter effects in general from observation of only five experimenters sampled at random.

The random effects model will be given a more formal statement in succeeding sections, and will be illustrated first for a one-way analysis of variance, and then extended to the two-way case. It will be seen that, computationally, analyses using the fixed and random effects models are identical, but that the inferences drawn are different, and that the actual form of the F test is different in the two-way analysis.

Parenthetically, the student is warned that designation of the random effects model as Model II, and the mixed model as Model III, though convenient for the organization adopted here, is not uniform across authors. It is probably better for the student to learn to think of these three models as "fixed effects," "random effects" (or "components of variance"), and "mixed" from the very outset, rather than relying on any number-designation alone.

13.1 MODEL II: RANDOM EFFECTS

In the study outlined in the section above, consider a particular experimenter, whom we will call, for the moment, experimenter v. When this particular experimenter is paired with a particular subject, say subject i, then some value of the dependent variable Y is observed. Let us denote the value of Y obtained for subject i under experimenter v by the symbol $Y_i(v)$. Over all possible subjects that might be paired with a particular experimenter v there is some hypothetical distribution of possible values of $Y_i(v)$. That is, we can think of a distribution of values for a population of potential subjects, each observed under the same experimenter v. For the moment let us call the mean of this population distribution $\mathbf{\mu}(v)$. Then the score associated with any particular subject drawn from this population for experimenter v is thought of as

$$Y_i(v) = \mathbf{\mu}(v) + e_i(v).$$

Here, $\mathbf{\mu}(v)$ is the mean of the population of all possible observations under this experimenter across subjects, $e_i(v)$ is the deviation from that mean, or the error, peculiar to subject i observed under experimenter v.

However, there are a great many possible experimenters who might carry out this particular experiment. Suppose that the number of such possible experimenters is some large, but perhaps finite, number V. We select a random sample of some J such experimenters for our particular study; note that this is tantamount to selecting some J distinct populations of observations for representation in our experiment, since each experimenter v that we might use is associated with a potential population of observations to be made. Any given experimenter either may or may not appear in our actual sample of J experimenters. This situation is unlike the fixed effects model, where if a treatment (a particular population of potential observations) is represented once in an experiment, it is thought of as always being represented in any complete repetition of that experiment. In this new, random effects, model, whether or not a particular treatment (experimenter) appears in any repetition of the experiment is purely a matter of chance. In short, only a random selection of some J of the total number V of possible experimenters is actually used in the performance of the experiment, and consequently only J of the V populations of ultimate interest actually are represented in the data. When $J = V$, so that every population of interest actually is represented by observations in the experiment, the fixed effects and the random effects models are identical. On the other hand, when J is less than V, so that only a subset of the treatments of interest (the experimenters) appear in any given repetition of the experiment, the fixed effects model no longer applies.

Now imagine a particular experimenter v who happens to be among those sampled for the experiment proper, and who is the jth such experimenter actually observed; we will designate this experimenter as v_j. Experimenter v_j is observed in association with a distinct random sample of some n subjects. The score for any observation i made in the experiment under the particular experimenter v_j will be denoted by $Y_i(v_j)$. This score can also be thought of as composed of two components:

$$Y_i(v_j) = \mathbf{\mu}(v_j) + \mathbf{e}_i(v_j).$$

However, the j subscript here indicates only that the particular experimenter v was the jth member of the sample of experimenters, and $\mathbf{\mu}(v_j)$ is merely the true mean associated with experimenter v, or $\mathbf{\mu}(v)$. If that experimenter were to be observed over all possible subjects, rather than for a limited sample of n subjects, the mean value for such observations should be $\mathbf{\mu}(v)$.

In general, the thing to be remembered is that not only is a random sample of observations made under each of a set of J different treatment populations, but also that the different treatment populations themselves are members of a random sample of a much larger set of *possible* treatment populations. The reason for the relatively complicated notation used here is this: any given treatment v is itself associated with a potential population of observations with mean $\mathbf{\mu}(v)$, and the particular observation i made under that treatment is sampled from that population. The treatments that actually are used in the experiment itself are, moreover, a random selection from a large set of possible such treatments. Any potential treatment v may or may not appear in the actual experiment carried out; this is purely a matter of chance. But wherever it does appear, a population of potential observations associated with that treatment, having a mean $\mathbf{\mu}(v)$ and a variance $\mathbf{\delta}^2(v)$, is represented by a sample in the experiment.

Now, over all of the V possible experimenters who might have been included among the J actually sampled, there is a general mean μ. That is,

$$\mu = \frac{\sum\limits_{v} \mathbf{\mu}(v)}{V}.$$

We may define the effect associated with a particular experimenter v as the deviation of $\mathbf{\mu}(v)$ from this grand over-all mean:

$$\text{effect of experimenter } v = \mathbf{a}(v) = \mathbf{\mu}(v) - \mu.$$

Then the score associated with any observation i that might be made under experimenter v can be thought of as

$$Y_i(v) = \mu + \mathbf{a}(v) + \mathbf{e}_i(v). \qquad [13.1.1]$$

That is, the score $Y_i(v)$ may be thought of as a sum consisting of the general mean μ, the effect associated with the particular experimenter v, and an error component associated uniquely with observation i made under experimenter v. Similarly, for any experimenter v included in the sample as the jth experimenter in the actual study, we can think of the effect $\mathbf{a}(v_j)$ associated with *that* experimenter. Any observation i made in association with that experimenter should then have a score with a composition given by

$$Y_i(v_j) = \mu + \mathbf{a}(v_j) + \mathbf{e}_i(v_j). \qquad [13.1.2]$$

Since experimenter v_j is a randomly chosen member of the population of V experimenters, then $\mathbf{a}(v_j)$ is a randomly chosen value from the distribution of effects over the various possible experimenters. The effect associated with any observation i made under the jth experimenter included is a value of a random variable, or a **random effect**. Recall that in the fixed effects model of Chapter 12, the composition of the score for individual observation i in condition j was thought of as

$$Y_{ij} = \mu + \alpha_j + e_{ij}$$

where α_j symbolizes the effect attributable to the jth treatment administered. In Model I, the particular treatment in the experiment represented by the subscript j is conceived as fixed, since in any repetition of the experiment that treatment appears and is associated with the subscript j. On the other hand, in the random effects model, or Model II, the value represented by $\mathbf{a}(v_j)$ need not be the same from repetition to repetition of the experiment. The value $\mathbf{a}(v_j)$ is a value of a random variable, which depends upon the particular, randomly selected, experimenter who happens to appear as the jth experimenter actually employed. This is the reason for the "random effects" nomenclature: the particular experimenter under whom the observation is made is most definitely thought of as exerting a systematic influence on the obtained score, via the effect $\mathbf{a}(v)$ associated with him, but the opportunity for this experimenter to appear in the actual data as experimenter v_j, exerting effect $\mathbf{a}(v_j) = \mathbf{a}(v)$, is governed purely by a random process. Thus, $\mathbf{a}(v_j)$ is spoken of as a random variable, or random effect.

Although this discussion has been carried on so far in terms of particular experimenters sampled from a population of V such persons, it should be obvious that the same argument applies to any set of V potential treatments or levels from which some J might be sampled for use in a given experiment. Any particular treatment v may or may not appear in the actual experiment, but if it does appear, then there is an effect $a(v)$ associated with that treatment which should be the same each and every time that particular treatment shows up in the sample of J such treatments.

In both the fixed effects and the random effects models, our first interest is in judging if there really are different effects associated with the different possible treatments. For the present study, we would like to see if "experimenter effects" exist. Recall that in the fixed effects analysis of variance the hypothesis to be tested can be stated as

$$H_0: \alpha_j = 0, \qquad \text{for all } j.$$

In a somewhat similar way the hypothesis for the random effects model can be stated:

$$H_0: a(v) = 0, \qquad \text{for all possible treatments } v.$$

In other words, we wish to see if any effects at all exist attributable to the treatments, including not only those actually sampled but also the potential treatments not appearing in the experiment actually carried out. In the fixed effects situation of Chapter 12, we found that a test of the hypothesis of no treatment effects was equivalent to a test that there was no variability among the effects of the J treatments: that is,

$$\text{variance of effects} = \frac{\sum\limits_{j} \alpha_j^2}{J} = 0$$

when the null hypothesis is true. In the random effects model, the population of potential treatments is thought of as generating a distribution of effects $a(v)$, with a mean of zero and a variance given by

$$\sigma_A^2 = E[a(v)]^2,$$

where the expectation is over all possible treatments v. The hypothesis of no treatment effects is true when, and only when, the value of σ_A^2 is zero. Our immediate task is to find a way to test the hypothesis that $\sigma_A^2 = 0$.

13.2   ASSUMPTIONS MADE IN THE RANDOM EFFECTS MODEL

Once the conceptual distinctions between Models I and II are clear, we can adopt a somewhat simpler notation. Let us define

$$a_j = a(v_j)$$
$$e_{ij} = e_i(v_j)$$
$$Y_{ij} = Y_i(v_j).$$

Then the basic model given by expression **13.1.2** becomes

$$Y_{ij} = \mu + a_j + e_{ij}. \qquad [13.2.1*]$$

Remember that a_j represents a random variable, the value of which depends upon which one of the basic set of treatments is selected at random and denoted by the subscript j. Furthermore, e_{ij} is a random variable that depends on the particular observation i that happens to be made under the treatment designated by v_j. It has become conventional to use Roman letters such as a_j to denote the random variables in Model II, in contrast to Greek letters such as α_j used for the comparable terms conceived as constants in the fixed effects model.

The assumptions to be made in deriving a test of the hypothesis of no treatment effects are as follows:

1. The possible values a_j represent a random variable having a distribution with a mean of zero, and a variance σ_A^2.
2. For any treatment j, the errors e_{ij} are normally distributed with a mean of zero and a variance σ_e^2, which is the same for each possible treatment j.
3. The J values of the random variable a_j occurring in the experiment are completely independent of each other.
4. The values of the random variable e_{ij} are completely independent.
5. Each pair of random variables a_j and e_{ij} are completely independent.

You will note that here we are making assumptions about two different kinds of distributions. First of all there is a distribution of possible values for the effects a_j that might appear in a given repetition of the experiment. For the experiment proper, a random sample is taken from among all possible such effects values when a random sample of treatments is drawn. For the time being, no special assumptions need be made about the distribution of effects from which the sample of effects is drawn, other than that the mean is zero and that the variance has some (finite) value σ_A^2. Later, when we discuss this model more generally, we will assume that the distribution of effects is normal. The second kind of distribution is that of errors within particular treatments. Just as in Model I, we assume errors to be normally and independently distributed, with the same variance irrespective of the particular treatment under which the observation is made. Notice that we are, at present, making no assumptions about the number V of potential treatments from which a sample of J is taken for the experiment, except that V is greater than J. We are, however, assuming that within any given treatment population an infinite number of potential units for observation exist.

13.3 THE MEAN SQUARES FOR MODEL II

In this section the mean squares for the one-way analysis will be examined. *The discussion will be restricted to the situation where exactly the same number of observations n are made under each treatment.* Although the treatments may have different numbers of sample observations in the one-way case for the fixed effects model, some difficulties arise in Model II unless the numbers of observations are equal.

The partition of the sum of squares for the random effects model is carried out in exactly the same way as for Model I. Furthermore, the mean squares are found just as before:

$$\text{MS between} = \frac{\text{SS between}}{J-1} = \frac{\sum_j n(M_j - M)^2}{J-1} \qquad [13.3.1\dagger]$$

and

$$\text{MS within} = \frac{\text{SS within}}{N-J} = \frac{\sum_j \sum_i (y_{ij} - M_j)^2}{N-J}. \qquad [13.3.2\dagger]$$

Bear in mind that for any random sample of J treatments or factor levels, there is a corresponding random sample of effects a_j. Let the sample mean of these effects be denoted by

$$M_a = \frac{\sum_j a_j}{J}.$$

Although the mean of the effects over all of the *possible* treatments must be zero, the value of M_a need not be zero, since this is but the mean of a sample of effects actually occurring in the experiment.

In the same way, sample means for error may be defined:

$$M_{ej} = \frac{\sum_i^n e_{ij}}{n}$$

and

$$M_e = \frac{\sum_j \sum_i e_{ij}}{N} = \frac{\sum_j M_{ej}}{J}.$$

In these terms, MS between groups becomes

$$\frac{\sum_j n(M_j - M)^2}{J-1} = \frac{\sum_j n(a_j + M_{ej} - M_a - M_e)^2}{J-1}.$$

Now suppose that we take the expectation of the mean square between groups, over all samples of subjects *and* over all samples of treatments:

$$E(\text{MS between}) = E \sum_j n \frac{(a_j + M_{ej} - M_a - M_e)^2}{J-1}$$

or

$$E(\text{MS between}) = nE \sum_j \frac{(a_j - M_a)^2}{J-1} + nE \sum_j \frac{(M_{ej} - M_e)^2}{J-1}$$

$$+ 2n \sum_j E \frac{(a_j - M_a)(M_{ej} - M_e)}{J-1}. \quad [13.3.3]$$

Consider the first term on the right in equation **13.3.3**. Since the J values a_j are a sample from a population of effects, this first term

$$\sum_j \frac{(a_j - M_a)^2}{J-1}$$

is an unbiased estimate of the *effects* variance σ_A^2, so that

$$nE \sum_j \frac{(a_j - M_a)^2}{J-1} = n\sigma_A^2. \quad [13.3.4*]$$

The last term on the right in expression **13.3.3** is zero because of the assumption that the a_j and the e_{ij} are independent. The middle term in **13.3.3** is simply

$$nE \sum_j \frac{(M_{ej} - M_e)^2}{J-1} = \sigma_e^2$$

since it involves the unbiased pooled estimate of the variance of the mean error, σ_e^2/n. Finally, we see that

$$E(\text{MS between}) = n\sigma_A^2 + \sigma_e^2. \quad [13.3.5*]$$

The expected mean square between is the weighted sum of two variances, that of the population of treatment effects, and that of the error.

Now we turn to the mean square within treatments:

$$E(\text{MS within}) = E \sum_j \sum_i \frac{(e_{ij} - M_{ej})^2}{J(n-1)}. \quad [13.3.6\dagger]$$

For any j, $\sum_i \frac{(e_{ij} - M_{ej})^2}{n-1}$ is an unbiased estimate of σ_e^2, the error variance, so that

$$E(\text{MS within}) = \sum_j \frac{\sigma_e^2}{J} = \sigma_e^2. \quad [13.3.7*]$$

Thus the mean square within is always an unbiased estimate of error variance alone.

13.4   THE NULL HYPOTHESIS FOR MODEL II

In Model I, the null hypothesis is that each effect α_j is zero, which corresponds to the equality of all of the treatment-population means. We saw in Section 13.1 that, in Model II, if there are no treatment effects at all, either for those represented in the experiment or for any other possible treatment in the set sampled, $a(v) = 0$ for each possible treatment v, and $\sigma_A^2 = 0$. Hence the null hypothesis of no treatment effects for Model II is usually written

$$H_0 \colon \sigma_A^2 = 0.$$

Now suppose that this null hypothesis were true. In this case

$$E(\text{MS between}) = \sigma_e^2 \qquad [13.4.1*]$$
$$E(\text{MS within}) = \sigma_e^2 \qquad [13.4.2*]$$

so that *both* mean squares are unbiased estimates of error variance alone.

The mean square between can be shown to be independent of the mean square within under the assumptions of Section 13.1, so that when H_0 is true the ratio

$$F = \frac{\text{MS between}/\sigma_e^2}{\text{MS within}/\sigma_e^2} = \frac{\text{MS between}}{\text{MS within}} \qquad [13.4.3\dagger]$$

can be referred to the F distribution with $J - 1$ and $N - J$ degrees of freedom, as a test of this null hypothesis. Significant values of F lead to a rejection of $H_0 \colon \sigma_A^2 = 0$ in favor of $H_1 \colon \sigma_A^2 > 0$.

13.5   AN EXAMPLE

Suppose that the experiment described in the introduction to this chapter were actually carried out. Five experimenters chosen at random conduct the same experiment, each on a different set of eight subjects randomly sampled and assigned at random among the experimental groups. The data are as shown in Table 13.5.1.

The hypothesis that $\sigma_A^2 = 0$ (that there are no experimenter effects) is to be tested using $\alpha = .01$.

The computations are:

$$\text{SS total} = 1455.94 - \frac{(240.8)^2}{40} = 6.32$$

$$\text{SS between} = \frac{(44.0)^2 + \cdots + (49.3)^2}{8} - \frac{(240.8)^2}{40}$$

$$= 1453.09 - 1449.62 = 3.47$$

$$\text{SS within} = \text{total} - \text{between} = 6.32 - 3.47 = 2.85.$$

Table 13.5.1

		Experimenter		
1	2	3	4	5
5.8	6.0	6.3	6.4	5.7
5.1	6.1	5.5	6.4	5.9
5.7	6.6	5.7	6.5	6.5
5.9	6.5	6.0	6.1	6.3
5.6	5.9	6.1	6.6	6.2
5.4	5.9	6.2	5.9	6.4
5.3	6.4	5.8	6.7	6.0
5.2	6.3	5.6	6.0	6.3
44.0	49.7	47.2	50.6	49.3

$$\sum_{j} \sum_{i} Y_{ij} = 240.8$$

Table 13.5.2 is the summary table. The F value required for rejection at the .01 level for 4 and 35 degrees of freedom is between 4.02 and 3.83 (that is, between the values for 30 and 40 degrees of freedom denominator). Accordingly, the

Table 13.5.2

Source	SS	df	MS	E(MS)	F
Between (experimenters)	3.47	4	.868	$8\sigma_A^2 + \sigma_e^2$	10.72
Within	2.85	35	.081	σ_e^2	
Total	6.32	39			

hypothesis of no experimenter effects may be rejected. There is sufficient evidence to say that experimenter effects exist and contribute to the variance of Y.

13.6   ESTIMATION OF VARIANCE COMPONENTS IN A ONE-WAY ANALYSIS

Instead of estimating effects directly by taking differences of the treatment means from the grand mean, as in the fixed effects model, in Model II we will estimate σ_A^2, the true variance due to treatments. For the foregoing experiment, this is the true variance attributable to experimenters.

Since the expectation of the mean square between is

$$E(\text{MS between}) = n\sigma_A^2 + \sigma_e^2, \qquad [13.6.1]$$

an unbiased estimate of σ_A^2 may be found by taking

$$\frac{\text{MS between} - \text{MS within}}{n} = \text{est. } \sigma_A^2. \qquad [13.6.2\dagger]$$

(Note: est. $\sigma_A^2 = 0$ when MS within $\geqq$ MS between.)

For the example, this estimate is

$$\text{est. } \sigma_A^2 = \frac{.868 - .081}{8} = .098.$$

The variance of Y_{ij} over the population of all possible potential observations is

$$\sigma_Y^2 = E(Y_{ij} - \mu)^2 = \sigma_A^2 + \sigma_e^2$$

so that the true variance consists of two independent parts or components: the variance due to treatments, and that due to error alone. The best estimate of the total variance σ_Y^2 is given by

$$\text{est. } \sigma_Y^2 = \text{est. } \sigma_A^2 + \text{est. } \sigma_e^2 = \frac{\text{MS between} + (n - 1) \text{ MS within}}{n}. \qquad [13.6.3\dagger]$$

For the example,

$$\text{est. } \sigma_Y^2 = .098 + .081$$
$$= .179.$$

This fact that the total variance must consist of two components allows one to make a somewhat more informative use of the estimates of σ_A^2 and σ_e^2. We can take the ratio of the estimated σ_A^2 to the estimated total variance to find the estimated proportion of variance accounted for by the treatments,

$$\text{est. proportion of variance accounted for} = \frac{\text{est. } \sigma_A^2}{\text{est. } \sigma_Y^2} \qquad [13.6.4\dagger]$$

$$\text{est. proportion of variance accounted for by experimenters} = \frac{(.098)}{.179}$$
$$= .55.$$

Here we estimate that over one half of the variance among observations seems to be due to experimenter differences. This would be a most important finding in a real experiment, as it would suggest that different experimenter's repetitions of this experiment would not necessarily be comparable.

At this point it is useful to consider what this particular experiment says about other, similar, experiments. For example, the experiment in the previous section might be repeated with two experimenters and twenty subjects each, or with twenty experimenters and two subjects each. The F values obtained under these two conditions will likely be very different. However, we should expect to find the estimates of the components of variance agreeing relatively well with each other across the experiments. It is the components of variance, rather than the mean squares or the F values, that we should expect to remain relatively the same across experiments of the same type.

In many contexts it is highly informative to estimate variance components and to turn these estimates into proportions of variance accounted for.

Such indices give one of the best ways to decide if a factor is a predictively important one relative to the dependent variable. We have already noted that it is entirely possible for a given factor to show up as statistically significant in a study, even though only a very small percentage of variance is attributable to that factor. This is most likely to happen if the sample n is very large, of course. On the other hand, when there is both significant evidence for effects of a factor *and* evidence that the factor accounts for a relatively large percentage of variance, then this information may be an important key in interpreting the experiment or in deciding how the experimental findings might be applied. Thus, in Model II experiments, in which levels of a factor are sampled, it is good practice to estimate the components of variance, and judge the "predictive" significance of the factor on this basis in addition to the result of the F test.

13.7 THE INTRACLASS CORRELATION COEFFICIENT

One way of expressing the idea that a factor accounts for a given amount of variance is by the index known as the population intraclass correlation coefficient:

$$\rho_I = \frac{\sigma_A^2}{\sigma_A^2 + \sigma_e^2}. \qquad [13.7.1*]$$

The intraclass correlation coefficient for the grand population will be zero only when σ_A^2 is zero, and will reach unity only when $\sigma_e^2 = 0$, given that $\sigma_Y^2 > 0$. Notice that this index is simply another way of expressing the proportion of variance attributable to the factor A. The population index ρ_I is identical to ω^2 in its general form and meaning, although ρ_I applies to the random effects and ω^2 to the fixed effects model so that slightly different estimation methods apply in the two situations.

Quite often the intraclass correlation is used to express the fact that observations in the same category are related, or tend on the average to be more like each other than observations in different categories. The larger the value of ρ_I, the more similar do observations in the same treatment category tend to be, relative to observations in different categories. For example, in a study of the similarity in intelligence of twins, a random sample of sets of twins might be taken, each pair of twins constituting a natural "treatment" with $n = 2$. An estimated intraclass correlation greater than zero would indicate that some of the variability in intelligence is accounted for variation among sets of twins, so that pairs of twins tend to be more alike in this respect than are pairs of nontwins. In general, for two or more classes of observations, the value of $1 - \rho_I$ can be interpreted as the ratio of the expected squared difference between two observations in the same class to that of two observations from different classes. The value of ρ_I is thus a measure of the homogeneity of observations within classes, relative to between classes.

Still another special application of this idea of an intraclass correlation occurs in the study of the reliability of repeated measurements of individuals. The reliability of a single measurement Y_{ij} for an individual j can be defined in a way equivalent to ρ_I for a population of such individuals. The estimate of ρ_I given by **13.6.4** is then an estimate of the true reliability for such measurements. A modern discussion of these methods for estimating reliability in terms of the random effects analysis of variance is given in Winer (1962, Chapter 4).

13.8   INTERVAL ESTIMATION FOR PROPORTION OF VARIANCE ACCOUNTED FOR

Under Model II, it is also possible to estimate intervals for the proportion of variance accounted for by a factor. Here, however, we must assume that the basic distribution of effects, $a(v)$, is normal. Given this assumption, the required limits can be found from the distribution of the quantity θ, where

$$\theta = \frac{\sigma_A^2}{\sigma_e^2} \qquad [13.8.1]$$

and

$$\text{true proportion of variance accounted for} = \frac{\theta}{1 + \theta} = \rho_I. \qquad [13.8.2]$$

The required confidence interval is found as follows: Let F' be the value in an F distribution with $J - 1$ and $N - J$ degrees of freedom, cutting off the *upper* $(\alpha/2)$ proportion in the distribution, and let F'' be the value cutting off the *lower* $(\alpha/2)$. For this distribution, it must be true that

$$\text{prob.}(F'' \leq F \leq F') = 1 - \alpha.$$

Furthermore, regardless of the true value of σ_A^2, it will be true that the ratio

$$\frac{\text{MS between}/(\sigma_e^2 + n\sigma_A^2)}{\text{MS within}/(\sigma_e^2)} \qquad [13.8.3]$$

is distributed as an F variable with $J - 1$ and $N - J$ degrees of freedom, since numerator and denominator *are* independent chi-square variables divided by degrees of freedom when Model II is true, and the basic distribution of effects is normal. This ratio is equivalent to

$$\left(\frac{\text{MS between}}{\text{MS within}}\right)\left(\frac{1}{1 + n\theta}\right), \qquad [13.8.4]$$

so that

$$\text{prob.}\left(F'' \leq \frac{\text{MS between}}{\text{MS within}} \frac{1}{1 + n\theta} \leq F'\right) = 1 - \alpha.$$

By algebraic operation on this inequality, it can be shown that the $100(1 - \alpha)$ percent confidence interval for θ is

$$\frac{1}{n}\left[\frac{\text{MS between}}{(\text{MS within})\ F'} - 1\right] \le \theta \le \frac{1}{n}\left[\frac{\text{MS between}}{(\text{MS within})\ F''} - 1\right]. \quad \text{[13.8.5†]}$$

This may be turned into a confidence interval for ρ_I very easily by the relation

$$\rho_I = \frac{\theta}{1 + \theta}. \quad \text{[13.8.6†]}$$

The value of F' can be found directly from the F tables for $(\alpha/2)$, and the degrees of freedom $J - 1$, and $N - J$. However, the value of F'' must be found by the methods of Section 11.12, for values on the *lower* tail of an F distribution. These confidence intervals, like those for σ^2 given in Chapter 11, need not be optimal.

As an example let us find the 95 percent confidence interval for θ and ρ_I for the data of Section 13.5. Here,

$$\frac{\text{MS between}}{\text{MS within}} = 10.5.$$

The first thing we need is the value of F', that value cutting off the *upper* 2.5 percent in an F distribution with 4 and 35 degrees of freedom. Table V shows this to be about 3.2. Next, we must find F'', the value on the lower tail of this same distribution cutting off the lower 2.5 percent. From the same table we take the value cutting off the *upper* 2.5 percent in a distribution *with 35 and 4 degrees of freedom*: this is about 8.4. Thus,

$$F'' = \frac{1}{8.4} = .119.$$

The approximate 95 percent confidence interval for θ is then

$$\frac{1}{8}\left[\frac{10.72}{3.2} - 1\right] \le \theta \le \frac{1}{8}\left[\frac{10.72}{.119} - 1\right]$$
$$.29 \le \theta \le 11.1.$$

The corresponding interval for ρ_I is

$$\frac{.29}{1 + .29} \le \rho_I \le \frac{11.1}{1 + 11.1}$$

or
$$.224 \le \rho_I \le .917.$$

Over random samples the probability is approximately .95 that an interval such as this covers the true value of ρ_I. Notice that for this example the value 0 *does not* fall into the interval, reflecting the significance of the F test carried out in Section 13.5. This procedure is applicable to the estimation of ρ_I, the proportion of variance accounted for, only in the random effects model, or Model II. *It does* *not apply to the estimation of ω^2 in Model I, the fixed effects situation.*

When the F ratio is small, the confidence intervals for θ and for ρ_I may show negative values for the lower limit. When this happens, the lower limit for the confidence interval can be regarded as equal to zero, since neither true ρ_I nor θ can be negative.

13.9 OTHER HYPOTHESES TESTABLE USING MODEL II

The hypothesis tested in the example above was that there exist no effects of the individual treatments, so that $\sigma_A^2 = 0$. The theory is not limited to this situation, however. When the distribution of effects can be assumed to be normal it is possible to test many other hypotheses about the ratio $\theta = \sigma_A^2/\sigma_e^2$, or, equivalently, about ρ_I.

In the general case, the hypothesis tested is

$$H_0: \theta \leq \theta_0 \text{ versus } H_1: \theta_0 < \theta$$

where θ_0 is any hypothetical value, $0 \leq \theta_0$.

This general hypothesis is tested as follows: after α is chosen, we determine the F value cutting off the upper α proportion of cases in a distribution with $J - 1$ and $N - J$ degrees of freedom. Call this value F_α. Now we find the F ratio for the sample in the usual way (Section 13.4). The hypothesis is rejected in favor of H_1 when

$$\frac{\text{MS between}}{\text{MS within}} \geq (1 + n\theta_0)F_\alpha. \qquad [13\ 9.1\dagger]$$

For example, suppose that in planning a study the experimenter was undecided about whether a given factor A should be controlled or allowed to vary. The experimenter made up his mind that he would control for A only if he could be confident that A accounted for at least 10 percent of the variance in the dependent variable, as judged from a preliminary experiment. This is equivalent to a true ρ_I of .10. The pilot experiment was carried out, with a random selection of seven levels of factor A and three observations chosen at random in each level. The F ratio for this experiment, with 6 and 14 degrees of freedom, turned out to be 4.0. The F_α value required for significance at the 5 percent level and for these degrees of freedom is 2.85.

Since the question involves a percentage of variance accounted for, or ρ_I, this must be turned into the equivalent value of θ_0. If the variance accounted for were exactly .10 of the total, then

$$\frac{\sigma_A^2}{\sigma_e^2 + \sigma_A^2} = \frac{\theta_0}{1 + \theta_0} = .10$$

$$\theta_0 = \frac{1}{9}.$$

Hence,

$$F_\alpha(1 + n\theta_0) = 2.85\left(1 + \frac{3}{9}\right) = 3.80.$$

The obtained value of F exceeds this, and so the experimenter concluded that more than 10 percent of the variance is attributable to A, and that the factor A should be controlled in the experiment.

13.10   THE POWER OF THE F TEST UNDER MODEL II

It was noted above that when there is a value of $\theta > 0$ which characterizes the experimental population, the random variable

$$\frac{\text{MS between}}{(\text{MS within})(1 + n\theta)}$$

actually is distributed as F with $J - 1$ and $J(n - 1)$ degrees of freedom. On the other hand, the ordinary ratio (MS between/MS within) is not distributed as F in this instance.

The problem in finding the power of the F test when a given value of θ is true is that of finding the probability

$$\Pr\left(\frac{\text{MS between}}{\text{MS within}} > F_\alpha;\, \theta,\, \nu_1,\, \nu_2\right).$$

That is, the power of the test corresponds to the probability of rejecting the null hypothesis at the α level, when there are ν_1 and ν_2 degrees of freedom, and the particular value of θ is indeed true.

Now since [MS between/(MS within)$(1 + n\theta)$] is distributed as F, it follows that

$$\Pr\left[\frac{\text{MS between}}{\text{MS within}} > F_\alpha\right] = \Pr[F(1 + n\theta) > F_\alpha]$$
$$= \Pr[F > F_\alpha/(1 + n\theta)].$$

This last probability can be calculated from the distribution of F, and thus the power can be determined. Bear in mind, however, that this method of determining power depends upon the assumption that effects are sampled randomly from a normal distribution of such effects. This method may also require fairly extensive tables of the distribution of F, and such tables are not always easy to find. [One reasonably extensive set of tables for F is found in Dixon and Massey (1957)]. On the other hand, the method of approximating F distributions as beta distributions, and by use of binomial tables, as discussed in Chapter 11, can be useful here.

As an example, suppose that it were required to find the power of an F test with 2 and 21 degrees of freedom, when the true value of $\rho_I = .20$. This implies that $\theta = .25$. The value of n in this experiment is 8. The .01 significance level is to be employed. Then we find that $F_{.01} = 5.78$, so that $F_{.01}/[1 + 8(.25)] = 1.928$. We want the probability that an F variable equals or exceeds this value in a distribution with 2 and 21 degrees of freedom.

If we let

$$x = \frac{[\nu_1 F/(1 + n\theta)]}{[\nu_1 F/(1 + n\theta)] + \nu_2},$$

then the value of x can be referred to a beta distribution with $r = \frac{1}{2}\nu_1$ and $N - r = \frac{1}{2}\nu_2$. Furthermore, the cumulative probability for x can be found from cumulative binomial probabilities as shown in Chapters 8 and 11. Thus, we find

$$x = \frac{2(1.928)}{2(1.928) + 21}$$
$$= .155.$$

The cumulative probability for x corresponds to one minus the cumulative binomial probability. However, since the power of the test will be one minus the cumulative probability for x, in this instance the cumulative binomial probability actually gives the power. We take

$$\text{cum.pr.}_{\text{bin}}[\tfrac{1}{2}\nu_1 - 1; \; p = x, \; N' = \tfrac{1}{2}(\nu_1 + \nu_2) - 1],$$

which can be calculated most closely from a binomial distribution with $p = .15$ and $N' = 11$. We require the cumulative probability for 0 successes in such a distribution, and this turns out to be .1673. Hence, when $\theta = .25$, and $J - 1 = 2$, $J(n - 1) = 21$, $n = 8$, and $\alpha = .01$, the F test has a power of .17, approximately.

The power of the F test can be used to determine the value of n, the size of each experimental group. Naturally, one must have an idea of the magnitude of ρ_I or of θ that he wants to be reasonably certain of detecting as a significant result. Then, as n is varied, it is a fairly simple matter to explore the consequences for the test's power against these alternatives.

13.11   IMPORTANCE OF THE ASSUMPTIONS IN MODEL II

By now it should surely be clear that the actual *arithmetic* of the simple analysis of variance is the same regardless of the model adopted. The analysis of variance is, strictly speaking, only a way of arranging this arithmetic. However, the inferences made from the sample values are really quite different, depending on the model invoked. All the inferences made under Model I concern means (and differences between means). On the other hand, the inferences made using Model II deal with variances: that is, *the basic inference has to do with the variance of the population of effects actually sampled by the experimenter.* This distinction between the two models has an influence on the importance of the assumptions made in each.

In the first place, the assumption of normality can be quite important in Model II. Provided that one is concerned only with a test of the hypothesis $\sigma_A^2 = 0$, then slight departures from normality among the error distributions should have only minor consequences for the conclusions reached using reasonably large sample size. Here, no assumption at all about the distribution of the a values need be made. On the other hand, for tests of the more general hypothesis given in Section 13.9, it is most important that the distributions *both* of effects and of errors be normal in form. In the same way, interval estimates for θ and for ρ_I depend heavily upon the assumptions of normal distributions of the a and the e values for their validity.

The assumption of equality of variance has a somewhat different status in the random effects than in the fixed effects model. If the random effects model applies, and if the effects a_j are independent of the errors e_{ij}, we need assume only one distribution of errors with variance σ_e^2, so that in a sense, the error variance for all possible observations given some treatment j must be the same as for any other treatment. However, the important assumptions are those involving independence: the errors *must* be independent both of the particular treatment effects and of each other. The random effects model is not really applicable to data where the *errors* in observations must be related to the effect of the treatment applied, or to each other. Examples of situations where errors are related are most often those in which subjects are given repeated trials at some task, and some trial-related change such as learning occurs, making the observed scores depend on each other in some serial fashion. In many learning situations, the occurrence of an error makes the occurrence of another error of the same magnitude or direction less likely than otherwise, also introducing such a statistical dependence.

However, as always, we must be very careful not to let a purely statistical consideration intrude on experimental practice. The mere fact that an experiment involves repeated observations or observations on matched subjects does not necessarily imply a serious violation of the statistical assumptions. For example, in some situations the careful experimenter can plan his study in such a way that an initial "warm-up" or practice session does away with this form of dependency among observations, and then an analysis of variance may be fully applicable to the experimental data. In other situations, still other experimental devices may exist for controlling for such dependencies. These purely statistical requirements must not be interpreted as prohibitions against certain kinds of experiments, but only as warnings that special care should be taken in planning the experiment itself if a particular form of analysis is desired.

Although statistical dependency among the error components of scores creates a special problem in any analysis of variance, it is not true that the *observations themselves* must be completely independent for Model II to apply. We have just seen that the intraclass correlation coefficient expresses the degree of similarity or dependency among observations given the same treatment. When σ_A^2 is greater than zero some statistical relation must exist between pairs of score values within the possible treatment groups. As we saw in Section 13.9, hypotheses about the degree of relatedness can be tested via Model II. However, the form of relatedness that makes the simple Model II analysis inapplicable is most often a trend or serial relation, ordinarily implying nonindependence of the error terms. It is this sort of nonindependence that makes a special problem for studies involving repeated measurements of the same individuals.

Finally, it should be emphasized that the application of Model II analysis of variance depends on two kinds of random sampling: the treatments or factor levels themselves must be sampled at random, and then within each treatment group a random sample of individual observations must appear. The inferences drawn under this model are not strictly valid unless *both* forms of random sampling are actually carried out.

All in all, the Model II analysis of variance is more heavily dependent upon the assumption of normal distributions, both of treatment effects and of errors, than is Model I. In a way, one gets more for his money in a Model II one-way analysis: an inference is permitted that goes beyond the factor levels actually observed, interval estimates of strength of association can be made, and a general hypothesis about ρ_I can be tested. None of these things have direct parallels in the fixed effects models. However, some of these "additional" inferences are paid for by relatively more restrictive assumptions.

13.12   Two-factor experiments with sampling of levels

Now we turn to the analysis of experiments involving two factors, each of which is represented in the experiment only by a sample of its levels.

For example, suppose that a projective test involves ten cards administered individually to a subject. The subject must respond by giving as many free-associations to each card as he is able. These responses are scored in a number of ways, but the total number of responses is an important index of overall performance. The developer of this test has some idea that the *order* in which cards are presented has a bearing on the total number of responses given, and so he would like to see if this factor of order is an important one in accounting for variation in test performance. If it should turn out that order is important, he will try to find an order of presentation for the cards that will be optimal in evoking responses. Furthermore, this psychologist has worked out a standard set of instructions for test administrators, in the hope that administrator effects on performance are thereby made negligible. However, he would like to see if this is the case.

Here are two factors, order of presentation and test administrator, that may account for variance in total response to the test. Obviously, neither factor can be represented at all levels in any experiment, since there are exactly 3,628,800 different ways of presenting the ten cards, and a very large number of persons who might be trained to administer the test. Hence, the experimenter decides to conduct a study in which each of these factors will be sampled. From a single set of data he will be able to answer the question of the relative contribution to variance of each of the factors, as well as the secondary question of possible interaction between test administrator and order of presentation.

This hypothetical experiment will be developed as an example in Section 13.15. For the time being we will turn our attention to extending Model II to cover such situations.

13.13   Model II for two-factor experiments

The two different factors in the experiment will be designated A and B. A random sample of C different levels of A will be drawn, and shown as columns in the data table, and a random sample of R different levels of B will appear as the rows in the table. Within each combination shown by a cell of the data table, n observations are to be made at random.

The score of individual i in column j and row k of the table will be thought of as a sum:

$$Y_{ijk} = \mu + a_j + b_k + c_{jk} + e_{ijk}.$$

Here, a_j is the random variable standing for the effect of the sample treatment appearing in the data table as column j, b_k the random variable indicating the effect of the sample treatment in row k, and c_{jk} is the random interaction effect associated with cell jk. The term e_{ijk} is a random variable standing for the error effect of the observation of individual i under the joint conditions indicated by column j and row k. Observe that in this linear model all components of a score except the grand mean μ are values of *random variables*, the sampled effects.

The assumptions made are:

1. The a_j are normally distributed random variables with mean zero and variance σ_A^2.
2. The b_k are normally distributed with mean zero and variance σ_B^2.
3. The c_{jk} have a bivariate normal distribution (Section 15.26) with mean zero and variance σ_{AB}^2.
4. The e_{ijk} are normally distributed with mean zero and variance σ_e^2.
5. The a_j, the b_k, the c_{jk}, and the e_{ijk} are pair-wise independent.

13.14 THE MEAN SQUARES

The computations for the two-way analysis under Model II are exactly the same as for Model I (Section 12.31). Thus, the total sum of squares is partitioned into a sum of squares for rows, a sum of squares for columns, a sum of squares for interaction, and an error sum of squares.

On examining the mean squares, however, we will find that the expectations are quite different from those for Model I. Consider the mean square for rows. In terms of the model, Section 13.13,

$$\text{MS rows} = \frac{\sum_k Cn(M_k - M)^2}{R - 1}$$

where

$$M_k = \frac{\sum_j \sum_i (\mu + a_j + b_k + c_{jk} + e_{ijk})}{Cn}$$

$$= \mu + M_a + b_k + c_k + M_{ek} \qquad [13.14.1]$$

and

$$M_a = \frac{\sum_j a_j}{C}; \qquad M_b = \frac{\sum_k b_k}{R}$$

$$b_k = \frac{\sum\limits_j b_k}{C}; \qquad a_j = \frac{\sum\limits_k a_j}{R}$$

$$c_k = \frac{\sum\limits_j c_{jk}}{C}; \qquad c_j = \frac{\sum\limits_k c_{jk}}{R}$$

$$M_{ek} = \frac{\sum\limits_j \sum\limits_i e_{ijk}}{Cn}; \; M_{ej} = \frac{\sum\limits_k \sum\limits_i e_{ijk}}{Rn}.$$

Furthermore,

$$M = \frac{\sum\limits_k \sum\limits_j \sum\limits_i (\mu + a_j + b_k + c_{jk} + e_{ijk})}{RCn}$$

$$= \mu + M_a + M_b + M_c + M_e \qquad [13.14.2]$$

where

$$M_c = \frac{\sum\limits_k \sum\limits_j c_{jk}}{RC} = \frac{\sum\limits_k c_k}{R} \qquad [13.14.3]$$

$$M_e = \frac{\sum\limits_k \sum\limits_j \sum\limits_i e_{ijk}}{RCn} = \frac{\sum\limits_k M_{ek}}{R}.$$

Then, for any row k,

$$(M_k - M)^2 = (b_k - M_b + c_k - M_c + M_{ek} - M_e)^2. \qquad [13.14.4*]$$

At this point, one of the really important differences between Models I and II appears. You may recall that in Section 12.26 it was pointed out that for the fixed effects model the following things are assumed true of the interaction effects γ_{jk}:

$$\sum_j \gamma_{jk} = 0, \sum_k \gamma_{jk} = 0, \text{ and } \sum_j \sum_k \gamma_{jk} = 0.$$

That is, in Model I we assume that over the column treatments the interaction effects sum to zero, that over the row treatments interaction effects sum to zero, and that over both rows and columns these effects sum to zero. The net result is that in the sum of squares for, say, rows, only row effects and error can possibly be included:

$$(M_k - M)^2 = (\beta_k + M_{ek} - M_e)^2.$$

For the fixed effects model, any interaction effect cannot possibly contribute to the sum of squares for rows because this sum of squares is itself based on the data *summed* over columns, automatically making the interaction effects sum to zero. For the same reason, the sum of squares for columns does not include any of the interaction effects.

On the other hand, in Model II the set of R row treatments represents a *sample* from a large set of possible such treatments, and the C column treatments another sample from a large set. Here it is reasonable to define interaction effects so that

$$E(c_{jk}) = 0$$
$$\phantom{E(c_{jk}) = 0}_{j}$$

$$E(c_{jk}) = 0$$
$$\phantom{E(c_{jk}) = 0}_{k}$$

$$E(c_{jk}) = 0.$$
$$\phantom{E(c_{jk}) = 0}_{jk}$$

That is, the expected value of the interaction term c_{jk} is zero over *all possible* treatments that might have been selected for column j with a fixed row treatment k. Furthermore, the expected value of the interaction term c_{jk} is zero over all possible treatments that might have been selected for row k with a fixed column treatment j. However, these requirements hold only when one considers *all possible* row or column treatments that might have been selected. Unlike the situation in the fixed effects model, *here there is no requirement at all that the interaction effects sum to zero over the particular set of R row treatments or over the particular set of C column treatments that just happened to appear in the sample.* For this reason, none of c_k or c_j or M_c need be zero in any given set of data. The important consequence of this fact is that *the sum of squares for rows reflects deviations due not only to row treatment effects, but also to interaction effects.* In the same way *the sum of squares for columns includes both column and interaction effects.* This has a most important bearing on estimates and tests made under Model II, as we shall see.

Resuming the argument for the expected mean square for rows, we note that in expression **13.14.4** no term a_j can appear in any one of the squares entering into the mean square, but that each such square includes a term $(c_k - M_c)$ due to interaction. As pointed out above, this is due to the fact that the values c_{jk} summed over j for any particular k *need not* be zero. For this reason it can be shown that

$$E(\text{MS between rows}) = \sigma_e^2 + n\sigma_{AB}^2 + Cn\sigma_B^2. \qquad [13.14.5^*]$$

Unlike the expectation for fixed effects, *here the mean square for rows estimates a weighted sum of the error variance, the variance due to row treatments, and the interaction variance,* σ_{AB}^2.

A parallel situation holds for factor A, represented by the columns :

$$E(\text{MS columns}) = E \frac{(\text{SS columns})}{C - 1}$$
$$= \sigma_e^2 + n\sigma_{AB}^2 + Rn\sigma_A^2. \qquad [13.14.6^*]$$

Turning to the expected value for the mean square interaction, we find that

$$E(\text{MS interaction}) = E \frac{(\text{SS interaction})}{(R - 1)(C - 1)}$$
$$= \sigma_e^2 + n\sigma_{AB}^2. \qquad [13.14.7^*]$$

The expectation of the mean square error is, as always,

$$E(\text{MS error}) = E \frac{\left[\sum_j \sum_k \sum_i (e_{ijk} - M_e)^2\right]}{RC(n - 1)} = \sigma_e^2.$$

[13.14.8*]

13.15 HYPOTHESIS TESTING IN THE TWO-WAY ANALYSIS UNDER MODEL II

We have just seen that the expected values both for the mean square for rows and the mean square for columns are different in Model II from those in Model I. The practical implication of this fact is that the hypotheses of no row and no column effects are tested in a different way for this model than for Model I. Consider the test of the hypothesis

$$H_0: \sigma_A^2 = 0.$$

When this hypothesis is true, then

$$E(\text{MS columns}) = \sigma_e^2 + n\sigma_{AB}^2$$

[13.15.1*]

which is *not* the same as the expectation of mean square error, but rather that of *mean square interaction:*

$$E(\text{MS interaction}) = \sigma_e^2 + n\sigma_{AB}^2.$$

[13.15.2*]

These two mean squares (MS columns and MS interaction) can be shown to be independent, and when each is divided by $\sigma_e^2 + n\sigma_{AB}^2$, each is distributed as χ^2 divided by degrees of freedom. Thus, the hypothesis of no column effects is usually tested by

$$F = \frac{\text{MS columns}}{\text{MS interaction}}$$

[13.15.3†]

with $C - 1$ and $(R - 1)(C - 1)$ degrees of freedom.

In the same way, the hypothesis

$$H_0: \sigma_B^2 = 0$$

is tested by a comparison of MS rows with the interaction MS, since if this hypothesis is true,

$$E(\text{MS rows}) = \sigma_e^2 + n\sigma_{AB}^2,$$

[13.15.4*]

which is the same as the expected value of the mean square for interaction. The hypothesis is tested by the ratio

$$F = \frac{\text{MS between rows}}{\text{MS interaction}},$$

[13.15.5†]

with $R - 1$ and $(R - 1)(C - 1)$ degrees of freedom. The hypothesis of no interaction is tested just as for Model I:

$$H_0: \sigma_{AB}^2 = 0$$

tested by

$$F = \frac{\text{MS interaction}}{\text{MS error}} \qquad [13.15.6\dagger]$$

with $(R - 1)(C - 1)$ and $RC(n - 1)$ degrees of freedom.

It should be emphasized that the test of row and column effects against interaction is really appropriate *only* when the factors have been randomly sampled, as in Model II (and in some instances of the mixed models introduced in sections to follow). When the experiment qualifies for Model I, the ratio of the mean square for a factor to the interaction mean square is *not necessarily* distributed as F, even when the null hypothesis is true. This is one of the major practical distinctions between the fixed effects and the random effects models.

There is one situation, however, when the main effects are not tested against interaction in Model II. This occurs when the experimenter has decided that the interaction variance σ_{AB}^2 is zero. There is either some theoretical or some empirical reason to believe that interaction effects should not occur, or there is *strong* evidence for no interaction in the data. Of course, there is a difficult problem involved in deciding when there really *is* strong evidence that interaction effects do not exist. When the mean square for interaction is equal to or less than the mean square for error, we are ordinarily quite justified in reaching this conclusion. However, a simple rule that applies to a broad class of situations is that due to Paull (1950): **when each of the mean squares for error and for interaction has a number of degrees of freedom greater than six, then pool these into a combined estimate of error when the F ratio for interaction is less than 2.00.**

When there is sufficient evidence that no interaction effects exist, the two sums of squares, error and interaction, are pooled as follows:

$$\text{pooled MS error} = \frac{\text{SS interaction} + \text{SS error}}{(R - 1)(C - 1) + RC(n - 1)}. \qquad [13.15.7\dagger]$$

When $\sigma_{AB}^2 = 0$, the expectation of this pooled MS error is σ_e^2. The expectation of MS rows is also σ_e^2 under the hypothesis of no row effects, so that the test may be carried out by use of the ratio

$$F = \frac{\text{MS rows}}{\text{pooled MS error}} \qquad [13.15.8\dagger]$$

with $R - 1$ and $RCn - R - C + 1$ degrees of freedom. In the same way a test of no column effects is carried out by

$$F = \frac{\text{MS columns}}{\text{pooled MS error}}, \qquad [13.15.9\dagger]$$

with $C - 1$ and $RCn - R - C + 1$ degrees of freedom. This procedure is *not* recommended, however, unless the evidence for no interaction effects is strong enough to satisfy a criterion such as Paull's.

As an example, suppose that developer of the projective test mentioned in Section 13.12 actually carried out the experiment. Recall that this study was designed to test both for the presence of effects associated with order of presentation and with particular administrators of the test. Four persons were selected at random from among psychologists and trained to administer the test. Also, four orders of presentation of the test were selected at random. Each administrator gave the test to a different pair of randomly selected normal adults under each one of the selected order conditions, so that a total of thirty-two different test performances were observed in all. The dependent variable was the total number of responses a subject gave to the test cards. The experimenter was prepared to assume normal distributions of order effects, of administrator effects, of interaction effects, and of error. The data were as shown in Table 13.15.1. The α level chosen for the test of each hypothesis was .05.

Table 13.15.1

Order	Administrator 1	2	3	4	Totals
I	26 25	30 33	25 23	28 30	220
II	26 24	25 33	27 17	27 26	205
III	33 27	26 32	30 24	31 26	229
IV	36 28	37 42	37 33	39 25	277
Totals	225	258	216	232	931

The analysis proceeds in the usual manner for two-way designs (Section 12.31):

$$\text{SS total} = \sum_j \sum_k \sum_i y_{ijk}^2 - \frac{\left(\sum_j \sum_k \sum_i y_{ijk}\right)^2}{N}$$

$$= 27965 - \frac{(931)^2}{32}$$

$$= 878.7$$

$$\text{SS rows} = \frac{\sum_k \left(\sum_j \sum_i y_{ijk}\right)^2}{Cn} - \frac{\left(\sum_j \sum_k \sum_i y_{ijk}\right)^2}{N}$$

$$= \frac{(220)^2 + \cdots + (277)^2}{8} - \frac{(931)^2}{32} = 363.0$$

$$\text{SS columns} = \frac{\sum_j \left(\sum_k \sum_i y_{ijk}\right)^2}{Rn} - \frac{\left(\sum_j \sum_k \sum_i y_{ijk}\right)^2}{N}$$

$$= \frac{(225)^2 + \cdots + (232)^2}{8} - \frac{(931)^2}{32} = 122.3$$

$$\text{SS error} = \sum_j \sum_k \sum_i y_{ijk}^2 - \frac{\sum_j \sum_k \left(\sum_i y_{ijk}\right)^2}{n}$$

$$= 27965 - \frac{(51)^2 + (50)^2 + \cdots + (64)^2}{2}$$

$$= 310.5$$

$$\text{SS interaction} = \text{SS total} - \text{SS rows} - \text{SS columns} - \text{SS error}$$

$$= 878.7 - 363.0 - 122.3 - 310.5 = 82.9.$$

Table 13.15.2 summarizes the analysis.

Table 13.15.2

Source	SS	df	MS	E(MS)	F
Rows (orders)	363.0	3	121.0	$\sigma_e^2 + 2\sigma_{AB}^2 + 8\sigma_B^2$	$\left(\dfrac{121.0}{15.7} = 7.7\right)$
Columns (admin.)	122.3	3	40.8	$\sigma_e^2 + 2\sigma_{AB}^2 + 8\sigma_A^2$	$\left(\dfrac{40.8}{15.7} = 2.6\right)$
Interaction	82.9	9	9.2	$\sigma_e^2 + 2\sigma_{AB}^2$	
Error	310.5	16	19.4	σ_e^2	
Totals	878.7	31			

Ordinarily, the tests both for row and for column effects would be carried out by dividing the mean square for rows or columns by mean square interaction. However, it is immediately apparent that here the mean square for error is larger than mean square for interaction, so that the estimate of σ_{AB}^2 is zero. Notice that the Paull criterion is also satisfied. Thus, we find

$$\text{pooled MS error} = \frac{82.9 + 310.5}{9 + 16} = 15.7,$$

which has $9 + 16$ or 25 degrees of freedom.

We test for row effects by finding

$$F = \frac{121.0}{15.7} = 7.7.$$

For 3 and 25 degrees of freedom, this value exceeds that required for the .05

level, and so the hypothesis of no row effects is rejected. The same procedure is carried out for columns, and here

$$F = \frac{\text{MS columns}}{\text{pooled MS error}} = 2.6$$

for 3 and 25 degrees of freedom. This fails to reach the value of 2.99 required to reject the null hypothesis. (These F values are enclosed in parenthesis in the summary table to show that they are not obtained in the usual way. In reports of such data, the F tests would usually be accompanied by an explanatory footnote to the summary table.)

The conclusions from the experiment are, then, as follows:

1. Order does seem to have an effect on the total number of responses given in the test.
2. There is insufficient evidence to determine if true administrator differences exist.
3. There is virtually no evidence for administrator-order interaction.

Incidentally, the comment in Section 12.36 about the statistical dependency of F tests based on the same denominator mean square applies to these results. The prior probability that at least one of the three significance tests carried out leads to a Type I error is not equal to the nominal α value of .05, but rather is some undetermined value between .05 and .15. Similarly, the proportion of significant results representing Type I error need not be .05. There is no simple way of determining how much the significance level given by one of these tests depends on the levels given by the others.

13.16   Point estimation of variance components

The four expressions in the $E(\text{MS})$ column of the summary table allow us to estimate the components of variance, as follows:

$$\text{est. } \sigma_B^2 = \frac{\text{MS rows} - \text{MS interaction}}{Cn} \qquad \text{[13.16.1†]}$$

$$\text{est. } \sigma_A^2 = \frac{\text{MS columns} - \text{MS interaction}}{Rn} \qquad \text{[13.16.2†]}$$

$$\text{est. } \sigma_{AB}^2 = \frac{\text{MS interaction} - \text{MS error}}{n} \qquad \text{[13.16.3†]}$$

$$\text{est. } \sigma_e^2 = \text{MS error.} \qquad \text{[13.16.4†]}$$

If any of the first three estimates turn out to be negative in sign, then that component is estimated to be zero.

In the example above, the estimate of σ_{AB}^2 is zero, and the pooled MS error was taken to represent σ_e^2. Thus, to be consistent, one should estimate the other components as follows:

$$\text{est. } \sigma_B^2 = \frac{\text{MS rows} - \text{pooled MS error}}{Cn}$$

$$= \frac{121.0 - 15.7}{8} = 13.16$$

$$\text{est. } \sigma_A^2 = \frac{40.8 - 15.7}{8} = 3.14$$

$$\text{est. } \sigma_e^2 = 15.7.$$

The proportion of variance accounted for by factor B (rows) is estimated from

$$\frac{\text{est. } \sigma_B^2}{\text{est. } (\sigma_e^2 + \sigma_{AB}^2 + \sigma_A^2 + \sigma_B^2)} = \frac{13.16}{15.7 + 0 + 13.16 + 3.14} = .41.$$

An estimated 41 percent of the variance in total response is attributable to order of presentation of this projective test.

For factor A (columns) the estimated proportion of variance accounted for is

$$\frac{3.14}{15.7 + 0 + 13.16 + 3.14} = .098$$

so that if factor A accounts for any variance at all, our best guess is that this is less than 10 percent of the total variance.

It is possible to find a confidence interval for

$$\theta = \frac{\sigma_A^2}{\sigma_e^2 + n\sigma_{AB}^2}$$

by the method of Section 13.8. Unfortunately, this confidence interval cannot be turned directly into a confidence interval for proportion of variance accounted for, as in the one-way case. Thus, its use is rather more limited. Approximate confidence intervals for the variance components σ_A^2, σ_B^2, and so forth, can be found by methods outlined in Scheffé (1959, pp. 231–235).

13.17 MODEL III: A MIXED MODEL

Multifactor experiments involving Model II are relatively rare in psychological research. It is far more common to encounter experiments where one or more factors have fixed levels and the remaining factors are sampled. This situation calls for a third model of data, in which each individual observation results in a score that is a sum of *both* fixed and random effects. Obviously, mixed models such as this apply only to experiments where two or more factors are under study.

As an example of an experiment fitting a mixed model, suppose that a study is concerned with the muscular tension induced in subjects by three different varieties of task. The subject performs using pencil and paper with the preferred hand, meanwhile holding the bulb of a sensitive pressure recording gauge in the other. The mean reading on this gauge during the performance provides the dependent variable score. Three separate kinds of tasks are administered: in one type the subject solves relatively complicated problems in arithmetic; in a second type he writes a short, imaginative, composition; and in the third he must make a careful tracing of a line drawing.

Now it is obvious that individual differences may exist among subjects with respect to the characteristic pressure each will exert with the nonpreferred hand. Furthermore, over the three tasks, some systematic, trial-related, change might occur. For these reasons the experimenter decides to use the same subjects over all three tasks, and to give each type of task twice to each subject, making a total of six performances to be carried out by any subject. The actual order of presentation for these six performances is chosen at random and separately for each subject. After a preliminary period of practice on a neutral task designed to accustom him to the experimental arrangement, each subject carries out the six performances, two of each type, in the random order previously selected for him. Some six subjects are selected at random for participation in this experiment.

This experiment illustrates several things. In the first place, the experimenter is interested in three, and only three, types of task that might be given to the subject. Therefore the factor of "tasks" is properly considered to be composed of three fixed levels; this is a fixed effects factor. On the other hand, the experimenter is not at all interested in any particular subject or small set of subjects, but rather in the large set of potential subjects who might be observed under these task conditions. Granted that the only subjects about whom an inference is to be made are those of a certain age, sex, ability to understand instructions, and so forth, the experimenter would, nevertheless, like to extend his inference to all possible such subjects. The factor of "subjects" consists of a great many possible levels, from which only a few are selected at random for representation in the experiment. Hence, "subjects" corresponds to a random effects factor.

Notice that by giving each type of problem to each subject, a form of control over individual differences among subjects is introduced, since a possibility such that all of the relaxed, easy-going subjects in the sample might accidentally pile up in one experimental condition, and all of the nervous wrecks in another, is thereby eliminated. We will have more to say about this form of experimental control in succeeding sections.

Finally, the experimenter knows that the order in which a subject receives the various tasks and repetitions of tasks might also account for variance in the experiment. Since he cannot use enough subjects to introduce this factor as a systematic feature of the experiment, he does the next best thing by giving each subject a different, randomly selected, order of performance of the tasks and their repetitions. In this way, the possible effects of order are effectively "mixed up" with the possible effects of individual subjects; the sampled factor actually may be regarded as "subject-order" combinations. However, this is irrelevant to the purposes of the experimenter, who is most interested in possible effects associated with the tasks, rather than in the study of individual differences among subjects.

This is a fairly straight-forward and typical example of an experiment calling for a mixed model of the analysis of variance. Next we will turn to a study of the model itself before pursuing this example further. We shall find that the combination of fixed and random effects in the same experiment requires no change at all in the computational procedures for a simple two-factor design such as this. However, it does make an important difference in the F ratios employed to test for the different effects.

13.18 A MIXED MODEL FOR A TWO-FACTOR EXPERIMENT

Let the factor having fixed levels be labeled A, and represented by the columns of the table, and let the randomly sampled factor be B, and shown by the rows of the data table. Now it is assumed that

$$Y_{ijk} = \mu + \alpha_j + b_k + c_{jk} + e_{ijk} \qquad [13.18.1*]$$

where α_j is the fixed effect of the treatment indicated by the column j, b_k is the random variable associated with the kth row, c_{jk} the random interaction effect operating in the cell jk, and e_{ijk} is the random error associated with observation i in the cell jk. We make the following assumptions.

1. The b_k and the c_{jk} are jointly normal, each with mean of zero and with variances σ_B^2 and σ_{AB}^2 respectively.
2. The e_{ijk} are normally distributed, with mean zero and variance σ_e^2.
3. The e_{ijk} are independent of the b_k and the c_{jk}.
4. Over the possible rows k, the degree of dependence between any pair of observations in different columns j and j' is the same for all pairs of columns.

This last assumption requires some elaboration. It will not, in general, be true that the observations within any *sampled* level (row) will be independent in this model. Observations in the same sample level from factor B but in different levels on factor A (columns) may be dependent or related to some extent. Here we are assuming that over the possible levels of B, the degree of statistical relation is the same for observations in any pair of levels on A. This assumption is very questionable for many kinds of data, especially in psychological research, but it must be made to justify the test of A effects given below. A much more advanced method of testing this hypothesis exactly without assumption 4 is given in Scheffé (1959, pp. 270–273). However, our discussion will be restricted to the case where assumption 4 is justified. When this assumption is very questionable, approximate, though conservative, procedures outlined in Winer (1962, p. 123) may be employed.

13.19 THE EXPECTED MEAN SQUARES IN A MIXED MODEL

The partition of the sum of squares and the calculation of the mean squares proceed exactly as for any two-way analysis of variance. However, let us look into the composition of the mean square between columns (the fixed effects factor) under the model outlined in Section 13.18. Basically, the mean square between columns is found from the squared deviations of column means from the general mean of the data. In the mixed model, the squared difference between M_j, the mean of a particular column j, and M, the general mean, can be thought of as having the following composition:

$$(M_j - M)^2 = (\alpha_j + M_b + c_j + M_{ej} - M_b - M_c - M_e)^2$$
$$= (\alpha_j + c_j - M_c + M_{ej} - M_e)^2.$$

Here, M_b, c_j, M_{ej}, and M_e are all defined just as in Section 13.14. Notice that **any deviation of a column mean from the grand mean includes not only the column effect and a mean error deviation, but also a deviation due to inter-action effects,** $c_j - M_c$. The reason for this is exactly the same as given for the occurrence of interaction effects in expression **13.14.4**: there is no requirement at all that interaction effects must sum to zero across rows when the row treatment effects, and thus the interaction effects, are only a random sample of such effects. Thus column means, formed by summing observations across rows, include not only the fixed column effects, but also some of the random interaction effects as well.

For this reason, given the mean square for columns calculated in the usual way, it can be shown that the expectation for this mean square is

$$E(\text{MS columns}) = \sigma_e^2 + n\sigma_{AB}^2 + \frac{Rn \left(\sum_j \alpha_j^2 \right)}{C - 1}. \qquad [13.19.1^*]$$

In Model III, the mean square for columns (the fixed effects factor) is an estimate of a weighted sum of the error variance, the sum of the squared effects associated with the column factor itself, *and* the interaction variance.

This often strikes students as a violation of intuition, but a little reflection should convince you that this is reasonable. Consider the example of Section 13.17 once again. Why should the means corresponding to the different tasks (columns) in the experiment tend to differ? In the first place there actually may be systematic differences between the population means, or effects, connected with the different tasks. Furthermore, since we deal only with sample means rather than with population means, a certain amount of random error may enter into the average differences we observe between the tasks. However, there is still another reason why the column means might differ in any sample. Any given subject might find one or two of the tasks quite stressful and tension-producing and the others quite easy. Furthermore, the pattern of such hard and easy tasks might vary over subjects, so that the effect that a task has *within* a particular subject might be different across subjects. Subject 1 finds arithmetic problems extremely challenging, although imaginative composition is a breeze for him. On the other hand, subject 2 finds arithmetic problems no challenge at all, but is brought to a standstill when faced with a writing task. Such systematic differences among the tasks *within* particular subjects, but differing from one subject to another, is what we mean by "subject by task interaction." Since subjects were sampled for the experiment, it is entirely possible for such interaction effects to tend to "pile up" in one or another of the columns of the data table, and thus to contribute to the apparent differences between means for the tasks; whether or not this happens depends on the particular sample of subjects we obtain. Such apparent differences among means need not be due to the main effects α_j of the tasks themselves, since interaction effects representing differences within *particular* subjects need not have any special connection with the differences between means for the tasks across *all possible* subjects. On the other hand, this is not the same as random error, since these differences do reflect systematic effects occurring in the experiment. The presence of interaction effects in the deviation of a column mean M_j from the general mean M is actually due to the circumstance that one set of factor levels (subjects) is sampled, so that the interaction effects are under no necessity to "cancel out" in the particular sample drawn.

Now let us examine the composition of the mean square for rows (the sampled factor). This mean square is based on the squared deviation of each row mean, M_k, from the general sample mean M. Applying the model of Section 13.18 once again we find that

$$(M_k - M)^2 = (b_k + M_{ek} - M_b - M_e)^2.$$

Here, the deviation depends only on row effects and upon error deviation; the interaction effects do not appear in any such deviation making up the mean square for rows.

The reason for the absence of interaction effects in the mean square for rows is really very simple. You may recall from Section 13.13 that when the interaction effects c_{jk} were regarded as sampled, we assumed that the expected values for the interaction effects over rows, over columns, and row and column combinations, should be zero. The same assumption holds here, and in particular we are assuming that for a given row k

$$\underset{j}{E}(c_{jk}) = 0,$$

the expectation of c_{jk} over all possible levels that might appear as column j is zero. However, all possible column treatments under consideration are actually included among those used in the experiment. This implies that

$$\underset{j}{E}(c_{jk}) = \frac{\sum\limits_{j} c_{jk}}{C} = 0.$$

Since this is true, then within any given row the sum of the c_{jk} values taken over columns must be zero. Because the columns factor is *not* sampled, and each possible column treatment is actually represented in the experiment, interaction effects within rows *do* cancel out and sum to zero. Thus, $c_k = 0$ and $M_c = 0$ in the mixed model.

As a result it can be shown that

$$E(\text{MS rows}) = \sigma_e^2 + Cn\sigma_B^2, \qquad\qquad \text{[13.19.2*]}$$

the mean square for rows estimates a weighted sum of the error variance and the variance attributable to the sampled factor B.

The interaction variance is based on the squared difference between the deviation from M for the mean of each cell jk in the data table, and the corresponding row mean's and column mean's deviations:

$$[(M_{jk} - M) - (M_k - M) - (M_j - M)]^2 = (M_{jk} - M_k - M_j + M)^2.$$

In terms of our mixed model:

$$
\begin{aligned}
(M_{jk} - M_k - M_j + M)^2 &= (\alpha_j + b_k + c_{jk} + M_{ejk} - b_k - M_{ek} - \alpha_j - c_j \\
&\qquad - M_b - M_{ej} + M_b + M_e)^2 \\
&= (c_{jk} - c_j + M_{ejk} - M_{ek} - M_{ej} + M_e)^2.
\end{aligned}
$$

Observe that any such deviation depends only on interaction effects and on error. Then it can be shown that

$$E(\text{MS interaction}) = \sigma_e^2 + n\sigma_{AB}^2. \qquad\qquad \text{[13.19.3*]}$$

That is, the interaction mean square is an estimate of a weighted sum of the error variance and the interaction variance.

When the hypothesis of no column effects is true, so that $\alpha_j = 0$ for each of the column treatments j, then

$$E(\text{MS columns}) = \sigma_e^2 + n\sigma_{AB}^2. \qquad\qquad \text{[13.19.4*]}$$

Notice that this is exactly the same as the expectation for the MS interaction. Under our assumptions (particularly assumption 4) the ratio

$$F = \frac{\text{MS columns}}{\text{MS interaction}} \qquad\qquad \text{[13.19.5†]}$$

with $C - 1$ and $(R - 1)(C - 1)$ degrees of freedom provides an appropriate test of the hypothesis of no column effects. *Be careful to notice that it is the fixed effects factor that is tested against interaction in this mixed model.*

On the other hand, when the hypothesis of no row effects is true,

$$E(\text{MS rows}) = \sigma_e^2, \qquad\qquad \text{[13.19.6*]}$$

which is the same as

$$E(\text{MS error}) = \sigma_e^2.$$

Then the test for the existence of row effects (that is a test of $H_0: \sigma_B^2 = 0$) is given by

$$F = \frac{\text{MS rows}}{\text{MS error}} \qquad\qquad \text{[13.19.7†]}$$

with $R - 1$ and $RC(n - 1)$ degrees of freedom.

Thus, we see that there are two distinctly different kinds of F tests employed in the analysis of variance for this mixed model: the hypothesis that all the fixed effects are zero is tested by comparing the mean square for that factor against the mean square for interaction. On the other hand, the hypothesis associated with the random effects factor is tested by comparing its associated mean square with the mean square for error. *The important procedural differences among these three models thus have to do with the way in which hypotheses are tested, and especially with the denominators used in the various F ratios.*

The test for interaction in a two-way analysis is the same for all three models:

$$F = \frac{\text{MS interaction}}{\text{MS error}}, \qquad\qquad [13\ 19.8\dagger]$$

with $(R - 1)(C - 1)$ and $RC(n - 1)$ degrees of freedom.

When there is good reason to believe that interaction effects do not exist, there is some advantage in pooling the interaction and the error sums of squares to get a new estimate of the error variance. This is done exactly as outlined in Section 13.15, and Paull's rule is a handy device for deciding when a pooled estimated of error should be used in this model as well.

13.20   AN EXAMPLE FITTING MODEL III

Suppose that the experiment outlined in Section 13.17 is carried out, and that the data turn out as shown in Table 13.20.1. In this table, the two values appearing for each combination of a subject with a task represent the two performances of that task by the subject. Although the task on each of these two trials is the same, the concrete problem given to the subject is different on the repetition of the task. Furthermore, since the order of all six performances is randomized individually for each subject, so that the two distinct problems within a task appear in a randomly chosen order, the experimenter is prepared to regard these scores as two independent observations under the same task condition. That is, the repetition of the task by a subject is regarded as constituting an independent replication of the experiment.

The .05 level is chosen for α, and the analysis proceeds as follows:

$$\text{SS total} = (7.8)^2 + \cdots + (10.5)^2 - \frac{(358.5)^2}{36} = 123.57$$

$$\text{SS columns} = \frac{(103.5)^2 + (136.0)^2 + (119.0)^2}{12} - \frac{(358.5)^2}{36} = 44.04$$

$$\text{SS rows} = \frac{(61.3)^2 + \cdots + (62.9)^2}{6} - \frac{(358.5)^2}{36} = 6.79$$

$$\text{SS error} = (7.8)^2 + \cdots + (10.5)^2 - \frac{(16.5)^2 + \cdots + (19.1)^2}{2}$$

$$= 14.53$$

$$\text{SS interaction} = 123.57 - 44.04 - 6.79 - 14.53 \doteq 58.21.$$

Table 13.20.2 is the summary table for this analysis.

For 2 and 10 degrees of freedom an F value of 4.10 is required for significance, so that the hypothesis of no column effects is *not* rejected. Notice that the F ratio for columns is formed by dividing the column mean square by

Table 13.20.1

Subjects	Tasks			Total
	I	II	III	
1	7.8	11.1	11.7	
	8.7	12.0	10.0	
	16.5	23.1	21.7	61.3
2	8.0	11.3	9.8	
	9.2	10.6	11.9	
	17.2	21.9	21.7	60.8
3	4.0	9.8	11.7	
	6.9	10.1	12.6	
	10.9	19.9	24.3	55.1
4	10.3	11.4	7.9	
	9.4	10.5	8.1	
	19.7	21.9	16.0	57.6
5	9.3	13.0	8.3	
	10.6	11.7	7.9	
	19.9	24.7	16.2	60.8
6	9.5	12.2	8.6	
	9.8	12.3	10.5	
	19.3	24.5	19.1	62.9
Totals	103.5	136.0	119.0	358.5

the mean square for interaction. If this experiment had been incorrectly analyzed under Model I, quite a different conclusion would have been reached; this illustrates the importance of using the proper model in testing hypotheses.

For 5 and 18 degrees of freedom an F value of 2.77 is required in order to reject the null hypothesis at the 5 percent level. The F ratio formed by the mean square for rows over the mean square error gives a value smaller than this, so that the hypothesis of no subject effects is also not rejected.

On the other hand, the F test for interaction exceeds the required F of 2.41, and the hypothesis of no interaction *is* rejected.

In summary, there is insufficient evidence to permit us to conclude

Table 13.20.2

Source	SS	df	MS	E(MS)	F
Columns (tasks)	44.04	2	22.02	$\sigma_e^2 + 2\sigma_{AB}^2 + \dfrac{12\sum\limits_j \alpha_j^2}{2}$	3.78
Rows (subjects)	6.79	5	1.36	$\sigma_e^2 + 6\sigma_B^2$	1.68
Interaction (tasks by subjects)	58.21	10	5.82	$\sigma_e^2 + 2\sigma_{AB}^2$	7.19
Error	14.53	18	.81	σ_e^2	
Total	123.57	35			

either that there are task effects or that subject effects exist in this experiment. There is, however, fairly strong evidence for the presence of interaction effects. It stands to reason that there is something about the combination of a particular subject with a particular task that accounts for variance in the data. Thus, within subjects, task differences apparently exist, but these tend to be different for different subjects. Similarly, subject differences may exist within tasks, but these intersubject differences tend to vary across the three tasks. In this light, the findings of this study seem very trivial: they suggest that if you want to say how these tasks differ, you must specify the subject to whom the tasks are given. Of course, the usual qualification holds: this result might reflect only sampling error, or the influence of some uncontrolled aspect of the study.

13.21 VARIANCE ESTIMATION IN MODEL III

Just as in Model I, the best estimate of any one of the fixed effects is given by the difference between the treatment mean and the grand mean in the sample:

$$E(M_j - M) = \alpha_j.$$

Notice, however, that in the example given above, since interaction effects seem to exist with the *sampled* factor, such estimates of main effects have little or no practical use. In a two-way design such as the example the variance component associated with the random effects can be estimated by:

$$\text{est. } \sigma_B^2 = \frac{MS \text{ rows} - MS \text{ error}}{Cn} \qquad [13.21.1\dagger]$$

and that for interaction by:

$$\text{est. } \sigma_{AB}^2 = \frac{MS \text{ interaction} - MS \text{ error}}{n}. \qquad [13.21.2\dagger]$$

13.22   SOME CONNECTIONS AMONG THE THREE MODELS

Up to this point, the question of the number of possible treatments or levels from which we sample has been skimmed over rather lightly. However, the time has come to consider the situation where one actually knows the number of distinctly different levels of the factor from which he samples. This question was unimportant for our development of a test for the simple one-way analysis under the random effects model, although we later found it necessary to assume a normal distribution of effects, implying an infinite number of factor levels, in the tests and interval-estimation procedures of Sections 13.8 and 13.9. Similar normality assumptions are necessary in the two-factor situations for both Models II and III.

Now we will consider a possible situation where the two factors of the experiment are each represented by a sampling of levels, and where something is known of the number of the possible such levels for each. Suppose that the factor shown as columns in the data table is represented by some C randomly selected levels from among a total of some V possible levels. The factor shown as rows is represented by some R levels selected at random from among some U possibilities. Both U and V are here thought of as known, finite, numbers. Let any particular level that might be selected for inclusion among the columns be denoted by v, and any particular one of the possible levels for rows by u. Returning to the notation of Section 13.1 once again, a possible column effect can be symbolized by $a(v)$, and a possible row effect by $b(u)$. A possible interaction effect associated with a combination (v,u) from among the VU possible such combinations can be denoted by $c(v,u)$. Now suppose that we define the true variance due to the columns factor A as

$$\sigma_A^2 = \frac{\sum_v [a(v)]^2}{V - 1},$$

that for the factor B represented by the rows as

$$\sigma_B^2 = \frac{\sum_u [b(u)]^2}{U - 1},$$

and the true interaction variance as

$$\sigma_{AB}^2 = \frac{\sum_v \sum_u [c(v,u)]^2}{(U - 1)(V - 1)}.$$

Under these conditions the expectations of the mean squares become:

$$E(\text{MS columns}) = \sigma_e^2 + n\left(1 - \frac{R}{U}\right)\sigma_{AB}^2 + nR\sigma_A^2$$

$$E(\text{MS rows}) = \sigma_e^2 + n\left(1 - \frac{C}{V}\right)\sigma_{AB}^2 + nC\sigma_B^2$$

$$E(\text{MS interaction}) = \sigma_e^2 + n\sigma_{AB}^2$$
$$E(\text{MS error}) = \sigma_e^2$$

Viewing the expected mean squares in this way lets us show the connections between Models I, II, and III very simply. Suppose first of all that $V = C$ and $U = R$, so that the sample of C column levels completely exhausts the set of possible levels for that factor, and the sample of R row levels exhausts the possible levels for factor B. Then, letting

$$\boldsymbol{a}(v) = \boldsymbol{a}(v_j) = \alpha_j,$$

and

$$\boldsymbol{b}(u) = \boldsymbol{b}(u_k) = \beta_k,$$

the expectations given above become

$$E(\text{MS columns}) = \sigma_e^2 + \frac{nR \sum\limits_{j} \alpha_j^2}{C - 1}$$

$$E(\text{MS rows}) = \sigma_e^2 + \frac{nC \sum\limits_{k} \beta_k^2}{R - 1}$$

$$E(\text{MS interaction}) = \sigma_e^2 + \frac{n \sum\limits_{j} \sum\limits_{k} \gamma_{jk}^2}{(R - 1)(C - 1)}$$

$$E(\text{MS error}) = \sigma_e^2.$$

These are precisely the expected mean squares for the fixed effect Model I found in Chapter 12. This illustrates once again that when both sets of factor levels exhaust the possible levels that might appear in the experiment, Model I applies.

On the other hand, suppose that both U and V are thought of as infinitely large numbers, so that infinitely many possibilities exist both for the row levels and for the column levels. Here the expected mean squares become:

$$E(\text{MS columns}) = \sigma_e^2 + n\sigma_{AB}^2 + nR\sigma_A^2$$
$$E(\text{MS rows}) = \sigma_e^2 + n\sigma_{AB}^2 + nC\sigma_B^2$$
$$E(\text{MS interaction}) = \sigma_e^2 + n\sigma_{AB}^2$$
$$E(\text{MS error}) = \sigma_e^2.$$

These are the mean square expectations found in Section 13.14 for Model II, where we assumed that both rows and columns represented samples from possible sets of levels of (practically) infinite size.

Finally, suppose that the levels of factor A (the columns) exhaust the possible levels for this factor, but that the levels for factor B are a sample from among an infinite number of such levels. Then the expected mean squares are:

$$E(\text{MS columns}) = \sigma_e^2 + n\sigma_{AB}^2 + \frac{nR \sum\limits_{j} \alpha_j^2}{C - 1}$$

$$E(\text{MS rows}) = \sigma_e^2 + nC\sigma_B^2$$
$$E(\text{MS interaction}) = \sigma_e^2 + n\sigma_{AB}^2$$
$$E(\text{MS error}) = \sigma_e^2.$$

These are the expectations for mean squares found in Section 13.19 under Model III, a mixed model.

This is a useful way to visualize the similarities and differences between these three models for the analysis of variance. The essential differences between these models depend on the conception each embodies of number of possible treatments or factor-levels that might be selected for inclusion in the actual experiment.

13.23   Randomization in experiments

The basic statistical tools of the experimenter are randomization and control. Randomization enters into the experiment in at least two different ways. The actual individuals observed in any sample may differ in any number of ways from the "typical" individual in the population as a whole. It is entirely possible to obtain a sample of N individuals who are "peculiar" as compared with the population they represent, and the conclusions reached from this sample need not be indicative of what the population as a whole tends to be like. By sampling at random from the population, the experimenter is able to identify the peculiarities of his sample with random error, and to allow for the possibility of an atypical sample in his conclusions. Furthermore, given some sample it will always be true that factors other than the ones manipulated by the experimenter will contribute to the observed differences between subjects in the particular situation. If it should happen that some extraneous factor operates unevenly over several treatment groups or over different subjects, this can create spurious differences, or mask true effects in the data. Such a factor is a "nuisance-factor," playing a role analogous to "noise" in communication. Randomization of subjects over treatments is one device for "scattering" the effects of these nuisance factors through the data. Often, when particular nuisance factors are known, levels of these factors are scattered at random throughout the design. In order to randomize his experiment, the researcher uses some scheme such as a table of random numbers to allot individual subjects to experimental groups or nuisance levels to subgroups in a purely random, unsystematic, "chancy" way. By randomization, the possibility of "pile-ups" of nuisance effects in particular treatment groups is identified with random error, and the experimenter can rest assured that overall repetitions of his experiment under the same conditions, true effects will eventually emerge if they exist.

Every experiment is randomized to some extent. A study having one or more experimental factors and certain constant controls where every other factor contributing to variance is randomized is ordinarily called a **fully randomized** design: usually a random sample is drawn, and then subjects are allotted to treatments or treatment combinations purely at random. In addition, levels of one or more nuisance factors may be assigned at random. Most of the experiments used as examples in Chapter 12 were of this type. Furthermore, many other experiments fitting either the fixed effects or the random effects models can also be thought of in this way if nature or some other agency is conceived as carrying out the random assignment of individuals to groups.

On the other hand, sometimes it is advantageous to the experimenter *not* to randomize the effects of particular nuisance factors, but rather to represent them systematically in his experiment. These nuisance factors are treated as though they were experimental factors, when actually the purpose of representing them in the experiment in the first place is to control these factors and thereby reduce error variance. It should be obvious from our discussion of the indices ω^2 and ρ_I that deliberate introduction of any systematic effects that account for a portion of the variance σ_Y^2 must also *reduce* the error variance σ_e^2. That is, the true variance of any dependent variable is thought of as a sum of components:

$\sigma_Y^2 =$ (the variance attributable to systematic features of the experiment)

+ (the variance attributable to unsystematically represented factors).

The more *relevant* factors that can be introduced into the experiment in a systematic way, the smaller will be the variance considered as error. Furthermore, the smaller the error variance, then the more precise will be the experiment, in that confidence intervals will be smaller and true effects more likely to be detected if present.

It follows that in multifactor experiments the introduction of each factor can serve two purposes: not only does the representation of the factor permit inferences to be drawn about *its* effects, but also it permits inferences to be made about other factors with relatively greater precision than otherwise. Often there is genuine interest in the possible effects and interaction effects of each factor. On the other hand, it is very common to find one or more factors introduced only as a device for reducing error variance, when little or no interest is taken in that factor per se. The experimenter somehow knows beforehand that certain nuisance factors do contribute to variance, and that introducing one or more of them systematically into the experiment will increase the precision of his inferences about the other factors of primary interest.

Finally, the principle of randomization is important to statistics and experimental design in an even more basic way. There is a growing body of evidence that procedures such as the analysis of variance and F test may be justified by a different kind of mathematical rationale, based on the actual randomization performed in the experiment. This rationale is rather different from the "classical" or "normal-theory" models presented here. The probability statements made under such randomization models do not refer specifically to distributions over all samples, but rather to distributions over **all possible randomizations of the same sample** (an example of this general idea will be given in Chapter 17). The connection between the idea of randomization in the experiment and the use of the F test is not a particularly simple one, and space and the mathematical level assumed do not permit our going further into this rationale here. Suffice it to say that in some instances this randomization model does permit one to analyze data by the analysis of variance and to make inferences using the F test, when Models I, II, and the mixed model are, for various reasons, inappropriate. A modern discussion of randomization models is given in Scheffé (1959, Chapter 9), although this presentation is not elementary and the student is, as usual, advised to seek help in the use of such a model for his data.

13.24   SIMPLE MATCHED-GROUPS EXPERIMENTS UNDER THE MIXED MODEL: RANDOMIZED BLOCKS

We are going to conclude our discussion of mixed models by considering one of the most common of the simple experimental designs involving matching as a device for increased precision. In this design there is one factor A that is actually the experimental variable of interest. However, there is another factor B that is primarily of a nuisance character. *The experimenter knows that factor B accounts for some variance in Y.* Each level or category of the nuisance factor consists of a group of subjects who are "matched," each showing the same status on this nuisance factor. Then *within* each such matched group, all the different levels of factor A are represented.

The experiment analyzed in Section 13.20 represents one fairly extreme form of a matched-groups design. Here, each subject was effectively matched with himself over the conditions of the experiment. The nuisance factor was "subjects" (or, more properly, "subject-order combinations") while the experimental factor of real interest was "task." The experimenter really did not care about the possible effects attributable to subjects (or to subject-order combinations) in and of themselves. He was, however, interested in controlling for such effects, both as generators of between-treatment differences in the data and as contributors to his estimate of error variance. He hoped thereby to increase both the interpretability and precision of his experiment. Although the "subjects" factor was tested in the analysis given above, this was really for illustrative purposes, and the test for the sampled factor is sometimes omitted altogether when such a nuisance factor is involved. It is most important, however, that the sum of squares for such a factor be extracted from the over-all sum of squares; otherwise, the statistical advantage gained from matching or repeating observations is lost.

The purpose of the experimenter in introducing various factors into his design must be considered in the analysis and interpretation of the experiment. It is often a temptation to think of the shape of the data table as dictating the tests and interpretations to be made. However, factors introduced solely for the purpose of statistical control have quite a different status in the over-all evaluation of the results from that of the experimental factors of genuine concern to the experimenter. The fact that evidence exists for effects or interactions of some nuisance variable may contribute absolutely nothing to the interpretation of the experiment. The researcher had a pretty good idea that the nuisance factor contributed to variance in the first place, or otherwise he would not have troubled to introduce it into the experiment in a systematic way. Nevertheless, whether or not one intends to test for the effects of such deliberately introduced nuisance factors, he must allow for their presence in the over-all partition of the sum of squares.

Consider an experiment in which some J different treatments are to be compared. Each treatment corresponds to one level on the real experimental factor of interest. These effects are regarded as fixed. However, one or more nuisance factors exist, and suppose that previous work has shown considerable statistical association between these factors and the variable Y. The experimenter would like to represent levels or combinations of levels of those factors in his experiment. If all levels or combinations of levels for one or more nuisance factors can be represented, then the model he uses is that of the fixed effects analysis described in Chapter 12: the experimental treatments are observed under each of the nuisance levels, with subjects within each nuisance level being allotted at random to the treatments.

However, it most often happens that there are very many levels to the nuisance factor, or many combinations of factors, and the experimenter must sample from among these levels for a representation in his experiment. He draws a sample of K nuisance levels, each containing some nJ subjects. Each group representing a qualitatively different category on the nuisance factor or factors is called a **matched group,** or a **block** of subjects. Then within each block, subjects are allotted at random to the J different treatments. Such a matched group design, with sampling of the nuisance factor levels, is often called a **randomized blocks design** (especially when there are J observations per block, or one subject per treatment within a matched group).

It must be emphasized that whatever the model adopted there is no requirement that the various levels of the nuisance or matching factor be stages on some single continuum or attribute. Levels on the matching factor may represent qualitative differences of any kind, or combinations of factor levels. *The important point is that the factor or factors underlying the groupings must actually contribute to variance.* If the factors leading to the matching of subjects have nothing at all to do with the dependent variable, then this is not an economical way to design an experiment, since formation of matched groups is troublesome and "costly," and the experimenter gets nothing in return (or may actually lose something) for his extra effort.

For an example of a randomized blocks design, suppose that in an experiment on human performance at some motor task, the experimenter knows that sex, age, and type of occupation all contribute to the variance of scores. Therefore, he wishes to use subjects matched in these respects. However, it would manifestly be impossible to introduce each of these things in the experiment at all levels or combinations of levels. Instead, the experimenter samples at random from among age-sex-occupation groups, finding J subjects who are alike in all three respects for each such combination sampled. Some K different groups or blocks are sampled, each containing J subjects. Then, within any age-sex-occupation grouping, the subjects are assigned *at random* to the J experimental treatments, each treatment getting one and only one subject. Here, $N = JK$ for the experiment.

A randomized blocks experiment such as this is usually analyzed as **a mixed model design without replication.** The analysis of variance proceeds just as for any two-way analysis, except that the partition of the sum of squares contains only three terms:

$$\text{SS total} = \text{SS rows} + \text{SS columns} + \text{SS interaction}.$$

Notice that a separate SS error cannot be calculated, since there is only one observation per cell. Table 13.24.1 is the summary table for the analysis. Since there is only one individual per cell in this instance $(n = 1)$ the only F test to be made is that for treatments, which involves the MS rows divided by the MS interaction. The rows (blocks or matched groups effects) cannot be tested in the usual way unless at least 2 observations are made in each block and treatment combination. On the other hand, a conservative form of test is possible even in this situation, since under the null hypothesis the ratio (MS blocks)/(MS interaction) has a distribution which is stochastically smaller (that is, this ratio is likely to be smaller) than the corresponding F ratio. This could be useful in deciding whether or not there is any advantage in continuing to use randomized blocks.

Table 13.24.1

Source	SS	df	MS	E(MS)	F
Columns (treatments)		$J - 1$	$\dfrac{\text{SS columns}}{J - 1}$	$\sigma_e^2 + \sigma_{AB}^2 + \dfrac{K \sum_j \alpha_j^2}{J - 1}$	$\dfrac{\text{MS col.}}{\text{MS int.}}$
Rows (blocks of matched individuals)		$K - 1$	$\dfrac{\text{SS rows}}{K - 1}$	$\sigma_e^2 + J\sigma_B^2$	
Interaction		$(J - 1)(K - 1)$	$\dfrac{\text{SS int.}}{(J - 1)(K - 1)}$	$\sigma_e^2 + \sigma_{AB}^2$	
Total		$JK - 1$			

Matched group designs of the randomized-blocks type (one observation per cell) are extremely common in any number of research areas and especially in agriculture (hence the name, each "block" being a separate set of plots of ground). It is a highly efficient way of proceeding, provided that one can be reasonably sure that the factors on which individuals are matched actually are associated with the dependent variable; presumably this kind of information is available to the experimenter from his knowledge of the research area or from pilot studies for his experiment. (The student may already have recognized that the matched pairs design, analyzed by use of the t test in Chapter 10, is an example of a randomized blocks design with $J = 2$.)

When blocks represents a randomly sampled factor, the assumptions of the mixed model must of course be met. It often makes much more sense to regard the matching variable as sampled than to limit the generality of the experiment only to certain levels actually represented in the experiment. However, a troublesome point then emerges: if the matching factor actually is sampled, then the expectation of MS between columns (the fixed effects factor) contains interaction variance, and the test for treatment effects should be carried out against the MS interaction. If there is a rather small variance attributable to the nuisance factor, but a considerable interaction variance component, then it may still take enormous treatment effects to show up as significant. This situation is illustrated in the example of Section 13.20. When variance due to blocks is small, but large variance due to interaction is present, one of the basic advantages usually gained by introducing matching in the first place is, in effect, lost, since the experiment may be relatively less precise for detecting over-all effects of the treatments than it would have been if the nuisance variable had simply been randomized over subjects. Consequently when the levels of the nuisance variable are to be sampled, the experimenter does well to think about possible interactions of this nuisance variable. If for some reason or other he can conclude that interaction effects will not exist, then the unreplicated matching scheme may be quite effective for increasing the precision of the experiment. Otherwise, if he anticipates quite sizable interaction variance he should either replicate the experiment and regard the matching factor levels as fixed, gaining some precision at the risk of losing some generality, or he should abandon the matching and be content with randomizing the nuisance variable. However, if his interest is in the *uniformity* of the treatments over the various nuisance-factor levels, the tendency for general effects to emerge in spite of the deliberately introduced interaction between treatments and levels of the nuisance factor, then the randomized blocks type of design is sufficient to the purpose, and the nuisance factor levels should be sampled.

Occasionally, the levels of the matching factor are not sampled, but rather are chosen deliberately to represent extreme degrees of the matching variable, or large qualitative differences among groups in one or more ways. In such designs the mixed model is really not appropriate, and the experiment should either be replicated (two or more subjects or observations per treatment in each matched group) and analyzed by Model I, or the experimenter must either know or be prepared to assume that no interaction effects exist in order to achieve a fully valid F test. A test for interaction or "nonadditivity" exists for fixed effects unreplicated designs (Tukey, 1949), and this provides a basis for judgment about the appropriate test for treatment effects in this situation.

This is an area in which the beginner in experimentation may need considerable help in designing his particular study. The concerns just raised lie very close to the core of experimental psychology. Matched-group, and particularly repeated-observation, designs either of the replicated or unreplicated variety are enormously useful in many areas such as learning and psychophysics. Such designs are basic to studies where one is trying to establish the form of functional relation between the experimental and the dependent variable. But here the problem of interaction with a nuisance variable such as "subjects" becomes a serious one. How do you interpret a function that changes its form in an essential way from subject to subject? The answer seems to lie in the study of such interaction effects in and of themselves, and particularly in the search for ways to transform or rescale dependent variables so as to eliminate such interactions. In some situations interaction effects can be eliminated by an appropriate transformation of the dependent variable, and then general, interpretable, functional relations may appear. In other situations the problem of nuisance interactions may be resolved by the extraction of other components of variance in the analysis. These are not elementary considerations, however, and a text such as this is not the place to pursue them further. The essential point is that there is no way to design an experiment using matched groups, repeated observations, or any other form of control, that is guaranteed to make the results nice, neat, and interpretable unless the experimenter knows something about the relations of the nuisance variables themselves to the dependent variable.

This is also not the place to go into the actual mechanics for matching subjects for an experiment. There are a number of pitfalls to be avoided, and, as usual, the beginner does well to seek help. An introductory exposition of the problem is given in Lindquist (1953, Chapter 5).

The really important point to remember is that matching is often a good device for introducing controls into the experiment and for reducing error-variance estimates, thereby for achieving greater sensitivity to the presence of real treatment effects. However, matching accomplishes this aim only if the matching variable or variables actually *do* show a relatively strong association with the dependent variable in the experiment. To reiterate: here is one place where the experimenter must make full use of his knowledge about the area of study and previous findings about the strengths of association or proportions of variance accounted for among the several relevant variables. Knowledge of possible interaction effects can be particularly important. "Blind" matching, without the use of such information, can actually impair the efficiency of the experimental design.

13.25 REPEATED OBSERVATION OF THE SAME SUBJECTS

The practical limit to the strategy of matching subjects in an experiment is reached when each subject is matched with himself. Each subject is given each of the J experimental treatments, with each individual being assigned to the treatments in some different, randomly chosen, order. Conceptually, each group of J observations of a single subject is like a matched group of observations: any subject represents some constellation of "nuisance factor levels" that he brings to the experiment, and that are presumably constant over the treatments for him. A sampling of subjects is formally like a sampling of nuisance-factor combinations, just as when matched groups are sampled.

In the example of Section 13.20 the same subjects were exposed to a number of different experimental treatments, and, in addition, each subject produced two score values under each treatment. A simpler, somewhat more typical plan is one in which a group of n subjects is selected at random, and each of J fixed treatments is given once to each subject. Each subject-treatment combination yields exactly one score value. This arrangement is frequently encountered in practice, and it seems worthwhile to give it a little elaboration here. Actually, this design can be viewed as a special case of the mixed model, or Model III, and can be analyzed accordingly. However, it is worth going through the simplest of these cases, not only to show how the arithmetic is often arranged into a pattern somewhat different from that of Section 13.20, but also to emphasize how this design handles a particular sort of systematic dependency among observations.

Consider some J experimental treatments once again. This set of experimental treatments is assumed to be fixed. Then, a sample of some n subjects is selected at random, and each subject i is given each of the experimental treatments j. The score value produced by subject i under treatment j is symbolized by Y_{ij}.

Obviously, this situation falls under the mixed model, or Model III. That is, we are dealing with a fixed set of experimental treatments, and a randomly selected set of subjects, where each subject is given each treatment. This suggests the following linear model:

$$Y_{ij} = \mu + \alpha_j + a_i + e_{ij}$$

where

$$\alpha_j = \mu_j - \mu, \qquad \sum_i \alpha_j = 0,$$

and a_i is the random effect associated with subject i. Then e_{ij} is the random error associated with subject i under treatment j. In this interpretation of the model, we are dealing only with treatment effects and subject effects. No treatment by subject interaction is assumed to exist.

We assume further that

$E(e_{ij}) = 0$	over all i and for every j,
$E(a_i) = 0$	over all i,
$\text{Cov}(a_i, e_{ij}) = 0$	over all i and for every j, (Appendix B)
$\text{Cov}(e_{ij}, e_{ik}) = 0$	over all i and for every treatment pair j and k.

In other words, we are assuming that over the entire population of individuals the mean of the error term for any treatment j is zero. We also assume that the mean of the individual effects a_i is zero in the population of individuals i, and that the individual effects a_i and the error effects e_{ij} are independent. The error in treatment j is assumed to be independent of the error in treatment k over all individuals i, and for any two treatments j and k. As usual for such a model, we assume that the distribution of effects a_i is normal, and that the distribution of errors e_{ij} is also normal, for any j. Finally, the error terms, e_{ij}, are assumed to have the same variance, or σ_e^2, within any treatment j.

It is important to note that we do *not* assume that two scores produced by the same individual in two different treatments are independent; that is, we are not assuming that $\text{Cov}(Y_{ij}, Y_{ik})$ is equal to zero (see Appendix B). This model then does allow for a certain kind of dependency among observations due to the same individual subjects. Let us proceed to find an expression for this covariance between Y_{ij} and Y_{ik}. From Appendix B we have

$$\text{Cov}(Y_{ij}, Y_{ik}) = E(Y_{ij}, Y_{ik}) - E(Y_{ij})E(Y_{ik})$$
$$= E(\mu_j + a_i + e_{ij})(\mu_k + a_i + e_{ik}) - E(\mu_j + a_i + e_{ij})E(\mu_k + a_i + e_{ik}).$$

On carrying out these multiplications, and distributing the expectations, we find that the assumptions given above permit us to simplify this to

$$\text{Cov}(Y_{ij}, Y_{ik}) = \mu_j\mu_k + E(a_i^2) - \mu_j\mu_k$$
$$= E(a_i^2).$$

But, since $E(a_i) = 0$, $E(a_i^2)$ is simply the variance of the individual effects over the whole population of individuals. Hence

$$\text{Cov}(Y_{ij}, Y_{ik}) = \sigma_a^2.$$

In a similar way we can investigate the variance of score values Y_{ij} within any treatment j:

$$\sigma_{Y_j}^2 = E(\mu_j + a_i + e_{ij})^2 - [E(\mu_j + a_i + e_{ij})]^2,$$

which, under our assumptions, is equal to

$$\sigma_{Y_j}^2 = E(a_i^2) + E(e_{ij}^2).$$

The first term on the right is simply σ_a^2, and the second term is σ_e^2. Since neither of these terms depends on j, then for any j we have

$$\sigma_Y^2 = \sigma_a^2 + \sigma_e^2.$$

Putting our findings to this point together, we obtain

$$\sigma_e^2 = \sigma_Y^2 - \sigma_a^2,$$

or, since $\sigma_a^2 = \text{Cov}(Y_{ij}, Y_{ik})$,

$$\sigma_e^2 = \sigma_Y^2 - \text{Cov}(Y_{ij}, Y_{ik}).$$

Since the covariance must be the same, or σ_a^2 for any pair of treatments j and k, we can denote the covariance by the symbol for the average covariance, or $\overline{\text{Cov}}$. Thus,

$$\sigma_e^2 = \sigma_Y^2 - \overline{\text{Cov}}.$$

The error variance is the variance of score values within any treatment, minus the average covariance across treatments. Bear this equation in mind, as it will be used to show why this form of dependency among score values makes the ordinary, one-way, analysis of variance inappropriate for this design.

The sum of squares between treatments has the usual form:

$$\text{SS between treatments} = \sum_j n(M_j - M)^2,$$

so that the expectation is, as usual,

$$E(\text{SS between treatments}) = (J - 1)\sigma_e^2 + n \sum_j \alpha_j^2.$$

Now consider the sum of squares between subjects:

$$\text{SS between subjects} = \sum_i J(M_i - M)^2$$

$$= \sum_i J(\mu + a_i + M_{ei} - \mu - M_a - M_e)^2$$

$$= \sum_i J[(a_i - M_a) + (M_{ei} - M_e)]^2.$$

From this it follows that the expectation of SS between subjects is

$$E(\text{SS between subjects}) = (n - 1)[\sigma_e^2 + \sigma_a^2]$$
$$= (n - 1)(\sigma_Y^2).$$

When the sum of squares within subjects is considered, it turns out that

$$\text{SS within subjects} = \sum_i \sum_j (y_{in} - M_i)^2$$

$$= \sum_j n\alpha_j^2 + \sum_i (e_{ij} - M_{ei})^2,$$

so that the expectation of this sum of squares is

$$E(\text{SS within subjects}) = \sum_j n\alpha_j^2 + n(J - 1)\sigma_e^2.$$

We must find an estimate of σ_e^2 against which to compare MS between treatments. If we take

$$\text{SS within subjects} - \text{SS between treatments} = \text{SS residual},$$

we have the basis for such an estimate, since

$$E(\text{SS within subjects}) - E(\text{SS within treatments})$$
$$= \sum_j n\alpha_j^2 + n(J - 1)\sigma_e^2 - \sum_j n\alpha_j^2 - (J - 1)\sigma_e^2$$
$$= (n - 1)(J - 1)\sigma_e^2.$$

Hence

$$E(\text{MS residual}) = \frac{E(\text{SS residual})}{(n - 1)(J - 1)} = \sigma_e^2.$$

Now notice that if we had mistakenly computed the sum of squares within, as for a one-way analysis of variance with independent groups of observations, this mean square would not have given a "pure" estimate of the error variance:

$$E(\text{SS within treatments}) = J(n - 1)[\sigma_e^2 + \sigma_a^2]$$

so that

$$E(\text{MS within treatments}) = \sigma_a^2 + \sigma_e^2 \geq \sigma_e^2.$$

Any variability at all attributable to subjects, or to dependency of scores within the same subject, will make the estimate of error taken from within treatments too large. Equivalently,

$$E(\text{SS within treatments}) = J(n - 1)[\overline{\text{Cov}} + \sigma_e^2].$$

This particular kind of dependency among score values, due to the repeated observation of the same subjects, makes it necessary to calculate the estimate of error from SS residual (or its equivalent) rather than from SS within treatments.

The basic computational analysis for such a simple repeated measurements design can be outlined as follows:

$$\text{SS between subjects} = \frac{\sum_i \left(\sum_j y_{ij}\right)^2}{J} - \frac{\left(\sum_i \sum_j y_{ij}\right)^2}{nJ},$$

$$\text{SS within subjects} = \sum_i \sum_j y_{ij}^2 - \frac{\sum_i \left(\sum_j y_{ij}\right)^2}{J},$$

$$\text{SS between treatments} = \frac{\sum_j \left(\sum_i y_{ij}\right)^2}{n} - \frac{\left(\sum_i \sum_j y_{jj}\right)^2}{nJ},$$

$$\text{SS residual} = \text{SS within subjects} - \text{SS between treatments}.$$

The number of degrees of freedom for SS between subjects is $n - 1$, for within subjects is $n(J - 1)$, and for SS residual is $(n - 1)(J - 1)$. An approximate F test for the hypothesis of no treatment effects is provided by

$$F = \text{MS treatments}/\text{MS residual}$$

with $J - 1$ and $(n - 1)(J - 1)$ degrees of freedom.

Although this form of arranging the arithmetic looks very different from the analysis carried out in the example of Section 13.20, it is, essentially, the same. In order to demonstrate this, let us for the moment ignore the fact that in the previous example each individual provided two scores under each condition, and let us deal only with the sum of the scores in any condition for each individual. This makes eighteen basic observations in all, as though six subjects provided scores in each of three treatments.

Now the total SS would be

$$\text{SS total} = (16.5)^2 + (23.1)^2 + \cdots + (19.1)^2 - \frac{(358.5)^2}{18} = 218.08.$$

The sum of squares between subjects is found from

$$\text{SS between subjects} = \frac{(61.3)^2 + \cdots + (62.9)}{3} - \frac{(358.5)^2}{18} = 13.58$$

and

$$\text{SS within subjects} = (16.5)^2 + \cdots + (19.1)^2 - \frac{(61.3)^2 + \cdots + (62.9)^2}{3}.$$

We also have

$$\text{SS between treatments} = \frac{(103.5)^2 + \cdots + (119.0)^2}{6} - \frac{(358.5)^2}{18}$$

$$= 88.08.$$

Completing the analysis, we have

Source	SS	df	MS	F
Between subjects	13.58	5		
Within subjects	204.50	12		
Treatments	88.08	2	44.04	3.78
Residual	116.42	10	11.64	
Total[a]	218.08	17		

[a] Between subjects + within subjects.

On the other hand, if we had arranged the arithmetic as in Section 13.20, the summary would look this way:

Source	SS	df	MS	F
Subjects	13.58	5		
Treatments	88.08	2	44.04	3.78
Interaction (residual)	116.42	10	11.64	
Total	218.08			

Since interaction effects are assumed not to exist, then the F ratio in each table provides a test of the hypothesis that treatment effects do not exist. The pattern of analysis shown in the first table has the advantage of drawing attention to the fact that this is a repeated measurements design, and thinking about the analysis in this way may help to keep things straight in more complicated situations; otherwise, there is no particular advantage to doing the analysis one way rather than the other. When different individuals are given different treatments, even though the individuals are matched, it is still reasonable to assume the error portions of their scores to be independent. In the random and mixed models the possibility of a particular kind of statistical association or dependence between scores of individuals in the same group of repeated observations is allowed, but the error portions of those scores must be independent, both of each other and of the random effects in the model. Texts in statistics and experimental design sometimes leave the impression that all the problems introduced by repeated observations are solved when "subjects" appears as a factor in the experimental analysis, and the mixed or random effects model is employed. This is not necessarily true. It is true that the analysis of variance for these models can be applied when a *particular form* of dependency exists among the observations (strictly speaking, when the dependency involves equal *correlations* among pairs of observations within subjects). However, we have no good reason to believe that most repeated observation data have this character, and we have ample reason to believe that in some such data the error components of observations are dependent, particularly when the material is such that learning or other serial changes occur.

Nevertheless, these problems are not insurmountable for the experienced and competent researcher. As mentioned before, a well-designed experiment can eliminate some of the sources of dependency among observations, and various experimental devices exist for handling this problem in different situations. As usual, these statistical considerations should not be interpreted as prohibitions against repeated observation experiments. Quite the contrary; such experimental designs are especially well suited to the problems of the research psychologist. On the other hand, these statistical requirements should serve as guideposts in the process of matching a well-thought-out experiment with an appropriate analysis. The experimenter may have to exercise considerable thought and ingenuity in planning the experiment if he wants the analysis of variance and F tests to be appropriate to the data. But there is nothing sacred or mandatory about the use of analysis of variance; the real concern of the experimenter must always be for the logic of the experiment and the interpretability of the data.

Incidentally, these problems of dependency do not ordinarily arise when the basic scores assigned to individuals are means or other summary measures over repeated trials under the *same* condition, and each individual is represented by one such score. So long as each subject is paired with only one score in the data to be analyzed, and errors for different subjects are independent, then one of the methods already discussed is usually applicable.

13.26   THE GENERAL PROBLEM OF EXPERIMENTAL DESIGN

The factorial design mentioned in the last chapter and the randomized blocks design just discussed are but two of the many ways that the experimenter may choose to design his study. The literature of experimental design is full of ways for collecting and analyzing data for particular experimental situations. Many of these designs in common use in psychology are found in references such as Edwards (1960), Lindquist (1953), and particularly in Winer (1962). A more advanced treatment of the subject is given in Cochran and Cox (1957). An informative, nonmathematical introduction to design is given in D. R. Cox (1958).

From a broad point of view, the problem of choosing a design for an experiment is a problem in economics. The experimenter has some question or questions that he wants to answer. He wants to be sure of the following points.

1. The actual data collected will contain all the information that he needs to

make inferences, and that this information can be extracted from the data.
2. The important hypotheses can be tested validly and separately.
3. The level of precision reached in estimation, and the power of his statistical tests, will be satisfactory for his purposes.

Consideration number 1 involves the actual selection of treatments and treatment combinations that the experimenter will observe. What are the factors of interest? How many levels of a factor will be observed? Will these levels be sampled or regarded as fixed? Which factors should be crossed and which nested in the design. Consideration 2 is closely related to the first: must all combinations of factors to be observed, or is there interest in only some treatment combinations? If only part of the possible set of treatment combinations can be observed, is it possible to make separate inferences about the various factors and combinations? Consideration 3 involves the choice of sample size and of experimental controls. Is the sample size contemplated large enough to give the precision of estimation or power (or both) that the experimenter feels is necessary to have for his inferences?

However, attendant to each of these considerations are parallel considerations of cost.

4. The more different treatments administered, necessitated by the more questions asked of the data, the more the experiment will cost in time, subjects, effort, and other expenses.
5. The more kinds of information the experimenter wants to gain the larger the set of assumptions he usually must make to obtain valid inferences.
6. The more hypotheses that can be tested validly and separately, the greater the number of treatment combinations necessary, and the larger may be the required number of subjects. Furthermore, the clarity of the statistical findings may be lessened, and the experiment as a whole may be harder to interpret.
7. The experimenter can increase precision and power by larger samples, or by exercising additional controls in his experiment, either as constant control, or by a matching procedure. Each possibility has its real costs in time, effort, and, perhaps, money.

Other things being equal, the experimenter would like to get by with as few subjects as possible. Even "approximately" random samples are extremely difficult to obtain, and often the sheer cost of the experiment in time and effort goes up with each slight increase in sample size. Furthermore, the experiment may not be carried out as carefully with a large number as with only a few subjects. Throwing in lots of ill-considered treatments just to see what will happen can be an expensive pastime for the serious experimenter. The number of treatment combinations can increase very quickly when many factors are added, and some of these combinations may be of no interest at all to the experimenter or add little or nothing to precision, so that he may be paying a high price for discovering "garbage" effects in his data. If a large sample is out of the question, power and precision can be bought at the price of constant or matching controls. However, remember that constant control reduces the generality of the conclusions the experimenter can reach from a set of data, and matching may be extremely hard to carry out.

All in all, there are several things the experimenter wants from his experiment and several ways to get them. Each desideratum has its price, however, and the experimenter must somehow decide if the gain in designing the experiment in a particular way is offset by the loss he may incur in so doing. This is why the problem of experimental design has strong economic overtones.

Texts in experimental design present ways of laying out the experiment so as to get "the most for the least" in a given situation. Various designs emphasize one aspect or another of the considerations and costs involved in getting and analyzing experimental data. Texts in design can give only a few standard types or layouts that study and experience have shown optimal in one or more ways, with, nevertheless, some price paid for using each design. Obviously, the best design for every conceivable experiment does not exist "canned" somewhere in a book, and experienced researchers and statisticians often come up with novel ways of designing an experiment for a special purpose. On the other hand, a study of the standard experimental designs is very instructive for any experimenter, if only to let him appreciate the strengths and weaknesses of different ways of laying out an experiment.

We can go no further into the question of experimental design here. However, a principle for the beginner to remember is, *keep it simple!* Concentrate on how well and carefully you can carry out a few *meaningful* manipulations on a few subjects, chosen, *if possible*, randomly from some well-defined population and *randomized* among treatments. The experiment and its meaning is the thing to keep in mind, and not some fancy way of setting up the experiment that has no real connection with the basic problem or the economics of the situation. Then, when the novice finally knows his way around in his experimental area, the refinements of design are open to him.

EXERCISES

1. As part of an experiment on psychogalvanic reactions of dogs to unexpected visual stimuli, "artificial" human faces were constructed by combining elements of actual human features. There were six positions in which such elements might appear in a given face, and six possible elements in each position. Thus there were 6^6 or 46,656 different faces possible. As part of a pretest of the main experiment, eight of these possible stimulus faces were chosen at random, and each presented to a different group of two dogs. The results were as follows:

1	2	3	4	5	6	7	8
18	12	20	15	16	19	17	22
19	14	21	13	18	20	19	25

Test the hypothesis of no true between-stimulus variance.

2. From the results of problem 1, what proportion of the total variance in dogs' responses would you judge to be due to stimulus differences?

3. In a certain one-way random effects design, the mean square between treatments was 28.9 and the mean square within was 3.7. The mean square between had 10 degrees of freedom, and the mean square within 33 degrees of freedom. Find the 95 percent confidence limits for ρ_I, variance accounted for.

4. What is the power of the test described in problem 3 if the alternative hypothesis that $\rho_I = .50$ is true?

5. A one-way, random-effects analysis of variance with 6 and 21 degrees of freedom gave an F-value just significant at the .01 level for the hypothesis that $\sigma_A^2 = 0$. Test the hypothesis that $\rho_I \leq .10$ against the alternative that $\rho_I > .10$.

6. A random sample of five supermarkets was selected from a very large chain, and within each supermarket the number of items purchased on a single day by each of 10 customers selected at random was noted. The entries in the following table are the numbers of items purchased, arranged by store:

		STORES			
1	2	3	4	5	6
13	11	18	45	26	18
16	14	27	48	29	36
19	17	51	27	32	14
16	17	57	33	32	14
19	25	9	18	38	28
22	26	18	27	41	31
19	24	19	36	18	14
19	26	28	40	21	25
13	17	30	28	12	28
6	23	34	10	25	30

1-way random

Test the hypothesis of no variance in number of items purchased attributable to stores. (Use $\alpha = .05$.)

7. What is the power of the test in problem 6 against the hypothesis that variance between stores accounts for 50 percent of the total variance in number of items purchased?

8. An investigator was interested in the possible effects of the measured body-build or "somatotype" of a person, together with the ethnic background of that person, on his tendency to be overweight. Thus, a sample of 5 different somatotypes was chosen, along with a sample of 4 different ethnic-geographical groupings. Two men were found in each somatotype and ethnic combination, and then a typical week's calorie consumption of each was found. The following table shows the average daily consumption in thousands of calories for each man.

ETHNIC-GEOGRAPHIC GROUP

		A	B	C	D
	I	4.21 3.38	4.37 3.29	2.91 3.35	3.22 2.46
	II	3.58 1.82	3.58 1.37	2.31 2.96	3.02 2.11
SOMATOTYPE	III	3.30 1.89	3.01 2.34	2.01 1.85	2.17 1.22
	IV	3.37 2.21	2.49 3.86	3.36 3.10	2.78 2.79
	V	3.56 2.58	3.70 3.48	3.44 2.78	3.74 2.35

Does there seem to be a relation between somatotype, ethnic-geographic background, or their combination, and average daily calorie consumption? (Use $\alpha = .05$.)

9. An automobile company was interested in the comparative efficiency of three different sales-approaches to one of their products. They selected a random sample of 10 different large cities, and then assigned the various selling approaches at random to three agencies within the same city. The results in terms of sales volume over a fixed period for each agency were as follows:

APPROACH

		A	B	C
	1	38	27	28
	2	47	45	48
	3	40	24	29
	4	32	23	33
	5	41	34	26
CITY	6	39	23	31
	7	38	29	34
	8	42	30	25
	9	45	31	26
	10	41	27	34

Does there seem to be a significant difference ($\alpha = .05$) among these three sales approaches?

10. Suppose that the data given in problem 16 for Chapter 12 had represented two factors, each with randomly selected levels. Carry out the appropriate analysis of variance and F-tests.

11. In a pilot study preliminary to a larger experiment there was interest in the proportion of variance attributable to a particular experimental factor. Four levels of this factor were selected at random, and three subjects were assigned at random to each. The data turned out as follows:

LEVEL 1	LEVEL 2	LEVEL 3	LEVEL 4
3.9	1.9	5.3	4.5
4.7	2.8	3.0	5.4
2.4	3.2	4.5	4.7

Establish 95 percent confidence limits for the true proportion of variance of the dependent variable accounted for by the experimental factor.

12. In a given experiment, twelve randomly selected pairs of children, each pair matched with respect to age, sex, and intelligence, were used. The members of each pair were randomly assigned to two fixed experimental conditions. The data in terms of the dependent variable are given below:

PAIR	CONDITION I	CONDITION II
1	18	23
2	19	34
3	27	26
4	25	23
5	28	29
6	15	30
7	17	14
8	29	41
9	36	37
10	25	24
11	46	45
12	31	32

Carry out an analysis of variance and F-test on these data.

13. Analyze the data of problem 12 via a t-test for matched pairs, and compare the result with that found in the preceding problem. Do the two methods agree as they should?

14. In an experiment on paired-associate learning, eight randomly chosen subjects were presented with three different lists of 35 pairs of words to learn. Each subject was successively given the three lists in some randomly chosen order. The score for a subject was the number of pairs correctly recalled on the first trial. Do these three lists seem to be significantly different ($\alpha = .05$) in their difficulty for subjects?

		LIST A	LIST B	LIST C
	1	22	15	18
	2	15	9	12
	3	16	13	10
SUBJECT	4	19	9	10
	5	20	12	13
	6	17	14	12
	7	14	13	10
	8	17	19	18

15. Below is an analysis of variance summary table. Carry out F-tests ($\alpha = .05$ throughout) and interpret the findings under
    (a) a fixed-effects model.
    (b) a random-effects model (both variables).
    (c) Factor A fixed, Factor B random.

SOURCE	d.f.	SS
Factor A	5	186.0
Factor B	9	1427.2
Interaction	45	699.6
Error	120	3109.3
Total	179	5422.1

16. Write out the expectations for each of the mean squares for the data in problem 15, under each of the three models. See if you can estimate the values of

$$\sigma_A^2, \sigma_B^2, \sigma_{AB}^2$$

for the situation where both factors are random.

17. In a study of human memory subjects were given lists of fifty words to memorize. *A priori*, however, the experimenter felt that the four lists chosen (regarded as fixed treatments in the experiment) might differ considerably in difficulty. Therefore each of ten subjects chosen at random was given all four lists in a randomly assigned order. The data, in terms of words correctly recalled turned out as follows. Are the lists significantly different from each other?

		LISTS			
		I	II	III	IV
	1	15	21	24	20
	2	27	35	32	34
	3	26	28	30	30
	4	38	41	40	39
	5	14	22	19	23
SUBJECTS	6	29	33	30	31
	7	32	37	34	35
	8	45	45	40	42
	9	28	38	35	32
	10	33	34	33	29

18. Suppose that the four lists used in problem 17 above had been a random sample from a very large set of lists that might be used in such a study. What difference, if any, would this have made in the analysis?

14 Individual Comparisons among Means

The true experimental scientist gains insights as well as pleasure from "snooping around" in his data. He is satisfied only when he feels that he understands what actually went on in the experiment and what the data actually show. In any sensibly planned experiment there will be clues that can widen understanding, and the good scientist searches them out.

Experimentation is almost never conducted in the hermetically sealed atmosphere that statistical reports often suggest. Most emphatically, good research requires all the care and control that the experimenter can provide in the actual experiment. But research hypotheses are not entertained or dismissed in the cavalier way that a statistics book may imply. These books set down the formal rules of the game, a set of prearranged signals from one scientist to another that he has found something worth looking into, or that his results contain sampling error. Research reports are written in this conventionalized language, and no doubt will continue to be until something better is devised. Nevertheless, the formal rules of the game should not be confused with the actual work of the scientist.

The novice in the use of statistics often feels that statistical tests somehow cut him off from the close examination of his data. There is something that seems so final about an analysis of variance summary table, for example, that one is tempted to regard this as some ultimate distillation of all meaning in the experiment. Nothing could be further from the truth. Statistical summaries and tests are certainly useful, but they are not *ends* of experimentation. The important part of data analysis often begins when the experimenter asks himself, "What accounts for the results I obtain?" and turns to the detailed exploration of the data.

Considered by themselves, all that the F tests in an analysis of variance can tell you is that *something* seems to have happened. If the F is significant, then some effects presumably exist that can be expected to occur again under similar circumstances; if the test is not significant, something notable still may have happened, but if treatment effects exist they are at least partially obscured by other variation. Other than this, a F test alone tells almost nothing. If all experimenters were satisfied with the general statement that something did (did not) happen, then science would progress by slow stages indeed. Yet, in psychology, it is remarkable that published research papers so often emphasize over-all significance of results, and give little evidence either of the experimenter's prior planning or of the combing through data for meaning and potential application that is the heart of research.

Although statistical methods and principles cannot do the experimenter's basic job for him, important statistical aids nevertheless exist at all stages of the research enterprise. When the experimenter only *looks* at his data, combining and recombining them in different ways, a basic statistical principle assists him: the best estimates he can make of the true experimental effects are those he actually observes in his data, regardless of statistical significance. However, even more potent statistical tools are available for the detailed analysis of data.

The material in this chapter deals with two devices for analyzing data in more detail than that provided by the ordinary analysis of variance. Both methods are designed for comparing means or groups of means in a variety of ways. First, the technique of **planned comparisons** or **contrasts** will be introduced. Here, instead of planning to analyze his data to see if any over-all experimental effects exist, the researcher at the outset has a number of particular questions he wants to answer separately. This technique of planned comparisons is used *instead of* the ordinary analysis of variance and F test. Indeed, as will be shown, the usual analysis of variance actually summarizes the evidence for many possible such individual questions that might be asked of the data.

Next, the even more important technique of **incidental** or **post-hoc comparisons** (or **contrasts**) will be discussed. Here, differences among means combined in any number of ways can be evaluated for significance, *after* the over-all F test has shown significance. This procedure of post-hoc comparisons is an important supplement to the usual analysis of variance, and is very useful for the further exploration of data after the initial analysis has suggested the existence of real effects in the data.

Both techniques, planned and post-hoc comparisons, apply only to fixed effects or mixed model experiments. The inferences to be made concern particular population means combined in particular ways, and these methods make sense only when applied to treatment groups representing *fixed level, non-sampled, factors.*

14.1 ASKING SPECIFIC QUESTIONS OF DATA

In Section 8.9 the idea of linear combinations of random variables was introduced, and the sampling distribution of such a linear combination was explored. Our concern with linear combinations was justified by the fact that the evidence pertaining to particular experimental questions can often be found from various ways of combining the data; specifically, by linear combinations of means. In this section we will pursue this idea further, and see how special linear combinations are useful to the experimenter.

Consider a study of the influence of the manner of persuasion upon the tendency of persons to change their attitude toward some institution or group. The particular attitude studied is that toward a minority group, and the experimental modes of persuasion used were:

1. a motion picture favorable to the minority group;
2. a lecture on the same topic, also favorable to the minority group;
3. a combination of the motion picture and the lecture.

Subjects were to be assigned at random to the different experimental groups, each having first been given a preliminary attitude test. Following the experimental treatment, each subject was to be given the test once again; the dependent variable here is the *change* in attitude score. It was recognized, however, that the mere repetition of a test may have some influence on the individual's changing his score, and so a randomly selected control group was also to be used, in which the subjects were to be given no special experimental treatment.

However, during the design of the experiment the question arose, "Would the subjects perhaps change as a result of seeing *any* movie or hearing *any* lecture?" Thus it was decided to introduce two more control groups, the first to be shown a movie completely unrelated to the minority group, and the second given a lecture by the same person, but on a quite different topic.

The final experimental design can be diagrammed as follows:

Experimental groups			Control groups		
I	II	III	IV	V	VI
Movie	Lecture	Movie and lecture	Nothing	Neutral movie	Neutral lecture

A total sample of thirty subjects was drawn at random, and subjects were assigned at random to the six conditions, with five subjects per condition.

In this study, the experimenter was not interested in the over-all existence of treatment effects. Rather, from the outset his interest lay in answering the following specific questions:

1. Do the experimental groups as a whole tend to differ from the control groups?
2. Is the effect of the experimental lecture-movie combination different from the average effect of either movie or lecture separately?
3. Is the effect of the experimental lecture different from the effect of the experimental movie?
4. Among the control groups, is there any effect of the neutral movie or lecture as compared with the group receiving no treatment?

In other words, the experimenter entertained a number of particular questions before the collection of the data, and he wished to analyze his data to answer those questions. An over-all analysis of variance and F test would give indication of the existence of *any* systematic effects. However, he was interested only in the particular differences among population means corresponding to answers to these questions.

Notice that the evidence pertaining to each question comes from the various sample means combined in some special way. For example, the evidence for question 3 involves only the difference between the means for groups I and II. On the other hand, the evidence for question 2 involves both the difference between mean I and mean III *and* the difference between mean II and mean III. Still other combinations of means pertain to the other questions asked. Can one, however, attach a significance level to each of these comparisons among means, permitting a statement about such differences among the *population* means?

This is a problem of planned comparisons among means, and the technique used will now be outlined. Let it be emphasized that this procedure applies only when the experimenter has specific questions to be asked *before* the data are collected, and this method is used *instead of* the ordinary analysis of variance and F test.

14.2 Planned comparisons

The basic theory underlying planned comparisons has already been outlined in Sections 8.9 and 8.10. There it was noted that given normally distributed variables sampled independently and at random, values from any linear combination of those random variables will also be normally distributed. Furthermore, if the mean and variance for each variable is known, then the mean and variance of the sampling distribution of the linear combination is also known. Both these principles were used in Section 10.12 to establish the rationale for the t test of independent differences. Now we will go far beyond the simple two-sample case and apply these principles for any linear combination of means.

First, we need to define a **population comparison or comparison among population means**: given the means of J distinct populations, $\mu_1, \cdots, \mu_J$, a comparison among those means is any linear combination or weighted sum, with weights c_j not all equal to zero:

$$\psi = c_1\mu_1 + c_2\mu_2 + \cdots + c_J\mu_J = \sum_j c_j\mu_j. \qquad [14.2.1*]$$

We will use the symbol ψ (small Greek psi) to stand for the value of some particular *population* comparison. The weights for a comparison ψ are *some set of real numbers* $(c_1, \cdots, c_J)$ *not all zero*.

In many applications of the theory, the requirement is made that the *sum* of the weights c_j equals zero:

$$\sum_j c_j = 0. \qquad [14.2.2*]$$

(Conventionally, a comparison in which the sum of the weights is zero is called a contrast.) In the following we will simply assume that this requirement is met, and defer giving the reason for this requirement until Section 14.9.

A **sample comparison** is defined exactly as for a population comparison, except that **sample means** are weighted and summed:

$$\hat{\psi} = c_1 M_1 + \cdots + c_J M_J = \sum_j c_j M_j \qquad [14.2.3*]$$

(the caret or "hat" over the psi will always indicate that this comparison is a sample value). Once again, we will require that $\sum_j c_j = 0$. Be sure to notice that the symbol for a comparison, either ψ or $\hat{\psi}$, stands for a single number, since it equals a *weighted sum of numbers*.

14.3   ESTIMATES OF POPULATION COMPARISON VALUES

Some of the statistical properties of comparisons will now be viewed in the light of what we know about linear combinations. We know from Section 8.10 that since each sample comparison is a linear combination, it is true that

$$E(\hat{\psi}) = E\left(\sum_j c_j M_j\right) = \sum_j c_j E(M_j) = \psi. \qquad [14.3.1*]$$

Any sample comparison is an unbiased estimate of the population comparison involving the same weights c_j. If we wish to evaluate the weighted sum of population means, our best estimate is the weighted sum of sample means. In the example above, each question asked by the experimenter can be answered by some particular comparison among population means. His best evidence for each question is some weighted sum based on sample means.

14.4   THE SAMPLING VARIANCE OF
PLANNED COMPARISONS

Suppose that each of the sample means M_j is based on a different sample of n_j cases drawn at random from a population with true mean μ_j. Furthermore, suppose that each population has the same variance σ_e^2, so that the sampling distribution of M_j values has a variance σ_e^2/n_j.

By the principle of Section 8.10, the sampling variance of a comparison $\hat{\psi}$ based on independent means must be

$$\text{var.}(\hat{\psi}) = \sum_j c_j^2 \, \text{var.}(M_j) = \sigma_e^2 \sum_j \frac{c_j^2}{n_j}. \qquad [14.4.1*]$$

Thus, for example, suppose that the weights in a sample comparison of five means were $(-2, -1, 3, 1, -1)$, so that

$$\hat{\psi} = (-2)M_1 + (-1)M_2 + (3)M_3 + (1)M_4 + (-1)M_5. \qquad [14.4.2]$$

Furthermore, suppose that each mean is based on 10 cases, so that $n = 10$ for all j. Provided that σ_e^2 is the variance of each population, then the sampling distribution of values of $\hat{\psi}$ must have a variance given by

$$\text{var.}(\hat{\psi}) = \sigma_e^2 \left[\frac{(-2)^2 + (-1)^2 + (3)^2 + (1)^2 + (-1)^2}{10} \right]$$

$$= \sigma_e^2 \left(\frac{16}{10} \right)$$

$$= \frac{8\sigma_e^2}{5}. \qquad [14.4.3]$$

Obviously, in no practical situation will we know the value of σ_e^2. However, *the value of σ_e^2 can be estimated from the data in exactly the same way that it is estimated in the analysis of variance:*

$$\text{est. } \sigma_e^2 = \text{MS error.}$$

(Here, σ_e^2 is the same as σ_e^2 in the expected mean squares for analysis of variance, and mean square error is the same as mean square within groups in the one-way analysis.) Thus

$$\text{est. var.}(\hat{\psi}) = (\text{MS error}) \sum_j \frac{c_j^2}{n_j}. \qquad [14.4.4\dagger]$$

Now we have the three essential ingredients for statistical inference: an unbiased estimate $\hat{\psi}$ of some population value ψ, an estimate of the sampling variance of $\hat{\psi}$, and we know the form of the sampling distribution of $\hat{\psi}$ (normal if populations are normal). These facts taken together suggest the use of t (Section 11.16). Next, the question of a test will be undertaken.

14.5 INTERVAL ESTIMATES AND TESTS FOR PLANNED COMPARISONS

Given the value of the planned sample comparison $\hat{\psi}$, a confidence interval for the population comparison value ψ can be found from

$$\hat{\psi} - t_{(\alpha/2;\,\nu)} \sqrt{\text{est. var.}(\hat{\psi})} \leq \psi \leq \hat{\psi} + t_{(\alpha/2;\,\nu)} \sqrt{\text{est. var.}(\hat{\psi})}. \qquad [14.5.1\dagger]$$

For the $100(1 - \alpha)$ percent confidence interval, the value of $t_{(\alpha/2;\,\nu)}$ represents the value cutting off the *upper* $\alpha/2$ proportion of sample values in a distribution of t with ν degrees of freedom. The number of degrees of freedom is the same as the degrees of freedom for the mean square error used to estimate σ_e^2. Such a confidence interval for ψ rests on the usual assumptions for a fixed effects analysis of variance: normally distributed populations, each with variance σ_e^2, and independent errors. As in the over-all F test for such an analysis, the assumption of normality is relatively innocuous for reasonably large samples.

On the other hand, violation of the assumption of homogeneity of error variances can have rather serious consequences in some instances, although such violations have their minimum effect when n is constant over all groups (*cf.* Section 12.16).

This confidence interval can be used to test any hypothetical value of ψ at all. If some hypothetical value of ψ fails to be covered by the confidence interval, then that hypothesis may be rejected beyond the α level (two-tailed).

In most instances, the only hypothesis of interest in planned comparisons is

$$H_0 : \psi = 0.$$

Then a test for this hypothesis is given by

$$t = \frac{\hat{\psi}}{\sqrt{\text{est. var.}(\hat{\psi})}}, \qquad [14.5.2\dagger]$$

distributed as t with degrees of freedom as for mean square error ($N - J$ in the one-factor experiment). If the hypothesis is to be tested against a directional alternative, then the sign of the t is considered, just as in the ordinary test for two means.

Notice that the t test for a difference between two independent means (Section 10.15) is merely a test of a comparison

$$\hat{\psi} = (1)M_1 + (-1)M_2$$

where the two weights are 1 and -1 respectively. However, the t test can be used to test any particular comparison, regardless of the number of groups involved. This is the exception in the use of t for means, previously mentioned in Section 12.14.

When the hypothesis to be tested is nondirectional, the t test for any planned comparison may be replaced by the equivalent F test:

$$F = \frac{(\hat{\psi})^2}{\text{MS error} \left(\sum_j c_j^2 / n_j \right)} \qquad [14.5.3\dagger]$$

with 1 and $N - J$ degrees of freedom. (Recall, once again, that t^2 is distributed as an F variable.)

Although this discussion has referred to comparisons planned among a set of J means for data in a one-factor experiment, exactly the same ideas apply when comparisons are planned in a two-factor (or higher) experiment fitting the fixed effects or mixed model. Thus, when the comparisons are to be made among the J column means in a two-factor experiment with K rows and n per cell in the data table,

$$\text{est. var.}(\hat{\psi}) = (\text{MS error}) \left(\sum_j \frac{c_j^2}{Kn} \right)$$

since each mean is based on Kn observations. The t or F value is computed just as in **14.5.2** or **14.5.3,** with $JK(n-1)$ degrees of freedom for the denominator. The test has this form, however, only when the fixed effects model applies.

When the second factor, represented by the rows, is sampled, so that the mixed model applies, then any comparison among the column means has an estimated variance given by

$$\text{est. var.}(\hat{\psi}) = (\text{MS interaction}) \left(\sum_j \frac{c_j^2}{Kn} \right)$$

and the number of degrees of freedom for t or the denominator of F is $(J-1)$ times $(K-1)$.

If, for some reason, cell means are to be compared in a two-way table, with n per cell, the estimated variance for the comparison is given by

$$\text{est. var.}(\hat{\psi}) = (\text{MS error}) \left(\frac{\sum_j \sum_k c_{jk}^2}{n} \right)$$

where the mean square error is the mean square within cells, and the degrees of freedom are found from $(JK)(n-1)$.

14.6 INDEPENDENCE OF PLANNED COMPARISONS

A single planned comparison among means is usually tested by a simple t ratio, as we have seen. However, seldom does an experimenter have interest in only one comparison on his data. Usually there are sets of questions he wants answered, each corresponding to some comparison among means. This brings up the very critical problem of the independence of comparisons.

Just as the possible answers to some questions may depend logically upon answers to others, so the values of some comparisons made on a given set of means may depend upon the values of other comparisons. This fact has serious consequences for estimates and tests of several comparisons, since the questions involved in the respective comparisons cannot be given separate and unrelated answers unless the comparisons are statistically independent of each other. Fortunately, a simple method exists for determining whether or not two comparisons are independent, given normal population distributions with equal variances σ_e^2. **The determination of the independence of two comparisons depends only on the weights each involve, and in no way on the means actually observed.** One can plan comparisons that will be independent *before* the data are collected.

The solution to the problem of independent comparisons rests on still another general principle having to do with linear combinations of variables. This principle is another instance of the extraordinary utility of the assumption of normal distributions in statistical theory. The principle may be stated as follows:

Given J independent random variables, $Y_1, \cdots, Y_J$, each normally distributed with variance σ_e^2, and two linear combinations,

$$\hat{\psi}_1 = c_{11}Y_1 + \cdots + c_{1J}Y_J$$

with weights $(c_{11}, \cdots, c_{1J})$, and

$$\hat{\psi}_2 = c_{21}Y_1 + \cdots + c_{2J}Y_J$$

with weights $(c_{21}, \cdots, c_{2J})$, then the random variables $\hat{\psi}_1$ and $\hat{\psi}_2$ are themselves independent *provided* that

$$\sum_j c_{1j}c_{2j} = 0. \qquad [14.6.1*]$$

In other words, **given independent and normally distributed random variables having the same variance, one can decide if two different linear combinations of those variables are independent simply by seeing if the products of the weights assigned to each variable sum to zero:**

$$c_{11}c_{21} + \cdots + c_{1J}c_{2J} = 0.$$

This principle is necessarily true only for normally distributed variables with equal variances.

Now suppose that we assume J normally distributed populations with equal variance, and consider independent random samples of size n from each population. Then from this principle, *two different sample comparisons $\hat{\psi}_1$ and $\hat{\psi}_2$ among means are independent only if*

$$\sum_j c_{1j}c_{2j} = 0, \qquad [14.6.2*]$$

where the c_{1j} are the weights used in the first comparison, $\hat{\psi}_1$, and c_{2j} are the weights used in the second, $\hat{\psi}_2$.

Two comparisons satisfying this condition are said to be **orthogonal** comparisons; orthogonality of two comparisons is equivalent to the statistical independence of sample comparisons only when the populations are normal, however. When two comparisons are statistically independent, the information each provides is actually nonredundant and unrelated to the information provided by the other. The estimate $\hat{\psi}_1$ is unrelated to the estimate $\hat{\psi}_2$. Thus, seeing if comparisons are orthogonal lets the experimenter judge whether or not he is gaining unrelated, nonoverlapping pieces of information about his experiment. On the other hand, although separate confidence intervals and tests can be constructed for such comparisons, these intervals and tests ordinarily are *not* statistically independent, due to the fact that the estimate of σ_e^2 each involves is based on the mean square error for the whole experiment. This is precisely the same problem encountered before with F ratios; the different values determining the numerators of several F ratios in an analysis of variance usually are independent, but the F ratios themselves are not independent if they involve the same denominator value. Only when the number of degrees of freedom for the mean square error is very large can the confidence intervals and tests for individual planned comparisons be regarded as independent. We will have more to say on this issue in the last section of this chapter.

When the J distinct samples have different sizes, symbolized by n_j, the criterion for orthogonality for two comparisons among the means becomes

$$\sum_j \frac{c_{1j}c_{2j}}{n_j} = 0,$$

so that the product of comparison weights for each sample is weighted inversely by n_j before the sum is taken. If this weighted sum of products is zero, then the comparisons may be regarded as orthogonal.

In somewhat more formal terms, given two or more mutually orthogonal comparisons with population values ψ_1, ψ_2, and so on, then

$$\frac{(\hat{\psi}_1 - \psi_1)^2}{\sigma^2 \sum_j c_{1j}^2/n_j}, \quad \frac{(\hat{\psi}_2 - \psi_2)^2}{\sigma^2 \sum_j c_{2j}^2/n_j}, \text{ etc.}$$

can be regarded as independent chi-square variables, each with 1 degree of freedom, under the usual assumptions of normality and equality of variances. This fact makes it possible to find separate confidence intervals for the various true ψ values, or to test separate hypotheses about the comparison values.

In the following, the terms "independent" and "orthogonal" will be used interchangeably. However, this is proper only because normal distributions with homogeneous variances are assumed throughout the discussion.

14.7 An illustration of independent and nonindependent planned comparisons

Four treatment-groups were observed in an experiment, with a sample of six randomly assigned subjects in each. The means of the four groups were

$$
\begin{array}{cccc}
I & II & III & IV \\
17 & 24 & 27 & 16.
\end{array}
$$

The mean square error, found by the usual method for a one-way analysis of variance, was 5.6, with $24 - 4$, or 20, degrees of freedom. Before the data were seen, the experimenter decided that he was interested basically in the following questions.

1. Does mean I differ from the average of means II, III, and IV?
2. Does mean II differ from the average of means III and IV?
3. Does mean III differ from mean IV?
4. Does the average of I and II differ from the average of III and IV?

It is assumed that each population is normally distributed, with the same variance in each. Now, each of the questions can be put into the form of comparisons among means by using the following sets of weights. In Table 14.7.1, the rows represent the various questions (comparisons) and the columns the various samples. In the cells the numbers are the weights c_j to be assigned to a given sample mean for a given comparison:

Table 14.7.1

	Means			
Question	I	II	III	IV
1	1	$-\frac{1}{3}$	$-\frac{1}{3}$	$-\frac{1}{3}$
2	0	1	$-\frac{1}{2}$	$-\frac{1}{2}$
3	0	0	1	-1
4	$\frac{1}{2}$	$\frac{1}{2}$	$-\frac{1}{2}$	$-\frac{1}{2}$

There is certainly nothing mysterious about the way these weights were chosen. The first comparison is designed to investigate a difference between mean I and the *average* of means II, III, and IV. This calls for a weighting of mean I by unity, and finding the average of the three other means, which is tantamount to weighting each one by a *negative* 1/3 in the comparison with mean I. Similarly, mean I does not figure in the second question, and so gets weight 0, while mean II is contrasted with the *average* for means III and IV. The other weights are found in similar ways. Note that the sum of weights for each row in the table is zero, as it should be for comparisons among means.

Are these four comparisons orthogonal, and thus independent of each other? Consider the weights in rows 1 and 2; the sum of their products across columns is

$$\sum_j c_{1j}c_{2j} = (1)(0) + \left(\frac{-1}{3}\right)(1) + \left(\frac{-1}{3}\right)\left(\frac{-1}{2}\right) + \left(\frac{-1}{3}\right)\left(\frac{-1}{2}\right) = 0,$$

so that comparison 1 and comparison 2 are orthogonal and hence independent. In the same way, by computing the sum of the products of weights we see that comparison 2 and comparison 3, and comparison 1 and comparison 3 are also independent.

Now look at comparison 1 and comparison 4. Here the sum of the products of weights is

$$(1)\left(\frac{1}{2}\right) + \left(\frac{-1}{3}\right)\left(\frac{1}{2}\right) + \left(\frac{-1}{3}\right)\left(\frac{-1}{2}\right) + \left(\frac{-1}{3}\right)\left(\frac{-1}{2}\right) = \frac{2}{3}$$

so that comparisons 1 and 4 are not orthogonal. Neither are comparisons 2 and 4. On the other hand, comparisons 3 and 4 are orthogonal.

These judgments about the orthogonality and consequent independence of comparisons depend, of course, upon the assumption that the means *themselves* are independent over samples, and that the sampling distribution represented by each mean is normal with variance σ_e^2/n.

The computation and tests for the first three comparisons will next be illustrated. Presumably, our experimenter would not test comparison 4 separately because of its nonindependence of the others. The information he gains from comparison 4 is redundant, depending on the outcomes of the first three comparisons.

The value of comparison 1 is

$$\hat{\psi}_1 = (1)(17) + \left(\frac{-1}{3}\right)(24 + 27 + 16) = -5.3.$$

The estimated variance of this comparison is, from **14.4.4,**

$$\text{est. var.}(\hat{\psi}_1) = \frac{(5.6)}{6}\left[(1)^2 + \left(\frac{-1}{3}\right)^2 + \left(\frac{-1}{3}\right)^2 + \left(\frac{-1}{3}\right)^2\right].$$
$$= .93(1.33) = 1.24.$$

Under the hypothesis that $\psi_1 = 0$, the t ratio is

$$t = \frac{-5.3}{\sqrt{1.24}} = -4.8.$$

For a nondirectional test, with 20 degrees of freedom, this result is significant beyond the 1 percent level, so that we reject the hypothesis that the true comparison value is zero. We can assert confidently that population mean I does differ from the average of means II, III, and IV.

In a similar fashion we find

$$\hat{\psi}_2 = (1)24 + \left(\frac{-1}{2}\right)(27) + \left(\frac{-1}{2}\right)(16) = 2.5$$

with

$$\text{est. var.}(\hat{\psi}_2) = \frac{5.6}{6}\left[(1)^2 + \left(\frac{-1}{2}\right)^2 + \left(\frac{-1}{2}\right)^2\right]$$
$$= 1.40.$$

Then

$$t = \frac{2.5}{\sqrt{1.40}} = 2.11.$$

This is just significant at the 5 percent level for a nondirectional test; mean II does differ significantly from means III and IV. Finally,

$$\hat{\psi}_3 = 27 - 16 = 11,$$

with

$$\text{est. var.}(\hat{\psi}_3) = \frac{5.6}{6}(1 + 1) = 1.87.$$

Here,

$$t = \frac{11}{\sqrt{1.87}} = 8.04,$$

which is very significant. We can say with confidence that population mean III is different from population mean IV.

Recall once again, however, that although these various comparisons' *values* may be regarded as independent estimates, the *tests* accompanying these estimates are not independent. There is no very satisfactory elementary way to determine how much our conclusion about any one of the comparisons depends upon our conclusions about any of the others. All we have really done by use of orthogonal, planned, comparisons is to package the data into nonoverlapping, informative, portions, affording separate tests.

14.8 MULTIPLE *t* TESTS ON A SET OF DATA

For the moment, let us return to the problem of many *t* tests carried out on the same set of data, first mentioned in Section 12.14. Now we can show more clearly why carrying out all *t* tests between pairs of means is not a very satisfactory way to package the data for maximum clarity of results.

Suppose that some *J* treatment groups exist, and the experimenter decides to carry out a *t* test in the ordinary way for each such pair of group means. The significance level for each test is reported just as though each were carried out in the absence of any of the others. Under these circumstances the comparisons tested by the *t* tests cannot be regarded as independent, and the various tests themselves refer to redundant, overlapping, aspects of the data. Some pattern of significant and nonsignificant results will undoubtedly result. How does one interpret this pattern? How much does the result of one such test dictate the findings on another? What do the significance levels mean, both alone and relative to each other? What correspondence does the pattern of significant results have to the significance level that would be found by use of one over-all *F* test for these means? None of these questions has a simple answer, it seems.

The interpretation of a set of such results is very difficult, particularly because of the nonindependence of various comparisons tested. This can be illustrated from the example of the preceding section. Each possible *t* ratio for a difference between means corresponds to a test of one of the comparisons in Table 14.8.1. Suppose that the *t* test corresponding to row 1 is carried out first,

Table 14.8.1

Comparison	*I*	*II*	*III*	*IV*
		Means		
1	1	−1	0	0
2	1	0	−1	0
3	1	0	0	−1
4	0	1	−1	0
5	0	1	0	−1
6	0	0	1	−1

that for row 2 second, and so on. It can be seen immediately by the rule of Section 14.6 that the comparison involved in the fourth t test is not independent either of the first or of the second. As an illustration of why this is true, suppose that you were given the information that the difference between means I and II is -7, and that the difference between I and III is -10. Then what do you know of the difference between II and III? The answer is that you know it to be exactly -3; the values of comparisons 1 and 2 actually determine the value of comparison 4.

Furthermore if by chance mean I is a little high, then both t values for comparisons 1 and 2 will tend to be more positive than they should; that is, t_1 and t_2 are not independent because they depend upon mean I. When one considers the entire set of six comparisons and tests it becomes clear that a fairly complex pattern of dependency runs throughout the possible t tests. It is absurd to regard each new comparison and test as somehow providing new information. Here, the information in the data is packaged in redundant ways, and this makes the determination of over-all significance levels for such results extremely difficult. One really has no simple way to tell how many of the significant results are due to chance alone, to the circumstance that several tests were carried out on the same data, or, most important, to the circumstance that some results dictate others. Hypothesis testing seems to lose most of its usual meaning when carried out in this "shotgun" fashion.

On the other hand, must the experimenter limit himself only to inferences corresponding to orthogonal comparisons in the data? Fortunately, the answer is "no," since there is a proper way of inspecting all differences or comparisons of interest, orthogonal or not, in a set of data, where the significance level attached to each comparison is quite meaningful. However, this method properly belongs under post-hoc rather than planned comparisons, and will be discussed in Section 14.15.

14.9   THE INDEPENDENCE OF SAMPLE COMPARISONS AND THE GRAND MEAN

Comparisons in which the sum of the weights is zero (that is, contrasts) represent weighted differences among sets of means. Such contrasts are usually of most interest to the experimenter. In principle, any set of weights (not all zero) could be used in carrying out a comparison among several sample means; however, as we have seen one usually uses only sets of weights such that $\sum_j c_j = 0$ for each comparison. Now one reason for this requirement can be shown: by insisting that the weights applied in any comparison sum to zero, we make each comparison value be independent of the value of the grand sample mean.

In Chapter 8 it was pointed out that the mean of any sample is a linear combination of random variables. Given N individuals divided among J sample groups, the grand mean over all individuals in all groups is a linear combination

$$M = \sum_j \sum_i \frac{y_{ij}}{N}$$

where each score gets a weight of $1/N$. Furthermore, again in terms of the individual scores, any sample comparison can be written variously as

$$\hat{\psi} = \sum_j c_j M_j = \sum_j c_j \sum_i \frac{y_{ij}}{n_j} = \sum_j \sum_i \left(\frac{c_j}{n_j}\right) y_{ij}.$$

Each *individual* score y_{ij} actually gets a weight (c_j/n_j), depending on the group j to which it belongs.

In most instances, we want to ask questions about combinations of population means that will be unrelated to any consideration of what the overall mean of the combined populations is estimated to be. Thus, we choose weights such that any comparison to be made is orthogonal to the linear combination standing for the grand mean. In terms of the basic scores, the principle of Section 14.6 implies that it should be true that

$$\sum_j \sum_i \left(\frac{1}{N}\right)\left(\frac{c_j}{n_j}\right) = 0,$$

which in turn implies that $\sum_j c_j = 0$.

The comparison for the mean can also be the basis of a test of a hypothesis about the population mean, such as H_0: $\mu = k$. One takes $\hat{\psi}_M = M$, and then $F_{(1,N-J)} = (\hat{\psi}_M - k)^2/(MSW/N) = t^2_{(N-J)}$.

14.10   THE NUMBER OF POSSIBLE INDEPENDENT COMPARISONS

It stands to reason that given any finite amount of data, only a finite set of questions may be asked of those data if one is to get nonredundant, nonoverlapping, answers. There is just so much information in any given set of data; once this information has been gained, asking further questions leads to answers that depend upon the answers already learned.

This idea of the amount of information in a set of data has a statistical parallel in the number of possible independent (orthogonal) comparisons to be made among J means:

Given J independent sample means, there can be no more than $J - 1$ comparisons, each comparison being independent both of the grand mean and of each of the others.

In other words, the experimenter can frame no more than $J - 1$ different comparisons standing for questions about his data, if he wishes these comparisons to be completely independent of each other and the grand mean. This does not say that many different sets of $J - 1$ mutually independent comparisons cannot be found for any set of data. It does say that once a set of $J - 1$ comparisons has been found where the comparisons are independent of each other and the grand mean, it is impossible to find one more comparison which is also independent both of the grand mean and all of the rest. Thus the number of questions the experimenter may ask his data as planned comparisons is limited, if the statistical answers are to be regarded as independent.

On the other hand, there are a great many *sets* of orthogonal comparisons that may be applied to any given set of data. In fact, there is no limit at all to the number of ways in which one might choose weights for the various comparisons, and such that a set of $J - 1$ such comparisons could be formed. The point is that in any set of such comparisons, that set may consist of no more than $J - 1$ which are mutually orthogonal.

This was the difficulty in the example of Section 14.7. There were only four sample means, and four comparisons were planned; it is impossible to find four mutually orthogonal comparisons *whatever* the weights used, if they all are to be orthogonal to the grand mean in addition. This principle also illuminates the dependencies among multiple t tests; given J samples, there will be $(J)(J - 1)/2$ possible t tests between pairs of means. If J is greater than two, this will always be more than the number of possible independent comparisons.

It is no accident that the number of independent comparisons, $J - 1$, is also the number of degrees of freedom for between groups in the analysis of variance. This connection will be clarified in the next section.

14.11   PLANNED COMPARISONS AND THE ANALYSIS OF VARIANCE

There is a very intimate connection between analysis of variance and the technique of planned comparisons. **Each and every degree of freedom associated with treatments in any fixed effects analysis of variance corresponds to some possible comparison of means.** The number of degrees of freedom for the mean square between is the number of possible *independent* comparisons to be made on the means. Any analysis of variance is equivalent to a break-down of the data into *sets* of orthogonal comparisons.

In order to show how this is true, we need to define the **sum of squares for a comparison:** the sum of squares for any comparison $\hat{\psi}_g$ is

$$\mathrm{SS}\,(\hat{\psi}_g) = \frac{(\hat{\psi}_g)^2}{w_g} \qquad [14.11.1\dagger]$$

where

$$w_g = \sum_j \frac{c_j^2}{n_j}. \qquad [14.11.2\dagger]$$

For any comparison $\hat{\psi}_g$ this sum of squares has *one* degree of freedom. It follows that

$$\mathrm{MS}\,(\hat{\psi}_g) = \mathrm{SS}\,(\hat{\psi}_g). \qquad [14.11.3]$$

Notice that the F ratio for a comparison $\hat{\psi}_g$ could be written as

$$F = \frac{\mathrm{MS}\,(\hat{\psi}_g)}{\mathrm{MS\ error}}, \qquad [14.11.4\dagger]$$

with 1 and $N - J$ degrees of freedom.

Now suppose that we have two *independent* comparisons $\hat{\psi}_g$ and $\hat{\psi}_h$ on the same data. If we find the values of SS $(\hat{\psi}_g)$ and SS $(\hat{\psi}_h)$ and add them up we have

$$\text{SS } (\hat{\psi}_g \text{ and } \hat{\psi}_h) = \text{SS } (\hat{\psi}_g) + \text{SS } (\hat{\psi}_h). \qquad [14.11.5]$$

Because $\hat{\psi}_g$ and $\hat{\psi}_h$ are independent, this new sum of squares has *two* degrees of freedom by the additive property of chi-square variables. Furthermore, we could use this sum of squares to test the hypothesis that *both* ψ_g and ψ_h (population values) simultaneously are zero:

$$H_0\colon \psi_g = 0, \psi_h = 0$$

The appropriate F test is given by

$$F = \frac{\text{MS } (\hat{\psi}_g \text{ and } \hat{\psi}_h)}{\text{MS error}} = \frac{[\text{SS } (\hat{\psi}_g) + \text{SS } (\psi_h)]/2}{\text{MS error}} \qquad [14.11.6\dagger]$$

with 2 and $N - J$ degrees of freedom. One can *either* test each comparison separately, *or* test both together via this F ratio; one does not do both kinds of tests, however, if the usual interpretation of a significance level is to be valid.

Extending this idea further, suppose that there were $J - 1$ independent comparisons on the data, and that the SS were calculated for each of them. Then

$$\text{SS } (\text{all } \hat{\psi}_g) = \text{SS } (\hat{\psi}_1) + \cdots + \text{SS } (\hat{\psi}_g) + \cdots + \text{SS } (\hat{\psi}_{J-1}) \qquad [14.11.7]$$

with $J - 1$ degrees of freedom.

If we wanted to test the hypothesis that *all* comparisons were simultaneously zero, in some set G consisting of $J - 1$ mutually orthogonal comparisons,

$$H_0\colon \psi_g = 0 \text{ for all } g \epsilon G,$$

then the F test would be

$$F = \frac{\text{MS } (\text{all } \hat{\psi}_g)}{\text{MS error}} = \frac{\sum\limits_{g} \text{SS } (\hat{\psi}_g)/(J - 1)}{\text{MS error}} \qquad [14.11.8\dagger]$$

with $J - 1$ and $N - J$ degrees of freedom. Again, a prior choice exists between separate or collective tests of these comparisons, but it makes no sense to test *both* the separate and the collective hypotheses in the usual way.

It can be shown that **for any set of $J - 1$ independent sample comparisons on any set of J means,**

$$\text{SS } (\text{all } \hat{\psi}_g) = \sum\limits_{g} \text{SS } (\hat{\psi}_g) = \text{SS between groups.} \qquad [14.11.9*]$$

The total of the sum of squares for any $J - 1$ independent comparisons on J means is always equal to the sum of squares between the J groups.

In the fixed effects analysis of variance, the F test between groups is always equivalent to a simultaneous test that all comparisons among means are zero. The F test is an "omnibus test" of all possible comparisons to be made among a particular set of means in the data. Each of an independent set of these comparisons can be tested separately instead of all together, if the experimenter has definite questions to ask about his data to begin with. If he has no such questions, then the overall F test still permits him to ask "did *anything* happen?" On the other hand, a much more powerful test of a particular hypothesis may be possible by testing one or more comparisons separately than by the overall F test. The important thing to remember is that for each degree of freedom in the sum of squares between groups or treatments, there is a potential prior question to be asked of the data. The F test gives evidence to let us judge if all of a set of $J - 1$ such orthogonal comparisons are simultaneously zero in the populations. For this reason, if planned orthogonal comparisons are tested separately, the overall F test is not carried out, and vice-versa.

The same idea extends to higher-order analyses of variance as well. Suppose that in a two-factor experiment, there are R rows and C columns in the data matrix. Then the SS rows is the sum of the $R - 1$ separate and independent sums of squares corresponding to orthogonal comparisons that might be made on row means. When the experimental design itself is orthogonal, each and every one of these $R - 1$ comparisons is itself orthogonal to any of the $C - 1$ independent comparisons that might be made on the column means. Thus, for an orthogonal design, the SS rows and SS columns are independent, since the values making up each sum are independent. For such a design, and given normal population distributions with equal variances, the mean square for rows and the mean square for columns are completely independent values. Furthermore, the $(R - 1)(C - 1)$ potential orthogonal comparisons summarized as mean square interaction are independent of those made on rows, of those made on columns, and of each other. A considerable part of the statistician's ability to formulate different kinds of experimental designs for different purposes comes from the general principle showing which potential comparisons are independent and which are not in a given way that data might be collected.

Incidentally, the same sort of argument used above can also shed some light on the concept of degrees of freedom in general. You should remember that the unbiased variance estimate s^2 is said to have $N - 1$ degrees of freedom when computed from a sample of N independent observations. Now the rationale given in Section 10.8 for this number of degrees of freedom is rather inadequate; although the N different observed values are independent, the N deviations from the sample mean going to make up s^2 are *not* independent. Nevertheless, it is always possible to find $N - 1$ linear combinations of those N values that will be orthogonal to each other and the sample mean. For normal populations, values of these $N - 1$ orthogonal linear combinations *are* independent. Furthermore, it can be shown that the value of s^2 depends only on the values of $N - 1$ such independent linear combinations. Each such linear combination of scores can be turned into a squared and weighted value, much like the sum of squares for a comparison among means, and s^2 is algebraically equivalent to the sum of these squared and weighted values divided by the number of degrees of freedom. Each single such squared value has 1 degree of freedom, since it was based on a single linear combination; the sum of these values has $N - 1$ degrees of freedom since the linear combinations themselves are orthogonal. In short, the degrees of freedom for s^2 is properly thought of as the number of linear combinations of scores, mutually orthogonal and orthogonal to the mean, sufficient to determine the value of s^2. Note the similarity of this formulation to that of mean square between as a value based on $J - 1$ linear combinations orthogonal to each other and to the grand sample mean; the underlying ideas are the same in both instances.

A simple example may help to clarify this point. Consider a random sample of three independent observations, yielding values y_1, y_2, and y_3. Three orthogonal linear combinations of these values are formed:

$$v_1 = \frac{y_1 + y_2 + y_3}{3}$$

$$v_2 = y_1 - y_2$$

and

$$v_3 = \frac{y_1 + y_2}{2} - y_3.$$

The first of these linear combinations is just the sample mean, and you can check for yourself that the other two combinations are orthogonal both to the sample mean and to each other. Now we take

$$SS = \frac{v_2^2}{2} + \frac{v_3^2}{1.5}$$

just as though these two v values were based on means. However, in terms of the original y values,

$$SS = \frac{(y_1 - y_2)^2}{2} + \frac{1}{1.5}\left(\frac{y_1 + y_2}{2} - y_3\right)^2,$$

so that on expanding this expression and turning SS into a mean square, we have

$$MS = \frac{SS}{2} = \frac{y_1^2 + y_2^2 + y_3^2 - y_1 y_2 - y_1 y_3 - y_2 y_3}{3}.$$

On the other hand, on taking s^2 in the usual way, we have

$$s^2 = \frac{\sum_i y_i^2 - \left(\sum_i y_i\right)^2 / 3}{2} = \frac{y_1^2 + y_2^2 + y_3^2 - y_1 y_2 - y_1 y_3 - y_2 y_3}{3}$$

$$= MS.$$

Thus, the mean square based on two orthogonal linear combinations, each orthogonal to the sample mean, is exactly equivalent to the unbiased estimate s^2 calculated from the values for these three observations. This mean square has two degrees of freedom, since it is based on two out of the set of three possible orthogonal linear combinations of values for three observations.

14.12 POOLING THE SUMS OF SQUARES FOR "OTHER COMPARISONS"

Quite often it happens that no set of prior questions exists that would use up all the possible $J - 1$ independent comparisons in the data. The experimenter may have only one or two questions he particularly wants answered, and is content to lump all other comparisons into a single test, corresponding to "are there any other effects?" Then a combination of planned comparisons and analysis of variance techniques is possible.

Suppose that out of $J - 1$ possible comparisons the experimenter decides that only two are of overriding interest. Let us call these comparisons $\hat{\psi}_1$ and $\hat{\psi}_2$. He calculates SS $(\hat{\psi}_1)$ and SS $(\hat{\psi}_2)$ and carries out the tests for each. Now what of all the remaining $J - 1 - 2$ independent comparisons he might make? A very simple way exists for finding a SS value for those remaining. First of all, by the analysis of variance he calculates SS between groups in the ordinary way. Then

SS (all $\hat{\psi}$ independent of $\hat{\psi}_1$ and $\hat{\psi}_2$)
$$= \text{SS between} - \text{SS } (\hat{\psi}_1) - \text{SS } (\hat{\psi}_2). \quad [14.12.1\dagger]$$

In other words, one simply subtracts the sum of squares for comparisons 1 and 2 from the sum of squares between groups to find the sum of squares for *all remaining* comparisons independent of the first two. Instead of only one F test, the experimenter now makes three tests: one for comparison 1, one for comparison 2, and one for all remaining comparisons, given by

$$F = \frac{\text{SS between} - \text{SS } (\hat{\psi}_1) - \text{SS } (\hat{\psi}_2)}{(J - 3) \text{ MS error}} \quad [14.12.2\dagger]$$

with $(J - 3)$ and $(N - J)$ degrees of freedom.

Any subset of some total possible set of independent comparisons may be planned for in this way: the comparisons of special interest are checked for independence and then tested. All remaining comparisons independent to those tested are embodied in the difference between the SS between and the sum of the SS for the special comparisons tested. The F test for "other comparisons" has its degrees of freedom reduced by one for each comparison tested separately. If this F value is significant, then these comparisons of secondary interest can be examined individually by post-hoc methods.

This is a strategy in the analysis of experiments that is quite versatile and is used all too seldom in psychological research; this is partly the result of an unfortunate confusion between planned and post-hoc comparisons that often appears in psychology. The method is particularly effective where various control groups are introduced into the experiment, and the experimenter's interest in the control groups is quite different from his concern with the experimental treatments. It should be emphasized, however, that the various comparisons to be tested separately must be planned before the data are seen for the inferences to be valid. Comparisons are tested *instead of*, rather than as a supplement to, the ordinary "omnibus" F test. Remember that it is essential to this method that the comparisons designed for separate testing be orthogonal according to the rule given in Section 14.6.

14.13 AN EXAMPLE USING PLANNED COMPARISONS

The experiment outlined in Section 14.1 was carried out with a total of 30 subjects, assigned at random into groups of 5. The weights for the comparisons representing the four basic questions are shown in Table 14.13.1. A check using the criterion for orthogonality (Section 14.6) shows that these comparisons can be regarded as independent. (Try to figure out for yourself what the remainder, or 5th, comparison must be if it is to be orthogonal to the first four.)

Table 14.13.1

	Treatments			Controls		
Comparison	Movie	Lecture	Movie & lecture	Nothing	Neutral movie	Neutral lecture
1	$\frac{1}{3}$	$\frac{1}{3}$	$\frac{1}{3}$	$-\frac{1}{3}$	$-\frac{1}{3}$	$-\frac{1}{3}$
2	$\frac{1}{2}$	$\frac{1}{2}$	-1	0	0	0
3	1	-1	0	0	0	0
4	0	0	0	1	$-\frac{1}{2}$	$-\frac{1}{2}$

It was decided to test each of these comparisons at the .05 level. The data turned out as shown in Table 14.13.2. The numbers in the table are changes in attitude score for each person.

Table 14.13.2

	Movie	Lecture	Movie & lecture	Nothing	Neutral movie	Neutral lecture	
	6	3	7	-6	5	-1	
	10	6	9	0	-5	3	
	1	-1	4	-5	3	2	
	6	5	9	2	-4	-1	
	4	2	3	2	5	-6	
Totals	27	15	32	-7	4	-3	68
Means	5.4	3	6.4	-1.4	.8	$-.6$	

The numerical computations for the ordinary one-way analysis of variance are first carried out; not only does this give the MS error needed for the comparisons tests, but also the SS between useful in testing any remaining comparisons.

$$\text{SS total} = 720 - \frac{(68)^2}{30} = 565.9$$

$$\text{SS between} = \frac{(27)^2 + \cdots + (-3)^2}{5} - \frac{(68)^2}{30} = 256.3$$

$$\text{SS error} = 309.6$$

$$\text{MS error} = \frac{309.6}{24} = 12.9.$$

Now for comparison 1:

$$\hat{\psi}_1 = \left(\frac{1}{3}\right)(5.4) + \left(\frac{1}{3}\right)(3) + \left(\frac{1}{3}\right)6.4 - \left(\frac{1}{3}\right)(-1.4) - \left(\frac{1}{3}\right)(.8) - \left(\frac{1}{3}\right)(-.6)$$

$$= \frac{16}{3} = 5.3$$

$$w_1 = \frac{1}{5}\left[\frac{1}{9} + \frac{1}{9} + \frac{1}{9} + \frac{1}{9} + \frac{1}{9} + \frac{1}{9}\right] = \frac{2}{15}$$

$$\text{SS}(\hat{\psi}_1) = \frac{15(5.3)^2}{2} = 210.7.$$

The test for comparison 1 is given by

$$F = \frac{210.7}{12.9} = 16.3$$

which is significant far beyond the 5 percent level for 1 and 24 degrees of freedom. There does seem to be a reliable difference between experimental and control groups, in general.

For comparison 2,

$$\hat{\psi}_2 = \left(\frac{1}{2}\right)(5.4) + \left(\frac{1}{2}\right)(3) - (1)(6.4) = -2.2$$

$$w_2 = \frac{1}{5}\left[\left(\frac{1}{2}\right)^2 + \left(\frac{1}{2}\right)^2 + 1\right] = .3$$

$$\text{SS}(\hat{\psi}_2) = \frac{(2.2)^2}{.3} = 16.1,$$

so that the F test for this comparison is

$$F = \frac{16.1}{12.9} = 1.2.$$

This is not significant. There is not enough evidence to say that the combined movie-lecture effect is different from the average of their separate effects.

Comparison 3 gives

$$\hat{\psi}_3 = (1)(5.4) - 1(3) = 2.4$$

with

$$w_3 = \frac{1}{5}(1 + 1) = .4$$

so that

$$SS\,(\hat{\psi}_3) = \frac{(2.4)^2}{.4} = 14.4.$$

The F test for comparison 3 is then

$$F = \frac{14.4}{12.9} = 1.1,$$

again not significant. There is not enough evidence to say that a movie-lecture difference exists in the populations.

The value for comparison 4 is

$$\hat{\psi}_4 = (1)(-1.4) + \left(\frac{-1}{2}\right)(.8) + \left(\frac{-1}{2}\right)(-.6) = -1.5$$

with

$$w_4 = \frac{1}{5}\left[1 + \frac{1}{4} + \frac{1}{4}\right] = .3.$$

The sum of squares

$$SS\,(\hat{\psi}_4) = \frac{(-1.5)^2}{.3} = 7.5$$

is less than MS error, so that the F test is definitely not significant.

Only four of the five possible independent comparisons have been made. The sum of squares for the fifth comparison can be found from

$$\text{SS between} - SS\,(\hat{\psi}_1) - SS\,(\hat{\psi}_2) - SS\,(\hat{\psi}_3) - SS\,(\hat{\psi}_4)$$
$$= 256.3 - 210.7 - 16.1 - 14.4 - 7.5 = 7.6.$$

This last comparison might be tested, but we can see that since its sum of squares is less than MS error, it cannot be significant.

The results of this analysis can be put into tabular form as shown in Table 14.13.3.

Table 14.13.3

Source	SS	df	MS	F
Between groups	256.3	5		
Comparison:				
1	210.7	1	210.7	16.3
2	16.1	1	16.1	1.2
3	14.4	1	14.4	1.1
4	7.5	1	7.5	—
Remainder	7.6	1	7.6	—
Error (within groups)	309.6	24	12.9	—
Totals	565.9	29		

As always, however, we should bear in mind that for any set of several significance tests, the probability that at least one of tests will give significant results can be relatively large, even though the nominal α level used for each separate test is conventionally small. A related problem in the evaluation of a set of results is the now familiar principle that the several F or t ratios, even for a set of independent comparisons, need not be independent. This point will be discussed further in Section 14.17. The over-all meaning of a set of statistical results such as ours is rather difficult to assess. We have, however, at least arranged the data into portions pertaining to hypotheses we believe important and meaningful for the over-all interpretation of the experiment. This was the contribution of the analysis by planned comparisons.

14.14   THE CHOICE OF THE PLANNED COMPARISONS

Given any J independent means, there are any number of ways to choose the $J - 1$ independent comparisons to be made among these means. The important thing is that the experimenter have definite prior questions that he wants answered, and that these questions can be framed as orthogonal, or independent comparisons. The experimenter must assure himself that he has collected data that will provide answers to his questions in terms of orthogonal comparisons. For example, in the experiment just analyzed, the experimenter might have had the following initial concerns:

1. Is there any effect of showing the experimental movie, as opposed to the initial lecture or nothing?
2. Is there any effect of the experimental lecture, as opposed to the experimental movie or nothing?
3. Is the effect of the experimental movie the same whether or not it is accompanied by a lecture?
4. Does the neutral movie have the same effect as the neutral lecture?

These four questions can be embodied in four comparisons quite different from those employed above (Table 14.14.1). A check shows that each of these

Table 14.14.1

Comparisons	Movie	Lecture	Movie & lecture	Nothing	Neutral movie	Neutral lecture
1	$\frac{1}{2}$	$-\frac{1}{2}$	$\frac{1}{2}$	$-\frac{1}{2}$	0	0
2	$-\frac{1}{2}$	$\frac{1}{2}$	$\frac{1}{2}$	$-\frac{1}{2}$	0	0
3	$-\frac{1}{2}$	$-\frac{1}{2}$	$\frac{1}{2}$	$\frac{1}{2}$	0	0
4	0	0	0	0	1	-1

comparisons is independent of each of the others. Notice, however, that some of these comparisons are *not* independent of some of the first set.

This new set of orthogonal comparisons also illustrates another point: If the experimenter had conceived the study as a two-factor design, in which "experimental movie versus no experimental movie" constituted the levels of the first factor, and "experimental lecture versus no experimental lecture" the levels of the second, then the first three comparisons outlined above correspond exactly to the three degrees of freedom associated with main effects and interaction in such a design. That is, comparison 1 provides a SS value that is exactly the same as the SS for the first factor, comparison 2 gives a SS value that corresponds to the SS for the second factor, and comparison 3 provides a SS value that is identical to the SS for interaction in a two-factor design. The fourth comparison is, however, based on two groups not included in the two-way design. This illustrates the fact mentioned in Section 14.11: the single degrees of freedom included in the SS between groups for an analysis of variance each correspond to some potential comparison among means.

For any given experiment, there will undoubtedly be many interesting questions that could be framed as planned comparisons. Only the particular experimenter can decide, of course, which comparisons are of interest to him. As a rule, he should not try to think up a sensible question to correspond to each of the $J - 1$ degrees of freedom associated with J means just because he can, in principle, do so. Rather, the technique of planned comparisons should be used only when there are a few important specific questions to ask of the data that will clarify the whole experiment if answered. Once these questions have been decided upon, they must be phrased so that the answer will be evident from some way of weighting and combining means. Usually, such questions resolve themselves into differences between groups of means, just as, in the last set of comparisons, the answer to question 1 depends upon the difference between the averages of two groups of means:

$$\frac{(\text{movie}) + (\text{movie \& lecture})}{2} - \frac{(\text{lecture}) + (\text{nothing})}{2}.$$

This difference weights the (movie) mean and the (movie & lecture) means each by $1/2$, and the (lecture) and (nothing) group means each by $-1/2$, so that this difference gives us our weights for that comparison. The same is true for each of the other comparisons discussed here. In general, the idea is to frame the question, determine what data are necessary for the answer, and then ask how the data may be combined to answer it. The questions of most importance should be formulated first, and then if other important questions can be framed as independent comparisons, well and good. However, if all the different questions cannot be framed as separate orthogonal comparisons, the more important may be tested directly and the others lumped into an over-all "other comparisons" F test.

14.15   INCIDENTAL OR POST-HOC COMPARISONS IN DATA

Even though tests for planned comparisons form a useful technique in experimentation, it is far more common for the experimenter to have no special questions to begin with. His initial concern is to establish only that some real effects or comparison differences do exist in his data. Given a significant over-all test, his task is then to explore the data to find the source of these effects, and to try to explain their meaning.

The technique for comparisons to be introduced now is strictly applicable only to the situation where a preliminary analysis of variance and F test has shown over-all significance. It is not a device for rescuing poor experiments by data-juggling. Instead, if the experimenter has found evidence for over-all significance among his experimental groups, he may use this method of post-hoc comparisons to evaluate *any* interesting comparisons among means.

This statement is not to be interpreted to mean that post hoc comparisons are somehow illegal or immoral if the original F test is not significant at the required α level. It means only that the probability statements that one makes about such comparisons are not necessarily true when F does not reach that level, and that going through the procedure may be something of a waste of time if one really is interested in accurate probability statements. In spite of considerable confusion in the psychological literature and elsewhere on this issue, you can investigate comparisons among means whenever and wherever you like. Such comparisons may be quite suggestive about what is going on in the data, or about new questions calling for new experiments. What one cannot do is to attach an unequivocal probability statement to such post hoc comparisons, unless the conditions underlying the method have been met. It is not correct to assert that some post hoc comparison is significant at the α level by this method unless the overall F is significant at the α level, but you can look at anything or say anything else you wish.

The idea of a comparison here is exactly the same as defined in Section 14.2. A sample comparison $\hat{\psi}$ is a linear combination of sample means

$$\hat{\psi} = \sum_j c_j M_j$$

where $\sum_j c_j = 0$.

After the over-all F has been found significant, then *any* comparison $\hat{\psi}$ may be made. *Unlike planned comparisons, there is no requirement that such post-hoc comparisons be independent.* Any and all comparisons of interest may be made. Most often the experimenter may be interested in examining all pairs of means, but any comparison $\hat{\psi}$ is legitimate.

There are a number of methods that have been devised for testing the significance of post-hoc comparisons, only one of which will be given here. This is the method due to Scheffé (1959), which has advantages of simplicity, applicability to groups of unequal sizes, and suitability for any comparison. This method is also known to be relatively insensitive to departures from normality and homogeneity of variance. The method due to Tukey (outlined in Winer, 1962) seems to be preferable to the Scheffé method when the various samples are of equal size and the experimenter's interest lies only in all possible pairwise differences between means. However, the Scheffé method is emphasized here because of its simplicity and versatility over a wide variety of situations.

Given any comparison g made on the data after a significant F has been found for the relevant factor, the significance of the comparison value $\hat{\psi}_g$ may be found by use of the following confidence interval:

$$\hat{\psi}_g - S\sqrt{V(\hat{\psi}_g)} \le \psi_g \le \hat{\psi}_g + S\sqrt{V(\hat{\psi}_g)} \qquad \text{[14.15.1†]}$$

where

$$\sqrt{V(\hat{\psi}_g)} = \sqrt{(\text{MS error})w_g} = \sqrt{\text{est. var.}(\hat{\psi})} \qquad \text{[14.15.2†]}$$

and

$$S = \sqrt{(J-1)F_\alpha}. \qquad \text{[14.15.3†]}$$

(Take care to note that S as used here, is *not* the sample standard deviation.) The w_g is defined in Section 14.11, and F_α is the value required for significance at the α level, with $J-1$ and $N-J$ degrees of freedom (that is, the F required for significance at the α level for an *over-all* test of the J means). For any α, this gives the $100(1-\alpha)$ percent confidence interval for ψ_g, the true value of the comparison. When the confidence interval fails to cover zero, the comparison is said to be significant, and identified as one possible contributor to the over-all significance of F.

The meaning of this confidence interval for post-hoc comparisons requires some special comment. If we consider all possible comparisons ψ_g that might be carried out on the true means of the J groups, then *the probability is $1 - \alpha$ that the statement* 14.15.1 *is true simultaneously for all* ψ_g. That is, if we could work out all possible comparisons on the data, and for each comparison calculate a 95 percent confidence interval, then the chances are 95 in 100 that every one of these confidence intervals would contain the true value for that comparison. There is only a 5 percent chance that one or more confidence intervals will not cover the corresponding true comparison value.

This is a somewhat different conception from the usual confidence-interval statement. When we make a 95 percent confidence interval for, say, a single mean μ, we are correct in saying that if we took all possible random samples of size N and calculated the interval for each, the interval would cover the true value of μ 95 percent of the time. For the post-hoc comparisons, however, we are referring to all possible comparisons that might be made on a given set of data, and the probability statement is about the event of all such intervals computed from a set of data covering the corresponding true values. For this reason it is important that the initial F be significant, giving us prior reason to believe that reliable departures from zero exist to be found among the possible comparisons.

Note that one says that a comparison is significant at the α level if the confidence interval for that comparison does not cover zero. Any confidence interval for a comparison ψ_g that includes zero within its limits is said to be nonsignificant.

If the over-all F test is significant at the α level, then some comparison $\hat{\psi}_g$ must be significant at or beyond the same level. Indeed, a significant F test can be interpreted as evidence that at least one true comparison value among all those possible is not zero. This does not mean that just because the over-all F was significant you will necessarily find the significant comparisons, but only that they exist to be found. Hence our interpretation of a significant F as a signal, "Something's here—start looking."

14.16   AN EXAMPLE OF POST-HOC COMPARISONS
FOLLOWING A ONE-WAY ANALYSIS OF VARIANCE

In a psychological study subjects were assigned randomly to five different groups, representing five different experimental treatments. Twelve subjects were used in each group. The means of the various groups were

I	II	III	IV	V
63	82	80	75	70

and the analysis of variance is summarized in Table 14.16.1. For an α of .05, the required F for 4 and 55 degrees of freedom is approximately 2.53, so that this obtained value is significant. Hence, post-hoc comparisons may be tested for significance. The .05 level will be used in these tests as well.

Table 14.16.1

Source	SS	df	MS	F
Between groups	2856	4	714	4.0
Error (within groups)	9801	55	178.2	
Totals	12652	59		

Now suppose that the experimenter wished to examine all pair-wise differences between means. It is convenient to put all ten such differences in a table, such as Table 14.16.2. (The difference in a cell is the mean represented by column subtracted from the mean represented by the row; thus $63 - 82 = -19$, and so on.)

Table 14.16.2

			Group		
	Mean	II	III	IV	V
Mean		82	80	75	70
Group					
I	63	-19	-17	-12	-7
II	82		2	7	12
III	80			5	10
IV	75				5

The 95 percent confidence interval will be found for each of these differences, as follows.

First we note that each of the comparisons being made uses the same numbers as weights, and that each mean is based on the same number of subjects, twelve; hence

$$w_g = \frac{1}{12}(1+1) = \frac{1}{6}$$

and $\qquad \sqrt{V(\hat{\psi}_g)} = \sqrt{\frac{178.2}{6}} = 5.45$ for each pair of means.

Also

$$S = \sqrt{(4)F_{(.05)}} = \sqrt{4(2.53)} = 3.18.$$

The 95 percent confidence interval for each comparison difference is thus

$$\hat{\psi}_g - (3.18)(5.45) \le \psi_g \le \hat{\psi}_g + (3.18)(5.45)$$

or $\qquad\qquad \hat{\psi}_g - 17.33 \le \psi_g \le \hat{\psi}_g + 17.33.$

For this interval to *exclude* zero for any of these differences, the obtained difference would have to be *greater* than 17.33 in absolute magnitude. One of the differences (between I and II) is this large, and so this pairwise comparison is significant (at the .05 level). We can say that this difference contributes to the over-all significance of F.

However, many other comparisons could be tested. Suppose that we compare the average of II and III with the average of I, IV, and V. Here the value of the comparison is

$$\hat{\psi}_g = \frac{82 + 80}{2} - \frac{63 + 75 + 70}{3}$$

$$= \left(\frac{1}{2}\right)(82) + \left(\frac{1}{2}\right)(80) - \left(\frac{1}{3}\right)63 - \left(\frac{1}{3}\right)75 - \left(\frac{1}{3}\right)(70) = 11.67.$$

Now the confidence interval is different, since

$$w_g = \frac{1}{12}\left(\frac{1}{4} + \frac{1}{4} + \frac{1}{9} + \frac{1}{9} + \frac{1}{9}\right) = \frac{5}{72} = .069,$$

and $\qquad\qquad \sqrt{V(\hat{\psi}_g)} = \sqrt{(178.2)(.069)} = 3.51.$

The value of S is still 3.18 for $\alpha = .05$. Therefore, the confidence interval for this comparison is

$$\hat{\psi}_g - (3.18)(3.51) \le \psi_g \le \hat{\psi}_g + (3.18)(3.51)$$

or

$$11.67 - 11.16 \le \psi_g \le 11.67 + 11.16.$$

This interval does *not* include zero, and, so this comparison also is significant beyond the .05 level.

The mere fact that one can find a significant comparison does not insure that the comparison is a meaningful one. It is definitely not profitable to work out every conceivable comparison among the means and test each for significance, in hopes that something of meaning will emerge. Just the reverse procedure should be used: inspecting the data, the experimenter comes to tentative conclusions about where the large and interpretable effects lie. These tentative conclusions are then tested.

14.17 Planned versus post-hoc comparisons

It is obvious that in any given experiment it is always possible either to plan comparisons to be tested in lieu of the over-all F test, or to perform post-hoc comparisons should the over-all F be significant. What are the arguments for and against these two ways of proceeding?

An important point in favor of planned comparisons is this: consider any true comparison ψ among J means ($J > 2$), such that $\psi \neq 0$. **The probability of a test's detecting that ψ is not zero is greater with a planned than with an unplanned comparison on the same sample means.** In other words, for any particular comparison, the test is more powerful when the comparison is planned than when it is post-hoc.

This can be seen in a slightly different way by considering confidence intervals. As we have just seen, the confidence interval for a post-hoc comparison is

$$\hat{\psi}_g - S \sqrt{V(\hat{\psi}_g)} \leq \psi_g \leq \hat{\psi}_g + S \sqrt{V(\hat{\psi}_g)}.$$

where

$$S = \sqrt{(J - 1)F_\alpha}.$$

The confidence interval for a planned comparison may be written as

$$\hat{\psi}_g - t_{(\alpha/2)} \sqrt{V(\hat{\psi}_g)} \leq \psi_g \leq \hat{\psi}_g + t_{(\alpha/2)} \sqrt{V(\hat{\psi}_g)}$$

where $t_{(\alpha/2)}$ is the value of t required for two-tailed significance at the α level with $N - J$ degrees of freedom. It can be shown that for more than two means

$$t_{(\alpha/2)} < S,$$

so that **the confidence interval for any given comparison is shorter when that comparison is planned than when it is post-hoc.**

The practical implication is that the *importance* of the comparison should dictate whether or not it is tested by the planned or the post-hoc procedure. If the experimental question represented by the comparison is an important one for the interpretation of the experiment, and it is essential that a Type II error not be made in the accompanying test, then a planned comparison should be carried out. On the other hand, if the question is a minor one, and a considerable risk of overlooking a true non-zero value can be tolerated, then the post-hoc method suffices.

However, we have also seen that the number and variety of planned comparisons to be made and tested is limited by the independence requirement. No such requirement holds for the post-hoc method. The data may be only partly explored by the planned method, but fully by the post-hoc method.

One final but very mportant point must be made before we leave the subject of planned and post-hoc comparisons. This point applies both to the method of planned comparisons and to any technique such as the multiway analysis of variance, where some K different significance tests are applied to the data. It is true that when comparisons are orthogonal or the experimental design itself is orthogonal these various tests may be carried out separately. Furthermore, they may be regarded as approximately independent when the degrees of freedom for error is very large. Under these circumstances, if each test is conducted with α set at some level, say .05, then one should expect about 5 percent of these tests to show significance by chance alone, even though each and every H_0 tested is true.

However, the temptation is very strong for the experimenter interpreting the entire set of significance tests and evaluating the experiment as a whole to focus on the several tests that did turn out significant, and to overlook the fact that many such tests were carried out. What is the probability that *at least one test* will show spurious significance out of K tests carried out? We have already seen that for K independent tests this can be found from

$$p(\text{one or more significant results, all } H_0 \text{ true}) = 1 - (1 - \alpha)^K$$

which for small α is approximately equal to $K\alpha$. Thus, as the number of different tests increases, the probability of at least one spuriously significant result tends to increase as well. Although this probability can be calculated exactly only for independent tests, this is a feature of *any* set of ordinary significance tests on the same set of data, whether of planned comparisons or for effects in a complicated analysis of variance. Perform enough tests and the probability is large that at least one will turn out significant by chance alone.

On the other hand, in carrying out post-hoc comparisons the situation is different. Here, the probability is α that at least one comparison will turn out to be spuriously significant at the α level, regardless of how many comparisons are made. It is true that, for any given comparison, the probability of *overlooking* a true difference from zero is greater in the post-hoc than in the planned method (in other words, the Type II error probability is greater). Nevertheless, the probability of committing a Type I error for *one or more* such post-hoc comparisons is exactly α.

This fact has two important implications for us: In the first place, the multiple results one gets from a complicated experiment are hard to interpret from a statistical point of view. Given enough significance tests, you are almost bound to get at least one significant result, even though the experiment is the purest nonsense. The probability that at least one of your over-all conclusions is wrong is considerably greater than α. This is not an argument against the use of planned comparisons any more than against the use of multifactor experimental designs; it is, however, an argument against the indiscriminate use of the α level from single significance tests as a means of evaluating the conclusions drawn from the experiment as a whole, and especially against the overinterpretation of single significant findings.

The second implication is that the importance of not committing a Type II error should have a great deal to do with the choice of planned versus post-hoc comparisons. Thus, casual inspection of the data seems to call for the post-hoc variety of comparison, whereas the "big guns" of planned comparisons should be reserved for the really important questions, where one wants to be almost sure to detect true differences from zero where they exist.

If, for some reason, the experimenter feels that he must have the probability equal α or less that *one or more* planned comparisons turn out significant by chance, then he may adopt the following procedure: Each of the K separate comparisons may be tested at the α/K level, and then it will be true that the probability of one or more significant results by chance alone is no greater than α. For example, suppose that given some 10 experimental groups, 5 comparisons are to be tested separately. The experimenter tests each at the .01 level. Then the probability that one or more turns out significant by chance alone is no more than 5(.01) or .05. When only a few out of relatively many possible comparisons are to be tested separately, this procedure gives the experimenter some control over the "error rate" for the separate tests, without completely sacrificing the gains in power and "neat packaging" of the data achieved by making planned rather than post-hoc comparisons.

Since the procedure outlined in the preceding paragraph will often require t values for percentage points not listed in the standard tables, the following approximation is useful. In order to find the approximate value of t cutting off the upper α proportion in a distribution with ν degrees of freedom, take

$$t_{(\alpha;\,\nu)} = z_\alpha + (z_\alpha^3 + z_\alpha)/4(\nu - 2),$$

where z_α is the corresponding value in a standardized normal distribution. (The literature on planned and post-hoc comparisons is really quite extensive, and this subject has received a great deal of attention from psychologists and other users of applied statistics in recent years. A brief review of some of the more interesting developments especially relevant to psychology may be found in Hays [1968].)

In summary, then, let it be said that the planned comparisons method is best suited for situations where a few overriding concerns dictate the interpretation of the whole experiment. Here one must have the most powerful tests possible for resolving these issues. The post-hoc method is suited for trying out hunches gained during the data analysis and for inferring the sources of the significant over-all F test. We will return to the subject of planned and post-hoc comparisons in Chapter 16, where their applicability to still another kind of experiment will be pointed out.

EXERCISES

1. A set of seven fixed treatments, each with the same number of subjects, is to be used in an experiment. Find the weights for six orthogonal comparisons among the resulting means.
2. Consider four groups, and let there be two orthogonal comparisons with weights as follows:

	I	II	III	IV
ψ_1	1	1	-1	-1
ψ_2	1	-1	-1	1

Show that the weights, c_{3j}, for a third orthogonal comparison must satisfy the simultaneous equations

$$c_{31} + c_{32} + c_{33} + c_{34} = 0$$
$$c_{31} + c_{32} - c_{33} - c_{34} = 0$$
$$c_{31} - c_{32} - c_{33} + c_{34} = 0.$$

Solve the equations for these weights. Does this suggest a general method for finding weights for sets of orthogonal comparisons? How would one go about this?

3. Consider the following set of means, each corresponding to one fixed experimental treatment on a separate group of 10 independent subjects:

I	II	III	IV	V
86	95	92	80	104

If the mean square error is 40, test the significance of the planned comparisons having the following weights:

I	II	III	IV	V
1	-1	0	0	0
0	0	0	1	-1
-1	-1	0	1	1
1	1	-4	1	1

4. Suppose that in problem 3 above there had been interest only in the first two comparisons individually. Make an analysis of variance summary table showing the resulting mean squares and tests of significance.

5. Given the overall significance level established for problem 3, carry out *post-hoc* tests for the difference between all pairs of means.

6. In the following table, the means for a three by three factorial experiment with n observations per cell are symbolized. If interactions among the cell means are to be independent of comparisons among row or among column means, see if you can work out at least 3 of the 4 orthogonal comparisons underlying the interaction sum of squares, and show that they are orthogonal to row and column interactions.

M_{11}	M_{12}	M_{13}
M_{21}	M_{22}	M_{23}
M_{31}	M_{32}	M_{33}

7. For problem 16 of chapter 12, treat the following comparisons among columns as planned, and test for significance:

I	II	III	IV
1	-1	-1	1
1	0	0	-1
0	1	-1	0

8. For problem 16, chapter 12, suppose that the comparisons listed in problem 7 had been unplanned. Carry out a *post-hoc* test for these comparisons.
9. For problem 12 of chapter 12 carry out a test of the planned comparisons corresponding to "male versus female cues" and to "cues versus no cues."
10. Construct the 95 percent confidence interval for the two comparisons made in problem 9 above.
11. For problem 12, chapter 12, carry out a test of the comparison corresponding to "same-sex subjects and cues versus different-sex subjects and cues." To which sum of squares does this comparison contribute?
12. For problem 16, chapter 12, carry out and test *post-hoc* comparisons on all differences of between column means.
13. For problem 16, chapter 12, devise and carry out a test of two planned comparisons among rows.
14. Suppose that planned comparisons between lists I and II and between lists III and IV had been planned for problem 17, chapter 13. Find the 95 percent confidence limits for these comparisons.
15. If it is appropriate to do so, carry out *post-hoc* comparisons between all "list" means for problem 17, chapter 13.
16. Why are comparison methods limited to designs corresponding to fixed or mixed models?
17. Below is an analysis of a variance summary table.

SOURCE	d.f.	s.s.
Between rows	(4)	
Comparison 1	1	203
Comparison 2	1	25
Other row comparisons	2	250
Between columns	(6)	
Comparison 1	1	78
Other column comparisons	5	507
Interaction	(24)	
Comparison 1	1	215
Other comparisons	23	
Error	105	3405
Total	139	5660

Assuming a fixed-effects model, complete the table and carry out the tests indicated ($\alpha = .01$).

18. See if you can suggest circumstances under which the theory of comparisons might extend to statistical analyses other than the comparison of means in the analysis of variance.

15 Problems in Linear Regression and Correlation

So far the emphasis in this book has been upon problems of inferring population characteristics from samples, and of finding evidence for statistical relations between variables. There is, however, quite another aspect to the theory of statistical inference: the problem of predicting a score for the individual case, given some prior information about him. How does one *use* the statistical relations found in data? For example, a personnel officer in an industrial setting must decide whether or not to hire an individual for a specific job. He has some information about him, perhaps a score on trade-skills test, and in the light of this information he must "bet" or predict how the person will work out on a particular job. A vocational counselor must give some advice to a prospective college student about the course of study in which he is likely to find success. The counselor has the student's high-school record, as well as his intelligence and interest test scores— what is the best thing to advise the student? A physician examines a patient who exhibits certain symptoms and laboratory test results. What kind of treatment is the best bet to relieve these symptoms? In the real work-a-day world, the problem of using the results of statistical studies to make a decision about an individual case is not only one of the most common, but perhaps the most important, single use of statistical findings.

Quite apart from any practical application, the problem of prediction arises in a purely scientific study as well. Several times in the preceding discussion we have emphasized that the scientist is looking for evidence of stable relationships in his data, and that the idea of strength of a statistical relation is based on the extent to which knowing individual status on variable X reduces uncertainty about status on Y, the thing predicted. The indices ω^2 and ρ_I both rest on this notion: statistical association is strong to the extent that specifying the event X reduces the variance of the possible Y values, so that, literally, information about X tells us something about Y. Furthermore, we have noted that if there is actually a functional relation between X and Y, perfect prediction is possible, so that X tells us *everything* about Y, other things being equal.

However, we have said very little about possible *rules* for prediction. There has been a very good reason for us to avoid the question of prediction rules: heretofore, X has been treated categorically, as standing for any one of a set of qualitatively distinct categories into which an observation might fall. The analysis of variance, for instance, deals only with a set of different samples representing different populations of potential observations. The actual experimental basis for distinguishing among the different samples is, formally, immaterial; the samples may represent either qualitative or quantitative distinctions of any sort. In the simple analysis of variance, we acknowledge only that the different samples are somehow different n one experimental respect, and then we look for systematic differences in the dependent variable Y. However, rules for going from the *qualitative* status of an observation to some quantitative prediction are ordinarily very difficult to formulate, and so we have had to be content with discussing evidence for the existence of a statistical relation, and with forming estimates of its strength according to indices such as ω^2.

Now, however, we are going to deal with situations where the independent variable X actually takes on *numerical* values. The variable X may stand for an amount of some treatment administered in an experiment, such as the dosage of a drug, or the number of trials given at a learning task, values fixed quite arbitrarily by the experimenter. Or, perhaps, X may stand for some measured property of the individual himself, such as his height, his age, or his intelligence quotient. Associated with each possible X value there is some distribution of Y scores, the conditional distribution of Y given X (compare Section 4.7). Perfect prediction of Y from X will be possible only if the relation between the two variables is functional, of course. For a functional relation, each conditional distribution of Y given X consists of a single value. Nevertheless, regardless of whether or not the relation is actually a function, we can ask three questions of any set of data in which each individual observed represents the occurrence of some joint *numerical* event (x, y):

1. Does a statistical relation affording some predictability appear between the random variables X and Y?
2. How strong is the apparent degree of the statistical relation, in the sense of possible predictive ability the relation affords?
3. Can a simple rule be formulated for predicting Y from X, and if so, how good is this rule?

The ordinary statistical techniques we have studied apply to the first two of these questions, but the third is a new feature. In this chapter we are going to study the possibility of applying a *rule* for the prediction of Y from X. We are going to act *as though* the true relation actually were a function, and, using a function rule, make predictions or "bets" about Y values from knowledge of X values. Then we are going to evaluate the *goodness* of this prediction rule in terms of how well one actually would do by predicting according to the rule. If the statistical relation actually is a function then some rule exists that affords perfect prediction; for the usual statistical relation, no rule permits perfect prediction, but some function rule may nevertheless provide a good "fit" to the relation under study. Proceeding in this fashion gives us two important advantages: quite often we are able to achieve a fair degree of predictive ability by adopting a particular function rule, even though the true relation itself is not really a function. Second, by studying our errors using this rule, we gain information about how the rule might be made better and how the general form of the relation might be specified more adequately.

The basic topic of this chapter is the study of prediction using a *linear* function rule. When pairs of (x, y) values fall exactly into a function that can be plotted as a straight line, the function is said to be linear. Such functions have extremely simple rules, which are always of the form

$$Y = bX + a$$

where b and a are two constant numbers. In this chapter we will see how prediction rules of this linear form can be applied, and how the fit of such a rule to a given relation can be evaluated.

The reasons for starting with linear rules for prediction are several: linear functions are the simplest to discuss and understand; such rules are often good approximations to other, much more complicated, rules; and we will find that in certain circumstances the only prediction rule that *can* apply is linear. However, do not jump to the conclusion that just because we deal first with linear prediction this is the only important way to predict, or that all real relationships must be more or less like linear functions. In the next chapter we will find that there are many other, nonlinear, function rules that might also be applied to a given problem.

15.1 REGRESSION AND CORRELATION PROBLEMS

At the very outset it must be emphasized that we are going to deal with two rather different sampling situations in this chapter. First, we will discuss techniques applicable to so-called **regression problems**. These are the problems and techniques that will ordinarily be of most interest to the experimental psychologist. The general ideas we will use are straightforward extensions of those underlying the analysis of variance, except now we will consider *numerical* experimental variables.

For example, a typical regression problem might be the following: an experimenter is interested in the effect of some tranquilizing drug on a subject's ability to solve reasoning problems. Individuals are assigned at random to different experimental groups, and the same drug is administered in different preassigned amounts to the various groups. These dosages were decided upon before the experiment, and the experimenter is concerned only with these fixed amounts. Then the groups are compared on the basis of scores on a reasoning task. The experimenter is interested not only in the existence of a possible statistical relation between the amount of drug administered and the reasoning score, but also in the extent to which this relation may be approximated by the use of a linear function rule.

Several things are important about this example. In the first place, there is a clearly identified independent variable X, each value of which is an amount of the drug administered. The possible values of this variable are fixed in advance by the experimenter. The dependent variable Y is the reasoning score for a subject, and within each experimental group this variable is free to take on different values. Furthermore, *both* the independent and dependent variables are *numerical:* each subject in a group j has the same X score, the amount of drug administered, although a particular subject i in group j has his own dependent variable score Y_{ij}. Finally, our interest is in the *form* of the relation that occurs, and specifically, on whether this relation can be approximated well by a linear function. In other words, can one predict the value of Y_{ij} from that of X_j using such a linear rule?

In a later section we will discuss the meaning of the term "regression" as used here. For the moment, suffice it to say that in a problem in regression, one variable is clearly the independent or predictor variable, the thing manipulated or known first by the experimenter. This variable X is represented at several arbitrary values in the experiment. Here the only interest is in the possibility and degree of linear prediction of Y from X.

The other application of the theory presented in this chapter occurs in **problems of correlation**. In a correlation problem, there is no clear-cut distinction between two kinds of scores as to which is the independent or predictor variable. Both variables are left completely free to take on any value for any observed individual. A sample of N individuals is drawn and each individual observed represents the occurrence of a joint (x,y) event. The basic question asked concerns the relation between the two variables: can Y be predicted from X or X predicted from Y using a linear rule?

For example, suppose that 100 children are observed, and the height (X) and weight (Y) of each child noted. The pair of values for each child is thus a representative of some joint (x,y) event. Any possible height, any possible weight, or any possible combination may occur in the data. There is no implication that height is somehow "responsible" for weight, or weight for height. The experimenter wants only to see how well the relation between the two kinds of scores can be thought of as a linear relation. Can *either* variable be predicted from the other by a linear rule?

This distinction between the two kinds of problems is an important one, and will be emphasized by our discussing the two types separately. Although there are close theoretical connections between the methods appropriate to these two kinds of problems, in practice the two situations differ in the way in which samples are drawn, the status of one variable as *the* independent or predictor variable, the assumptions one makes in significance testing, and in some circumstances, in the kinds of conclusions drawn. Thus, regression and correlation problems will be distinguished from each other much as we did the different models for analysis of variance, where a roughly parallel situation exists.

In the next few sections, however, this distinction between regression and correlation problems is not important. We are going to lay the groundwork for the descriptive statistics of regression, for how one actually finds a linear rule to use for prediction from some complete set of data. Then we will turn to problems of inference in the regression and the correlation situations, where the regression-correlation distinction *is* important.

15.2   The descriptive statistics of regression and correlation

Just as we discussed how one could find and interpret the mean and variance as descriptions of particular aspects of a set of data, we will now turn to the problem of finding a linear rule that "fits" a given set of data as well as possible. For the moment, our interest is only in a specific set of data, the scores for some particular set of N observed individuals.

Imagine this kind of situation: a teacher of a large introductory college course is interested in the possible relationship between the high-school preparation in mathematics that a student has and his. success in the course. In a particular semester the teacher has a class of 91 students, and at the outset he asks each student to tell him the mathematics courses he has taken in high school (four years). The teacher weights these courses in a routine way and assigns scores running from 2 through 8 to the students. Let us call these "mathematics scores" X.

The teacher, however, files these reports away and does not look at them until after the final examination in the course has been given. The actual raw scores on this examination will be called the variable Y. After both scores for each student are known, the teacher asks this question: "To what extent is there a linear relation between the X and the Y scores?" In other words, how well does a simple linear rule allow one to predict the Y score of a student drawn at random from this group, given the information about the X score? The problem is to find the best possible linear rule for predicting from these data, and then to evaluate the goodness of such a rule.

Actually, the teacher is not especially interested in predicting the raw Y score of a student so much as he is in the relative performance of the student in terms of Y. That is, he would like to be able to predict the standard score z_Y, given by

$$z_Y = \frac{y - M_Y}{s_Y}.$$

This prediction is to be based on the standard score z_X, where

$$z_X = \frac{x - M_X}{s_X}.$$

Since a linear rule is to be used for this prediction, this means a rule of the form

$$z'_Y = bz_X + a \qquad [15.2.1*]$$

where b and a are constants. The predicted score is labeled as z'_Y to indicate that it *need not* be the same as z_Y, the true standard score for any given individual.

Several individuals may have the same z_X standard score, but quite different z_Y scores; by use of the rule, however, *only one z'_Y or predicted standard score will be given for each z_X value.*

The problem is shown graphically in Figure 15.2.1. Here, the horizontal axis represents possible values for z_X, the vertical axis the possible values for z_Y, and any point within the plane defined by these two axes represents a pair of z scores, (z_X, z_Y), that might be associated with any individual observation. Points in the functional relation $z'_Y = bz_X + a$ lie along the straight line in the figure. For the particular value of z_X represented in the figure, the linear rule affords a *predicted* value z'_Y; this *need not* correspond to the actual value z_Y corresponding to any individual showing the particular value of z_X shown in the figure. The extent of "miss" or error between the predicted value z'_Y and z_Y for an individual is represented by the vertical distance between the two points

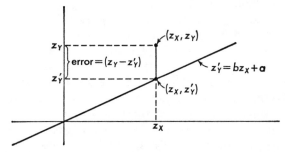

FIG. 15.2.1. Plot of a linear regression equation for the prediction of the standard score on Y from the standard score on X

(z_X, z_Y) and (z_X, z'_Y). We would like our prediction rule to be such that, across all individuals, the fit between predicted and actual standard scores on Y is as good as possible.

To recapitulate, the problem is to predict the z_Y from z_X *as though* these two numerical values were functionally related. The function rule to be used is a linear one, with constants chosen to give predictions that are, in some sense, "best." The user is perfectly well aware that the relation between the two kinds of scores is *not actually* a function (that is, any given value of z_X can occur in pairs with any number of values of z_Y), but he does want to see how well the use of such a linear function rule allows him to guess the standard score on variable Y given the standard score on X.

The reader may be puzzled by the use of the term "prediction" in this context; the teacher actually *has* the two scores for each of the 91 students. Why doesn't he merely look at the standardized Y score for any student? Actually the methods to be developed here will apply to situations where the user wants to go beyond the immediate data, and to forecast the Y or z_Y score for an individual for which this information is not already available. However, the basis for these methods is best seen if one deals only with one intact group of N cases, each having two scores, X and Y. For the moment, "prediction" consists of drawing one case at random from this particular group, noting the z_X, and then finding a predicted value of z'_Y by use of the linear rule.

The first problem is to find constants a and b that will make the linear rule give the "best possible" predictions. We will find these constants by the method of "least squares" in the next section.

15.3   THE REGRESSION EQUATION FOR PREDICTING z_Y FROM z_X

Given that we want to predict the relative status of an individual on variable Y, or z_Y, knowing only z_X, and that we must use the linear rule given by **15.2.1,**

$$z'_Y = bz_X + a,$$

how do we go about finding b and a?

Obviously, we want our rule to be *good*, in the sense that it gives the best or closest predictions possible. There are a number of criteria that might be chosen to define what makes a "good" prediction: hitting the actual score on the nose the largest proportion of times, least absolute error on the average, and so on, are some of the possibilities for defining good prediction. However, the theory of correlation and regression is based on the **least squares** criterion. For our purposes, this means that **we want to choose constants for our linear prediction equation in such a way that the average squared error in prediction will be as small as possible.** Thus, for any individual case i we will make some prediction, z'_Y; this need *not* be the same as the true value z_{Yi} for that individual, and so some error will exist,

$$\text{error} = (z'_{Yi} - z_{Yi}).$$

The least-squares criterion requires that we choose a and b in such a way that the average squared error over individual predictions be as small as possible. Thus, given N individuals i, we want to choose a and b so as to make

$$\frac{\sum_i (z'_{Yi} - z_{Yi})^2}{N} \qquad [15.3.1*]$$

have its minimum possible value.

Now, first of all we will show that by the least-squares criterion, *if we are predicting standard scores, the value of the constant a must be zero, so that the best linear rule is actually*

$$z'_Y = bz_X. \qquad [15.3.2*]$$

This can be shown as follows: Substituting **15.2.1** into **15.3.1** and rearranging terms we have

$$\frac{\sum_i (z'_{Yi} - z_{Yi})^2}{N} = \frac{\sum_i [(bz_{Xi} - z_{Yi}) + a]^2}{N}. \qquad [15.3.3]$$

On carrying out the square for each i and summing, we have

$$\frac{\sum_i [(bz_{Xi} - z_{Yi}) + a]^2}{N} = \frac{\sum_i (bz_{Xi} - z_{Yi})^2}{N} + 2a \frac{\sum_i (bz_X - z_Y)}{N} + \frac{\sum_i a^2}{N}$$

$$= \frac{\sum_i (bz_{Xi} - z_{Yi})^2}{N} + a^2 \qquad [15.3.4]$$

since a and b are constants, and the mean of each set of z scores must be zero.

Now, assuming b fixed, for what value of a can the expression on the right in **15.3.4** be at its smallest? The first term is a mean of squared numbers and hence must be positive, and a^2 must be positive as well; it follows that this expression can be at its smallest value *only* when a is zero. Thus, the value of a dictated by the least-squares criterion is zero.

Next, we will show that by the least-squares criterion the value of b for predicting z scores must be

$$b = \frac{\sum_i z_{Xi} z_{Yi}}{N}, \qquad [15.3.5*]$$

that is, the constant b must equal the average of the *products* of the standard scores across the N individuals. This value of b is actually the **correlation coefficient** (or Pearson product-moment correlation coefficient) r_{XY}, about which we will have much to say. By definition,

$$r_{XY} = \frac{\sum_i z_{Xi} z_{Yi}}{N}. \qquad [15.3.6*]$$

Thus, by the criterion of least squares, our prediction rule must be

$$z'_Y = r_{XY} z_X. \qquad [15.3.7*]$$

We can show as follows how the least-squares criterion dictates that $b = r_{XY}$ for the prediction of standard scores. Since we know that a must be zero, this time we start by substituting expression **15.3.2** into **15.3.1**:

$$\frac{\sum_i (z'_{Yi} - z_{Yi})^2}{N} = \frac{\sum_i (bz_{Xi} - z_{Yi})^2}{N}. \qquad [15.3.8]$$

Expanding the square on the right hand of **15.3.8** and summing, we have

$$\frac{\sum_i (z'_{Yi} - z_{Yi})^2}{N} = \frac{b^2 \sum_i z^2_{Xi}}{N} - \frac{2b \sum_i z_{Xi} z_{Yi}}{N} + \frac{\sum_i z^2_{Yi}}{N}$$

$$= b^2 - 2br + 1 \qquad [15.3.9]$$

since the variance of standard scores is always 1. Now suppose that b differed from r by some amount c, either a positive or negative number:

$$b = r + c.$$

Substituting $r + c$ for b in **15.3.9** above, we find

$$\frac{\sum_i (z'_{Yi} - z_{Yi})^2}{N} = (r + c)^2 - 2(r + c)r + 1$$

$$= r^2 + 2rc + c^2 - 2r^2 - 2rc + 1$$

$$= (1 - r^2) + c^2. \qquad\qquad [15.3.10]$$

When $b = r$, so that c is zero, the mean squared error must be at its smallest,

$$\frac{\sum_i (z'_{Yi} - z_{Yi})^2}{N} = (1 - r^2), \qquad\qquad [15.3.11*]$$

and for any $c \neq 0$, the value must be $(1 - r^2)$ *plus a positive number, c^2. Hence taking $b = r$ gives the least squared error in linear prediction, on the average, for standard scores.*

15.4   THE STANDARD ERROR OF ESTIMATE FOR STANDARD SCORES

In the preceding section, we used the idea of the mean squared error made in prediction of z_Y from z_X, and we found that this mean squared error is at its minimum when the regression equation is given by **15.3.7,** or

$$z'_Y = r_{XY}z_X.$$

This notion of the mean squared error is an important one in its own right, and we will give it a special symbol and name. Let

$$S^2_{z_Y \cdot z_X} = \frac{\sum_i (z'_{Yi} - z_{Yi})^2}{N}$$

$$= 1 - r^2_{XY} \qquad\qquad [15.4.1*]$$

be called the **sample variance of estimate for standard scores.** This variance of estimate reflects the *poorness* of the linear rule for prediction of standard scores, the extent to which squared error is, on the average, large.

Most often, however, this index is discussed in terms of its positive square root

$$S_{z_Y \cdot z_X} = \sqrt{1 - r^2_{XY}}, \qquad\qquad [15.4.2]$$

which is called the **sample standard error of estimate for predicting standard scores.**

Obviously, there is a close connection between the size of the standard error of estimate in a sample and the value of r in the regression equation: **the larger the absolute value of r_{XY}, the smaller is the standard error of estimate.** Now we turn to some interpretations of the index r_{XY}.

15.5 INTERPRETATIONS OF THE CORRELATION COEFFICIENT IN A SAMPLE

The connection between the variance of estimate and the correlation coefficient shows us at once that r_{XY} can take on values *only* between -1 and 1. Notice that the variance of estimate, being a weighted sum of squares, can be only a positive number or zero. If r_{XY} were less than -1, or greater than $+1$, then the variance of estimate could not be positive. Hence, $-1 \leq r_{XY} \leq 1$.

What does it mean when r_{XY} is exactly zero? When this is true, one predicts $z'_Y = 0$, corresponding to the mean Y, *regardless* of the value of X; for any X, the mean of Y is the best linear prediction when the correlation is zero. Furthermore,

$$S^2_{z_Y \cdot z_X} = 1$$

for $r_{XY} = 0$. This means that when the correlation coefficient is zero the variance of estimate for standard scores is exactly the same as the variance of the standard scores z_Y with X *unspecified*. Thus, when r_{XY} is zero, predicting by the linear rule *does not* reduce the variability of z_Y below the variability present when z_X is unknown. In short, the fact that r_{XY} is zero means that if a predictive statistical relation exists for the set of data, it is not linear, and the linear rule gives no predictive power.

On the other hand, when r_{XY} is either $+1$ or -1, the variance of error in prediction is zero, so that each prediction is exactly right. These maximum values for r_{XY} can occur only when X and Y *are* functionally related, *and* follow a linear rule.

All intermediate values of r_{XY} indicate that some prediction is possible using the linear rule, but that this prediction is not perfect, and some error in prediction exists. Any value of r_{XY} between 0 and 1 in absolute magnitude indicates either that the relationship is not functional or that if it is a function, the rule is not exactly linear, although a linear rule does afford some predictability.

In the regression equation for standard scores the correlation coefficient plays the role of converting a standard score in X into a predicted standard score in Y. Rather loosely, the correlation coefficient can be said to be "the rate of exchange," the value of a "standard deviation's worth" of X in terms of *predicted* standard deviation units of Y. This becomes clearer if we look on a plot of the function given by the regression equation, as shown in Figure 15.5.1. This plot shows the pairing of z_X and *predicted* z_Y, or z'_Y values (do not fall into the mistake of thinking that such a plot shows *actual* z_X and z_Y pairings). The regression function is shown by the straight line plot generated by the regression equation. Regardless of the relationship between X and Y, z_X and z'_Y will plot as a

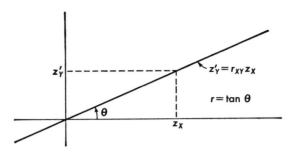

FIG. 15.5.1. The correlation coefficient as the slope of the regression line for predicting the value of Z_Y from Z_X

straight line. Given a particular z_X, as shown in the figure, then the ratio of the predicted value to the known standard score value for X,

$$\frac{z_Y'}{z_X} = \tan \theta, \qquad [15.5.1]$$

the trigonometric tangent of the angle θ. However, this ratio is also r_{XY}, so that

$$r_{XY} = \tan \theta, \qquad \left\{ \begin{array}{l} 0° \angle \theta \angle 45° \\ 135° \angle \theta \angle 180°. \end{array} \right. \qquad [15.5.2]$$

The tangent of the angle θ gives the *slope* of the regression line for predicting standard scores, and the value of r_{XY} corresponds to the slope of this line. When r_{XY} is between 0 and 1, the angle θ in such a plot must be between 0 and 45 degrees. For an r_{XY} between 0 and -1, the angle θ must be between 135 and 180 degrees, so that in this instance a high standing on X yields a prediction of low standing on Y, and vice versa.

15.6 The idea of regression toward the mean

The term *regression* has come to be applied to the general problem of prediction by use of a wide variety of rules, although the original application of this term had a very specific meaning, as we shall show. The term "regression" is a shortened form of **regression toward the mean in prediction.** The general idea is that **given any standard score** z_X, **the best linear prediction of the standard score** z_Y **is one relatively nearer the mean of zero than is** z_X.

This can be illustrated quite simply from our regression equation 15.3.7. Suppose that an individual has a standard score z_X of 2. Also suppose that the regression equation we have found for the group to which he belongs is

$$z_Y' = .5z_X.$$

Then we *predict* this individual to have a z_Y score of 1, since

$$z_Y' = .5(2) = 1.$$

Notice that we predict him to fall relatively *nearer* the mean on Y than he fell on X. That is, we predict in accordance with *regression toward the mean.* For another set of data, the regression equation might be

$$z'_Y = -.75z_X.$$

Now in this instance, suppose that the z_X for some randomly selected individual were 1.5. Then

$$z'_Y = (-.75)(1.5) = -1.125.$$

Since the correlation coefficient is negative, the prediction is that this individual falls *below* the mean of Y, given that he falls *above* the mean on X. However, in absolute terms, we again predict that he falls relatively *closer* to the mean on Y than he falls on X.

This principle of predicting relatively closer to the mean, or regression toward the mean, is a feature of any linear prediction rule that is best in the "least-squares" sense of Section 15.3. The idea is that if we are going to use such a linear rule for prediction, then it is always a good bet that an individual will fall *relatively closer to the group mean on the thing predicted than he does on the thing actually known*. This does *not* imply that an individual *must* fall relatively closer to the mean on Y than he does on X, however, but only that our best *bet* about him is that he will do so. Regression toward the mean is not some immutable law of nature, but rather a statistical consequence of our choosing to predict in this linear way.

The idea of regression, and, indeed, most of the foundations for the theory of correlation and regression equations, came from the work of Sir Francis Galton in the nineteenth century. In his studies of hereditary traits, Galton pointed out the apparent regression toward the mean in the prediction of natural characteristics. In some biological traits, regression toward the mean does seem to occur just as the linear theory implies. However, the modern point of view in statistics is that regression toward the mean is built into the statistical assumptions and methods we use for prediction, and is not necessarily a feature of the natural world. Given the value of z_X, the *true* value of z_Y for an individual observation can be anything, and regression toward the mean simply describes our best guess about the value of z_Y, when "best" is defined in terms of minimal squared error.

15.7 THE PROPORTION OF VARIANCE ACCOUNTED FOR BY LINEAR REGRESSION

In our discussion of the population indices of statistical relationship, ω^2 and ρ_I, we found it convenient to define the over-all strength of relationship in terms of a proportional reduction in variance. Exactly the same idea may be applied in judging just how good a linear rule is for predicting Y from X. In other words, we want an index to specify the strength of *linear* relationship in a set of data, showing how well a linear rule approximates the true form of the relation.

Because we are dealing with standard scores, the total variability of the z_Y scores for N individuals is 1.00. Furthermore, we have seen that the variance of estimate, $S^2_{z_Y \cdot z_X}$, is the average squared *error* in prediction using the linear rule. This can be thought of as representing the "left-over" variability, given the rule and the value of z_X. Thus, it makes sense to think of the proportional *reduction* in variance as

$$\frac{S^2_{z_Y} - S^2_{z_Y \cdot z_X}}{S^2_{z_Y}}$$

which is simply

$$\frac{1 - (1 - r^2)}{1} = r^2. \qquad [15.7.1*]$$

The proportional reduction in variance of Y given the linear rule and X is r^2_{XY}, sometimes called the coefficient of determination. You can always think of r^2_{XY} as representing the strength of *linear* relationship in a given set of data.

Thus, if the value of a correlation coefficient is .50 (positive or negative in sign) then some .25 of the variability in Y is accounted for by specifying the linear rule and X. If the correlation is .80, then 64 percent of the variance in Y is accounted for in this way. A correlation of positive or negative 1.00 means that 100 percent of variability in Y can be accounted for by the linear rule and X, but if $r_{XY} = 0$, none of the variability is thereby accounted for. All in all, not the correlation coefficient per se but the *square* of the correlation coefficient informs us of the "goodness" of the linear rule for prediction.

15.8   THE REGRESSION OF z_X ON z_Y

There is really nothing in our discussion so far that has made it necessary to think of X as the independent variable, or the value somehow known first or predicted from. It is entirely possible to consider a situation where one might want to predict z_X from knowing z_Y. What does this do to the linear prediction rule, the correlation coefficient, and so on?

In the first place, the same argument used in Section 15.3 shows that for predicting z_X from z_Y,

$$z'_X = r_{XY} z_Y, \qquad [15.8.1*]$$

where the correlation coefficient s just as before,

$$r_{XY} = \frac{\sum\limits_{i} z_{Xi} z_{Yi}}{N}.$$

This equation **15.8.1** is for regression of z_X on z_Y, or z_X predicted from z_Y.

You may be wondering why the regression equation for predicting z_X from z_Y is not found by solving equation **15.3.7** for z_X, to give

$$z_X = \frac{1}{r_{XY}} (z'_Y). \qquad [15.8.2]$$

However, recall what the symbols z'_Y and z'_X actually represent. These are *predicted* values and do not necessarily symbolize the actual values of z_Y and z_X at all. Solving the expression **15.8.2** for z_X *might* be useful if one wanted to know the value of z_X *known*, given that z'_Y were the predicted value, although it is hard to see why anyone would ordinarily want this information. The form of the regression equation used (**15.3.7** or **15.8.1**) depends strictly on which variable, X or Y, is designated as the independent variable, or the thing known first in a prediction situation.

For prediction of X from Y, the sample variance of estimate for standard scores is

$$S^2_{z_X \cdot z_Y} = 1 - r^2_{XY} \qquad [15.8.3*]$$

(notice the reversal in subscripts when z_X is predicted from z_Y). The proportional variance in X accounted for by Y is, once again, r^2_{XY}.

This brings up the point that the correlation coefficient is a *symmetric* measure of linear relationship. *So long as we are talking about the correlation coefficient alone, it is immaterial which we designate as the independent and which the dependent variable; the measure of possible linear prediction is the same.* However, when we deal with the actual regression equations themselves, this symmetry is not usually present. As we shall see in the next section, it does make a difference whether you are predicting Y from X or X from Y when it comes to finding the regression equations and errors of estimate for *raw* scores.

15.9 THE REGRESSION EQUATIONS FOR RAW SCORES

Up to this point, we have considered only the problem of predicting standard scores from standard scores. Introducing regression and correlation in terms of standard scores makes the algebra somewhat easier, and the essential ideas somewhat simpler. Nevertheless, each feature of correlation and regression shown for standard score prediction is also valid for the prediction of raw scores. For any given set of data, each standard score corresponds uniquely to some raw score, and vice versa, so that linear prediction which is optimal in standard score terms is optimal in raw score terms as well.

It is quite simple to put the regression equation for prediction z_Y from z_X into raw score form. We start with

$$z'_Y = r_{XY} z_X,$$

which is exactly the same as

$$\frac{y' - M_Y}{S_Y} = r_{XY} \frac{(x - M_X)}{S_X}$$

where Y' is the *predicted raw score* for the individual, and X is his *known* raw score on the independent variable. A little algebraic manipulation gives

$$y' = \frac{r_{XY} S_Y}{S_X} (x - M_X) + M_Y. \qquad [15.9.1\dagger]$$

This is the raw score form of the regression equation for prediction of Y from X.

It will be convenient to write this regression equation as

$$y' = b_{Y \cdot X}(x - M_X) + M_Y, \qquad [15.9.2\dagger]$$

where

$$b_{Y \cdot X} = \frac{r_{XY}S_Y}{S_X}. \qquad [15.9.3*]$$

The value $b_{Y \cdot X}$ is called the **sample regression coefficient** of Y on X.

In an identical way, we can turn the regression equation for z_X predicted from z_Y into

$$x' = b_{X \cdot Y}(y - M_Y) + M_X. \qquad [15.9.4\dagger]$$

This is the raw score form of the regression equation for predicting X from Y. Here

$$b_{X \cdot Y} = \frac{r_{XY}S_X}{S_Y} \qquad [15.9.5*]$$

is the sample regression coefficient for predicting X from Y.

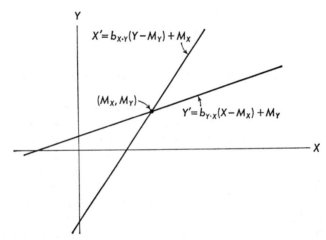

FIG. 15.9.1. Plot of the two regression lines for predicting Y from X and X from Y

Notice that when no specification is put on which is to be regarded as the independent or predictor variable, there are two possible sample regression coefficients, $b_{Y \cdot X}$ and $b_{X \cdot Y}$, and that

$$\sqrt{b_{Y \cdot X}b_{X \cdot Y}} = r_{XY}, \qquad [15.9.6]$$

the square root of the product of the two regression coefficients is the correlation coefficient.

Figure 15.9.1 shows the two raw score regression lines that might apply to a given set of data.

15.10 THE ERRORS OF ESTIMATE FOR RAW SCORES

When prediction of raw scores is to be carried out, the **sample variance of estimate** for predicting Y from X is

$$S_{Y \cdot X}^2 = S_Y^2 (1 - r_{XY}^2) \qquad \text{[15.10.1*]}$$

and the **sample standard error of estimate** is

$$S_{Y \cdot X} = S_Y \sqrt{1 - r_{XY}^2}. \qquad \text{[15.10.2*]}$$

Similarly, the sample variance of estimate for predicting X from Y is

$$S_{X \cdot Y}^2 = S_X^2 (1 - r_{XY}^2) \qquad \text{[15.10.3*]}$$

and the sample standard error of estimate is

$$S_{X \cdot Y} = S_X \sqrt{1 - r_{XY}^2}. \qquad \text{[15.10.4*]}$$

Finally, the proportion of variance accounted for by linear relationship (either in predicting Y from X or X from Y) is r_{XY}^2, just as for standardized scores.

15.11 COMPUTATIONAL FORMS FOR r_{XY} AND $b_{Y \cdot X}$

Although the **sample** correlation coefficient was actually defined in Section 15.3 as a summed product of standard scores (**15.3.6**), in practice the index is seldom computed in this way. An equivalent raw score computational form will now be found. We will show that the raw score form of the correlation coefficient is

$$r_{XY} = \frac{\left(\sum_i x_i y_i / N \right) - M_X M_Y}{S_X S_Y}. \qquad \text{[15.11.1†]}$$

Starting with the definition **15.3.6**, and substituting the raw score equivalents of z_X and z_Y, we have

$$r_{XY} = \frac{\sum_i (x_i - M_X)(y_i - M_Y)}{N S_X S_Y}$$

$$= \frac{1}{N S_X S_Y} \left[\sum_i x_i y_i - \sum_i x_i M_Y - \sum_i y_i M_X + N M_X M_Y \right]$$

$$= \frac{1}{N S_X S_Y} \left[\sum_i x_i y_i - 2N M_X M_Y + N M_X M_Y \right]$$

$$= \frac{\left(\sum_i x_i y_i / N \right) - M_X M_Y}{S_X S_Y}.$$

Still another computing form for r_{XY} that is useful when work is being done on a desk calculator is

$$r_{XY} = \frac{N \sum_i x_i y_i - \left(\sum_i x_i\right)\left(\sum_i y_i\right)}{\sqrt{\left[N \sum_i x_i^2 - \left(\sum_i x_i\right)^2\right]\left[N \sum_i y_i^2 - \left(\sum_i y_i\right)^2\right]}}.$$ [15.11.2†]

Various other methods for computing r_{XY} are available. One method especially popular in past years is designed for data grouped into a joint frequency distribution or "scatter plot," and this version is to be found in many elementary statistics texts. However, when desk calculators are available this method is not usually so efficient as a method dealing directly with the raw scores, and we will not consider it here.

Given the correlation coefficient, it is possible to find the sample regression coefficient $b_{Y \cdot X}$ directly from

$$b_{Y \cdot X} = r_{XY}\left(\frac{S_Y}{S_X}\right).$$

However, for many problems in regression it is desirable to calculate $b_{Y \cdot X}$ without first finding r_{XY}. This may be done most simply by taking

$$b_{Y \cdot X} = \frac{\sum_i x_i y_i - N M_X M_Y}{\sum_i x_i^2 - N M_X^2}$$ [15.11.3†]

or

$$b_{Y \cdot X} = \frac{N \sum_i x_i y_i - \left(\sum_i x_i\right)\left(\sum_i y_i\right)}{N \sum_i x_i^2 - \left(\sum_i x_i\right)^2}.$$ [15.11.4†]

Incidentally, it should be noted that the sample correlation coefficient is a *dimensionless* quantity, since it is not expressed directly either in units of X or in units of Y. As we saw, the basic definition of the correlation coefficient (15.3.6) depends upon the individual X and Y *standard* scores, which are themselves in standard units. This fact makes it possible to simplify the computation of the correlation coefficient by transforming *either X or Y or both* into new variables u and v by the linear rules

$$u_i = cX_i + g$$
$$v_i = dY_i + h$$

where c and d are any positive constants, and g and h are any constant numbers at all. In other words, instead of carrying out the computations for r_{XY} using X and Y values, we can carry these computations out using the transformed values u and v. If u and v are both such positive linear functions of X and Y, the standard scores for each individual are unchanged by this conversion to new scores, and thus the value of r_{XY} is the same as the correlation value between u and v. This principle is often used to cut down the labor in calculation for correlation coefficients, and will be illustrated in the problem of Section 15.28. Unlike r_{XY}, however, the value of $b_{Y \cdot X}$ is not invariant over such transformations of the scores.

15.12 AN EXAMPLE OF REGRESSION AND CORRELATION COMPUTATIONS FOR A SAMPLE

Consider once again the example of Section 15.3. The teacher collected data for a class of 91 students, obtaining for each a score X, based on the number of courses taken n high-school mathematics, and a score Y, the actual score on the final examination for the course. These data are shown in Table 15.12.1. The arrangement of data is similar to that used in a problem in regression: the columns of the table represent the various possible values of X, and within each column the different Y values are shown. We will use y_{ij} to stand for a Y value for individual i in grouping j, and x_j for the X value common to all individuals in grouping j.

Table 15.12.1
FINAL EXAM SCORES Y ARRANGED ACCORDING TO X SCORE OBTAINED

				36								
			34	35								
			26	30								
			25	27								
	26		23	25								
22	23	29	23	25								
17	18	26	22	24								
16	18	26	22	21								
14	16	24	19	20								
10	13	24	18	19				46				
9	12	23	17	19				38				
7	10	22	17	18		41		34	44		42	53
5	10	16	17	18	28	32	32	33	37	52	41	48
3	7	9	12	12	27	27	25	27	32	46	38	40
2	6	8	8	3	16	19	25	20	28	37	35	40

X

2	2.5	3	3.5	4	4.5	5	5.5	6	6.5	7	7.5	8

Freq.

10	11	10	14	15	3	4	3	6	4	3	4	4

The mean of the Y values is

$$\frac{\sum_j \sum_i y_{ij}}{N} = \frac{2169}{91} = 23.84$$

and that of the X values is

$$\frac{\sum_j \sum_i x_j}{N} = \frac{\sum_j x_j (\text{freq.})}{N} = \frac{381.5}{91} = 4.19.$$

The two standard deviations are

$$S_X = \sqrt{\frac{\sum_j x_j^2 (\text{freq.})}{N}} - M_X^2 = \sqrt{\frac{1874.75}{91}} - 17.56 = 1.74$$

$$S_Y = \sqrt{\frac{\sum_j \sum_i y_{ij}^2}{N}} - M_Y^2 = \sqrt{\frac{64043}{91}} - 568.34 = 11.64.$$

The correlation coefficient will be computed next:

$$\frac{\sum_j \sum_i x_j y_{ij}}{N} = \frac{(2)(22) + (2)(17) + \cdots + (8)(40) + (8)(40)}{91} = \frac{10580}{91} = 116.26$$

so that

$$r_{XY} = \frac{\sum_j \sum_i x_j y_{ij} / N - M_X M_Y}{S_X S_Y}$$

$$= \frac{116.26 - (4.19)(23.84)}{(1.74)(11.64)}$$

$$= .81.$$

Thus, the regression equation for predicting z_Y from z_X for these data is

$$z_Y' = .81 z_X.$$

The raw score regression coefficient $b_{Y \cdot X}$ is

$$b_{Y \cdot X} = r_{XY} \frac{S_Y}{S_X}$$

$$= (.81) \frac{11.64}{1.74}$$

$$= 5.42.$$

Using this regression coefficient we find that the raw score regression equation for predicting Y from X is

$$y_j' = (5.42)(x_j - 4.19) + 23.84.$$

For instance, given that an individual has a high-school mathematics score of 5, the teacher can predict that his score on the course examination is

$$y' = (5.42)(5 - 4.19) + 23.84$$
$$= 28.23.$$

Figure 15.12.1 shows a plot of the regression equation, together with the actual (x, y) pairs for these data. Notice that although the actual pairs of scores do tend to cluster about the predicted (x, y') pairs, there is nevertheless

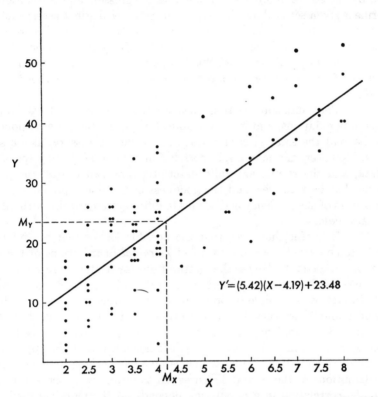

$$Y' = (5.42)(X - 4.19) + 23.48$$

FIG. 15.12.1. Plot of the data in Section 15.12, showing the regression line for predicting Y from X

some "scatter" of the actual Y scores about the predicted value for each X. This, of course, is reflected in the fact that the obtained r_{XY} is not 1.00.

For these data, the sample standard error of estimate is

$$S_{Y \cdot X} = S_y \sqrt{1 - r_{XY}^2}$$
$$= (11.64)(.586)$$
$$= 6.82.$$

15.13   ASSUMPTIONS MADE IN COMPUTING CORRELATION AND REGRESSION COEFFICIENTS FOR SAMPLE DATA

A few comments are in order about the propriety of computing correlations and regression equations for sample data. In some of the older literature in psychological research several misleading ideas appear about when it is proper to compute these indices and equations as *descriptive* statistics. The modern researcher needs to be clear about these matters. It is *not* necessary to make any assumptions at all about the form of the distribution, the variability of Y scores within X columns or "arrays," or the true level of measurement represented by the scores in order to employ linear regression and correlation indices to describe a given set of data. So long as there are N distinct cases, each having two numerical scores, X and Y, then the *descriptive* statistics of correlation and regression may be used. In so doing, we describe the data *as though* a linear rule were to be used for prediction, and this is a perfectly adequate way to talk about the tendency for *these* numerical scores to associate or "go together" in a linear way *in these data*.

The confusion has arisen because in inference about true linear relationships in populations, and in some applications of regression equations to predictions beyond the sample, assumptions do become necessary, as we shall see presently. However, one may apply correlation techniques to any set of paired-score data, and the results are valid descriptions of two things: the particular linear rule that best applies, and the goodness of the linear prediction rule as a summarization of the tendency of Y scores to differ systematically with differences in X *in these data*.

It is true, however, that the possible values that r_{XY} may assume depend to some extent upon the forms of *marginal* distributions of both X and Y in the joint data-table. Unless the distributions for X and Y are similar in form, it is not necessarily true that the obtained value of r_{XY} can range between -1 and $+1$. In fact, it is possible to produce examples where the forms of the distributions of X and Y are very different, and the maximum possible absolute value of r_{XY} is only .3 or less. The fact that the value of the correlation coefficient can, in principle, range from -1.00 to $+1.00$ does not mean that the opportunity for a linear predictive relation to appear in a sample has nothing to do with the marginal distributions of the X and Y scores. In the same way, the actual possible range of the correlation in a population depends on the marginal distributions. This fact has very important implications for those who study the patterns of correlations in multivariate studies, and particularly for those who must employ some variant on, or approximation to, the correlation coefficient in such studies. An informative discussion of these issues is given in Carroll (1961).

A related problem with the value of r_{XY} in a sample has to do with selection of cases to appear in the sample, and in particular with systematic restrictions on the range of X (or of Y) values that appear. This introduces a bias into the value of r_{XY} as an estimate of the correlation had this selection not been exercised. In particular, the absolute value of the correlation coefficient tends to be lowered by the introduction of such systematic selection reducing the possible range of one or both variables. These matters are also discussed by Carroll.

15.14 POPULATION CORRELATION AND REGRESSION

Imagine a population where each distinct elementary event qualifies for one and only one joint (x, y) event, where both X and Y are random variables.

Then indices of population correlation and regression can be defined in ways completely analogous to the sample indices.

In the first place, consider the expectation of the product of deviations from expectation:

$$E[(X - \mu_X)(Y - \mu_Y)] = \text{cov.}(X,Y). \qquad [15.14.1^*]$$

The expectation is taken over all individuals in the population, where

$$\mu_X = E(X)$$

and

$$\mu_Y = E(Y).$$

The expectation of this product, or cov.(X,Y), is called the **population covariance** of X and Y.

We have already used this idea of the population covariance several times. In rule 6 of Appendix B, it is stated that if two random variables are independent, then

$$E[(X - \mu_X)(Y - \mu_Y)] = 0,$$

so that cov.$(X,Y) = 0$. This rule has been applied a number of times, particularly in our discussion of the expectations for analysis of variance sums of squares. A corollary to this rule is that if

$$\text{cov.}(X,Y) \neq 0,$$

then X and Y are *not* statistically independent. It is obvious that the value of the covariance for two random variables must have something to do with possible association, or statistical relationship, between X and Y. The absence of a statistical relation between X and Y implies that the covariance must be zero; however, the reverse implication does not necessarily hold, and **the fact that cov.$(X,Y) = 0$ does not necessarily mean that X and Y are independent.**

Notice that the **sample** parallel to the population covariance is

$$\frac{\sum_i (x_i - M_X)(y_i - M_Y)}{N} = \frac{\sum_i x_i y_i}{N} - M_X M_Y = \text{sample covariance.} \quad [15.14.2^*]$$

This is the term we calculate for the numerator in finding the *sample* correlation coefficient, so that the correlation coefficient for a sample may also be defined as

$$r_{XY} = \frac{\text{sample covariance}}{S_X S_Y}. \qquad [15.14.3^*]$$

In precisely this way we define the **population** correlation coefficient,

$$\rho_{XY} = \frac{\text{cov.}(X,Y)}{\sigma_X \sigma_Y}. \qquad [15.14.4^*]$$

The population correlation coefficient is the covariance of X and Y, divided by the product of the population standard deviations for X and Y.

Just as r_{XY} must lie between -1 and 1 inclusive, so must it be true that

$$-1 \leq \rho_{XY} \leq 1.$$

The population regression equations for predicting Y from X and X from Y are also parallels to the sample forms:

$$Y' = \rho_{XY} \frac{\sigma_Y}{\sigma_X} (X - \mu_X) + \mu_Y \qquad [15.14.5*]$$

and

$$X' = \rho_{XY} \frac{\sigma_X}{\sigma_Y} (Y - \mu_Y) + \mu_X. \qquad [15.14.6*]$$

The **population regression coefficients** are

$$\beta_{Y \cdot X} = \rho_{XY} \frac{\sigma_Y}{\sigma_X} \qquad [15.14.7*]$$

and

$$\beta_{X \cdot Y} = \rho_{XY} \frac{\sigma_X}{\sigma_Y}. \qquad [15.14.8*]$$

Finally, for any such population, we can discuss the *true* variance of estimate for Y predicted from X:

$$\sigma_{Y \cdot X}^2 = \sigma_Y^2 (1 - \rho_{XY}^2) \qquad [15.14.9*]$$

as well as the true variance of estimate for X predicted from Y

$$\sigma_{X \cdot Y}^2 = \sigma_X^2 (1 - \rho_{XY}^2). \qquad [15.14.10*]$$

The two **true standard errors of estimate** are thus

$$\sigma_{Y \cdot X} = \sigma_Y \sqrt{1 - \rho_{XY}^2} \qquad [15.14.11*]$$

and

$$\sigma_{X \cdot Y} = \sigma_X \sqrt{1 - \rho_{XY}^2}. \qquad [15.14.12*]$$

Notice that just as for a sample, one can interpret the square of the correlation coefficient as **the proportion of variance accounted for by linear regression.** That is, $\sigma_{Y \cdot X}^2$ is "error" variance in the use of the linear rule, the variability *not* accounted for by linear regression. Thus,

$$\sigma_Y^2 - \sigma_{Y \cdot X}^2$$

is the variance *accounted* for by linear regression, or the reduction in variance accomplished by using a linear prediction rule. It follows that

$$\rho_{XY}^2 = \frac{\sigma_Y^2 - \sigma_{Y \cdot X}^2}{\sigma_Y^2}, \qquad [15.14.13*]$$

the square of the correlation coefficient is the relative reduction in variance accomplished by the use of a linear prediction rule. (Notice the very close similarity in meaning among ω^2, ρ_I, and ρ_{XY}^2; each is a relative reduction in population variance. However, ω^2 and ρ_I each stand for the reduction in variance of Y obtained by specifying the X **qualitative classification**, whereas ρ_{XY}^2 is the reduction in variance obtained by specifying the **numerical value** of X and a **linear** least-squares prediction rule.)

All in all, the features describing correlation and the ability to predict Y from X using a linear rule can also be defined for populations of potential observations, in which each element represents some joint (x, y) event. As usual, we want to make inferences about these population characteristics from a sample. Consider our college teacher, once again. The real motive for his study of the class of students was to find some way of forecasting a student's degree of success in his course ahead of time. A student wants admission to the course. He is found to have a score of 4 in terms of high-school credits in mathematics. How good a bet is this student to do well in the course? Unfortunately, the only evidence the teacher has comes from a *sample* of students. The new student is a sample of 1 observation presumably from the same population as the class of 91 students already observed. How can the teacher actually use the information he has about the sample to make inferences about population and correlation, and then use the sample information to predict about new students?

In order to carry out this form of prediction, the teacher needs to estimate the true, or population, regression equation. What information about the regression equation appropriate to the population is contained in the regression equation found for a sample? Next we will consider ways to estimate the population regression equation given only the sample regression equation.

15.15   ESTIMATES OF THE PARAMETERS OF THE POPULATION REGRESSION EQUATION, AND STANDARD ERRORS

Given a random sample, the value of the sample regression coefficient $b_{Y \cdot X}$ is our best available estimate of $\beta_{Y \cdot X}$, the population regression coefficient. Moreover, the best estimate of $\beta_{Y \cdot X}(\mu_X)$ is given by $b_{Y \cdot X}(M_X)$. As usual, our best estimate of μ_Y is simply M_Y.

Since each of these estimates corresponds to a term in the sample regression equation, we can use the sample equation itself as our best estimate of the population regression equation; that is, our estimate of the population regression equation is given by

$$Y' = b_{Y \cdot X}(X - M_X) + M_Y.$$

In the example, the teacher *is* justified in using his sample regression equation to predict for new students drawn from the same population as the original class, *provided* that the class used to find this regression equation is actually a random sample from a population of such students.

In addition, an unbiased estimate of the *true* variance of estimate for predicting Y from X is given by

$$\text{est. } \sigma^2_{Y \cdot X} = \frac{N}{N-2} s^2_{Y \cdot X}.$$

The reason for the appearance of $N-2$ in the denominator will be found in Section 15.19.

Before we consider the problem of confidence intervals for these estimates, and how a particular prediction might be made and itself qualified by a confidence interval, assumptions must be made about the nature of the population distribution of (x, y) pairs. This takes us directly into the study of problems in regression, which is our next topic. Here, for the first time, the essential differences between regression and correlation problems will be important in our discussion.

15.16 The Model for Simple Linear Regression

Imagine a study of this kind: in an experiment on transfer of training some N subjects were selected and assigned at random to J different groups. Each of the experimental groups was given a different number of trials at learning a particular task, say task A. Then, given that experience, each individual in each group was put to work learning another performance, task B. The number of trials required by a subject to reach a criterion level of performance on B was noted. Here, the independent experimental variable X is number of trials given on task A, and the dependent variable Y is number of trials required to learn task B.

This is clearly an example of a regression problem. The possible values of X to be represented in the experiment were chosen *in advance* by the experimenter, and each X value decided upon was made to appear some prearranged number of times in the data. On the other hand, within each value of X, the data show some obtained distribution of Y values. The experimenter's interest is primarily in evaluating the strength of linear relationship, if any, between X and Y, and in estimating the linear regression equation.

Just as various models can be adopted for various kinds of situations in the analysis of variance, so in regression analysis different models are possible, and the form of analysis and the assumptions made depend on which model is used. Now the simplest model for regression analysis will be stated; this is, however, not the only model possible for analyzing such an experiment. An alternative, and in many instances better, model will be given in the next chapter.

For the moment, we are going to assume that any raw score Y_{ij} has the following composition:

$$Y_{ij} = \mu_Y + \beta_{Y \cdot X}(X_j - \mu_X) + e_{ij} \qquad [15.16.1^*]$$

where Y_{ij} is the score of individual i in the jth treatment category, X_j is the quantitative X score associated with each individual in the jth category, and e_{ij} is the random error associated with individual i in category j.

Notice that within each category j, each and every individual exhibits the same value of X, or X_j. It will be assumed that the J treatment categories are fixed, just as in fixed effects analysis of variance. This means that *the same sample distribution of X_j values will appear in any exact repetition of the experiment for a different sample of N individuals, although the distribution of Y values within a category may differ from sample to sample.*

Notice also that this model is just another way of stating the linear regression equation for a population. Let

$$e_{ij} = Y_{ij} - Y'_j,$$

which is the miss or error in predicting Y_{ij} given the value of X_j and the *population* regression equation. Then

$$Y'_j = \mu_Y + \beta_{Y \cdot X}(X_j - \mu_X)$$

which is exactly the population regression equation, **15.14.5.**

The essential point, however, is that in adopting this model we are assuming that the linear regression equation is *true* as a description of the relation between X and Y, and that any apparent deviation from this function rule is random error, the influence of other and uncontrolled factors obscuring this true functional relation. This model asserts that, basically, Y *is* a linear function of X, could we but control the influence of all other factors. This need not really be true, of course, but in adopting this model we are assuming it true.

Given this linear model, and the fixed effects assumption for the categories j, then the true mean of X for each category j must be X_j, or

$$\mu_{X_j} = X_j.$$

This implies that the model may also be written

$$Y_{ij} = \mu_Y + \beta_{Y \cdot X}(\mu_{X_j} - \mu_X) + e_{ij}. \qquad [15.16.2^*]$$

Be sure to notice the close similarity to the fixed-effects analysis of variance model; if we identify

$$\beta_{Y \cdot X}(\mu_{X_j} - \mu_X) = \alpha_j$$

we see this to be only a special version of Model I for the one-way analysis of variance.

15.17 THE SUMS OF SQUARES FOR REGRESSION PROBLEMS

In any regression problem, the deviation of a score y_{ij} from the grand mean M_Y can be thought of as the sum of three parts:

$$(y_{ij} - M_Y) = (y_{ij} - M_{Y_j}) + (M_{Y_j} - y'_j) + (y'_j - M_Y). \qquad [15.17.1]$$

The first term $(y_{ij} - M_{Y_j})$ is simply the deviation of the particular score from the mean of its group, M_{Y_j}. The second, $(M_{Y_j} - y'_j)$, is the deviation of the mean of group j from the score *predicted* by the *sample* regression equation from the X_j for that group. The third part is the deviation of the predicted score itself from the grand mean.

By the now familiar argument used in Section 12.4 and elsewhere it can be shown that over all individuals in all groups, the total sum of squares can be partitioned into

$$\sum_j \sum_i (y_{ij} - M_Y)^2 = \sum_j \sum_i (y_{ij} - M_{Y_j})^2 + \sum_j n_j (M_{Y_j}^2 - y_j'^2)$$
$$+ \sum_j n_j (y_j' - M_Y)^2. \quad [15.17.2*]$$

The first of these parts is

$$\text{SS error} = \sum_j \sum_i (y_{ij} - M_{Y_j})^2, \quad\quad\quad [15.17.3]$$

which is just the ordinary *SS within* found as for a one-way analysis of variance.

The second sum of squares is

$$\text{SS deviations from linear regression} = \sum_j n_j (M_{Y_j}^2 - y_j'^2) \quad [15.17.4]$$

and the third part is

$$\text{SS linear regression} = \sum_j n_j (y_j' - M_Y)^2. \quad\quad [15.17.5]$$

In short, for any regression problem it is true that

SS total = (SS error) + (SS deviations from lin. reg.)
$$+ \text{ (SS lin. reg.).} \quad [15.17.6\dagger]$$

(Bear in mind that for any set of J distinct groups it is *also* true that

SS total = (SS error) + (SS between groups)

as shown in Section 12.4. This fact will be used later.)

Now let us examine the expectations of the various sums of squares, beginning with the sum of squares for linear regression.

15.18   THE EXPECTATION OF THE SS FOR LINEAR REGRESSION

The sum of squares for linear regression can be put into a slightly more recognizable form by substituting from **15.9.2** for y_j':

$$\sum_j n_j (y_j' - M_Y)^2 = \sum_j n_j [b_{Y \cdot X} (x_j - M_X) + M_Y - M_Y]^2$$
$$= b_{Y \cdot X}^2 \sum_j n_j (x_j - M_X)^2$$
$$= N b_{Y \cdot X}^2 s_X^2. \quad\quad [15.18.1*]$$

That is, for a given sample N, the sum of squares for linear regression actually depends only upon $b_{Y \cdot X}^2$, the *sample* regression coefficient squared, and s_X^2, the *sample* variance of X.

Furthermore, since

$$b_{Y \cdot X} = r_{XY} \frac{S_Y}{S_X}$$

then

$$\text{SS linear regression} = N r_{XY}^2 S_Y^2 \qquad [15.18.2\dagger]$$

and also

$$\text{SS linear regression} = \frac{N[\text{sample cov.}(X,Y)]^2}{S_X^2} \qquad [15.18.3\dagger]$$

by **15.14.3**.

Look for a moment at the sample covariance,

$$\text{sample cov.}(X,Y) = \frac{\sum_j \sum_i x_j y_{ij}}{N} - M_X M_Y. \qquad [15.18.4]$$

Since x_j is the same for each individual i in group j, the sample covariance in this instance is

$$\text{sample cov.}(X,Y) = \frac{\sum_j n_j (M_{Y_j})(x_j)}{N} - \frac{\sum_j n_j (M_{Y_j}) M_X}{N}$$

$$= \sum_j \frac{n_j (x_j - M_X)(M_{Y_j})}{N}. \qquad [15.18.5]$$

You are probably unaware of it, but we have just shown a most interesting fact: the sample covariance in a regression problem is a *linear combination* or comparison among the means M_{Y_j} for J samples, each mean being weighted by a value

$$c_j = \frac{n_j (x_j - M_X)}{N}. \qquad [15.18.6]$$

Furthermore,

$$\sum_j c_j = \sum_j \frac{n_j (x_j - M_X)}{N} = 0, \qquad [15.18.7]$$

since the sum of deviations of individual x_j values about the M_X must be zero for any sample. Thus, **formally, the sample covariance in a regression problem is simply a comparison among sample means.**

Now the information about comparisons in Section 14.11 can be used to find what the expectation of SS linear regression must be. For any sample comparison $\hat{\psi}$, we found that

$$\text{SS}(\hat{\psi}) = \frac{\hat{\psi}^2}{w}$$

is a sum of squares with 1 degree of freedom. For the covariance comparison

$$w = \sum_j \frac{c_j^2}{n_j} = \sum_j \frac{n_j^2 (x_j - M_X)^2}{n_j N^2}$$

$$= \frac{S_X^2}{N}.$$ [15.18.8]

The sum of squares for the covariance comparison is thus

$$\frac{[\text{sample cov.}(X,Y)]^2}{(S_X^2/N)} = \frac{N[\text{sample cov.}(X,Y)]^2}{S_X^2}$$
$$= \text{SS linear regression.}$$ [15.18.9*]

The sum of squares for linear regression has one degree of freedom, since it corresponds to the sum of squares for a single comparison among sample means.

If the true value for any comparison ψ is zero, then

$$E(\text{SS } \hat{\psi}) = E(\text{MS } \hat{\psi}) = \sigma_e^2,$$

which provides the principle used in testing comparisons. Now this principle will be applied to the SS linear regression. When $\beta_{Y \cdot X}$ is truly zero, so that the true cov.(X,Y) is zero, then

$$E(\text{SS linear regression}) = \sigma_e^2.$$ [15.18.10*]

Furthermore, when each of the populations represented by the fixed X_j values has a *normal* distribution of Y values, with equal variance for each such population, then

$$\frac{\text{MS linear regression}}{\sigma_e^2} = \chi_{(1)}^2,$$

the ratio of MS linear regression to σ_e^2 is a chi-square variable with 1 degree of freedom, given that $\beta_{Y \cdot X}$ is zero. In short, the same situation exists as in any test of a planned comparison when the true comparison value is zero.

For the regression model, the weighting factor w for the covariance comparison $\hat{\psi}$ is

$$w = s_X^2/N$$

so that

$$E(\text{MS linear regression}) = E(\hat{\psi}^2/w)$$
$$= \frac{N[\text{cov.}(X,Y)]^2}{s_X^2} + \sigma_e^2$$
$$= N\beta_{Y \cdot X}^2 s_X^2 + \sigma_e^2.$$

If we define

$$\rho^2 = \frac{\sigma_Y^2 - \sigma_{Y \cdot X}^2}{\sigma_Y^2},$$

the squared correlation coefficient as the proportion of variance accounted for by linear regression on the variable X in the population, then

$$E(\text{MS linear regression}) = \sigma_e^2 + N\rho^2\sigma_Y^2.$$

(It might be mentioned that in the regression model adopted here, the correlation coefficient for the population, ρ, and the covariance, $\mathrm{cov.}(X,Y)$, are not, strictly speaking, defined, since each is based on the notion of the expectation of the products of *two* random variables, X and Y. In this model, Y is a random variable, but X is treated as a fixed constant. Nevertheless, it is useful to think of $\mathrm{cov.}(X,Y)$ as a comparison ψ among the population means, and of ρ^2 as the proportion of σ_Y^2 accounted for by linear regression.)

15.19 The Expectations of the Other Sums of Squares

Given the linear model we adopted at the outset, it must follow that the true mean for each population j is the true value of Y_j' found from the population regression rule. Thus any departure of M_{Yj} from the sample Y_j' can reflect only error. For this reason it can be shown that

$$\frac{E[\text{SS deviations from linear regression}]}{\text{degrees of freedom}} = \sigma_e^2. \qquad [15.19.1]$$

However, what are the degrees of freedom here? Recall that for any J groups,

SS between = (SS linear regression) + (SS deviations from linear regression)

since the values of SS total and SS error remain unchanged regardless of whether we are doing an analysis of variance or a regression analysis on the same data. We saw in Section 14.11 that the SS between groups has only $J - 1$ degrees of freedom, and loses a degree of freedom for each particular orthogonal comparison sum of squares subtracted out. The sum of squares for deviations from regression must thus have only $J - 2$ degrees of freedom, and the expectation for this sum of squares is $(J - 2)\sigma_e^2$.

As usual, we assume equal variance σ_e^2 within each X population, so that

$$E(\text{SS error}) = \sigma_e^2(N - J). \qquad [15.19.2]$$

Since for this model both MS deviations and MS error estimate only σ_e^2, the two SS values are pooled, giving

SS deviations and error = (SS error) + (SS deviations from linear regression).

Therefore, the mean square for deviations *and* error is

$$\text{MS deviations and error} = \frac{\text{SS deviations and error}}{N - 2}. \qquad [15.19.3\dagger]$$

The degrees of freedom associated with this mean square come from the sum of the degrees of freedom for deviations from regression, $J - 2$, and those for error, $N - J$:

$$N - 2 = J - 2 + N - J.$$

Under the assumption that each X population has the same variance σ_e^2 (the assumption of *homoscedasticity*), then the population variance of estimate is the same as the value of σ_e^2:

$$
\begin{aligned}
\sigma_Y^2 &= \rho^2\sigma_Y^2 + \sigma_{Y \cdot X}^2 \\
&= \rho^2\sigma_Y^2 + \sigma_e^2
\end{aligned}
$$

and

$$
\sigma_e^2 = \sigma_Y^2(1 - \rho^2).
$$

Hence, an unbiased estimated of $\sigma_{Y \cdot X}^2$ is provided by the unbiased estimate of σ_e^2:

$$
\begin{aligned}
\text{est. } \sigma_{Y \cdot X}^2 &= \frac{\text{SS deviations and error}}{N - 2} \\
&= \frac{NS_Y^2(1 - r_{XY}^2)}{N - 2}.
\end{aligned}
$$

15.20   Assumptions underlying a test using this regression model

For a genuine regression problem, where the various X_j values are selected beforehand for observation, the following assumptions are made, in addition to the assumption of the general model:

$$
Y_{ij} = \mu_Y + \beta_{Y \cdot X}(X_j - \mu_X) + e_{ij}
$$

1. within each population j, the distribution of Y_{ij} values is normal;
2. within each population j, the variance σ_e^2 is the same;
3. the errors e_{ij} are completely independent.

In other words, the assumptions made here are identical to those for the fixed effects analysis of variance. Notice that absolutely no requirement is put on the distribution of X values in the grand population, although if the fixed effects idea is to apply, presumably the relative proportions of the different X_j values in the grand population is the same as in the sample. This need not trouble us unduly, since the main application of this procedure is to experimental situations where this question of the "actual" population distribution of X values is irrelevant anyway.

15.21   Tests for zero linear regression

The hypothesis to be tested is of the form

$$
H_0\text{: } \beta_{Y \cdot X} = 0
$$

or, equivalently,

$$
H_0\text{: } \rho_{Y \cdot X}^2 = 0.
$$

Since the SS for linear regression has only one degree of freedom, we have the option of using either the t or the F ratio:

$$F = \frac{\text{MS linear regression}}{\text{MS error and deviations}}$$ [15.21.1†]

with 1 and $N - 2$ degrees of freedom. However, the t ratio is ordinarily used, and can be put into a simple form: in terms of $b_{Y \cdot X}$,

$$t = \frac{b_{Y \cdot X} S_X \sqrt{N - 2}}{S_{Y \cdot X}}$$ [15.21.2†]

with $N - 2$ degrees of freedom.

In terms of r_{XY}, the test statistic is even simpler:

$$t = \frac{r_{XY} \sqrt{N - 2}}{\sqrt{1 - r_{XY}^2}}$$ [15.21.3†]

with $N - 2$ degrees of freedom. Either a directional or a nondirectional alternative may be used.

It is important to remember that in the model underlying this test it is assumed that if *any* possibility of prediction exists at all, then the linear rule completely specifies the predictive part of the relationship. Second, the hypothesis tested is that no prediction is possible even *using* the linear rule (that ρ_{XY}^2 or $\beta_{Y \cdot X}$ is zero). When we can reject H_0, we can feel confident that some prediction in the population is possible using a linear prediction rule. However, there is a difficulty with this model and the test it affords: the model itself may be wrong, as in a situation where almost perfect prediction is possible using *some* rule, but the linear rule gives very poor or no prediction. This leads to nonsignificant results which *do not* imply that no prediction is possible in the population. On the other hand, there may be a significant result, and the sample r^2 may be moderately large; this does not mean, however, that the r^2 obtained is an index of *possible* predictive power in the relationship, but only that there is a particular amount of predictive ability afforded by a *linear* rule. Some other rule may exist that greatly increases our ability to predict.

In the next chapter a somewhat different model for regression analysis will be explored, which permits one to evaluate both linear and other rules for describing and using a statistical relationship. For the moment, however, we will confine our attention to estimation methods within the linear regression model.

15.22   INTERVAL ESTIMATION IN REGRESSION PROBLEMS

Under the regression model, it is quite possible to form confidence intervals for $\beta_{Y \cdot X}$, the true regression coefficient. The $100(1 - \alpha)$ percent confidence interval is found from

$$b_{Y \cdot X} - \frac{\text{est. } \sigma_{Y \cdot X} t_{(\alpha/2)}}{S_X \sqrt{N}} \leq \beta_{Y \cdot X} \leq b_{Y \cdot X} + \frac{\text{est. } \sigma_{Y \cdot X} t_{(\alpha/2)}}{S_X \sqrt{N}} \qquad [15.22.1\dagger]$$

where

$$\text{est. } \sigma_{Y \cdot X} = \sqrt{\frac{NS_Y^2 - Nb_{Y \cdot X}^2 S_X^2}{N - 2}} = \sqrt{\frac{NS_Y^2(1 - r_{XY}^2)}{N - 2}},$$

or

$$\text{est. } \sigma_{Y \cdot X} = \sqrt{\text{MS deviations and error}} \qquad [15.22.2\dagger]$$

found from **15.19.3**. The $t_{(\alpha/2)}$ value is found for $N - 2$ degrees of freedom.

Occasionally, interval estimates are desired for the predicted values Y' using the *population* regression rule, and some specific X_j value. Remember that the predicted value Y' found using a *sample* regression equation does not necessarily agree with the Y' that would be found using the population regression equation. In a regression problem, there are *two* possible sources of disagreement between a sample Y' and the true value: the sample mean M_Y may be in error, and the sample estimate of $\beta_{Y \cdot X}$ may be wrong to some extent. Considering both of these sources of error, we have for a given score x_j the following confidence interval for *predicted* Y for population j:

$$y_j' - t_{(\alpha/2)} \text{ est. } \sigma_{Y \cdot X} \sqrt{\frac{1}{N} + \frac{(x_j - M_X)^2}{NS_X^2}} \leq \text{true } y_j'$$

$$\leq y_j' + t_{(\alpha/2)} \text{ est. } \sigma_{Y \cdot X} \sqrt{\frac{1}{N} + \frac{(x_j - M_X)^2}{NS_X^2}}. \qquad [15.22.3*]$$

The number of degrees of freedom is again $N - 2$.

Here, where $y_j' = M_Y + b_{Y \cdot X}(x_j - M_X)$, and where true

$$y_j' = \mu_Y + \beta_{Y \cdot X}(x_j - M_X),$$

there must be two kinds of variability in the distribution of Y_j' values over samples, given a constant value for $X_j - M_X$. The first source of variability is the difference between a given value of the mean M_Y and the true mean μ_Y. The second source of variability is the difference between the sample regression coefficient $b_{Y \cdot X}$ and the true coefficient $\beta_{Y \cdot X}$. The two terms under the radical sign in **15.22.3** reflect these two sources of variability.

Be sure to notice the interesting fact that **the regression equation found for a sample is not equally good as an approximation to the population rule over all the different values of** X_j. The sample rule is at its best as a substitute for the population rule when $X_j = M_X$, the mean of the X values, since the confidence interval is smallest at this point. However, as X_j values grow increasingly deviant from M_X in either direction, the confidence intervals grow wider. This indicates that for the more extreme values of X_j we cannot really be sure that the value Y_j' predicted for a sample for individuals each showing X_j comes anywhere near the value we would predict if we knew the regression equation for the population.

However, you will also notice that three things operate to make the predicted Y' values agree better with their population counterparts: large sample size; a *large* value of S_X^2, such as would be obtained by using a wide range of X_j values each with equal n; and a *small* value of S_Y^2, such as would be attained by controls operating in the experiment as a whole.

Finally, if we wish, we can establish a confidence interval for the *actual score* (not the predicted score) of a *single* individual given that the over-all relationship is linear. In this instance, the predicted score of an individual based upon a sample will differ from the prediction based upon population values, and even the "best" prediction, or true y_j', will still differ from the individual's actual score, y_{ij}. Hence in the distribution of predictions of actual scores there are three kinds of variability: of M_Y, over the various samples, of $b_{Y \cdot X}$, also over samples, and of the differences between y_{ij} and true y_j' within the population. These three kinds of variability are reflected in the confidence interval found from

$$y_j' - t_{(\alpha/2)} \text{ est. } \sigma_{Y \cdot X} S_j \leq y_{ij} \leq y_j' + t_{(\alpha/2)} \text{ est. } \sigma_{Y \cdot X} S_j \qquad \textbf{[15.22.4]}$$

where

$$S_j = \sqrt{1 + \frac{1}{N} + \frac{(x_j - M_X)^2}{N S_X^2}}.$$

For small samples, this may be a very wide interval indeed, particularly for extreme values of X_j.

Finally, for the case in which N approaches an infinite value, the confidence interval for the prediction of an individual score becomes simply

$$y_j' - z_{(\alpha/2)} \sigma_{Y \cdot X} \leq y_{ij} \leq y_j' + z_{(\alpha/2)} \sigma_{Y \cdot X} \qquad \textbf{[15.22.5†]}$$

since the variability of M_Y and $b_{Y \cdot X}$ over samples approaches zero, and since we have assumed that each of the populations j is normally distributed with the same variance $\sigma_{Y \cdot X}^2$.

The assumptions of normal distributions within each X_j population and equality of variance within populations are quite important when one is using these interval estimation methods. *It is possible to make an estimate of the regression equation, and to use this equation for prediction without assuming anything except random sampling, but interval estimates depend heavily on these assumptions for their validity.*

On the other hand, the *test* of the hypothesis that $\beta_{Y \cdot X}$ is zero apparently shares the features of the fixed effects model, in that the assumptions of normality and of homogeneous variances are somewhat less important than in other situations. In particular, arranging to have an equal number of cases per X_j grouping is good policy if one suspects that the equal variances assumption is not a reasonable one.

15.23   ESTIMATING THE STRENGTH OF LINEAR ASSOCIATION IN A REGRESSION PROBLEM

As mentioned in the preceding sections, the sample regression co-efficient $b_{Y \cdot X}$ is an unbiased estimate of $\beta_{Y \cdot X}$, the population regression coefficient, and confidence limits may be established for the value of $\beta_{Y \cdot X}$ as well. On the other hand, in a *regression* problem, it is not so simple to estimate ρ_{XY}^2, the true proportion of variance accounted for by linear regression. Indeed, the correlation coefficient ρ for a population is not even well defined in the regression model. In the population we can rather arbitrarily state that

$$\beta_{X \cdot Y}^2 = \frac{\rho_{XY}^2 \sigma_Y^2}{\sigma_X^2}$$

so that the point estimate of $\beta_{Y \cdot X}$ certainly has some bearing on an estimated value of ρ_{XY}^2. The value of σ_X^2 is presumably known, since this is given by s_X^2 for the distribution of X values in the experiment; but the joker is σ_Y^2, which need not be the same as S_Y^2. However, a point estimate of sorts for ρ_{XY}^2 can be made much as for ω^2 in the analysis of variance, by taking

$$\begin{aligned}
\text{est. } \rho_{XY}^2 &= \frac{(\text{MS regression}) - (\text{MS error and deviations})}{(\text{MS regression}) + (N-1)(\text{MS error and deviations})} \\
&= \frac{(N-1)r_{XY}^2 - 1}{(N-1) - r_{XY}^2}.
\end{aligned} \qquad [15.23.1]$$

For very large N this is simply r^2. When MS regression is less than the pooled error mean square (as when the F value is less than 1.00) then ρ_{XY}^2 is estimated to be zero. For regression problems using this model, simple interval estimates for ρ_{XY}^2 are apparently not possible.

15.24   AN EXAMPLE OF A REGRESSION PROBLEM

A social psychologist was interested in problem-solving carried out cooperatively by small groups of individuals. The theory upon which his experiments were based suggested that within a particular range of possible group sizes, the relationship between group size and average time to solution for particular kind of problem should be linear and negative: the larger the group, the less time on the average should it take for the problem to be solved. In order to check on this theory, the psychologist decided to form a set of experimental groups, ranging in size from groups consisting of a single individual to groups consisting of six individuals. Thus, 105 individuals were chosen at random from some specific population, and then formed at random into five groups consisting of 1 individual each, five groups each consisting of 2 individuals, five groups of 3, and so on until there were six different and nonoverlapping sets of five groups each, the last consisting of five groups each of size 6. Each individual subject participated in one and only one group in the study.

Each experimental group was given the same set of problems to solve, and given a score Y based on the average time to solve the problems. (The normal distribution of such scores might be open to some question, but we'll simply ignore this problem.) Thus, the experimental unit of observation was a group, and the experimental design can be represented in terms of the following six samples of 5 groups each:

	5 groups	5 groups	5 groups	5 groups	5 groups	5 groups
Group size	1	2	3	4	5	6

Notice that in this example $N = 30$, the number of distinct experimental *groups*.

Now the independent variable X in this experiment is the group size; X can take on only the whole values 1 through 6 in this experiment, and each value occurs with frequency 5. The dependent variable Y is the score attached to each distinct group. Does a linear relation appear to exist between X and Y in this range of group sizes?

Table 15.24.1

GROUP SCORES ARRANGED ACCORDING TO GROUP SIZE

		x			
1	*2*	*3*	*4*	*5*	*6*
32	30	27	19	23	20
29	30	26	21	19	19
30	26	24	20	20	16
28	27	24	20	22	18
26	26	23	18	18	16
145	139	124	98	102	89

The raw data were as shown in Table 15.24.1. Preliminary computations show that

$$\sum_j \sum_i x_j = (1)(5) + (2)(5) + (3)(5) + (4)(5) + (5)(5) + (6)(5) = 105$$

$$\sum_j \sum_i x_j^2 = (1)(5) + (4)(5) + \cdots + (36)(5) = 455$$

$$\sum_j \sum_i y_{ij}^2 = (32)^2 + \cdots + (16)^2 = 16813$$

$$\sum_j \sum_i y_{ij} = 697$$

$$\sum_j \sum_i y_{ij}x_j = (32)(1) + (29)(1) + \cdots + (18)(6) + (16)(6)$$
$$= 2231.$$

Given these values, a simple substitution into **15.11.4** gives the sample value of $b_{Y \cdot X}$:

$$b_{Y \cdot X} = \frac{30(2231) - (105)(697)}{(30)(455) - (105)^2}$$
$$= \frac{-4024}{2625} = -1.53.$$

Notice the negative sign of $b_{Y \cdot X}$, indicating that the apparent linear relationship *is* negative.

Now S_X^2 is simply

$$S_X^2 = \frac{\sum_j \sum_i X_j^2}{30} - M_X^2$$
$$= \frac{455}{30} - (3.5)^2$$
$$= 2.92$$

so that from **15.18.1**,

$$\text{SS linear regression} = 30(1.53)^2(2.92)$$
$$= 205.06.$$

Since

$$\text{SS total} = 16813 - \frac{(697)^2}{30}$$
$$= 619.37$$

then

$$\text{SS deviations and error} = 619.37 - 205.06 = 414.31$$

and

$$\text{MS deviations and error} = \frac{414.31}{30 - 2} = 14.79.$$

Under the hypothesis of no linear regression the F test (**15.21.1**) is given by

$$F = \frac{\text{MS linear regression}}{\text{MS deviations and error}}$$
$$= \frac{205.06}{14.79} = 13.86.$$

For 1 and 28 degrees of freedom this value is significant beyond the .01 level. On considering the sign of the regression coefficient one can say with considerable assurance that the null hypothesis of no linear regression is *not* true, and accept the alternative that some *negative* linear regression exists. Although we carried out the test in terms of F, we could just as well have used the (directional) t test given by **15.21.2** or **15.21.3**.

Now suppose that we wished to find confidence limits for $\beta_{Y\cdot X}$, the true regression coefficient for this problem. First of all we find

$$\text{est. } \sigma_{Y\cdot X} = \sqrt{\text{MS deviations and error}}$$
$$= 3.85.$$

Then, by **15.22.1**, the 95 percent confidence interval for $\beta_{Y\cdot X}$ is given by

$$-1.53 - \frac{(3.85)(2.048)}{\sqrt{(2.92)(30)}} \leq \beta_{Y\cdot X} \leq -1.53 + \frac{(3.85)(2.048)}{\sqrt{(2.92)(30)}}$$

or about

$$-2.37 \leq \beta_{Y\cdot X} \leq -.69.$$

In terms of the *sample* regression equation, the predicted y' score for a group with $X = 6$ is found from **15.9.2** to be

$$y' = -1.53(6 - 3.5) + 23.23$$
$$= 19.41.$$

Now suppose that we wished to find the 95 percent confidence interval for the value to be predicted for this group of size 6 using the *population* regression equation; we would take confidence limits given by **15.22.3**:

$$19.41 \pm (2.048)(3.85) \sqrt{\frac{1}{30} + \frac{(6 - 3.5)^2}{(2.92)(30)}}$$

or

$$19.41 \pm 2.55.$$

In a similar way, we can form confidence limits for Y' for any other value of X represented in the experiment.

The sample correlation coefficient value obtained for these data is about $-.56$. In terms of population ρ_{XY}^2, we can estimate the strength of linear association in the way represented by **15.23.1**, which gives

$$\text{est. } \rho_{XY}^2 = \frac{205.06 - 14.79}{205.06 + (29)(14.79)}$$
$$= .31.$$

This estimate, that about 31 percent of the variance in Y is accounted for by X and the linear rule, lends assurance that a moderately strong *linear* relationship exists in the population represented by these data.

On the other hand, all these inferences about the existence and strength of linear relationship in the population strictly apply *only to this particular distribution of X values*. The experimenter has no assurance at all that the linear predictive relationship will hold this strongly, or at all, over a different range of X possibilities. Consequently, his inferences do not necessarily extend to all possible distributions of X values, but only to a distribution of X values paralleling that actually used in the experiment.

There is one additional problem connected with the distribution of X values employed in regression analysis. Even when the values of the quantitative experimental variable are chosen in advance, there is the possibility of some error in the assignment of a value to an individual observation. In this instance, the distribution of the "true" values of the X variable will differ to some unknown extent from the apparent values represented in the experiment. For example, the experimental variable may be a measure of finger dexterity, and the experiment concerns the relationship of this variable to typing speed after a period of training. Individuals are selected with specific scores on finger dexterity, and groups of individuals having the same score represent the experimental variable X. The measure of finger dexterity is known to have some error of measurement, so that the score of an individual on any given occasion may be different from his "true" or long-run average score. Individuals who are assigned to the same X value group may actually be different in their real ability. These actual differences are obscured by errors of measurement or by the crudity of the measuring device. What effect will this have on inferences about linear regression of Y on X?

In general, the existence of such errors associated with the X values in a regression problem creates very serious problems, and in some such situations the regression analysis itself may be inappropriate. Fortunately, however, there is one circumstance in which the analysis may still be carried out in the usual way, in spite of the fact of the existence of errors in the values of X represented in the experiment. When the values of the experimental, quantitative, variable are selected in advance, and when these values themselves are subject to error, such as error of measurement, the regression analysis outlined above may still be applied without change, provided that such errors are themselves independently and normally distributed, as well as independent of the values of the X and the Y, and have equal variances and expectations of zero. In the measurement context, the expectation of zero for such errors means that there is no systematic bias in measurement of X. Such unsystematic errors in the quantitative experimental variable X do not affect the validity of the regression analysis, provided that the conditions just stated are true, and so long as the various values to be represented are selected in advance, rather than sampled.

On the other hand, errors in the X values create difficult problems when these conditions are not met, or when the X values are themselves sampled. Furthermore, the problem becomes even more complicated when the regression is not linear. In the analysis of nonlinear regression, to be discussed in the next chapter, the results can be seriously affected by such errors, even though all of the conditions permitting an analysis of linear regression are met. A discussion of this problem and its ramifications is contained in Scheffé (1959, Chapter 6).

15.25   THE ANALYSIS OF COVARIANCE

Although at this point only a brief treatment of the analysis of covariance is possible, it does deserve mention. Whereas the analysis of variance deals with experimental factors treated qualitatively, and regression analysis deals with quantitative factors, the analysis of covariance represents a link between these two approaches. In the analysis of covariance, some of the experimental factors are qualitative and others are quantitative. The theory of the analysis of covariance thus has elements both of analysis of variance and of regression theory. In the following it will be useful to distinguish between the qualitative experimental factor (the independent variable) and the quantitative factor. The first we will call simply the experimental factor; the second we will call the "concomitant variable." The dependent variable, Y, is of course quantitative.

One of the principal uses of the analysis of covariance is exemplified in the following situation: Suppose that an experiment were planned on the relationship between different methods of teaching students how to program for the computer and their performance at the end of a year's instruction. Although originally none of the students knows how to program, it is known that they vary in general intelligence. It might be possible to match the students on intelligence prior to the experiment, and then analyze the results accordingly. An alternate strategy is not to match the students, but rather to measure the general intelligence of each and to adjust for this factor by use of the analysis of covariance. This technique permits the experimenter to adjust the results after the fact, in such a way that performance differences among the different treatment groups, due to the linear relationship of performance and intelligence, are effectively removed from consideration.

Generally speaking, the analysis of covariance permits a post-hoc, statistical control for one or more concomitant variables, removing their influence from the comparison of groups on the main experimental variable. Naturally, there are some important qualifications to the use of this procedure, and it is not so universally useful as it might at first appear. In particular, this method is effective only where the relationship between the concomitant and the dependent variable is linear, and where the degree of this relationship does not itself depend upon the experimental variable. Nevertheless, in situations in which its requirements are satisfied, the use of the analysis of covariance can be quite effective in helping to clarify statistical relationships.

The model for simple analysis of covariance is a direct extension of the model of linear regression. Let A be the main, qualitative experimental factor, represented by J distinct levels. Each observation i within any group j yields two score values. The first is a value y_{ij} on the dependent variable; the second is a value x_{ij} on the concomitant variable. Then the linear model that is assumed to hold is

$$y_{ij} = \mu + \alpha_j + \beta_{Y.X}(x_{ij} - \mu_X) + e_{ij}.$$

That is, y_{ij} depends upon the general mean μ, plus an effect due to treatment j, and it also depends upon the value of x_{ij}, as weighted by the linear regression coefficient $\beta_{Y.X}$. There is also a random error component e_{ij}. We assume further that only a linear relationship exists between X and Y, and that the value of $\beta_{Y.X}$ does not depend upon j. In other words, within each treatment population, the relationship between X and Y is linear and has the same regression coefficient $\beta_{Y.X}$. The other, usual, analysis-of-variance assumptions are also made.

Now notice that this model allows for the possibility of examining the differences between Y means in terms that rule out the effects of X. The mean

Y value for any group j is

$$M_{Y_j} = \mu + \alpha_j + \beta_{Y.x}(M_{X_j} - \mu_X) + M_{e_j},$$

so that an adjusted mean, free of the effects of X, can be found from

$$M'_{Y_j} = M_{Y_j} - \beta_{Y.x}(M_{X_j} - \mu_X) = \mu + \alpha_j + M_{e_j}.$$

The adjusted mean M'_{Y_j} depends only on the grand population mean, the effect of treatment j, and error. Basically, the analysis of covariance is a procedure for estimating the value of $\beta_{Y.x}$, adjusting the Y means so as to remove the linear effects of X, and then comparing the adjusted means for evidence of treatment effects.

A simple analysis of covariance requires three sets of calculations. First, for the concomitant variable X, one finds the usual sums of squares:

$$SS_X \text{ between} = \frac{\sum_j \left(\sum_i x_{ij}\right)^2}{n_j} - \frac{\left(\sum_j \sum_i x_{ij}\right)^2}{N},$$

$$SS_X \text{ total} = \sum_j \sum_i x_{ij}^2 - \frac{\left(\sum_j \sum_i x_{ij}\right)^2}{N},$$

$$SS_X \text{ within} = SS_X \text{ total} - SS_X \text{ between}.$$

The same sorts of computations are then carried out for the Y variable:

$$SS_Y \text{ between} = \frac{\sum_j \left(\sum_i y_{ij}\right)^2}{n_j} - \frac{\left(\sum_j \sum_i y_{ij}\right)^2}{N},$$

$$SS_Y \text{ total} = \sum_j \sum_i y_{ij}^2 - \frac{\left(\sum_j \sum_i y_{ij}\right)^2}{N},$$

$$SS_Y \text{ within} = SS_Y \text{ total} - SS_Y \text{ between}.$$

Finally, we must calculate **sums of products** of X and Y, both within and between groups. These sums of products form the basis for estimates of the regression coefficient. These calculations are

$$SP_{XY} \text{ between} = \frac{\sum_j \left(\sum_i x_{ij}\right)\left(\sum_i y_{ij}\right)}{n_j} - \frac{\left(\sum_j \sum_i x_{ij}\right)\left(\sum_j \sum_i y_{ij}\right)}{N},$$

$$SP_{XY} \text{ total} = \sum_j \sum_i x_{ij} y_{ij} - \frac{\left(\sum_j \sum_i x_{ij}\right)\left(\sum_j \sum_i y_{ij}\right)}{N},$$

$$SP_{XY} \text{ within} = SP_{XY} \text{ total} - SP_{XY} \text{ between}.$$

Sums of squares for the adjusted means and the adjusted error are found as follows:

$$\text{SS}'_Y \text{ between} = \text{SS}_Y \text{ between} + \frac{(\text{SP}_{XY} \text{ within})^2}{\text{SP}_X \text{ within}} - \frac{(\text{SP}_{XY} \text{ total})^2}{\text{SS}_X \text{ total}}, \quad [15.25.1]$$

$$\text{SS}'_Y \text{ error} = \text{SS}_Y \text{ within} - \frac{(\text{SP}_{XY} \text{ within})^2}{\text{SS}_X \text{ within}}. \quad [15.25.2]$$

Then the analysis of adjusted means can be summarized thus:

Source	SS	df	MS	F
Adjusted means	SS'_Y between	$J - 1$	$\dfrac{\text{SS}'_Y \text{ between}}{J - 1}$	
Error	SS'_Y error	$N - J - 1$	$\dfrac{\text{SS}'_Y \text{ error}}{N - J - 1}$	$\dfrac{\text{MS}'_Y \text{ between}}{\text{MS}'_Y \text{ error}}$
Adjusted total		$N - 2$		

Note that the F test here has $J - 1$ and $N - J - 1$ degrees of freedom. One pays a slight price (one degree of freedom in the denominator of F) for the ability to adjust for linear effects of X on Y. The hypothesis being tested is, of course, the usual one: $H_0: \alpha_j = 0$ for all j.

Perhaps the biggest hitch in the application of the analysis of covariance is the assumption that the regression coefficient is constant within each of the treatment populations. Furthermore, it is necessary to assume that the regression of X on Y is linear within each such population. Fortunately, one can examine the data in order to see if the first of these assumptions is at least reasonable. Let β_j be the true regression coefficient within population j. Then we wish to test the hypothesis $H_0: \beta_j = \beta$ for every j.

In order to carry out this test of the homogeneity of within-treatment regression coefficients, we calculate for each group j the quantities

$$\text{SP}_{XY_j} = \sum_i x_{ij} y_{ij} - \frac{\left(\sum_i x_{ij}\right)\left(\sum_i y_{ij}\right)}{n_j}$$

and

$$\text{SS}_{X_j} = \sum_i x_{ij}^2 - \frac{\left(\sum_j x_{ij}\right)^2}{n_j}.$$

Then one takes

$$A = \text{SS}_Y \text{ within} - \sum_j \frac{\text{SP}_{XY_j}^2}{\text{SS}_{X_j}}$$

$$B = \text{SS}'_Y \text{ error} - A.$$

This leads to the F ratio

$$F = \frac{B/(J-1)}{A/(N-2J)} \qquad [15.25.3]$$

with $J - 1$ and $N - 2J$ degrees of freedom. Large values of F are evidence against the hypothesis of equal regression within groups, and thus cast doubt on the validity of this assumption in the analysis of covariance. Although the usual reason for testing the hypothesis of equal regression among treatments is a desire to check on the assumptions underlying the analysis of covariance, there may also be interest in this hypothesis in its own right. The text by Winer (1962) also gives several other tests growing out of the analysis of covariance.

Employed with discretion, the analysis of covariance can be a very useful statistical tool. It does provide a way to introduce, "after the fact," statistical controls for the linear regression of one variable in the analysis of another. However, as usual, one's ability to achieve this advantage comes about at the price of some rather restrictive assumptions. The differences between the means on the dependent variable are free of the influence of the concomitant variable *only* if the relationship between concomitant and dependent variables is linear. If appreciable nonlinear regression exists, then this may have the effect of inflating the error term in the analysis of covariance, and thereby lower the efficiency of the experiment. Furthermore, when the concomitant variable is itself affected by the experimental treatment, adjusting the means of the dependent variable may actually remove some portion of the treatment effects themselves. If this situation is suspected to exist, it may be worthwhile to carry out a prior analysis of variance on the concomitant variable, to see if there is an appreciable influence of the experimental variable upon it. The requirement of equal regression within treatment groups is also not trivial, and the interpretation of the analysis of covariance is open to serious question when this assumption is not met. In view of these potential problems, it is probably best for the beginner to use the analysis of covariance only with considerable caution, or with expert assistance.

Although only the simplest case has been considered here, analysis of covariance is not limited to simple one-factor designs with one concomitant variable. The method can be extended to quite complex designs with a number of experimental factors, and with two or more concomitant variables. The same idea may also be applied to repeated measures and similar designs.

For example, if there is one experimental variable and two concomitant variables, the model can be written as

$$y_{ij} = \mu + \alpha_j + \beta_X(x_{ij} - \mu_X) + \beta_W(w_{ij} - \mu_W) + e_{ij}.$$

Here, β_X is a regression coefficient for the linear relationship of X and Y, where the relationships of W to X and W to Y have been ruled out; similarly, β_W is a regression coefficient reflecting the linear relationship of W and Y, where the relationships of X and Y and W and X have been ruled out. Such multiple regression coefficients will be considered in the next chapter. The same general sort of model can be stated for any number of experimental factors and for any number of quantitative concomitant variables. Indeed, it is even possible to extend the model to the situation where there are several experimental factors, several concomitant variables, and several dependent variables, providing the basis for a multivariate analysis of variance and covariance. Given modern computer technology, such complicated analyses are quite feasible, although a discussion of such techniques is far beyond our scope. Many of these advanced techniques are described in standard texts in multivariate analysis such as that by Rao (1952). Furthermore, good brief introductions to some of these methods are given in the book by Cooley and Lohnes (1963).

Even though we will not be able to go into the multivariate extensions of the basic regression model, along the lines suggested above, we are still not quite through with this model. In the next chapter the model will be extended to allow for nonlinear as well as linear relationships between the concomitant and the experimental variable.

15.26 PROBLEMS IN CORRELATION IN BIVARIATE NORMAL POPULATIONS

In a regression problem, the values of X are selected in advance of the observation of the Y values. However, quite frequently in psychological research a sample of N individuals is drawn from some population where each individual "brings along" a pair of values (x, y). The experimenter exerts no direct control over either X or Y values for a given subject. Such a study is usually called a problem in correlation. The main interest actually focuses on the value of r_{XY} itself, especially as an estimate of ρ_{XY} for the population. Given a particular assumption about the population distribution of joint (x, y) events we can test not only if X and Y are linearly related, but also if any systematic relationship at all exists between the two variables.

In Chapter 4 it was pointed out that in a joint distribution of discrete random variables a probability is associated with each possible X and Y pair. Obviously, a similar conception holds when X and Y are continuous variables, and a probability is associated with any joint interval of values. Such joint distributions are called **bivariate.**

Although any number of theoretical bivariate distributions are possible in principle, by far the most studied is the **bivariate normal distribution.** The density function for this joint distribution has a rather elaborate-looking rule. However, for standardized variables, this can be condensed to

$$f(z_X, z_Y) = \frac{1}{K} e^{-G}$$

where

$$G = \frac{(z_X^2 + z_Y^2 - 2\rho z_X z_Y)}{2(1 - \rho^2)}$$

and

$$K = 2\pi \sqrt{(1 - \rho^2)}.$$

Notice that in a bivariate normal distribution, the population correlation coefficient ρ appears as a parameter in the rule for the density function. Thus, even though z_X and z_Y are both standardized variables, the particular bivariate distribution cannot be specified unless the value of the correlation ρ is known.

In a bivariate normal distribution, the marginal distribution of X over all observations is itself a normal distribution, and the marginal distribution of Y is also normal. Furthermore, given any X value, the *conditional* distribution of Y is normal; given any Y, the conditional distribution of X is normal. In other words, if a bivariate normal distribution is conceived in terms of a table of joint events (as in Section 4.10), where the number of possible X values, of Y values, and of possible joint (x,y) events is infinite, then within any possible row of the table one would find a normal distribution, a normal distribution also exists within any possible column, and the marginals of the table also exhibit normal distributions.

For our purposes, however, the feature of most importance in a bivariate normal distribution is this: **given that densities for joint (x,y) events follow the bivariate normal rule, then X and Y are independent if and only if** $\rho_{XY} = 0$. For *any* joint distribution of (x,y) the independence of X and Y implies that $\rho_{XY} = 0$, but it may happen that $\rho_{XY} = 0$ even though X and Y are *not* independent. However, *for this special joint distribution, the bivariate normal,* $\rho_{XY} = 0$ *both implies and is implied by the statistical independence of X and Y.* The only predictability possible in a bivariate normal distribution is that based on a *linear* rule.

On the other hand, just because the distribution of X and the distribution of Y both happen to be normal, when considered as marginal distributions, this does *not* necessarily mean that the joint distribution of (x,y) values is bivariate normal. Hence, it is entirely possible for a nonlinear statistical relation to exist even though both X and Y are normally distributed when considered separately. It is not, however, possible for any but a linear relation to exist when X and Y jointly follow the bivariate normal law.

Most of the classical theory of inference about correlation and regression was developed in terms of the bivariate normal distribution. **If one can assume such a joint population distribution, inferences about correlation are equivalent to inferences about independence or dependence between two random variables.** For the kinds of problems here called **correlation problems,** the assumption of a bivariate normal distribution is usually made. When this assumption is valid, any inference about the value of ρ is equivalent to an inference about the *independence,* or *degree of dependence* between two variables; this is not, however, a feature of regression problems, where the bivariate normal assumption need not be made. As always, by adopting more stringent assumptions about the form of the population distribution, one is able to make much more positive statements from sample results.

The general notion of the bivariate normal distribution has also been extended to the multivariate situation where any finite number of random variables are considered, and each observation represents a joint event such as $(x, y, \ldots, z, w)$. The same general form of density function is assumed to exist as for a bivariate normal distribution, except that in the density function rule for a multivariate normal distribution, the correlations ρ_{XY} between each *pair* of variables appear as parameters. In a multivariate normal distribution, fixing one or more of the random variables at constant value still results in a multivariate normal, bivariate normal, or normal distribution of the "free" variables. This distribution will be important to our discussion in the second half of the next chapter; for the moment, we will confine our attention to the bivariate normal distribution, where there are only two random variables, X and Y.

15.27 TESTS AND INTERVAL ESTIMATES IN CORRELATION PROBLEMS

For many correlation problems the hypothesis of interest is

$$H_0: \rho_{XY} = 0.$$

When this hypothesis is true and the population can be assumed to be bivariate normal in form, the distribution of the sample correlation coefficient tends, rather slowly, toward a normal distribution for increasing N. For N extremely large, a test of the hypothesis that $\rho = 0$ might be made in terms of the normal distribution. On the other hand, it can be shown that for any sample size N, a test of this hypothesis in a bivariate normal population is given by the t ratio:

$$t = \frac{r_{XY}\sqrt{N-2}}{\sqrt{1-r_{XY}^2}} \qquad [15.27.1\dagger]$$

with $N - 2$ degrees of freedom. This is precisely the same test statistic used in a regression problem to test the hypothesis that $\beta_{Y \cdot X}$ is zero, even though the rationale underlying these two tests is different, depending on the regression or correlation model adopted.

The various methods of interval estimation for $\beta_{Y \cdot X}$, for the mean predicted value of Y for a fixed value of X, and for the value of a given observation in terms of Y, originally presented in Section 15.22, are also justified in a problem in correlation. Notice that when a bivariate normal distribution is assumed, the requirement of a normal conditional distribution of Y values *given* any specific X value is automatically met. The basic reasoning behind these methods of interval estimation is somewhat different in the regression and correlation problem models, but the form of interval estimation is the same in either instance.

Under the assumptions made in a problem in correlation, the value of r_{XY} may be used directly as an estimator of ρ_{XY} for the population. Although it is a sufficient and consistent estimator for ρ_{XY}, the sample correlation is slightly biased; however, the amount of bias involves terms of the order of $1/N$, and for most practical purposes can be ignored.

As mentioned earlier, for very large samples the distribution of the sample correlation coefficient may be regarded as approximately normal when $\rho_{XY} = 0$. Even for relatively small samples ($N > 4$) this sampling distribution is unimodal and symmetric. However, when ρ_{XY} is other than zero, the distribution of r_{XY} tends to be very skewed. The more that ρ differs from zero, the greater is the skewness. When ρ_{XY} is greater than zero, the skewness tends to be toward the left, with intervals of high values of r_{XY} relatively more probable than similar intervals of negative values. When ρ_{XY} is negative, this situation is just reversed, and the distribution is skewed in the opposite direction. The fact that the particular form of the sampling distribution depends upon the value of ρ_{XY} makes it impossible to use the t test for other hypotheses about the value of the population correlation, or to set up confidence intervals for this value in some direct elementary way. Although the sampling distribution of r_{XY} for $\rho_{XY} \neq 0$ has been fairly extensively tabled, it is much simpler to employ the following method.

R. A. Fisher showed that tests of hypotheses about ρ_{XY}, as well as confidence intervals, can be made from moderately large samples from a bivariate normal population if one uses a particular *function* of r_{XY}, rather than the sample correlation coefficient itself. The function used is known as the Fisher r to Z transformation, given by the rule

$$Z = \frac{1}{2} \log_e \left(\frac{1 + r_{XY}}{1 - r_{XY}} \right). \qquad [15.27.2*]$$

The function is of the type called "one to one": for each possible value of r there can exist one and only one value of Z, and for each Z, one and only one value of r. This fact makes it possible to convert a sample r value to a Z value, make inferences in terms of Z, and then to turn those inferences back into statements about correlation once again.

Fisher showed that for virtually any value of ρ_{XY}, for samples of moderate size the sampling distribution of Z values is *approximately normal*, with an expectation given approximately by

$$E(Z) = \zeta = \frac{1}{2} \log_e \left(\frac{1 + \rho_{XY}}{1 - \rho_{XY}} \right). \qquad [15.27.3]$$

(The population value of Z, corresponding to ρ, is denoted by ζ, small Greek zeta.) The sampling variance of Z is approximately

$$\text{var.} (Z) = \frac{1}{N - 3}. \qquad [15.27.4*]$$

The goodness of these approximations increases the *smaller* the absolute value of ρ, and the *larger* the sample size. For moderately large samples the hypothesis that ρ_{XY} is equal to any value ρ_0 (not too close to 1 or -1) can be tested. This is done in terms of the test statistic

$$\frac{Z - \zeta}{\sqrt{1/(N - 3)}} \qquad [15.27.5\dagger]$$

referred to a *normal* distribution. The value taken for $E(Z)$ or ζ depends on the value given for ρ_0 by the null hypothesis:

$$\zeta = E(Z) = \frac{1}{2} \log_e \left(\frac{1 + \rho_0}{1 - \rho_0} \right)$$

and the sample value of Z is taken from the sample correlation,

$$Z = \frac{1}{2} \log_e \left(\frac{1 + r_{XY}}{1 - r_{XY}} \right).$$

It should be emphasized that the use of this r to Z transformation *does* require the assumption that the (x, y) events have a bivariate normal distribution in the population. On the surface, this assumption seems to be a very stringent one, which may not be reasonable in some situations, though there is some evidence that the assumption may be relatively innocuous in others. However, the consequences of this assumption's not being met seem largely to be unknown. Perhaps the safest course is to require rather large samples in uses of this test when the assumption of a bivariate normal population is very questionable.

Table VI in Appendix C gives the Z values corresponding to various values of r. This table is quite easy to use, and makes carrying out the test itself extremely simple. Only positive r and Z values are shown, since if r is negative, the sign of the Z value is taken as negative also.

For example, suppose that we wanted to test the hypothesis that $\rho_{XY} = .50$ in some bivariate normal population. A sample of 100 cases drawn at random gives a correlation r_{XY} of .35. The hypothesis is to be tested with $\alpha = .05$, two-tailed.

Then, from the Table VI, we find that for $r_{XY} = .35$,

$$Z = .3654.$$

For $\rho_{XY} = .50$, we find

$$\zeta = E(Z) = .5493.$$

The test statistic is then

$$\frac{.3654 - .5493}{\sqrt{1/97}} = -1.81.$$

In a normal sampling distribution, a standard score of 1.96 in absolute value is required for rejecting the hypothesis at the .05 level, two-tailed. Thus, we do not reject the hypothesis that $\rho_{XY} = .50$ on the basis of this evidence. Observe that the test made in terms of Z leads to an inference in terms of ρ.

Occasionally one has two *independent* samples of N_1 and N_2 cases respectively, where each is regarded as drawn from a bivariate normal distribution, and he computes a correlation coefficient for each. The question to be asked is, "Do both of these correlation coefficients represent populations having the *same* true value of ρ?" Then a test of the hypothesis that the two populations show equal correlation is provided by the ratio

$$\frac{Z_1 - Z_2}{\sigma_{(Z_1 - Z_2)}}, \qquad \text{[15.27.6†]}$$

where Z_1 represents the transformed value of the correlation coefficient for the first sample, Z_2 the transformed value for the second, and

$$\sigma_{(Z_1 - Z_2)} = \sqrt{\frac{1}{N_1 - 3} + \frac{1}{N_2 - 3}}. \qquad [15.27.7\dagger]$$

For reasonably large samples, this ratio can be referred to the normal distribution. Remember, however, that the two samples must be independent (in particular, not involving the same or matched subjects) and the population represented by each must be bivariate normal in form.

More generally, suppose that there are J independent samples, each drawn from a bivariate normal distribution of (x, y) pairs. Each sample j yields a sample correlation r_j between X and Y. Then the hypothesis that the true value ρ_{XY} is the same for all of the populations can be tested by the statistic

$$V = \sum_j (n_j - 3)(Z_j - U)^2 \qquad [15.27.8\dagger]$$

which is distributed as chi-square with $J - 1$ degrees of freedom when the null hypothesis that $\rho_1 = \rho_2 = \cdots = \rho_J$ is true. Here, n_j is the number of observations in the sample j, and

$$U = \frac{\sum_j (n_j - 3)Z_j}{\sum_j (n_j - 3)}.$$

15.28   CONFIDENCE INTERVALS FOR ρ_{XY}

If the population has a bivariate normal distribution of (x, y) events, then the r to Z transformation can be used to find confidence intervals, very much as for a mean of a large sample. It is approximately true that for random samples of size N, an interval such as

$$Z - z_{(\alpha/2)} \sqrt{\frac{1}{N - 3}} \leq \zeta \leq Z + z_{(\alpha/2)} \sqrt{\frac{1}{N - 3}} \qquad [15.28.1\dagger]$$

will cover the true value of ζ with probability $1 - \alpha$. Here, Z is the sample value corresponding to r_{XY}, ζ is the Z value corresponding to ρ_{XY}, and $z_{(\alpha/2)}$ (*definitely to be distinguished from* Z) is the value cutting of the upper $\alpha/2$ proportion in a normal distribution. Thus, the expression **15.28.1** above gives the $100(1 - \alpha)$ percent confidence interval for ζ. On changing the limiting values of Z back into correlation values, we have a confidence interval for ρ_{XY}.

In the example given in the preceding section,

$$Z = .3654$$
$$N = 100,$$

so that

$$\sqrt{\frac{1}{N-3}} = .1.$$

For $\alpha = .05$, $z_{(\alpha/2)} = 1.96$, so that the 95 percent confidence interval for ζ is given approximately by

$$.3654 - (1.96)(.1) \leq \zeta \leq .3654 + (1.96)(.1)$$

or

$$.1694 \leq \zeta \leq .5614.$$

The corresponding interval for ρ_{XY} is then approximately

$$.168 \leq \rho_{XY} \leq .510$$

(the correlation values here are taken to correspond to the nearest tabled Z values). We can assert that the probability is about .95 that sample intervals such as this cover the true value of ρ_{XY}.

15.29   An Example of a Correlation Problem

A study was made of the tendency of the height of a wife to be linearly related to that of her husband, and it was desired to find a sample correlation between husband and wife's heights, and to use this to test the hypothesis of no linear relationship.

A sample of 15 American *couples* was drawn at random, and the data are shown in Table 15.29.1.

Table 15.29.1

	Heights in inches	
Couple	X (Wife's height)	Y (Husband's height)
1	70	75
2	67	72
3	70	75
4	71	76
5	67	70
6	64	68
7	71	72
8	63	67
9	65	67
10	65	68
11	65	68
12	65	71
13	66	68
14	65	71
15	61	62

Table 15.29.2

Couple			Transformed scores		
	u	v	uv	u^2	v^2
1	10	5	50	100	25
2	7	2	14	49	4
3	10	5	50	100	25
4	11	6	66	121	36
5	7	0	0	49	0
6	4	−2	−8	16	4
7	11	2	22	121	4
8	3	−3	−9	9	9
9	5	−3	−15	25	9
10	4	−2	−8	16	4
11	5	−2	−10	25	4
12	5	1	5	25	1
13	6	−2	−12	36	4
14	5	1	5	25	1
15	1	−8	−8	1	64
	94	0	142	718	194

For these data the computations for the correlation coefficient can be simplified by subtracting 60 from the height of each wife, and 70 from the height of each husband; this does not alter the value of r_{XY} obtained (Section 15.11). Then the new scores are as shown in Table 15.29.2.

The correlation coefficient computed by the formula **15.11.2** turns out to be

$$r_{XY} = \frac{(15)(142) - (94)(0)}{\sqrt{[(15)(718) - (94)^2][(15)(194) - (0)^2]}}$$

$$= \frac{2130}{\sqrt{(1934)(2910)}}$$

$$= .89.$$

On the evidence of the sample, we conclude that there is a very strong linear relation between the heights of wives and husbands.

If we wished only to test the hypothesis that the true correlation is zero, we would employ the t test given in Section 15.27:

$$t = \frac{(.89)\sqrt{15 - 2}}{\sqrt{1 - (.89)^2}}$$

$$= \frac{3.209}{.456}$$

$$= 7.04.$$

This greatly exceeds both the values required for $\alpha = .05$ and for $\alpha = .01$ for a t with 13 degrees of freedom (two-tailed).

Suppose, however, that the question had been, "Does the height of the wife and the linear relation account for more than 50 percent of the variance in the observed heights of husbands?" That is, we actually want to test the hypothesis

$$\rho^2_{XY} \leq .50$$

against the alternative

$$\rho^2_{XY} > .50.$$

Given the assumption that ρ is positive, this is equivalent to the test of

$$H_0: \rho_{XY} \leq \sqrt{.50} \text{ or } H_0: \rho_{XY} \leq .707$$

against

$$H_1: \rho_{XY} > .707.$$

Here, the Fisher r to Z transformation is used to find the test statistic

$$\frac{Z - E(Z)}{\sqrt{1/(N - 3)}} = \frac{1.42 - .88}{\sqrt{1/12}}$$
$$= 1.87.$$

In a normal sampling distribution, this exceeds the value required for the 5 percent significance level, one-tailed. Thus we may safely conclude from this sample that more than 50 percent of the variance in Y is accounted for by the apparent linear relation.

This example is made up, of course, and correlations this large are not usually found in psychological work. It does illustrate one thing, however. Even though the correlation found is sizable, it makes no sense at all to think of the height of the wife as "causing" the height of the husband, or that of the husband the height of the wife. These are simply two numerical measurements that happen to occur together in a more or less linear way, according to the evidence of this sample. The reason *why* this linear relation exists is completely out of the realm of statistics, and the correlation coefficient and tests shed absolutely no light on this problem. In this example, it is perfectly obvious that personal preferences and current standards of society cause *some* selection to occur in the process of mating, and these factors in turn underlie our observations that (x, y) pairs to occur in a particular kind of relationship. As a description of a population situation, our inferences may very well be valid, but this fact alone gives us no license to talk about the cause of the apparent linear relation.

15.30 RETROSPECT: REGRESSION AND CORRELATION PROBLEMS

By now it must be obvious that although both problems in correlation and those in regression involve the same machinery of *descriptive* statistics, the sampling procedures, the assumptions made, and the inferences drawn tend to be somewhat different in the two situations. In regression problems, the methods usual in any fixed-effects experiment are employed: the various values of X to be observed are selected in advance, just as one selects factor levels or treatments for an experiment. Indeed, quite often the values of X actually are amounts of some quantitative experimental treatment. No requirement is put on the experimental distribution of X, and only the usual Model I analysis of variance assumptions are made about the Y values within each X array. Then inferences are possible about *linear* regression, the ability to predict Y from X using a linear rule. It is not necessarily true, however, that *all* the ability to predict Y from X resides in some linear rule; small or zero regression coefficients are *not necessarily* indicators of independence between X and Y values.

In correlation problems, no explicit restrictions are put on the possible X values to be observed in the sample. A sample of N cases is drawn, and their (x, y) pairings noted. In making inferences, a bivariate normal distribution is assumed, and when this assumption *is valid* inferences about the population correlation are equivalent to inferences about the entire *strength* of predictive relationship possible in the population.

It is undoubtedly repetitious to say so, but the significance level obtained in either kind of problem is of small moment as compared with the estimate of our ability to predict, as given by an estimate of ρ_{XY}^2 such as the sample r_{XY}^2. Very tiny estimated values of ρ_{XY}^2 can come from extremely significant results, showing that the ability to predict X from Y linearly is not completely absent, but that there very well may be no practical advantage to be gained from knowing X in the linear prediction of Y. Results deserving serious attention are those that are *both* significant and give estimated values of ρ_{XY}^2 indicating a really important percentage reduction in variance of Y, given X and the regression equation.

In psychology, there is often a tendency to regard regression problems as somehow "better" than problems in correlation. This is largely because a problem in regression carries along all of the paraphernalia of the usual controlled experiment, and a problem in correlation, ordinarily, does not. This distinction is not altogether fair, either to the methods or to the people who use them. The important thing is the research question, and the selection of the best technique to shed light on that question. The theory of normal correlation, involving the bivariate normal assumption, grew out of biological problems and related applied situations. In problems where the variables X and Y symbolize measurements of biologically determined traits, the bivariate normal distribution assumption often makes sense. It was mentioned in Section 8.6 that the normal distribution of natural trait measurements is usually accounted for by an argument similar to that for the limiting form of the binomial distribution. It also happens that the bivariate or multivariate normal distribution can be shown to be the limiting form of a *multinomial* distribution; exactly the same kind of argument for normality can thus be applied to joint distributions of natural traits as to univariate distributions, and in this light the bivariate normal assumption is not quite so arbitrary as it first appears.

In one way, the correlational approach of drawing a random sample of (x, y) pairings is superior to the regression approach using fixed values, particularly if the main reason for the study is an intended use of a linear rule for prediction in some *existent* population. Notice that in the regression model, nothing was said about the distribution of X values in some population; presumably, the relative frequencies for X in the population are exactly like those in the sample (with the reservation, noted above, that such X values may contain error, such as error of measurement). Now suppose that a study is done using the regression-problem procedure for sampling, where X values are selected arbitrarily, but then the regression equation is actually used to make predictions for individuals in some existing population. *We can have no confidence that the sample regression equation is even remotely like the best way to predict for some population unless the proportional representation of X values in the population is like that in the sample, or unless X values themselves are sampled at random.* In most practical prediction situations, we simply do not know the probability of occurrence of any X value, but we suspect that it is not the same as the relative frequency of X in the sample. For a population of "natural" (x, y) events, where each individual sampled brings both an X and a Y value for observation, the regression sort of experiment ordinarily does not make much sense as a way to find a prediction rule for use outside the experimental situation itself.

However, in the correlation type of problem, the distribution of X values in the sample need not have any particular connection with the distribution of X in the population, other than describing a random sample from that population distribution. For correlation problems, the regression equation arrived at from the random sample is the best approximation one can make to the true linear regression equation for optimal prediction, *regardless* of the distribution of X in the population. Hence, it seems reasonable then that the correlational approach should be used when interest lies in *actually predicting* an individual's status on one trait from his status on another in some "natural" population where the independent variable represents a natural or previously acquired characteristic of the individual.

Nevertheless, the regression approach is also useful. When some experimental treatment is given in quantity, such as the dosage of a drug, an amount of shock administered, amount of practice, and so on, the various X values do not really "occur" in some real population of individuals, and it makes no particular sense to think of some existent distribution of (x, y) values. In this instance, where X represents an amount of some experimental treatment administered, both the nature of the population (x, y) distribution and the character of the regression equation change. The population sampled in the experiment is potential or hypothetical. The population distribution of X is whatever the experimenter chooses to make it; in estimating the regression equation he is saying, literally, "If I repeated this experiment, not on J samples of individuals, but on J distinct populations having a marginal distribution of X like that in the sample, then the best rule for linear prediction of an individual Y value given the X value for the population should be thus and so." The predictions intended apply only to members of this potential grand population. In short, when X is an amount of something actually *done* to a subject by the experimenter, and not a measure of something that the subject brings to the study as a natural or acquired trait, the regression scheme for sampling does lead to a sensible rule for linear prediction. Predictions can be made about an individual so treated, *within this hypothetical population.*

In summary, there is no value judgment to be put on regression as opposed to correlation problems. Each is appropriate to a particular kind of research enterprise: the correlation scheme applies especially to problems of linear prediction of natural traits, and the regression procedures to studies of the ability to predict results of quantitative *experimental* treatments using a linear function rule.

EXERCISES

1. List the assumptions made in a test of significance in a linear regression problem; compare them with the assumptions made in fixed-effects analysis of variance. Compare the assumptions made in significance tests for correlation problems with the assumptions made in random-effects analysis of variance.
2. Under what circumstances must uncorrelated variables also be independent variables? Under what circumstances might two independent variables be correlated?
3. Make sketches showing the regress on line $z'_Y = r z_X$ when $r = .30, r = -.60, r = -1.00$, $r = +1.00$, and $r = 0$.
4. Suppose that in some population

$$Y_{ij} = \mu_Y + \beta_{Y \cdot X}(X_j - \mu_X) + e_{ij}$$

where $E(e_{ij}) = 0$.
Let $E(Y|X)$ be the mean of Y values for individuals all of whom have the same X value. If $E(Y|X)$ were plotted for all of the different values of X, what should the result look like? What must the mean value of $E(Y|X)$ over all X values be?
5. Suppose that for the population described in problem 4 above, each mean $E(Y|X)$ was the same, regardless of the X value. What would the value of $\beta_{Y \cdot X}$ be? What would the correlation ρ_{XY} be? Furthermore, suppose that the variance, $\text{Var}(Y|X)$, of Y values for individuals all of whom had the same value of X were different for different X. Would X and Y then be independent? (See chapter 4, and the discussion of independence.)
6. An experimenter was interested in the possible linear relation between the time spent per day in practicing a foreign language and the ability of the person to speak the language at the end of a six-week period. Some fifty students were assigned at random among five experimental conditions ranged from 15 minutes practice daily to 3 hours practice per day. Then at the end of six weeks, each student was scored for proficiency in the language. The data follow:

PROFICIENCY SCORES, BY DAILY PRACTICE TIME

117	106	86	140	105
85	81	98	128	149
112	74	125	108	110
81	79	123	104	144
105	118	118	132	137
109	110	94	133	151
80	82	93	96	117
73	86	91	101	113
110	111	122	103	142
78	113	130	135	112
.25	.50	1	2	3

X = PRACTICE, IN HOURS

Find the linear regression equation for predicting Y, the proficiency of a student, from X, the practice time per day. Plot the obtained data and the straight line representing the regression equation.

7. Test the hypothesis that, for the population of students sampled in problem 6, the true regression coefficient is zero. Furthermore, estimate the true proportion of variance in Y-values accounted for by linear regression on X.

8. From the data of problem 6, find the 95 percent confidence limits for the value of the regression coefficient $\beta_{Y \cdot X}$ in the population. Find the 95 percent confidence limits for the value of Y that *should* be predicted when $X = 2$ if one knew the *population* regression equation.

9. An experimenter was interested in the possible linear relationship between the measure of finger dexterity X, and another measure representing general muscular coordination Y. A random sample of 25 persons showed the following scores:

PERSON	X-VALUE	Y-VALUE
1	75	84
2	77	94
3	75	90
4	76	90
5	75	91
6	76	86
7	73	87
8	75	95
9	74	83
10	75	85
11	76	88
12	74	91
13	72	80
14	75	85
15	73	87
16	75	82
17	78	86
18	76	83
19	74	85
20	74	88
21	77	100
22	75	98
23	76	89
24	74	91
25	75	99

Compute the correlation coefficient, and test its significance ($\alpha = .05$).

10. Based on the data of problem 9 above, find the regression equation for predicting X from Y. Plot this regression equation along with the raw data. What is the appropriate measure of the "scatter" or horizontal deviations of the obtained points in this plot about the regression line?

11. For the data of problem 9, find the 99 percent confidence limits for the true value of ρ, the population correlation. (Hint: use the r to Z transformation.) What are we assuming when we find these confidence limits?

12. An investigator was interested in the ability of teachers to judge a child's age from his drawings. In order to gather data on this, he selected at random 12 drawings by children aged 4, 12 by children aged 5, 12 from age 6, and 12 from age 7. Then each of a panel of teachers assigned a "guessed-age" to each drawing. The data below show the four age groups together with the average age guessed for each child from his drawings.

AVERAGE AGES GUESSED FOR CHILDREN,
ARRANGED ACCORDING TO ACTUAL AGE.

5.87	6.46	6.01	6.09
4.49	4.80	5.03	5.87
5.83	4.12	6.10	7.25
4.83	4.13	5.13	5.77
5.60	5.25	6.75	6.51
5.18	5.78	5.27	6.69
4.85	5.59	5.99	6.40
4.77	4.64	6.19	6.10
5.79	4.56	5.90	7.40
4.41	4.66	5.60	5.59
4.23	6.09	6.90	6.09
5.62	5.87	5.09	5.87
4	5	6	7

X = ACTUAL AGE

Is there a significant linear relation between a child's actual age X, and the average age guessed for him by the panel of teachers Y?

13. Use the data or problem 12 to set up a regression equation for predicting Y from X. What is the value of the sample standard error, $S_{Y \cdot X}$? Set up the 95 percent confidence limits for the value of the regression coefficient for a population of such drawings judged by the teachers.

14. In a study of the relation between the age at which study of the piano was begun and the eventual proficiency of the student after five years practice, a random sample of 100 students, each just completing his fifth year of piano-study, was selected. Each student was given the same piece to play, and a panel of judges rated his performance. Summary statistics emerging from the final data are given below:

$$\Sigma X = 1475 \qquad \Sigma Y = 12890$$
$$\Sigma X^2 = 24459 \qquad \Sigma Y^2 = 1714421$$
$$\Sigma XY = 186659$$

Establish the 95 percent confidence limits for the true correlation between age at beginning piano lessons X, and fifth-year proficiency Y. Test the hypothesis that the true correlation is $-.50$.

15. Suppose that a person began piano lessons at age 35. On the basis of the data in problem 14, predict his proficiency at age 40. Given that proficiency after five years of study is normally distributed, establish the 95 percent confidence limits for this person's actual proficiency.

16. Comment on the assumptions made in establishing confidence limits in problems 14 and 15. Do these assumptions appear particularly reasonable in this example? Explain.

17. In a study of factors related to success in medical school, the dependent variable Y was

an overall numerical rating of success during the four years of study. The independent variable was a measure of the "authoritarian personality" characteristics of the individual student. Students were arranged into four groups, from very low to very high, on this personality trait. Finally a concomitant variable X was used, representing the average anxiety level of the student. Carry out an analysis of covariance on the resulting data.

AUTHORITARIANISM							
VERY LOW		LOW		HIGH		VERY HIGH	
X	Y	X	Y	X	Y	X	Y
30	23	22	17	27	21	19	23
30	20	26	20	26	20	24	28
24	20	24	19	30	24	28	32
27	19	28	23	27	19	30	36
23	15	26	18	28	20	29	33
28	20	25	22	27	19	24	30

18. For the data of problem 17, test the hypothesis of equal regression among the four experimental groups.

19. In an investigation of the relation between the height of an adolescent boy and a measure of his physical stamina, three random samples of 40 boys each were used. In the first sample, all boys were 15 years old; in the second, all were 16 years old, and in the third, the boys were 17 years of age. The following correlations were obtained:

SAMPLE 1	SAMPLE 2	SAMPLE 3
$r = .11$	$r = .23$	$r = .19$
$N_1 = 40$	$N_2 = 40$	$N_3 = 40$

Test the hypothesis that the correlation between height and stamina is the same for the three populations of boys represented by these samples. (Hint: use the r to Z transformation.)

20. Sometimes data are presented in the form of a bivariate frequency distribution, in which one variable, X, is grouped into class-intervals and represented as columns in the data table. Similarly, the other variable, Y, is grouped into class-intervals and represented by the rows in the data table. Each observation in the sample then qualifies for one class-interval for X, and one class interval for Y, thus falling into one and only one cell in the table. Each cell in the completed table shows the frequency in some pair of (X,Y) intervals. How would you go about computing the correlation between X and Y from a bivariate frequency distribution such as that given below? Remember that the X distribution can be found by summing the frequencies contained in each column, and the frequency distribution for Y by summing within the rows. Hence, M_X and S_X can be found in terms of the X midpoints, and M_Y and S_Y can be found from the Y midpoints (see chapter 3). Furthermore, the frequency of a cell, multiplied by the product of midpoints for that cell, can be found and the result summed over all cells. See if you can complete the calculation of r for this bivariate table:

FREQUENCIES OF JOINT (X,Y) INTERVALS

Y MIDPOINTS						f
16	0	0	1	2	3	6
14	0	2	3	6	2	13
12	4	3	10	13	7	37
10	4	8	17	13	4	46
8	5	11	5	4	2	27
6	3	6	2	1	0	12
	25	30	35	40	45	X midpoints
	16	30	38	39	18	f

Y (left side label)

X (bottom label)

21. Test the significance of the correlation coefficient found from the data of problem 20.
22. Find the two regression equations for predicting Y from X, and X from Y, using the results of problem 20. Then calculate the sample standard errors of estimate accompanying these two regression equations.

16 Other Topics in Regression and Correlation

16.1 CURVILINEAR REGRESSION

All the discussion in Chapter 15 was centered on linear regression, the use of a linear function rule for the prediction of Y from X. However, the theory of "regression" is much more extensive than the preceding discussion of linear regression might suggest. Indeed, the linear rule for prediction is only the simplest of a large number of such rules that might apply to a given statistical relation. Linear regression equations may serve quite well to describe many statistical relations that are roughly like linear functions, or that may be treated as linear as a first approximation. Nevertheless, there is no law of nature requiring all important relationships between variables to have a linear form. It thus becomes important to extend the idea of regression equations to the situation where the relation is *not* best described by a linear rule. Now we are going to consider problems of curvilinear regression, problems where the best rule for prediction need not specify a simple linear function.

Our study of nonlinear prediction rules will be confined to *problems of regression*, as defined in the last chapter. An independent variable X is identified, and various values of X are chosen in advance to be represented in the experiment. Ordinarily, each X value corresponds to an experimental treatment administered in some specific quantity. A regression problem where a curvilinear rule for prediction might make sense is the following: there is an investigation of the effect of environmental noise on the human subject's ability to perform a complex task. Several experimental treatments are planned, each treatment differing from the others only in the intensity of background noise present while the subject performs the task. Each treatment represents a one-step interval in the intensity of noise within a particular range. A group of some N subjects chosen at random are assigned at random to the various groups, with n subjects per group. In the experiment proper, each subject works on the same problem individually in the presence of the assigned intensity of background noise. The dependent variable is thus a subject's score on the task, and the independent variable is the noise intensity. The experimenter is interested not only in the possible existence of a linear relation between noise intensity and performance, but also a possible curvilinear relationship between X and Y.

In the next section, a general model for dealing with curvilinear as well as linear regression will be explored. Using this model we will see first how a test may be made for the existence of a nonlinear relationship. Then we will deal with the problem of inferring the form of the statistical relation by use of planned comparisons among means. Finally, inferences about form of relation made from post hoc comparisons will also be considered.

16.2   THE MODEL FOR LINEAR AND CURVILINEAR REGRESSION

In Chapter 15, inferences about the existence and extent of linear regression were made in terms of the model

$$Y_{ij} = \mu_Y + \beta_{Y \cdot X}(X_j - \mu_X) + e_{ij},$$

which we saw to be an instance of the fixed effects model for analysis of variance. In adopting this model, we assume that the only systematic tendency of Y to vary with X is due to the linear regression of Y on X.

Now we will extend this model to allow for the existence of other kinds of systematic dependence of Y on X. The model we will use can be stated as follows:

$$Y_{ij} = \beta_0 + \beta_1(X_j - \mu_X) + \beta_2(X_j - \mu_X)^2 + \cdots$$
$$+ \beta_{J-1}(X_j - \mu_X)^{J-1} + e_{ij}. \quad [16.2.1*]$$

That is, we assume that the value of Y might depend on the value of $(X_j - \mu_X)$, of $(X_j - \mu_X)^2$, and, indeed, on any power of $(X_j - \mu_X)$ up to and including the $J - 1$ power, each term weighted by a constant β_1, β_2, and so on.

Some intuitive feel for this model may be gained as follows. Suppose that it were really true that the best possible prediction of Y_{ij} were afforded by

$$Y'_{ij} = \beta_0 + \beta_1(X_j - \mu_X) + \beta_2(X_j - \mu_X)^2.$$

However, the experimenter does not know this, and he applies the simple linear rule:

$$Y'_{ij} = \beta_1(X_j - \mu_X) + \mu_Y.$$

Does this describe all the predictability possible in the relationship? No, since if β_2 is not zero, the *deviations* of the predicted from the obtained values of Y should be predictable using the rule

$$d_{ij} = (Y_{ij} - Y'_{ij}) = \beta_2(X_j - \mu_X)^2 + e_{ij} + (\beta_0 - \mu_Y).$$

In short, the deviations from linear regression here represent not only error, but also some further predictability depending on the *squared* value of $(X_j - \mu_X)$. We should expect that, when plotted, the deviations from linear regression for the various sample groups should tend to lie along a parabola, much as in Figure 16.2.1. The constant β_2 is a coefficient of **second-degree or quadratic regression.**

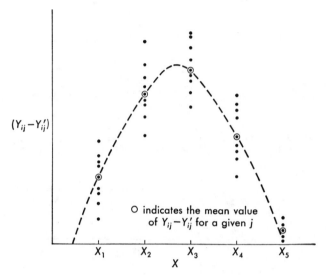

FIG. 16.2.1. Plot of deviations from linear regression, suggesting the presence of second-degree or quadratic regression

We need a way to decide if β_2 really is zero if we are going to tell whether or not the best prediction rule possible is at least of the second degree.

Extending this idea, suppose that the true relation between X and Y is given by the rule

$$Y_{ij} = \beta_0 + \beta_1(X_j - \mu_X) + \beta_2(X_j - \mu_X)^2 + \beta_3(X_j - \mu_X)^3 + e_{ij}.$$

However, the experimenter actually predicts using either a first- or a second-degree regression equation. Suppose that he uses an equation of the second degree. Then, nevertheless, the deviations from the predicted Y_{ij} values will tend to be systematically related to X values, since

$$(Y_{ij} - Y_{ij}') = \beta_3(X_j - \mu_X)^3 + e_{ij} + \beta_0'.$$

Unless β_3 is zero, cubic trends exist between X and Y, and the plot of deviations from prediction for the various populations should tend to fall into a cubic function, something like that shown in Figure 16.2.2.

The same general ideas hold for any number of experimental groups each associated with a fixed value X_j. A regression equation of the first, second, third, and so on up to the degree $J - 1$ is assumed possible. It can be shown, however, that for any finite number J of groupings, each associated with some fixed value of X_j, the best prediction equation possible in this model is of *no higher*

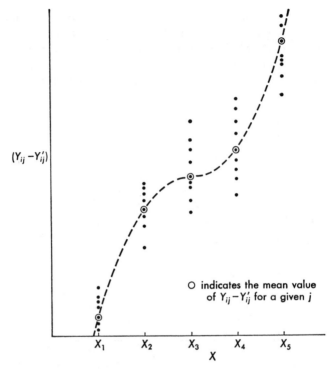

$(Y_{ij} - Y'_{ij})$

O indicates the mean value
of $Y_{ij} - Y'_{ij}$ for a given j

X_1 X_2 X_3 X_4 X_5

X

FIG. 16.2.2. Deviations from linear or quadratic regression, suggesting the presence of third-degree or cubic regression

degree than $J - 1$. Thus, a first-degree equation always suffices to describe the relation between X values and *mean* Y values for two groups, a second-degree equation suffices to describe the relation between X and mean Y for three groups, a third-degree equation for four groups, and so on.

Incidentally, the student who has had elementary calculus may recognize that this model corresponds to the expansion of a function $y = f(x)$ by a Taylor series. That is, for a large class of functions, the value of y associated with any x may be found from a sum of the form

$$f(x) = c_0 + c_1(x - a) + c_2(x - a)^2 + c_3(x - a)^3 + \cdots$$

where a, c_0, c_1, and so on are appropriately chosen constants. This fact has a bearing on how we interpret the results of an analysis under this model: when we infer the existence of a trend of a certain degree under this model, all we are really able to say is that a power series or expansion carried to a particular degree is capable of describing the relation obtained from the data. We are not really entitled to talk of the "real" function rule that may, in some sense, underlie what we observe, but only that when the function rule is expressed as a weighted sum of powers in X, an equation of a certain degree is sufficient to describe the relation we find. It is not necessarily true that the underlying process, whatever it may be, that generated the data is somehow conceptually equivalent to a sum of powers. This point will be important when we begin to discuss the various trends that appear to exist in a set of data, as in Section 16.8.

Regardless of the number J of different X_j groups, this model can still be regarded as a special version of the fixed effects analysis of variance model. Here, the effect ξ (small Greek xi) associated with group j can be thought of as

$$\xi_j = \sum_{k=0}^{J-1} \beta_k (X_j - \mu_X)^k$$

and

$$Y_{ij} = \mu_Y + \xi_j + e_{ij}.$$

This suggests that the machinery of fixed effects analysis of variance applies to problems of curvilinear as well as linear regression, and we will proceed in this way. However, an adequate justification of the F tests here is somewhat beyond our scope, and we will have to be content with an analogy to the simple fixed-effects analysis of variance.

16.3 ANOTHER LOOK AT THE PARTITION OF THE SUM OF SQUARES FOR REGRESSION PROBLEMS

In Section 15.17 it was mentioned that for any regression problem at all,

SS total = (SS error) + (SS deviations from linear regression)
+ (SS linear regression).

Under the model used in the last chapter, nothing could contribute either to SS error or to SS deviations from linear regression except random errors, e_{ij}. Consequently, these two sums of squares were pooled and used to estimate σ_e^2.

Now, however, we have a different model, which permits other, nonlinear, kinds of relations to exist. Nevertheless, the SS error still should reflect only error variance in the data. Furthermore, the SS linear regression depends only on the true regression coefficient β_1 and error, just as before. On the other hand, the status of the SS deviations from linear regression changes in this new model: this sum of squares reflects all the possible *nonlinear* relations as well as random error.

We saw in Section 15.18 that the SS linear regression can be thought of as the sum of squares for a comparison among the means of the J different groups. Now notice once again that for any set of distinct J groups

SS between groups = (SS linear regression)
+ (SS deviations from linear regression)   [16.3.1†]

since SS total and SS error are found in the ordinary way regardless of whether the problem is one of regression or a simple analysis of variance. Since SS linear regression is based upon a comparison among means, SS deviations from regression represents all the $J - 2$ remaining comparisons among means *orthogonal* to the comparison for linear regression. In this model, the systematic differences between group means can reflect either linear or curvilinear association, and SS deviations from linear regression can reflect only *curvilinear* regression and error.

This can be shown in terms of the expected mean squares: in the discussion of the analysis of variance, we defined the index of *total* strength of relationship as

$$\omega^2 = \frac{\sigma_Y^2 - \sigma_e^2}{\sigma_Y^2} = \frac{\displaystyle\sum_j n_j \alpha_j^2}{N\sigma_Y^2}.$$

Furthermore, in Chapter 15, we found that the strength of linear relationship is indexed by

$$\rho_{XY}^2 = \frac{\sigma_Y^2 - \sigma_{Y \cdot X}^2}{\sigma_Y^2}.$$

It will always be true that $\omega^2 \geqq \rho_{XY}^2$.

The strength of curvilinear relationship must then be indexed by the difference

$$\omega^2 - \rho_{XY}^2, \qquad\qquad\qquad [16.3.2*]$$

the proportion of variance in Y accounted for by X, but not by linear regression. From the results of Section 15.18 we can deduce that

$$E(\text{SS linear regression}) = \sigma_e^2 + N\rho_{XY}^2 \sigma_Y^2. \qquad [16.3.3]$$

We already know from Section 12.6 that

$$E(\text{SS between groups}) = (J - 1)\sigma_e^2 + \sum_j n_j \alpha_j^2, \qquad [16.3.4]$$

or, in terms of ω^2,

$$E(\text{SS between groups}) = (J - 1)\sigma_e^2 + N\omega^2 \sigma_Y^2. \qquad [16.3.5]$$

Putting these two results, **16.3.3** and **16.3.5,** together we have

$$
\begin{aligned}
E(\text{SS deviations from linear regression}) &= E(\text{SS between} \\
- \text{SS linear regression}) &= (J - 2)\sigma_e^2 + (\omega^2 - \rho_{XY}^2)N\sigma_Y^2. \quad [16.3.6]
\end{aligned}
$$

The SS deviations from linear regression reflects only systematic differences between population group means not due to linear regression. We should expect SS deviations from linear regression to reflect only error when β_2, β_3, and so on are *all* zero. If no curvilinear regression exists in the population, then

$$E(\text{MS deviations}) = \frac{E(\text{SS deviations})}{J - 2} = \sigma_e^2. \qquad [16.3.7*]$$

16.4 Testing for linear and nonlinear regression

Given the usual assumptions for the one-way fixed effects analysis of variance it can be shown that an F test for the hypothesis

$$H_0\colon \omega^2 - \rho_{XY}^2 = 0$$

becomes possible. In terms of the foregoing argument, this is the hypothesis of no *curvilinear* regression. This hypothesis may be tested apart from the hypothesis

$$H_0: \rho_{XY}^2 = 0,$$

so that one can ask *both* if any linear regression exists *and* if any curvilinear regression exists in the population (that is, predictability not afforded by a linear rule).

The actual F test employed to test for linear regression is different in this model from that in Section 15.21. Here, we take

$$F = \frac{\text{MS linear regression}}{\text{MS error}} \qquad [16.4.1\dagger]$$

with 1 and $N - J$ degrees of freedom to test for the existence of linear regression ($\rho_{XY}^2 \neq 0$, or $\beta_1 \neq 0$).

The test for curvilinear regression is given by

$$F = \frac{\text{MS deviations from linear regression}}{\text{MS error}} \qquad [16.4.2\dagger]$$

with $J - 2$ and $N - J$ degrees of freedom.

A convenient way to find the SS linear regression directly is

$$\text{SS lin. reg.} = \frac{N \left[\sum_j \sum_i x_j y_{ij} - \left(\sum_j n_j x_j \right) \left(\sum_j \sum_i y_{ij} \right) / N \right]^2}{N \left(\sum_j n_j x_j^2 \right) - \left(\sum_j n_j x_j \right)^2} \qquad [16.4.3\dagger]$$

although the method of Section 15.24 can be used. The separate calculation of the SS deviations from linear regression is actually the only new computational feature. The quantities needed for the analysis are

1. SS total (formula **12.10.1**)
2. SS between groups (formula **12.10.2**)
3. SS linear regression (formula **16.4.3**)
4. SS error = SS total − SS between
5. SS deviations from linear reg. = SS between − SS linear reg.

The analysis ordinarily is summarized in a table such as the following:

Source	*SS*	*df*	*MS*	*F*
Between groups		$J - 1$	—	—
Linear regression		1	$\dfrac{\text{SS lin.}}{1}$	$\dfrac{\text{MS lin.}}{\text{MS error}}$
Deviations from linear regression		$J - 2$	$\dfrac{\text{SS dev.}}{J - 2}$	$\dfrac{\text{MS dev.}}{\text{MS error}}$
Error		$N - J$	$\dfrac{\text{SS error}}{N - J}$	
Total		$N - 1$		

A significant value for the first of these F tests indicates that it is safe to conclude that some predictability is afforded by a linear rule. In the same way, a significant value for F for the second test lets one conclude that some further prediction is possible using a curvilinear rule. However, neither of these tests guarantees that there exists either a strong linear nor a strong curvilinear relationship in the population sampled, but only that from the evidence, $\rho_{XY}^2 > 0$, and that $\omega^2 - \rho_{XY}^2 > 0$. Nevertheless, for regression problems we can form at least rough estimates for both of these population indices, as shown in Section 16.5.

Naturally, the simplest and most readily available estimate of the form of relation in the data is given by a plot of the Y scores against the X values. In many instances a regression curve drawn "by eye" on the scatter plot suffices to describe the general form of predictive relation that appears to exist. One can regard the test for curvilinear regression as a license to infer that the curvilinear regression line that appears to fit the data also is a good bet about the population. However, much more sophisticated methods exist for judging the form of the prediction equation after the test for curvilinear regression has proved significant. One of the simplest of these methods will be outlined in Section 16.12. A more general treatment of the problem of fitting curvilinear regression lines will be found in Lewis (1960).

16.5   ESTIMATION OF THE STRENGTH OF LINEAR AND CURVILINEAR RELATIONSHIP FROM DATA

Using the model of Section 16.2, we have seen that

$$E(\text{SS lin. regression}) = \sigma_e^2 + N\rho_{XY}^2\sigma_Y^2$$

and

$$E(\text{MS error}) = \sigma_e^2,$$

so that

$$E(\text{SS lin. reg.} - \text{MS error}) = N\rho_{XY}^2\sigma_Y^2.$$

Then an estimate of ρ_{XY}^2 may be found from

$$\text{est. } \rho_{XY}^2 = \frac{\text{SS lin. reg.} - \text{MS error}}{\text{SS total} + \text{MS error}}, \qquad [16.5.1\dagger]$$

since

$$E(\text{SS total} + \text{MS error}) = N(\sigma_e^2 + \omega^2\sigma_Y^2) = N\sigma_Y^2.$$

Furthermore, by the same line of argument, a reasonable estimate of $\omega^2 - \rho_{XY}^2$, the proportion of variance accounted for by curvilinear regression, is given by

$$\text{est. } (\omega^2 - \rho_{XY}^2) = \frac{\text{SS deviations from lin. reg.} - (J - 2) \text{ MS error}}{\text{SS total} + \text{MS error}}. \qquad [16.5.2\dagger]$$

Once again, when MS deviation is less than MS error, so that the F for curvilinear regression is less than 1.00, the estimate is taken to be zero.

Notice that

$$\text{est. } (\omega^2 - \rho^2_{XY}) + \text{est. } \rho^2_{XY} = \text{est. } \omega^2, \qquad [16.5.3\dagger]$$

the estimated total strength of relationship.

16.6 THE CORRELATION RATIOS $\eta^2_{Y \cdot X}$ AND $\eta^2_{X \cdot Y}$

Our index ω^2 has been defined only in terms of the strength of association between the independent and dependent variables in some population. Similarly, the difference $(\omega^2 - \rho^2_{XY})$ reflects the tendency toward a curvilinear relationship in the population considered.

However, for some purposes it is convenient to have a descriptive index of total relationship *in a given set of data*. For this purpose the index $\eta^2_{Y \cdot X}$ is sometimes used (Greek eta, squared). The definition of $\eta^2_{Y \cdot X}$ is

$$\eta^2_{Y \cdot X} = \frac{\sum_j n_j (M_{Yj} - M_Y)^2}{\sum_j \sum_i (y_{ij} - M_Y)^2} \qquad [16.6.1^*]$$

which is exactly the same as

$$\eta^2_{Y \cdot X} = \frac{\text{SS between groups}}{\text{SS total}} \qquad [16.6.2\dagger]$$

where each group corresponds to one X_j value. The index $\eta^2_{X \cdot Y}$ is often called the **correlation ratio** for the relation of Y to X.

If Y is regarded as the independent variable, and the X values in the data are arranged by Y groupings, then another index $\eta^2_{X \cdot Y}$ may be found,

$$\eta^2_{X \cdot Y} = \frac{\sum_k n_k (M_{Xk} - M_X)^2}{\sum_k \sum_i (x_{ik} - M_X)^2} \qquad [16.6.3^*]$$

where X_{ik} is the value of the ith individual in the kth category, corresponding to a Y value Y_k. Ordinarily, these two indices will give different values for a given set of data, unless $r^2_{XY} = 1$, so that $r^2_{XY} = \eta^2_{Y \cdot X} = \eta^2_{X \cdot Y} = 1$.

So long as the group we are dealing with is, in fact, the population, the two indices ω^2 and $\eta^2_{X \cdot Y}$ are absolutely identical. However, the utility of the correlation ratio is rather limited if one desires more than a descriptive statistic for a sample. This is the reason that the symbol ω^2 has been used to discuss the *population* strength of total association in all of the foregoing: the correlation ratio $\eta^2_{Y \cdot X}$ usually applies only to a sample, and is usually discussed only when both variables X and Y are numerical. Applications of the two different η^2 indices make sense only when the sample is drawn as for a correlation problem. On the other hand the index ω^2 applies only to a population, and can be discussed even when the independent variable is categorical in character.

The correlation ratio has a direct connection with the F test for the presence of any association (i.e., H_0: true $\eta^2_{Y \cdot X} = 0$). Under this hypothesis,

$$F = \left(\frac{N - J}{J - 1}\right)\left(\frac{\eta^2_{Y \cdot X}}{1 - \eta^2_{Y \cdot X}}\right) \qquad [16.6.4^*]$$

actually is distributed as F with $J - 1$ and $N - J$ degrees of freedom. However, for the fixed effects or regression models, this is not true for any other hypothesis about the true correlation ratio.

Furthermore, the sample value of the correlation ratio is not an especially good way to estimate the true correlation ratio, and about the best one can do is to estimate ω^2 as we have shown. In general, our estimate of ω^2 gives a lower estimate of total association than does η^2; considering the error in the sample, this is not a bad feature of our estimate to have. Since $\eta^2_{Y \cdot X}$ makes most sense as a descriptive statistic, and since talking about degree of dependence *within a sample* has rather limited utility, we will continue to deal here only with the population quantity ω^2.

16.7 AN EXAMPLE OF TESTS FOR LINEAR AND CURVILINEAR REGRESSION

Consider the example mentioned in Section 16.1. Six different levels of noise intensity are employed. Each level represents a one-step interval in a scale of intensity, although this need not be true in general for this procedure. The exact units of noise intensity need not bother us here, and we can represent the levels of X as 1, 2, 3, and so on, since the different groupings represent equal intervals in noise intensity. The dependent variable is the score Y_{ij} of a subject in a complex performance given the noise intensity level X_j. It is desired to test both for linear and for curvilinear regression, using $\alpha = .01$. The scores are shown in Table 16.7.1.

Table 16.7.1

		Noise intensity levels, X_j			
1	*2*	*3*	*4*	*5*	*6*
18	34	39	37	15	14
24	36	41	32	18	19
20	39	35	25	27	5
26	43	48	28	22	25
23	48	44	29	28	7
29	28	38	31	24	13
27	30	42	34	21	10
33	33	47	38	19	16
32	37	53	43	13	20
38	42	33	23	33	11
270	370	420	320	220	140

The usual computations for a one-way analysis of variance are carried out first:

$$\text{SS total} = (18)^2 + (24)^2 + \cdots + (11)^2 - \frac{(1740)^2}{60}$$

$$= 7252$$

$$\text{SS between} = \frac{(270)^2 + \cdots + (140)^2}{10} - \frac{(1740)^2}{60}$$

$$= 5200$$

$$\text{SS error} = \text{SS total} - \text{SS between} = 7252 - 5200 = 2052.$$

Next the SS for linear regression is found from equation **16.4.3**. Since $n_j = 10$ for each group, this becomes

$$\text{SS lin. regression} = \frac{60 \left[\sum_j \sum_i x_j y_{ij} - \left(\sum_j 10 x_j \right) \left(\sum_j \sum_i y_{ij} \right) / 60 \right]^2}{60 \left(\sum_j 10 x_j^2 \right) - \left(\sum_j 10 x_j \right)^2}.$$

Here

$$\sum_j \sum_i x_j y_{ij} = \sum_j x_j \left(\sum_i y_{ij} \right) = 1(270) + 2(370) + \cdots + 6(140) = 5490$$

$$\sum_j 10 x_j = 10(1) + \cdots + (10)(6) = 210$$

$$\sum_j 10 x_j^2 = 10(1 + 4 + \cdots + 36) = 910$$

so that

$$\text{SS lin. reg.} = \frac{60[5490 - (210)(1740)/60]^2}{60(910) - (210)^2}$$

$$= 2057.1.$$

Then

$$\text{SS dev. from lin. reg.} = \text{SS between} - \text{SS lin. reg.}$$
$$= 5200 - 2057.1$$
$$= 3142.9.$$

The completed summary table is thus

Source	SS	df	MS	F
Between groups	5200	5	—	—
Linear reg.	2057.1	1	2057.1	54.1
Dev. from lin.	3142.9	4	785.7	20.7
Error	2052	54	38	
Totals	7252	59		

On evaluating these two F tests, we find that each is significant far beyond the .01 level. In short, we can reject both the hypothesis

$$H_0: \rho_{XY} = 0,$$

and the hypothesis

$$H_0: \omega^2 - \rho_{XY}^2 = 0,$$

although we should recall from Section 14.26 that the probability of error in one or both of these conclusions is not necessarily .01. Nevertheless, here we may say with some confidence that both linear and curvilinear regression exist. The best rule for population prediction apparently involves both X_j and powers of X_j, in terms of this model.

A plot of the Y means for these data supports our findings (Figure 16.11.1). The sample means tend, roughly, to fall along a straight line with negative slope (the dotted line in the figure). However, they would cluster even more closely along a curved line suggesting a parabolic arc. The experimenter would be likely to conclude that, in the range studied, relatively low levels of noise intensity may actually facilitate the performance of subjects in this task, but that beyond a particular point (about level 3) average performance falls off rapidly with further increases in noise intensity.

We may apply our estimation methods (**16.5.1** and **16.5.2**) to obtain some idea of the strength of linear and curvilinear relationship in the population sampled. Using **16.5.1**, we find

$$\text{est. } \rho_{XY}^2 = \frac{2057.1 - 38}{7252 + 38}$$

$$= \frac{2019.1}{7290}$$

$$= .28.$$

From the evidence at hand, it appears that about 28 percent of the variance in Y may be attributable to *linear* relationship with X, the noise levels.

Furthermore, using **16.5.2,** we have

$$\text{est. } (\omega^2 - \rho_{XY}^2) = \frac{3142.9 - (4)(38)}{7290}$$

$$= \frac{2990.9}{7290}$$

$$= .41.$$

Something on the order of 41 percent of the variance of Y is attributable to *curvilinear* relationship with X. Notice that we estimate a considerably stronger curvilinear than linear relationship in the population, even though the F tests were both very significant.

Finally, the total contribution of X to the variance of Y, or ω^2, is estimated to be

$$\text{est. } \omega^2 = \text{est. } \rho_{XY}^2 + \text{est. } (\omega^2 - \rho_{XY}^2) = .69$$

so that we infer quite a strong statistical association to exist between X and Y.

16.8 PLANNED COMPARISONS FOR TREND: ORTHOGONAL POLYNOMIALS

The separation of sums of squares into components representing linear and nonlinear regression is but the simplest of the procedures that may be applied to problems of linear and curvilinear regression. Many times an experiment is set up to answer certain specific questions about the *form* of the relation existing between two variables. In general when the experimenter wants to look into the question of trend, or form of relationship, he has some prior ideas about what the population relation should be like. These hunches about trend often come directly from theory, or they may come from the extrapolation of established findings into new areas. At any rate, specific questions are to be asked about the *degree* of the regression equation sufficient to permit prediction.

The technique of planned comparisons studied in Chapter 14 lends itself directly to this problem. Given J groups, each of which is associated with a value X_j of the experimental variable, each of the $J - 1$ degrees of freedom for comparisons can be allotted to the study of *one question* about the form or degree of the predictive relation. In this discussion, we will restrict our attention to the situation where the following conditions are satisfied.

1. The various X_j values represent *equally spaced unit intervals* on the X dimension or continuum. The units may be anything, but the difference between each X_j and X_{j+1} is precisely one of these units.
2. The number of observations made, n_j, is the same for each value of X_j.
3. Inferences are to be made about a hypothetical population in which *only possible* values of X_j are those actually represented in the experiment.

The first two of these requirements are not especially severe limitations, since this technique is ordinarily applied to experimental studies where one is free to choose the X_j values and the n per group quite arbitrarily. In such studies it is usually more convenient than otherwise to form equal spacings of X_j values, and to assign individuals at random and equally among the groups. The third restriction is quite serious, however. In uses of this method, **really nothing can be said about values of** X_j **that are not directly represented in the experiment, in so far as statistical inference is concerned.** The form of the relation may be quite different for X outside of the range represented, or even for potential X_j values intermediate to two actually represented. It is up to the experimenter to exercise caution and good scientific judgment in extrapolating beyond the X_j values he has actually observed. The statistician cannot really help him here, since this is a question of scientific, not statistical, inference. On the other hand, the question of errors in the X variable is potentially quite serious, as was pointed out in Section 15.24. This method is really not applicable when such errors exist, unless their magnitude is so small as to be negligible. In the following we will simply assume that such errors do not exist.

We have already seen (Section 15.18) that in the regression model the sum of squares for *linear* regression corresponds to the SS for one possible comparison among the sample means. Extending this idea, an orthogonal comparison can also be made reflecting second-degree or quadratic relationship. Still another orthogonal comparison can be carried out representing the third-degree, or cubic, relationship, and so on, until the experimenter has either investigated all the specific questions of interest, or $J - 1$ orthogonal comparisons have been made (whichever happens first). As for any set of *planned* comparisons, particular values from among the $J - 1$ possible comparisons are tested separately instead of the over all F test of between group means. If any one of these comparisons is significant, then the experimenter has relatively strong evidence that the corresponding β value in the population is not zero, and that the best prediction rule is at least of the degree represented by that comparison.

The weights which each trend comparison involves are completely dictated by J, the number of different X_j groups, and by the degree of the particular trend investigated. That is, given any number of groups J, one standard set of weights exists for investigating linear, or first-degree, trends, an orthogonal set exists for second-degree, or quadratic, trends, another for cubic, or third-degree, trends, and so on, up to and including trends of degree $J - 1$. These standard sets of comparison weights are called **coefficients of orthogonal polynomials,** and the method itself is sometimes called **the method of orthogonal polynomials.**

The general theory underlying orthogonal polynomials is much too advanced for us to go into here. Suffice it to say that the **orthogonal polynomial coefficients are so derived that the particular comparisons among means each represent one and only one kind of possible trend or form of relationship in the data.** This sounds much more complicated than it actually is. Perhaps a simple example may clarify this point. For four groups, equally spaced with regard to the X_j values, the three sets of standard orthogonal polynomial weights are

Comparison	M_1	M_2	M_3	M_4
linear	-3	-1	1	3
quadratic	1	-1	-1	1
cubic	-1	3	-3	1

The symbol M_1 denotes the mean Y value for the group with the *lowest* X value, or X_1, M_2 the mean Y for the next lowest value of X, or X_2, and so on. Applying the weights given in the first row to the means, we get the first, or *linear* comparison:

$$\hat{\psi}_1 = -3(M_1) - 1(M_2) + 1(M_3) + 3(M_4).$$

The sample value of $\hat{\psi}_1$ is an unbiased estimate of the *population* comparison

$$\psi_1 = -3(\mu_1) - 1(\mu_2) + 1(\mu_3) + 3(\mu_4).$$

If this population comparison is not equal to zero, then there is at least a linear trend in the relationship. Higher order trends may be present, however.

By applying the weights of the second comparison to the sample means, we get

$$\hat{\psi}_2 = 1(M_1) - 1(M_2) - 1(M_3) + 1(M_4)$$

which estimates ψ_2, the same linear combination applied to the population means. This comparison value in the population can be zero only if $\beta_2 = 0$, so that no quadratic trend exists in the statistical relationship. If this comparison value in the population is other than zero, then at least a quadratic trend exists in the statistical relationship.

Finally, the comparison for cubic trends is

$$\hat{\psi}_3 = -1(M_1) + 3(M_2) - 3(M_3) + 1(M_4)$$

This comparison estimates ψ_3. If the population value is other than zero, a cubic trend exists in the relationship.

If all three comparisons, ψ_1, ψ_2, ψ_3, are zero when applied to population values, then $\omega^2 = 0$, and no predictive relationship at all exists in the data.

Now suppose that we apply these comparisons to some hypothetical population situations. First of all, suppose that it were really true that

$$Y_{ij} = \beta_0 + \beta_1(X_j - \mu_X) + e_{ij}$$

which implies that the predictive relationship is truly linear, and that the population means must fall on a straight line when plotted against the corresponding

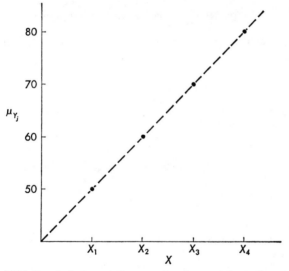

FIG. 16.8.1. Hypothetical population means in a linear relation to the value of X

X_j values. Suppose that the true plot of (X_j, μ_{Y_j}) pairs were as shown in Figure 16.8.1. Applying the standard weights for the linear comparison to the population means we have

$$\psi_1 = -(3)(50) - 1(60) + 1(70) + 3(80) = 100,$$

giving a large comparison value. However, if we apply the other two sets of weights to form the quadratic and the cubic comparisons, we find

$$\psi_2 = 1(50) - 1(60) - 1(70) + 1(80) = 0$$

and

$$\psi_3 = -1(50) + 3(60) - 3(70) + 1(80) = 0$$

reflecting the fact that the *only* predictive relation is linear.

However, suppose that it is really true that only a quadratic relation exists, following the rule

$$Y_{ij} = \beta_0 + \beta_2(X_j - \mu_X)^2 + e_{ij}.$$

Here, the true means must all lie on a *parabola* when plotted against X_j values (Figure 16.8.2).

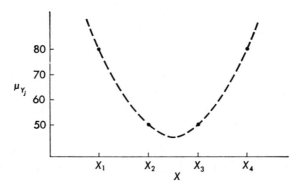

FIG. 16.8.2. Population means in a quadratic relation to the value of X

Suppose that the four population means were 80, 50, 50, and 80, as shown in the figure. Then the *linear* comparison gives

$$\psi_1 = -3(80) - 1(50) + 1(50) + 3(80) = 0$$

and the *cubic* gives

$$\psi_3 = -1(80) + 3(50) - 3(50) + 1(80) = 0.$$

However, in this instance, the *quadratic* comparison on the population means is

$$\psi_2 = 1(80) - 1(50) - 1(50) + 1(80) = 60,$$

showing that the predictive relation is purely of the second degree.

Finally, imagine that the relation is such that prediction from X to Y is afforded by a rule of the third or cubic degree, such as

$$Y_{ij} = \beta_0 + \beta_3(X_j - \mu_X)^3 + e_{ij}.$$

As a fairly extreme example, suppose that the plot of population means against X_j values showed the form pictured in Figure 16.8.3. In this figure the population means are 60, 80, 50, and 70. Then we have

$$\psi_1 = -3(60) - 1(80) + 1(50) + 3(70) = 0$$
$$\psi_2 = 1(60) - 1(80) - 1(50) + 1(70) = 0$$
$$\psi_3 = -1(60) + 3(80) - 3(50) + 1(70) = 100$$

reflecting only the cubic trend in the relationship.

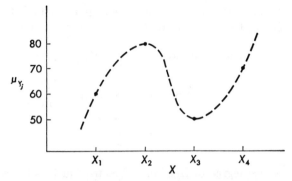

FIG. 16.8.3. A pronounced cubic trend among population means

This example may give you some feel for how these particular comparisons among sample means, used as estimates for population comparisons, give information about the general form of the predictive relationship, if any, between X and Y. Large and significant sample comparison values are evidence that a trend of at least that particular degree exists in the population relation.

One more thing should be pointed out: these weights used in the example are bona fide comparison weights, since the weights for each comparison sum to zero. Furthermore, the comparisons are all orthogonal, since the summed products of weights for any two comparisons is also zero. Thus the theory we have studied for any set of planned orthogonal comparisons applies perfectly well to trend comparisons of this type.

Table VII in Appendix C gives the required weights for trend comparisons for $J = 3$ through $J = 15$, and for trends through the sixth degree, which certainly should cover most psychological research on such problems. One simply finds the appropriate section of the table, and reads off the required weights. W thin each section, the rows represent the X_j groupings, with the lowest value of X_j shown as the first or topmost row. The columns represent the *degree* of trend to be examined. Then the entry in any row j and column h gives the weight applied to the mean of the jth group in the comparison for a trend of degree h. A larger table of required weights is given in *Biometrika Tables for Statisticians* (1967).

16.9 COMPUTATIONS FOR TREND ANALYSIS USING ORTHOGONAL POLYNOMIALS

In planned comparisons for trend made on data from single-factor experiments, one proceeds just as for any problem involving planned comparisons. First, an ordinary one-way analysis of variance is performed, although the over-all F test for the analysis is not carried out. The SS error is needed for testing the individual trend comparisons, and the SS between is useful either as an over-all check or for finding the pooled SS for all comparisons not of special interest by themselves.

If the linear trend ($h = 1$) is to be tested, one finds the set of weights c_{1j} from Table VII and the appropriate J, or number of groups. Then

$$\text{linear comparison} = \sum_j c_{1j} M_j = \hat{\psi}_1 \qquad [16.9.1\dagger]$$

and

$$\text{SS} (\hat{\psi}_1) = \text{SS linear} = \frac{\hat{\psi}_1^2}{w_1} \qquad [16.9.2\dagger]$$

where

$$w_1 = \sum_j \frac{c_{1j}^2}{n}. \qquad [16.9.3\dagger]$$

The SS linear has *one* degree of freedom, of course, just as for any comparison among means.

Then if there is interest in testing for quadratic trends ($h = 2$) one finds the set of weights c_{2j} from Table VII. The comparison is

$$\text{quadratic comparison} = \sum_j c_{2j} M_j = \hat{\psi}_2 \qquad [16.9.4\dagger]$$

$$\text{SS} (\hat{\psi}_2) = \text{SS quadratic} = \frac{\hat{\psi}_2^2}{w_2} \qquad [16.9.5\dagger]$$

where

$$w_2 = \sum_j \frac{c_{2j}^2}{n}. \qquad [16.9.6\dagger]$$

The number of degrees of freedom is again 1.

This general procedure is followed until as many of the comparisons as have a-priori interest for the experimenter have been made. These comparisons should actually be *planned* before the data are seen, and should really correspond to questions that are germane to the experimental problem and will clarify the interpretation of the experiment by their separate examination. For more than four or five experimental groups, it is seldom practical or useful to work out all the possible $J - 1$ trend comparisons. Usually, there is little understanding to be gained by finding, for example, that a sixth-degree or higher trend is significant. Ordinarily one stops after the cubic or perhaps quartic comparison, and then relegates all other trend effects to the pooled sum of squares:

$$\text{SS other trends} = \text{SS between} - \sum_{h=1}^{G} \text{SS} (\hat{\psi}_h) \qquad [16.9.7*]$$

where h symbolizes any from the G comparisons made separately. Thus, for five groups, one might be interested only in linear and quadratic trends. He would then find the pooled SS for other trends from

$$\text{SS (cubic and quartic)} = \text{SS between} - \text{SS lin.} - \text{SS quad.}$$

which has $4 - 2$ or 2 degrees of freedom.

In some applications of this method to problems of curve-fitting, in which one desires to find the trend of the lowest possible degree fitting the data satisfactorily, each successive comparison is tested against pooled SS for all comparisons representing trends of higher degree than the one tested. This method is presented quite clearly in Graybill (1961, page 182) but we shall not go into it here. In our example the value obtained for SS error will be used in the test for each trend comparison.

16.10   An example of planned comparisons for trend

Suppose that the experiment on noise intensity-level and performance had actually been planned as a study of trend. The experimenter is especially interested in judging if linear and quadratic trends exist in the relation between X and Y. We will pretend that the data in Section 16.7 had been collected specifically for this purpose.

The SS total, the SS between, and the SS error are found just as before, of course. However, let us list the group means and find the linear and

Table 16.10.1

	Group					
	1	2	3	4	5	6
Means	27	37	42	32	22	14
Weights						
linear	−5	−3	−1	1	3	5
quadratic	5	−1	−4	−4	−1	5

quadratic comparison values. Table 16.10.1 gives the six means and the weights that apply to each for linear and quadratic comparisons, as found from Table VII. The comparison values themselves are

$$\hat{\psi}_1 = -5(27) - 3(37) - 1(42) + 1(32) + 3(22) + 5(14)$$
$$= -120,$$

and

$$\hat{\psi}_2 = 5(27) - 1(37) - 4(42) - 4(32) - 1(22) + 5(14)$$
$$= -150.$$

The w values for the corresponding sums of squares are found from

$$w_1 = \frac{25 + 9 + 1 + 1 + 9 + 25}{10} = 7$$

and

$$w_2 = \frac{25 + 1 + 16 + 16 + 1 + 25}{10} = 8.4$$

so that

$$\text{SS linear} = \frac{(-120)^2}{7} = 2057.1$$

$$\text{SS quad.} = \frac{(-150)^2}{8.4} = 2678.6$$

(Notice that SS linear agrees exactly with the SS linear regression found in Section 16.7.) Then

$$\text{SS other trends} = 5200 - 2057.1 - 2678.6 = 464.3$$

with $\qquad J - 1 - 2 = 3$ degrees of freedom.

The summary table for this analysis is

Source	SS	df	MS	F
Between	5200	5	—	—
Linear	2057.1	1	2057.1	54.1
Quadratic	2678.6	1	2678.6	70.5
Other trends	464.3	3	154.8	4.0
Error	2052	54	38	
Totals	7252	59		

The F tests for both linear and quadratic trend greatly exceed the value required for the .01 level of significance, but the over-all test for other trends does not reach the .01 level. Thus, we can be fairly sure that in the population predictive association does exist between X and Y, and that the general rule for prediction includes both linear and quadratic components. We cannot, however, really be sure that the population prediction rule does not include higher order components as well. Even so, we infer from these data that a curvilinear regression equation affording at least some prediction has the general form

$$Y'_{ij} = \beta_0 + \beta_1(X_j - \mu_X) + \beta_2(X_j - \mu_X)^2$$

where the β_1 and β_2 are the linear and quadratic regression coefficients in the population.

16.11   ESTIMATION OF A CURVILINEAR PREDICTION FUNCTION

Sometimes it is convenient to find the actual pairings of X_j and predicted Y'_j values given by the curvilinear regression equation estimated from a sample. The experimenter may be interested in examining the departures his data tend to show from this estimated function, or perhaps in displaying in graphic form the relation suggested by the significant trends in his data. This is rather tedious to do in terms of the original X_j values, but when the X_j values are equally spaced a very simple substitute exists, involving the orthogonal polynomial coefficients once again. Instead of finding the predicted Y directly as a function of X_j, we can represent Y' as a function of the *polynomial weights* given to group X, and then use these predicted Y values to find and plot all (x_j, y'_j) pairs.

A rule that describes the systematic relation between X and mean Y in a set of data is

$$y'_j = M_Y + A_1c_{1j} + A_2c_{2j} + \cdots + A_{J-1}c_{(J-1,j)} = M_Y + \sum_{h=1}^{J-1} A_h c_{hj} \quad [16.11.1]$$

where c_{hj} is the orthogonal polynomial weight given to the jth group in the comparison for the trend of degree h. The constants A_h are found from

$$A_h = \frac{\hat{\psi}_h}{nw_h} \quad [16.11.2]$$

where $\hat{\psi}_h$ is the actual value of the h-degree comparison, and w_h is found just as in the calculation of the corresponding sum of squares (for example, **16.9.3** and **16.9.6**). Given that the values for each and every one of the $J - 1$ possible comparisons have been calculated correctly for the data, then by using this rule we will always find that

$$y'_j = M_{Yj}$$

the sample mean of the jth group is the value predicted by the rule. That is, the best possible estimate of the Y'_j value accompanying X_j is simply the *mean* of the jth group. This rule completely describes the function relating each *set* of c_j values for an X_j value to the corresponding sample M_{Yj} value. With a little more work, the rule itself could be put into terms of X_j, rather than c_{hj} values, although we shall not go into this here.

For most practical purposes, however, the experimenter wants to construct a function relating X_j to Y'_j, exhibiting only the trends he believes actually to exist in the population; that is, those corresponding to significant trend comparisons. Thus, the rule actually used has the form

$$y'_j = M_Y + \sum_{h \in G} A_h c_{hj} \quad [16.11.3\dagger]$$

where each h is one of the set G of *significant* trends in the data. With a table of orthogonal comparisons handy, the predicted Y'_j are very easy to find. Next we will illustrate this procedure for the data of Sections 16.7 and 16.10.

Recall first of all that for this example we concluded that *both* linear and quadratic trends exist. From the evidence at hand, these trends should be relatively strong in the population, and so we will construct a function relating Y'_j to X_j which reflects *only* linear and quadratic trends. First of all, we take the value already found for the linear comparison,

$$\hat{\psi}_1 = -120,$$

and divide by

$$nw_1 = (10)(7)$$

to find

$$A_1 = \frac{-120}{70} = -1.71.$$

Next we take the value for the quadratic comparison,

$$\hat{\psi}_2 = -150$$

and
$$nw_2 = (10)(8.4) = 84,$$

to find
$$A_2 = \frac{-150}{84} = -1.79.$$

Then our prediction rule for Y'_j, in terms of the **polynomial weights** c_{hj} is

$$y'_j = M_Y + A_1 c_{1j} + A_2 c_{2j}$$
$$= 29 + (-1.71)c_{1j} + (-1.78)c_{2j}.$$

Now suppose that we want to predict for the group with the *lowest* value of X or X_1. A look at Table VII for $J = 6$ shows that in the comparison for linear trends the mean of this group is weighted by -5, and in the comparison for quadratic by $+5$, so that

$$c_{11} = -5$$
$$c_{21} = 5.$$

For this group, the predicted value of Y'_j in terms of linear and quadratic trends is thus

$$y'_1 = 29 - (1.71)(-5) - (1.78)(5)$$
$$= 28.65.$$

Our best guess is that an observation in the lowest X-value group will have a Y score of about 28.65.

Now we predict for X_2, the second lowest value of X. Here, in the comparison for linear trends the weight assigned is -3, and for quadratic, -1. Thus

$$y'_2 = 29 - (1.71)(-3) - (1.78)(-1)$$
$$= 35.91.$$

Therefore, we predict the mean of the second group to be 35.91.

Going on in the same way for the other groups, we find the following functional pairings of X_j and *predicted* Y'_j values:

$$\text{for } x_1, \; y'_1 = 28.65$$
$$\text{for } x_2, \; y'_2 = 35.91$$
$$\text{for } x_3, \; y'_3 = 37.83$$
$$\text{for } x_4, \; y'_4 = 34.41$$
$$\text{for } x_5, \; y'_5 = 25.65$$
$$\text{for } x_6, \; y'_6 = 11.55.$$

These (x_j, y'_j) are plotted in Figure 16.11.1, together with the actual (x_j, M_{Yj}) pairings for the sample. The actual data points are shown as the heavy dots, and the fitted curve is drawn as *though* this function obtained over the whole range of X values. Notice that although the fit between the predicted values and the sample means is not perfect, since each mean point departs somewhat from its predicted position on the regression curve, the general form of the curve does

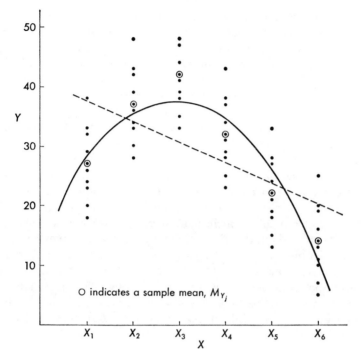

more or less parallel the systematic differences among the sample means. By way of contrast, the predictions using the linear regression equation alone correspond to points falling on the dotted line; it is apparent that adding a quadratic component to the prediction equation greatly improves our description of the relationship between X and Y exhibited in this sample, since the various means depart considerably from the regression line. On the other hand, if we added cubic, quartic, and quintic terms to our prediction equation, then we would find a curve on which the mean points for each group would fall exactly. There is not much profit in this, however, as the best description of a set of data is a curve fitting the points most nearly with a *simple* function rule (one of the lowest possible degree). Any set of means from J sampled groups can be fitted with a curve of at most degree $J - 1$.

This device for actually obtaining predicted points using a curvilinear regression equation is quite easy to apply after a trend analysis, since most of the computations have already been done. It is useful as a way to visualize how the kind of trend suggested by the significant comparisons might actually appear in a plot, as well as a device to dramatize the predictive strength, or the weakness, of the relation inferred from the data. Remember, however, that such an obtained regression curve is actually only a description of *these* data. *It can be regarded as an estimate of the population regression function only for hypothetical populations showing exactly the same representation of X values as used in the sample.* Free and easy extrapolation outside these values is done at the user's peril!

This example of a simple analysis for trend has been carried out in terms of a single set of distinctly different samples, each corresponding to one fixed value of the experimental variable X. Obviously, however, in many psychological settings this sort of question is studied more naturally within a more complex design, perhaps one in which repeated observations on the same subjects are made. This requires some special considerations that we will not be able to pursue here, and the reader is referred to the book by Winer (1962) for details.

A word of caution: curve fitting by use of orthogonal polynomials is but the simplest approach to the problem of fitting a function to a set of data. A number of other approaches to this problem exist. In particular, it must be borne in mind that the curvilinear regression function determined by this method depends on our ability to represent a functional relation by a power series, or Taylor expansion. Merely because we can reproduce the general form of relation in the data by, say, a function of the third degree, it does not follow that the true relation is best expressed in these terms. It may well be that another, and perhaps simpler, form of function rule best describes the relation, with parameters that can be linked to the psychological factors involved. In particular, when one has definite ideas about the mathematical form of the relation, and he wishes to estimate parameters for this function, other methods are called for. The book by Lewis (1960) gives a useful introduction to the general subject of curve-fitting for various purposes.

16.12   Trend analysis in post-hoc comparisons

It should be mentioned that the trend comparisons can perfectly well be made *after* the over-all analysis of variance has shown significant evidence for effects of the experimental variable. In particular, it may be informative to investigate trends of various degrees after significant evidence for over-all curvilinear regression has been found. The Scheffé method discussed in Chapter 14 applies to any post-hoc comparison among means, and one can use the orthogonal polynomial weights in the usual way to make post-hoc trend comparisons. However, just as for any set of post-hoc comparisons, the probability of our detecting a true trend as a significant result is ordinarily less in the test for a post-hoc than in the test for a planned comparison for that trend. Even so, post-hoc trend comparisons provide a most useful device for exploring data after the over-all F for between groups has shown significance, when the independent variable groupings are equally spaced quantitative distinctions. Unless the experimenter has definite questions about particular trends prior to the data, he ordinarily tests for trend by the post-hoc method. Estimation of the function can then be carried out as in Section 16.11.

16.13   Other uses of regression theory

We have considered only the barest rudiments of the theory of regression in this and the preceding chapter. The ideas of linear and curvilinear regression are capable of great extension and application to a wide variety of experimental problems. A very few of the directions of application of this general theory will be mentioned in this section.

In the first place, we have considered only relations following a linear prediction rule or those representable by rules involving powers of X (a *power series* in X). Obviously, many forms of relation may exist among variables, such as those predictable from the rules

$$Y = m[v^X]$$

or
$$Y = B \log (X) + a, \qquad X > 0,$$

to take only two simple examples. The methods of regression analysis can often be extended to the study of such relations as well. Usually, some transformation is made of the Y values to make the linear or power models fit: thus

$$\log_e Y = X(\log_e v) + \log_e m$$

may be studied by ordinary linear regression methods, even though

$$Y = m[v^X]$$

cannot. The basic methods of regression analysis can be extended in this way to cover many kinds of possible relations in data.

Another very important use of regression theory is also beginning to appear in psychological research. This involves the problem of testing a quantitative theory about two or more variables. Suppose that some theory exists which permits the deduction that the value of Y should be a function of X, or $y = g(x)$, other things being equal. The theory is sufficiently advanced for the quantitative rule for this function to be stated, and predicted values of Y obtained. The experimenter wants to check this deduction, but, obviously, other things will not be equal in his experiment, and so random error will exist, making his predicted and obtained values disagree. However, if the theory is true, and the uncontrolled variation is random, there should appear a strong *linear* relation between theoretical values of $g(x)$ and the obtained values of Y. Each experimental group has some X value, and hence affords a known value of $g(x)$ within the theory. Instead of testing for linear regression between Y and X, the experimenter tests for a linear relation between Y and the values given by the function of X, or $g(x)$ values. Then, a significant result and an estimated strong degree of association in the data is direct evidence for the theory's giving a good fit to data. Estimation is also applied to suggest ways in which the theory might be improved to give an even better fit. Evidence for curvilinear relationship between Y and $g(x)$ sheds even further light on the range within which the theoretical deduction seems to apply, and gives evidence for changes that might improve the goodness of the theory as a predictor of actual human or animal behavior. When dealing with an actual quantitative theory capable of precise deductions about Y values, this often is a far more sensible way for the experimenter to proceed than to cast the theory in the role of the null hypothesis. When the theoretical deduction itself corresponds to the null hypothesis a positive statement usually comes from significant results only if the theory is poor; proceeding by regression analysis between the predicted and obtained values makes significant linear association support the theory, and the experimenter can say something definite following a significance test and estimation procedure.

Nothing has been said so far about the comparisons of trends *among* populations. It may well be that the form of relation is different among different populations of experimental subjects, and the experimenter wants to discover this fact, if true. Especially important is the comparison of the form of apparent relationship between X and Y obtained in different variations of the same experiment. We have noted several times that the discovery and mathematical specification of functional relations involves weeding out variables that are irrelevant to the true relationship; if the form of obtained relation remains exactly the same over levels of some other variable, presumably that variable is irrelevant to the general form of relationship of X and Y. On the other hand, if the relation between X and Y shows a different form when another factor W is varied as well, then W *is* relevant, and the ability to predict Y in general may require knowledge of X *and* W. Observing the difference in form of relation across experiments or across populations is an important step in building up and testing a quantitative theory. Fortunately, the theory of planned comparisons among trends extends directly to this problem of intersample comparisons of trends, and so this question may also be studied statistically. This problem of cross-level comparison of trend is essentially that of *interaction* in a two-way or higher analysis of variance with quantitative factors. An introduction to this problem and elementary methods for its solution are given in Lewis (1960). Furthermore, it is entirely possible to extend the methods of the analysis of covariance, discussed in Chapter 15, so as to cover not only the linear regression of the dependent variable on one or more concomitant variables but also to consider nonlinear regression as well.

All of these extensions rest ultimately on the general ideas introduced here, however, and all were developed basically for the kinds of experiments we have called **problems in regression**. The novice experimental psychologist owes it to himself to develop some feel for regression techniques, and to become acquainted with some of the resources along these lines offered by modern statistics, although he always does well to seek expert help in applying the more advanced techniques. The book by Lewis (*op. cit.*) provides a good departure for the study of more advanced techniques in regression analysis and curve-fitting, and a fairly extensive survey of regression methods appears in Bennett and Franklin (1954), and in Anderson and Bancroft (1952). A thorough, though advanced, treatment of topics in regression theory is given by Graybill (1961, Chapters 8, 9, and 10).

16.14   Linear versus curvilinear regression

Past work in psychology dealing with statistical relations and the possibility of prediction has been cast largely in terms of linear regression. There

have been several reasons for this. As a practical first approximation to a complicated relationship, a linear function rule often serves quite well. Psychologists
have often been satisfied with assuming only linear regression because of the
crude state of many of our measurement techniques; it seems rather pointless
to read much meaning into the form of the relationship when the X and Y scores
themselves have some undetermined relationship to the quantities the psychologist hopes to measure. The needs of practical prediction may be met quite well
by a simple linear regression equation. In fact, the psychologist may even be
lucky if he can answer the question "Is any prediction at all possible?" without
going into the niceties of the form of the function rule that might serve best to
predict Y from X.

Another reason for the emphasis on linear prediction is that the
theory of correlation is extremely well developed along certain lines, some of
which will be explored in the next few sections. As we have seen, under the bivariate or multivariate normal assumptions, the absence of a linear relation is
the absence of any systematic relation. In the theory of normal correlation, linear
relations are really all one needs to study. Unfortunately, the distinction between
regression and correlation studies is not always a sharp one, and many psychologists seem unaware of the dependence of the classical theory of correlation on
normal theory. Some psychologists seem to operate on the principle that the
study of correlations must lead to the discovery of all important predictive relations running through data; in some areas this may well be true, but it is not
necessarily true in general, and seems an especially inappropriate assumption in
many experimental studies.

Finally, the theory of correlation and linear regression has important ties to several branches of psychological methodology not considered here,
such as test theory and factor analysis. This fact alone is sufficient to keep psychologists preoccupied with linear regression and correlation for some time to
come.

Nevertheless, the beginner in psychological research should not discount the possibility of important and predictive nonlinear relations between
variables. Current practice and practical necessity should not be confused with
the way nature or man behaves. There is no guarantee that all psychological
relationships of theoretical or applied interest must be linear in form. Presumably,
as our science emerges from its rather primitive beginnings into a truly quantitative experimental discipline, we will turn increasingly away from the crude
"Did anything happen?" approach to experimentation and toward problems of
form of relation and refinement of prediction. This seems to have been true in
the evolution of the other experimental sciences, and the author, for one, believes
that it is beginning to happen in psychology. The time seems to be approaching
for more estimation and quantitative specification of results in psychology, with
less emphasis upon the "Gee Whiz" or "shotgun" sort of study where everything
turns out both significant and meaningless.

On the other hand, at this stage of the game it is well for the experimenter to have good a-priori reasons for expecting particular kinds of predictive

relations in the data before he spends much time and effort in analyzing trends. Especially, the importance of a trend in a single set of data must not be over-emphasized. The same comment applies with even more force to the practice of fitting functions to data; these are nice ways to display findings, and the function obtained does describe *those data*, but means for any old set of data can always be fit perfectly with *some* function rule. Even the fact that some or all trends are found significant does not guarantee that the best description of the data is even remotely like the true form of relation. Functions fitted to data, like all other statistical devices, are only as good as the data to which they are applied. Careful study must be given to theory and great rigor must be exercised in the conduct of the experiment if the application of these relatively advanced methods is to uncover valid and predictive relations. Nonrandom samples and experimental errors can make monumental differences in conclusions reached about the form of a relationship from data. Furthermore, even with the most carefully collected data, the experimenter must be very leery about extending his findings beyond the domain actually studied in the sample. Functions that fit the values obtained perfectly can be wildly inappropriate over a larger range of X values, or even a more finely divided set of X intervals in the same range. Only if the experimenter has both good empirical evidence and sound theoretical reasons for doing so should he attempt to generalize beyond the range of X studied. Finally, the "true" form and the parameters entering into a relation are not always apparent from the data. A great deal of evidence must accumulate before firm statements about the precise form of relation and the relevant parameters of a predictive rule can be made. And yet, the situation is not hopeless; evidence *does* cumulate and theories *do* exist. Used with prudence, these methods for studying the possibility of prediction and functional form can give us hunches and insights to pursue, and from just such hunches and insights a quantitative and predictive science will eventually grow.

16.15 SOME RUDIMENTS OF MULTIPLE REGRESSION THEORY

In practical prediction situations, it is seldom that only one item of prior information is known about the individual subject. Usually several tests are given in an employment interview before the decision to hire or not to hire an applicant is made. The admissions officer in a college may have college entrance scores, high-school achievement records, educational level of parents, and a great many other items of information about a college applicant each of which may figure in the prediction made about his probable success in college. In other words, there are often several variables, with known values, each of which may contribute to the prediction about the value of the dependent variable.

In the same way, in multifactor experiments where each experimental factor is quantitative it may be desirable to investigate the extent to which all the factors considered together account for variance in Y. Furthermore, it may be important to be able to predict the value of Y given the *combination* of quantitative factor levels or experimental treatments administered to a subject.

The ability to predict from several variables considered simultaneously is studied in the theory of **multiple regression.** In this section we will consider only the barest rudiments of this theory, especially as it applies to a special kind of rule for predicting Y from values of several variables; we are going to consider only simple extensions of linear regression equations, and so this part of the theory is appropriately called **multiple linear regression.** In principle, multiple curvilinear regression can be studied as well, but this is far beyond our scope.

Some changes in notation will be necessary in discussing multiple regression. Imagine a set of K different variables, $\{X_k\}$, distinguished by the subscripts

$$(1, \; \cdots \;, k, \; \cdots \;, K).$$

The first of these variables, X_1, will be thought of, quite arbitrarily, as the dependent variable, and the others, $X_2, \; \cdots \;, X_K$, as the experimental or predictor variables for some particular problem. Similarly, relative to some fixed set of data, the standard score corresponding to X_1 will be denoted by z_1, that for X_2 by z_2, and so on.

Now we want to generalize the simple linear regression equation for one predictor variable to the situation where there are some $K - 1$ predictor variables. In our new notation, the linear regression equation found in Section 15.3 for standard scores becomes

$$z_1' = b_{1 \cdot 2} z_2, \tag{16.15.1}$$

where the constant $b_{1 \cdot 2}$ turns out to be r_{12}.

Suppose, however, that instead of two variables we were dealing with three: X_1, the dependent variable, and the two other variables X_2 and X_3. Then the linear rule for predicting the standard score on variable 1 from both the standard scores on 2 and 3 is simply

$$z_1' = b_{12 \cdot 3} z_2 + b_{13 \cdot 2} z_3. \tag{16.15.2*}$$

Here there are two constants, $b_{12 \cdot 3}$, reflecting the weight given the z_2 in predicting z_1, and $b_{13 \cdot 2}$, reflecting the weight given to z_3.

Just as in the two-variable situation, our job is to find values of these constants that will give the best predictions in the "least-squares" sense, making the average squared error in prediction a minimum. It is difficult to show the rationale for the required constants in terms of elementary methods alone, but we can at least suggest in a fairly crude way how these "best" values of $b_{12 \cdot 3}$ and $b_{13 \cdot 2}$ are determined.

Suppose that this prediction rule exists and that we find the covariance between the predicted values z_1' and the obtained values of z_2 for all individuals i; or

$$\text{cov.}(z'_1, z_2) = \frac{\sum_i z'_{1i} z_{2i}}{N}. \qquad [16.15.3]$$

Each z'_1 is thought of as a true z_1 plus some error: if these errors themselves have mean zero and are not correlated with z_2, then it must be true from **16.15.3** that

$$\text{cov.}(z'_1, z_2) = \frac{\sum_i (z_{1i} + e_{1i})(z_{2i})}{N}$$

$$= \frac{\sum_i z_{1i} z_{2i}}{N}$$

or

$$\text{cov.}(z'_1, z_2) = r_{12}, \qquad [16.15.4]$$

the covariance between the predicted z_1 scores and z_2 equals the correlation between z_1 and z_2. However, taken together with **16.15.2**, this implies that across individuals i,

$$r_{12} = \frac{\sum_i z'_{1i} z_{2i}}{N} = \frac{\sum_i z_{2i}(b_{12 \cdot 3} z_{2i} + b_{13 \cdot 2} z_{3i})}{N}$$

$$= \frac{\sum_i (b_{12 \cdot 3}) z_{2i}^2}{N} + \frac{\sum_i (b_{13 \cdot 2})(z_{2i} z_{3i})}{N}. \qquad [16.15.5]$$

Expression **16.15.5** reduces to

$$r_{12} = b_{12 \cdot 3} + b_{13 \cdot 2}(r_{23}), \qquad [16.15.6]$$

since the variance of a set of z scores is 1, and $\sum_i z_{2i} z_{3i}/N$ actually defines r_{23}, the correlation between X_2 and X_3 for this group of individuals.

Applying this same argument to the correlation r_{13}, we find that

$$r_{13} = b_{12 \cdot 3}(r_{23}) + b_{13 \cdot 2}. \qquad [16.15.7]$$

Given **16.15.6** and **16.15.7**, and the actual values of the three correlation coefficients, r_{12}, r_{13}, and r_{23}, found for any given set of data, the required b values can be found. That is, we have two equations in two unknowns:

$$b_{12 \cdot 3} \qquad + b_{13 \cdot 2}(r_{23}) = r_{12}. \qquad [16.15.8^*]$$
$$b_{12 \cdot 3}(r_{23}) + b_{13 \cdot 2} \qquad = r_{13}$$

Solving these equations for $b_{12 \cdot 3}$ and $b_{13 \cdot 2}$ gives us the constants we need for the actual regression equation. *These constants guarantee a least-squares fit between the predicted and the actual X_1 values for a given set of data.*

It is very simple to illustrate the way in which values of these constants are found for three-variable problems. Suppose that a personnel officer in an industrial plant has two items of information about each and every applicant for some speciälized job; his score on a standard intelligence test (X_2) and his score on a particular trade-skills test (X_3). Each person employed can, after a time, be given a proficiency score (X_1) indicating how he works out on the job. It is desired to find a regression equation for predicting standing on job proficiency from intelligence level and the trade-skills score. From the set of N people who have already been employed, the personnel officer finds that

$$r_{12} = .20$$
$$r_{13} = .40$$
$$r_{23} = .30.$$

Thus, the weights entering into the regression equation must be the solutions to the simultaneous linear equations represented by **16.15.8:**

$$b_{12 \cdot 3} + .30(b_{13 \cdot 2}) = .20$$
$$.30(b_{12 \cdot 3}) + b_{13 \cdot 2} = .40.$$

On solving these equations we have

$$3.333b_{12 \cdot 3} + b_{13 \cdot 2} = .667$$
$$.300b_{12 \cdot 3} + b_{13 \cdot 2} = .400$$

$$3.033b_{12 \cdot 3} \qquad\quad = .267$$
$$b_{12 \cdot 3} \qquad\quad = .088$$

and

$$b_{13 \cdot 2} \qquad\quad = .374.$$

Thus, the regression equation is found to be

$$z_1' = (.088)z_2 + (.374)z_3.$$

The regression weights b entering into this equation suggest that for optimal linear prediction of the standard score in work proficiency, the standard score on intelligence (z_2) should be weighted by much less $(.088)$ than that (z_1) for trade skills $(.374)$. Should a man show a standard score in intelligence of 1.5, and a trade-skills standard score of $-.3$, then the predicted standard score on job performance is found to be

$$z_1' = (.088)(1.5) + (.374)(-.3)$$
$$= .02,$$

almost at the mean. On the other hand, when $z_2 = -.3$ and $z_3 = 1.5$, $z_1' = .53$; the predicted score is considerably higher. The relative standing on trade skills is literally weighted more in prediction than is intelligence standing.

In general, for any finite number of variables K, the same general scheme is employed. An equation for predicting standard scores on X_1 is desired, of the form

$$z_1' = z_2(b_{12 \cdot 3 \cdots K}) + z_3(b_{13 \cdot 2 \cdots K}) + \cdots + z_K(b_{1K \cdot 2 \cdots K-1}). \quad [\textbf{16.15.9*}]$$

The predicted value of z_1 is to be a weighted sum of z values for variables 2 through K. The required weights are found from the simultaneous solutions to the $K - 1$ equations beginning with

$$(b_{12\cdot3\cdots K}) + r_{23}(b_{13\cdot2\cdots K}) + \cdots + r_{2K}(b_{1K\cdot2\cdots K-1}) = r_{12}$$

and ending with

$$r_{2K}(b_{12\cdot3\cdots K}) + r_{3K}(b_{13\cdot2\cdots K}) + \cdots + (b_{1K\cdot2\cdots K-1}) = r_{1K}.$$

The big computational chore in multiple regression problems with several variables consists of first finding all of the $\binom{K}{2}$ correlations needed, and then solving the $K - 1$ equations. Most older statistics books give considerable space to ways of solving these equations by hand. However, all signs point to such problems being handled by electronic computers in the future (as they already are in many research settings). Thus, we will not devote more space to the actual mechanics of solving large sets of simultaneous equations.

16.16   THE COEFFICIENT OF MULTIPLE CORRELATION

For a two-variable problem, we saw that r_{XY}^2 describes the proportion of variance in Y accounted for by linear regression on X. It is also possible to give r_{XY} a slightly different interpretation, however. Suppose that we found the correlation between the predicted Y' scores and the *actual* values of Y across all N subjects i:

$$r_{(Y'Y)} = \frac{\sum\limits_i z'_{Yi} z_{Yi}}{N S_{z_{Y'}}} = \frac{\sum\limits_i (r_{XY} z_{Xi} z_{Yi})}{N \sqrt{r_{XY}^2}}$$

$$= r_{XY}. \qquad [16.16.1^*]$$

Thus the correlation coefficient between X and Y is also the correlation between predicted and obtained values Y' and Y of the dependent variable.

Now we can extend this idea to define a **multiple-correlation coefficient.** For three variables, X_1, X_2, and X_3, suppose that predictions of X_1 were made according to **16.15.2.** Now we take the correlation between the predicted value z'_1 and the actual value z_1 over all individuals i:

$$r_{(z_1'z_1)} = \frac{\sum\limits_i z'_{1i} z_{1i}}{N S_{z_1'}} = \frac{\sum\limits_i (b_{12\cdot3} z_{2i} + b_{13\cdot2} z_{3i}) z_{1i}}{N S_{z_1'}}.$$

Although we shall not prove this result, it can be shown that

$$S_{z_1'} = \sqrt{b_{12\cdot3} r_{12} + b_{13\cdot2} r_{13}}$$

so that

$$r_{(z_1'z_1)} = \left(\sum_i b_{12\cdot3}(z_{2i} z_{1i}) + \sum_i b_{13\cdot2}(z_{3i} z_{1i}) \right) \Big/ N \sqrt{b_{12\cdot3} r_{12} + b_{13\cdot2} r_{13}}$$

$$= \sqrt{b_{12\cdot3} r_{12} + b_{13\cdot2} r_{13}}$$

$$= R_{1\cdot23}. \qquad [16.16.2^*]$$

The correlation of predicted and obtained scores for variable X_1 is the multiple-correlation coefficient, denoted by $R_{1 \cdot 23}$. The value of the multiple-correlation coefficient itself is found from **16.16.2** to be

$$R_{1 \cdot 23} = \sqrt{b_{12 \cdot 3} r_{12} + b_{13 \cdot 2} r_{13}}. \qquad [16.16.3^*]$$

For our three variable examples of Section 16.15, the multiple-correlation coefficient is

$$
\begin{aligned}
R_{1 \cdot 23} &= \sqrt{(.088)(.20) + (.373)(.40)} \\
&= \sqrt{.0176 + .1492} \\
&= \sqrt{.1668} \\
&= .408.
\end{aligned}
$$

In general, in K-variable problems, the squared value of the multiple-correlation coefficient turns out to be

$$R^2_{1 \cdot 2 \ldots K} = (b_{12 \cdot 3 \ldots K}) r_{12} + \cdots + (b_{1K \cdot 2 \ldots K-1}) r_{1K} \qquad [16.16.4^*]$$

the weighted sum of the correlation coefficients of each variable with variable 1, each coefficient weighted by the regression weight applied to that variable in the regression equation. Then the value of R is merely the positive square root of this R^2 value.

16.17 THE STANDARD ERROR OF ESTIMATE AND THE INTERPRETATION OF R^2

In a two-variable problem the sample standard error of estimate for predicting the z score of the dependent variable is, in our new notation,

$$S^2_{z_1 \cdot z_2} = \frac{\sum_i (z'_{1i} - z_{1i})^2}{N} = 1 - r^2_{12}. \qquad [16.17.1]$$

If we applied precisely the same idea to predictions using a multiple regression equation, we could find

$$
\begin{aligned}
S^2_{z_1 \cdot z_2 \ldots z_K} &= \frac{\sum_i (z'_{1i} - z_1)^2}{N} \\
&= 1 - R^2_{1 \cdot 2 \ldots K}. \qquad [16.17.2^*]
\end{aligned}
$$

This is the **sample variance of estimate for z scores**. The corresponding **sample standard error of estimate** is thus

$$S_{z_1 \cdot z_2 \ldots z_K} = \sqrt{1 - R^2_{1 \cdot 2 \ldots K}}. \qquad [16.17.3^*]$$

Just as in the two-variable situation, the value of the variance of estimate reflects the "poorness" of prediction, or *the proportion of variance in variable 1 not accounted for by variables 2, 3, and so on in terms of a linear prediction rule.* Hence, it must be true that the proportion of variance accounted for is given by

$$1 - S^2_{z_1 \cdot z_2 \cdots z_K} = 1 - (1 - R^2_{1 \cdot 2 \cdots K}) = R^2_{1 \cdot 2 \cdots K}. \qquad [16.17.4*]$$

Just as for r^2, **the squared multiple correlation coefficient indicates the proportion of variance in X_1 accounted for by the set of $K - 1$ remaining variables, in terms of predictions using this form of linear rule.** The index R^2 is sometimes called the **coefficient of multiple determination.**

We can never do less well in linear predictions using several variables than the best we can do using a single variable. This means that the value of R^2 is always greater than or equal to the largest value of r^2 between any single independent variable and the dependent variable. That is,

$$R^2_{1 \cdot 2 \cdots K} \geqq r^2_{1k} \qquad [16.17.5*]$$

for any variable k in the set of $K - 1$ *predictor* variables.

This fact lets us evaluate how much we have gained in ability to predict by the addition of more items of prior information. For example, in Section 16.15 we found that the value of $R_{1 \cdot 2 \cdots K}$ was only .408, so that the squared value is close to .17. However, the original correlation between the job-proficiency and trade-skills scores was .40, giving a value of .16 to r^2_{13}. This says that almost nothing is contributed to the ability to predict by the addition of an intelligence-test score to the trade-skills score. The personnel officer would apparently do about as well using the trade-skills score alone for prediction as he would using the multiple-regression equation.

16.18   MULTIPLE REGRESSION EQUATIONS
FOR RAW SCORES

Although the multiple-regression equation and the multiple-correlation coefficient have been presented here only in standard scores form, the same idea applies to raw scores. The raw score equivalent of formula **16.15.9** is

$$x'_1 = M_1 + b_{12 \cdot 3 \cdots K} \left(\frac{S_1}{S_2} \right) (x_2 - M_2) + b_{13 \cdot 2 \cdots K} \left(\frac{S_1}{S_3} \right) (x_3 - M_3)$$

$$+ \cdots + b_{1K \cdot 2 \cdots K-1} \left(\frac{S_1}{S_K} \right) (x_K - M_K). \qquad [16.18.1*]$$

For *raw* scores, the sample standard error of estimate is

$$S_{1 \cdot 2 \cdots K} = s_1 \sqrt{1 - R^2_{1 \cdot 2 \cdots K}}. \qquad [16.18.2*]$$

The value of the multiple-correlation coefficient is exactly the same, however, regardless of whether raw scores or standard scores are involved.

16.19   INFERENCES ABOUT MULTIPLE REGRESSION AND CORRELATION

Exactly as in our discussion of the descriptive statistics of simple correlation and regression, *one intact set of data are presumed in definitions of the descriptive statistics of multiple regression and correlation*. The regression equation, the value of R, and the standard error of estimate all refer to *this* set of data. The same ideas can be applied to a population, however, and we could define the true multiple-regression equation and the true R in terms of this population. The status of the sample regression equation relative to the population is the same as in two-variable problems: the sample regression equation is usually the best guess we can make about the particular rule that applies in the population, but we have no assurance that the multiple linear predictive relationship (or lack thereof) in the sample is not a chance result rather than a true reflection of the population situation. Thus, considerations of sampling become important, and under certain assumptions we can make inferences about the population.

Just as for simple regression theory, different assumptions and different kinds of inferences can be made, depending on whether the problem is one of regression or of correlation. The general fixed effects model of analysis of variance extends directly to multiple-regression problems, and permits a test of the hypothesis that true R is zero. It is far more usual, however, to regard the N cases as a sample from a *multivariate normal distribution*, each case representing an occurrence of some joint event $(x_1, x_2, \cdots, x_K)$. Under this assumption it can be shown that the ratio

$$F = \left(\frac{R^2}{1 - R^2}\right)\left(\frac{N - K}{K - 1}\right) \qquad\qquad [16.19.1\dagger]$$

is distributed as F with $K - 1$ and $N - K$ degrees of freedom, provided that true R^2 is zero. Here, N is the number of subjects, and K is the number of variables. This ratio tests only the hypothesis that the true value of R^2 is zero, however, which is ordinarily not a hypothesis of overwhelming interest. Methods exist for testing some other hypotheses about multiple regression, especially those having to do with increase in R^2 with the addition of more independent variables, but these methods are beyond our scope.

It must be emphasized that population R^2, like ρ^2, represents the strength only of a particular kind of relation among variables. If each and every variable from X_2 through X_K is statistically independent of X_1, then true R^2 must be zero. However, if the population distribution is not normal it is entirely possible for true R^2 to be zero even though some statistical relationship, and thus predictability, exists. Statements about association derived from multiple correlation and regression analyses apply only to this particular linear way of predicting X_1 from the other variables, and do not necessarily imply that all possible predictability is thereby summarized, unless the multivariate normal assumption holds.

16.20   Partial correlation

Complementary to the notion of multiple correlation is that of the partial correlation between two variables, with the effects of some third variable being held constant. That is, consider three variables X_1, X_2, and X_3. The three correlation coefficients r_{12}, r_{13}, and r_{23} are presumably known for the sample. However, in the computation of r_{12}, the value of X_3 is left completely free to vary; given the X_1 and X_2 values for an individual his X_3 score could, in principle, be anything.

Suppose, however, that the value of X_3 were fixed at some constant level. What would happen to the correlation between X_1 and X_2? This correlation might be different, because some of the apparent linear predictability of X_1 from X_2 (and vice versa) may be due to the association of each with X_3. For example, there is undoubtedly some positive correlation between body weight and the ability to read among normal school children. However, it is patently true that the tendency for such scores to covary arises in part from another factor strongly related to each, the child's chronological age. Could we hold chronological age constant, then this correlation between weight and reading ability should vanish, or at least be appreciably lowered. For studies centered on the influence of some extraneous variable (or variables) on the tendency of two other variables either to correlate or fail to correlate, the partial correlation coefficient is a very useful descriptive device.

The partial correlation coefficient $r_{12\cdot3}$ is the correlation between X_1 and X_2 adjusted for the linear regression of each on X_3. The partial correlation can be defined in this way: Suppose that z_1 were predicted from z_3, by

$$z_1' = r_{13}z_3.$$

Then the error in that prediction must represent the operation of all factors associated with variance in X_1 other than the linear association with factor 3. The variance of such errors is the variance in X_1 unaccounted for by the correlation r_{13}. In standard score terms, this variance is just

$$S_{z_1\cdot z_3}^2 = 1 - r_{13}^2.$$

Similarly, the error in prediction of z_2 from z_3 must reflect factors other than linear association with factor 3.

Then the tendency for linear association between variables 1 and 2, free of the linear association each has with variable 3, must be represented in the correlation of these errors in predicting each variable from variable 3. Using **15.11.1**, we can define the partial correlation as a correlation between errors in prediction by writing

$$r_{12\cdot3} = \frac{\sum_i (z_{1i}' - z_{1i})(z_{2i}' - z_{2i})}{N(s_{z_1\cdot z_3})(s_{z_2\cdot z_3})} \qquad [16.20.1*]$$

where both z_{1i}' and z_{2i}' are predicted from the individual's standard score z_{3i}. A little algebra lets us put this partial correlation coefficient in terms of the original correlations:

$$r_{12\cdot3} = \frac{\sum_i r_{13}r_{23}z_{3i}^2 - \sum_i r_{13}z_{3i}z_{2i} - \sum_i r_{23}z_{3i}z_{1i} + \sum_i z_{1i}z_{2i}}{N\sqrt{(1-r_{13}^2)(1-r_{23}^2)}}$$

$$= \frac{(r_{13}r_{23}) - (r_{13}r_{23}) - (r_{13}r_{23}) + r_{12}}{\sqrt{(1-r_{13}^2)(1-r_{23}^2)}} \qquad [16.20.2]$$

or

$$r_{12\cdot3} = \frac{r_{12} - (r_{13}r_{23})}{\sqrt{(1-r_{13}^2)(1-r_{23}^2)}}. \qquad [16.20.3\dagger]$$

This last expression is the usual computational form for a partial correlation coefficient based on the three original correlations. Obviously, we can define $r_{13\cdot2}$ and $r_{23\cdot1}$ in an analogous way.

Be sure to notice that the interpretation of the partial correlation coefficient as reflecting the extent of linear association between 1 and 2 with 3 held constant makes sense only if all the predictive relations among the variables are linear to begin with. Quite different relations between X_1 and X_2 may exist for different constant values of X_3 when the basic interrelations among the variables are not linear.

The partial correlation coefficient may be extended to four or more variables in the same general way. However, we shall not take the space to do this here, since these formulas can be found in many statistics books.

There is an intimate connection between partial correlation and multiple regression. The multiple-regression equation for three variables in Section 16.15 can also be written as

$$z_1' = (r_{12\cdot3})\left(\frac{S_{z_1\cdot z_3}}{S_{z_2\cdot z_3}}\right)(z_2) + (r_{13\cdot2})\left(\frac{S_{z_1\cdot z_2}}{S_{z_3\cdot z_2}}\right)(z_3). \qquad [16.20.4]$$

Observe that the b weight that a score z_2 receives depends upon the partial correlation between 1 and 2 with 3 held constant, and that for z_3 depends on the partial correlation $r_{13\cdot2}$. Each variable is weighted in terms of its unique linear contribution to the variance of X_1. Furthermore, it follows that partial correlations can be found directly from particular multiple-regression weights. For three variables, once again,

$$r_{12\cdot3} = \sqrt{(b_{12\cdot3})(b_{21\cdot3})} \qquad [16.20.5]$$

where $b_{12\cdot3}$ is the weight applied to z_2 in the multiple-regression equation for predicting z_1, and $b_{21\cdot3}$ is the weight applied to z_1 in the equation for predicting z_2. (Notice the similarity to the expression **15.9.6.**)

When a random sample of N cases can be assumed to come from a multivariate normal distribution, then it is permissible to use Fisher's r to Z transformation just as in Section 15.26 to test hypotheses and find confidence limits for the true value of the partial correlation. However, the standard error of Z becomes

$$\sigma_Z = \frac{1}{\sqrt{N - 3 - (K - 2)}},$$

where K is the total number of variables considered. Thus, for three variables $(K = 3)$, when the partial correlation $r_{12 \cdot 3}$ is used to test the hypothesis that $\rho_{12 \cdot 3}$ is, say 0, the value of $r_{12 \cdot 3}$ is changed to a corresponding Z value via Table VI, and the test is carried out using the normal distribution of the statistic

$$\frac{Z - 0}{[1/\sqrt{N - 3 - (3 - 2)}]} = Z[\sqrt{(N - 4)}].$$

If this statistic shows a value significant at the α level, then the hypothesis that $\rho_{12 \cdot 3} = 0$ can be rejected at the α level. All in all, when the problem is one of *correlation*, and the assumption of a multivariate normal population is tenable, any procedure using the r to Z transformation applicable to a simple correlation also applies to a partial correlation, provided that the standard error of Z is adjusted. However, the assumption of a multivariate normal distribution (all variables normally distributed, all pairwise joint events such as (x_1, x_2) bivariate normal, and so on) is very unreasonable in many instances, and such tests should not be undertaken lightly.

16.21   TESTING SIGNIFICANCE FOR INTERCORRELATIONS

Before we leave the topic of correlation, a word must be said about significance tests for intercorrelations. It is quite common to find research in psychology where a number of different variables are studied in the same sample, and all sample intercorrelations are found among these variables. For example, a study may concern three variables, X_1, X_2, and X_3, and values are found for r_{12}, r_{13}, and r_{23}. This in itself is fine as a description of linear relations in the data, and is the first step in virtually any multivariate analysis, such as finding a multiple-regression equation or carrying out a factor analysis.

However, one often finds the experimenter testing the significance of each one of these $\binom{K}{2}$ intercorrelations by the method of Section 15.26, as though each one were based on a different sample. The resulting significance levels are largely meaningless, for reasons much like those making t tests for all differences among a set of means a dubious procedure. In the first place, even for independent tests of significance, when so many tests are carried out the probability that some Type I errors are being made may be very high. Even worse, the t tests for correlations are quite redundant and are not statistically independent when carried out on a table of intercorrelations. Consequently, the set of results can be grossly misleading. In particular, one should ordinarily expect *more* than $\binom{K}{2}\alpha$ such tests to show significance by chance alone.

It is simple to illustrate that dependencies must exist among inter-correlations. Imagine a sample of N cases, each of which provides three score values X_1, X_2, and X_3. Imagine that r_{12} turns out to be $-.80$ and that r_{13} is also equal to $-.80$. What is the smallest value that r_{23} may be? We can determine this by examining the partial correlation coefficient for $r_{23 \cdot 1}$. Since we know that this partial correlation coefficient must be greater than or equal to -1, we take

$$r_{23 \cdot 1} = \frac{r_{23} - r_{12}r_{13}}{\sqrt{(1 - r_{12}^2)(1 - r_{13}^2)}}$$

$$= \frac{r_{23} - (-.80)^2}{[1 - (-.80)^2]} \geq -1.00,$$

and then

$$r_{23} - .64 \geq -1 + .64,$$

so that

$$r_{23} \geq .28.$$

The smallest value that r_{23} can possibly have, and still be consistent with r_{12} and r_{13} is thus .28. On the other hand, if $r_{12} = -.5$ and $r_{13} = -.5$, then r_{23} can also be as low as $-.5$. This value would make the partial correlation $r_{23 \cdot 1}$ be equal to exactly -1. Fixing the value of two of the correlations determines the necessary *lower* limit of the third. Higher-order patterns of dependency exist among any set of intercorrelations, and these patterns may be used to fix a lower bound for any subset of the intercorrelations. Such lower bounds may be calculated in a variety of ways. However, the following rule gives an absolute lower bound for the average of the intercorrelations.

In general, for K variables, the average of the $\binom{K}{2}$ intercorrela-tions among these variables must be greater than (or equal to) $-1/(K - 1)$. It follows that given the values of some of the intercorrelations, the average lower limit for all the other correlations is not -1, but some number greater than -1. The larger K is, the closer this lower limit comes to 0. Hence, it is somewhat pointless to treat each of the correlations in turn as though the sampling distri-bution of values could extend from -1 through $+1$, when with each successive value of r known from the sample the possible lower limit to the next set of values is raised. One should either not test for significance in the ordinary way in dealing with intercorrelations found for a single sample, or he should interpret the signifi-cance levels with *considerable* latitude.

It is rather hard to see why anyone would want to know if all the true intercorrelations among a set of variables are zero anyway. If these variables are to be used to predict some other variable, then a test of significance for R^2 is much more meaningful. If some other regression method is contemplated, appro-priate tests may also exist for the results of applying this method. Tracing rela-tionships among variables is the legitimate business of the scientist, but simply asking if *anything* relates linearly to anything else in a large set of variables is a pretty crude way to do business.

EXERCISES

1. Why is the discussion of the statistical theory of curvilinear regression limited to "regression problems" (as defined in chapter 15) rather than "problems of correlation?"
2. Show that the polynomial weights for linear, quadratic, and cubic comparisons are orthogonal when, say, the number of experimental groups is 7. Attempt to explain why the highest possible trend is of the sixth degree when there are only seven sample means.
3. Use expression 16.20.4 to prove the following:

$$1 - R^2_{1 \cdot 23} = (1 - r^2_{12})(1 - r^2_{13 \cdot 2})$$

From this result, prove that $R^2_{1 \cdot 23}$ cannot be less than either r^2_{12} or r^2_{13}.
4. Suppose that the intercorrelation among three variables could be $-.75$, $-.75$, and $-.75$. What would this imply about the partial correlation between the first and second variables, holding the third constant? Can this situation actually exist? What are the *smallest* intercorrelations that can exist among three variables, if all three intercorrelations are the same?
5. A study was carried out concerning the influence upon a subject's reaction time of the time-interval between warning signal and stimulus. Subjects were randomly assigned to six groups of nine subjects each and each group received a different time-interval between the warning buzzer and the stimulus. Then the mean reaction time of each subject was found. These data are shown in the table below:

TEST GROUPS

REACTION TIMES OF SUBJECTS, GROUPED BY EXPERIMENTAL CONDITIONS

.14	.19	.15	.12	.28	.29
.26	.19	.09	.22	.18	.23
.14	.08	.18	.07	.10	.17
.25	.14	.12	.23	.26	.25
.20	.10	.15	.17	.20	.18
.27	.17	.06	.07	.16	.17
.21	.16	.06	.13	.10	.27
.29	.24	.14	.15	.25	.20
.15	.09	.07	.25	.15	.27
0	.1	.2	.3	.4	.5

X = TIME INTERVAL, IN SECONDS

Test time significance of the linear trend in these data ($\alpha = .05$). Is there a significant curvilinear trend?

6. Estimate the relative proportions of variance in reaction time accounted for by a linear relation to signal-stimulus interval. Then estimate the proportion of variance accounted for by curvilinear relationship to the experimental variable. What total proportion of variance in reaction time seems to be accounted for by the experimental variable?
7. Carry out *post-hoc* comparisons and significance tests for trends of 2nd, 3rd, and 4th

degree in the data of problem 5, granted that this is justified by the F-test for overall curvilinear trends. (Hint: use orthogonal polynomials to make the comparisons, and then by the Scheffé method given in chapter 14.)

8. In a study of factors in alcoholism, samples of United States cities were taken, and the rate of verified alcoholism over the past ten years in each found. The sizes of the cities ranged from 10,000 through 109,000, and the cities were classified into five class-intervals according to population-size, as shown below. Five cities were selected at random from each population-grouping. The average alcoholism rate per 1000 population for each city over the past ten years is shown below:

ALCOHOLISM RATE, PER THOUSAND

17.57	19.08	18.75	20.07	24.74
16.35	18.02	17.99	19.01	23.30
17.37	18.89	18.78	20.16	24.09
16.26	18.03	18.08	19.29	22.86
14.19	17.26	18.59	18.48	23.90
10–29	30–49	50–69	70–89	90–109

X = POPULATION, IN THOUSANDS

Test for linear, quadratic, cubic, and quartic trends. What would you conclude about the relation of alcoholism-rate and size of city, based on these samples?

9. Based on the significant trends found in problem 8, find the curvilinear regression equation. Use the method of orthogonal polynomials to find the required equation. Plot the regression equation, along with the obtained data points for problem 8.

10. Make a freehand extrapolation of the regression curve found in problem 9, in order to guess at the alcoholism rate for a city of 1,000,000 population. Does this rate seem particularly reasonable? What does this illustrate about extrapolating findings beyond the range actually sampled in such studies?

11. For the data of problem 17, Chapter 15, let the group labeled "very low" be given the value 0, the "low" group the value 1, the value 2 for "high", and the value 3 for "very high." Consider these as values of a variable W. Now find the partial correlation between Y and W holding X constant. Does X appear to have an appreciable effect on the relationship between Y and W? Can one test the significance of this partial correlation according to the procedures given in Section 16.20? Why?

12. Once again assign values to the data of problem 17, Chapter 15, as in the problem above. Now find the multiple regression equation for predicting Y from X and W. How much is the predictive power improved by the addition of X to W, as contrasted with W alone?

13. In a study of 50 cases sampled at random, the correlations among three variables, X_1, X_2, and X_3, were as follows:
$$r_{12} = .38$$
$$r_{13} = .45$$
$$r_{23} = -.17$$

Find the multiple regression equation for the standardized score of an individual on variable X_1, given the standardized values on X_2 and X_3. If an individual had a standardized score of 1.9 on X_2 and -1.2 on variable X_3, what would you predict his standardized score to be on X_1?

14. Using the data of problem 13, find the multiple correlation and the proportion of variance in X_1 accounted for by multiple regression on X_2 and X_3. Test the significance of this multiple correlation.

15. Using the data of problem 13, find the partial correlation of variables X_1 and X_2, holding X_3 constant. Find the partial correlation of variable X_2 and X_3, holding X_1 constant.

16. Test the significance of these partial correlations found in problem 15 above.

17. Test the data of problem 6, chapter 15, for curvilinear regression. Estimate the true proportion of variance in Y due to curvilinear regression in these data.

17 Comparing Entire Distributions: Chi-Square Tests

Heretofore, problems of hypothesis testing and interval estimation have centered very largely on summary characteristics of one or more population distributions. That is, we have been concerned with the value of a population mean, the differences among two or more population means, the equality of two population variances, the value of a population regression coefficient, and so on. In almost every instance the hypothesis tested was composite, so that the null hypothesis itself did not really specify the population distribution or distributions exactly. Of course, it was necessary to make assumptions about the population distribution in order to arrive at the appropriate test statistics and sampling distributions for these various hypotheses. Nevertheless, our interest was really in comparing population values with hypothetical values, or several populations in particular ways. With the exception of problems dealing with a single proportion (the two-class "Bernoulli" situation) we have not really considered hypotheses about the **identity** of two or more population distributions.

You may recall, however, that in Chapter 4 the basic ideas of the independence of variables and of attributes were introduced in terms of the *absolute identity* of two or more distributions. Most of the statistical tests we have discussed for experimental problems really do pertain to this question of independence of variables or attributes, but they actually do so because of assumptions such as normality and homogeneous variances; the assumptions make the hypotheses tested about, say, means, be equivalent to hypotheses about identical distributions. If the populations can be said to differ in any summary characteristic, then their distributions cannot be identical.

There are research problems in which one wants to make direct inferences about two or more distributions, either by asking if a population distribution has some particular specifiable form, or by asking if two or more population distributions are identical. These questions occur most often when *both* variables in some experiment are qualitative in character, making it impossible to carry out the usual inferences in terms of means or variances. In these instances we need methods for studying independence or association from categorical data. Other situations exist, however, when we wish to ask if a population distribution of a random variable has some precise theoretical form, such as the normal distribution, without having any special interest in summary properties such as mean and variance.

The methods considered in this chapter all pertain in one way or another to this central problem: how does one make inferences about a population *distribution* in terms of the distribution obtained in the sample? Remember that population distributions may have some random variable as the domain, as in all the examples in Chapters 10 through 16, or the distribution may consist of a probability assigned to each of a set of mutually exclusive and exhaustive qualitative classes. As suggested in Chapter 4, such a set of mutually exclusive and exhaustive *qualitative* events is often called an **attribute.** The methods of this chapter were originally developed for the study of theoretical distributions on attributes, and especially for the problem of independence or association of attributes. However, as we shall see, even the distribution of a random variable may be studied by these methods if the domain of the distribution is thought of as divided into a set of distinct class intervals.

It stands to reason that the best evidence one has about a population distribution grouped into qualitative classes is the sample distribution, grouped in the same way. Presumably the discrepancy between sample and theoretical distribution should have some bearing on the "goodness" of the theory in the light of the evidence. Furthermore, the comment made at the end of Chapter 5 was not altogether casual: in principle, exact probabilities for various possible sample distributions can always be found by use of the multinomial or hypergeometric rules, given some discrete population distribution. As we shall see, the rationale for the tests in this chapter is based essentially on this idea.

The first topic of this chapter is the comparison of a sample with a hypothetical population distribution. We would like to infer whether or not the sample result actually does represent some particular population distribution. We will deal only with discrete or grouped population distributions, and our inference will be made through an *approximation* to the exact multinomial probabilities. Such problems are said to involve "goodness of fit" between a single sample and a single population distribution.

Next, we will extend this idea to the simultaneous comparison of several discrete distributions. Ordinarily, the reason for comparing such distributions in the first place is to find evidence for **association** between two qualitative attributes. In short, we are going to employ a test for independence between attributes, which can be regarded as based on the comparison of *sample* distributions.

Finally, we will take up the problem of measuring the strength of association between two attributes from sample data. Tests and measures of association for qualitative data are very important for psychology, where many of the most important distinctions made are, essentially, categorical or qualitative in character. The methods in this chapter are widely used, both because of the kinds of data psychologists collect, and because of the computational simplicity of their application. However, the theory underlying these tests is not simple, and misapplication of these tests is very common. For this reason a good deal of space will be devoted to discussing some of the basic ideas underlying these methods, and some pains will be taken to emphasize their inherent limitations.

17.1 Comparing sample and population distributions: goodness of fit

Suppose that a study of educational achievement of American men were being carried on. The population sampled is the set of all normal American males who are twenty-five years old at the time of the study. Each subject observed can be put into one and only one of the following categories, based on his *maximum* formal educational achievement:

1. college graduate;
2. some college;
3. high-school or preparatory-school graduate;
4. some high school or preparatory school;
5. finished eighth grade;
6. did not finish eighth grade.

These categories are mutually exclusive and exhaustive: each man observed must fall into one and only one classification.

The experimenter happens to know that ten years ago the distribution of educational achievement on this scale for twenty-five-year-old men was:

Category	Relative frequency
1	.18
2	.17
3	.32
4	.13
5	.17
6	.03

He would like to ask if the present population distribution on this scale is exactly like that of ten years ago. Therefore the hypothesis of "no change" in the distribution for the present population specifies the exact distribution given above. The alternative hypothesis is that the present population does differ from the distribution given above, in some unspecified way.

A random sample of 200 subjects is drawn from the current population of twenty-five-year-old males, and the following frequency distribution obtained:

Category	f_{oj} (obtained frequency)	f_{ej} (expected frequency)
1	35	36
2	40	34
3	83	64
4	16	26
5	26	34
6	0	6
	200	200

(These figures *are* hypothetical!) The last column on the right gives the *expected* frequencies under the hypothesis that the population has the same distribution as ten years ago. For each category, the expected frequency is

$$Np_j = f_{ej} = \text{expected frequency}$$

where p_j is the relative frequency for category j dictated by the hypothesis.

How well do these two distributions, the obtained and the expected, agree? At first blush, you might think that the difference in frequency obtained and expected across the categories, or

$$\sum_j (f_{oj} - f_{ej}),$$

would describe the difference in the two distributions. However, it must be true that

$$\sum_j (f_j - f_{ej}) = \sum_j f_{oj} - \sum_j f_{ej}$$
$$= N - N$$
$$= 0,$$

so that this is definitely not a satisfactory index of disagreement.

On the other hand, the sum of the *squared* differences in observed and expected frequencies does begin to reflect the extent of disagreement:

$$\sum_j (f_{oj} - f_{ej})^2.$$

This quantity can be zero only when the fit between the obtained and expected distributions is perfect, and must be large when the two distributions are quite different.

An even better index might be

$$\sum_j \frac{(f_{oj} - f_{ej})^2}{f_{ej}},$$

[17.1.1]

where each squared difference in frequency is weighted inversely by the frequency expected in that category. This weighting makes sense if we consider that a departure from expectation should get relatively more weight if we expect rather few individuals in that category than if we expect a great many. Somehow, we are more "surprised" to get many individuals where we expected to get few or none, than when we get few or none where we expected many; thus, the departure from expectation is appropriately weighted in terms of the frequency expected in the first place, when an index of over-all departure from expectation is desired.

Remember, however, the real purpose in the comparison of these distributions is to test the hypothesis that the expectations are correct, and that the current distribution actually is the same as ten years ago. One might proceed in this way: given the probabilities shown as relative frequencies for the hypothetical population distribution, the exact probability of any sample distribution can be found. That is, given the hypothesis and the assumption of independent random sampling of individuals (with replacement), the exact probability of a particular sample distribution can be found from the **multinomial** rule (Section 5.23). Thus, in terms of the hypothetical population distribution, the probability of a sample distribution exactly like the one observed is

p (obtained distribution$|H_0$)

$$= \frac{200!}{35!\ 40!\ 83!\ 16!\ 26!\ 0!}\ (.18)^{35}(.17)^{40}(.32)^{83}(.13)^{16}(.17)^{26}(.03)^0.$$

With some effort we could work this value out exactly. However, we are not really interested in the probability of exactly this sort of obtained distribution, but rather in all possible sample results this much or more deviant from expectation according to an index such as expression **17.1.1**. The idea of working out a multinomial probability for each possible such sample result is ridiculous for N and J this large, as an absolutely staggering amount of calculation would be involved.

When the theoretical statistician finds himself in this kind of impasse he usually begins looking around for an approximation device. In this particular instance, it turns out that the multivariate normal distribution provides an approximation to the multinomial distribution for very large N, and thus the problem can be solved. We will not go into this derivation here, as it is far beyond our capabilities; suffice it to say that the basic rationale for this test does depend on the possibility of this approximation, and that the approximation itself is really good only for very large N. Then the following procedure is justified:

We form the statistic

$$\chi^2 = \sum \frac{(f_{oj} - f_{ej})^2}{f_{ej}} \qquad\qquad \text{[17.1.2†]}$$

which is known as the Pearson χ^2 statistic (after its inventor, Karl Pearson). Given that the exact probabilities for samples follow a multinomial distribution, and given a very large N, **when H_0 is true this statistic χ^2 is distributed approximately as chi-square with $J - 1$ degrees of freedom.** Probabilities arrived at using this statistic are *approximately* the same as the exact multinomial probabilities we would like to be able to find for samples as much or more deviant from expectation as the sample obtained. The larger the sample N, the better should this approximation be.

Note that the number of degrees of freedom here is $J - 1$, the number of distinct categories in the sample distribution, or J, minus 1. You may have anticipated this from the fact that the sum of the differences between observed and expected frequencies is zero; given any $J - 1$ such differences, the remaining difference is fixed. This is very similar to the situation for deviations from a sample mean, and the mathematical argument for degrees of freedom here would be much the same as for degrees of freedom in a variance estimate.

To return to our example, the value of the χ^2 statistic is

$$\chi^2 = \sum_j \frac{(f_{oj} - f_{ej})^2}{f_{ej}}$$

$$= \frac{(35 - 36)^2}{36} + \frac{(40 - 34)^2}{34} + \frac{(83 - 64)^2}{64}$$

$$+ \frac{(16 - 26)^2}{26} + \frac{(26 - 34)^2}{34} + \frac{(0 - 6)^2}{6}$$

$$= 18.30.$$

This value is referred to in the chi-square table (Table IV) for $J - 1 = 6 - 1 = 5$ degrees of freedom. We are interested only in the upper tail of the chi-square distribution in such problems, because the only reasonable alternative hypothesis (disagreement between true population and hypothetical distribution) must be reflected in *large values* of χ^2. Table IV shows that for 5 degrees of freedom, a value of 11.07 cuts off the upper .05 of the distribution, and a value of 15.09 corresponds to the upper .01. The hypothesis that the current distribution of educational achievement is exactly like that of ten years ago may thus be rejected, either at the .05 or at the .01 level, as we choose.

Tests such as that in the example, based on a single sample distribution, are called "goodness-of-fit" tests.

Chi-square tests of goodness of fit may be carried out for any hypothetical population distribution we might specify, provided that the population distribution is discrete, or is thought of as grouped into some relatively small set of class intervals. However, in the use of the Pearson χ^2 statistic to approximate multinomial probabilities, it *must* be true that:

1. each and every sample observation falls into one and only one category or class interval;
2. the outcomes for the N respective observations in the sample are independent;
3. sample N is large.

The first two requirements stem from the multinomial sampling distribution itself: the multinomial rule for probability holds only for mutually exclusive and exhaustive categories, and for independent observations in a sample (random sampling with replacement). The third requirement comes from the use of the chi-square distribution to approximate these exact multinomial probabilities: this approximation is good only for large sample size. Furthermore, unless N is infinitely large, the Pearson χ^2 itself is not distributed exactly as the chi-square variable.

The fact that when H_0 is true the Pearson χ^2 statistic for goodness of fit is not distributed exactly as the random chi-square variable can be seen from the expected value and the variance of the Pearson χ^2, where ν symbolizes the degrees of freedom, $J - 1$:

$$E(\chi^2) = \nu \qquad\qquad\qquad [17.1.3]$$

$$\text{var.}\,(\chi^2) = 2\nu + \frac{1}{N}\left(\sum_j \frac{1}{p_j} - \nu^2 - 4\nu - 1\right). \qquad [17.1.4]$$

Recall that we learned in Chapter 11 that the expected value of a chi-square variable with ν degrees of freedom is ν, and that the variance is 2ν. Although the expected value of the Pearson χ^2 statistic is also ν, the variance of this statistic need not equal the variance of the random variable chi-square, unless N is infinitely large. This implies that the Pearson χ^2 is not ordinarily distributed as the chi-square variable for samples of finite size. Note that the expression for the variance of the Pearson χ^2 indicates that the goodness of the chi-square distribution as an approximation to the distribution of the Pearson statistic depends on several things, including not only the size of N but also the true probabilities p_j associated with the various categories and the number of degrees of freedom.

How large should sample size be in order to permit the use of the Pearson χ^2 goodness-of-fit tests? Opinions vary on this question, and some fairly sharp debate has been raised by this issue over the years. Many rules of thumb exist, but as a conservative rule one is usually safe in using this chi-square test for goodness of fit if each *expected* frequency, f_{ej}, is 10 or more when the number of degrees of freedom is 1 (that is, two categories), or if the expected frequencies are each 5 or more where the number of degrees of freedom is greater than 1 (more than two categories). We will have more to say about sample size and Pearson χ^2 tests in Section 17.7.

Be sure to notice, however, that *these rules of thumb apply to expected, not observed, frequencies per category.*

17.2 THE RELATION TO LARGE-SAMPLE TESTS OF A SINGLE PROPORTION

The goodness-of-fit test with 1 degree of freedom is formally equivalent to the large-sample test of a proportion, based on the normal approximation to the binomial. That is, imagine a distribution with only two categories.

Category j	Expected frequency	Obtained frequency	Expected proportion	Obtained proportion
1	f_{e1}	f_{o1}	p	P
2	f_{e2}	f_{o2}	q	Q
	$\overline{}$	$\overline{}$	$\overline{}$	$\overline{}$
	N	N	1.0	1.0

Suppose that the normal approximation to the binomial is to be used to test the hypothesis that the true population proportion in category 1 is p. Then we would form the test statistic

$$ z = \frac{NP - Np}{\sqrt{Npq}} = \frac{-(NQ - Nq)}{\sqrt{Npq}} \qquad [17.2.1]$$

or

$$ z = \frac{f_{o1} - f_{e1}}{\sqrt{f_{e1}(N - f_{e1})/N}}. \qquad [17.2.2]$$

For very large N, this can be regarded as a standardized normal variable. Now consider the *square* of this standardized variable z:

$$ z^2 = \frac{N(f_{o1} - f_{e1})^2}{f_{e1}(N - f_{e1})} $$

$$ = \frac{(f_{o1} - f_{e1})^2}{f_{e1}} + \frac{(f_{o1} - f_{e1})^2}{N - f_{e1}}. $$

Since $(f_{o1} - f_{e1}) = -(f_{o2} - f_{e2})$, we may write

$$ z^2 = \frac{(f_{o1} - f_{e1})^2}{f_{e1}} + \frac{(f_{o2} - f_{e2})^2}{f_{e2}} $$

$$ = \chi^2. \qquad [17.2.3*]$$

When the frequency distribution has only two categories, the Pearson χ^2 statistic has exactly the same value as the square of the standardized variable used in testing for a single proportion, using the normal approximation to the binomial. From the definition of a chi-square variable with 1 degree of freedom (Section 11.1), it can be seen that if $E(P) = p$, and if N is very large, then sample values of χ^2 should be distributed approximately as a chi-square variable with 1 degree of freedom. For a large sample and so long as a two-tailed test is desired, it is immaterial whether we use the normal-distribution test for a single proportion or the Pearson χ^2 test for a two-category problem. Furthermore, the square root of this χ^2 value gives the equivalent z value if one does desire a one-tailed test. This direct equivalence between z and $\sqrt{\chi^2}$ holds only for 1 degree of freedom, however.

Again, for 1 degree of freedom, the Pearson χ^2 test may be improved somewhat if the test statistic is found by taking

$$ \chi^2 = \frac{(|f_{o1} - f_{e1}| - .5)^2}{f_{e1}} + \frac{(|f_{o2} - f_{e2}| - .5)^2}{f_{e2}} \qquad [17.2.4\dagger]$$

so that the *absolute value* of the difference between observed and expected frequencies is reduced by .5 for each category before the squaring is carried out. This is known as **Yates' correction** and depends on the fact that the binomial is a discrete, and the normal a continuous, distribution. However, Yates' correction applies only when there is 1 degree of freedom.

17.3 A SPECIAL PROBLEM: A GOODNESS-OF-FIT TEST FOR A NORMAL DISTRIBUTION

One use of the Pearson goodness-of-fit χ^2 test is in deciding if a continuous population distribution has a particular form. That is, we might be interested in seeing if a sample distribution of scores might have arisen from some theoretical form such as the normal distribution. It is important to recall that like most theoretical distributions useful in statistics, the normal is a family of distributions, particular distributions differing in the parameters entering into the rule as constants; for the normal distribution, these parameters are μ and σ, of course. It is also important to remember that although such a distribution is continuous, it is necessary to think of the population as grouped into a finite number of distinct class intervals if the Pearson χ^2 test is to be applied.

For example, suppose that there is good reason to believe that intelligence scores on some test are normally distributed among American men in general. However, we are curious about the distribution of such intelligence scores for men serving time in prison. Is it reasonable to believe that this distribution is also approximately normal? The question refers not to the mean, nor to the variance of this distribution, but rather to the distribution's form. We decide to draw a random sample of 400 men in prison to test the hypothesis of a normal population distribution.

However, the population distribution must be thought of as grouped into class intervals. Furthermore, it is necessary that the number expected in each class interval be relatively large. Therefore, the experimenter must first decide on the number of class intervals he will use to describe both theoretical and obtained distributions. He would like to have intervals insuring a fairly "fine" description, as well as a sizable number expected in each interval. Suppose that for our problem we decide to think of the population distribution as divided into eight class intervals, in such a way that each interval should include exactly one eighth of the population. What would the limits of these eight intervals be? From Table I we find the limits shown in Table 17.3.1. Notice that this arrangement into class intervals refers to the *population* distribution, assumed to be normal under the null hypothesis, and that this choice of class intervals is made before the data are seen. This arrangement is quite arbitrary, and some other number of class intervals might have been chosen; in fact, with a sample size this large, an experimenter would be quite safe in taking many more intervals with a much smaller probability associated with each. Notice also that the population is thought of as divided into class intervals of unequal size, in order to give equal probability of intervals. On the other hand, it is perfectly possible to decide on some arbitrary

Table 17.3.1

Class limits in terms of z for class interval j	Approx. p_j	f_{ej}
1.15 and above	1/8	50
.68 to 1.15	1/8	50
.32 to .68	1/8	50
.00 to .32	1/8	50
−.32 to .00	1/8	50
−.68 to −.32	1/8	50
−1.15 to −.68	1/8	50
below −1.15	1/8	50
		400

class-interval size in z score terms, and then allow the various probabilities to be unequal. Our way of proceeding has two possible advantages: departures from normality in the middle of the score range are relatively more likely to be detected in this way than otherwise, and computations are simplified by having equal expectations for each class interval in the distribution. Furthermore, when the population is conceived as divided into class intervals having equal probability of occurrence, we can see from expression **17.1.4** that the sampling variance of the Pearson χ^2 statistic should be relatively closer to that of a chi-square variable; thus, when $\nu = J - 1$ and $p_j = 1/J$, we find that

$$\text{var.}(\chi^2) = \frac{2\nu(N - 1)}{N}.$$

A relatively better approximation should be afforded when intervals giving equal expected frequencies are chosen, other things being equal. However, grouping in this way may not be especially advantageous in some studies, and other ways of thinking of the population as grouped may be more desirable. The point is that this step is *arbitrary*, even though the conclusions reached from the analysis may depend heavily upon how the population is regarded as grouped. A rejection of the normal-distribution hypothesis based on a certain grouping leads to the conclusion that the population is not normally distributed; however, a failure to reject, based on a particular grouping, does not imply that some other grouping would not have led to a rejection of this hypothesis. The point should be clear: in testing for a normal population distribution, the experimenter must give considerable prior thought to how he wants to regard the population as grouped, within the limitation that he must have a sizable expectation for frequencies within each category.

      Now to proceed with the example. Before the sample distribution can be grouped into the same categories as the population, something must be known about the population mean and standard deviation. Our best evidence comes from the sample estimates, M and s, and so these are used in place of the unknown μ and σ. In the actual data for this example suppose that it turned out that the sample mean, M, is 98, and the estimate s is 8.4. Then these values are used as estimates of the true values of μ and σ for this population.

Using these estimates, we turn each score for an individual in the sample into a standard score. In terms of the arbitrary class intervals decided upon, the distribution of these *sample* standard scores is shown in Table 17.3.2.

Table 17.3.2

Interval	f_{oj}	f_{ej}
1.15 and above	14	50
.68 to 1.15	17	50
.32 to .68	76	50
.00 to .32	105	50
$-$ .32 to .00	71	50
$-$ .68 to $-$.32	76	50
-1.15 to $-$.68	31	50
below -1.15	10	50

Now the χ^2 test for goodness of fit is carried out:

$$\chi^2 = \frac{(14 - 50)^2}{50} + \frac{(17 - 50)^2}{50} + \frac{(76 - 50)^2}{50} + \frac{(105 - 50)^2}{50}$$
$$+ \frac{(71 - 50)^2}{50} + \frac{(76 - 50)^2}{50} + \frac{(31 - 50)^2}{50} + \frac{(10 - 50)^2}{50}$$
$$= 183.3.$$

Before the significance level is ascertained, however, an adjustment must be made in the degrees of freedom. Two parameters had to be estimated in order to carry out this test: the mean and the standard deviation of the population. *One degree of freedom is subtracted from $J - 1$ for each separate parameter estimated in such a test.* Therefore, the correct number of degrees of freedom here is

$$\nu = J - 1 - 2 \text{ or } 5.$$

For 5 degrees of freedom, Table III shows that this obtained χ^2 value far exceeds that required for significance at either the .05 or .01 level. The experimenter can feel quite confident in saying that scores in this population are not normally distributed.

Once again, let it be repeated that **the arrangement into population class intervals is arbitrary.** It may seem more reasonable to group the obtained distribution into convenient class intervals, and then to ask the question about the population distribution grouped in the same way. However, remember that the expected frequency in each interval must be relatively large, say 5 or more. In most instances this will require some combining of extreme class intervals to make the expected frequencies large enough to permit the test and *this combining operation amounts to a tinkering with the randomness of the sample.* Combining was not necessary in our example because we determined the required grouping ahead of time so as to guarantee large expected frequencies in each class.

17.4   Pearson χ^2 tests of association

The general rationale for testing goodness of fit also extends to tests of independence (or lack of statistical association) between categorical attributes. Quite often situations arise where N independent observations are made, and each and every observation is classified in two qualitative ways. One set of mutually exclusive and exhaustive classes can be called the A attribute. Any one of the C distinct classes making up this attribute can be labeled A_j, where j runs from 1 through C. Thus $A = \{A_1, \cdots, A_j, \cdots A_C\}$. Furthermore, on some other attribute B, there are some R mutually exclusive and exhaustive classes.

$$B = \{B_1, \cdots, B_k, \cdots, B_R\}.$$

Each observation then represents the occurrences of one joint event (A_j, B_k). The entire set of data can be shown as a **contingency table,** with the C classes of attribute A making up the columns, and the R classes of attribute B the rows; each and every possible (A_j, B_k) joint event thus is shown by a cell in the table. Be sure to notice that each distinct observation can represent only one joint event, and thus *each observation qualifies for one and only one cell in the table.*

For example, suppose that a random sample of 100 school children is drawn. Each child is classified in two ways: the first attribute is the sex of the child, with two possible categories:

$$\text{male} \ = A_1$$
$$\text{female} = A_2.$$

The second attribute, B, is the stated preference of a child for two kinds of reading materials:

$$\text{fiction} = B_1$$
$$\text{nonfiction} = B_2$$

(we will assume that each child can be given one and only one preference-classification). The entire set of data could be arranged into a joint-frequency distribution, or contingency table, symbolized by

	A_1	A_2
B_1	(A_1,B_1)	(A_2,B_1)
B_2	(A_1,B_2)	(A_2,B_2)

N

Each cell of this table contains the frequency among the N children for the joint event represented by that cell. The data might, for example, turn out to be

	A_1	A_2	
B_1	19	32	51
B_2	29	20	49
	48	52	$100 = N$

This table shows that exactly 19 observations represent the joint event (male, prefers fiction), 32 the joint event (female, prefers fiction) and so on.

In general, for some C classes making up the attribute A and some R classes the attribute B, the joint-frequency distribution is represented in an $R \times C$ table, each cell of which shows the frequency for one possible joint event in the data. The sum of frequencies over cells must be N, and each possible observation qualifies for one and only one cell in the table.

Corresponding to the joint-frequency distribution for a sample, there is a joint *probability* distribution in the population. For the example

	A_1	A_2	
B_1	$p(A_1,B_1)$	$p(A_2,B_1)$	$p(B_1)$
B_2	$p(A_1,B_2)$	$p(A_2,B_2)$	$p(B_2)$
	$p(A_1)$	$p(A_2)$	1.00

Suppose that we had some hypothesis where the probability of each possible joint event is specified. We might want to ask how this hypothetical *joint* distribution actually fits the data. Just as for the simple frequency distribution considered in Section 17.1, the hypothesis about the joint distribution tells us what to expect for each *joint* event's frequency. That is, for cell (A_1,B_1) we should expect exactly $Np(A_1,B_1)$ cases to occur in a random sample; for cell (A_1,B_1) our expectation is $Np(A_1,B_2)$ and so on for the other cells. Given this *complete specification* of the population joint distribution, tnen for sufficiently large N we could apply the Pearson χ^2 test of goodness of fit. Just as for any goodness-of-fit test where no parameters are estimated, the number of degrees of freedom would be the number of distinct event classes minus 1. Since there are RC *joint* events in this instance, $RC - 1$ is the degrees of freedom for a goodness-of-fit test for such a joint distribution.

However, exact hypothesis about joint distributions are quite rare in psychology. Very seldom would one want to carry out such a test even though it is possible. Instead, the usual null hypothesis is that the two attributes A and B are independent. If the hypothesis of independence can be rejected, then we say that the attributes A and B are statistically related or associated.

In Section 4.7 it was stated that two discrete attributes are considered independent *if and only if*

$$p(A_j,B_k) = p(A_j)p(B_k)$$

for *all possible* joint events (A_j,B_k). Given that the hypothesis of independence is true, and given the *marginal* distributions showing $p(A_j)$ and $p(B_k)$, we know what the joint probabilities *must* be.

On the other hand, this fact alone does us little good, because independence is defined in terms of the *population* probabilities, $p(A_j)$ and $p(B_k)$, and these we do not know. What can we use instead? **The best estimates we can make of the unknown marginal probabilities are the sample marginal proportions:**

$$\text{est. } p(A_j) = \frac{\text{freq. of } A_j}{N}$$

and

$$\text{est. } p(B_k) = \frac{\text{freq. of } B_k}{N},$$

for each A_j and B_k. Given these estimates of the true probabilities, then we expect that the frequency of the joint event (A_j, B_k) will be

$$f_{ejk} = \text{expected frequency of } (A_j, B_k) = N[\text{est. } p(A_j)][\text{est. } p(B_k)].$$

However, since these probability estimates are based on *sample* relative frequencies, it must then be true that

$$f_{ejk} = \frac{(\text{freq. } A_j)(\text{freq. } B_k)}{N}. \qquad [17.4.1\dagger]$$

In tests for independence, the expected frequency in any cell is taken to be the product of the frequency in the column times the frequency in the row, divided by total N. Using these expected frequencies, the Pearson χ^2 statistic in a test for association is simply

$$\chi^2 = \sum_j \sum_k \frac{(f_{ojk} - f_{ejk})^2}{f_{ejk}} \qquad [17.4.2\dagger]$$

where f_{ojk} is the frequency actually observed in cell (A_j, B_k), f_{ejk} is the expected frequency for that cell under the hypothesis of independence, and the sum is taken over all of the RC cells.

The number of degrees of freedom for such a test, where sample estimates of the marginal probabilities are made, differs from the degrees of freedom for a goodness-of-fit test. As we saw, if each joint probability actually were completely specified by the hypothesis, then a goodness-of-fit χ^2 could be carried out over the cells: this would have $RC - 1$ degrees of freedom. However, in the last section there was an instance of the principle that *1 degree of freedom is subtracted for each estimate made.* How many different estimates must we actually make in order to carry out the χ^2 test for association? Since there are C categories for attribute A, we must actually estimate $C - 1$ probabilities for this attribute, since given the first $C - 1$ probabilities, the last value is determined by the fact that $\sum_j p(A_j) = 1.00$. Furthermore, we also must estimate $R - 1$ probabilities for attribute B. In all some $(C - 1) + (R - 1)$ estimates are made and this number must be subtracted out of the total degrees of freedom. Therefore, the degrees of freedom for a Pearson χ^2 test of association is

$$\nu = RC - 1 - (C - 1) - (R - 1) = (R - 1)(C - 1).$$

In summary, given a joint-frequency table with C columns and R rows, the hypothesis

$$H_0: p(A_j, B_k) = p(A_j)p(B_k)$$

can be tested by the Pearson χ^2 statistic with $(R - 1)(C - 1)$ degrees of freedom, *provided* that

1. each and every observation is independent of each other observation;
2. each observation qualifies for one and only one cell in the table;
3. sample size N is large.

For our example, the expected frequency for cell (A_1, B_1) is found to be

$$f_{o11} = \frac{(\text{freq. } A_1)(\text{freq. } B_1)}{N} = \frac{(48)(51)}{100} = 24.48,$$

that for cell (A_2, B_1) is

$$f_{o21} = \frac{(\text{freq. } A_2)(\text{freq. } B_1)}{N} = \frac{(52)(51)}{100} = 26.52,$$

and so on, until the following set of expected frequencies is found for the table as a whole:

	A_1	A_2	
B_1	24.48	26.52	51
B_2	23.52	25.48	49
	48	52	$100 = N$

Notice that in any row the sum of the expected frequencies must equal the obtained marginal frequency for that row, and the sum of the expected frequencies in any column must also equal the obtained frequency for that column. In computations of expected frequencies, it is wise to carry a sizable number of decimal places, and not round to a few places until the final result.

Given the expected frequencies, the χ^2 test for this example is based on

$$\chi^2 = \frac{(19 - 24.48)^2}{24.48} + \frac{(32 - 26.52)^2}{26.52}$$
$$+ \frac{(29 - 23.52)^2}{23.52} + \frac{(20 - 25.48)^2}{25.48}$$
$$= 4.83.$$

In this 2×2 table, the degrees of freedom are $(2 - 1)(2 - 1) = 1$. For 1 degree of freedom, the χ^2 value needed to reject the hypothesis at the .05 level is 3.84, and so the hypothesis of independence can be rejected (however, if α were set at .01, then the hypothesis would not be rejected). Rejection of the hypothesis of independence lets one say that some statistical association does exist between the two attributes, A and B. In this instance, the experimenter might say that rejection of the hypothesis lets him feel safe in concluding that the sex and the reading preference of a child are in some way related.

17.5   AN EXAMPLE OF A TEST FOR INDEPENDENCE IN A LARGER TABLE

There is no reason at all why the number of categories in either A or B must be only 2, as in the preceding example. Any number of rows and any number of columns can be used for classifying observations into a contingency table. Provided that there is a sufficiently large random sample of independent observations where each occurs in one and only one cell, a χ^2 test can be carried out. However, for larger tables sample N must be rather large if a sufficiently large expected frequency is to be associated with each cell (this problem will be discussed in the next section).

Not only will the example in this section involve a larger table, but also it will illustrate that the row or column classes may be quantitative in their original character, *if* the experimenter is willing to treat the data in terms of class intervals. That is, either the attribute A or the attribute B may represent numerical measurements grouped into class intervals. Under these conditions the χ^2 test is still an adequate way to answer the question of association between A and B, but in using a χ^2 test the experimenter is actually ignoring the detailed numerical information in the data and is treating the different class intervals simply as qualitative distinctions.

Consider an experiment where a psychologist is studying the possible inheritance of patterns of emotional response in rats. He has noticed that four strains of rats tend to differ in their tendency toward emotional response in a new situation. These four strains are each isolated and carefully inbred until, after several generations, a large number of representatives of four pure strains are available for study. Each rat is put individually into a new and presumably fear-inducing situation, and left for a fixed amount of time. By some standard procedure, each individual rat is given a "score" on emotional response, based on a composite of a number of behavior indices. The experimenter is looking for evidence of association between the emotionality score and the strain the rat belongs to. Some 25 rats are selected at random from each of the four pure strains, giving a total of 100 individuals in all. Within each strain used in the experiment

Table 17.5.1

Scores	Group				Total
	1	2	3	4	
9 and over	5	8	6	4	23
6–8	10	7	8	6	31
3–5	5	4	7	8	24
0–2	5	6	4	7	22
	25	25	25	25	100

a grouped frequency distribution of "emotionality scores" is found, the same class intervals being used for each distribution. These frequency distributions for the four groups are shown in Table 17.5.1. *Because of his previous experience with such scores, the experimenter was able to set up these four class intervals in advance of his seeing the data.*

Basically, the experimenter is interested in comparing the four populations in terms of their distributions of scores. Looked at in this way, the problem is the comparison of four grouped frequency distributions, each based on the same grouping scheme. However, this can also be framed as a problem in testing the independence of two attributes.

Let us call rat population (pure breed) j the class A_j, and let us call any class interval of scores B_k. Now if all the populations were exactly *alike* in their score distributions, it should be true that

$$p(B_k|A_j) = p(B_k).$$

That is, the probability of any rat's showing a score in the class interval B_k given that he belongs to the population A_j should be the same as the probability in general associated with that class interval. This is tantamount to the independence of the attribute A (breeds) and the attribute B (scores arranged in class intervals). Therefore the problem of the simultaneous comparison of a number of frequency distributions is basically the same as the problem of testing for independence.

Going ahead on this basis, we carry out the χ^2 test for this table. The expected frequencies, each found by multiplying the column frequency by the row frequency and dividing by N, are shown in Table 17.5.2.

Table 17.5.2

Scores	Group 1	2	3	4	Total
9 and over	5.75	5.75	5.75	5.75	23
6–8	7.75	7.75	7.75	7.75	31
3–5	6.00	6.00	6.00	6.00	24
0–2	5.50	5.50	5.50	5.50	22
	25.00	25.00	25.00	25.00	100

Observe that for these *expected* frequencies, the relative frequency distribution in each column is precisely the same as for the column marginal total. If there are no differences between the population distributions, we expect each sample to show exactly the same frequency distribution. The estimate of this expected distribution within a column is provided by the pooled sample frequency distribution (the B_k marginals).

The χ^2 test is given by

$$\chi^2 = \frac{(5 - 5.75)^2}{5.75} + \frac{(8 - 5.75)^2}{5.75} + \cdots + \frac{(7 - 5.5)^2}{5.5}$$

$$= 4.97.$$

Here there are $(4 - 1)(4 - 1) = 9$ degrees of freedom, and the value required for significance at the .05 level is 16.92, making the result clearly not significant. The experimenter does not have enough evidence to say whether or not the breed is associated with the score a rat gets on emotionality level.

This example should illustrate the fact that χ^2 tests can be used to compare entire frequency distributions grouped in the same way. The grouping should, however, be decided upon in advance of the actual data. The only inference drawn from a significant result is that the different samples do not represent the same population distribution. Be very sure to notice, however, that this is rather different from the question asked in the analysis of variance, where several samples are also compared. In the analysis of variance, the numerical scores themselves are used as the basis for comparison in terms of means; in the χ^2 analysis, the scores play a role only in the grouping of observations into qualitatively distinct categories. The fact that these groupings are class intervals of a numerical score distribution is quite immaterial in the test itself. Population differences of any sort may show up as a significant result, and the experimenter can conclude only that the population distributions are not identical when the test is significant. Furthermore, given that the underlying assumptions are met, the conclusions reached in analysis of variance are about population means; the experimenter can conclude not only that the populations differ, but also how and by how much they differ. The introduction of the scores themselves into the analysis makes for relatively stronger conclusions. Nevertheless, there are situations where several frequency distributions are to be compared directly, class intervals can be formed in advance of the data, and the sample size is relatively large. In this instance it is possible to use the χ^2 test for independence, provided that the experimenter realizes that he is not using all of the information in the data and that his conclusions will not be directly comparable to those of the analysis of variance. In general, the Pearson χ^2 analysis is best reserved for the truly qualitative data for which it is most appropriate.

17.6   THE SPECIAL CASE OF A FOURFOLD TABLE

Two-by-two contingency or joint-frequency tables are especially common in psychological research. For such fourfold tables, computations for the Pearson χ^2 test can be put into a very simple form. Consider the following table:

a	b	$a + b$
c	d	$c + d$
$a + c$	$b + d$	N

Here, the small letters a, b, c, and d represent the frequencies in the four cells, respectively. Then, the value of χ^2 can be found quite easily from

$$\chi^2 = \frac{N(ad - bc)^2}{(a + b)(c + d)(a + c)(b + d)}$$ [17.6.1†]

with, of course, one degree of freedom.

This value of χ^2 is usually corrected to give a somewhat better approximation to the exact multinomial probability. With the correction, the value is found from

$$\chi^2 = \frac{N(|ad - bc| - N/2)^2}{(a + b)(c + d)(a + c)(b + d)}$$ [17.6.2†]

which is compared with Table IV (1 degree of freedom) in the usual way. This is another instance of **Yates' correction for continuity,** which we encountered first in Section 17.2. This correction should be applied only when the number of degrees of freedom is one, however.

17.7   THE ASSUMPTIONS IN χ^2 TESTS FOR ASSOCIATION

Chi-square tests are among the easiest for the novice in statistics to carry out, and they lend themselves to a wide variety of psychological data. This computational simplicity is deceptive, however, as the use of the chi-square approximation to find multinomial probabilities is based on a fairly elaborate mathematical rationale, requiring a number of very important assumptions. This rationale and the importance of these assumptions has not always been understood, even by experienced researchers, and there is probably no other statistical method that has been so widely misapplied.

In the first place, since the exact probabilities to be approximated are assumed to follow the multinomial rule, each and every observation categorized should be independent of each other observation. In particular, this means that caution may be required in the application of χ^2 tests to data where dependency among observations may be present, as is sometimes the case in repeated observations of the same individuals. As always, however, it is not the mere fact that observations were repeated, but rather the nature of the experiment and the type of data, that let one judge the credibility of the assumption of independent observations. Nevertheless, the novice user of statistical methods does well to avoid the application of Pearson χ^2 tests to data where each individual observed contributes more than a single entry to the joint frequency table.

In the second place, the joint-frequency table must be complete, in the sense that each and every observation made must represent one and only one joint-event possibility. This means that each distinct observation made must qualify for one and only one row, one and only one column, and one and only one cell in the contingency table.

The stickiest question of all concerns sample size and the minimum size of expected frequency in each cell. Probabilities found from the chi-square tables for such tests are always approximate. Only when the sample size is infinite must these probabilities be exact. The larger the sample size, the better this approximation generally is, but the goodness of the approximation also depends on such things as true marginal distributions of events, the number of cells in the contingency table, and the significance level employed. *Furthermore, there are no hard and fast rules that are sufficient to cover all the things which can influence the goodness of the chi-square approximation.* Rules of thumb for sample size do exist, but even these vary in different statistics texts; the simple reason for this is that statisticians themselves vary in their standards for a "good" approximation. This makes it very hard for the user of statistics, since a rule of thumb that may be fine for one kind of problem may not be advisable for another.

Without going further into the complexities of the matter, we will simply state a rule that is at least current, fairly widely endorsed, and generally conservative. **For tables with more than a single degree of freedom, a minimum expected frequency of 5 can be regarded as adequate, although when there is only a single degree of freedom a minimum expected frequency of 10 is much safer.** This rule of thumb is ordinarily conservative, however, and circumstances may arise where smaller expected frequencies can be tolerated. In particular, if the number of degrees of freedom is large, then it is fairly safe to use the χ^2 test for association even if the minimum expected frequency is as small as 1, provided that there are only a few cells with small expected frequencies (such as one out of five or fewer). On the whole, however, it may be wise for the beginner using this technique to err, if he must, on the conservative side, and apply Pearson χ^2 tests only to data having fairly large expected frequencies. When you find yourself in a situation where relatively many expected frequencies tend to be small and sample size cannot be made large, ask for expert advice. In some instances it may be possible to carry out an exact test of the hypothesis of independence, as suggested in Section 17.9.

A word must be said about the practice of pooling categories to attain large expected frequencies after the data are seen. This has been done routinely for many years, and many statistics texts advise this as a way out of the problem. However, this may amount to trading the devil for the witch! The whole rationale for the chi-square approximation rests on the randomness of the sample, and that the categories into which observations may fall are chosen in advance. When one starts pooling categories after the data are seen he is doing something to the randomness of the sample, with unknown consequences for his inferences. The manner in which categories are pooled can have an important effect on the inferences one draws. This practice is to be avoided if at all possible; better the risk of a poor approximation to the exact probabilities than a result without any statistical interpretation at all.

17.8 LIKELIHOOD RATIO TESTS FOR CATEGORICAL DATA

The method of the likelihood ratio mentioned in Chapter 9 has been applied to the problems of tests both of goodness of fit and of association for categorical data. Like the χ^2 tests, the likelihood ratio methods are based ultimately on the multinomial rule for the distribution of sample results, given some hypothesis about a discrete distribution. However, the actual form of the test provided by the likelihood ratio method differs somewhat from the familiar χ^2 statistic. For a goodness-of-fit test, the test statistic for large N is

$$\chi^2 = 2 \sum_{j=1}^{J} n_j(\log_e n_j - \log_e p_j) - 2N \log_e N \qquad [17.8.1]$$

where n_j is the observed number in category j, p_j is the probability for that category as specified by the null hypothesis, and the logarithmns are to the base e. The value of this statistic is referred to chi-square table for $J - 1$ degrees of freedom.

For tests of association, the likelihood ratio test is given by

$$\chi^2 = 2N \log_e N + 2 \sum_{j}^{C} \sum_{k}^{R} n_{jk} \log_e n_{jk} - 2 \sum_{j}^{C} n_j \log_e n_j - 2 \sum_{k}^{R} n_k \log_e n_k \qquad [17.8.2]$$

where n_j is the number observed in column j, n_k is the number observed in row k, n_{jk} is the observed frequency in cell j,k, and the logarithmns are again taken to the base e. This statistic is referred to in the distribution of chi-square with $(R - 1)(C - 1)$ degrees of freedom.

For very large N, these likelihood ratio tests become equivalent to the ordinary Pearson χ^2 tests introduced earlier. However, there is some reason to believe that these tests may be somewhat less affected by small sample size than are the Pearson χ^2 tests, particularly when the number of degrees of freedom is greater than one. Thus, for moderately large samples these procedures may be slightly superior. For very large samples, these methods and the χ^2 tests presented earlier should give virtually identical results. The assumptions having to do with the completeness of the data and of the independence of observations are, however, just as essential for these methods based on the likelihood ratio as for the others. A discussion of these methods is given in Mood and Graybill (1963).

17.9   THE POSSIBILITY OF EXACT TESTS FOR GOODNESS OF FIT AND FOR ASSOCIATION

As we have mentioned, the Pearson χ^2 tests of association and of goodness of fit give approximations to exact probabilities, which, in principle, may be found using the multinomial (or in some instances, the hypergeometric) rule. The basic reason for using the chi-square approximation is that actual computation of these exact probabilities is extremely laborious or downright impossible. However, in some situations where the sample size is so small that the use of the χ^2 tests is ruled out, it may be practicable to compute probabilities exactly. This will be illustrated only for a simple test of association, although other possibilities exist.

The test we will discuss is commonly known as **Fisher's exact test** for a 2×2 contingency table. This is not appropriately called a χ^2 test, since it does not use the chi-square approximation at all. Instead, the exact probability is computed for a sample's showing as much or more evidence for association than that obtained, given only the operation of chance.

There are at least two somewhat different rationales one might advance for this exact test, but the one we will adopt is based on the principle of randomization. This general idea has been referred to already in Section 13.25.

Suppose that some N subjects are categorized into the following 2×2 table:

	A_1	A_2	
B_1	a	b	$a+b$
B_2	c	d	$c+d$
	$a+c$	$b+d$	N

Now suppose for a moment that the N subjects actually make up the population, and that the distribution in the population shows $a + c$ individuals in the first *column* category, A_1, and $b + d$ individuals in the second column category, A_2. Some $n = a + b$ individuals are sampled at random and *without* replacement. What is the probability that in *this sample* exactly a individuals fall into A_1 and b into A_2? In Section 5.23 it was mentioned that the probability of a sample drawn *without* replacement from a finite population can be found by the *hypergeometric* rule. Applying the rule, we find that the probability of a in A_1 and b in A_2 *within the sample represented by row B_1* is just

$$\frac{\binom{a + c}{a}\binom{b + d}{b}}{\binom{N}{a + b}}$$

which is the same as

$$\frac{(a + b)! \,(c + d)! \,(a + c)! \,(b + d)!}{N! \,a! \,b! \,c! \,d!}.$$

Now consider the marginals of the sample table as fixed, so that regardless of the arrangement within the table we know the totals in rows and columns. Imagine that occurrences of the events represented by the rows and columns have absolutely nothing to do with each other, so that the two attributes are independent. Any sample result in the cells of the table occurs as though individuals in the columns were assigned to the rows at random. Then the probability of any particular random arrangement is given by the use of the hypergeometric rule, just as for the probability found above. *If one finds the probability of the arrangement actually obtained, as well as every other arrangement giving as much or more evidence for association, then one can test the hypothesis that the obtained result is purely a product of chance by taking this probability as the significance level.* This amounts to finding both the probability of the obtained table, and every other table (with the same marginals) showing more disproportion between cells a and c than in the table obtained.

An example may make this idea seem more reasonable. Imagine a study involving 10 individuals. Each individual observation fell into one and only one cell of the following contingency table:

	A_1	A_2	
B_1	1	4	5
B_2	3	2	5
	4	6	10

Now assume that regardless of how else the individuals were categorized we would have got 4 cases in A_1, 6 cases in A_2, 5 cases in B_1, and 5 cases in B_2. Also assume that the arrangement obtained in the cells of the table is purely a result of chance. What is the probability of this result (the arrangement actually obtained)? Since

$$a = 1$$
$$b = 4$$
$$c = 3$$
$$d = 2$$

this is

$$\frac{4! \; 6! \; 5! \; 5!}{10! \; 1! \; 4! \; 3! \; 2!}$$

or

$$p(\text{obtained arrangement}) = .238.$$

Actually, we could stop at this point, since for $\alpha = .05$ or less, we definitely could not reject the hypothesis of a random arrangement, since the probability is at least .238 of a sample this "systematic" or more so by chance alone. However, we will continue in order to show the method.

An even more systematic-looking result that might have occurred is

	A_1	A_2
B_1	0	5
B_2	4	1

where $a = 0$, $b = 5$, $c = 4$, $d = 1$. Notice that if this result had occurred, then there would be even more discrepancy between the obtained distributions in rows B_1 and B_2. In other words, the relative frequencies in cells a and c are even more different than before. The probability of this result is

$$\frac{4! \; 6! \; 5! \; 5!}{10! \; 0! \; 5! \; 4! \; 1!} = .024.$$

Thus, the probability of our result or one *more* suggestive of association is

$$.238 + .024 = .262.$$

This is not as small as the conventional levels set for α, and so the hypothesis that the apparent association is the product of chance is *not* rejected. Notice, however, that if the actual result had corresponded to the table given last, we could have rejected the hypothesis for any $\alpha \geq .024$.

Unlike the χ^2 test, the Fisher exact test is essentially one-tailed. The probabilities are calculated for all possible results departing as much or more in a specific direction from the marginal distribution of A as does our sample. When the rows (or columns) have equal frequencies (as in the example) the final probability can be doubled to arrive at the two-tailed significance level. However, when both the two rows and the two columns have unequal marginal frequencies, the probability should be found for each possible table where the absolute difference

$$\left| \frac{a}{a+b} - \frac{c}{c+d} \right|$$

is as great or greater than in the table actually obtained.

Convenient tables are available for the Fisher exact test, so that it is not usually necessary to carry out all the computations given here in order to perform this test. These tables are available in the *Biometrika Tables for Statisticians* (1967) and in Siegel (1956) among other places.

It must be emphasized that the reasoning underlying this test is rather different from that for χ^2 tests. Here, one actually refers his sample result to the distribution over possible randomizations of that sample. The hypothesis is that chance is the determiner of the apparent association in the table. This hypothesis is rejected as the actual table obtained falls among those unlikely to occur over all random assignments of subjects in columns to the rows. This rationale makes the test especially suitable for experiments where subjects actually *are* assigned to groups at random. Small sample size is no restriction for this test; indeed, small samples actually make the computations quite simple, although very strong association must exist to show up as significant when sample size is quite small.

Conceptually, no special difficulty exists in extending the exact test idea to larger tables, although there are two obvious problems that arise: the computations required mount up very rapidly as the size of the table increases, and one must have some sort of index (such as the Pearson χ^2 statistic) to evaluate the apparent evidence for association in each possible sample. Thus, exact tests are not in common use for larger tables. However, if it should be very important to have a test for a small sample in some problem, it is well to remember that these exact procedures are possible, even for larger tables.

17.10   A TEST FOR CORRELATED PROPORTIONS IN A TWO-BY-TWO TABLE

A problem that often arises in psychological research is somewhat different from the problem of association between attributes. As an illustration of this problem, suppose that some N individual subjects are each observed by two independent judges. Each judge places each subject into one of two mutually exclusive and exhaustive categories, such as "high leadership potential" versus "low leadership potential." It is assumed that a judge's ratings of different individuals are independent. Let us call these categories simply "H" and "L" for the moment. We would like to ask if these two judges, given all possible subjects in the population, would show the same true proportion of individuals rated in category "H." In other words, in the population of all subjects to be rated, does $p_1(H) = p_2(H)$, where $p_1(H)$ is the proportion rated in category H by judge 1, and $p_2(H)$ is the proportion rated in that category by judge 2?

This is a problem of **correlated proportions,** since each of the two sample proportions will be based in part on the same individuals. A test due to McNemar (1955) applies to this situation. Suppose that the sample of N individuals were arranged into the following 2×2 table:

JUDGE 1

		H	L
JUDGE 2	H	a	b
	L	c	d

An *exact* test of the hypothesis that $p_1(H) = p_2(H)$ is possible using binomial distribution. Under this hypothesis, the probability of a given sample result, showing a particular pair of cell frequencies b and c, is just

$$\binom{b+c}{b} (.5)^{b+c}. \qquad [17.10.1*]$$

To carry out the exact test, let g equal the smaller of the two frequencies, b or c. Then one takes the sum of probabilities

$$2 \sum_{h=0}^{g} \binom{b+c}{h} (.5)^{b+c}. \qquad [17.10.2*]$$

If this number is less than or equal to the value chosen for α, then the null hypothesis may be rejected (two-tailed).

When N is relatively large, the exact probability may be approximated by use of χ^2, where

$$\chi^2 = \frac{(|b - c| - 1)^2}{b + c} \qquad [17.10.3\dagger]$$

with one degree of freedom.

For our example, a significant result would let one conclude that the *true* distributions of judgments for the two judges differ. Be sure to notice that this is not an ordinary test of association for a contingency table, but rather a test of the equality of two proportions where each sample proportion involves some of the same observations, making the two sample proportions dependent.

Obviously, this test can be applied in contexts other than two judges' categorizations of subjects, although this is a fairly common problem in psychology; another common example is the difference in difficulty of two test items administered to the same subjects. This test has been extended to more complicated situations by Cochran; the procedure is discussed in Section 18.5.

17.11   MEASURES OF ASSOCIATION IN CONTINGENCY TABLES

One of the oldest problems in descriptive statistics is that of indexing the strength of statistical association between qualitative attributes. Although a number of simple and meaningful indices exist to describe association in a fourfold table, this problem grows more complex for larger tables, and has perhaps never been solved to everyone's real satisfaction.

Why should there be any special problem in indexing the strength of association between qualitative or categorical attributes? As we have seen, most of our notions of the strength of a statistical association rest on the concept of the variance of a random variable. Thus, indices such as ρ^2, ρ_I, η^2, and our index ω^2, all rest on the idea of a proportional reduction in variance in the dependent variable afforded by specifying the value of the independent variable. However, when the independent and dependent variables are each categorical in nature, the variance per se is not defined. Something else must be used in specifying how knowledge of the A category to which an observation belongs increases our ability to predict the B category.

Three somewhat different approaches to this problem will be discussed here. The first rests directly on the notion of statistical independence between two attributes, defined by $p(A_j, B_k) = p(A_j)p(B_k)$. In this approach, the strength of association is measured basically in terms of the difference

$$|p(A_j, B_k) - p(A_j)p(B_k)|,$$

the extent to which the probability of a joint occurrence differs from the probability that would be true if the attributes A and B were independent. As we shall see, this conception is adequate from a statistical point of view but seems to lack a simple interpretation in terms of how one *uses* the statistical relation.

Another and much more recent approach deals with **predictive association.** Association between categorical attributes is indexed by the reduction in the probability of error in prediction afforded by knowing the status of the individual on one of the attributes. This way of defining association makes intuitive good sense, but is not as directly tied to tests of association as the first approach.

Finally, a brief mention will be made of still another point of view on this problem, based on concepts from information theory. Here an analogue to the variance does exist for categorical data. This makes it possible to define the strength of association in contingency tables in a way very similar to the usual indices for numerical data.

17.12   The phi-coefficient and indices of contingency

Before we go into the problem of describing statistical association in a sample, a general way of viewing statistical association in a population will be introduced. This is the **index of mean square contingency,** originally suggested by Karl Pearson, the originator of the χ^2 test for association. Imagine a *discrete joint probability distribution* represented in a table with C columns and R rows. The columns represent the qualitative attribute A, and the rows B. Then the mean square contingency is defined to be

$$\varphi^2 = \sum_j \sum_k \frac{p(A_j,B_k)^2}{p(A_j)p(B_k)} - 1 \qquad [17.12.1*]$$

This population index φ^2 (small Greek phi, squared) can be zero only when there is complete independence, so that

$$p(A_j,B_k) = p(A_j)p(B_k)$$

for each joint event (A_j,B_k). However, when there is *complete association* in the table, the value of φ^2 is given by

$$\text{max. } \varphi^2 = L - 1,$$

where L is the *smaller* of the two numbers R or C (number of rows or columns in the table). Thus, a convenient index of strength of association in a population is provided by

$$\varphi' = \sqrt{\frac{\varphi^2}{L - 1}}, \qquad [17.12.2*]$$

which will always lie between the values 0 and 1.

For the special case in which $R = C = 2$ (a 2×2 table) then φ is itself an index of association, since $\varphi = \varphi'$. Except in a 2×2 table the sign of φ is always taken as positive, and even in 2×2 tables the sign is meaningless unless the categories are regarded as ordered.

Now consider a sample of data arranged into a fourfold contingency table, as in Section 17.6. The *sample* value of φ is given by

$$\varphi = \frac{(bc - ad)}{\sqrt{(a + b)(c + d)(a + c)(b + d)}}. \qquad [17.12.3\dagger]$$

Notice that this is almost exactly the square root of the expression for χ^2 in a 2×2 table, given by **16.6.1.** In fact

$$\chi^2 = N\varphi^2 \qquad [17.12.4\dagger]$$

and

$$\varphi = \sqrt{\frac{\chi^2}{N}}. \qquad [17.12.5\dagger]$$

Since both χ^2 and population φ reflect the degree to which there is nonindependence between A and B, a test for the hypothesis

$$H_0: \text{true } \varphi = 0$$

is provided by the ordinary χ^2 test for association in a 2×2 table.

In a 2×2 table there is an interesting link between sample φ and the correlation coefficient r. Let the categories within each attribute A and B be thought of as *ordered*. Suppose that the individuals i in the categories are assigned numerical scores as follows:

$X_i = 1$ if i falls in the higher category of A
$X_i = 0$ if i falls in the lower category of A
$Y_i = 1$ if i falls in the higher category of B
$Y_i = 0$ if i falls in the lower category of B.

The data in the table would then be of this form

		a	b
Y	1		
	0	c	d
		0	1
			X

Suppose that the correlation between these scores were computed across the N individuals i. We would find that

$$
\begin{aligned}
r_{XY} &= \frac{\sum_i x_i y_i / N - M_X M_Y}{S_X S_Y} \\
&= \frac{Nb - (b + d)(a + b)}{\sqrt{(a + b)(c + d)(a + c)(b + d)}} \\
&= \frac{ab + b^2 + bc + bd - ab - b^2 - ad - bd}{\sqrt{(a + b)(c + d)(a + c)(b + d)}} \\
&= \varphi.
\end{aligned}
$$

What we have shown is that the coefficient φ may be regarded as the correlation between the attributes A and B when the categories are associated with "scores" of 0 and 1. If the categories are ordered for each attribute, and 1 means a higher category than does 0 on A, and similarly on B, then the sign of the φ coefficient becomes meaningful: a positive sign implies a tendency for the high category on A to be associated with the high category of B, and vice-versa. On the other hand, a negative value implies a tendency for a high category on one attribute to be associated with a low category on the other. Because of this connection with r, φ is often called the **fourfold point correlation**.

The idea of φ^2 extends to samples in larger contingency tables as well. For a set of data arranged into an $R \times C$ table, the *sample* value of φ^2 is simply

$$\varphi^2 = \frac{\chi^2}{N}.$$

A convenient way to describe the apparent strength of association in a sample is to find

$$\varphi' = \sqrt{\frac{\varphi^2}{L-1}} = \sqrt{\frac{\chi^2}{N(L-1)}} \qquad [17.12.6\dagger]$$

which must lie between 0, reflecting complete independence, and 1, showing complete dependence, of the attributes. The Pearson χ^2 is computed in the ordinary way as for a test of association (**17.4.2**) and L is the *smaller* of R, the number of rows, or C, the number of columns.

This index φ' (Cramér's statistic) is not to be confused with the ordinary *coefficient of contingency*, sometimes used for the same purpose. The coefficient of contingency is defined by

$$C_{AB} = \sqrt{\frac{\chi^2}{N+\chi^2}}.$$

This last index has the disadvantage that it cannot attain an upper limit of 1.00 unless the number of categories for A and B is infinite. Obviously, this limits the usefulness of C_{AB} as a descriptive statistic, and the index given by φ' is superior.

The sample coefficient φ' gives a way to discuss the apparent strength of statistical association in any contingency table, and there is a direct connection with χ^2 tests making it possible to test the significance of any obtained φ' value from a sufficiently large sample. However, it is rather hard to put the meaning of φ' in common-sense terms, particularly for larger tables. Our other indices of association such as ω^2 do have such an interpretation in terms of reduction in variance, or variance accounted for, but this idea is not directly applicable to the φ' indices.

Now we turn to an index of association in the *predictive* sense. How much does knowing the classification A improve one's ability to predict the classification on B?

17.13   A MEASURE OF PREDICTIVE ASSOCIATION FOR CATEGORICAL DATA

Suppose that for some population, the joint-probability distribution of (A_j, B_k) events were as follows:

	A_1	A_2	
B_1	.20	.15	.35
B_2	.10	.30	.40
B_3	.10	.15	.25
	.40	.60	1.00

That is, $p(A_1, B_1) = .20$, $p(A_2, B_1) = .15$, and so on.

Now suppose that, knowing these probabilities, you were asked to predict the B category for some case drawn at random. You know *absolutely nothing* about which A category this particular case belongs to. Which B category should you bet on? Your probability of being exactly right is largest if you bet on B_2, since this category has the largest probability of occurrence. Let us symbolize this largest probability in the marginal distribution for B by $\max_{k} p(B_k)$, the maximum $p(B_k)$ over all the possible events B_k. In this instance,

$$\max_{k} p(B_k) = .40.$$

In this way of predicting, *not knowing* the A classification, the probability of an *error* in prediction is

$$p(\text{error}|A_j \text{ unknown}) = 1 - \max_{k} p(B_k).$$

For this particular example,

$$p(\text{error}|A_j \text{ unknown}) = .60.$$

Now, however, suppose that a case is drawn at random and you were *told* the A group into which he falls. Given this information, you must predict the B class. Assume that the case came from group A_1; what should you predict? The largest *conditional* probability, $\max_{k} p(B_k|A_1)$, occurs for category B_1:

$$\max_{k} p(B_k|A_1) = \frac{.20}{.40} = .50.$$

Thus, given A_1, category B_1 should be predicted, and the probability of an error in this instance is

$$p(\text{error}|A_1) = 1 - \max_{k} p(B_k|A_1) = .50.$$

On the other hand, if the information were that the case belongs to A_2, then category B_2 would be predicted, since

$$\max_{k} p(B_k|A_2) = p(B_2|A_2)$$
$$= \frac{.30}{.60} = .50.$$

An error in this prediction has probability of

$$p(\text{error}|A_2) = 1 - \max_{p} p(B_k|A_2).$$

Under this model of prediction, *on the average*, over all cases, the probability of error is, then

$$
\begin{aligned}
p(\text{error}|\text{given } A) &= p(\text{error}|A_1)p(A_1) + p(\text{error}|A_2)p(A_2) \\
&= 1 - \max_{k} p(A_1 B_k) - \max_{k} p(A_2 B_k) \\
&= 1 - .20 - .30 \\
&= 1 - .50 \\
&= .50.
\end{aligned}
$$

Notice that when A is not specified, then the probability of an error in prediction is .60, but when A is specified, this average probability of an error is only .50. This shows that there is *predictive* association between A and B; the A category quite literally tells one something about how to bet on B, since the probability of error is reduced when the particular A category is known.

This idea is the basis for an **index of predictive association.** This index, which was developed by Goodman and Kruskal (1954), will be called λ_B, (small Greek lambda, sub B):

$$\lambda_B = \frac{p(\text{error}|A_j \text{ unknown}) - p(\text{error}|A_j \text{ known})}{p(\text{error}|A_j \text{ unknown})}. \qquad [17.13.1^*]$$

This index shows the proportional reduction in the *probability* of error afforded by specifying A_j. If the information about the A category does not reduce the probability of error at all, the index is zero, and one can say that there is no predictive association. On the other hand, if the index is 1.00, no error is made given the A_j classification, and there is complete predictive association.

It must be emphasized that this idea is not completely equivalent to independence and association as reflected in χ^2 and φ'. It is quite possible for some statistical association to exist even though the value of λ_B is zero. In this situation, A and B are not independent, but the relationship is not such that giving A_j causes one to change his bet about B_k; the index λ_B is other than 0 only when *different* B_k categories would be predicted for different A_j information.

On the other hand, if there is complete proportionality throughout the table, so that φ' is zero, then λ_B must be zero. Furthermore, when there is complete association, so that perfect prediction is possible, both λ_B and φ' must be 1.00.

Sample values of λ_B can be calculated quite easily from a contingency table. Here, the sample is regarded as though it were the population, and probabilities are taken from the relative frequencies in the sample. Thus we interpret λ_B as the proportional reduction in the probability of error in prediction for cases drawn at random from *this* sample, or, if you will, a population exactly like this sample in its joint distribution.

In terms of the frequencies in the sample, we find

$$\lambda_B = \frac{\sum_j \max_k f_{jk} - \max_k f_{.k}}{N - \max_k f_{.k}} \qquad [17.13.2\dagger]$$

where

f_{jk} is the frequency observed in cell (A_j, B_k)

$\max_k f_{jk}$ is the *largest* frequency in column A_j

$\max_k f_{.k}$ is the largest *marginal* frequency among the rows B_k.

When two or more cells in any column have the same frequency, larger than any others in that column, then the frequency belonging to any single one of those cells is used as the maximum value for the column. Similarly, if several row marginals each exhibit the same frequency, which is largest among the rows, that frequency is used.

As an example, suppose that some 40 observations were grouped into the following 3×4 contingency table:

	A_1	A_2	A_3	A_4	
B_1	0	2	3	5	10
B_2	7	6	1	1	15
B_3	3	2	6	4	15
	10	10	10	10	40

The value of λ_B, for predictions of B from A, is found to be

$$\lambda_B = \frac{7 + 6 + 6 + 5 - 15}{40 - 15}$$

$$= .36.$$

By way of contrast, consider the values of χ^2 and of φ':

$$\chi^2 = 17.97$$

which for 6 degrees of freedom is significant beyond the .01 level. The sample value of φ' is

$$\varphi' = \sqrt{\frac{17.97}{(40)(2)}}$$

$$= .48$$

which apparently indicates a considerable degree of association.

However, the index of predictive association is .36, apparently somewhat less than indicated by φ'. Although it is rather difficult to give an exact meaning to φ', the meaning of λ_B is quite clear: in predictions of B from A, information about the A category reduces the probability of error by some 36 percent on the average.

The index λ_B is an *assymetric* measure, much like ω^2 and η^2. It applies when A is *the* independent variable, or the thing ordinarily known first, and B is the thing predicted. However, for the same set of data, it is entirely possible to reverse the roles of A and B, and obtain the index

$$\lambda_A = \frac{p(\text{error}|B_k \text{ unknown}) - p(\text{error}|B_k \text{ known})}{p(\text{error}|B_k \text{ unknown})}, \qquad \text{[17.13.3*]}$$

which is suitable for predictions of A from B. In terms of frequencies:

$$\lambda_A = \frac{\sum_k \max_j f_{jk} - \max_j f_{j.}}{N - \max_j f_{j.}}. \qquad \text{[17.13.4†]}$$

In general, the two indices λ_B and λ_A will not be identical; it is entirely possible to have situations where B may be quite predictable from A, but not A from B.

Finally, in some contexts it may be desirable to have a *symmetric* measure of the power to predict, where neither A nor B is specially designated as the thing predicted from or known first. Rather, we act as though sometimes the A and sometimes the B information is given beforehand. In this circumstance the index λ_{AB} can be computed from

$$\lambda_{AB} = \frac{\sum\limits_{j} \max\limits_{k} . f_{jk} + \sum\limits_{k} \max\limits_{j} . f_{jk} - \max\limits_{k} . f_{.k} - \max\limits_{j} . f_{j.}}{2N - \max\limits_{k} . f_{.k} - \max\limits_{j} . f_{j.}}. \qquad [17.13.5\dagger]$$

In the example, the symmetric measure is

$$\lambda_{AB} = \frac{7 + 6 + 6 + 5 + 5 + 7 + 6 - 15 - 10}{2(40) - 15 - 10}$$

$$= .31.$$

This says that knowing *either* the A or B classification considerably improves our ability to predict the other category, in that the probability of error is reduced by about 31 percent. The value of λ_{AB} will, incidentally, always lie between λ_A and λ_B. (A basis for a sampling theory for such measures can be found in Goodman and Kruskal [1963].)

These measures of *predictive* association form a valuable adjunct to the tests given by χ^2 methods. When the value of χ^2 turns out significant one can say with confidence that the attributes A and B are not independent. Nevertheless, the significance level alone tells almost nothing about the strength of association. Usually we want to say something about the predictive strength of the relation as well. If there is the remotest interest in actual predictions using the relation studied, then the λ measures are worthwhile. Statistical relations so small as to be almost nonexistent can show up as highly significant χ^2 results, and this is especially likely to occur when sample size is large. All too often the experimenter then "kids himself" into thinking that he has discovered some relationship observable to the "naked eye," which will be applicable in some real-world situation. Plainly, this is not necessarily true. The λ indices do, however, suggest just how much the relationship found implies about real predictions, and how much one attribute actually does tell us about the other. Such indices are a most important corrective to the experimenter's tendency to confuse statistical significance with the importance of results for actual prediction. Virtually any statistical relation will show up as highly significant given a sufficient sample size, but it takes a relation of considerable strength to enhance our ability to predict in real, uncontrolled, situations. It can happen that even though a χ^2 test is significant, the predictor's *behavior* is not changed one whit by this new information. The λ measures show how one is led to predict *differentially* in the light of the relationship.

17.14   INFORMATION THEORY AND THE ANALYSIS
OF CONTINGENCY TABLES

A few words must be said about still another development for the description of the relation exhibited in a contingency table. These ideas come from information theory, and suggest analogies to the classical statistics based on variance, but applicable to the qualitative situation. Unfortunately, space does not permit a thorough discussion of these notions, and so only the barest sketch will be given.

Imagine a discrete probability distribution based on, say, four categories

$$
\begin{array}{ll}
A_1 & p(A_1) \\
A_2 & p(A_2) \\
A_3 & p(A_3) \\
A_4 & p(A_4).
\end{array}
$$

Consider a particular case x drawn at random from this distribution. The probability that this case falls in category A_1 is $p(A_1)$; in category A_2, $p(A_2)$; and so on. Now, however, you find out that the particular case x belongs to category A_1. *How much information have you gained?* Before this information was supplied, some probability value less than 1 was attached to each possible category, but afterward, these probabilities are

$$
\begin{array}{l}
p(x \text{ in } A_1|\text{data}) = 1 \\
p(x \text{ in } A_2|\text{data}) = 0 \\
p(x \text{ in } A_3|\text{data}) = 0 \\
p(x \text{ in } A_4|\text{data}) = 0.
\end{array}
$$

Your distribution of probabilities relative to the occurrence of that case, x, has changed. There was some amount of uncertainty before the exact category was known, but afterward there was no uncertainty.

In information theory, the *average prior amount of uncertainty* over all possible such cases x is defined to be

$$
H(A) = - \sum_j p(A_j) \log p(A_j). \tag{17.14.1}
$$

Thus, given any prior distribution for the possible cases x, the average uncertainty can be calculated. However, after the information about the category for any particular case is given, there is *no* uncertainty, and

$$
\text{posterior } H(A) = 0.
$$

Thus, the average amount of information gained by the occurrence of particular cases is the *reduction* in average uncertainty, which is

$$
\text{average amount of information} = H(A) - \text{posterior } H(A) = H(A).
$$

From a purely statistical point of view, the interesting thing about this formulation is that the index $H(A)$ is a very close analogue to the *variance* σ^2 of a distribution. When $p(A_j) = 1.00$ for some j, then $H(A) = 0$; similarly, if there is only one possible value for a random variable, $\sigma^2 = 0$. On the other hand, when there is a wide range of possible events, $H(A)$ tends to be large; indeed, the more evenly spread are the probabilities over the various possible events, the larger will $H(A)$ be. In the same way, σ^2 is large when the distribution of a random variable is "spread-out." In short, $H(A)$ is a variance-like index that can be computed for any discrete distribution, even though the event-classes are purely qualitative.

Furthermore, for any joint-probability distribution, one can define the average *joint* uncertainty

$$H(A,B) = -\sum_j \sum_k p(A_j,B_k) \log p(A_j,B_k) \qquad [17.14.2]$$

and the average conditional uncertainties

$$H(A|B) = -\sum_k \sum_j p(A_j,B_k) \log p(A_j|B_k) \qquad [17.14.3]$$

and

$$H(B|A) = -\sum_j \sum_k p(A_j,B_k) \log p(B_k|A_j). \qquad [17.14.4]$$

Now given these variance-like measures for *qualitative* data, indices of strength of association much like ρ^2 and ω^2 may be formed:

relative reduction in uncertainty in B given $A = \dfrac{H(B) - H(B|A)}{H(B)}.$ [17.14.5*]

Notice the similarity between this index and ω^2: if B is thought of as the dependent variable, this index gives the proportion by which knowing the A category reduces uncertainty about B, just as ω^2 tells the extent to which fixing X reduces the variance in Y.

A symmetric measure of association highly analogous to ρ^2 is

$$\frac{H(A) + H(B) - H(A,B)}{\text{minimum } [H(A), H(B)]} \qquad [17.14.6*]$$

indicating the relative strength of association between both variables.

These indices can be computed from the relative frequencies in sample contingency tables, and used purely as descriptive statistics if one so desires. Furthermore, a sampling theory exists for these statistics, and tests for zero association are possible. This sampling theory is highly related to the likelihood ratio tests of Section 17.8. Their main contribution to statistical method lies, however, in the possibility of extending the familiar notions having to do with variance and factors accounting for variance to qualitative data situations. We will not go further into this topic, but the interested student is advised to look into the books by Attneave (1959) and Garner (1962), where these possibilities are discussed in detail. It is safe to say that in the future this information-analysis approach will provide many new links between methods developed for numerical data, and those based on qualitative distinctions; some important ideas for extending the standard experimental designs to cover the qualitative measurement situation have already come from this approach.

17.15 RETROSPECT: THE χ^2 TESTS AND MEASURES OF ASSOCIATION

The problems to which the Pearson χ^2 tests apply are basically those of comparing distributions: in tests of goodness of fit, the null hypothesis states that some theoretical distribution obtains, and the question itself is one of "fit" between the hypothetical and the obtained distributions in terms of the same event categories or groupings. Tests for association may be regarded in a similar way: the hypothesis of independence between two attributes dictates a particular relationship we should expect to hold between the cell frequencies and the marginal frequencies in the obtained joint distribution. Divergence between the expected and obtained frequencies is regarded as evidence against independence.

On the face of the matter, χ^2 tests are simple and appealing. They are almost the only methods most psychologists know for handling qualitative data. On the other hand, χ^2 tests are always approximate, and the evidence at hand suggests that the goodness of the approximation varies with a number of factors, not all of which can be taken into account in a simple rule of thumb. Most prominent among the requirements for a satisfactory use of χ^2 tests is a large sample size, but even here statisticians are in disagreement about a sufficient number of cases to permit use of these tests. About the only advice that can be given the beginner in statistics in using these tests is caution: the Pearson χ^2 tests look easy, but probably nowhere in statistical inference is it more important to recognize the "if–then" nature of the conclusions.

Chi-square tests have been used wholesale in many areas in psychology, and in studies of the most varied kinds of problems. In the light of the statistical requirements for these tests, a large proportion of these applications are doubtless unjustified. However, even granting that the test is justified in the first place, what does the use of a χ^2 test for association actually tell the psychologist? If N is very large, as it should be for the best application of the test, virtually any "degree" of true statistical relationship between attributes will show up as a significant result. The test detects virtually any departure from strict independence between the attributes for these large sample sizes. Given the significant result, the experimenter can say that the two attributes are not independent, but is that his real interest? It has been said before, but it bears saying again: surely nothing on earth is completely independent of anything else. Given a large enough sample size, the chances are very good that the psychologist can demonstrate the association of any two qualitative attributes via a χ^2 test.

It seems to this author that the really important thing is some measure of the *strength* of association between the attributes studied. Such measures do take account of sample size, and do give some indication of how knowledge of one attribute can contribute to prediction of status on the other. Especially useful in this regard are the measures of predictive association and the information measures mentioned above. Such measures may actually tell the psychologist much more about the possible importance and meaning of a given relationship than can any χ^2 test alone. Thus, perhaps the emphasis in many such studies should be shifted from the sheer significance of the test to an appraisal of the strength of relationship represented. How much of a real increase in ability to *predict* does this finding seem to represent?

Admittedly, this is not always the experimenter's problem. There are circumstances where for theoretical or other reasons the experimenter wants merely a comparison between distributions, either sample and theoretical, or for two samples. Provided that the distributions considered are of some random variable, so that the obtained distributions can be put into cumulative form, then methods superior to χ^2 tests exist. Most notable are the so-called Kolmogorov-Smirnoff tests, either for one or two samples. These tests provide a direct comparison between distributions, and thus handle one aspect of the problem to which a Pearson χ^2 test is often applied, without some of the objectionable features of such a test. A description of the Kolmogorov-Smirnoff tests is given in Siegel (1956). Other tests are also possible when data are numerical in form, as will be shown in the next chapter.

All in all, the use of χ^2 tests is not a panacea for the problems of the psychologist; far from it! Granted that psychologists do want to study qualitative data, and do want to make inferences about possible relations, the price paid in using χ^2 tests for finding out only that "something" goes with "something" is fairly high, and may not be worth it. On the other hand, regardless of sample size, there *is* information in qualitative data about the *strength* of statistical relationship, and in many problems this is where the emphasis in data analysis should fall. In the light of its somewhat complicated statistical character, a significant Pearson χ^2 test may mean next to nothing, but an apparent *predictive* relationship in the data is usually worth looking into.

EXERCISES

1. In the Midwest, a large number of sportscasters make predictions each Thursday about the outcomes of Saturday's football games during the season. On a certain Thursday, each of 50 sportscasters made predictions about the same eight games. The numbers of correct predictions of the winner are as follows:

NUMBER CORRECT	NUMBER OF SPORTSCASTERS
8	1
7	3
6	5
5	11
4	15
3	8
2	5
1	1
0	1

Test the hypothesis that the correct predictions of any given sportscaster are outcomes of a stable and independent binomial process with $p = .5$.

2. On a simple test of arithmetic, with items all of equal difficulty, a teacher recorded the item number on which each child made his first error. For 100 children, the item numbers on which the first error occurred were distributed as follows:

ITEMS WITH FIRST ERROR	NUMBERS OF CHILDREN
6 or more	9
5	7
4	10
3	15
2	29
1	30
	100

If an error on any item for any child is like the outcome of a stable and independent Bernoulli process, test the hypothesis that p(error) = 1/3. (Hint: recall the Pascal and geometric distributions.)

3. In a study of the possibility of a sex-linkage with the occurrence of identical twins, a random sample was taken of records of normal deliveries from a particular set of hospitals over a one year period. These records revealed the following:

	MALE	FEMALE
SINGLE BIRTHS	658	688
IDENTICAL TWINS	39	34

Is there significant evidence that identical twins are more likely to be a given sex than are infants born singly? What does one assume about the respective deliveries recorded in this table?

4. In a study of parents' and teachers' perceptions of children, members of a random sample of normal first-grade children were rated separately by their parents and by their teachers on muscular coordination. Is there a significant relationship between the ratings of parents and teachers? How would you describe this relationship, if such exists?

		PARENTS' RATINGS			
		POOR	FAIR	GOOD	EXCELLENT
	POOR	33	48	113	209
TEACHERS'	FAIR	41	100	202	255
RATINGS	GOOD	39	58	70	61
	EXCELLENT	17	13	22	10

5. Compute the Goodman-Kruskal indices of predictive association for the data of problem 4, for predictions of teachers' ratings from those of parents, and parents' ratings

from those of teachers. Comment on the meaning of the values obtained from these indices.

6. A large horse show employed two judges to admit entrants into the final competition. There were 106 initial entrants. The results were as follows:

		JUDGE I	
		ADMIT	DON'T ADMIT
JUDGE II	ADMIT	28	14
	DON'T ADMIT	22	42

Given that these entries are a random sample from some large population, and that the respective judgements were independent, would you say that the two judges have the same probability of admitting an entrant into the horse show finals?

7. An experimenter constructed a six item multiple-choice test, each item having four possible answers. Suppose that when a subject is simply guessing, the probability of getting the right answer on any given item is exactly 1/4. Furthermore, suppose that the answer guessed on any item can be assumed to be independent of the answer guessed on any other item. Now the experimenter gave the test to 420 subjects, and found the following frequency distribution of numbers of items correct:

NUMBER CORRECT	FREQUENCY
6	32
5	10
4	34
3	62
2	108
1	121
0	53
	420

Test the hypothesis ($\alpha = .01$) that each subject was guessing independently on each item.

8. According to Mendelian genetics, if a parent having two dominant characteristics A and B is mated with another parent having the two recessive characteristics a and b, the offspring should show combinations of dominant and recessive characteristics with the following relative frequencies:

TYPE	RELATIVE FREQUENCY
Ab	9/18
aB	4/18
Ab	4/18
ab	1/18

In an actual experiment, when an AB parent was mated with an ab parent, the following frequency distribution of offspring resulted:

TYPE	RELATIVE FREQUENCY
Ab	39
aB	19
Ab	16
ab	1
	$\overline{75} = N$

Test the hypothesis that the Mendelian theory holds for these dominant and recessive characteristics ($\alpha = .05$).

9. In a comparison of child-rearing practices within two cultures, a researcher drew a random sample of 100 families representing Culture I, and another sample of 100 families representing Culture II. Each family was classified according to whether the family was father-dominant or mother-dominant, in terms of administration of discipline. The results follow:

	CULTURE 1	CULTURE II
FATHER-DOMINANT	53	37
MOTHER-DOMINANT	47	63

At approximately what α level can one reject the hypothesis of no association between culture and the dominant parent in a family?

10. For the data of problem 9, find the coefficient of predictive association (λ) for predicting parent-dominance of a family from the culture. Do these data suggest the presence of a very strong predictive association here? Explain.

11. Four large midwestern universities were compared with respect to the fields in which graduate degrees are given. The graduation rolls for last year from each university were taken, and the results put into the following contingency table:

UNIVERSITY	LAW	MEDICINE	FIELD SCIENCE	HUMANITIES	OTHER
A	29	43	81	87	73
B	31	59	128	100	87
C	35	51	167	112	252
D	30	49	152	98	215

Is there significant association ($\alpha = .05$) between the university and the fields in which it awards graduate degrees? What are we assuming when we carry out this test?

12. In a study of the possible relationship between the number of years of nursery and kindergarten-school training experienced by a child, and his rated deportment in the first grade, a random sample of 150 first-graders was obtained, and each was rated in terms of behavior:

	DEPORTMENT IN 1ST GRADE		
PRIOR EXPERIENCE	POORLY BEHAVED	MODERATELY WELL-BEHAVED	VERY WELL-BEHAVED
2 YRS. + KINDERGARTEN	6	12	0
1 YR. + KINDERGARTEN	12	25	6
KINDERGARTEN ONLY	14	31	12
NO KINDERGARTEN	2	23	7

Is there significant association ($\alpha = .05$) between amount of nursery school and kindergarten and rated deportment in the first grade, according to these data?

13. As judged from the data of problem 12, to what extent does knowing the number of years in pre-school and kindergarten permit us to predict the behavior rating of a child in first-grade? (Hint: compute the index of predictive association.) Comment on this result in the light of the finding in problem 12 above.

14. Four random samples of 44 subjects each were drawn, and each sample assigned to a different experimental condition. Each subject was given the same problem-solving test, with a possible score of 0 through 12. The results yielded four frequency distributions, as follows. Test the hypothesis that these four samples represent identical population distributions ($\alpha = .05$):

SCORES	SAMPLE 1	SAMPLE 2	SAMPLE 3	SAMPLE 4
12	1	5	3	2
11	1	2	3	6
10	1	2	3	5
9	2	6	3	6
8	3	2	3	1
7	8	4	4	1
6	12	2	6	2
5	8	4	4	1
4	3	2	3	1
3	2	6	3	6
2	1	2	3	5
1	1	2	3	6
0	1	5	3	2
	44	44	44	44

15. By inspecting the data table in problem 14, see if you can conclude what the F-value would have been if an analysis of variance had been used. Would this be significant?

Why? How do these sample distributions differ? What does this show about a comparison of samples via a chi-square test as opposed to a comparison via the F-test in analysis of variance?

16. A psychological study yielded the following data for 50 randomly-chosen subjects:

66	78	82	75	94	77	69	74	68	60
96	78	89	61	75	95	60	79	83	71
79	62	67	97	78	85	76	65	71	75
86	84	75	81	68	63	62	75	76	77
73	65	88	87	60	62	71	78	85	72

Test the hypothesis that the population distribution of such scores is normal. (Hint: use 10 class intervals, each of which has probability 1/10 in a normal distribution.)

17. In a study of the effect of a particular kind of cortical lesion upon the ability of a monkey to learn a discrimination problem, two groups of six monkeys each were used. The following data were found:

	SOLVED PROBLEM	DID NOT SOLVE PROBLEM
EXPERIMENTAL GROUP	1	5
CONTROL GROUP	5	1

Is there a significant difference ($\alpha = .05$) between these two groups of monkeys, so that one may say that the experimental group is less likely to solve the problem?

18. A researcher was interested in the stability of political preference among American women voters. A random sample of 80 women who voted in the elections of 1956 and 1960 showed the following results:

		1960 VOTE		
		REPUBLICAN	DEMOCRAT	
1956 VOTE	REPUBLICAN	34	11	45
	DEMOCRAT	5	30	35

Do these data afford significant evidence ($\alpha = .05$) that the true proportion of women who voted Republican in 1960 was different from the true proportion who voted Republican in 1956? (Hint: this problem involves *overlapping* or *correlated* proportions.)

19. In a study of the possible relationship between a military officer's own confidential judgment of his effectiveness as a leader, and the judgment of him by his immediate superior, a sample of 112 officers was drawn at random. Each officer rated himself with respect to his leadership ability, and then his immediate superior was asked to rate him. These ratings were made independently. The data turned out as follows:

RATING BY SUPERIOR

	LOW	MOD LOW	MOD. HIGH	VERY HIGH
VERY HIGH	1	9	7	6
HIGH	2	5	8	12
MOD. LOW	4	12	15	3
LOW	5	10	8	5

SELF-RATING

Does there seem to be significant ($\alpha = .05$) association between an officer's own judged leadership ability and the judgment of his immediate superior? Compute the Cramér coefficient showing the relative degree of association between self and superior's ratings.

20. Two pairs of dice are under study. One pair of dice is black, and the other pair white. Each pair of dice is tossed 360 times, and the following distributions of results obtained:

NUMBER OF SPOTS	BLACK DICE	WHITE DICE
12	11	13
11	21	17
10	29	35
9	41	35
8	48	51
7	62	60
6	51	49
5	39	45
4	31	25
3	19	23
2	8	7

Can we say ($\alpha = .05$) that
   (a) the black dice are fair (probability of each side of each die equal to 1/6)?
   (b) the white dice are fair?
   (c) the true distribution of results is the same for the black and the white dice?

18 Some Order Statistics

So far in our discussion of statistical inference we have dealt with three different kinds of experimental situations. In the first situation the experimental factor is thought of as a set of qualitative distinctions, and the dependent variable as some numerical measure. Tacitly we assumed that each Y score represented some interval-scale measurement of a property of interest, and that the basic question had to do with the possible effects of the experimental manipulation on the underlying property under study. However, as pointed out in Chapter 3, the question of level of measurement represented by the numerical values Y has to do with the *interpretation* put on the experiment, and statistical considerations enter in terms of the population distribution of the random variable Y, whatever these values actually represent. The classic methods, such as t and F tests, require that assumptions such as normality of parent distributions and homogeneity of variances be made; for this reason, such tests are often called "parametric," since their derivation involves explicit assumptions about population distributions and parameters. However, it is entirely possible to have an experimental situation fitting this general pattern where it is quite unreasonable to make these assumptions. Here, the study of association between experimental and dependent variables clearly must be approached in some other way.

In the second kind of experimental situation, both variables X and Y are numerical scores: these are the problems of regression and of correlation. Once again, the statistical methods per se are neutral about what these scores really represent, but the interpretation of the experiment may depend very heavily on whether interval scale or some other level of measurement holds for X and Y. For instance, it may well be true that, considered simply as numerical scores, the values of X and Y exhibit a more or less linear relationship. However, if either X or Y is *not* an interval-scale representation of the magnitude of some underlying property, we have no basis at all for saying that the "real" psychological variables they represent have this particular form of relationship. Furthermore, significance tests and confidence intervals for regression and correlation are *parametric* methods, since they derive their validity from assumptions having to do with populations, such as the assumptions of homogeneous variances and bivariate normal distributions. Nevertheless, sometimes an experimenter wants to know if there *is* a statistical relation between two psychological factors, even though he can commit himself neither about the interval-scale character of the numerical values themselves nor about the population distribution of the scores.

Finally, in the preceding chapter we discussed situations where both variables in an experiment are treated qualitatively. Each observation is thought of as a representative of some qualitative *joint* event, and the essential question concerns the statistical association between two qualitative attributes (that is, nominal scales). Here, it is necessary to assume very little about the population sampled, but a great deal is necessary in terms of sample size, independence of observations, and so on, if we are to get a valid and workable test of the hypothesis of nonassociation or independence. Tests such as χ^2 are appropriately called "nonparametric" since they require almost no "gratuitous" assumptions about the population sampled. However, such tests are *not assumption-free;* far from it! Rather, tests such as χ^2 are actually based on discrete sampling distributions, which, in principle, can be worked out exactly by application of elementary rules of probability. When discrete sampling distributions are involved, and exact probabilities can be found for sample results, it is usually possible to dispense with some of the assumptions made in deriving tests based on continuous sampling distributions. Even when the actual test statistic, such as χ^2, gives only an approximation to the exact results, the basic rationale nevertheless involves rather simple probability theory. As we saw in the preceding chapter, difficulties with the Pearson χ^2 statistic arise mostly from its use to *approximate* multinomial probabilities; the argument for the multinomial distribution of possible sample results is straightforward. In Chapter 17 it was also pointed out that in some situations tests such as χ^2 use only part of the potential information in the data, and the user may pay for this waste of information by being unable to make particular kinds of inferences from his result. Furthermore, the relative freedom from assumptions in such tests is offset in part by the large sample size necessary to justify the use of approximate methods.

In this chapter we are going to discuss methods based primarily on the **order relations** among observations in a set of data. The reasoning behind each of these tests involves relatively simple applications of probability theory; it happens that discrete sampling distributions often can be found in particularly simple ways if only the order features of the data are considered. In particular, many of these tests rest on the idea of the possible randomizations of the data. Perhaps the preceding discussion has suggested two reasons why we should be interested in analyzing data in terms of ordinal properties: In the first place, the only relevant information in a set of numerical scores may actually be ordinal. That is, it may well be true that the operation used to measure some psychological property actually is valid only at the ordinal level, so that the numerical scores

obtained actually give information only about relative magnitudes of the underlying property, and arithmetic differences between scores have no particular
meaning in terms of this underlying property. This is a common situation in
psychological measurement. When the experimenter knows this to be true of his
measurement operation, he wants only the relevant order information in the data
to figure in the analysis. Even if he can assume only that the observed score is
some unknown function of the underlying variable he hoped to represent, he may
still be able to reach a judgment about independence or association of underlying
variables by an ordinal analysis of the data (see Section 4.9).

An equally common reason for using order methods is that one or
more assumptions about the population distributions, strictly necessary for the
parametric method to apply, may be quite unreasonable. Rather than use the
parametric method anyway and wonder about the validity of his conclusion, the
experimenter prefers to change the question in such a way that another method
does apply. Ordinarily, adoption of such a method will require that only certain
features of the raw numerical data be considered if the objectionable assumptions
are to be avoided, so that all of the numerical information in the scores is not used.
Consequently, the experimenter may lose something by deciding to use the nonparametric method: the question answered by the nonparametric method is seldom exactly the same as that answered by the corresponding parametric method,
and for a given sample size the nonparametric test may represent a considerably
"weaker" use of the evidence.

Because of their direct dependency on elementary probability
theory, and their comparative freedom from assumptions about population distributions and parameters, order techniques are usually classified among the
nonparametric methods. However, not all nonparametric techniques involve considerations of order, nor are all order methods completely free of assumptions
about distributions and parameters. Therefore, in this chapter, the author prefers to limit discussion to a few methods based on order, and more or less beg
the complicated question of "parametric" versus "nonparametric." For this, and
virtually every other question concerned with such statistical methods, the reader
is referred to the monumental *Handbook* by Walsh (1962, 1965). This chapter
simply shows some techniques that are often found useful, and that happen to
involve order. Nevertheless, a few comments will be made apposite to the use of
these techniques in situations where other techniques such as t and F might
also apply.

18.1   ORDER TECHNIQUES AS SUBSTITUTES FOR
THE CLASSICAL METHODS

In the past few years, a great deal of study has been given to relative merits of parametric and nonparametric tests in situations where both types
of methods apply. These studies show clearly that advantages and drawbacks
exist in the use of any of these methods. However, research workers, and especially psychologists, sometimes gain the impression that they "get away with
something" by using an order method or some other nonparametric technique

in preference to one of the classic statistical tests. In some miraculous way, using such a technique is supposed to solve all the problems raised by unknown measurement level, objectionable assumptions, and so on (including, some apparently believe, sloppy data). If this is true, this is the only known example of something for nothing in statistics or anywhere else. Just as with any statistical method, there are both potential gains and losses in the decision to use a nonparametric technique, and the choice among methods can be evaluated only in the light of what the experimenter wants to do and the price he is willing to pay.

Clearly a word of warning is called for in the use of order methods as stand-ins for the classical methods in situations where both kinds of methods are appropriate. At least two things must be borne in mind:

First, the actual hypothesis tested by a given order method is seldom exactly equivalent to the hypothesis tested by a parametric technique. For example, when the usual assumptions for the simple analysis of variance are true, then the hypothesis that the means of the populations are equal *both implies and is implied by* the absolute identity of the population distributions. Under these assumptions, the test statistic is distributed as F when and only when the null hypothesis is true. On the other hand, in order methods designed for the comparison of J experimental groups, the actual null hypothesis ordinarily is that all possible orderings of observations by their scores in the data are equally likely. This is implied by the hypothesis of identical populations, so that if the equal-likelihood hypothesis is rejected, the hypothesis of identical populations can also be rejected. Thus, these order tests can be regarded essentially as testing the hypothesis of identical population distributions. Regardless of how well the actual test statistic agrees with expectation, however, the population distributions still *might* be different in particular ways. In most order tests the sampling distribution implied by a true H_0 can also obtain when H_0 is not true in various ways. Departures from strict identity among the population distributions may exist to which the order methods are very insensitive. If one is willing to make only minimal assumptions about the population distributions, then the kinds of true differences among populations that the test fails to detect may be quite unknown. Of course, various additional assumptions can be made about the population distributions, and then the sensitivity of these tests to various alternatives can be studied. Nevertheless, these gratuitous assumptions may be fully as offensive as the assumptions the test was designed to avoid in the first place. It is sad but true: the specificity of our final conclusion is more or less bought in terms of what we already know or can at least assume to be true. If we do not know or assume anything, we cannot conclude very much.

Second, when both the order method and a parametric method actually do apply (that is, when the parametric assumptions are true), the power of the two kinds of tests may be compared, given α, the sample N, and the true situation. Order techniques share with other nonparametric methods the disadvantage of being relatively low-powered as compared with parametric tests. This means that, other things such as α and N being equal, one is taking more risk of a Type II error in using the order method. If Type II errors are to be avoided, then a relatively larger sample size (or a larger α value) is required in the use of the order technique as compared to the parametric method.

In connection with the second point, a useful concept in the comparison of tests of hypotheses is that of "power-efficiency." The general idea may be given in this way: Suppose that there is some null hypothesis to be tested, and that either of two tests (methods of testing H_0), test U and test V, might appropriately be applied. The power of test U and of test V against any alternative to H_0 will depend upon several things, of course, one of which is sample size. For a given degree of power against a specific true alternative, test U may require N_U cases, whereas test V may require N_V cases. In general, for different tests, N_U and N_V will be different for the same power level.

Now suppose that U is the more powerful test, in the sense that it requires fewer cases to detect a true alternative to H_0 for some fixed α and $1 - \beta$ probabilities. Then the **power-efficiency** of test V relative to test U is

$$\text{power-efficiency of } V = \frac{(100)N_U}{N_V}.$$

The more cases N_V that test V requires to attain the same power as test U with N_U cases, the smaller is the power-efficiency of test V relative to test U.

For instance, if test U requires 20 cases to reach power of .95 for a given true situation and $\alpha = .05$, and test V requires 40 cases to reach the same power under the same conditions, then the power efficiency of V relative to U is

$$\frac{100(20)}{40} = 50 \text{ percent.}$$

In this way nonparametric statistics (such as order methods) may be compared with parametric methods such as the t test and the analysis of variance. Results of such comparisons show that **order methods generally have less than 100 percent power-efficiency when used in situations where the most powerful parametric tests such as t and F apply.** Order methods require more evidence than parametric methods to yield comparable conclusions.

These remarks should not be construed as holding against all applications of order statistics, however. The comments on power hold only where the appropriate "high-powered" methods can appropriately be applied. If the data are collected as order data in their own right, perhaps because no higher-level measurement operation is available, then this objection does not necessarily hold; the concept of comparative power between parametric and nonparametric tests is useless here since the experimenter really has no choice to make. Similarly, comparative power is difficult or impossible to study when the assumptions underlying parametric tests are not true. For data that are essentially ordinal, or when assumptions are manifestly untrue, some of the tests described here may be about as powerful as can be devised. Furthermore, since order methods apply to all sorts of population distributions, their applicability is considerably more general than the parametric methods. Note well, however: this refers to the generality of *application* of the method, and not to the *generalizability* of the results from a sample, which depends on how the sample was drawn and randomized. It is not clear at all that the use of a nonparametric technique makes *conclusions* more general.

The decision to use or not to use order methods in a given problem cannot be given a simple prescription. This is but another place where the experimenter has to think about what he wants to accomplish and how. It is wrong to conclude that statistical assumptions are bad, that only bona fide interval-scale data can be subjected to the classical statistical treatment, that tests with relatively low power-efficiency are useless, and so on. None of these statements is necessarily true in all situations. Through practice and all the help he can get the experimenter must learn to pick and choose among all of the various methods available, finding the one that most clearly, economically, and reasonably sheds light on the particular question he wants answered. This is not a simple task, and a brief discussion such as this can only begin to suggest some of these issues.

In the sections to follow, several types of order statistics and tests will be discussed. First of all, tests will be mentioned that are appropriate to experimental data where the experimental factor is categorical and the dependent variable is treated at the ordinal level. Then some correlation-like methods for ordinal data will be surveyed, and finally an index of association for ordered classes will be discussed.

18.2   Comparing Two or More Independent Groups: The Median Tests

In Chapter 5 the median test was introduced as an example of the use of the binomial distribution. Now we will go into this method in more detail. The median test, which is one of the simplest of the order methods, involves the comparison of several samples on the basis of deviations from the median rather than from the mean. We assume that the underlying variable on which the populations are to be compared is continuous, and that the probability of a tie between two observations in actual value of the underlying variable is, in effect, zero.

The null hypothesis to be tested is that the J different populations are absolutely identical in terms of their distributions. The alternative is simply the contrary of H_0. If we wish to add the assumption that whatever the differences in central tendency that may exist among the distributions, they are at least identical in *form*, then the test actually becomes one of central tendency; however, it is hard to see why such an assumption would ordinarily be justified in a situation where an order method such as this is called for. Thus, in this method, as well as in most of the methods to follow, the null hypothesis will state only that the distributions are identical. The median test will be sensitive to differences in central tendency for the various population distributions, but failure to reject H_0 does not necessarily imply that the distributions are, in fact, identical, unless one is willing to make other assumptions.

The method is as follows: The J different sample groups are combined into a single distribution and the grand median for the sample, Md, is obtained. Now each score in each group is compared with Md. If the particular score is above the grand median, the observation is assigned to a "plus" category; if the particular score is not above the grand median, the observation is assigned to a "minus" category. Let a_j be the number of "plus" observations in group j, and let $n_j - a_j$ represent the number of "minus" observations. Then the data are arranged into a $2 \times J$ table such as the following:

		GROUP				
	1	...	j	...	J	*Total*
PLUS	a_1		a_j		a_j	a
MINUS	$n_1 - a_1$		$n_j - a_j$		$n_J - a_J$	$N - a$
	n_1	...	n_j	...	n_J	N

Notice that this is simply a joint-frequency or contingency table, where one attribute is "group" and the other is "plus or minus."

Now, if the value Md actually divides each of the populations in exactly the same way there should be a binomial distribution of the sample numbers a_j for random samples from population j. That is, the probability of exactly a_j "plus" observations in group j is

$$\binom{n_j}{a_j} p^{a_j} q^{n_j - a_j}$$

where p is the probability of an observation's falling into the "plus" category and $q = 1 - p$. Under the null hypothesis, the probability p should be the same for each and every population, since any value Md should divide the population distributions identically.

However, we want the probability of the obtained *sample* result, conditional to the fact that the *marginal* frequency of "plus" observations is a. Under the null hypothesis, the probability of exactly a observations above the value Md is $\binom{N}{a} p^a q^{n-a}$. Then the conditional probability for a particular arrangement in the table works out to be

$$\frac{\binom{n_1}{a_1}\binom{n_2}{a_2} \cdots \binom{n_j}{a_j} \cdots \binom{n_J}{a_J}}{\binom{N}{a}}.$$

This is the probability of this *particular* sample result conditional to this particular value of a. Since this probability can be found for any possible sample result, the significance level can be found by exact methods, involving finding all possible sample results differing this much or more among the J groups. Such exact probabilities may be very laborious to work out, of course, unless J is only 2 or 3.

For large samples a fairly good approximation to the exact significance level is found from the statistic

$$\chi^2 = \frac{(N-1)}{a(N-a)} \sum_{j=1}^{J} \frac{(Na_j - n_j a)^2}{N n_j}. \qquad [18.2.1\dagger]$$

For reasonably large samples ($N \geq 20$, $n_j \geq 5$ for each j), this statistic is distributed *approximately* as chi-square with $J - 1$ degrees of freedom. Rejection of the null hypothesis lets us assert that the populations represented are *not* identical.

One difficulty with this test is that it is based on the assumption that ties in the data will not occur. Naturally, this is most unreasonable since tied scores ordinarily will occur. This is really a problem only if several scores are tied at the over-all median, Md, since other ties have no effect on the test statistic itself. When ties occur at the median, several things may be done to remove this difficulty, but the safest general procedure seems to be to allot the tied scores (those tied with the grand median) within each group in such a way that a_j is as close as possible to $n_j - a_j$. This at least makes the test relatively conservative. Remember that this χ^2 method gives an approximate test, and really should be used only when the sample size within each group is fairly large. In principle, exact probabilities may be computed, as suggested above, when sample N is small.

As an example, consider the following problem. An experimenter was interested in the effect which the difficulty of admission to a club has on the desire of a person to become a member. Thus, he decided to form four experimental "social clubs" at a large college. A sample of 100 girls was drawn and each was assigned at random to one of four experimental groups. Each girl was sent a letter asking her to come for an interview as a prospective member of a service club having secret membership. In experimental treatment I, the goals of the club were outlined, and quite easy membership requirements stated. In treatment II, exactly the same information was given, but with somewhat harder requirements to be met. Treatments III and IV included severe and truly formidable entrance requirements respectively. After a standard "dummy" interview, each girl was asked to rate on a 12-point scale just how eager she was to join such a group. On this scale 1 represented *most* eager, and 12 *least* eager to join. The data are shown in Table 18.2.1.

Table 18.2.1

	Group			
Rating	I	II	III	IV
12	0	0	0	0
11	3	2	1	0
10	7	3	4	1
9	5	5	4	4
8	5	5	1	0
7	3	6	3	10
6	2	2	5	8
5	0	2	4	1
4	0	0	3	1
3	0	0	0	0
	25	25	25	25 $N = 100$

The experimenter wanted to test the hypothesis that these groups represented four potential populations having identical distributions of ratings. The combined distribution of ratings was as follows:

	f	cf
12	0	100
11	6	100
10	15	94
9	18	79
8	11	61
7	22	50
6	17	28
5	7	11
4	4	4

$$Md = 7.5$$

When each rating was compared to the value of Md, the following results emerged:

	I	II	III	IV	
plus	20	15	10	5	$50 = a$
minus	5	10	15	20	$50 = N - a$

$$\chi^2 = \frac{(99)}{(50)(50)} \left\{ \frac{[(100)(20) - (25)(50)]^2}{2500} + \cdots \right.$$
$$\left. + \frac{[(100)(5) - (25)(50)]^2}{2500} \right\}$$

$$= 19.8$$

For 3 degrees of freedom, this value exceeds the χ^2 value for the .01 level, and so the experimenter can say that the populations are not identical. The hypothesis tested is not about any particular characteristic of the populations considered, but rather about the absolute identity of their distributions. Some departure from strict identity of population distributions is inferred from this significant result, and the experimenter has evidence that the experimental treatment is associated with the rated attraction of a club for a girl.

18.3   THE MEDIAN TEST FOR MATCHED GROUPS

It is possible to extend the idea of the median test to situations where the various groups are not independent random samples, but rather are matched in terms of one or more factors. The method outlined below is especially appropriate to the kind of experiment referred to as a randomized blocks design in Chapter 13. That is, some J experimental treatments are to be compared, and there are some K blocks or levels of a nuisance factor. Within each of the K blocks, some J subjects are chosen and assigned at random to the treatments. Ordinarily, these K blocks or levels are a random sample of all possible such levels of the nuisance factor, so that if treated by the analysis of variance this would correspond to a mixed-model design.

In these circumstances, the data are arranged into a table with J columns and K rows, just as for the corresponding analysis of variance. The median of each row, or Md_k, is found. Next, each observation within a row is given a categorization as "plus" or "minus," depending on whether or not its value falls above the median for its row. Now the number of plus observations occurring in column j is found.

Let

$$a_j = \text{number of "plus" observations in column } j.$$

Having done this for each column j, we have a table exactly like that in Section 18.2. However, here we let

$$a = \frac{J}{2} \qquad \text{for } J \text{ even}$$

and

$$a = \frac{J - 1}{2} \qquad \text{for } J \text{ odd.}$$

Then, under the assumption that each and every row population has the same form of distribution, we can test the hypothesis that the distributions represented by the columns are identical. Actually, one is here doing something analogous to a test of column effects against interaction in an analysis of variance; if interest lies in the extent to which populations differ in terms of central tendency, either he should be prepared to assume that interaction effects do not exist, or the data should correspond to a mixed or random model experiment.

The actual test statistic is given by

$$\chi^2 = \frac{J - 1}{Ka(J - a)} \frac{\displaystyle\sum_{j=1}^{J} (Ja_j - Ka)^2}{J} \qquad [18.3.1\dagger]$$

for $J - 1$ degrees of freedom. For a relatively large number of rows, this test is satisfactory even though there is only one observation per cell in the data table. Mood (1963), one of the developers of this test, suggests that the large sample approximation should be fairly good when the number of *cells* in the data table is at least 20.

For example, suppose that the experiment of the preceding section had been carried out, but that 10 *matched groups* of 4 girls each had been used. The girls were matched in terms of social and service activities on the campus. Suppose that the data were as shown in Table 18.3.1. Be sure to notice that the

Table 18.3.1

Group	I	II	III	IV	Row median
		Treatment			
1	9(+)	5(−)	8(+)	4(−)	6.5
2	10(+)	6(+)	5(−)	3(−)	5.5
3	11(+)	8(+)	3(−)	4(−)	6.0
4	10(+)	7(+)	6(−)	5(−)	6.5
5	8(+)	5(−)	6(−)	7(+)	6.5
6	9(+)	8(+)	4(−)	2(−)	6.0
7	9(+)	7(+)	4(−)	3(−)	5.5
8	9(+)	10(+)	5(−)	4(−)	7.0
9	11(+)	10(+)	9(−)	8(−)	9.5
10	11(+)	6(+)	5(−)	4(−)	5.5
a_j	10	8	1	1	

rating value in each cell is compared with the median of the *row* for that cell. The over-all table of frequencies of plus and minus categories within columns is thus:

	I	II	III	IV	
plus	10	8	1	1	$20 = Ka$
minus	0	2	9	9	$20 = N - Ka$

Here, the test statistic has a value given by

$$\chi^2 = \frac{3}{20(2)} \left\{ \frac{[4(10) - 10(2)]^2}{4} + \cdots + \frac{[4(1) - 10(2)]^2}{4} \right\} = 19.8.$$

For 3 degrees of freedom, this greatly exceeds the χ^2 value required for significance at the .01 level.

This same general idea can be extended to two-way designs with replication, where a test of both main effects and interaction becomes possible. However, the number of required assumptions goes up in this situation, and the tests, especially that for interaction, are quite laborious to carry out. These methods are outlined in Mood (1950).

18.4   THE SIGN TEST FOR MATCHED PAIRS

In Chapter 5 we encountered a test applicable when the number of experimental treatments is only two, and *pairs* of observations are matched, the so-called sign test, which is based on the binomial distribution. We will now examine this test somewhat more closely.

Let N be the number of pairs of observations, where one member of each pair belongs to experimental (or natural) treatment 1 and the other to treatment 2. Here, the only relevant information given by the two scores for pair is taken to be the **sign of the difference** between them. If the two treatments actually represent identical populations, and chance is the only determiner of which member of a pair falls into which treatment, we should expect an equal number of differences of plus and of minus sign. The theoretical probability of a "plus" sign is .5, and so the probability of a particular number of plus (or minus) signs can be found by the binomial rule with $p = .5$ and N. Notice, however, that we must assume either that the population distributions are continuous so that exact equality between scores has probability zero, or that ties are otherwise impossible. Ordinarily, pairs showing zero differences are simply dropped from the sample, although this makes the final conclusion have a conditional character, "In a population of untied pairs . . . ," and so on. The test is carried out as follows:

First, the direction of the difference (that is, the sign of the difference) between the two observations in each pair is noted, with the same order of subtraction always maintained. Thus each *pair* is given a classification. of Plus or Minus, according to the sign of the difference between scores. If the null hypothesis of no difference between the two matched populations were true, one would expect half the nonzero differences to show a positive sign, and half to show a negative sign. Thus, one may simply take the *proportion* of plus differences, and test the hypothesis that the sample proportion arose from a true proportion of .50.

The normal approximation to the binomial can be used if the num ber of sample pairs is large (10 or more):

$$z = \frac{|P - p| - 1/(2N)}{\sqrt{pq/N}} = \frac{|P - .50| - 1/(2N)}{\sqrt{(.50)(.50)/N}}. \qquad \textbf{[18.4.1†]}$$

Here, $1/2N$ is a correction for continuity, as in Section 17.2. Although this form holds for a two-tailed test, either a one- or a two-tailed test may be carried out, depending on the alternative hypothesis appropriate to the particular problem. Naturally, in the one-tailed test the sign of the difference between P and .5 is considered.

If the sample size is fewer than 10 pairs, the binomial distribution should be used to give an exact probability. That is, the binomial table with $p = .5$ should be used to find the probability that the *obtained frequency* of the *more frequent sign* should be equaled or exceeded by chance give a true p of .50. These tables of exact probabilities are given in Siegel (1956) as well as many other places, although the binomial probabilities are not hard to work out. For a one-tailed test, one also checks that the more frequent sign accords with the alternative hypothesis before carrying out the test, of course. For a two-tailed test, this one-tailed binomial probability is doubled.

As an example of the sign test, consider this situation: It was desired to see the effect that a frustrating experience might have upon the "social age" of a child. Each child could be age-rated from his play by a trained observer.

Each of a random sample of preschool children was first rated by the observer according to the social-age level of his play during a free-play period. Pairs of children were then formed having the same rated "social age." One randomly chosen member of each pair was frustrated by being allowed to play with a desirable toy that was then taken away, while the pair-mates were not frustrated. The observer did not know which children were frustrated, and the children were given these experimental conditions separately. Finally, in a postexperimental session the children were again rated at free play. Suppose that twenty pairs were used, and that the postexperimental ratings were as shown in Table 18.4.1.

Table 18.4.1

Pair	Frustrated	Not frustrated	Sign of diff.
1	32	36	+
2	35	34	−
3	33	34	+
4	36	40	+
5	44	42	−
6	41	40	−
7	32	35	+
8	38	40	+
9	37	38	+
10	35	35	0
11	29	35	+
12	34	32	−
13	50	51	+
14	40	38	−
15	39	42	+
16	31	33	+
17	47	46	−
18	41	42	+
19	30	29	−
20	35	35	0

The final rating of the frustrated child was subtracted from that of his pair-mate in order to find the sign of the difference between them. Eleven pairs showed a positive sign, seven showed a negative sign, and two showed a zero difference. Did presence or absence of frustration seem to be related to the difference in rated social age? The null and alternative hypotheses are

$$H_0: p = .5$$
$$H_1: p \neq .5$$

where p is the population proportion of plus changes among pairs (plus being the more frequent sign of change in the sample). The statistic for this test is thus

$$z = \frac{|11/18 - .50| - 1/36}{\sqrt{(.50)(.50)/18}} = \frac{.611 - .500 - 028}{.118} = .70.$$

Note that the two pairs showing no difference were excluded from the number of pairs figuring in the test, so that N is reduced to 18. When referred to a normal distribution, this z value leads to the conclusion of no significant difference. There is not enough evidence to conclude that frustration did lead to a difference in social-age rating.

18.5 COCHRAN'S TEST

A test that seems to stand midway between methods designed for contingency-table data and methods based on order is due to Cochran (1950). This test can be viewed as a generalization of the McNemar two-sample test mentioned in Section 17.10, and it is appropriate in an experiment involving repeated observations (or matched groups) where the dependent variable can take on only two values:

$Y_{jk} = 1$ (for "success," "pass," and so on, recorded for individual k in treatment j).

$Y_{jk} = 0$ (for "fail," and so on, recorded for individual k in treatment j).

For example, suppose that in some experiment K subjects were observed in a standard situation where each subject performed individually under each of J different experimental conditions. Each subject was assigned the conditions in some random order. In each condition, the task of the subject was to solve one of a set of simple reasoning problems. If the problem was solved correctly within one minute, the performance was recorded as a "success," and as a "failure" otherwise. Suppose that the interest of the experimenter was in seeing if the problems had equal difficulty for the subjects; that is, if the true proportion of successes was constant over the problems. In this situation the Cochran test would apply.

As usual, let the experimental treatments (problems, in the example) be shown as columns in the data table. The subjects are shown as the rows. Then the entry in the cell formed by column j and row k contains Y, which is 1 for a success and 0 for a failure. Let

$$y_k = \sum_j y_{jk}$$

be the marginal total for row k and let

$$y_j = \sum_k y_{jk}$$

be the marginal total for column j. Finally, let

$$\bar{T} = \frac{\sum_j y_j}{J}.$$

Then the statistic for Cochran's test is given by

$$Q = \frac{J(J-1) \sum\limits_{j=1}^{J} (y_j - \bar{T})^2}{J \left(\sum\limits_{k} y_K \right) - \left(\sum\limits_{k} y_k^2 \right)}. \qquad [18.5.1\dagger]$$

For relatively large K, this is distributed approximately as chi-square with $J - 1$ degrees of freedom, when the hypothesis is true that the probability of a "success" is constant over all treatments J.

Table 18.5.1

	Problem				
Subject	1	2	3	4	Y_k
1	1	1	1	0	3
2	0	1	1	1	3
3	0	0	1	0	1
4	1	1	1	1	4
5	0	1	0	0	1
6	0	0	1	0	1
7	1	0	0	0	1
8	0	0	1	1	2
9	0	0	0	0	0
10	1	0	0	0	1
11	1	0	1	0	2
12	0	0	1	1	2
13	0	1	0	1	2
14	1	0	0	0	1
15	0	1	0	0	1
16	1	0	1	1	3
17	0	1	0	0	1
18	0	0	1	0	1
19	0	1	1	0	2
20	0	0	1	1	2
Y_j	7	8	12	7	34

$\bar{T} = \dfrac{34}{4} = 8.5$

$\sum\limits_{k} y_k^2 = 76$

$$Q = \frac{(4)(3)[(7 - 8.5)^2 + \cdots + (7 - 8.5)^2]}{4(34) - 76}$$

$$= 3.40$$

To continue the example, suppose that the experiment outlined above had been carried out. Twenty randomly selected subjects were given each of four problems, and each subject got the problems in some different, randomly chosen, order. A "1" was recorded for a successful solution, and a "0" for a failure. The data turned out to be as shown in Table 18.5.1. For 3 degrees of freedom, the χ^2 table shows this value not significant at either the .05 or .01 level. Thus, the hypothesis of no difference between problems is not rejected.

So far, all the tests discussed in this chapter have actually depended on some way of arranging the observations into only two ordered classes: above or below the median in the median tests, plus or minus differences in the sign test, 1 and 0 categories in the Cochran test. The only role of the numerical scores has been to assign individuals to one of these two categories. When numerical data are reduced to only two categories, it is obvious that much of the possible information in the data is sacrificed. However, other techniques exist which use somewhat more of the information in the scores themselves; that is, observations are rank-ordered in terms of their scores in these methods. Next we turn to one of the simplest of the tests where the numerical score serves to give each observation a place in order.

18.6 THE WALD-WOLFOWITZ "RUNS" TEST FOR TWO SAMPLES

This test applies to the situation where two unmatched samples are to be compared, and each observation is paired with a numerical score. The underlying variable that these scores represent is assumed to be continuously distributed.

Suppose that the numbers of observations in the two experimental groups are N_1 and N_2 respectively. All of the $N_1 + N_2$ observations in these samples are drawn independently and at random. For convenience, we will call any observation appearing in sample 1 an "A" and any observation in sample 2 a "B." Now suppose that all these sample observations, irrespective of group, are arranged in order according to the magnitude of the scores shown. Then there will be some *arrangement* or pattern of A's and B's in order. In particular there will be *runs* or "clusterings" of the A's and B's. This is easily illustrated by an example.

Suppose that in some two-sample experiment the data turned out as shown in Table 18.6.1. When these scores are combined into a single set and arranged in order of magnitude, we get the following

B B B   A A A   B B   A A   B   A A   B B   A A   B B   A
\1 1 2/\3 4 4/\5 5/\6 6/\7/\8 8/\12 13/\14 15/\18 19/\20/

1   2   3   4   5   6   7   8   9   10

Above each observation's score is an A or a B, denoting the group to which that observation belongs. Now notice that there are runs of A's and B's. That is, the

Table 18.6.1

Sample 1(A)	Sample 2(B)
8	12
6	13
8	19
4	18
14	7
4	2
15	1
20	1
3	5
6	5

ordering starts off with a run of three B's . . . this run is underlined and numbered 1. Then there is a run of three A's, which is run number 2. Proceeding in this way, and counting the beginning of a new run whenever an A is succeeded by a B or vice versa, we find that there are 10 runs.

It should be obvious that there must be at least two runs in any ordering of scores from two groups. If the groups are of equal size N there can be no more than $2N$ runs in all. In general, the number of runs cannot exceed $N_1 + N_2$.

Now suppose that the two groups were random samples from *absolutely identical population distributions*. In this instance we should expect many runs, since the values for the two samples should be well "mixed up" when put in order. On the other hand, if the populations differ, particularly in central tendency, we should expect there to be less tendency for runs to occur in the sample ordering. This principle provides the basis for a test based on fairly simple probability considerations.

For the moment, let R symbolize the total number of runs appearing for the samples. Then it can be shown that if R is any *odd* number, $2g + 1$, the probability for that number of runs is

$$\text{prob.}(R = 2g + 1) = \frac{\binom{N_1 - 1}{g - 1}\binom{N_2 - 1}{g} + \binom{N_1 - 1}{g}\binom{N_2 - 1}{g - 1}}{\binom{N_1 + N_2}{N_1}}$$

when all arrangements in order are equally likely.

If R is an *even* number, $2g$, then

$$\text{prob.}(R = 2g) = \frac{2\binom{N_1 - 1}{g - 1}\binom{N_2 - 1}{g - 1}}{\binom{N_1 + N_2}{N_1}}.$$

On this basis, the exact sampling distribution of R can be worked out, given equal probability for all possible arrangements of A and B observations. It turns out that for fairly large samples the distribution of R can be approximated by a normal distribution with

$$E(R) = \frac{2N_1N_2}{N_1 + N_2} + 1$$

and

$$\sigma_R^2 = \frac{2N_1N_2(2N_1N_2 - N_1 - N_2)}{(N_1 + N_2)^2(N_1 + N_2 - 1)}.$$

Thus, an approximate large sample test is given by

$$z = \frac{R - E(R)}{\sigma_R}$$

referred to a normal distribution. Since the usual alternative hypothesis entails "too few" runs, the test is ordinarily *one-tailed*, with only negative values of z leading to rejection of the hypothesis of identical distributions.

For sample size less than or equal to 20 in *either* sample, exact values of R required for significance are given in Siegel (1956).

There is some reason to believe that the runs test generally has rather low power-efficiency, compared either with a t test for means or with other order tests for identical populations. However, it is mentioned here because of its utility in various problems where other methods may not apply. Actually, A and B may designate any dichotomy within a *single* sample, and any principle at all that gives an ordering to A's and B's may be used. For example, it may be of interest to see if there is a time-related trend such as learning in a *single* set of data. In this instance, we might find the over-all median for the set of scores, calling above the median A and below B. Here time is used as the ordering principle, and the occurrence of few runs is treated as evidence that time trends do exist. In any problem where the data may be given a dichotomous classification in one respect and then simply ordered in another respect, the runs test gives a way to answer the question of possible association between the basis for the ordering and the categorization. This makes the runs test useful in a variety of problems where other tests do not apply directly, and especially to problems where the *experimental* variable is ordinal and the *dependent* variable categorical.

A major technical problem with this test is the treatment of ties. In principle ties should not occur if the scores themselves represent a continuous random variable. But, of course, ties do occur in actual practice, since we seldom represent the underlying variable directly or precisely in the data. If tied scores all occur among observations in the *same* group (as in the example above), then there is no problem; the value of R is unchanged by any method of breaking these ties. However, if members of *different* groups show tied scores, the number of runs depends upon how these ties are resolved in the final ordering. One way of meeting this problem is to break all ties in a way *least* conducive to rejecting the null hypothesis (so that the number of runs is made as large as possible). This at least makes for a conservative test of H_0. However, if cross-group ties are very numerous, this test is really inapplicable.

18.7   THE MANN-WHITNEY TEST FOR TWO
INDEPENDENT SAMPLES

Unlike the runs test, this test employs the actual *ranks* of the various observations directly as a device for testing hypotheses about the identity of two population distributions. It is apparently a good and relatively powerful alternative to the usual *t* test for equality of means.

The Mann-Whitney test can best be justified in a way somewhat different from the tests previously discussed. The rationale for this test can be based directly on the method of randomization, alluded to in our discussion of the Fisher exact test (Section 17.9), and in Chapter 13. That is, probability statements actually refer directly to all possible randomizations of the *same* sample of *N* subjects among the various treatment. However, we will not go further into this rationale here; suffice it to say that one can evaluate Type I error probability by referring to the sample space of all possible results of randomizing the same set of data provided that the original random sample is itself representative of the population to be investigated. The Mann-Whitney test, the Wilcoxen test (18.8), and, indeed, most of the order tests can be justified in terms of possible randomizations of the same sample.

We assume that the underlying variable on which two groups are to be compared is continuously distributed. The null hypothesis to be tested is that the two population distributions are identical. Then we proceed as follows:

The scores from the combined samples are arranged in order (much as in the runs test). However, now we assign a *rank* to each of the observations, in terms of the magnitude of the original score. That is, the lowest score gets rank 1, the next lowest 2, and so on. Now choose one of the samples, say sample 1, and find the *sum* of the ranks associated with observations in that sample. Call this T_1. Then find

$$U = N_1 N_2 + \frac{N_1(N_1 + 1)}{2} - T_1. \qquad [18.7.1\dagger]$$

If the resulting value of U is *larger* than $N_1 N_2/2$, take

$$U' = N_1 N_2 - U.$$

The statistic used is the *smaller* of U or U'. (Incidentally, this choice of the smaller of the two values for U is important in using tables to find significance for small samples, but is really immaterial in the large sample test to be described later.)

As an example, consider the following data:

Sample 1(A)	Sample 2(B)
8	1
3	7
4	9
6	10
	12

Arranged in order and ranked, these data become

	B	A	A	A	B	A	B	B	B
Score	1	3	4	6	7	8	9	10	12
Rank	1	2	3	4	5	6	7	8	9

The sum of the ranks for the A observations (group 1) is

$$T_1 = 2 + 3 + 4 + 6 = 15.$$

This is turned into a value of U by taking

$$U = 4(5) + \frac{4(5)}{2} - 15$$

$$= 15.$$

This is larger than $\frac{(4)(5)}{2}$ or 10, and so we take

$$U' = 20 - 15 = 5$$

as the value we will use.

Now notice that given these 9 *scores*, the value of U depends only on how the A's and B's happen to be arranged over the rank order. The number of possible random arrangements is just $\binom{N_1 + N_2}{N_1}$, and if the hypothesis of completely identical populations is true, the random assignment of individuals to groups should be the only factor entering into variation among obtained U values. Under the null hypothesis, all arrangements should be equally likely, and this gives the way for finding the probability associated with various values of U. For large samples, this sampling distribution of U is approximately normal, with

$$E(U) = \frac{N_1 N_2}{2}$$

and

$$\sigma_U^2 = \frac{N_1 N_2 (N_1 + N_2 + 1)}{12}.$$

Thus, for large samples, the hypothesis of no difference in the population distributions is tested by

$$z = \frac{U - E(U)}{\sigma_U}.$$

For a two-tailed test, either U or U' may be used, since the absolute value of z will be the same for either. However, if the alternative hypothesis is such that one of the populations should tend to have a lower average than the other (assuming distributions of similar *form*), then a one-tailed test is appropriate, and the sign of the z should be considered.

For situations where the larger of the two samples is 20 or more and the samples are not too different in size, the normal approximation given above should suffice. However, when the larger sample contains fewer than 20 observations, tables given in Siegel (1956) should be used to evaluate significance of U.

This test is one of the best of the nonparametric techniques with respect to power and power-efficiency. It seems to be very superior to the median test in this respect, and compares quite well with t when assumptions for both tests are met. For some special situations, it is even superior to t. This makes it an extremely useful device for the comparison of two independent groups.

Ordinarily, ties are treated in the Mann-Whitney test by giving each of a set of tied scores the *average* rank for that set. Thus, if three scores are tied for fourth, fifth, and sixth place in order, each of the scores gets rank 5. If two scores are tied for ninth and tenth place in order, each gets rank 9.5, and so on. This introduces no particular problem for large sample size when the normal approximation is used and ties are relatively infrequent. However, when ties exist σ_U^2 becomes

$$\sigma_U^2 = \frac{N_1 N_2}{12} \left[N_1 + N_2 + 1 - \frac{\sum_{i=1}^{G} (b_i^3 - t_i)}{(N_1 + N_2)(N_1 + N_2 - 1)} \right],$$

where there are some G distinct *sets* of tied observations, i represents any *one* such set, and b_i is the number of observations tied in set i. For a small number of ties and for large $N_1 + N_2$, this correction to σ_U^2 can safely be ignored.

18.8   THE WILCOXEN TEST FOR TWO
MATCHED SAMPLES

As we saw in Section 18.4, the problem of comparing two matched samples can be treated by the sign test if the only feature of the data considered is the sign of the difference between each pair. However, this still overlooks one other important property of any pair of scores: not only does a difference have a direction, but also a size that can be ranked in order among the set of all such differences. The Wilcoxen test takes account of both features in the data, and thus uses somewhat more of the available information in paired scores than does the sign test. This procedure has very close ties with the Mann-Whitney test; both are based essentially on the randomization idea. The Wilcoxen test also has very high power-efficiency compared to other methods designed specifically for the matched-pair situation.

The mechanics of the test are very simple: the signed difference between each pair of observations is found, just as for the t test for matched groups (Section 10.23). Then these differences are rank-ordered in terms of their absolute size. Finally, the sign of the difference is attached to the rank for that difference. The test statistic is T, the *sum of the ranks with the less frequent sign.*

Suppose that in some experiment involving a single treatment and one control group, subjects were first matched pairwise, and then one member of each pair was assigned to the experimental group at random. In the experiment proper, each subject received some Y score. Perhaps the data turned out as shown in Table 18.8.1. Here, the differences are found, their absolute size ranked, and then the sign of the difference attached to the rank. The less frequent sign is minus, and so

$$T = 3.5 + 1 + 3.5 + 7 = 15.$$

Table 18.8.1

Pair	Treatment	Control	Difference	Rank	Signed rank
1	83	75	8	8	8
2	80	78	2	2	2
3	81	66	15	10	10
4	74	77	−3	3.5	−3.5
5	79	80	−1	1	−1
6	78	68	10	9	9
7	72	75	−3	3.5	−3.5
8	84	90	−6	7	−7
9	85	81	4	5	5
10	88	83	5	6	6

The hypothesis tested by the Wilcoxen test is that the two populations represented by the respective members of matched pairs are identical. When this hypothesis is true, then each of the 2^N possible sets of *signed* ranks obtained by arbitrarily assigning $+$ or $-$ signs to the ranks 1 through N is *equally* likely. The random assignment of subjects to experimental versus control group in the example is tantamount to such a random assignment of signs to ranks when the null hypothesis is true. On this basis, the exact distribution of T over all possible randomizations can be worked out. For large N (number of pairs), the sampling distribution is approximately normal with

$$E(T) = \frac{N(N + 1)}{4}$$

and

$$\sigma_T^2 = \frac{N(N + 1)(2N + 1)}{24}$$

so that a large sample test is given by

$$z = \frac{T - E(T)}{\sigma_T}.$$

This test can be either directional or nondirectional, depending on the alternative hypothesis. However, a directional test usually makes sense only if one is prepared to assume that the distributions have the same form, and that a signed deviation of T from $E(T)$ is equivalent to a particular difference in central tendency between the two populations. For samples larger than about $N = 8$, this normal approximation is adequate. For very small samples, a table given in the book by Siegel (1956) should be used. Since only one set of differences is ranked, ties present no special problem unless they occur for zero differences. If an *even* number of zero differences occur, each zero difference is assigned the average rank for the set (zero differences, of course, rank lowest in absolute size), and then half are arbitrarily given positive and half negative sign. If an odd number of zeros occur, one randomly chosen zero difference is discarded from the data, and the procedure for an even number of zeros followed, except that N is reduced by 1, of course. For other kinds of tied differences, the method used in the example may be followed. Be sure to notice that even when several pairs are tied in absolute size so that they all receive the midrank for that set, the sign given to that midrank for different pairs may be different. For fairly large samples with relatively few ties, this procedure of assigning average ranks introduces negligible error.

All in all, the Mann-Whitney and the Wilcoxen tests are generally regarded as the best of the order tests for two samples. They both compare favorably with t in the appropriate circumstances, and when the assumptions for t are not met they may even be superior to this classical method. However, each is fully equivalent to a classical test of the hypothesis that the *means* of two groups are equal only when the assumptions appropriate to t are true. Unless additional assumptions are made, these tests refer to the hypothesis that two population distributions of unspecified form are *exactly* alike. In many instances, this is the hypothesis that the experimenter wishes to test, especially if he is interested only in the possibility of statistical independence or of association between experimental and dependent variables. However, if he wants to make particular kinds of inferences, particularly about population *means*, then other assumptions become necessary. Without these assumptions, the rejection of H_0 implies only that the populations differ in *some* way, but the test need not be equally sensitive to all ways that population distributions might differ.

18.9   THE KRUSKAL-WALLIS "ANALYSIS OF VARIANCE" BY RANKS

The same general argument for the Mann-Whitney test may be extended to the situation where J independent groups are being compared. The version of a J-sample rank test given here is due to Kruskal and Wallis (1952). This test has very close ties to the Mann-Whitney and Wilcoxen tests just discussed, and can properly be regarded as a generalized version of the Mann-Whitney method.

Imagine some J experimental groups in which each observation is associated with a numerical score. As usual, we assume that the underlying variable is continuously distributed. Now, just as in the Mann-Whitney test, the scores from all groups are pooled, arranged in order of size, and ranked. Then the rank-sum attached to *each separate group* is found. Let us denote this sum of ranks for group j by the symbol T_j

$$T_j = \text{sum of ranks for group } j.$$

For example, suppose that three groups of small children were given the task of learning to discriminate between pairs of stimuli. Each child was given a series of pairs of stimuli, in which each pair differed in a variety of ways. However, attached to the choice of one member of a pair was a reward, and within an experimental condition, the cue for the rewarded stimulus was always the same. On the other hand, the experimental treatments themselves differed in the *relevant* cue for discrimination: in treatment I, the cue was form, in treatment II, color, and in treatment III, size. Some 36 children of the same sex and age were chosen at random and assigned at random to the three groups, with 12 children per group. The dependent variable was the number of trials to a fixed criterion of learning. Suppose that the data turned out to be as shown in Table 18.9.1.

Table 18.9.1

	Treatment	
I	II	III
6 (1)	31 (34.5)	13 (10)
11 (7)	7 (2)	32 (36)
12 (9)	9 (4)	31 (34.5)
20 (19)	11 (7)	30 (33)
24 (23)	16 (14)	28 (31)
21 (20)	19 (17.5)	29 (32)
18 (16)	17 (15)	25 (24)
15 (13)	11 (7)	26 (26.5)
14 (11.5)	22 (21)	26 (26.5)
10 (5)	23 (22)	27 (29.5)
8 (3)	27 (29.5)	26 (26.5)
14 (11.5)	26 (26.5)	19 (17.5)
T_j 139.0	200.0	327.0

$T = 666$

Here, the numbers in parentheses are the ranks assigned to the various score values in the entire set of 36 cases. Then the sum of the ranks for each particular group j is found, and designated T_j. The value of T is the sum of these rank sums; if the ranking has been done correctly, it will be true that

$$T = \frac{N(N + 1)}{2}.$$

Note here that

$$T = \frac{(36)(37)}{2} = 666.$$

For large samples, a fairly good approximate test for identical populations is given by

$$H = \frac{12}{N(N + 1)} \left[\sum_j \frac{T_j^2}{n_j} \right] - 3(N + 1).$$

This value of H can be referred to the chi-square distribution with $J - 1$ degrees of freedom for a test of the hypothesis that all J population distributions are identical.

For the example,

$$H = \frac{12}{36(37)} \left[\frac{(139)^2 + (200)^2 + (327)^2}{12} \right] - 3(37)$$
$$= 13.81.$$

However, since there were ties involved in the ranking, this value of H really should be corrected by dividing through by a value found from

$$C = 1 - \left(\frac{\sum_{i}^{G} (t_i^3 - t_i)}{N^3 - N} \right)$$

where G is the number of sets of tied observations, and t_i is the number tied in any set i. For the example, there are 4 sets of two tied observations, 1 set of three ties, and one set of four tied observations. Thus,

$$C = 1 - \left(\frac{4[(2)^3 - 2] + [(3)^3 - 3] + [(4)^3 - 4]}{(36)^3 - 36} \right)$$

so that

$$C = .997.$$

Finally, the corrected value of H is

$$H' = \frac{H}{C} = \frac{13.81}{.997} = 13.85.$$

Unless N is small, or unless the number of tied observations is very large, relative to N, this correction will make very little difference in the value of H. Certainly, this was true here. Furthermore, when each set of tied observations lie within the same experimental group, the correction becomes unnecessary.

In terms of Table IV, for 2 degrees of freedom the value of H exceeds that required for the .01 level, and so the experimenter can be quite confident in saying that the population distributions are not identical. Apparently there is some association between the type of cue given in the discrimination problem and the number of trials to criterion.

It is somewhat difficult to specify the class of alternative hypotheses appropriate to the J sample rank test, and so the question of power is somewhat more obscure than for the Mann-Whitney test. However, there is reason to believe that this test is about the best of the J-sample order methods. Certainly it should be superior in most situations to the median test discussed in Section 18.2. In comparisons with F from the analysis of variance, the Kruskal-Wallis test shows up extremely well.

When sample size is relatively small within the groups, the tables given by Siegel should be consulted.

18.10   THE FRIEDMAN TEST FOR J MATCHED GROUPS

Much as the Kruskal-Wallis test represents an extension of the Mann-Whitney test, so the Friedman test is related to the Wilcoxen matched-pairs procedure. Furthermore, we shall see that this test is related to the methods of Section 18.17. The Friedman test is appropriate when some K sets of matched individuals are used, where each set contains J individuals assigned at random to the J experimental treatments. It also applies when each of K individuals is observed under each of J treatments in random order. Thus, it is useful in situations where data are collected much as in the randomized-blocks experiments mentioned in Section 13.24.

The data are set up in a table as for a two-way analysis of variance with one observation per cell. The experimental treatments are shown by the respective columns, and the matched sets of individuals by the rows. Within each row (matched group) a rank order of the J scores is found. Then the resulting ranks are summed by columns, to give values of T_j.

For example, in an experiment with four experimental treatments $(J = 4)$, 11 groups of 4 matched subjects apiece were used. Within each matched group the four subjects were assigned at random to the four treatments, one subject per treatment. The data can be represented as in Table 18.10.1.

Table 18.10.1

Groups	I	II	III	IV
1	1 (2)	4 (3)	8 (4)	0 (1)
2	2 (2)	3 (3)	13 (4)	1 (1)
3	10 (3)	0 (1)	11 (4)	3 (2)
4	12 (3)	11 (2)	13 (4)	10 (1)
5	1 (2)	3 (3)	10 (4)	0 (1)
6	10 (3)	3 (1)	11 (4)	9 (2)
7	4 (1)	12 (4)	10 (2)	11 (3)
8	10 (4)	4 (2)	5 (3)	3 (1)
9	10 (4)	4 (2)	9 (3)	3 (1)
10	14 (4)	4 (2)	7 (3)	2 (1)
11	3 (2)	2 (1)	4 (3)	13 (4)
T_j	30	24	38	18

$$T = \frac{K(J)(J + 1)}{2}$$
$$= 110$$

It is important to remember that the ranks are given to the scores *within* rows. Then the T_j values are simply the sums of those ranks within columns.

The rationale for this test is really very simple. Suppose that within the population represented by a row, the distribution of values for all the *treatment* populations were identical. Then, under random sampling and randomization of observations, the probability for any given permutation of the ranks 1 through J within a given row should be the same as for any other permutation. Furthermore, across rows, each and every one of the $J!$ possible permutations of ranks across columns should be equally probable. This implies that we should expect the column sums of ranks to be identical under the null hypothesis. However, if there tend to be pile-ups of high or low ranks in particular columns, this is evidence against equal probability for the various permutations, and thus against the null hypothesis.

The test statistic for large samples is given by

$$\chi_r^2 = \frac{12}{KJ(J+1)} \left[\sum_j T_j^2 \right] - 3K(J+1)$$

distributed approximately as chi-square with $J - 1$ degrees of freedom.

For the example, we would take

$$\chi_r^2 = \frac{12}{11(4)(5)} [(30)^2 + (24)^2 + (38)^2 + (18)^2] - 3(11)(5)$$
$$= 11.79.$$

For 3 degrees of freedom, this is just significant at the .01 level, and so the experimenter may say with some assurance that the treatment populations differ.

If ties in ranks within rows should occur, a conservative procedure is to break the ties so that the T_j values are as close together as possible.

This test may well be the best alternative to the ordinary two-way (matched-groups) analysis of variance. Once again, there is every reason to believe this test, much like the Kruskal-Wallis test for one-factor experiments, is superior to the corresponding median test, and from the little evidence available, the result should compare well with F when both the classical and order methods apply.

The chi-square approximation given above is good only for fairly large K. However, the test should be satisfactory when $J \gtrsim 4$ and $K \gtrsim 10$. As usual, tables giving significance levels for small samples can be found in the book by Siegel.

18.11 RANK-ORDER CORRELATION METHODS

The tests discussed so far in this chapter are designed for situations where the experimental variable is categorical and the dependent variable is in ordinal terms, as given either by the original measurement procedure or by a transformation of the scores into ordered classes or ranks. In this section, measures and tests of association will be described for situations where *both* variables are represented in ordinal terms. First of all, two somewhat different measures of "agreement" or association between rank orders will be introduced. Next, a measure of association will be given for the special situation where both variables take the form of ordered classes. Finally, the problem of indexing simultaneous agreement among several rank orders will be treated.

It is customary to call some of these rank-order statistics "correlations," but this usage deserves some qualification. The Spearman rank correlation to be introduced next actually *is* a correlation coefficient computed for numerical values that happen to be ranks. However, the next index to be considered, Kendall's tau, is not a correlation coefficient at all. Neither of these indices is closely connected with the classical theory of *linear* regression: as we shall see, under some circumstances the Spearman correlation can be used as an estimator of the population correlation coefficient, but one is seldom interested in the possibility of linear regression equations for predicting rank on Y from rank on X, particularly since the degree of possible linear relationship between ranks tells little about the form of relationship between the *underlying* variables. If it is true that scores X and Y have a strong linear relationship, and both X and Y are interval-scale representations of their respective underlying variables, then we can also infer that the underlying variables show this same degree of linear relationship. However, if the numerical values analyzed are ranks, the correlation between the ranks may tell relatively little about the degree of *linear* relationship between the underlying variables. Especially the *square* of a correlation-like index on ranks is not to be interpreted in the usual way as a proportion of variance accounted for in the underlying variables.

Instead, it is somewhat better to think of both the Spearman and the Kendall indices only as showing "concordance" or "agreement," the tendency of two rank orders to be similar. As descriptive statistics, both indices serve this purpose very well, although the definition of "disagree" is somewhat different for these two statistics.

Still another interpretation can also be put on the use of these rank-order measures of association. Even though the Spearman and the Kendall statistics cannot ordinarily be thought of as showing the extent of *linear* relation between the variables underlying the ranks, they can be considered as indices of the general "monotonicity" of the underlying relation. A function $(Y = f(X))$ is said to be **monotone-increasing** if an increase in the value of X in the domain of the function always is accompanied by an increase in the corresponding value of Y. A **monotone-decreasing** function has the opposite property: an increase in the value of X in the domain is accompanied by a decrease in the value of Y. Linear functions are always monotone, but so are many other functions that are definitely not linear ($Y = X^3$, $Y = \log X$, and so on). On the other hand, a function that plots as a parabola is **nonmonotone;** in part of the domain of X, increases in X lead to increases in Y, and in another part, increases in X lead to decreases in Y by the function rule. In general, any functional relation between numbers with a plot showing one or more distinct "peaks" or "valleys" is nonmonotone.

The size of either the Spearman or the Kendall coefficient does tell something about the tendency of the underlying scores to relate in a *monotone* way. High absolute values of either index give evidence that the basic form of the relation between Y and X is monotone. Positive values suggest that the relation tends to be monotone-increasing, and negative values monotone-decreasing. On the other hand, much like r_{XY}, small absolute or zero values of these indices suggest *either* that the two variables are not related *at all*, or that the form of the relation is nonmonotone. In short, rank-order measures of association between variables do not reflect *exactly* the same characteristics as r_{XY} (or more properly r_{XY}^2). They do not necessarily show the tendency toward linear regression, per se, for either of the numerical variables underlying the ranks, but rather show a more general characteristic, which includes linear regression as a special case: these indices reflect the **tendency toward monotonicity,** and the **direction of relationship** that appears to exist. Thus, rank-order measures of association stand somewhere between indices such as φ and r_{XY}: φ gives evidence for *some unspecified departure from independence*, rank-order measures tell something of the *monotonicity of the relationship*, and r_{XY} reflects *the tendency toward linear relationship* between variables. Unfortunately, values given by these different statistics are not directly comparable with each other, but this difference among the statistics is a point to bear in mind when thinking about what the different rank-order indices actually "tell" about the underlying relationship. These various indices of "correlation" imply different things about the relation between variables, and are seldom interchangeable.

18.12 THE SPEARMAN RANK CORRELATION COEFFICIENT

Imagine a group of N cases drawn at random as for a problem in correlation. However, instead of having an X score for each individual, we have only his *rank* in the group on variable X (say for low to high, the ranks 1 through N). In the same way, for each individual we have his rank in the group on variable Y. The question to be asked is "How much does the ranking on variable X tend to agree with the ranking on variable Y?" and a measure is desired to show the extent of the agreement. Or, perhaps, we have two judges who each rank the same set of N objects. We wish to ask, "How much does judge A agree with judge B?" In either instance, there are two distinct rank orders of the same N things, and these rank orders are to be compared for their agreement with each other. Furthermore, when these objects or individuals constitute a random sample from some population, we may wish to test the hypothesis that the true agreement in ranks is zero.

Two simple ways for comparison of two rank orders for agreement have already been mentioned: the first, and older, method is the Spearman rank correlation, commonly symbolized r_S (although ρ is sometimes used, this symbol r_S will be used here to avoid confusion with the population correlation); the second method is the Kendall "tau" statistic, which will be discussed in the following section. Regardless of whether the data are two rank orders representing scores shown by individuals in a sample, or rankings of objects given by two judges, we can apply r_S. Suppose that in either circumstance we call the things ranked "individuals," and the two bases for ranking, the "variables." We can take the point of view that if rank orders agree, the ranks assigned to individuals should *correlate* positively with each other, whereas disagreement should be reflected by a negative correlation. A zero correlation represents an intermediate condition; no particular connection between the rank of an individual on one variable and his rank on the other.

For a descriptive index of agreement between ranks, the ordinary correlation coefficient can be computed on the *ranks* just as for any numerical scores, and this is how the Spearman rank correlation for a sample is defined:

r_S = correlation between ranks over individuals.

However, since the numerical values entering into the computation of the correlation coefficient actually are ranks in this instance, r_S can be given a very simple computational form when no ties in rank exist:

$$r_S = 1 - \left[\frac{6 \left(\sum_i D_i^2 \right)}{N(N^2 - 1)} \right]. \qquad [18.12.1\dagger]$$

Here D_i is the *difference* between ranks associated with the particular individual i, and N is the number of individuals observed.

The basis for this way of calculating r_S can be shown very easily, given two simple mathematical facts:

1. Given the whole numbers $1, 2, \cdots, N$, the sum of this set of numbers is $N(N + 1)/2$.
2. Given the whole numbers $1, 2, \cdots, N$, the sum of *squares* of these numbers is $(N)(N + 1)(2N + 1)/6$.

It follows quite easily that for *each* of the two sets of ranks, the mean rank is

$$M = \frac{N(N + 1)}{2N} = \frac{(N + 1)}{2}, \qquad [18.12.2]$$

and the variance of the ranks is

$$\begin{aligned} s^2 &= \frac{N(N + 1)(2N + 1)}{6N} - \frac{(N + 1)^2}{4} \\ &= \frac{N^2 - 1}{12}. \end{aligned} \qquad [18.12.3]$$

For the moment, let us symbolize the rank of individual i on variable X by x_i, and the rank of i on variable Y by y_i. Furthermore, let

$$D_i = (x_i - y_i).$$

Now, if we took the sum of the squares of these differences, we would find

$$\sum_i D_i^2 = \sum_i (x_i - y_i)^2 = \sum_i x_i^2 + \sum_i y_i^2 - 2 \sum_i x_i y_i. \qquad [18.12.4]$$

Using expression **18.12.4** and the second of the mathematical principles listed above, we have

$$\frac{\sum_i x_i y_i}{N} = \frac{\sum_i x_i^2}{2N} + \frac{\sum_i y_i^2}{2N} - \frac{\sum_i D_i^2}{2N}$$

$$= \frac{(N+1)(2N+1)}{6} - \frac{\sum_i D_i^2}{2N}.$$ [18.12.5]

On substituting **18.12.2**, **18.12.3**, and **18.12.5** into the equation for the correlation coefficient and carrying out a little algebra, we have

$$r_S = \frac{\sum_i x_i y_i / N - M_x M_y}{s_x s_y}$$

$$= \frac{(N^2 - 1)/12 - \sum_i D_i^2 / 2N}{(N^2 - 1)/12} = 1 - \left(\frac{6 \sum_i D_i^2}{N(N^2 - 1)} \right).$$

The Spearman rank correlation is thus very simple to compute when ranks are untied. All one needs to know is N, the number of individuals ranked, and D_i, the difference in ranking for each individual. In spite of the different computations involved, r_S is only an ordinary correlation coefficient calculated on ranks.

The computation of r_S will be illustrated in a problem dealing with agreement between judge's rankings of objects. In a test of fine weight discrimination, two judges each ranked 10 small objects in order of their judged heaviness. The results are shown in Table 18.12.1. Did the judges tend to agree?

Table 18.12.1

Object	Judge I	Judge II	D_i	D_i^2
1	6	4	2	4
2	4	1	3	9
3	3	6	3	9
4	1	7	6	36
5	2	5	3	9
6	7	8	1	1
7	9	10	1	1
8	8	9	1	1
9	10	3	7	49
10	5	2	3	9
				128

$$r_S = 1 - \frac{6(128)}{10(10^2 - 1)} = .224.$$

The Spearman rank correlation is only .224, so that agreement between the two judges was not very high, although there was some slight tendency for similar ranks to be given to the same objects by the judges.

Incidentally, notice that if the relative true weights of the objects are known, we could also find the agreement of each judge with the true ranking. When some criterion ranking is known, it may be useful to compare a judged ranking with this criterion to evaluate the accuracy or "goodness" of the judgments—this is not the same as the agreement of judges with each other, of course.

Formula **18.12.1** may not be used if there are ties in either or both rankings, since the means and variances of the ranks then no longer have the simple relationship to N present in the no-tie case. When ties exist, perhaps the simplest procedure is to assign mean ranks to sets of tied individuals; that is, when two or more individuals are tied in order, each is assigned the mean of the ranks they would otherwise occupy. Next an ordinary correlation coefficient is computed, using the ranks as though they were simply numerical scores. The result is a Spearman rank correlation that can be regarded as corrected for ties. On the other hand, if a test of the significance of r_S is the main object of the analysis, a conservative course of action is to find a way to break the ties that will make the absolute value of r_S as small as possible.

When no ties in rank exist, the exact sampling distribution of r_S can be worked out for small samples. Exact tests of significance for r_S are based on the idea that if one of the two rank orders is known, and the two underlying variables are independent, then each and every permutation in order of the individuals is equally likely for the other ranking. On this basis, the exact distribution of $\sum_i D_i^2$ can be found, and this can be converted into a distribution of r_S. Exact probability tables for $\sum_i D_i^2$ and r_S based on small N are to be found in books by Kendall (1955), Siegel (1956), and others.

The exact distribution of r_S is rather peculiar. Although the distribution is unimodal and symmetric when the rankings are independent, the plot of the distribution has a curious jagged or serrated appearance due to particular constraints on the possible values of $\sum_i D_i^2$ for a given N. With N large, the distribution does approach a normal form, but relatively slowly, so that for samples of small to moderate size the normal approximation is not very good.

The hypothesis of the independence of the two variables represented by rankings can also be given an approximate, large-sample, test in terms of r_S. This test has a form very similar to that for r_{XY}:

$$t = \frac{r_S \sqrt{N-2}}{\sqrt{1-r_S^2}}$$

with $N - 2$ degrees of freedom. This test is really satisfactory, however, only when N is fairly large; N should be at least greater than or equal to 10.

Remember that if inferences are to be made about the form of relation holding between two continuous underlying variables, one can fail to reject the null hypothesis either because the variables are independent or because the relation is nonmonotone. Furthermore, the rejection of the hypothesis does not necessarily let one conclude that linear association exists between the underlying variables, but only that some more or less monotone relation holds.

Under particular assumptions, especially that of a bivariate normal distribution, the value of r_S from a large sample can be treated as an estimate of the value of ρ for the variables underlying the ranks. When the population is bivariate normal with $\rho = 0$, values of r_S and r_{XY} correlate very highly over samples. On the other hand, these assumptions are rather special, and the status of r_S as an estimator of ρ *in general* is open to considerable question.

18.13   THE KENDALL TAU COEFFICIENT

A somewhat different approach to the problem of agreement between two rankings is given by the τ coefficient (small Greek tau) due to M. G. Kendall (1955). Instead of treating the ranks themselves as though they were scores and finding a correlation coefficient, as in r_S, in the computation of τ we depend only on the number of *inversions* in order for pairs of individuals in the two rankings. A single inversion in order exists between *any pair* of individuals b and c when b > c in one ranking and c > b in the other. When two rankings are *identical*, no inversions in order exist. On the other hand, when one ranking is exactly the reverse of the other, an inversion exists for *each pair* of individuals; this means that complete disagreement corresponds to $\binom{N}{2}$ inversions. If the two rankings *agree* (show noninversion) for as many pairs as they disagree about (show inversion) the tendency for the two rank orders to agree or disagree should be exactly zero.

This leads to the following definition of the τ statistic:

$$\tau = 1 - \left[\frac{2(\text{number of inversions})}{\text{number of pairs of objects}} \right]. \qquad \text{[18.13.1†]}$$

This is equivalent to

$$\tau = \frac{(\text{number of times rankings agree about a pair}) - (\text{number of times rankings disagree})}{\text{total number of pairs}}.$$

It follows that the τ statistic is essentially a difference between two proportions: the proportion of pairs having the *same* relative order in both rankings minus the proportion of pairs showing *different* relative order in the two rankings.

Viewed as coefficients of agreement, r_S and τ thus rest on somewhat different conceptions of "disagree." In the computation of r_S, a disagreement in ranking appears as the *squared* difference between the ranks themselves over the individuals. In τ, an inversion in order for any *pair* of objects is treated in the same way as evidence for disagreement. Although these two conceptions are related, they are not identical: the process of squaring differences between rank values in r_S places somewhat different weight on *particular* inversions in order, whereas in τ all inversions are weighted equally by a simple frequency count. Values of the statistics r_S and τ are correlated over successive random samples from the same population, but the extent of the correlation depends on a number of things, including sample size and the character of the relation between the underlying variables in the population. Nevertheless, the two statistics are closely connected, and a number of mathematical inequalities must be satisfied by the values of the two statistics. For example,

$$-1 \leq 3\tau - 2r_S \leq 1.$$

It will be convenient to discuss the numerator term in the τ coefficient separately, and thus we will define

$S = $ (number of agreements in order) $-$ (number of disagreements in order)

$$= \binom{N}{2} - 2(\text{number of inversions}).$$

Various methods exist for the computation of S and τ, but the simplest is a graphic method. In this method, all one does is to list the individuals or objects ranked, once in the order given by the first ranking, and again in the order given by the second. For example, suppose that in some problem there were seven objects {a,b,c,d,e,f,g} ranked by each of two judges, and that the rankings came out like this:

				Rank			
	1	*2*	*3*	*4*	*5*	*6*	*7*
Judge 1	c	a	b	e	d	g	f
Judge 2	a	c	e	b	f	d	g

Now straight lines are drawn connecting the same objects in the two parallel rankings, thus:

Then *the number of times that pairs of lines cross is the number of inversions in order*. Here, the number of crossings is 4, and so

$$S = \binom{7}{2} - (2)(4)$$
$$= 21 - 8$$
$$= 13.$$

The sample value of τ is

$$\tau = \frac{13}{21}$$
$$= .62.$$

Although r_S from a sample has a rather artificial interpretation as a correlation coefficient, the interpretation of the obtained value of τ is quite straightforward: if a *pair* of objects is drawn at random from among those ranked, the probability that these two objects show the *same* relative order in both rankings is .62 *more* than the probability that they would show different order. In other words from the evidence at hand it is a considerably better bet that the two judges will tend to order a randomly selected pair in the *same* way than in a different way.

This graphic method of computing τ is satisfactory only when no ties in ranking exist. For nontied rankings, however, it is very simple to carry out, even when moderately large numbers of individuals have been ranked. Notice that although both the examples of computations for r_S and τ featured judges' rankings of objects, exactly the same methods apply when the ranking principle is provided by scores shown by individuals on each of two variables.

Still another way of computing S will be mentioned, since we will find an extension of this method very convenient when ties exist in one or both rankings. The data are organized as in Table 18.13.1. Notice that this is really just a joint-frequency table based on ranks. Each distinct observation (object) has a pair of rank numbers, and this determines the cell in the table in which that observation falls. As a convention for the procedure described below, the ranks along the margins of the table are listed with the lowest ranks in the upper left-hand corner, as shown here.

Table 18.13.1

JUDGE 1

JUDGE 2	1	2	3	4	5	6	7
1		1(a)					
2	1(c)						
3				1(e)			
4			1(b)				
5							1(f)
6					1(d)		
7						1(g)	

Now S is found as follows: we compute a value S_+ by first taking any cell with nonzero frequency and, ignoring its row and column, *counting the number of entries to the right and below that cell.* Thus, for the cell containing the object c, there are five entries to the right and below. For the cell with object b, there are three entries to the right and below, and so on. Then S_+ is the sum of these numbers over cells. Over all nonzero cells, the value of S_+ is found to be:

$$
\begin{aligned}
&5 \text{ (for cell a)}\\
&5 \text{ (for cell c)}\\
&3 \text{ (for cell e)}\\
&3 \text{ (for cell b)}\\
&1 \text{ (for cell d)}\\
&0 \text{ (for cell f)}\\
&0 \text{ (for cell g)}\\
\hline
&S_+ = 17.
\end{aligned}
$$

Then

$$
S = 2(S_+) - \binom{N}{2} = 34 - 21 = 13
$$

and

$$
\tau = .62.
$$

As we shall see, this general method is advantageous when ties exist in the data, or when the data are put into ordered classes rather than ranks.

Notice that we might also compute a value S_- by taking the sum of the frequencies to the left and below the various nonzero cells. Then

$$
S = \binom{N}{2} - 2(S_-)
$$

since

$$
(S_+) + (S_-) = \binom{N}{2},
$$

when there are no tied ranks.

The exact test of significance for τ is based on the assumption that the variables underlying the ranks are continuously distributed, so that ties are impossible. Alternatively, one can imagine random sampling of complete rank orders of N things from a potential population of such rankings. In either instance, one assumes that if there is no true relationship between the pairs of rank orders observed, then given the first rank order, all possible permutations in order are equally likely to occur as the second rank order. This makes finding the exact sampling distribution of S or of τ for fixed N relatively simple.

However, for our purposes, the fact of importance is that the exact sampling distribution of τ approaches a normal distribution very quickly with successive increases in the size of N. Even for fairly small values of N, the distribution of τ is approximated relatively well by the normal distribution. Of course, this is true only when H_0 is true, so that the rankings are each equally likely to show any of the $N!$ permutations in order, and $E(\tau) = 0$. The distribution of τ is not simple to discuss when other conditions hold, and so we will test only the hypothesis of independence between rankings (implying equal probability of occurrence for each and every possible ordering of N observations on the second variable given their ordering on the first).

For N of about 10 or more, the test is given by

$$z = \frac{\tau}{\sigma_\tau}$$

referred to a normal distribution, where

$$\sigma_\tau^2 = \frac{2(2N + 5)}{9N(N - 1)}.$$

Equivalently, in terms of S,

$$z = \frac{S}{\sigma_S}$$

where

$$\sigma_S^2 = \frac{N(N - 1)(2N + 5)}{18}.$$

For small N, the exact tables in terms of S given by Kendall (1948) or Siegel (1956) can be used.

This approximate test can be improved if a correction for continuity is made. The correction for continuity involves subtraction of $1 \Big/ \binom{N}{2}$ from the absolute value of τ, or 1 from the absolute value for S, before the z statistic is formed.

18.14   KENDALL'S TAU VERSUS THE SPEARMAN RANK CORRELATION

The Spearman rank correlation is the traditional way to treat the problem of association between ranks. However, there is reason to believe that the τ coefficient may be superior for many purposes.

In the first place, τ can be given a very simple interpretation as a descriptive statistic, whereas the Spearman coefficient is meaningful, at least at an elementary level, only by analogy with the ordinary correlation coefficient. By the graphic method shown above, the value of τ is certainly as easy to compute as that of r_S, although either index can be troublesome for large samples.

From the definition of sample τ as a difference between two proportions, of the general form $P - Q$, it is easy to see that τ for a population can be defined as a corresponding difference between probabilities. For this reason, sample τ provides an unbiased estimate of its population counterpart. On the other hand, r_S is usually regarded as an estimator of the population correlation ρ, although studies have shown that its properties as an estimator of ρ vary considerably with the form of the population distribution and with the true value of ρ. Recently, however, Kruskal (1958) showed that ρ_S, the Spearman coefficient, can also be defined for a population, and has a very reasonable interpretation as an index of strict monotonicity of relation. Under this view, r_S can be shown to be biased as an estimator of ρ_S, and the curious circumstance emerges that an unbiased estimate of ρ_S involves *both* sample r_S and τ:

$$\text{unbiased estimate of } \rho_S = \frac{N+1}{N-2} r_S - \frac{3}{N-2} \tau.$$

For large N, however, this is virtually the same value as r_S. At any rate, it is becoming evident that each index supplies information of importance about the general monotone relation that may exist between variables, quite irrespective of the true value of the correlation coefficient ρ.

The big argument in favor of the use of τ in the test of the hypothesis of independence is the fairly rapid convergence of its sampling distribution to a normal form, as opposed to the somewhat slower convergence and other peculiarities of the distribution of r_S. For moderately large samples, where the exact tables of the sampling distributions cannot be used, τ seems to provide a better test of the hypothesis of no-association than does r_S.

18.15 KENDALL'S TAU WITH TIES

The determination of the exact sampling distribution of S (or τ) depends on the assumption of *no* ties. However, if the underlying model is altered in a particular way, τ can be computed and tested even though ties occur in one or both rankings. First of all, we will consider the method for finding S when ties occur, and then turn to the problem of a significance test.

The simplest method for finding S and τ when ties occur is to use the joint-frequency–table method suggested in Section 18.13. However, here it will occur that more than one entry will fall into a particular row, column, or cell. This method is best shown by example.

Suppose that some 12 subjects were observed, and ranked on each of two variables, so that the ranked data were as shown in Table 18.15.1. A table is formed in which the first ranking is represented by the columns and the second

Table 18.15.1

Subject	X ranking	Y ranking
1	5	8.5
2	7	11
3	5	12
4	1	8.5
5	2	8.5
6	3	6
7	5	5
8	11.5	8.5
9	11.5	2
10	9.5	2
11	9.5	2
12	8	4

ranking by the rows. (Actually, the score values themselves could have been used to form this table, since the actual rank values are not used at all after the data are tabled.) Table 18.15.2 shows these data.

Table 18.15.2

		X RANKING								
		1	2	3	5	7	8	9.5	11.5	Total
	2							2	1	3
	4						1			1
	5				1					1
Y	6			1						1
RANKING	8.5	1	1		1				1	4
	11					1				1
	12				1					1
	Total	1	1	1	3	1	1	2	2	12

Now we proceed to find S_+ just as above, except that the number of cases to the right and below a given cell is weighted by the number of cases in that cell. For example, for the cell given by the row labeled 2 and the column labeled 9.5, there is only 1 case to the right and below; however, since there are 2 cases *in* the cell, in the sum for S_+ we enter 2(1) or 2 for this cell. Proceeding in this way over all of the cells with nonzero frequencies, we have

$$\begin{aligned} S_+ &= 2(1) + 1 + 1(2) + 1(4) + 1(2) + 1(2) + 1(1) \\ &= 14. \end{aligned}$$

Now we find a value S_-, computed exactly as for S_+ except that we find the number of cases to the *left and below* a given cell, and weight this number by the cell frequency. Here

$$\begin{aligned} S_- &= 2(8) + 1(8) + 1(7) + 1(3) + 1(2) + 1(2) + 1(1) \\ &= 39. \end{aligned}$$

Then,

$$S = S_+ - S_-$$

so that for this example

$$S = 14 - 39 = -25.$$

Kendall suggests calculating the denominator of τ by taking

$$\sqrt{\left(\frac{N(N-1)}{2} - T_1\right)\left(\frac{N(N-1)}{2} - T_2\right)}$$

where

$$T_1 = \frac{\sum_j n_j(n_j - 1)}{2}$$

$$n_j = \text{the marginal total of column } j$$

$$T_2 = \frac{\sum_k n_k(n_k - 1)}{2}$$

$$n_k = \text{the marginal total for row } k.$$

(A simpler method for finding a "tau-like" index will be given in the next section, however.)

For this example,

$$T_1 = \frac{1}{2}[3(2) + 2(1) + 2(1)]$$
$$= 5$$

and

$$T_2 = \frac{1}{2}[3(2) + 4(3)]$$
$$= 9.$$

Then

$$\tau = \frac{S}{\sqrt{\left(\frac{N(N-1)}{2} - T_1\right)\left(\frac{N(N-1)}{2} - T_2\right)}}$$

$$= \frac{-25}{\sqrt{(66-5)(66-9)}}$$

$$= \frac{-25}{59}$$

$$= -.42.$$

The apparent degree of agreement between these two rankings is a *negative* .42.

A large-sample test of significance for tau with ties can be constructed along the lines given in Section 18.13. That is, one forms a ratio

$$z = \frac{S}{\sigma_S}$$

and refers this to a normal distribution. However, it must be emphasized that when ties exist in either or both rankings, this test is *conditional* to the distribution of ties, and the significance level refers to a probability of occurrence among samples each showing *exactly* the same distribution of ties as appeared in the sample actually obtained. When ties exist, the sampling variance of S, or σ_S^2, is found from

$$\sigma_S^2 = \frac{N(N-1)(2N+5) - \sum_j n_j(n_j-1)(2n_j+5) - \sum_k n_k(n_k-1)(2n_k+5)}{18}$$

$$+ \frac{\left[\sum_j n_j(n_j-1)(n_j-2)\right]\left[\sum_k n_k(n_k-1)(n_k-2)\right]}{9(N)(N-1)(N-2)}$$

$$+ \frac{\left[\sum_j n_j(n_j-1)\right]\left[\sum_k (n_k)(n_k-1)\right]}{2(N)(N-1)}.$$

After this rather forbidding-looking calculation for σ_S^2 has been carried out, the value of z may be found and referred to a normal table, provided that N is rather sizable. Unless N is quite large and ties are infrequent, a correction for continuity is required; this is given in Kendall (1955).

It is interesting to observe that even when ties exist, one really does not have to assign ranks at all in order to compute τ; we are really dealing with ordered categories of observations, and when we assign the midrank to a set of observations we are, in effect, allotting them to the same ordered category. Thus, in principle this procedure may be carried out on a joint *grouped* frequency distribution; the actual computations, however, really involve only the *ordinal* properties of the scores on which the distributions were based, and the class intervals are treated as ordered classes.

Furthermore, even when data are grouped in this way, the τ index can be regarded as reflecting monotonicity in the relationship between the underlying scores. Large values of τ lead to the conclusion that the relationship tends to be monotone, and small absolute values may indicate either that there is no statistical association, or that the form of the relationship tends to be nonmonotone.

18.16   A measure of association in ordered classes

It has just be suggested that the method for computing S when the rankings contain ties is tantamount to arranging the data into ordered classes. Unfortunately, the value of the τ statistic itself does not seem to have a very simple interpretation when ties are present in either ranking. This difficulty is removed if one uses the γ (Greek gamma) statistic suggested by Goodman and Kruskal (1954) specifically for data arranged into ordered classes. Actually, this statistic γ has the same numerator term as τ, and the S value is computed in exactly the same way. However, its denominator differs from that of τ, and permits γ to have a simpler interpretation.

In terms of the quantities S_+ and S_- computed above, the statistic γ is just

$$\gamma = \frac{S_+ - S_-}{S_+ + S_-}.$$ [18.16.1†]

For the example above, this is

$$\gamma = \frac{-25}{14 + 39}$$
$$= -.47.$$

This value can be interpreted as follows: suppose that a pair of subjects were drawn at random from the 12 actually observed. Given that these subjects were tied in neither of the rankings, is it a better bet that they show the same or a different ordering on X and Y? The value of γ shows that it is a much better bet that an untied pair has different ordering on the two variables, since the probability of finding a pair with a different ordering is .47 *more* than the probability of a pair with the same ordering, among all possible untied pairs we might draw. Put more formally,

$$\gamma = p(\text{same ordering|untied pairs}) - p(\text{different ordering|untied pairs})$$

for a pair chosen at random with replacement from among these 12 subjects.

If there are no tied ranks, then

$$\gamma = \tau.$$

The index γ has the same basic interpretation as τ: a difference in probability for same versus different ordering on the underlying variables for a randomly selected pair. The only difference is that γ is conditional to the set of *untied* such pairs. The index γ is much like the λ measures for categorical data (Section 17.13), since it has a simple interpretation in a predictive sense, although γ refers to *ordered* categories and thus reflects the general tendency toward monotonicity in the relationship, while λ reflects *any* form of predictive relationship.

18.17 KENDALL'S COEFFICIENT OF CONCORDANCE

Sometimes we want to know the extent to which members of a set of m distinct rank orderings of N things tend to be similar. For example, in a beauty contest each of 7 judges ($m = 7$) gives a simple rank order of the 10 contestants ($N = 10$). How much do these rank orders tend to agree, or show "concordance"?

This problem is usually handled by application of Kendall's statistic, W, the "coefficient of concordance." As we shall see, the coefficient W is closely related to the average r_S among the m rank orders.

The coefficient W is computed by putting the data into a table with m rows and N columns. In the cell for column j and row k appears the rank number assigned to individual object j by judge k. Table 18.17.1 might show the data for the judges and the beauty contestants. It is quite clear that the judges did not agree perfectly in their rankings of these contestants. However, what should the column totals of ranks, T_j, have been if the judges had agreed exactly? If each judge had given exactly the same rank to the same girl, then one column should total to 7(1), another to 7(2), and so on, until the largest sum should be 7(10). On the other hand, suppose that there were complete disagreement among the judges, so that there was no tendency for high or low rankings to pile up in particular columns. Then we should expect each column sum to be about the same.

In this example, the column sums of ranks are not identical, so that apparently some agreement exists, but neither are the sums as different as they should be when absolutely perfect agreement exists.

Table 18.17.1

	Contestants									
Judges	*1*	*2*	*3*	*4*	*5*	*6*	*7*	*8*	*9*	*10*
1	8	7	5	6	1	3	2	4	10	9
2	7	6	8	3	2	1	5	4	9	10
3	5	4	7	6	3	2	1	8	10	9
4	8	6	7	4	1	3	5	2	10	9
5	5	4	3	2	6	1	9	10	7	8
6	4	5	6	3	2	1	9	10	8	7
7	8	6	7	5	1	2	3	4	10	9
T_j	45	38	43	29	16	13	34	42	64	61

$$T = \frac{m(N)(N+1)}{2}$$

$$= 385$$

This idea of the extent of variability among the respective sums of ranks is the basis for Kendall's W statistic. Basically,

$$W = \frac{\text{variance of rank sums}}{\text{maximum possible variance of rank sums}}.$$

Because the mean rank and the variance of the ranks each depend only on N and m, this reduces to

$$W = \left(\frac{12 \sum_j T_j^2}{m^2 N(N^2-1)} \right) - \frac{3(N+1)}{N-1}.$$

For the example, we find

$$W = \left(\frac{12[(45)^2 + \cdots + (61)^2]}{49(10)(99)} \right) - \frac{3(11)}{9}$$
$$= 4.28 - 3.66$$
$$= .62.$$

There is apparently a moderately high degree of "concordance" among the judges, since the variance of the rank sums is 62 percent of the maximum possible.

Note that by its definition, W cannot be negative, and its maximum value is 1.

The value of the concordance coefficient is somewhat hard to interpret directly in terms of the tendency for the rankings to agree, but an interpretation can be given in terms of the average value of r_S over all possible *pairs* of rank orders. That is

$$\text{average } r_S = \frac{mW - 1}{m - 1}.$$

For the example,

$$\text{average } r_S = \frac{7(.62) - 1}{7 - 1}$$
$$= .56.$$

If we took all of the possible $\binom{7}{2}$ or 21 *pairs* of judges, and found r_S for each such pair, the average rank correlation would be about .56. Thus, on the average, judge-pairs do tend to give relatively similar rankings. The advantage of reporting this finding in terms of W rather than average r_S is that

$$\frac{-1}{m - 1} \leq \text{average } r_S \leq 1$$

whereas, regardless of the values for N or m,

$$0 \leq W \leq 1.$$

This makes W values more immediately comparable across different sets of data. Nevertheless, the clearest interpretation of W seems to be in terms of average r_S.

Recall that this is essentially the idea employed in the Friedman test for matched groups. In the Friedman procedure a matched group takes the place of a judge, and the rank order of scores for different treatments within a group is like an ordering of objects of judgment. Then, if the scores for different treatments tended to show up in substantially the same order for the various groups, a true difference in treatments is inferred.

An exact test is possible for the hypothesis that there is no actual agreement among judges (see Kendall, 1955; and Siegel, 1956, for tables). For m of at least 8, an approximate test is given by

$$\chi^2 = m(N - 1)W$$

referred to the chi-square distribution with $N - 1$ degrees of freedom. This test is really a good one only for fairly sizable m and N, however.

Exercises

1. In a study of engaged couples' preferences with respect to size of family, some twenty-six engaged couples were selected at random and were asked to state, independently, the ideal number of children they would like to have. Responses of men and women from each couple are listed below:

COUPLE	MEN	WOMEN
1	3	2
2	0	1
3	1	0
4	2	2
5	0	3
6	2	3
7	1	2
8	2	3
9	2	3
10	1	3
11	2	4
12	0	1
13	3	4
14	5	2
15	7	2
16	1	2
17	0	3
18	2	4
19	10	3
20	5	3
21	2	4
22	0	2
23	1	3
24	3	2
25	5	2
26	2	1

Use the sign test to decide if husbands and wives differ significantly on this issue.
2. A psychologist was interested in the verb-adjective ratio as an index of the habitual pattern of verbal expression for an individual. In a comparison of science majors and English majors in terms of verb-adjective ratio, 10 science majors and 12 English majors were sampled at random, and a selection of the free-writing of each subject taken. Each such selection was scored according to the ratio of number of verbs used relative to the number of adjectives. In terms of the resulting data, use the Mann-Whitney test to describe if science majors and English majors differ significantly in their relative usage of verb and adjectives.

VERB-ADJECTIVE RATIOS	
SCIENCE MAJORS	ENGLISH MAJORS
1.32	1.04
2.30	.93
1.98	.75
.59	.33
1.02	1.62
.88	.76
.92	.97
1.39	1.21
1.95	.80
1.25	1.16
	.71
	.96

3. Compare the two groups of problem 2 above in terms of the *runs* test.
4. In a psychological study, twenty subjects were used, each subject being tested on a different day. Although each subject was cautioned not to tell anyone about the experiment, the experimenter felt that the later subjects were performing systematically better than the earlier, and suspected that information about the experiment was being gradually "leaked." Test the hypothesis that there is no correlation between the day on which a subject was tested and his score. (Hint: compute a Spearman rank correlation between days and scores, and test its significance.)

DAY	SCORE	DAY	SCORE
1	120	11	132
2	124	12	136
3	123	13	133
4	128	14	119
5	125	15	140
6	127	16	138
7	134	17	139
8	129	18	161
9	130	19	142
10	137	20	145

5. Use the runs test to answer the question of problem 4. (Hint: divide the data according to position above and below the median score, and then look for runs above and below the median in terms of days.)
6. In an experiment, five groups of 10 randomly-selected cases each were used. Use the median test to decide if there are significant ($\alpha = .05$) differences among the populations sampled.

	GROUPS			
1	2	3	4	5
87	41	31	60	55
35	19	18	8	67
67	70	64	14	46
44	62	7	49	95
51	43	13	16	79
49	46	22	28	63
18	6	38	16	66
98	13	31	70	82
97	22	18	64	56
84	38	64	30	67

7. Analyze the data of problem 1 above by means of the Wilcoxen test for matched samples.
8. In the table 12.33.1 treat the college men and the professional athletes as two independent groups of subjects, and compare these groups by use of the runs test.
9. In chapter 12, take the data from table 12.13.1 and test the hypothesis that the four populations are identical by use of the median test.
10. Suppose that the data shown in table 13.15.1 were based on 20 matched groups of three subjects each. Use the median test to compare the three groups.
11. Suppose that the data of table 13.5.1 were actually the results of five repeated observations on each of eight subjects. Furthermore, let a score value greater than or equal to 6.00 be counted a success, and a value less than or equal to 6.00 a failure. Compare the five sets of repeated observations by use of Cochran's test.
12. Apply the Kruskal Wallis "analysis of variance by ranks" to the data of problem 10, Chapter 12.
13. In a study of upward mobility trends in American society, a random sample of 107 U.S. families with at least one male child was taken. The occupational level of the father and that of his oldest son was taken, and the data put into the following table:

		FATHER			
		UNSKILLED LABOR	SKILLED LABOR	WHITE COLLAR	PROFESSIONAL- EXECUTIVE
SON	UNSKILLED LABOR	2	3	7	3
	SKILLED LABOR	3	4	20	9
	WHITE COLLAR	1	2	14	19
	PROFESSIONAL- EXECUTIVE	0	4	2	14

Index the extent to which a *monotone* relationship seems to exist between occupational level of father and of son.
14. Test the hypothesis that no monotone relationship exists in the data of problem 13.
15. Using the data of table 12.33.1 once again, compare the college men and professional athletes by use of the Mann Whitney test.

16. Use the Kruskal Wallis analysis of variance by ranks to compare the groups represented by the columns in table 12.33.1.

17. Two judges each ranked 10 pictures (labeled A, B, C, etc) in order of beauty. The rankings are given below. Use Kendall's tau-coefficient to describe the agreement between the judges.

JUDGE 1	JUDGE 2
C	A
A	C
B	B
D	G
F	D
E	F
G	E
J	H
I	J
H	I

18. Use the data of problem 17 to test the hypothesis of zero true-agreement between the judges.

19. Compute and test the significance of the tau-coefficient representing the relationship shown in table 15.28.1.

20. Eighteen subjects were asked to respond to very faint tones by pushing a key specific to each tone. The task was carried out under five different conditions of noise. In each condition 100 tones were presented, and the number of correct responses recorded. The data follow:

		CONDITION			
	1	2	3	4	5
1	73	75	71	74	64
2	91	89	83	80	82
3	92	94	90	93	83
4	84	82	76	73	75
5	56	58	54	57	47
6	60	58	52	49	51
7	73	75	69	72	62
8	70	68	64	61	63
9	87	89	83	86	76
10	75	73	69	66	68
11	77	75	69	72	62
12	68	70	64	61	63
13	73	75	69	72	62
14	75	73	68	65	67
15	93	95	89	92	82
16	90	89	85	82	84
17	84	86	80	83	73
18	69	67	63	60	62

SUBJECT (label for rows, positioned at rows 9–10)

Are there significant differences among conditions ($\alpha = .05$)? Analyze these data using the Friedman matched-groups analysis by ranks.

21. In a study of attitudes toward international affairs, each of a group of 15 subjects ranked 10 countries according to their "responsibility" in international affairs. The data follow:

		COUNTRY								
	A	B	C	D	E	F	G	H	I	J
1	2	1	3	5	4	6	7	8	9	10
2	8	7	6	4	9	1	2	10	5	3
3	3	4	2	1	6	5	7	8	9	10
4	9	10	8	6	7	1	3	4	5	2
5	5	2	3	4	6	1	7	8	9	10
6	2	4	3	1	5	8	9	10	7	6
7	4	3	2	5	1	6	9	10	8	7
SUBJECT 8	2	4	1	6	5	3	8	7	9	10
9	9	10	6	7	8	1	4	3	5	2
10	5	4	1	10	2	3	7	8	6	9
11	2	4	1	5	3	6	7	9	10	8
12	3	4	2	5	1	10	9	6	8	7
13	4	3	5	2	10	1	8	6	9	7
14	5	3	4	2	1	8	9	6	10	7
15	4	5	10	1	3	2	8	9	6	7

Find the coefficient of concordance and the average rank correlation for this group of subjects. Assuming that the 15 subjects constitute a random sample, test the hypothesis of zero true-concordance among subjects in terms of their rankings of countries.

22. Using the figures given in table 16.7.1, construct a table in which the columns represent the variable Y, in a grouped distribution with $i = 5$, and the rows the values of X. Then, on the basis of this table, calculate Kendall's tau.

19 Some Elementary Bayesian Methods

In almost all of the methods of statistical inference discussed so far, the results gained from a sample are treated as the only source of relevant information about a population state of affairs. It is true, of course, that the choice and use of a particular statistical method will depend upon the meeting of certain assumptions. The theory rests on assumptions about such things as the form of the basic population distribution, the method of sampling, the independence or nonindependence of observations, and the like. In this sense the experimenter does bring some prior knowledge or prior opinions to the evaluation of the sample results; he does so whenever he decides that a given set of assumptions do or do not apply to the situation in question. However, seldom in traditional statistical analysis is there a place for the prior knowledge or opinion of the experimenter to enter more explicitly into the evaluation of data. These methods tend to treat each sample as though it were the first of its sort ever taken, and each estimate of a parameter the first ever made. Prior opinion and information are generally ruled out as belonging to a different domain of discourse from inferences drawn directly from a sample.

In this chapter we will be concerned with some examples of an alternate approach. Unlike the methods introduced so far, the approach that has come to be called "Bayesian" utilizes not only sample information, but also other relevant information. Under this view, in drawing conclusions about some true state of affairs, the experimenter can use not only the usual information derived from a sample but also any prior information he may have available. Such prior information may come from previous samples, from theoretical considerations, or simply from the experimenter's own opinions and beliefs about a particular state of affairs.

The Bayesian approach is particularly appealing to those concerned with practical decision-making on the basis of statistical evidence. It has been pointed out before that evidence from a single sample or set of samples is often quite inadequate as the basis for a choice among actions. Even given sample evidence, other considerations almost always enter in. Bayesian theory represents one attempt to bring in some of these other factors explicitly, so that, insofar as possible, all relevant information will be brought into play as the basis for inference and decision-making.

At the outset it is important to understand that Bayesian methods represent only an extension of the classical methods. All of the machinery of mathematical statistics is used in Bayesian statistics. However, the usual concerns of classical statistics are supplemented by an additional set of concerns: the treatment of the available prior information along with present sample information. Since the language of all of statistical theory is the language of probability, any information to be included in the shaping of a statistical inference must be in probabilistic terms. This has led most (though not all) statisticians working in the Bayesian area to adopt the personal or subjective interpretation of probability (see Section 2.10). Recall that in this interpretation of the formal concept of probability, a probability value is a measure of strength of an individual's opinion or belief about the existence of some situation or the occurrence of some event. Furthermore, recall that this approach to probability can be given a behavioral as well as a theoretical basis. Given the personal interpretation of probability, a language becomes available that can encompass the information contained in sample results as well as the prior information the experimenter has available. It has been emphasized repeatedly that the theory of probability and statistics is really quite neutral about the ultimate interpretation to be placed on the concept of probability, so long as the purely mathematical axioms of probability are satisfied.

The methods to follow lend themselves to the subjective, rather than exclusively to the relative-frequency, interpretation of this concept. Largely for this reason, Bayesian methods remain somewhat controversial and have not found universal acceptance among statisticians and the users of statistics. However, if Bayesian methods are to be criticized, the criticism should be based on their intuitive reasonableness and appeal, the reality of their assumptions about human behavior, their pragmatic value, their place within empirical science, and so on. They are not properly criticized on mathematical grounds, since they employ exactly the same mathematical machinery as other methods. There is no question of the formal validity of the probability statements such methods yield, so long as we play within the formal rules of the game. It is the interpretation of these statements that must be at issue.

It must be emphasized that the full machinery of probability theory and mathematical statistics is available, regardless of whether we choose to interpret a probability as a long-run relative frequency or, more broadly, as a measure of the degree of belief held by one or more individuals about an event. Quite often, much of the available information about an experimental or natural situation is reflected in choices or judgments made by one or more individuals. Under the personal interpretation of probability, the results of such judgments can be formally included in statistical analysis. In one sense Bayesian methods are techniques for showing how prior beliefs, expressed as probabilities, are modified in the light of new information, also expressed in probabilistic terms.

This approach to inference and decision-making is not new. Even in the early development of probability theory in the seventeenth and eighteenth centuries there was much concern with the question of how probabilities change in the light of new evidence. The Bayesian approach is named for the Reverend Thomas Bayes, an English clergyman who worked in the theory of probability during the mid-eighteenth century. It was he who first suggested and proved the result known as Bayes' theorem, a cornerstone of this whole approach to the revision of probability through evidence. However, over the years there has been much debate, sometimes acrimonious, over the applicability of Bayes' theorem. Some rather ridiculous uses of the theorem brought the entire Bayesian approach into disrepute in the late nineteenth and early twentieth centuries.

In the present century there has been a renewal of interest in methods involving Bayes' theorem, owing in part to a great surge of interest in the theory of decision-making. Increasingly complex problems are being encountered in economics, in business management, in natural resource development, and in a wide variety of other areas, each of which involves decision-making in the face of uncertainty. We cannot attack such problems without a solid theoretical basis for understanding the process of decision-making itself. And, as the theory of decision-making developed, more and more attention was directed to the Bayesian approach to the revision of probability. In recent years the theory of personal probability has been brought to a high state of development by Bruno deFinetti and Leonard J. Savage, among others, and practical Bayesian methods have been worked out by a number of statisticians, most notably Howard Raiffa and Robert Schlaifer. (An excellent introduction to the problem of decision-making under uncertainty is given by Raiffa [1968].)

A brief word of warning: Although Bayesian methods are often appealing in their apparent simplicity, it is a mistake to think that they are "short cuts" or in some sense "easy." Just the opposite is true. Development of the rationale for a Bayesian result often requires mathematical and statistical sophistication of a very high order. We will not even attempt here to show the mathematical rationale for most of these methods. Second, there is a great deal of similarity between the results of Bayesian techniques and results gained from the application of the more familiar techniques to the same data. The actual data are very powerful in shaping our final conclusions, regardless of whether Bayesian or classical methods are used. This is as it should be, of course. Nevertheless, in spite of the similarity in appearance of a Bayesian and a classical result, these methods may differ quite widely in the interpretation of the result. It is most important to remember that the two approaches to statistical inference start from different interpretations of probability, so that the final interpretations of the statistical conclusions will differ even though the conclusions may be almost identical in format. Finally, unlike the body of classical statistical methods, which, while still growing, is pretty well mapped out, the body of practical Bayesian methods is still developing and is far from complete. Bayesian parallels may not yet be available for some of the statistical methods in common use.

In spite of these limitations, the author believes that an introduction to some of the simpler Bayesian methods is worthwhile at this point. The basic point of view is appealing, particularly to behavioral scientists, since this point of view is essentially behavioral. Also, there is every reason to believe that such methods will become steadily more prominent in scientific and technical reports of the future. Indeed, the Bayesian point of view and the methods growing out of it may well be the basis of the "classical" statistics for future generations.

19.1   Bayes' theorem applied to random variables

The Bayesian approach is first concerned with a set of beliefs or opinions, expressed in the language of probability. Such probabilities change as the result of new information. The mathematical device for showing how such probabilities change is Bayes' theorem, which has already been introduced in Sections 4.5 and 9.28. In its earlier appearances Bayes' theorem was restricted to probabilities associated with qualitative events. Now we are ready to restate Bayes' theorem as it applies to random variables.

Let X be a discrete random variable that may take on any of the J values $A_1, \cdots, A_j, \cdots, A_J$, and let Y be a second random variable that can take on any of the K values $B_1, \cdots, B_k, \cdots, B_K$. Then the conditional probability that X takes on the value A_j, given that Y takes on the value B_k, is symbolized by

$$p(X = A_j | Y = B_k),$$

or, more simply by

$$p(A_j | B_k).$$

Bayes' theorem can now be written as follows:

$$p(A_j | B_k) = \frac{p(B_k | A_j) p(A_j)}{\sum_{i=1}^{J} p(B_k | A_i) p(A_i)},$$

$$1, 2, \cdots, i, j, \cdots, J; 1, 2, \cdots, k, \cdots, K.$$

The probability that X takes on the particular value A_j, given that Y takes on the particular value B_k, is equal to the conditional probability that $Y = B_k$ given that $X = A_j$, multiplied by the unconditional probability that $X = A_j$, and divided by the probability that $Y = B_k$ over all possible values of X. Be careful to notice the different kinds of probability distributions for the random variable X that are involved here. The first is the **unconditional distribution of** X, in which each possible value A_j is paired with its probability $p(A_j)$, not conditional on Y. On the other hand, there is the **conditional distribution of** X, **given some value for** Y, such as $Y = B_k$. In this conditional distribution, each possible value of X, such as $X = A_j$, is paired with a conditional probability $p(A_j | B_k)$, the probability that $X = A_j$ given that $Y = B_k$. There will be a distinct conditional distribution of X for each possible value of Y. That is, there will be a conditional distribution of X given $Y = B_1$, another conditional distribution of X given

$Y = B_2$, and so on through $Y = B_K$. As we learned in Chapter 4, if X and Y are independent, then each conditional distribution of X given Y will be identical to the unconditional distribution of X. In this situation, the specification of Y has no effect on the probability of X. On the other hand, if X and Y are not independent, then some of the conditional distributions of X given Y must differ from the unconditional distribution of X.

This form of Bayes' theorem, and all of the remarks above, also hold true if X and Y, rather than being discrete, are actually continuous variables, but each scale of possible values is divided into class intervals. Let a set of mutually exclusive and exhaustive class intervals for X be designated by $A_1, \cdots, A_j, \cdots, A_J$ and a similar set of class intervals for Y by $B_1, \cdots, B_k, \cdots, B_K$. Then the expression $p(X \,\epsilon\, A_j | Y \,\epsilon\, B_k)$, symbolizing the probability that the value of X falls in the interval A_j, given that the value of Y falls in the interval B_k, could be used in place of $p(X = A_j | Y = B_k)$. Analogous changes would, of course, be made in the other terms of Bayes' theorem.

Since it is very important to keep straight the various kinds of probabilities figuring in Bayes' theorem, some new terminology will be useful at this point. Let the distribution associating the unconditional probability $p(X = A_j)$ with each possible value or interval A_j be called the **prior** distribution of X. The prior distribution will be used to represent our set of beliefs about X, **prior** to the accumulation of additional evidence. It will be convenient for us to designate the prior probability for any event A by $p'(A)$.

Next, let the conditional distribution of X, given Y, be called the **posterior distribution** of X. In Bayesian analysis the posterior distribution represents our beliefs about the various values of X following the new evidence represented by Y. Any difference between the prior distribution and the posterior distribution is then a reflection of the change in our beliefs brought about by the evidence Y. In the following, the posterior probability for an event A will be designated by $p''(A)$.

Finally, the effective agent in this change is, by Bayes' theorem, $p(Y = B_k | X = A_j)$, the probability that Y takes on the value B_k, given that X has the value A_j. This set of conditional probabilities for Y given X will be called the **likelihood function**. In applications of Bayes' theorem, this will represent the likelihood of the occurrence of the sample information $Y = B_k$, given that X has the value A_j. These three kinds of probabilities—the prior probability, the posterior probability, and the likelihood—actually make up the "working parts" of Bayes' theorem. The other expressions, such as $\sum_i p(Y = B_k | X = A_i) p'(X = A_i)$,

really serve only in a standardizing role, guaranteeing that the posterior probabilities will each lie between 0 and 1, and will sum to 1.00.

In order to illustrate a prior distribution, a likelihood function, and a posterior distribution, let us turn to a simple example. Suppose that we are concerned about the actual academic ability possessed by a young child. We assume that he has some "true amount" of academic ability, and that this can, in principle, be measured. The amount of true academic ability will be symbolized by the random variable X. Although X is regarded as continuous, we can consider the scale of possible values of X as divided into a set of mutually exclusive and exhaustive class intervals. Unfortunately, we cannot measure X directly; we can only infer its value from other evidence. Even so, we have reason to believe that for children of this age and background, the distribution of X should have the following form:

Interval A_j	$p'(x \in A_j)$
130 and up	.02
115–129	.14
100–114	.34
85–99	.34
70–84	.14
up to 69	.02

In the absence of any other information about the X value for this particular child, let us say that this distribution accurately reflects our prior opinions about him. That is, we feel most confident that his true ability measure falls either in the interval 85–99 or in the interval 100–114, next most confident that the value of X falls either in the interval 70–84 or the interval 115–129, and so on. This is the **prior** distribution.

Now, however, we gain some further information that is relevant to the value of X. This information should have some impact on our opinions about the ability of the child in question. Suppose that a simple test of reasoning exists, such as one of the items on the Stanford-Binet test. Performance on this item is known to be related to the academic ability of a child of this age and background. Let us call the score of a child on this test Y, and let us further suppose that only three values of Y are possible: 0, 1, and 2. This test has been extensively studied, and the distribution of Y scores, given each possible level of X, is known. These conditional distributions of Y given X are shown as the rows in the following table (the apparent limits should be interpreted as real limits in the usual way):

Interval A_j	$p(y \mid x \in A_j)$		
	$y = 0$	$y = 1$	$y = 2$
$130 \leq x$	.05	.15	.80
115–129	.15	.35	.50
100–114	.30	.30	.40
85–99	.40	.30	.30
70–84	.50	.35	.15
$x \leq 69$	.80	.15	.05

In other words, for children whose ability scores fall at or above 130, 80 percent have Y scores of 2, 15 percent have Y scores of 1, and only 5 percent have Y scores of 0.

Each such conditional distribution of Y, given some value or interval of values of X, is a likelihood function. A likelihood function may be thought of as the sampling distribution of Y, given the value or interval of values for X.

Now suppose the child in our example is given the reasoning test and his Y score turns out to be 2. What is the resultant posterior distribution, representing our changed beliefs about his true X score? We apply Bayes' theorem to find the probability associated with each interval of X given that $Y = 2$. Starting with $130 \leq x$, we have

$$p''(130 \leq X | Y = 2) = \frac{p(Y = 2 | 130 \leq X)p'(130 \leq X)}{\sum_{i=1} p(Y = 2 | X \,\epsilon\, A_i)p'(X \,\epsilon\, A_i)}$$

or

$$\frac{(.80)(.02)}{(.80)(.02) + (.50)(.14) + (.40)(.34) + (.30)(.34) + (.15)(.14) + (.05)(.02)}$$
$$= .016/.346 = .046.$$

In this same way we find

$$p''(115 \leq X \leq 129 | Y = 2) = \frac{(.50)(.14)}{.346} = .202,$$

$$p''(100 \leq X \leq 114 | Y = 2) = \frac{(.40)(.34)}{.346} = .393,$$

$$p''(85 \leq X \leq 99 | Y = 2) = \frac{(.30)(.34)}{.346} = .295,$$

$$p''(70 \leq X \leq 84 | Y = 2) = \frac{(.15)(.14)}{.346} = .061,$$

$$p''(X \leq 69 | Y = 2) = \frac{(.05)(.02)}{.346} = .003.$$

Here we have the posterior distribution for X, given that $Y = 2$. This distribution reflects the impact of the information that $Y = 2$ on our opinions about the value of X. You will note that as compared with the prior distribution, the posterior distribution shows a marked shift upward, so that whereas the prior distribution was symmetric about a value for X of 100, the posterior distribution is skewed negatively. The interval 100–114 now represents the mode of the distribution, and the probability for that interval is considerably higher than that for any other interval. The probability is almost .60 that X equals or exceeds 100, whereas the corresponding probability was .50 in the prior distribution.

The terms "prior distribution" and "posterior distribution" are relative, depending upon where one is in the process of gathering information. Thus, in our example, if a second test of ability were administered to the child, the posterior distribution just calculated would serve as the prior distribution in the new set of calculations. Although this distribution was "posterior" to the first item of information, it was "prior" to the second. The amount of information collected, and the consistency of the information in pointing a certain way, show up as changes in the successive posterior distributions.

19.2   Parameters treated as random variables

Basically a parameter is a constant that enters into the rule for a particular probability or probability density function. As such, a parameter either has a certain value or it does not. We have seen repeatedly in the foregoing that questions about a particular population or process are often cast most conveniently in terms of the values of one or more parameters. Even though such parameters are properly regarded as fixed constants, we very seldom know these values exactly. However, we do often have reason to believe more strongly in some possible values than in others as the true value of some parameter. The strength of such beliefs varies over the possible values that might be true of a parameter. In this sense, Bayesian analysis often deals with parameters *as though* they were random variables.

Associated with each possible value of a parameter is a probability or probability density, representing the strength of belief that the parameter has that value. In such uses of the concept of a parameter it is important to keep in mind that it is not the true parameter value itself that varies, but rather our beliefs and opinions about the true value of such a parameter, when considered over all of the possibilities that might be true. Thus, suppose that u is the true mean height of American women at this moment in time. The parameter u has some true value, which is, for us, unknown. However, this does not keep us from having opinions about the value of u. Most of us would think it unlikely that the mean height of American women is about 6 feet or below 5 feet. We would be far more confident that u actually lies in an interval of values somewhere about 5.6 feet. Perhaps our various opinions about the true value of u could be represented as a normal distribution, with an expected value or $E(u)$, of about 5.6 feet, and some variance σ_u^2. The value of the normal density at any possible value for u would then reflect the strength of our belief that u actually does have that particular value. The point is that a set of opinions about the value of any parameter can be represented by a probability distribution or probability density function, wherein the parameter is treated as though it were a random variable, even though we are well aware of the fact that the actual parameter in question is a constant.

In Bayesian statistics, as in other statistical methods, the focus of many problems is on the value of one or more parameters. In consequence, the prior and posterior distributions in most Bayesian methods are distributions of some parameter. This is perfectly legitimate, and it need cause no confusion so long as one remembers the interpretation to be given to such distributions. Thus, consider some parameter symbolized by θ (theta). When we discuss the prior distribution of θ we really mean the distribution reflecting our own prior opinions about the true value of θ, over all the possibilities that might be true, and prior to additional information. The posterior distribution then reflects our opinions about θ after some relevant information has been acquired. We will symbolize the prior probability or probability density function for θ by $p'(\theta)$ or $f'(\theta)$, respectively. Then the posterior probability or probability density function, given the information $Y = B_k$, can be symbolized by $p''(\theta|Y = B_k)$ or by $f''(\theta|Y = B_k)$, or, where the context is clear, by $p''(\theta)$ or $f''(\theta)$.

For the moment, assume that the prior distribution of θ is discrete and finite. In this distribution θ is represented as a random variable that can take on only the J distinct values represented by $\theta_1, \theta_2, \cdots, \theta_J$. Let the random variable Y take on K distinct values, just as before. Then Bayes' theorem can be written as

$$p''(\theta = \theta_j|Y = B_k) = \frac{p(Y = B_k|\theta = \theta_j)p'(\theta = \theta_j)}{\sum_i p(Y = B_k|\theta = \theta_i)p'(\theta = \theta_i)},$$

or, more succinctly, as

$$p''(\theta_j|B_k) = \frac{p(B_k|\theta_j)p'(\theta_j)}{\sum_i p(B_k|\theta_i)p'(\theta_i)}.$$

Here $p'(\theta_j)$ is the prior probability that θ has the value θ_j, $p''(\theta_j|B_k)$ is the posterior probability that θ has the value θ_j given that Y has the value B_k, and $p(B_k|\theta_j)$ is the likelihood that Y takes on the value B_k given that θ has the value θ_j. Bayes' theorem in this form permits us to transform the discrete prior distribution of θ into the posterior distribution, which in this instance is also discrete.

One example of Bayesian inference about the value of a parameter has already been given in Chapter 9. We will now show the application of the method to a somewhat more complex problem.

Suppose that an experimenter is working with a colony of laboratory mice. For various reasons that need not concern us here, he is not certain about the genetic strain to which this colony belongs. He does know that the colony belongs either to genetic strain I, or to strain II, or is of a mixed genetic strain. Now suppose further that there is some physical characteristic that each mouse may or may not possess. Among strain I this characteristic should occur about 70 percent of the time, but only about 10 percent of the time in strain II. In a mixed strain, the physical trait should occur in about 30 percent of mice. If we symbolize the true proportion among the mice by the parameter π (pi), then π must have one of the values .10, .30, or .70. (In order to avoid confusion among the various probabilities, we will henceforth use π rather than p as the parameter of a Bernoulli process.)

The experimenter believes that the colony is just as likely to be mixed as to be of pure strain. Furthermore, if the strain is pure, he believes it equally likely that the strain is I or that the strain is II. In short, the prior set of beliefs on the part of the experimenter can be represented by the following table:

Strain	Value of π	Prior probability $p'(\pi)$
I	.70	.25
II	.10	.25
Mixed	.30	.50

Here the three genetic possibilities correspond exactly to three possible values of the parameter π, so that the prior opinions of the experimenter can be represented by a probability distribution over these three possible values of the parameter π.

Now suppose the presence of the physical property for any mouse corresponds to the outcome of a stable and independent Bernoulli process with parameter π. Then for a sample of N mice selected independently and at random (the colony is assumed very large), the number of mice showing the characteristic should follow a binomial sampling distribution. This means that the likelihood of r mice showing the characteristic, given N sampled, should be defined by the binomial rule, with parameter π. The experimenter actually takes such a sample and finds that of 10 mice sampled, 8 show the characteristic. Thus $N = 10$ and $r = 8$. The likelihood of such a sample result is given by

$$p(r = 8 | N = 10, \pi) = \frac{10!}{8!2!} \pi^8 (1 - \pi)^2.$$

(This likelihood, although in form identical to a binomial probability, is written as a conditional probability, in order to emphasize its dependence upon the values of N and π.)

We apply Bayes' theorem to obtain the posterior distribution. Consider first the possibility that the animals belong to strain I, so that $\pi = .70$. Then, by Bayes' theorem,

$$p''(\pi = .70 | r = 8, N = 10) = \frac{\binom{10}{8}(.70)^8(.30)^2(.25)}{p(r = 8 | N = 10)},$$

where

$$p(r = 8 | N = 10) = \binom{10}{8}(.70)^8(.30)^2(.25) + \binom{10}{8}(.10)^8(.90)^2(.25)$$
$$+ \binom{10}{8}(.30)^8(.70)^2(.50)$$
$$= (.2334)(.25) + (.0000)(.25) + (.0014)(.50)$$
$$= .059075.$$

(The binomial probabilities are found directly from Table II, Appendix C.) Consequently, the posterior probability for $\pi = .70$ is

$$p''(.70 | r = 8, N = 10) = \frac{.058375}{.059075} = .99, \quad \text{approximately.}$$

In the same manner we find

and

$$p''(.10|r = 8, N = 10) = 0, \quad \text{approximately}$$

$$p''(.30|r = 8, N = 10) = \frac{.0007}{.059075} = .01, \quad \text{approximately.}$$

In summary, the posterior distribution is

Strain I: $\pi = .70$; $p''(.70|r = 8, N = 10) = .99,$
Strain II: $\pi = .10$; $p''(.10|r = 8, N = 10) = .00,$
Mixed: $\pi = .30$; $p''(.30|r = 8, N = 10) = .01.$

The probability for strain I has jumped very dramatically from a prior probability of .25 to a posterior probability of .99. In the light of this evidence, the experimenter is almost certain that the colony of mice must belong to strain I.

It is fairly obvious why 8 out of 10 successes is sufficient to create such a change between prior and posterior probabilities. The likelihood of 8 successes out of 10 is very small given either $\pi = .10$ or $\pi = .30$, so that these possibilities are given almost no weight in the posterior distribution. On the other hand, a very large relative weight is given to $p = .70$, since this value of the parameter makes the occurrence of 8 out of 10 relatively likely.

This little example is quite contrived, of course, and it would be rare to find a real example in which the possible values of the parameter could be restricted in this way. Ordinarily, the prior distribution of the parameter is continuous, since the true value may be any one of an infinite number of possibilities. Next we will see how Bayes' theorem can be extended to meet this situation.

19.3 CONTINUOUS PRIOR AND POSTERIOR DISTRIBUTIONS

In many situations of practical interest, a parameter θ can have as its value any of the real numbers within a particular interval. Thus, the parameter π in a Bernoulli process might have as its value any of the real numbers between 0 and 1.00, or the mean μ for some population might have as its value any of the real numbers, $-\infty < \mu < \infty$. If we are to represent our beliefs about the various possible values of such a parameter, we must do so in terms of a continuous function. Bayes' theorem is accordingly modified to involve probability densities rather than probabilities. Consider a likelihood function for y that depends only on the value of a single parameter θ. Then

$$f''(\theta = g|y) = \frac{f(y|\theta = g)f'(\theta = g)}{\int_a^b f(y|\theta)[f'(\theta)]}, \quad a \leq \theta \leq b.$$

Here, $f'(\theta = g)$ is the prior probability density at some particular value $\theta = g$ in the prior distribution, $f(y|\theta = g)$ is the likelihood for the obtained value of Y given the value $\theta = g$, and $f''(\theta = g|y)$ is the posterior density for the particular value of θ given $Y = y$. Rather than a sum in the denominator, as for a discrete prior distribution, we now have an integral taken over the entire range of values within which θ must lie, or $a \leq \theta \leq b$. It should be apparent that Bayes' theorem for continuous prior and posterior distributions has a form completely analogous to the form for discrete distributions.

Quite often in statistics a transition from discrete to continuous random variables signals a shift from rather tedious computational methods to simpler and more routine procedures. In part this is true in Bayesian statistics as well. It is always possible in principle to calculate posterior distributions directly from prior distributions by means of the likelihood function and Bayes' theorem. In the case of discrete prior and posterior distributions one often has no choice but to apply Bayes' theorem directly. However, when the parameter under study is treated as a continuous random variable, a number of routine methods become available for use. Even so, the availability of such methods depends upon certain assumptions being met about the form of the prior distribution. One set of possibilities that lead to relatively simple procedures will be introduced next.

19.4 CONJUGATE PRIOR DISTRIBUTIONS

Very difficult mathematical complexities can sometimes arise when Bayes' theorem is applied directly. Not only can the computations be laborious even for fairly simple discrete prior and posterior distributions; in the continuous case the statistician is confronted by the necessity to evaluate the integral that appears in the denominator of Bayes' ratio. Doing so may require quite advanced methods. On the other hand, there is an approach that allows one to bypass many of these difficulties, and that does yield rather simple and routine computational procedures.

It has been found that given a likelihood function of a particular form, prior distributions that are members of certain families fit "naturally" with the likelihood function. Furthermore, these priors yield posterior distributions belonging to the same family. This has the result of greatly simplifying the mathematical complexities of Bayes' theorem, and of yielding a routine procedure in many instances. Such a family of distributions, whose members go along naturally with a given likelihood function, is called the **natural conjugate** of the likelihood function. The choice of a prior distribution that is conjugate to the likelihood function, implying that both the prior and the posterior distributions will belong to the same natural conjugate family, reduces the computational burden enormously. (Lest the idea of the *choice* of a prior distribution prove unsettling to the reader, in view of the foregoing discussion of states of belief, let it be said that the priors that are available as natural conjugates are so flexible that one's own prior beliefs can usually be represented among the families of distributions available. This will be illustrated in the following.)

In the next few sections we will discuss and illustrate several such families of conjugate distributions, each appropriate to a different kind of likelihood function. For instance, when samples of N independent observations are taken from a stationary Bernoulli process, the likelihood function is binomial, as has just been illustrated. Now it happens that *the natural conjugate of the binomial likelihood function is the family of beta distributions,* introduced in Chapter 8. When one believes the binomial to be the appropriate likelihood function, the choice of a prior distribution belonging to the beta family will greatly simplify the process of finding the posterior distribution, as we shall shortly see. *The posterior distribution will then also belong to the beta family.* For other sorts of likelihood functions, still other families of distributions serve as natural conjugates. Choosing a member of the natural conjugate family as the prior distribution greatly simplifies matters in these instances as well.

In the following it is most important to remember that **if the prior distribution is a member of the natural conjugate family for a particular likelihood function, the posterior distribution will also be a member of that family.** Thus, after the initial conjugate distribution is determined, the same basic methods may be applied over and over again in a routine way as more information accumulates.

It is desirable that the families of distributions used as priors be sufficiently flexible and "rich" in form that an experimenter's actual state of opinion or belief can be readily represented. To a large extent the conjugate families to be discussed here do have this property. The family of distributions employed for the prior and the posterior distributions should also be rather simple and easy to interpret. At the very least, it should be possible to calculate and interpret expectations easily from the distributions. Once again, the conjugate distributions to be discussed here tend to have these advantages. It is also helpful if distributions serving in the prior and posterior roles have been extensively studied and tabulated; by and large this also is true of the conjugate distributions that we will employ.

One does not *have* to adopt a conjugate prior distribution in order to carry out a Bayesian analysis. Other alternatives may exist that describe the prior state of opinion better than the available conjugate family, or direct computation may be preferable for other reasons. However, as we will see in the following, the use of conjugate prior distributions can reduce Bayesian methods to a set of very simple and routine computations.

19.5 BAYESIAN INFERENCE FOR A BERNOULLI PROCESS

Suppose now that we are concerned with data generated by a stationary and independent Bernoulli process. Lest we forget: this means that only the two events "success" and "failure" can occur, that occurrence of each event is independent of every other, and that the probability π of a success is constant on every trial. When N independent observations or trials are made from such a stable Bernoulli process, the sampling distribution of r, the number of successes, is binomial. The likelihood of r successes, given N and π, is given by the binomial rule.

Although we cannot even begin to do so here, it can be shown that *the natural conjugate of the binomial likelihood function is the beta family of distributions.* That is, in the method to follow, we will assume that the prior distribution of the parameter π follows the rule for a beta function with parameters r' and N':

$$f'(\pi|r', N') = \frac{(N' - 1)!}{(r' - 1)!(N' - r' - 1)!} (\pi)^{r'-1}(1 - \pi)^{N'-r'-1}, \; N' > r' > 0.$$

(Remember that although the binomial function is discrete, the beta function is continuous, and $0 \leq \pi \leq 1$. This is appropriate, since π *could* be any value whatever in the interval from 0 to 1, inclusive.) Such a beta prior distribution of π will have an expectation given by

$$E'(\pi) = \frac{r'}{N'}$$

and a variance

$$\text{Var}'(\pi) = \frac{r'(N' - r')}{N'^2(N + 1)}.$$

Because the beta family forms the natural conjugate to the binomial likelihood function, the following principle holds: **If the prior distribution of π is beta, with parameters r' and N', and N trials are taken from a stationary independent Bernoulli process with parameter π, then the posterior distribution of π will be beta with parameters $r'' = r' + r$, and $N'' = N' + N$, where r is the number of successes observed in the sample of N trials.** In short, when one is concerned with the parameter π of a stable Bernoulli process, if π is assumed to have a beta distribution prior to the sample of N independent trials, the posterior distribution of π will also be beta. The posterior parameters are found simply as sums of the prior parameter values r' and N', and the r and N values from the sample.

In order to see this principle in operation, take the following example. A psychologist constructing a test has a certain item he wishes to include. However, he does not know the difficulty of this item. In this instance he expresses the difficulty as the proportion of children of a certain age who are able to pass the item. Because of what he knows about similar items, he believes that $\pi = .30$ might well turn out to be the true value of π. However, he is not completely certain, and he feels that some range of values to either side of .30 might well include the true value of π. Let us suppose that his prior opinions about the value of π can be represented by a beta distribution with $r' = 3$ and $N' = 10$. The reasons for this choice of prior parameter values will be examined a little later. A graph of this prior distribution is shown in Figure 19.5.1.

The experimenter decides to administer the item to 50 children of the relevant age group, chosen at random. Each child either passes or fails the test, and the performance of each child is independent of that for any other. Under these conditions, the testing of each child on the item can be regarded as an independent trial from a stationary Bernoulli process with parameter π. It turns out that of the 50 children, 20 pass the item. Then the posterior distribution of π is of beta form, with parameters

$$r'' = 3 + 20 = 23$$

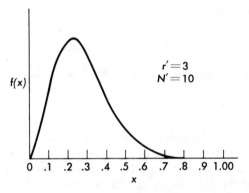

FIG. 19.5.1. A beta distribution with $r' = 3$ and $N' = 10$

and
$$N'' = 10 + 50 = 60.$$

This distribution is also pictured in Figure 19.5.2.

Notice the difference between the prior and the posterior distributions: Originally the mean of the prior distribution of was r'/N' or .30. In the posterior distribution the mean is $r''/N'' = 23/60 = .383$. Moreover, the posterior distribution tends to be concentrated in a smaller range of values than was the prior distribution. The variance of the prior distribution was

$$\frac{r'(N' - r')}{N'^2(N' + 1)} = \frac{(3)(7)}{(10)^2(11)},$$

or about .019, giving a standard deviation of about .138. On the other hand, the

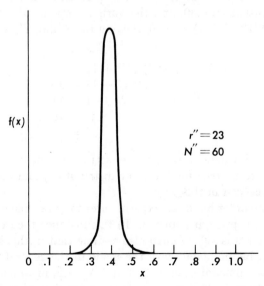

FIG. 19.5.2. A beta distribution with $r'' = 23$ and $N'' = 60$

variance of the posterior distribution is

$$\frac{r''(N'' - r'')}{(N'')^2(N'' + 1)} = \frac{23(37)}{(60)^2 61}.$$

or about .00387, corresponding to a standard deviation of about .0622. This reflects the extent to which the new information has decreased the experimenter's uncertainty about the value of π.

Notice also that the mean of the posterior distribution, or r''/N'', lies between the mean of the posterior distribution, or r'/N', and the proportion of successes found from the sample, or r/N. This will always be true in problems of this sort, since the mean of the posterior distribution can be regarded as the weighted average of the prior mean and the actual sample proportion:

$$\text{posterior mean} = \frac{r''}{N''} = \frac{\dfrac{r'}{N'}(N') + \dfrac{r}{N}(N)}{N' + N}.$$

The posterior distribution has a mean that is closer to the sample proportion of .40 than it is to the mean of the prior distribution, or .30. The reason is that the size of the sample, $N = 50$, is large relative to the prior parameter N', which was only 10. The sample information thus gains considerably more weight than does the prior distribution in the determination of the posterior distribution.

This method is obviously simple enough to execute, once the prior distribution has been specified. How might one decide upon the appropriate parameters of the prior distribution? The beta form of the prior is determined by the choice of the beta conjugate distribution to the binomial likelihood function. However, one must decide upon the values to be given to r' and to N'. If the experimenter can specify the mean and the variance of his prior distribution, the problem is essentially solved. That is, if the experimenter had held that the mean of his prior distribution was .30, and the variance about .019, then $r'/N' = .30$ and $r'(N' - r')/(N')^2(N' + 1) = .019$. Solving for r' and N', we have

$$r' = .3N',$$
$$.3N'(.7N')/(N')^2(N' + 1) = .019,$$
$$.21/(N' + 1) = .019,$$
$$N' + 1 = 11$$
$$N' = 10$$
$$r' = 3.$$

Experimental methods can be worked out for settling on the value of the mean and variance one holds to be true for the prior distribution, and our example might very well have proceeded in this way.

On the other hand, an experimenter may be able only to specify the expected value for his prior distribution. In this instance the expected value of π for this experimenter was .30. We can also assume that in this example the prior distribution should be unimodal, with the mode in the general vicinity of .30. A beta distribution is unimodal when $r' > 1$ and $N' > 2$, in which case the mode is at the value equal to $(r' - 1)/(N' - 2)$. This too sets boundary conditions on the choice of r' and N'. Finally, suppose that we insist that r' and N' be whole numbers. This is not absolutely necessary, but can be a conceptual and computational convenience. If r' and N' are both to be whole numbers, then $r' = 3$ and $N' = 10$ are the two smallest whole numbers that will produce a unimodal beta distribution with a mean of .30. A little exploration of the consequences of various choices for r' and N', such as demonstrated here, may well lead to a prior distribution fitting the prior beliefs of the experimenter adequately. In any case, the specification of the prior parameters is not so much a life-and-death matter as it may at first appear; for reanobly large sample N minor variations in the choice of r' and N' will have very little effect on the posterior distribution. You may very easily convince yourself that this is true by calculating the posterior distribution in the example on the basis of other, slightly different, values of r' and N'.

Perhaps the experimenter originally believes that any value of π in the interval from 0 to 1 might equally well be the true value. Can this situation be represented by a beta distribution? Yes—since setting $r' = 1$ and $N' = 2$ produces a beta distribution that is perfectly uniform over the interval from 0 to 1 inclusive. What would the effect on the posterior distribution have been if this uniform prior had been used? Here, given that the sample showed 20 successes out of 50 trials, the posterior distribution would have had parameters

$$r'' = 1 + 20, \qquad N'' = 2 + 50.$$

This posterior distribution, still beta in form, has a mean of $r''/N'' = 21/52$ or .408, and a variance of .0045, yielding a standard deviation of about .067. The posterior distribution does differ from the posterior distribution found previously, though not radically so. The size of the sample is such in this example that very little weight is given to the prior distribution. We will discuss this question of "flat" or uniform prior distributions at more length in a later section.

Before leaving this example, we should also note that the prior distribution might have come about empirically. That is, suppose the experimenter originally had no particular opinion about π, and that this state of affairs was represented by the uniform or "flat" beta with $r' = 1$ and $N' = 2$. Then the experimenter took a sample of say, $N = 8$, and found that $r = 2$ was the number of successes. His posterior distribution would then have been exactly the same as the prior distribution originally used in this example, where $r' = 3$ and $N' = 10$.

Perhaps this gives a clue as to one interpretation to be put on the prior parameters r' and N': these parameters function as though the experimenter had already taken a sample of $N' - 2$ cases and found $r' - 1$ successes, so that r' and N' can be thought of as the product of a prior sample, given a uniform prior distribution. When the experimenter has reason to say that his beliefs about the value of π are best described by a beta distribution with parameters r' and N', he is acting *as though* his belief were based on a previous sample of $N' - 2$ cases of which $r' - 1$ were successes. Such a prior sample may be real, or it may be purely fictitious; in either case r' and N' fit within the same conceptual framework as r and N for the sample actually taken.

Bayesian methods are really quite indifferent about the source of the prior distributions that are used. Prior distributions are treated in exactly the same way regardless of their source. As more and more information is accumulated, however, the details of the original prior distribution become less and less important. Even though you and I start from very different prior positions, if we are given exactly the same information and treat it in the same way, in the long run our posterior distributions will be identical. The weight of the sample information will finally overwhelm any initial differences in the priors. This is an enormous advantage of the Bayesian approach, and is really the main defense of these methods against arguments based on their "subjectivity." This theme will be elaborated as we turn to Bayesian methods designed for other data-generating processes.

19.6   Normal processes with known variance

Next we will illustrate the application of the concept of a conjugate family of distributions when the likelihood function is normal. For the sake of simplicity, this will be discussed first under the assumption that the normal processes generating the data has a variance σ^2 that is known. Suppose that we are engaged in observing some process in which any value of the random variable X has a likelihood given by the normal density function:

$$f(x) = \frac{1}{\sqrt{2\pi}\,\sigma}\, e^{-(x-\mu)^2/2\sigma^2}.$$

Typically in such a situation one is interested in making inferences about the value of the mean μ (the value of σ^2 is assumed known). Such inferences are most often based upon the value of M, the sample mean, as calculated from a sample of N observations made independently and at random from the basic normal process. The sampling distribution of the sample mean is then itself normal, with expectation μ and variance σ^2/N.

When the likelihood for the sample mean is normal, then the conjugate family of distributions is also normal. That is, given a basic normal process underlying the random variable X, and given N independent observations, the conjugate prior distribution of μ will be taken to be normal. The mean of the prior distribution will be designated by M', and the variance of the prior distribution by V'. Then, given the sample mean M, based on N independent observations, the posterior distribution of μ is normal with expectation

$$M'' = \frac{\left(\dfrac{1}{V'}\right) M' + \left(\dfrac{N}{\sigma^2}\right) M}{\dfrac{1}{V'} + \dfrac{N}{\sigma^2}}.$$

The variance of the posterior distribution is found from

$$\frac{1}{V''} = \frac{1}{V'} + \left(\frac{N}{\sigma^2}\right).$$

Notice that the posterior mean is a weighted average of the prior mean M' and the sample mean M. The weights are provided by the reciprocals of the prior variance and the variance of the mean, respectively.

The logic of this method becomes somewhat more apparent if we define the prior variance in a slightly different form. An inspection of the equations given above shows the V' plays the role of a reciprocal weight in the transformation of the prior to the posterior distribution. The other weight is provided by the variance of the sampling distribution of the mean, $\sigma_M^2 = \sigma^2/N$. The larger the value of N, the larger is the weight given to the sample evidence. In a similar way we can talk about the weight given to the prior distribution, just as though the prior actually represented a previous sample based on N' prior cases. We can define

$$N' = \frac{\sigma^2}{V'} \quad \text{so that} \quad V' = \frac{\sigma^2}{N'}.$$

Then the variance of the prior distribution becomes analogous to the variance of the sampling distribution of the mean, for samples of N' cases. Now the mean of the posterior distribution can be written as

$$M'' = \frac{\left(\dfrac{N'}{\sigma^2}\right) M' + \left(\dfrac{N}{\sigma^2}\right) M}{\dfrac{N'}{\sigma^2} + \dfrac{N}{\sigma^2}}$$

$$= \frac{N'M' + NM}{N' + N}.$$

The mean of the posterior distribution is simply a weighted or pooled mean, calculated as though it came from the pooling of two samples, of size N' and N respectively. Furthermore, for the posterior distribution,

$$N'' = N' + N \quad \text{and} \quad V'' = \frac{\sigma^2}{N''}.$$

Compare this with the situation where cases are drawn independently from a stationary Bernoulli process, and the beta conjugate family is employed. There, prior information was combined with sample information in the form of a weighted average. Here, prior information is also treated as though it represented the result of a previous sample of N' independent observations of the process, and prior information is combined with sample information through weighted averages.

Let us look at an example of the application of this method. In a classical experiment in visual perception, the stimulus is the well-known "Muller-Lyer illusion." The subject is called upon to match the length of a line tipped with "feathers" at either end by adjusting the length of a line tipped with arrowheads at either end.

A measurement is made of the error that the subject makes in matching the length of the standard line. Generally, subjects tend to overestimate the length of the standard. Let us suppose that under these experimental conditions, a considerable amount of evidence exists that subjects overestimate the standard stimulus by 6.7 units on the average. Furthermore, there is evidence that the distribution of such errors over subjects tends to follow a normal distribution.

Now our experimenter decides to modify the response procedure of the experiment in a certain way. He does this in the belief that this new procedure will systematically reduce the error in estimates of the length of the stimulus. Indeed, his hunch is that this procedure will reduce the average size of the error to zero. On the other hand, he sees no reason to believe that the variance of the estimates given across subjects will change. Past experience has shown that the standard deviation of such estimates is 3.5 units.

Here, the experimenter's prior beliefs about the value of μ under the new procedure can be represented by a normal distribution. The mean of the prior distribution should be zero, since this is the value that the experimenter feels μ will most likely turn out to have. He is also willing to bet with odds of 9 to 1 that the value of μ will not lie more than 10 units to either side of 0. For a normal distribution, this is equivalent to the statement that the value 10 or the value -10 will lie about 1.65 standard deviations from the mean of the distribution. Hence, he takes $1.65 \sqrt{V'} = 10$ and solves to obtain $V' = 36.07$ as the value of the prior variance. Since the process or population variance is assumed to be $\sigma^2 = (3.5)^2 = 12.25$, it must then be true that

$$
\begin{aligned}
N' &= \sigma^2/V' \\
&= 12.25/36.07 \\
&= .34, \quad \text{approximately.}
\end{aligned}
$$

Now the experiment is actually carried out on 25 subjects, drawn at random and working independently. For these subjects, the mean error in their estimates of line length is 4.8 units. What then is the posterior distribution of μ, given this information?

Since the normal prior distribution is a member of the conjugate family, given that the likelihood of μ is normal, the posterior distribution will be normal as well. The mean of the posterior distribution is found to be

$$
\begin{aligned}
M'' &= \frac{N'M' + NM}{N' + N} \\
&= \frac{0(.34) + 4.8(25)}{.34 + 25} \\
&= 4.73.
\end{aligned}
$$

The variance V'' of the posterior distribution is found from

$$
\begin{aligned}
N'' &= .34 + 25, \\
V'' &= \sigma^2/N'' \\
&= 12.25/25.34 \\
&= .483,
\end{aligned}
$$

so that the standard deviation is $\sigma = \sqrt{.483}$ or .695.

Once again, the posterior distribution for μ is decidedly different from the prior distribution. Now the posterior distribution, while still normal, has a mean of 4.73, in contrast to the mean of the prior distribution, which was zero. The fact that the experimenter is far less uncertain about the value of μ than formerly is reflected in the smaller standard deviation of the posterior distribution. This is now only .695, in contrast to the standard deviation of 6.06 assumed for the prior distribution. The values that the experimenter should believe to be good bets about μ are concentrated in a much narrower interval than formerly. Notice that $\mu = 0$ now has a much smaller probability density associated with it than formerly. The experimenter has sharply revised his opinions.

Even though the weight given to the prior distribution is very small in this example, it still exerts some influence. The posterior mean is 4.73, somewhat less than the value 4.8 as found from the sample mean. Nevertheless, in the posterior distribution the mean is very much closer to the sample value than to the prior mean, since relative to N' the actual sample N is quite large. Just as in the binomial example given above, the prior distribution has been treated as the result of a small "fictitious" sample. The actual sample information literally swamps this prior information in the final analysis.

Incidentally, the value of .34 for N' need cause no confusion. Remember that we are treating the prior distribution *as though* it reflected information from a prior sample. Such a sample *is* fictitious in this example, and there is no reason whatsoever for N' to be limited to a whole number, as N necessarily must be. The value of N' is actually only a weight, reflecting the original degree of certainty about the value of μ. In the posterior distribution the sample information is given 75 times as much weight ($N/N' = 25/.34$) as was the original information reflected in the prior.

On the basis of this posterior distribution, let us ask a few additional questions. First of all, within what interval of possible values for μ does the experimenter have a personal probability equal to 90 percent? That is, for what range of values, centered about $E''(\mu)$ in the posterior distribution, is the probability .90? This is easily found. Since a z value of 1.65 cuts off the upper 5 percent in a normal distribution, and a value of -1.65 the lower 5 percent, we take

$$M'' + 1.65 \sqrt{V''} = 4.73 + (1.65)(.695)$$
$$= 4.73 + 1.15$$
$$= 5.88,$$

and

$$M'' - 1.65 \sqrt{V''} = 4.73 - 1.15$$
$$= 3.58.$$

The probability in the posterior distribution is then .90 that μ lies between the values 3.58 and 5.88. (Contrast this *carefully* with the interpretation put upon confidence intervals in Chapter 9. These interpretations are not the same, although the calculations involved appear to be similar).

Above what value is the posterior probability only .01 that the true value of μ lies? In this instance we take

$$M'' + 2.326(\sqrt{V''}) = 4.73 + 2.326(.695)$$
$$= 4.73 + 1.62$$
$$= 6.35.$$

Hence, the probability that the true value of μ is at or above 6.35 is .01 in the posterior distribution.

Finally, how does this compare with result of a straightforward test of the hypothesis that $\mu \leq 0$, against the hypothesis that $\mu > 0$, by the classical methods? If the hypothesis were tested in the ordinary way (remember that σ^2 is assumed known), we would find that

$$z = \frac{M - 0}{3.5/5} = \frac{4.8}{.7} = 6.85,$$

which more than exceeds the usual standards for significance. On the other hand, what is the probability that the value of μ lies at or below zero in the posterior distribution? This we find by taking

$$z = \frac{0 - M''}{(3.5/\sqrt{N''})} = \frac{-4.73}{.695} = -6.80.$$

The area lying below this z value is extremely small. The experimenter is almost certain that the true value of μ exceeds zero, and hence he could reject the hypothesis that $\mu \leq 0$ in favor of the hypothesis that $\mu > 0$ with great confidence.

What if the prior distribution had been almost uniform, with an expected value at zero? That is, suppose that the value of V' had been almost infinitely large; this is tantamount to a value of N' that is almost zero. Then,

$$M'' = M, \qquad N'' = N,$$

so that the probability that μ lies at or below zero is found from the posterior distribution by the area cut off by the z value:

$$z = \frac{0 - M}{3.5/5} = -6.85.$$

Given an (almost) uniform prior distribution, the significance level found from the ordinary test of the hypothesis H_0: $\mu \leq 0$ against H_1: $\mu > 0$, is the same as the area cut off in the posterior distribution below the value $\mu = 0$. Later, we will find that probabilities involved in this type of significance testing usually correspond to areas cut off by such particular values in the posterior distribution, provided that the prior distribution is assumed to be uniform, or nearly so. There is a close connection between ordinary hypothesis testing and Bayesian analysis given uniform priors; this will be discussed in more detail later on.

19.7   THE VARIANCE OF A NORMAL PROCESS, μ KNOWN

Occasionally, interest focuses on the variance σ^2 of some normal process, rather than upon the mean μ. Now we will discuss briefly the Bayesian analysis of a normal process with a known value of μ, but unknown variance σ^2.

Although σ^2 is actually the parameter of interest, it is generally more convenient to deal with the reciprocal of the variance, or $h = 1/\sigma^2$. We shall call this quantity h the **precision** of the process; just as the size of σ^2 is a reflection of the amount of uncertainty that attends any observation drawn from the population or process, so h is a reflection of the certainty attending any observation made at random. Dealing with h rather than with σ^2 permits one to adopt once again the theory of a conjugate family of distributions.

Given a normal process or population, with mean μ and variance σ^2, let us assume that μ is known, and that our interest focuses on the value of h. Some N independent observations are to be made from this normal process. Then the gamma family of distributions forms the natural conjugate to the likelihood function for the sample variance. That is, the prior distribution of h is assumed to follow the gamma law

$$f(h) = \frac{e^{-(1/2)\,\nu'V'h}\big/\big(\tfrac{1}{2}\nu'V'h\big)^{(1/2)\,\nu'-1}\big(\tfrac{1}{2}\nu'V'\big)}{\big(\tfrac{1}{2}\nu'-1\big)!}$$

with parameters ν' and V'. This is simply a gamma function as discussed in Section 8.14, with the parameter $m = \tfrac{1}{2}\nu'V'$ and the parameter $r = \tfrac{1}{2}\nu'$. The posterior distribution of h is also gamma in form, with parameters ν'' and V''.

Since the mean of a gamma distribution is

$$E(X) = \frac{r}{m},$$

the mean of the prior distribution of h is

$$E'(h) = \frac{\tfrac{1}{2}\nu'}{\tfrac{1}{2}\nu'V'} = \frac{1}{V'}.$$

Furthermore, since the variance of a gamma distribution with parameters r and m is equal to

$$\mathrm{Var}(X) = \frac{r}{m^2},$$

the variance of the prior distribution of h is

$$\mathrm{Var}'(h) = \frac{\tfrac{1}{2}\nu'}{\big(\tfrac{1}{2}\nu'V'\big)^2} = \frac{2}{\nu'V'^2}.$$

Given the mean and the variance of the prior distribution, the two parameters ν' and V can be determined. The parameter V' can be thought of conveniently either as the reciprocal of the expectation of h, $V' = 1/E'(h)$, or as the variance determined from a prior, and perhaps fictitious, sample of N' cases. In the same way, ν' can be thought of as the degrees of freedom in a prior sample; in the present instance $\nu' = N'$, since the value of μ is assumed known.

Now a sample is drawn consisting of N independent observations from the normal process, and an estimate of the population variance is calculated, *based upon deviations of the sample values from the known value of μ:*

$$V = \frac{\sum_i (x_i - \mu)^2}{N},$$

The posterior distribution of h follows a gamma law, with parameters ν'' and V'' given by

$$\begin{aligned} \nu'' &= \nu' + N \\ &= N' + N \end{aligned}$$

and

$$\begin{aligned} V'' &= \frac{V'\nu' + VN}{\nu''} \\ &= \frac{V'N' + VN}{N' + N}. \end{aligned}$$

Observe that the posterior distribution of h also has parameters that can be thought of as derived from pooling prior and sample information. The parameter V'' has the form of a variance found from pooling two samples: a (fictitious) prior sample based on N' cases, and the actual sample of N cases. The mean of the posterior distribution of h is then

$$\text{posterior } E''(h) = \frac{1}{V''},$$

and the variance of the posterior distribution is

$$\text{posterior } \mathrm{Var}''(h) = \frac{2}{\nu'' V''^2}.$$

Remember that a gamma distribution is generally quite skewed to the right, or positively. For a value of ν' greater than 2.00, this distribution will be unimodal, and for increasing values of ν', there will be reduced skewness. Figure 19.7.1 pictures the form of the gamma function for certain values of ν' and V'.

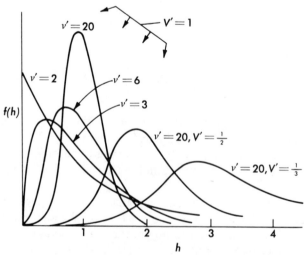

FIG. 19.7.1. Prior distributions of gamma form, with $V' = 1, \frac{1}{2}, \frac{1}{3}$ and various values of ν'

Bear in mind that this is the prior distribution for the parameter h, or $1/\sigma^2$. Hence there is some intuitive justification for the choice of a skewed function such as gamma, which shows relatively high probabilities for values near zero (i.e:, large values of σ^2) and lower probabilities for intervals of values far from zero (that is, small values of σ^2). In many situations we really do expect considerable variability among observations to be the rule, given no other information to go on.

Obviously, the requirement that μ be known is a considerable handicap. However, we will see that the foregoing discussion will be of aid in the next section when we discuss Bayesian methods applied when neither μ nor σ^2 is known.

19.8 NORMAL PROCESS WITH BOTH μ AND σ^2 UNKNOWN

In the typical situation where a normal process is believed to generate the data, neither μ nor σ^2 is known. Fortunately, a Bayesian analysis is also available for this situation. The theory underlying this method is somewhat more complicated than for the previous methods, although the actual mechanics of application are very similar. A brief sketch of this theory follows.

Suppose that a sample of N independent observations is to be drawn from a normal process with mean μ and variance σ^2, where the value of neither parameter is known. Then it can be shown that for the normal likelihood function that applies here, the natural conjugate must be a *joint* distribution of the parameters μ and h (equivalent to $1/\sigma^2$). Furthermore, this conjugate distribution must have a form known as a normal-gamma density function:

$$f'(\mu, h) = K[e^{-(1/2)hN'(\mu-M')^2}h^{1/2}][e^{-(1/2)V'\nu'h}h^{(\nu'/2)-1}].$$

This joint density function for μ and h has four parameters: M', V', N', and ν'. The constant K in the expression need not concern us here, nor should one worry overmuch about the complexity of this function rule. At this point only the results it provides need interest us.

When the prior joint distribution of μ and h has the form just given, and the likelihood function for a sample of N independent observations is normal, then the posterior joint distribution of μ and h will also be normal-gamma in form, with parameters M'', V'', N'', and ν''.

Even though the basic theory thus does involve a natural conjugate distribution which is joint for the parameters μ and h, it is nevertheless possible to deal with the distribution of μ alone, or of h alone. This is done in terms of the marginal distributions of μ and of h, based on the joint distribution of μ and h. That is, for any joint distribution of μ and h, it is possible to consider the marginal distribution of μ taken over all possible values of h, and the marginal distribution of h taken over all values of μ. Let us first consider the marginal distribution of μ.

Given the normal-gamma prior distribution for μ and h, then the marginal prior distribution for μ has the form of a **Student distribution** (Chapter 10), expressed by the density function

$$f(\mu|M', N', V', \nu') = G\left[\nu' + \frac{N'}{V'}(\mu - M')^2\right]^{-(\nu'+1)/2}\sqrt{\frac{N'}{V'}}$$

with parameters M', V', N', and ν'. The constant G depends only on the parameter ν'. This is exactly the same as the Student t distribution discussed in Chapter 10, with $t = \sqrt{(N'/V')}(\mu - M')$ and $\nu' = N' - 1$ degrees of freedom.

Computationally this method is rather simple, in spite of the apparent complexity of the underlying distributions. A sample of N independent observations is drawn from the normal process, and the mean and unbiased estimate of the variance calculated:

$$M = \frac{\sum_i x_i}{N}$$

and

$$V = s^2 = \frac{\sum_i (x_i - M)^2}{N - 1}.$$

Then the posterior marginal distribution of μ is a Student distribution with parameters

$$M'' = \frac{N'M' + NM}{N' + N},$$

$$N'' = N' + N,$$

$$V'' = \frac{(\nu'V' + N'M'^2) + [(N - 1)V + NM^2] - N''(M'')^2}{\nu' + N - 1}$$

$$\nu'' = \nu' + N - 1.$$

In other words, the posterior parameter M'' is simply the weighted average of the prior M' and the sample M. The posterior parameter V'' is analogous to the pooled variance from samples with ν' and $N - 1$ degrees of freedom, respectively. As usual, the posterior parameters are found as though there had been a prior sample of N' observations, giving a mean M' and a variance V', and these prior statistics had been pooled with the present sample to yield the posterior parameters.

In practical terms, the fact that the marginal distribution of μ is a Student distribution means that percentage points for this distribution can be found from tables of Student's t, such as Table *III* of Appendix C. That is, for the posterior distribution, one can define a statistic t, where

$$t = (\mu - M'') \sqrt{N''/V''}$$

and $\nu'' = N'' - 1$. Then the ordinary t table can be used to find percentage points for any value of μ that happens to be of interest. Exactly the same procedure can be used for the prior distribution, given N', V', and ν'.

Although the theory underlying this method *is* rather complicated, in practice it is no more difficult than the familiar t-test for small samples, *provided* that the prior values of M', V', and N' are somehow given. For the time being, let us simply assume that these values are known.

An example should clarify the method considerably. Imagine an experiment on social perception, where an individual subject is asked to rate another individual on his "effectiveness" in a given social interaction. Unknown to the subjects in the experiment, the person rated is a "stooge." Across a given population of subjects, the experimenter believes that ratings given by subjects should be approximately normally distributed. However, he does not know the mean of the ratings across such a population. A small pretest of the experiment involving 4 subjects did yield a mean of 75 and a variance of 25. The experiment proper is to be carried out with 20 subjects. Within what pair of possible values of μ should the experimenter have 50 percent confidence that the true value of μ lies, prior to the experiment? Within what pair of values should he be 50 percent confident that μ lies given the experimental results?

Since the experimenter's only available prior evidence appears to come from the small pretest, let us construct his prior distribution on that basis. The marginal prior distribution of μ thus has a Student form with parameters

$$N' = 4, \qquad M' = 75,$$
$$V' = 25 \qquad \nu' = N' - 1 = 3.$$

This implies that this marginal prior distribution will be symmetric about a mean value of 75, with a variance equal to $(V'/N')[(N' - 1)/(N' - 3)]$ or $(3V'/4) = 75/4$. If we form the statistic $t = (\mu - 75)\sqrt{4/25}$, we may find the 50 percent percentage points for the distribution. In Table III of Appendix C, we see that for $\nu' = 3$, the upper .25 of the distribution is cut off by $t = .765$, and the lower .25 by a value of $t = -.765$, since the distribution of t is symmetric. Then, if we solve for the μ values equivalent to these two t values, we have the two values cutting off the middle 50 percent in the prior distribution:

$$t = (\mu - 75)\frac{2}{5} = .765$$

$$\mu = 76.91,$$

$$t = (\mu - 75)\frac{2}{5} = -.765$$

$$\mu = 73.09.$$

Hence, prior to the experiment proper, the experimenter should believe that the probability is .50 that the true value of μ lies between 73.09 and 76.91. This conclusion is based solely on the small prior sample and the assumption of the conjugate prior distribution for μ and h jointly.

Now the actual experiment, providing ratings from 20 subjects independently tested, gives a mean rating of 76.5, and a (corrected) sample variance $s^2 = V = 32$. On this basis what can we say about the middle 50 percent region of the experimenter's posterior distribution for μ? Here, the new parameter values are:

$$M'' = \frac{4(75) + 20(76.5)}{24}$$

$$= 76.25,$$

$$V'' = \frac{3(25) + 4(75)^2 + 19(32) + 20(76.5)^2 - 24(76.25)^2}{20 + 3 - 1}$$

$$= \frac{22575 + 117653 - 139537.50}{22}$$

$$= 31.4,$$
$$N'' = 4 + 20 = 24,$$
$$\nu'' = \nu' + N - 1 = 22.$$

The posterior distribution is thus of Student form with parameters

$$M'' = 76.25$$
$$V'' = 31.4,$$

$N'' = 24$, and $\nu'' = 22$. This means that the posterior marginal distribution for μ is unimodal and symmetric about $M'' = 76.25$ with a variance equal to $(V''/N'')(\nu''/\nu'' - 2)$ or $(31.4/24)(22/20)$.

Now in order to find the interval of values cutting off the middle 50 percent in this posterior distribution, we proceed just as before. For $\nu'' = 22$, the required t values, cutting off the middle 50 percent of the Student t distribution, are $t = \pm .686$. Then we solve for the corresponding μ values by taking

$$(\mu - 76.25) \sqrt{24/31.4} = \pm .686,$$

which gives $\mu = 75.46$ and $\mu = 77.04$, approximately. Thus, in the marginal posterior distribution of μ, the probability is .50 that μ lies in the interval between 75.46 and 77.04. The interval representing the mid-50 percent is now somewhat higher up the scale of possible values of μ, representing the higher value of the sample mean as compared to the pretest mean. Furthermore, the interval of values is shorter than before, representing the increase in N'' as compared with N'. In this same way we could investigate any other percentage points in the prior and posterior marginal distributions of μ.

A parallel method exists for finding the prior and posterior marginal distributions of h (or $1/\sigma^2$) given that μ is unknown. This turns out to be very similar to the method for finding the prior and posterior distribution of h given that μ is known, as discussed above. Given the normal-gamma joint prior distribution of μ and h, then the marginal prior distribution of h has a gamma form, with parameters V' and ν'. Given N independent observations from a normal process, we calculate M and $s^2 = V$, the usual unbiased estimate of σ^2. (Notice that unlike the situation when μ is known, here we calculate the unbiased estimate of σ^2 by use of the sample mean M.) Then the posterior distribution of h is gamma with parameters

$$V'' = \frac{(\nu'V' + N'M'^2) + (\nu V + N M^2) - N''M''^2}{\nu' + \nu}$$
$$N'' = N' + N,$$
$$\nu'' = \nu' + \nu,$$

just as for the posterior marginal distribution of μ.

For our example, the prior marginal distribution of h would be gamma with parameters $V' = 25$ and $\nu' = 4 - 1 = 3$, as given by the pretest. The posterior marginal distribution of h would be gamma with parameters $V'' = 31.4$ and $\nu'' = 3 + 19 = 22$. Percentage points for these gamma distributions can be found by use of the chi-square table, Appendix C, as described in Section 8.14. Here, we would take the gamma parameter, $m = \frac{1}{2}V'\nu'$ and

$r = \frac{1}{2}\nu'$ for the prior distribution, and $m = \frac{1}{2}V''\nu''$ and $r = \frac{1}{2}\nu''$ for the posterior distribution.

19.9 Other natural conjugate families

Although we cannot go much further into the theory of natural conjugates to particular likelihood functions, it is worth noting that the theory has been extended to other kinds of data-generating processes and other sorts of parameters. For example, if a Poisson process with parameter m underlies the data, then the natural conjugate to the likelihood function for number of successes, r, is the family of gamma distributions. That is, the prior distribution for the parameter m is assumed to have the form

$$f'(m|r', g') = \frac{(g'm)^{r'-1}e^{-g'm}g'}{(r'-1)!}$$

so that $E'(m) = r'/g'$ and $\text{Var}'(m) = r'/(g')^2$. Then, if r successes are observed as outcomes of a stable and independent Poisson process during a unit time, the posterior distribution of m has the gamma form with parameters

$$r'' = r' + r, \qquad g'' = g' + 1.$$

The mean of posterior distribution of the parameter m is then

$$E''(m) = \frac{r' + r}{g' + 1}.$$

(Here, g' is assumed to be measured in the same units of time, length, or other dimension as that observed in the sample.) Other likelihood functions and assumed data-generating processes call for still other families of distributions as natural conjugates. The method has been extended in a variety of ways, such as to problems in regression. The student is referred to the volumes by D. V. Lindley (1965) for a discussion of some of these methods.

19.10 Diffuse prior distributions

So far we have assumed that the user of Bayesian inferential methods has a well-defined set of prior opinions about one or more parameters, and that these opinions are capable of being reflected in the form of a prior distribution. Then, either Bayes' theorem is applied directly, or by use of the theory of the natural conjugate of the likelihood function the parameters of the posterior distribution are worked out.

On the other hand, it may be that the prior opinions of the experimenter are not well defined. Perhaps there is little or no empirical evidence in the area of his interest, or perhaps there are no theoretical reasons to favor some values of the parameter over others as candidates for the true value. In this situation we say that the experimenter has a "diffuse" or "informationless" prior state. Even though prior information may not be altogether lacking, the experimenter feels that it should carry little or no weight relative to the sample evidence. It is not necessary for information to be totally lacking prior to the experiment in order for the prior state to be diffuse. Rather, such a state requires only that there be relatively little or relatively incoherent prior information, so that opinions based on such information should be easily "swamped" by information derived from the sample.

A completely informationless or diffuse prior state may be represented by a prior distribution that is perfectly "flat" over the possible values of the parameter. More typically, the experimenter's prior state of opinion is best represented by a distribution that is only relatively flat as compared with the likelihood function for the data given the value of the parameter. There are, after all, very few things about which no information at all exists. For almost any parameter, it is reasonable to believe that some intervals have nonzero probability, even though the prior distribution may be virtually uniform over such intervals. If you were asked to picture your prior distribution for the average weight of full-grown giraffes, surely you would rule out values such as 2 ounces and 200 tons. However, somewhere in the possible range of values there is bound to be at least one interval with almost uniform probability density, unless you happen to have an extraordinary fund of information about giraffes. Whether or not the prior distribution is uniform, or only "locally" uniform over intervals most likely to contain the true parameter value, the prior distribution will carry rather little weight in the determination of the posterior distribution.

Let us first consider the situation where the prior distribution is absolutely uniform over all possible values of the parameter θ. This means that the prior density function $f'(\theta)$ is a constant over the entire range of θ: $f'(\theta) = k$, for all values of θ, within some limits $a \leq \theta \leq b$. What does this imply about Bayes' theorem? The essence of Bayes' theorem can be written as follows:

$$f''(\theta|y) \propto f(y|\theta)f'(\theta),$$

where the symbol $\propto$ is read as "is proportional to." That is, the posterior density of θ given y is proportional to the likelihood function for y given θ, times the prior density of θ. (Remember that the denominator in Bayes' theorem really plays only the role of standardizing the density or probability values, so that the posterior distribution sums to 1.00 over all values of θ.)

Now, if the prior density is uniform over all possible values of θ, as in a diffuse state, then it must be true that

$$f''(\theta|y) \propto f(y|\theta)k$$

or

$$f''(\theta|y) \propto f(y|\theta).$$

When the prior distribution is uniform, the posterior density is proportional to the likelihood function. This means that for uniform prior distributions the form of the posterior distribution is actually dictated by the likelihood function.

Even when the prior distribution is not absolutely uniform, but only relatively flat compared with the likelihood function, the principle stated above is approximately true. Thus, when the prior distribution is diffuse, relative to the likelihood function, the posterior distribution depends almost exclusively on the likelihood function. Most important, it is not necessary for the prior distribution to be diffuse over the entire conceivable range of values for θ, but only over those values relatively close to the sample value. The principle holds when the prior distribution is only *locally* diffuse. Thus, if we are interested in the average height of American women, we can quite reasonably set the prior density at or near zero for values such as 40 centimeters or 5 yards. On the other hand, for an interval of values in the general vicinity of 5.6 feet, our prior distribution might well be almost flat; we really have no more reason to believe that the mean will turn out to be 5.55 feet than 5.60 feeet or 5.62 feet. Within some such interval our prior distribution might well be locally diffuse.

The upshot of all of this is that when little or nothing is known about the distribution of some parameter θ prior to the sample evidence, then the posterior distribution will be dictated by the likelihood function governing the occurrence of the data. This is true either for continuous prior distributions, with density functions that are almost uniform, or for discrete distributions, having probability functions that are constant.

Even when the prior state of opinion is diffuse, there is still an advantage in choosing a member of the conjugate family to represent this diffuse state. Some of the conjugate families do have members that are uniform, or very nearly uniform. Other conjugate families are not sufficiently flexible to permit showing a diffuse prior state by the use of one of their members. Let us first consider the beta family, the natural conjugate to a binomial likelihood function.

We have already seen that in the beta family of distributions when the parameters $r' = 1$ and $N' = 2$, a perfectly flat or uniform distribution results. The density function for this distribution is a constant for all possible values of the parameter π in the interval from zero to 1.00. The mean of the distribution is $1/2$. This uniform beta distribution can be used to represent an informationless or diffuse state of opinion about π.

However, it may be reasonable to imagine the prior N' and r' as having even smaller values. Indeed, some Bayesian statisticians argue that $N' = r' = 0$ are the proper values in the diffuse situation. This means that no weight is to be given to prior opinion. Unfortunately, the resulting prior beta distribution is not quite "proper" when these values are used; the graph becomes U shaped, with a minimum at $1/2$, and the function is not even defined for $\pi = 0$ and $\pi = 1$. This means that the total area under such a curve cannot be equal to 1.00 in the ordinary sense. Even so, this is a minor technicality as compared with the fact that $N' = 0$ does seem to reflect a total absence of prior information, so that the posterior distribution of π depends exclusively upon the sample results. The use of the technically faulty prior has no influence whatsoever on the further applications of the method given additional information. It seems to this author that the use of $r' = N' = 0$ actually is the best choice when there really is no information at all about π prior to the sample. Then the posterior parameters of the beta distribution are

$$r'' = r + r' = r,$$
$$N'' = N + N' = N.$$

In the case of a normal likelihood function, the normal family of distributions are conjugate in inferences about μ. It is really not possible to have a normal distribution that is perfectly uniform, of course. However, it is possible to define $N' = \sigma^2/V'$ in such a way that the prior distribution, though normal, will be almost uniform. This can be accomplished by setting N' as very nearly zero, which means that the diffuse prior represents the information in a fictitious prior sample of almost no observations at all. Technically, if N' is actually zero, the resulting distribution is not well defined; thus we make N' *almost* zero so that the prior distribution will be proper, but so that we can effectively ignore N' as negligible in further computations. For this diffuse prior situation,

$$N'' = N' + N = N, \qquad \text{approximately,}$$
$$M'' = \frac{N'M' + NM}{N' + N} = M, \quad \text{approximately.}$$

Other families of distributions serving as natural conjugates to particular likelihood functions cannot always be made to represent uniform priors. Thus, for example, the gamma family of distributions of h, which are conjugate to a normal likelihood function, tend to be very skewed to the right for small values of ν', or N'. In fact, as ν' approaches zero, the probability is concentrated more and more around $h = 0$, but the variance of h becomes infinitely large. In a way this is quite appropriate to a diffuse state; chances are that the variability in a given process will be large rather than small, given no special information about that process, and this is reflected in a high probability for intervals of possible h values close to 0. The diffuse character of the distribution is also reflected in the very large variance of h. Thus, setting ν' very close to zero means that a diffuse situation exists, in the sense of giving very little weight to the prior distribution, even though the prior is far from a uniform distribution.

In summary, when prior information is scanty or nonexistent, the prior state of the experimenter may be described as diffuse. Then, the prior distributions of parameters such as π in the Bernoulli situation or μ in the normal situation are represented by diffuse distributions. In spite of certain mathematical technicalities that may be raised as objections, it seems best to represent such a diffuse state by setting prior N' equal or very nearly equal to 0, so that the prior state has no influence on the posterior distribution. Similarly, although a diffuse prior may take a different form when σ^2 or h is the parameter of concern, it still makes sense to set N' or ν' at very nearly zero when a diffuse state is to be represented.

A diffuse prior state represents only the most basic starting place for Bayesian inference, of course. Setting N' or ν' equal or very nearly equal to zero is but a way of indicating that little or no prior information about the parameter of interest is available. The assumption of no prior information is mainly useful in "getting the ball rolling" in the continuing process of Bayesian inference. As additional information accumulates, further distributions are no longer diffuse, but rather reflect the information gained up to each successive point in time. Although somewhat "improper" prior distributions do result from $N' = 0$ or $\nu' = 0$ under the theory of conjugate distributions, such distributions can nevertheless be used to start off the continuing inferential process. The technical deficiencies of these prior distributions in no wise impair their usefulness.

On the other hand, if the experimenter has good reason to believe more strongly in certain values or ranges of values of the parameter than in others, then these beliefs should be represented in his prior distribution. Such prior states of opinion may come directly from prior empirical evidence, they may be based on theory, or they may come about by analogy with other and better known situations. The point is that when such prior information and prior opinions exist, Bayesian methods permit their incorporation in the formation of posterior distributions.

19.11 ON THE SUBJECTIVITY OF BAYESIAN METHODS

Perhaps the most frequent criticism of the Bayesian point of view is that it is essentially subjective. That is, some statisticians object to the inclusion of prior opinion about a parameter in the treatment of empirical data, in part on the grounds that the source of such opinions, and the weight to be given to them, are nowhere dictated by the theory. We have already run into the argument that a parameter is a constant in a given situation, and hence is not legitimately treated as a random variable capable of taking on various values with various probabilities. Finally a common criticism is that since the end product of a particular analysis depends upon what a particular experimenter happened to think at the outset, each set of results is peculiar to the experimenter or statistician who happens to be doing the analysis. This, such critics claim, violates one of the canons of empirical science: the open and "objective" treatment of results.

Although it is not the author's purpose to attempt an extended defense of Bayesian methods against these charges, some observations are worth making in this connection. In the first place, there are plenty of reasons to include the prior opinion of an experimenter in the evaluation of empirical data, not the least of which is that this prior opinion may itself be empirical in its source. Any experimenter worth his salt should have digested a good deal of prior evidence relevant to the problem at hand. This empirically based, "expert" opinion has a recognized place in the design of the experiment, the choice of a sampling method, and the various assumptions made about the data-generating process, regardless of the statistical method employed. Thus, it may also deserve a formal place in the statistical analysis. No statistical method is free of the influence of prior opinion, as anyone can attest who has pondered the many more or less gratuitous choices and assumptions made in gathering data.

Satisfaction or dissatisfaction with the formal treatment of parameters as random variables is perhaps as much a matter of taste and intuition as anything else. For some people this seems like a natural thing to do, and for others it does not. Although the Bayesian is just as aware of the true nature of a population parameter as is the most classical statistician, he still feels comfortable in discussing the distribution of a parameter, since he is willing to extend the concept of probability to cover possible states of opinion or belief. This leads him to make probability statements about parameter values that his more classically inclined brethren decline to make. Neither party can be proved right or wrong, since both rest their interpretations on formally correct systems of mathematical probability.

The last criticism deserves the most serious consideration. At first blush it may appear that Bayesian methods do result in a kind of statistical solipsism, in which each experimenter erects his own little world of private beliefs. However, this point overlooks the power and the durability of data. We have already suggested that prior distributions are relatively unimportant in the face of steadily increasing amounts of data. Even given very different prior distributions, two experimenters confronted with the same large volumes of data, steadily accumulating under the same conditions, will finally arrive at identical posterior distributions. The initial subjective element in Bayesian methods is extremely minor as compared with the weight of data. In the long run the Bayesian approach differs not at all from any other standard way of treating empirical data. (Remember, however, that although the classical and the Bayesian statisticians may arrive at the same final conclusions, the approach each uses is essentially different from the other.)

How can one report a Bayesian result so as to make absolutely sure that the reader understands the role of prior opinion as against the role played by the data? A very simple option is usually available. Not only can one report the posterior distribution, which reflects both data and prior state; for very little additional effort he can report the effect of the data given a diffuse or informationless prior state. As we have seen, this depends essentially on the likelihood function for the results actually obtained. Such a report of results is tantamount to a traditional statistical analysis. If the reader has his own prior opinions about the situation before the data were collected, with a consequently different weighting of the data in determination of a posterior distribution, he can carry out such an analysis himself. He may even adopt a more classical view of what the data imply, since all the raw materials for such an inference will be provided. The likelihood function depends upon the sampling distribution underlying the particular sample results. The outcome of a classical statistical method is a probability statement, giving the likelihood of such and such a sample result if so and so is true of the population parameter in question. When prior distributions are diffuse, the results of Bayesian methods are really quite similar to classical results, even though the Bayesian persists (perhaps perversely, according to some) in interpreting these results in terms of the posterior distribution of the parameter. One of the clear-cut examples of this difference in interpretation lies in the area of interval estimation. This will be discussed next.

19.12   Credible intervals

Estimation of the value of a parameter is a key problem in both the classical and the Bayesian approaches to inference. The issues raised in Chapters 7 and 9 are important under both points of view. How does one choose a single value to best represent the unknown parameter? These issues largely have to do with the kind and amount of information that a sample can supply about a population or process; for the sake of brevity, these issues will not be reviewed here. Suffice it to say that the Bayesian not only shares all of the concerns of the classical statistician in point estimation; he also has some additional choices before him not generally considered under the classical point of view. For example, the posterior distribution of a parameter θ will have a mean or expectation, as well as a mode and a median, and these are not always identical. Is the Bayesian's best bet about a parameter value the expectation, or one of these other measures of central tendency? In most work the expectation of the posterior distribution is taken as the best estimate of θ, but this is not invariably true. The mode of the posterior distribution is of special interest to the Bayesian, since this represents the most likely value of θ, given the prior distribution and the evidence. Furthermore, the experimenter using the Bayesian methods typically employs the standard deviation of the posterior distribution to reflect his uncertainty about the value of θ. Even so, other measures of the spread of the posterior distribution are available for use as well.

When the prior distribution is diffuse, Bayesian point estimators tend to be identical to the estimators ordinarily employed in classical work. Thus, given a diffuse prior distribution, the mode of the posterior distribution corresponds in value to the maximum likelihood estimator of classical statistics, though it is not interpreted in the classical way. The classical statistician would think of the maximum likelihood estimator of θ as the value that maximizes the likelihood of obtaining the sample results that actually occurred. The Bayesian would say simply that the mode of the posterior distribution is the most likely value of θ. Thus, even though the actual value each estimates for θ is the same, the interpretation given this value is different.

In Chapter 9 we considered interval estimation and saw that confidence intervals can be constructed for a parameter such as μ by use of the sampling distribution of the mean. In that chapter some pains were taken to interpret a confidence interval in the approved, traditional way. A confidence interval was defined as a range of values with a given high probability of covering the true population value. Now we shall explore a very similar-appearing method from the Bayesian point of view. The interpretation given such an interval estimate will, however, be quite different.

Suppose that the posterior distribution of the parameter μ is normal, with parameters M'' and N''. The value of the variance σ^2 is here assumed known. Now we want the interval of possible values of μ such that the posterior probability is .95 that μ falls in that interval. Furthermore, we want this to be a central interval of the distribution, so that the expectation of μ, or M'' forms the midpoint of the interval. Since the distribution is normal, this interval must be bounded by the two values

$$M'' - 1.96 \sqrt{\sigma^2/N''} \quad \text{and} \quad M'' + 1.96 \sqrt{\sigma^2/N''}.$$

The interval bounded by these two values is called the **95 percent credible interval** for μ. The limiting values themselves are called the **95 percent**

credible limits for μ. One says the probability is .95 that μ lies in the interval.

Does this interval look familiar? It should, since in form it appears identical to the 95 percent confidence interval for μ found in the usual way, if M'' is replaced by M and N'' by N. Indeed, if the prior is diffuse, so that N' is approximately zero, then M'' is approximately equal to M and N'' to N, and the two kinds of intervals are virtually identical, given that σ^2 is known. However, remember the important difference in interpretation: for the confidence interval there is no suggestion of a distribution of μ itself, whereas in the credible interval μ is treated quite explicitly as a random variable. (Recall the example in Section 19.6.)

Naturally, any other credible interval of interest can be calculated in a similar way. For instance, the 99 percent credible limits are

$$M'' - 2.58 \sqrt{\sigma^2/N''} \quad \text{and} \quad M'' + 2.58 \sqrt{\sigma^2/N''},$$

and these form the boundary values for the 99 percent credible interval for μ. When the sample size is relatively large, the unknown value of σ^2 ceases to be a particular problem. One simply takes the unbiased estimate s^2 from the sample as the substitute for σ^2 in the credible interval limits.

When the sample size is relatively small, and σ^2 is unknown, then a credible interval may be calculated for μ through the use of the t distribution. Once again, the method is very similar to that for a confidence interval involving t. If one wishes the $100(1 - \alpha)$ percent credible limits, in a posterior distribution of μ with parameters M'', V'', N'', ν'', he simply locates the value of $t_{(1/2)\alpha}$ in Table III, Appendix C, under the degrees-of-freedom entry equal to ν''. Then the credible limits are given by

$$\mu = M'' - t_{(1/2)\alpha} \sqrt{V''/N''}$$

and

$$\mu = M'' + t_{(1/2)\alpha} \sqrt{V''/N''}.$$

(This is illustrated in the example of Section 19.8, where the 50 percent credible interval is calculated for both the prior and the posterior distribution.) When the prior distribution is diffuse, and N' and ν' are both set at, or very nearly at, zero, then these credible limits become

$$\mu = M - t_{(1/2)\alpha} \sqrt{V/N}$$

and

$$\mu = M + t_{(1/2)\alpha} \sqrt{V/N}.$$

These are the same values that one calculates as the confidence limits for using the t distribution in the ordinary way. However, once again care should be taken to distinguish between the interpretation of the credible interval and that of the confidence interval.

Mutatis mutandis, credible intervals may also be found from the posterior distributions of other parameters such as π or h. Thus, the 95 percent credible interval for π is found from the beta distribution with parameter r'' and N''. This 95 percent credible limit for π can be found from

$$\pi = \frac{\nu_1 F_{.975, \nu_1, \nu_2}}{\nu_1 F_{.975, \nu_1, \nu_2} + \nu_2}$$

and

$$\pi = \frac{\nu_1 F_{.025, \nu_1, \nu_2}}{\nu_1 F_{.025, \nu_1, \nu_2} + \nu_2},$$

where $F_{.975, \nu_1, \nu_2}$ is the value corresponding to the upper 2.5 percent point in the F distribution with $\nu_1 = 2r''$ and $\nu_2 = 2(N'' - r'')$ degrees of freedom, and $F_{.025, \nu_1, \nu_2}$ is the corresponding lower 2.5 percent value, found as shown in Section 11.11. Thus, if $r'' = 5$ and $N'' = 25$, $\nu_1 = 2(5) = 10$, and $\nu_2 = 2(20) = 40$; then $F_{.975, \nu_1, \nu_2}$ is found from Table V to be 2.39, and $F_{.025, \nu_1, \nu_2}$ is found to be .307. Thus, the credible limits for π are worked out to be

$$\pi = \frac{2.39(10)}{[(2.39(10) + 40)]} = .374$$

and

$$\pi = \frac{(.307)(10)}{[(.307)(10) + 40]} = .071.$$

In this posterior distribution of π the probability is .95 that the value of π lies between .071 and .374. (Recall the connection between F and the beta distribution discussed in Chapter 11.)

19.13   ODDS RATIOS AND HYPOTHESIS TESTING

The main steps in Bayesian inference are: specifying a prior distribution of one or more states or parameters, determining the likelihood of the sample result given certain of the states or parameter values, and then determining the posterior distribution. Any decisions to be made depend upon the posterior distribution. Strictly speaking, it is not really necessary to adopt very much of the machinery of conventional hypothesis testing when one is operating within the Bayesian framework. Nevertheless, since so much of the discussion in previous chapters has been devoted to the testing of hypotheses, it may be helpful at this point to consider a Bayesian approach to these questions. We begin by considering the notions of prior and posterior odds ratios.

Consider a situation in which exactly one of two possible states must be true: state A or state B. State A has some prior probability $P'(A)$ and B has some prior probability $P'(B)$. Then we may define the **prior odds ratio of A to B** as

$$\Omega'_{AB} = P'(A)/P'(B).$$

The prior odds ratio Ω'_{AB} is simply the ratio of the prior probabilities.

Thus, for example, suppose that a psychologist constructing a test believes that a certain item either has a difficulty of .70 (which we will here interpret to mean that 70 percent of pupils of a certain age will fail the item) or a difficulty of .30. Let the state A be a true difficulty equal to .70 and the state B a true difficulty of .30. We will assume that states A and B are mutually exclusive and exhaustive. Now suppose that the prior probability the psychologist attaches to state A is $P'(A) = P'(.70) = .60$. The prior probability that he attaches to state B is $P'(B) = P'(.30) = .40$. Then the prior odds ratio for A over B is

$$\Omega'_{AB} = P'(A)/P'(B) = .60/.40 = 1.5.$$

We can also discuss the **posterior odds ratio** Ω'' for the states A and B. This is defined by

$$\Omega''_{AB} = P''(A)/P''(B),$$

which is the ratio of the posterior probabilities. Thus, if our psychologist took a sample of children and administered the item, and as a result his posterior probabilities were $P'(A) = .25$ and $P'(B) = .75$, the posterior odds ratio would be

$$\Omega''_{AB} = P''(A)/P''(B) = .33.$$

Given a particular sample result y, we can also consider the **likelihood ratio,** LR_{AB}, which is simply the likelihood of the sample result y given state A, divided by the likelihood of the sample result given state B. That is,

$$LR_{AB} = f(y|A)/f(y|B).$$

Thus, if our psychologist took a sample of 20 children and found that 8 of them failed the test, he could immediately find the likelihood of this result given the state A, or $\pi = .70$, by use of the binomial rule. Table II shows that the binomial probability for 8 "successes" out of 20, given $\pi = .70$, is .0039, (i.e., $12 = r$, $N = 20$, $p = .30$) so that the likelihood of this sample result given state A is .0039. On the other hand, given the truth of B, or $\pi = .30$, the binomial probability of 8 "successes" is .1144. Hence the likelihood ratio is

$$
\begin{aligned}
LR_{AB} &= P(8 \text{ out of } 20|\pi = .70)/P(8 \text{ out of } 20|\pi = .30) \\
&= .0039/.1144 \\
&= .034.
\end{aligned}
$$

Obviously, the data obtained are much more likely under state B than under state A.

Given these concepts of prior and posterior odds ratios and the likelihood ratio, then by Bayes' theorem the following must be true:

$$\Omega''_{AB} = \Omega'_{AB}(LR_{AB}),$$

the posterior odds ratio must be the prior odds ratio multiplied by the likelihood ratio. Thus, if the prior odds ratio is equal to 1.5, and the likelihood ratio is equal to .034, then the posterior odds ratio must be equal to $(1.5)(.034)$ or .051:

$$
\begin{aligned}
\Omega''_{AB} &= \Omega'_{AB}(LR_{AB}) \\
&= (1.5)(.034) \\
&= .051.
\end{aligned}
$$

By now a parallel with hypothesis-testing may begin to emerge. State A can be interpreted as H_0: $\pi = .70$, and state B as H_1: $\pi = .30$. Prior probabilities are given to each of these hypotheses: $P'(H_0) = .60$ and $P'(H_1) = .40$. Then, by the likelihood ratio for the data, given the truth respectively of each of the hypotheses, the posterior odds ratio Ω'' is found. The choice between the two hypotheses rests upon the magnitude of the odds ratio, and upon which of the two hypotheses is favored in this ratio, other things being equal. (A little later we will consider another factor in this choice, but for the time being let us assume that the larger the odds ratio, the more willing one is to accept that hypothesis favored by the ratio.)

Although in this example the two likelihoods were binomial probabilities, in other instances they may correspond to probability densities. Thus, if our psychologist had been comparing two values of μ, then the likelihood of a sample mean M would be probability density in the sampling distribution of M, given the particular value of μ and of the standard error of the mean. For samples drawn from a normal population, or for large samples, drawn independently and at random, this would be the normal probability density $f(z)$ associated with $z = (M - \mu)/\sigma_M$.

Now consider the situation in which the prior distribution is diffuse. This situation implies that the prior probability for H_0 is the same as the prior probability for H_1. Then

$$\Omega''_{AB} = LR_{AB},$$

since the prior odds ratio must be 1.00. The choice between the two hypotheses depends only upon the likelihood ratio when the prior distribution is diffuse.

The concerns of classical statistics in hypothesis-testing are bound up with the likelihood ratio. This fits in with the fact, noted above, that Bayesian methods based on diffuse priors also depend almost directly on the likelihood of the sample. In effect, in defining a rejection region for H_0 the classical statistician chooses a value c such that the hypothesis is rejected when $LR \leq c$. The actual value of c depends on the value of α chosen. When both H_0 and H_1 are exact, and all parameters are specified, LR is uniquely determined, and the statistician rejects or fails to reject H_0 on the basis of LR. The methods of classical statistics for dealing with inexact hypotheses can also be cast in this form, although the likelihood ratio is then replaced by the concept of a **maximum likelihood ratio,** where the denominator contains the maximum likelihood of y over all values of the parameter covered by H_0, and the denominator the maximum likelihood over all possible values of the parameter, including those covered by the alternative hypothesis. Although we cannot explore this way of looking at classical tests in detail, let it be said that all of the classical tests of hypotheses with which we have been concerned can be interpreted in terms of such a likelihood ratio. [A good discussion of the likelihood ratio as an approach to hypothesis testing is given in Mood and Graybill (1963).]

The Bayesian statistician also depends upon the likelihood ratio in making such choices. However, he is prepared to go considerably beyond the classical approach to the treatment of two hypotheses, and to deal not only with the likelihood ratio but also with prior and posterior probabilities for the hypotheses themselves. He does not necessarily assume that prior probabilities are equal or diffuse. We shall now examine simple hypothesis testing about the mean in this light.

19.14   TESTING HYPOTHESES ABOUT THE MEAN

For the sake of simplicity, we will deal here only with the situation where the hypothesis to be tested concerns the mean μ of a normal process or population. We will also simplify matters by assuming initially that the population or process is normal, and that the variance σ^2 is known (or that N is sufficiently large so that s^2 may be substituted for σ^2). The method extends directly to other situations, however.

The method for comparing two inexact hypotheses will be considered first. This is the same as the familiar one-tailed test of a hypothesis about μ. That is, given some value μ_0, the two hypotheses are

$$H_0: \mu \leqq \mu_0 \quad \text{and } H_1: \mu > \mu_0,$$

or the direction of the inequality may be reversed:

$$H_0: \mu \geqq \mu_0, \qquad H_1: \mu < \mu_0.$$

The prior distribution of μ is specified as having parameters M' and N'. Then the prior probabilities and prior odds ratio for H_0 versus H_1 may be calculated:

$$P'(H_0) = P'(\mu \leqq \mu_0) = \text{prior probability that } \mu \leqq \mu_0$$

and

$$P'(H_1) = 1 - P'(H_0) = P'(\mu > \mu_0).$$

It follows that

$$\Omega' = P'(H_0)/P'(H_1).$$

A sample of N cases, drawn independently and at random from the normal process, is taken and the sample mean M is computed. Then, as we have already learned above, for the posterior distribution

$$M'' = \frac{N'M' + NM}{N' + N}$$

and

$$N'' = N' + N.$$

Since the two hypotheses define two mutually exclusive and exhaustive regions in the distribution of μ, the posterior probabilities for the hypotheses are found by taking

$$z = \frac{M'' - \mu_0}{\sqrt{\sigma^2/N''}}$$

and referring it to a normal distribution. We have

$$P''(H_0) = P''(\mu \leqq \mu_0) = \text{posterior probability that } \mu \leqq \mu_0$$

and

$$P''(H_1) = 1 - P''(H_0).$$

If $P''(H_0)$ is less than or equal to some predetermined value α, or if Ω'' is less than or equal to some predetermined value c, where

$$\Omega'' = \frac{P''(H_0)}{P''(H_1)} \quad \text{and} \quad c = \frac{\alpha}{1 - \alpha},$$

the hypothesis H_0 is rejected in favor of H_1. For the moment let us take α to be one of the conventional values such as .05, .01, etc.

For example, suppose that in some experimental situation we are interested in whether or not the mean μ of a normal process is less than or greater than 40.5. That is

$$H_0: \mu \leq 40.5, \qquad H_1: \mu > 40.5.$$

We happen to know that the variance of this normal process is $\sigma^2 = 20$. Prior to the experiment we are willing to bet with odds of 4 to 1 that H_0 is true. This means that $P'(H_0) = .80$ and $P'(H_1) = .20$. On the other hand, we believe the most likely value of μ to be 38.8, so this is taken as M'. Now, assuming a normal distribution of μ as the prior, we can solve to find N'. Since the probability is .80 that μ falls at or below 40.5, we see from a normal table that

$$z = \frac{\mu_0 - M'}{\sqrt{\sigma^2/N'}} = .84.$$

Thus,

$$\frac{40.5 - 38.8}{\sqrt{20/N'}} = .84,$$

$$\frac{\sqrt{N'}\,(1.7)}{4.4721} = .84,$$

$$N' = 4.9, \quad \text{approximately.}$$

Now a random sample of 50 cases is taken independently at random. The sample mean turns out to be 45. Consequently,

$$M'' = \frac{(38.8)(4.9) + (45)(50)}{4.9 + 50}$$

$$= 44.45$$

with

$$N'' = 4.9 + 50 = 54.9$$

In order to find the posterior probabilities for H_0 and H_1, we first find the z value corresponding to 40.5 in this new distribution with $M'' = 44.45$ and $N'' = 54.9$:

$$z = \frac{40.5 - 44.45}{\sqrt{20/54.90}}$$

$$= -6.545.$$

The normal table shows that far less than .0001 proportion of the normal distribution lies at or below such a z value. Hence $P''(H_0) < .0001$ and $P''(H_1) > .9999$. Thus $\Omega'' < .0001$, and if $c = \alpha/(1 - \alpha) = .05/.95$, it follows that $\Omega'' < c$. The null hypothesis is so improbable, given this evidence, that it certainly can be rejected. Note that this conclusion is reached even in the face of the prior probability of .80 for H_0. Once again, the weight of the evidence has almost completely overcome the prior opinion held by the experimenter.

Now we will repeat this example with a diffuse prior distribution. We will let N' be very nearly zero and proceed to ignore it in the calculations. Then $M'' = M = 45$, and $N'' = N = 50$. In order to find $P''(H_0)$ we once again calculate a z value:

$$z = \frac{40.5 - 45}{\sqrt{20/50}}$$
$$= -7.143.$$

Here, $P''(H_0)$ is even smaller than before, and H_0 may definitely be rejected in favor of H_1.

The next example will show the procedure in the case of a smaller sample, so that the Student t distribution is used. Here, of course, it is not necessary to assume that σ^2 is already known. We will proceed on the basis of a diffuse prior once again. This time the two hypotheses to be compared are

$$H_0: \mu \leq 102, \qquad H_1: \mu > 102.$$

The underlying process is assumed to be normal. Some eight cases are drawn independently and at random. The sample mean M turns out to be 100 and the unbiased estimate of the variance, s^2, is 16. Then

$$M'' = M = 100,$$
$$V'' = s^2 = 16,$$
$$N'' = N = 8.$$

In order to find the posterior probability of H_0, we take the usual t statistic:

$$t = \frac{\mu_0 - M''}{\sqrt{V''/N''}}$$
$$= \frac{102 - 100}{\sqrt{16/8}}$$
$$= 2/\sqrt{2}$$
$$= 1.414.$$

Referring this value to the t table with 7 degrees of freedom, we see that such a t cuts off the lower .90 of the distribution with 7 degrees of freedom. Hence $P''(H_0) = .90$, $P''(H_1) = .10$, and the value of $\Omega'' = 9$. The hypothesis H_0 is thus very probable, given these data, and is not to be rejected. The posterior odds in favor of the null hypothesis are 9 to 1.

Although we have treated the probability of H_0 very much as we treated α in preceding chapters, this is not strictly appropriate here. Other considerations beyond the sheer probability of H_0 must enter into a decision between the hypotheses. These will be mentioned briefly a little later.

19.15 TWO-TAILED TESTS FOR THE MEAN

The testing of an inexact null hypothesis against an inexact alternative is quite simple under Bayesian procedures, provided that the two hypotheses are mutually exclusive and exhaustive. However, an interesting complication arises when the null hypothesis is exact and the alternative inexact. This is the typical situation in a two-tailed test of a mean. For the mean, a two-tailed test generally involves the two hypotheses

$$H_0: \mu = \mu_0, \qquad H_1: \mu \neq \mu_0.$$

This convention is satisfactory (and even necessary) from the classical point of view, but it becomes rather absurd in a Bayesian framework. The main difficulty is that if we regard the prior distribution of μ as continuous, then the probability of the *exact* value μ_0 is, in effect, zero. This implies that regardless of the value of the value of μ_0, $P'(H_0)$ must be regarded as zero. Even worse, the posterior probability $P''(H_0)$ must in consequence be zero.

Does it really make sense to test an exact hypothesis, which specifies a precise point on a continuum of possible values? Practically speaking, one is seldom interested in such an exact value. Rather, we generally deal with an interval of values, in which, for all practical purposes, each possible value is treated as equivalent to every other. We are not so much interested in μ_0 correct to an infinite number of decimal places as we are in μ_0 give or take a few units. Even the process of measurement never yields an exact value, regardless of how precise it may superficially appear. Rather, a measurement operation always yields a value that must be qualified by a small range of values to either side, reflecting at least the error of measurement.

In this light it is not unreasonable to revise an exact hypothesis as follows:

$$H_0: \mu \in A, \qquad H_1: \mu \notin A,$$

where $p(A) + p(\bar{A}) = 1.00$. Here, the null hypothesis is that μ lies in some interval of values A, as opposed to the alternative hypothesis that μ does not lie in the interval A. Generally, A will be an interval centered on the value μ_0, with boundaries $\mu_0 - d$ and $\mu_0 + d$, where d represents a difference from μ_0 having some practical or theoretical significance for the experimenter. For example, a psychologist might well know the standard error of measurement associated with some test of ability. He wishes to test the hypothesis that the mean score on this test is μ_0. On the other hand, he is willing to regard differences from μ_0 that fall within one standard error to either side as negligible in his conclusion. Then his null hypothesis would be

$$H_0: \mu_0 - d \leq \mu \leq \mu_0 + d,$$

where d is one standard error of measurement for the ability test.

Pursuing this example somewhat further, let us suppose that the psychologist has reason to believe that when the test is given to a new population, other than the one for which it was constructed, a different mean μ should result. The test was originally constructed so as to give a μ value of 500, with a standard error of measurement equal to 3.7. Although the new population might yield a new mean, there is no reason to believe that the standard error of measurement will be different from 3.7. He thus frames his question, "Is the mean of the test still 500 for the new population?" as

$$H_0: 500 - 3.7 \leq \mu \leq 500 + 3.7,$$
$$H_1: \text{not } H_0,$$

where the null hypothesis actually stands for "μ is within one standard error of 500."

Our experimenter draws a sample of 35 children independently and at random from the new population. He assumes that scores for this test should constitute an approximately normal distribution. He also assumes that the known variance of the old population, say $\sigma^2 = 50$, will also hold true of this new population.

The prior distribution for μ is assumed to be normal, with mean equal to 500 and with a variance equal to σ^2/N', or $50/N'$. Prior to the sample, the experimenter is willing to give 1-to-4 odds that the true value μ lies in the interval defined by one standard error of measurement to either side of 500, or 496.3 to 503.7. These odds can be used to determine N'. If $P'(H_0)$ is equal to .20, then the area cut off in the prior distribution by the interval between 496.3 and 503.7 must also be .20. Given a variance equal to $50/N'$, this implies that

$$z = \frac{(503.7 - 500)}{\sqrt{\sigma^2/N'}} = .26,$$

so that when we solve for N', we obtain $N' = .247$.

Now suppose that the sample mean turns out to be 507. Then, for the posterior distribution, the mean is

$$M'' = \frac{(.247)(500) + (35)(507)}{.247 + 35}$$
$$= 506.95.$$

The variance of the posterior distribution is $50/(N' + N)$ or 1.419. Thus, the interval corresponding to H_0 has an upper limit corresponding to the z value

$$z = \frac{503.7 - 506.95}{\sqrt{1.419}}$$
$$= \frac{-3.25}{1.19}$$
$$= -2.73.$$

Similarly, the z value for the lower limit to the interval, or 496.3, is found to be -8.95. The area defined by this interval in the posterior normal distribution must then be only about .003. Consequently, the posterior odds ratio Ω'' is only about $.003/(1 - .003)$ or .003, approximately. The experimenter almost surely rejects the null hypothesis in favor of the alternative hypothesis.

In short, the Bayesian approach to hypothesis testing can be extended to a case analogous to the test of an exact hypothesis, provided that the exact null hypothesis is reinterpreted to represent a region of values about the exact value of interest. In some contexts this region of values may have practical importance, as in the example just given involving the standard error of measurement. In other instances the region of values may have to be more or less arbitrarily chosen. In any case, this requirement of the Bayesian approach to hypothesis testing points up the fact that one is very seldom interested in an exact value of any given parameter; at best one is interested in a small range of values to either side of the putative "exact" value stated in the null hypothesis. In a field such as psychology it might be a good thing to focus more attention on the question of the meaning of an exact hypothesis, within a given scientific context. Should we really pretend to an exactitude of measurement that we do not, in fact, possess? Alternatively, if one *must* test an exact null hypothesis, then perhaps some attention should be paid to the choice of an exact alternative hypothesis. This alternative might well be a value of the parameter in question that has a special practical or theoretical significance if the null hypothesis happens not to be true. Given two exact hypotheses, their posterior likelihoods (densities) form the basis of the likelihood ratio LR, and this ratio together with Ω' is the basis for the choice between the hypotheses. In this connection it may be useful to recall some of the comments made in Chapter 9 with reference to the power of a test.

Although we shall not illustrate it here, the procedure involving the t distribution can be applied to an exact null hypothesis in very much this same way. Also, we will not demonstrate hypothesis testing for Bernoulli and other processes, simply because of limitations of space. Suffice it to say that the same general format applies when hypotheses are entertained about other parameters such as π or σ^2.

19.16 BAYESIAN INFERENCE AND DECISION-MAKING

A more complete introduction to the theory of Bayesian inference would now proceed to discuss the important role that the results of Bayesian procedures can play in decision-making. Some hint of this has already been given in Chapter 9. Here, we shall have to be content with some comments about hypothesis testing in the decision-making context, and the uses to which posterior probabilities may be put in decision-making. The connections between Bayesian inferential methods and the theory of decision-making are numerous, and the bulk of this topic is beyond our limited scope.

In the Bayesian approach to hypothesis testing, the end product is either a probability for each of the two competing hypotheses, or a posterior odds ratio, Ω''. Although we have acted as though the conventions about "small probabilities" such as α are still relevant here, in actuality the posterior probabilities obtained can be put into a larger context than that usually associated with the classical methods.

Let us once again consider errors of Type I, committed when we incorrectly reject H_0. Associated with any such error is a loss, which we can symbolize by $u(H_1; H_0)$ (see Section 9.7). On the other hand, for an error of Type II, committed when we incorrectly fail to reject H_0 in favor of H_1, there is a loss symbolized by $u(H_0; H_1)$. The respective losses, contingent on what we decide and what is actually true, can be shown in the following table:

TRUE STATE

	H_0 true	H_1 true
Accept H_0	0	$u(H_0; H_1)$
Reject H_0	$u(H_1; H_0)$	0

ACTION

Since under a Bayesian procedure we have posterior probabilities $P''(H_0)$ and $P''(H_1)$ available, we may adopt a decision rule that will minimize the expected loss. That is, the expected loss if one decides to accept H_0 is

$$E(u; H_0) = 0P''(H_0) + u(H_0; H_1)P''(H_1).$$

The expected loss in rejecting H_0 is found from

$$E(u; H_1) = u(H_1; H_0)P''(H_0) + 0P''(H_1).$$

Then if $E(u; H_0) < E(u; H_1)$, we accept H_0. If the reverse is true, we reject H_0 in favor of H_1.

What does this imply about the posterior odds ratios? We can form a loss ratio R by taking $R = u(H_1; H_0)/u(H_0; H_1)$. Then the decision rule can be written as

Reject H_0 if $\Omega'' \times R < 1.00$,
Accept H_0 if $\Omega'' \times R > 1.00$.

It is not the odds ratio alone that dictates the choice, but rather the odds ratio weighted by the loss ratio. In other words, the decision to reject or to accept H_0 depends both upon the posterior odds ratio Ω'' of H_0 to H_1, and upon the loss ratio R for Type I and Type II errors. This implies that if the odds in favor of H_0 are sufficiently small, H_0 may still be rejected even though a Type I error may be very serious as compared with a Type II error. On the other hand, it may be that H_0 will be rejected even when the odds favor it, simply because a Type II is so much more costly than a Type I error.

Use of the ordinary α values for rejection of the null hypothesis is tantamount to requiring that H_0 be extremely unlikely (from the Bayesian point of view) or that the sample results be extremely unlikely given H_0 (from the classical point of view). If the conventions about α are examined in the light of the decision rule just given, we can see they imply that Type I errors are more serious than Type II errors. By insisting that $P''(H_0)$ be less than .05 before we reject the null hypothesis, we are requiring that the odds ratio Ω'' be no more than .05/.95 or 1/19. This is like asking for insurance against a Type I error with a loss value 19 times that of a Type II error. However, it just might be that Type II errors are vastly more important, and sometimes it may be desirable to reject the null hypothesis even when the posterior odds favor it, simply to avoid a large expected loss. Bayesian methods can, of course, supply the posterior odds for such decision rules. However, R is a vital, and usually missing, ingredient in such a decision-theoretic approach to hypothesis-testing. The relative weights to be given to Type I and Type II errors are very seldom specified in real examples. Hence, we continue our reliance on conventionalized standards for judging the tenability of a hypothesis.

Let us return to the example of the preceding section in order to illustrate how the value of R might enter into a choice of action following the test of a hypothesis. Remember that the psychologist must decide whether the mean for a new population is the same, within one standard error of measurement, as that for an old population. Suppose further that he is confronted with the following possibilities for action. If the test administered to the new population turns out to have the same mean as the old, no alteration in the test is required. However, if the new population turns out to have a new mean, then the test must be restandardized. This implies a real cost in time and effort. In effect, H_0, if true, means "don't restandardize," while a true H_1 means "restandardize the test." A Type I error here implies that the test is reworked needlessly; a Type II error means that the test is not reworked when it should have been. Since the actual process of restandardization would tend to reveal that an error had been made if H_0 were really true, a Type I error is partially correctable. On the other hand, doing nothing to the test would only perpetuate the Type II error. Under this perspective it seems reasonable that a Type II error actually is more serious than an error of Type I, and that a Type II error should involve the larger loss.

The decision situation can be represented in the following table:

CONSEQUENCE

		True state H_0	True state H_1
	Do nothing	Acceptable test	Poor test; loss $= u(H_0; H_1)$
ACTION	*Rework test*	Poor test, but correctable; loss $= u(H_1; H_0)$	Acceptable test

Now suppose that the experimenter, pondering over the consequences of his possible actions, concludes that the loss ratio, R, is equal to 1/2, so that the loss entailed in a Type II error is twice as great as that for a Type I error. What does this imply about the experimenter's decision rule? It means that H_0 should be rejected if the posterior odds ratio Ω'' is less than 2, so that $\Omega'' \times R$ would be less than 1.00. In this way he chooses the action with the lower expected loss.

In the example worked out above, the posterior odds ratio for H_0 relative to H_1 was .003. Then, $\Omega'' \times LR$ is about .0015, and certainly less than 1.00. The experimenter would reject H_0 and proceed to rework the test.

On the other hand, what posterior odds ratio would lead him to accept H_0 and therefore do nothing to the ability test? If we take $(1/2)\Omega'' > 1$ and solve for Ω'', then Ω'' muxt be greater than 2. This means that H_0 will be accepted only if $P''(H_0) > 2/3$. In the view of one accustomed to classical hypothesis testing, it may seem strange that H_0 will still be rejected even when it has a probability of .4 or .5 or .6. Nevertheless, if the experimenter wishes a minimal expected loss, that is exactly what he should do in this circumstance. He should restandardize the ability test unless he is *extremely* confident that H_0 is true. Although this example may seem extreme, we could go on to construct one in which the experimenter would always reject H_0, regardless of how probable he believed it to be, in the face of an infinitely large loss value for a Type II error. Conversely, a situation can be imagined where the experimenter will never reject H_0, so terrified is he of a Type I error.

To reiterate: In the real world, and particularly in scientific work, losses and loss ratios are often extremely difficult to specify. Hence, most users of hypothesis-testing in either the Bayesian or the classical way will likely retain the familiar conventions of small values. Such values reflect either the small probability of the null hypothesis (Bayesian) or the small likelihood of the sample result given the hypothesis (classical). Even so, it is worth remembering that this decision-making machinery is ready for use when needed, and that the Bayesian approach does give one of the ingredients of this form of decision-making.

19.17   SOME FINAL REMARKS

This chapter has outlined some of the most elementary aspects of the Bayesian approach to statistical inference. The so-called classical approaches to statistical inference tend to limit their concerns to likelihood of the sample data, and do not explicitly admit consideration of prior and posterior probabilities. Inferences are based upon the likelihood of a sample result, given the value of one or more parameters and the form of the sampling distribution. On the other hand, Bayesian methods represent an extension of the classical methods, in that they employ not only the likelihood function for the sample result, but also prior and posterior distributions for the parameter or parameters under study. Such prior and posterior distributions are held to represent the prior and posterior states of opinion or belief on the part of the experimenter. The personal interpretation of probability is often invoked to justify the formal inclusion of such states of belief in the statistical analysis.

Although Bayesian and classical methods often require identical computations, they usually put different interpretations upon the end result of a statistical analysis. In interpreting prior and posterior distributions, the Bayesian is willing to treat a parameter as though it were a random variable; the classical statistician is not.

Bayesian methods offer advantages. The steady accumulation and systematic combination of information from a variety of sources is encouraged under the Bayesian approach. Although in classical statistics a number of ways do exist for combining results from different experimental occasions or settings, these tend to be less routine and more suited to special problems than is a thoroughgoing Bayesian approach. The student of classical statistics may begin to believe that the sequence "problem-sample-computation-inference" is all there is to statistical inference. It is easy to get the feeling that no one has ever drawn a sample before and no one ever will again. Although this is a parody of the true nature of statistical inference, textbooks, including this one, quite successfully leave this impression. In contrast, no one can look into Bayesian methods, however superficially, without carrying away the impression that statistical inference is an on-going process that is never completed.

Although we have not explored the matter fully here, Bayesian inference does lead naturally into a related theory of decision-making. The end results of a Bayesian analysis are in a form that lends itself easily to the calculation of expected values and expected losses, among other things. These are, of course, highly important features of the theory of decision-making.

Some of the objections to Bayesian methods are well founded: In at least one stage of their use, Bayesian methods do demand major participation by the experimenter. At the very outset the experimenter must face the question of the prior distribution. He must examine what he knows or believes to be true and then find a way to represent this state by a prior distribution. If nothing else, he must decide that he knows sufficiently little that a diffuse prior is appropriate. At this stage, the very beginning of a series of Bayesian inferences, the choice of a prior may be important in the determination of the posterior distribution. It is here, at the very outset of such a series of inferences, that the charge that the Bayesian methods are subjective is most telling.

If one really takes the continuity of a research effort seriously, and information does accumulate systematically, the problem of the prior distribution becomes negligible. If the Bayesian analysis is one of a continuing series, in which the mass of the data gradually becomes considerable, then the original prior distribution becomes irrelevant. Any effect of the original choice of prior must be totally swamped by the accumulated mass of the evidence. In the long run, experimenters with very different prior distributions will wind up with identical posterior distributions.

There are other drawbacks to the Bayesian approach as well, at least as of this writing. Not least is the fact that classical inferential methods are widely understood and used by workers in a great variety of fields. The introduction of a Bayesian analysis into a research report, or any other communication having to do with the evaluation of data, may very well invite misunderstanding. There is presently somewhat less danger of this in applied fields such as business, where the practical decision-theoretic aspects of the Bayesian approach are very important. The risk of misunderstanding can be considerable in other areas such as the sciences. Another disadvantage is the rather limited number of situations for which standard Bayesian methods exist, already worked out and ready for routine application. This condition is changing rapidly, however, and should cease to be a serious problem.

The Bayesian actually faces no serious obstacle in the communication of statistical results. It is easy enough to supply a diffuse prior to go along with any other prior distribution that might have been entertained. Since the raw materials are almost always the same as for a classical procedure, it is also fairly simple for the reader to see for himself what a classical analysis would show, even if he has only modest statistical skill. So far as this author can determine, there is no law against the inclusion of *both* kinds of analysis in a research report.

There is also another danger. Because these methods *look* somewhat subjective, with some latitude explicitly allowed the experimenter, the inexperienced or (Heaven forbid) slightly disingenuous experimenter might try to tinker with priors and things until the results come out in a certain way. Nothing will ever keep people from being stupid or dishonest if they are made that way, and the most pristine classical statistical analysis may be "adjusted." But Bayesian statistics must not be viewed as an open invitation to such practice. In time this danger will largely be averted as standard experimental and empirical routines are developed for designating prior distributions. There ought to be standard procedures by which the experimenter sets out to survey his knowledge and opinions prior to the collection of data, and converts these into prior probabilities. This is certainly possible in principle, as witness the connections pointed out in Chapter 2 between choice behavior, odds, and prior probability. The point is that these procedures are not yet part of the routine approach to an empirical investigation.

All in all, Bayesian methods are interesting, they shed a different light on many familiar problems, and they have already shown considerable utility in certain fields. Most of all, there is an intuitive "feel" about these methods that many, though by no means all, workers in statistics find satisfying. It thus behooves a potential user of statistical inference to know about Bayesian approaches, not as replacements for the methods that he has so laboriously learned, but as supplements to them.

EXERCISES

1. A gambler has two identical pairs of dice. One pair is loaded in such a way that a "seven" should appear with probability of 2/3, and "not a seven" with probability 1/3. The other pair of dice is fair. One pair of dice is selected. Your prior opinion is such that you believe it equally likely that the loaded pair or the fair pair of dice is being used. If the dice are tossed four times, and exactly one seven appears, what should your posterior probability for "loaded dice" be?

2. Suppose that a discrete random variable follows either

   Rule 1:

$$p(x) = \begin{cases} \dfrac{k}{(x+2)(x+1)} & x = 0, 1, 2 \\ 0 & \text{elsewhere} \end{cases}$$

   or

   Rule 2:

$$p(x) = \begin{cases} \dfrac{g(x+1)}{(x+2)} & x = 0, 1, 2 \\ 0 & \text{elsewhere} \end{cases}$$

Suppose that your prior opinion is that either rule is equally likely to hold true. One and only one rule must be true throughout the sequence of observations. Two observations are made, and the values obtained turn out to be 0 and 2 respectively. What is your posterior probability that the random variable follows Rule 1. You may leave your answer in terms of k and g.

3. Continuing problem 2, suppose that two additional observations were made, and that each of these observations yielded the value 0. Now what is your posterior probability that the random variable follows Rule 1? How does this differ from the posterior probability that would have obtained if four independent observations had originally been made, and had yielded the values 0, 2, 0, 0? Once again, you may leave your answer in terms of k and g.

4. A certain continuous random variable follows a rule which depends upon a single parameter θ. In a given experimental situation, exactly one of three possible values of θ must be true. Call these θ_1, θ_2, and θ_3. The prior probabilities are such that $p(\theta_1) = 2p(\theta_2)$ and $p(\theta_2) = p(\theta_3)$. Following observation of a set of data D, the posterior probabilities were all equal. How much more likely is D given that θ_2 is true than given θ_1 true? If the posterior probabilities had been .30 for θ_1, .20 for θ_2, and .50 for θ_3, what must the likelihood ratio for D given θ_1 as compared to D given θ_2 have been?

5. Suppose that in a stable Bernoulli process the probability π of a success had one of the three possible values: .20, .30, .40. If, prior to an experiment, each of these values is viewed as equally likely, what are the posterior probabilities given that fifteen independent observations would show exactly five successes?

6. An experimenter holds the prior probability of .5 that the mean of a certain population lies in the interval 140–160. He believes that the most likely value for the mean is 150. The standard deviation for the population is known to be 40. Assuming a normal prior distribution, what is the posterior distribution after a sample of 50 independent observations yields a mean of 143?

7. In assessing his prior distribution for μ in a particular population, an experimenter was willing to give odds of 3 to 1 that $\mu \leq 90$, and odds of 9 to 1 that $\mu \leq 130$. Given a normal prior distribution, what odds should the experimenter be willing to give that $\mu \leq 90$ and that $\mu \leq 130$ after a sample mean based on 35 independent observations yields a value of 86?

8. Suppose that the population involved in problem 7 is normal in its distribution, and that the prior odds were as stated at the beginning of that problem. Following a sample of N independent observations, the experimenter was willing to give odds of 4 to 1 that $\mu \leq 66$, and odds of 9 to 1 that $\mu \leq 72$. If N' was equal to 1, how big was N?

9. A certain normally distributed population is known to have a mean of 500. However, the variance is not known exactly. If the prior distribution of h is gamma, with an expected value $E(h) = 1/V'$ equal to .36, what is the posterior distribution produced by a sample estimate of the variance $V = 24$, given that the sample contained 19 independent observations. (For convenience, assume that $N' = 1$.)

10. In a certain competitive foot-racing event, a coach believed that the most likely mean time for high-school girls over a standard course should be about 2.800 minutes. He was also willing to give odds of 2 to 1 that this mean value was at or below 2.934 minutes. A sample of fifty high-school girls was selected at random and each ran the standard course. Their mean time for the event turned out to be 2.771 minutes. If the coach's prior distribution was normal, describe his posterior distribution.

11. For the posterior distribution found in problem 10, find the 99 percent credible interval for the mean μ.

12. An archeologist investigating a new stratum at a site where a great stone artifacts are to be found believes that about 30 percent of these artifacts should belong to a cultural Type A and the remainder to other cultural types. However, the distribution of his prior opinion on this matter is rather diffuse. From the large number of artifacts discovered in this stratum, he takes a random sample of twenty, and notes that some eight are of Type A, while the remainder belongs to other types. If the occurrence of such an artifact on a given sampling can be interpreted as the outcome of a stable and independent Bernoulli process, describe the posterior distribution for this archeologist.

13. In a large American city, statistics show that 25 years ago only one person in ten could identify at least one of the members of the city council by name. A modern researcher believes that this is still the most likely proportion, but he places rather little weight on his prior beliefs. A random sample of thirty individuals was drawn independently from this city. Of these, only four could identify one or more city councilmen. What is the researcher's posterior distribution concerning the true proportion? (Hint: for $r' > 1$ and $N' - r' > 1$, the mode of a beta distribution occurs at $(r' - 1)/(N' - r')$. If r' is the smallest whole number yielding a mode and otherwise fitting the qualifications of the problem, what is N'?)

14. For problem 1, chapter 9, assume a diffuse normal prior, and test the hypothesis $H_0: \mu \leq 28.6$ against the alternative $H_1: \mu > 28.6$.

15. For the data of problem 1, chapter 9, test the hypothesis $H_0: \mu = 28.6 \pm 1.0$ against the alternative H_1: not H_0. Assume a diffuse prior distribution.

16. In problem 2 chapter 9, assume a normal prior with $N' = 10$. Then test the hypothesis $H_0: \mu \geq 500$ against $H_1: \mu < 500$.

17. Using the circumstances of problem 16, and the data of problem 2, chapter 9, form the 95 percent credible interval for μ.

18. For problem 3, chapter 9, assume a diffuse prior distribution and test the stated hypothesis (Hint: use the F-distribution to find probabilities in the posterior beta distribution.)

19. In problem 6, chapter 9, let the prior distribution be normal with a mean of 15, and with $N' = 5$. Then test the hypothesis.

20. For the data of problem 1, chapter 10, assume a diffuse normal prior with mean of 98.6. Then test $H_0: \mu = 98.6 \pm .05$ against H_1: not H_0.

21. For the data of problem 20, form the 99 percent credible interval.

22. Frame problem 5, chapter 10, in Bayesian terms, with a diffuse normal prior, and then test the null hypothesis.

23. Find the 95 percent credible interval for the mean for problem 22.

24. Suppose that in problem 22, ν' were so small as to be negligible. Describe the posterior (marginal) distribution of h. How might this be used to find a credible interval for the population variance?

Appendix A:

Rules of Summation

SUBSCRIPTS

In the discussion to follow, let X and Y stand for numerical varia-bles. Ordinarily, the numerical values for these variables are scores in some given set of data, where each individual observed in a group of N cases is paired either with a single score, x, or with a pair of scores (x, y).

The *subscript* notation is useful in keeping track of the various score values across the different individuals or observations. For the moment, we will consider N distinct individuals, arbitrarily indexed by the set of numbers $(1, \cdots, i, \cdots, N)$. The "running subscript" i denotes any one of these individuals. Then, X_i denotes the particular value of X associated with any particular individual i. That is, the variable X_i ranges only over the set of par-ticular values

$$(x_1, \ x_2, \cdots, \ x_N).$$

Now we need a way to discuss the *sum* of these particular values of X over the entire set of N different individuals i. The symbol Σ (capital Greek sigma) stands for the operation "the sum of," referring to the values symbolized immediately after the summation sign. Thus,

$$\sum_{i=1}^{N} x_i$$

is read as "the sum of the N values x_i, beginning with $i = 1$ and ending with $i = N$." That is,

$$\sum_{i=1}^{N} x_i = x_1 + x_2 + \cdots + x_N.$$

Suppose, for example, that $N = 4$, and the actual values associated with the four different individuals were

$$6, 8, 3, 11.$$

Then,

$$\sum_{i=1}^{N} x_i = 6 + 8 + 3 + 11 = 28.$$

In many simple contexts where it is clear that the sum extends across *all* of a set of N individuals, the subscripts and superscripts on the summation sign are simplified or omitted. Thus, we often write:

$$\sum_{i} x_i$$

to indicate the sum of values across all N individuals i, or even

$$\Sigma x_i$$

to stand for this sum. However, the subscript is usually omitted only in very simple expressions where the set of values to be included in the sum is perfectly clear.

Quite frequently, symbols for values corresponding to individual observations are given two (or more) subscripts. That is, it happens that sometimes we want to discuss individual observations arranged into several groupings or cross-classifications. Then two (or more) sets of index numbers are used. For example, suppose that there were some J different *groups* of observations; these different groups are identified by being paired arbitrarily with the numbers

$$(1, \cdots, j, \cdots, J).$$

Furthermore, within any particular group, j, there is some number n_j of distinct individuals or observations, each of which is associated with an X score. Then, the particular value of the score for the ith individual in the group j is symbolized by x_{ij}. If we wanted to symbolize the sum of the scores for the particular group j, then we would write

$$\sum_{i=1}^{n_j} x_{ij}$$

which stands for "the sum of values for all of the n_j different individuals in the particular group labeled j." Where it is clear that all individuals scores in a particular group j are to be summed, this can be abbreviated to

$$\sum_{i} x_{ij}.$$

Notice that this quite literally is the symbol for "sum over i" for a set of scores x_{ij}; the fact that j appears as a subscript for x but that i alone appears below the sigma is a cue that all the scores summed in this particular operation *must* have the same subscript j (that is, must be in the same group of observations).

However, often one wants to indicate the sum of X_{ij} over *all* individuals in *all* groups. Written out in full, this would be symbolized by

$$\sum_{j=1}^{j} \sum_{i=1}^{n_j} x_{ij} = (x_{11} + x_{21} + \cdots + x_{n_j-1,J} + x_{n_j,J}).$$

This *double* summation is interpreted as "Take a particular group j, and sum scores over all the n_j different observations in that group; then, having done this for each of the J different groups, sum the results over all of these groups." Once again, this can be simplified to

$$\sum_{j} \sum_{i} x_{ij}$$

when the meaning of the symbol is perfectly clear in terms of which values are to be summed.

This idea can be extended, although in our discussions a *triple* summation is the largest such symbol used. In some problems, each individual i belongs simultaneously to *two* groupings. That is, suppose that all the scores in a set of data were arranged into a table with K rows, J columns, and JK cells, and that each individual observation belongs to one and only one row, one and only one column, and one and only one cell in the data table. Now let the various *column* groupings be symbolized by the numbers

$$(1, \cdots, j, \cdots, J).$$

The *rows* of the data table are indexed by the set of numbers

$$(1, \cdots, k, \cdots, K).$$

Then any cell of the table is indexed by a *pair* of numbers jk. The score of the ith individual in the cell jk of the data table is thus x_{ijk}. The number of individuals in the cell jk is shown by n_{jk}.

Now, when there are three subscripts like this,

$$\sum_{j=1}^{J} \sum_{k=1}^{K} \sum_{i=1}^{n_{jk}} x_{ijk},$$

the following is symbolized: "Take a particular cell, jk, and sum the scores of all individuals i in that cell; then, still within a particular column j, sum the results of doing this for each row k over the K respective rows; finally, sum the results for each column j over the J respective columns." A little thought should convince you that this is simply the sum of all individual scores in all cells, or the sum of all scores in the table.

In this instance, it is immaterial whether we write the sum for the rows or the sum for the columns first, since the total number symbolized is the same. However, this is not universally true, and order of the summation signs can be important.

Sometimes we want to symbolize only the sum for a particular cell, row, or column: First of all, letting n_{jk} stand for the number in cell j,k, we write

$$\sum_{i=1}^{n_{jk}} x_{ijk} = (x_{1jk} + \cdots + x_{njk}).$$

This denotes the sum of all individual scores in the *single* cell jk. If we write

$$\sum_{j=1}^{J} \sum_{i=1}^{n_{jk}} x_{ijk} = (x_{11k} + x_{21k} + \cdots + x_{nJk}),$$

we intend that first the sum of individuals in a particular cell in column j and row k shall be taken, and then, still within the particular row k, the result for all the different columns j shall be summed. This amounts to the sum of all scores within the particular row k. On the other hand, if we want the sum of all the scores in a particular column j, we could write

$$\sum_{k=1}^{K} \sum_{i=1}^{n_{jk}} x_{ijk} = (x_{1j1} + \cdots + x_{njK})$$

indicating the sum for a particular cell jk, and then the sum over the results for all rows k within the particular column j. In short, *unless a subscript appears both in the symbol for the x values and under the summation sign, the total value symbolized pertains to some particular set of observations, represented by that subscript not appearing under the summation symbol.* Notice that when the sum of all observations in a particular row k of the data table was intended, the sum was indicated over i and over j, but *not* over k. On the other hand, when the sum of observations in some column j was desired, the sum was indicated over i and over k, but not over j.

In learning to work with summation signs, it is well to start the habit of reading the summation symbols from *right to left*, and from *inside the punctuation outward*. The *innermost* summation indicated in a mathematical expression is ordinarily to be carried out first, then the next summation symbolized to the left, and then finally the summation appearing closest to the margin on the left.

RULES OF SUMMATION

A few simple rules can help one learn to interpret and carry out the computations symbolized by statistical expressions involving summations. Furthermore, elementary statistical derivations almost always involve one or more of these summation rules used as an algebraic equivalence. First of all, some simple rules will be given that apply when values for only one variable are being summed, and then we will consider situations involving two or more variables.

RULE 1. If a is some constant value over the N different observations i, then

$$\sum_{i=1}^{N} a = Na.$$

For example, suppose that 5 observations are considered, and paired with each observation is the constant number 10. Then

$$\sum_{i} a = 10 + 10 + 10 + 10 + 10 = (5)(10).$$

Any time that, in the particular sum indicated, each and every individual observation entering into the sum is paired with exactly the same constant value a, then the sum is equal to the number of individuals considered times the value of a.

Extending this rule to double or triple summations, we have

$$\sum_{j=1}^{J} \sum_{i=1}^{n_j} a = \sum_{j=1}^{J} (n_j)a$$

$$\sum_{k=1}^{K} \sum_{j=1}^{J} \sum_{i=1}^{n_{jk}} a = \sum_{k=1}^{K} \sum_{j=1}^{J} (n_{jk})a$$

and so on.

RULE 2. Given the value a, which is constant over all individuals entering into the summation, then

$$\sum_{i=1}^{N} ax_i = a \sum_{i=1}^{N} x_i.$$

For example, consider the five individual scores

$$6, 7, 3, 9, 20.$$

Suppose that each of these scores is multiplied by the constant 2. Then

$$\sum_{i} 2x_i = 2(6) + 2(7) + 2(3) + 2(9) + 2(20)$$

$$= 2(6 + 7 + 3 + 9 + 20) = 2 \sum_{i} x_i.$$

Once again, this rule can be applied to several sums:

$$\sum_{j=1}^{J} \sum_{i=1}^{n_j} ax_{ij} = a \sum_{j=1}^{J} \sum_{i=1}^{n_j} x_{ij}.$$

Notice here that the a symbol has no subscript, and thus may be regarded as a constant over the sum. However, it might happen that the sum is such that a_j appears instead of a, meaning that a_j is a constant for all observations i in the *same* group j, but is different for different groups. Then,

$$\sum_{j=1}^{J} \sum_{i=1}^{n_j} a_j x_{ij} = \sum_{j=1}^{J} a_j \left(\sum_{i=1}^{n_j} x_{ij} \right).$$

One must be sure to notice the set of observations for which some value *is* constant in multiple summations.

RULE 3. If some operation is to be carried out on the individual values of X before the summation, this is indicated by mathematical punctuation, such as parentheses, or equivalent symbols having the force of punctuation. Unless the summation symbol is included within this punctuation, the summation is to be done after the other operation: for example

$$\sum_i (x_i)^2 = \sum_i x_i^2 = (x_1^2 + x_2^2 + \cdots + x_N^2)$$

$$\sum_i \sqrt{x_i} = \sqrt{x_i} + \sqrt{x_2} + \cdots + \sqrt{x_N}.$$

However,

$$\left(\sum_i x_i \right)^2 = (x_1 + \cdots + x_N)^2$$

$$\sqrt{\sum_i x_i} = \sqrt{x_1 + \cdots + x_N}.$$

Be sure to notice that where the parentheses (or other symbols such as exponent and radical signs having the same force of parentheses) happen to appear makes a big difference in the sum symbolized. In general,

$$\sum_i x_i^2 \neq \left(\sum_i x_i \right)^2$$

$$\sum_i \sqrt{x_i} \neq \sqrt{\sum_i x_i}$$

$$\log \left(\sum_i x_i \right) \neq \sum_i \log x_i,$$

and so on.

This is especially important when there is multiple summation. For example,

$$\sum_{j=1}^{n} \left(\sum_i x_{ij} \right)^2 = \sum_j (x_{1j} + \cdots + x_{nj})^2.$$

The notation here tells us *first* to take the sum of all the individual values in a single group j; then, having done this for each group j, we square the *sum* for each group; finally, the sum of the *squared* sums for the various groups is found. Squaring at the wrong place in this sequence of steps will give quite a different result.

RULE 4. If the only operation to be carried out before a sum is taken is itself a sum (or difference), then the summation may be distributed. Thus,

$$\sum_i (x_i - 4) = \sum_i x_i - \sum_i 4$$

$$\sum_i (x_i^2 + 3x_i + 10) = \sum_i x_i^2 + \sum_i 3x_i + \sum_i 10.$$

This rule is very important in algebraic manipulations on expressions involving sums. For example, using this rule we can make the following algebraic changes:

$$\sum_{j=1}^{J} \sum_{i=1}^{n_j} (x_{ij} - m)^2 = \sum_{j=1}^{J} \sum_{i=1}^{n_j} (x_{ij}^2 - 2mx_{ij} + m^2)$$

$$= \sum_{j=1}^{J} \sum_{i=1}^{n_j} x_{ij}^2 - \sum_{j=1}^{J} \sum_{i=1}^{n_j} 2mx_{ij} + \sum_{j=1}^{J} \sum_{i=1}^{n_j} m^2.$$

Invoking rules 1 and 2, we can reduce this expression to

$$\sum_j \sum_i x_{ij}^2 - 2m \sum_j \sum_i x_{ij} + m^2 \sum_j n_j.$$

The value represented by this last expression is algebraically equivalent to the value of the first expression, but it may be far more convenient actually to compute this number in the second way, or to use this second expression in a mathematical argument.

SUMMATION RULES FOR TWO OR MORE VARIABLES

So far we have acted as though each distinct individual or observation were associated with one and only one numerical score. However, it may be that two distinct values x_i and y_i are paired with each individual or observation i. Then the rules given above may be extended:

RULE 5. If each of N distinct individuals i has two scores, x_i and y_i, then

$$\sum_{i=1}^{N} x_i y_i = x_1 y_1 + x_2 y_2 + \cdots + x_N y_N.$$

Notice that this symbol means that *the product of the pair of scores belonging to any given individual i is to be found first,* and then when this has been done for the entire set of individuals, *the sum of these products is to be found.* Be sure to notice that *in general*

$$\sum_i x_i y_i \neq \left(\sum_i x_i \right) \left(\sum_i y_i \right).$$

The other rules extend quite easily to the situation where two (or more) variables are considered:

RULE 6. $\displaystyle\sum_i a x_i y_i = a \sum_i x_i y_i.$

RULE 7. Given the constants a and b,

$$\sum_i (a x_i + b y_i) = a \sum_i x_i + b \sum_i y_i.$$

Sometimes it happens that when observations are arranged into groups, all observations i in the same group j will have the same value x_j on one variable, but different values y_{ij} on the other. Then, the following rule applies:

RULE 8. $\displaystyle\sum_{j=1}^{J} \sum_{i=1}^{n_j} x_j y_{ij} = \sum_j x_j \left(\sum_i y_{ij} \right).$

All of these rules may be extended by obvious analogy when there are more than three variables involved, or when there is multiple summation.

Like most of the mechanical skills of mathematics, real facility in algebraic manipulation of summations comes only with practice. However, when puzzled or in doubt about the equivalence to two sums, you can usually check this equivalence by actually writing out the terms in the sums symbolized.

Exercises—Rules of Summation

Write out in extended form the sums represented by each of the following expressions:

1. $\displaystyle\sum_{i=1}^{5} x_i$

2. $\displaystyle\sum_{i=1}^{6} x_i^2$

3. $\displaystyle\sum_{i=1}^{3} x_i^3 + \sum_{i=5}^{8} x_i^2$

4. $\displaystyle\sum_{i=2}^{4} 10(x_i + y_i)$

5. $\displaystyle\sum_{i=1}^{3} 4 \left(\sum_{j=1}^{2} x_{ij} \right)$

6. $\displaystyle\sum_{j=1}^{4} \left(\sum_{i=1}^{3} x_{ij} \right)^2$

7. $\displaystyle\left(\sum_{i=1}^{4} x_i^2 g_i \right) \bigg/ \left(\sum_{i=1}^{4} g_i \right)$

8. $\displaystyle\sum_{j=2}^{5} n_j \left(\sum_{i=1}^{3} x_{ij} \right)^2$

9. $\displaystyle\sum_{i=1}^{5} \sum_{j=2}^{4} (x_{ij}^2 + y_i)$

10. $\displaystyle\sum_{i=4}^{6} \sqrt{x_i + 10}$

Express each of the following extended sums in the appropriate summation notation:

11. $(x_1 + x_2 + x_3 + x_4)/N$

12. $x_1y_1 + x_2y_2 + x_3y_3 + \cdots + x_Ny_N$

13. $x_2f(x_2) + x_3f(x_3) + x_4f(x_4)$

14. $p(x_1) + p(x_2) + p(x_3) + p(x_{10}) + p(x_{11}) + p(x_{12})$

15. $3(x_1 + y_1)^2 + 3(x_1 + y_2)^2 + \cdots + 3(x_1 + y_N)^2$

16. $12(x_1^2 - 2x_1 + 3) + 12(x_2^2 - 2x_2 + 3) + 12(x_3^2 - 2x_3 + 3)$

17. $[3x_1 + 3x_2 + 3x_3 + 3x_4][5x_1 + 5x_2 + 5x_3 + 5x_4]$

18. $\sqrt{(x_1 - 2)(y_1 + 5) + (x_2 - 2)(y_2 + 5) + \cdots + (x_N - 2)(y_N + 5)}$

19. $[4(x_2 - 5)^2/y_2] + [4(x_3 - 5)^2/y_3] + [4(x_4 - 5)^2/y_4]$

20. $(x_{11} - M_1)^2 + (x_{12} - M_1)^2 + (x_{13} - M_1)^2 + (x_{21} - M_2)^2$
$$+ (x_{22} - M_2)^2 + (x_{23} - M_2)^2$$

21. $(x_1 + x_2 + x_3 + x_4)^2(y_1^2 + y_2^2 + y_3^2 + y_4^2)$

Reduce each of the following expressions into simplest form by application of the rules of summation:

22. $\displaystyle\sum_{i=1}^{10} (5x_i - 3)$

23. $\displaystyle\left(\sum_{i=1}^{5} (x_i + 2c) \right) \left(\sum_{j=1}^{3} 7 \right)$

24. $\displaystyle\sum_{i=4}^{6} (x_i - 2a)^2 + \sum_{i=1}^{3} (x_i + 2a)^2$

25. $\displaystyle\sum_{i=1}^{N} \left[x_i - \left(\sum_{i=1}^{N} x_i/N \right) \right]$

26. $\displaystyle\sum_{i=1}^{N} (x_i - 3)(x_i + 3)/N$

27. $\displaystyle\sqrt{\sum_{i=1}^{N} (a + b)} \ \sqrt{\sum_{i=1}^{N} c}$

28. $\displaystyle\left[\left(\sum_{i=1}^{N} x_i \right)^2 - \left(\sum_{i=1}^{N} y_i \right)^2 \right] \Big/ \sum_{i=1}^{N} (x_i + y_i)$

29. $\displaystyle\frac{\sum_{i=1}^{N} \left[x_i - \left(\sum_{i=1}^{N} x_i/N \right) \right]^2}{N}$

30. $\displaystyle(1/N) \sum_{i=1}^{N} \left[x_i - \left(\sum_{i=1}^{N} x_i/N \right) \right]\left[y_i - \left(\sum_{i=1}^{N} y_i/N \right) \right]$

Consider the following values:

$$
\begin{array}{llll}
x_1 = 7 & x_4 = 4 & x_7 = 2 & x_{10} = 5 \\
x_2 = 3 & x_5 = 0 & x_8 = 8 & \\
x_3 = 9 & x_6 = 1 & x_9 = 8 &
\end{array}
$$

In terms of these values for x_i, evaluate the following sums:

31. $\displaystyle\sum_{i=1}^{10} (x_i - 5)$

32. $\displaystyle\sum_{i=1}^{10} (x_i^2 - 3x_i + 1)$

33. $\displaystyle\sum_{i=1}^{5} x_i - \sum_{i=6}^{10} (x_i + 5)$

34. $\displaystyle\sum_{i=1}^{10} \left[x_i - \left(\sum_{i=1}^{10} x_i/10 \right) \right] \Big/ 10$

35. $\displaystyle\left(\sum_{i=1}^{10} x_i \right) \Big/ \left(\sum_{i=1}^{5} x_i \right)$

36. $\displaystyle\left(\sum_{i=1}^{10} x^2/10 \right) - \left(\sum_{i=1}^{10} x_i/10 \right)^2$

Appendix B:

The Algebra of Expectations

A very prominent place in theoretical statistics is occupied by the concept of the mathematical expectation of a random variable X. If the distribution of X is discrete, then the expectation (or expected value) of X is defined to be

$$E(X) = \sum_x xp(x)$$

where the sum is taken over all of the different values that the variable X can assume, and

$$\sum_x p(x) = 1.00.$$

For a continuous random variable X ranging over all the real numbers, the expectation is defined by

$$E(X) = \int_{-\infty}^{\infty} xf(x)\, d(x)$$

where
$$\int_{-\infty}^{\infty} f(x)\, d(x) = 1.00.$$

In essence, the expectation defined in either of these ways is a kind of weighted sum of values, and thus the rules of summation have very close parallels in the rules for the algebraic treatment of expectations. These rules apply either to discrete or to continuous random variables if particular boundary conditions exist; for our purposes these rules can be used without our going further into these special qualifications.

RULE 1. If a is some constant number, then

$$E(a) = a.$$

That is, if the same constant value a were associated with each and every elementary event in some sample space, the expectation or mean of the values would most certainly be a.

RULE 2. If a is some constant real number and X is a random variable with expectation $E(X)$, then

$$E(aX) = aE(X).$$

Suppose that a new random variable is formed by multiplying each value of X by the constant number a. Then the expectation of the new random variable is just a times the expectation of X. This is very simple to show for a discrete random variable X: By definition

$$E(aX) = \sum_x axp(ax).$$

However, the probability of any value aX must be exactly equal to the probability of the corresponding X value, and so

$$E(aX) = \sum_x axp(x) = a\sum_x xp(x) = aE(X).$$

RULE 3. If a is a constant real number and X is a random variable, then

$$E(X + a) = E(X) + a.$$

This can be shown very simply for a discrete variable. Here,

$$E(X + a) = \sum_x (x + a)p(x + a)$$

$$= \sum_X xp(x + a) + a\sum_X p(x + a).$$

However, $p(X + a) = p(X)$ for each value of X, so that

$$E(X + a) = E(X) + a\sum_x p(x) = E(X) + a.$$

The expectations of functions of random variables, such as

$$E[(X + 2)^2]$$
$$E(\sqrt{X + b})$$
$$E(b^X),$$

to give only a few examples, are subject to the same algebraic rules as summations. That is, the operation indicated within the punctuation is to be carried out *before* the expectation is taken. It is most important that this be kept in mind during any algebraic argument involving summations. In general,

$$E[(X + 2)^2] \neq [E(X) + E(2)]^2$$
$$E(\sqrt{X}) \neq (\sqrt{E(X)})$$
$$E(b^X) \neq b^{E(X)}$$

and so forth.

The next few rules concern two (or more) random variables, symbolized by X and Y.

RULE 4. If X is a random variable with expectation $E(X)$, and Y is a random variable with expectation $E(Y)$, then

$$E(X + Y) = E(X) + E(Y).$$

Verbally, this rule says that the expectation of a sum of two random variables is the sum of their expectations. Once again, the proof is simple for two discrete variables X and Y. Consider the new random variable $(X + Y)$. The probability of a value of $(X + Y)$ involving a *particular* X value and a *particular* Y value is the joint probability $p(x, y)$. Thus,

$$E(X + Y) = \sum_x \sum_y (x + y)p(x,y).$$

Notice that here the expectation involves the sum over all possible *joint* events (x, y). This could be written as

$$E(X + Y) = \sum_x \sum_y (x + y)p(x,y)$$
$$= \sum_x \sum_y xp(x,y) + \sum_x \sum_y yp(x,y).$$

However, for any fixed x,

$$\sum_y p(x,y) = p(x)$$

and for any fixed y,

$$\sum_x p(x,y) = p(y).$$

Thus,

$$p(x + y) = \sum_x xp(x) + \sum_y yp(y) = E(X) + E(Y).$$

In particular, one of the random variables may be in a functional relation to the other. For example, let $Y = 3X^2$. Then

$$E(X + Y) = E(X + 3X^2)$$
$$= E(X) + E(3X^2)$$
$$= E(X) + 3E(X^2).$$

This principle lets one *distribute* the expectation over an expression which itself has the form of a sum. We will make a great deal of use of this principle.

This rule may also be extended to any finite number of random variables:

RULE 5. Given some finite number of random variables, the expectation of the sum of those variables is the sum of their individual expectations. Thus,

$$E(X + Y + Z) = E(X) + E(Y) + E(Z)$$

and so on.

In particular, some of the random variables may be in functional relations to others. Let $Y = 6X^4$, and let $Z = \sqrt{2X}$. Then

$$E(X + 6X^4 + \sqrt{2X}) = E(X) + 6E(X^4) + E(\sqrt{2X}).$$

The next rule is a most important one that applies only to *independent* random variables (see Chapter 4):

RULE 6. Given random variable X with expectation $E(X)$ and the random variable Y with expectation $E(Y)$, then if X and Y are independent

$$E(XY) = E(X)E(Y).$$

This rule states that if random variables are *statistically independent*, the expectation of the product of these variables is the product of their separate expectations. An important corollary to this principle is:

If $E(XY) \neq E(X)E(Y)$, the variables X and Y are not independent.

The basis for rule 6 can also be shown fairly simply for discrete variables. Since X and Y are independent, $p(x,y) = p(x)p(y)$. Then,

$$E(XY) = \sum_x \sum_y (xy)p(x)p(y) = \sum_x \sum_y xp(x)yp(y).$$

However, for any fixed x, $yp(y)$ is perfectly free to be any value, so that

$$E(XY) = \sum_x xp(x) \sum_y yp(y) = E(X)E(Y).$$

By extension of rule 6 to any finite number of random variables we have:

RULE 7. Given any finite number of random variables, if all the variables are independent of each other, the expectation of their product is the product of their separate expectations: thus,

$$E(XYZ) = E(X)E(Y)E(Z),$$

and so on.

Incidentally, we will have occasion to use the idea of the *covariance* of two random variables, defined by

$$\text{cov.}(X,Y) = E[(X - E(X))(Y - E(Y))].$$

When X and Y are independent,

$$\text{cov.}(X,Y) = E[X - E(X)]E[Y - E(Y)]$$

by the rule given above. Since it is obvious that

$$E[X - E(X)] = E(X) - E(X) = 0$$

from rule 1 and rule 3, it follows that *when random variables are independent; their covariance is zero.* However, it is not necessarily true that zero covariance implies that the variables are independent.

Two more useful rules involve the variance of a random variable. The variance is defined by

$$\text{var.}(X) = \sigma_X^2 = E[X - E(X)]^2$$
$$\sigma_X^2 = E(X^2) - [E(X)]^2.$$

or

RULE 8. If a is some constant real number, and if X is a random variable with expectation $E(X)$ and variance σ_X^2, then the random variable $(X + a)$ has variance σ_X^2.

This can be shown as follows:

$$\text{var.}(X + a) = E[(X + a)^2] - [E(X + a)]^2.$$

By rule 3 above, and expanding the squares, we have

$$E[(X + a)^2] - [E(X + a)]^2 = E[(X^2 + 2Xa + a^2)] - [E(X) + a]^2$$
$$= E(X^2 + 2aX + a^2) - [E(X)]^2 - 2aE(X) - a^2.$$

Then by rules 4 and 1, we have

$$\text{var.}(X + a) = E(X^2) + 2aE(X) + a^2 - [E(X)]^2 - 2aE(X) - a^2$$
$$= E(X^2) - [E(X)]^2$$
$$= \sigma_X^2.$$

RULE 9. If a is some constant real number, and if X is a random variable with variance σ_X^2, the variance of the random variable aX is

$$\text{var.}(aX) = a^2\sigma_X^2.$$

In order to show this, we take

$$\text{var.}(aX) = E[(aX)^2] - [E(aX)]^2$$
$$= a^2E(X^2) - a^2[E(X)]^2$$
$$= a^2(E(X^2) - [E(X)]^2)$$
$$= a^2\sigma_X^2.$$

In short, adding a constant value to each value of a random variable leaves the variance unchanged, but multiplying each value by a constant multiplies the variance by the square of the constant.

EXERCISES—THE ALGEBRA OF EXPECTATIONS

Consider the following probability distributions of discrete random variables. Find the expectation, $E(X)$, for each.

1. x	$p(x)$
1	.75
0	.25
	1.00

2. x	$p(x)$
5	4/18
0	12/18
-5	2/18
	18/18

3. x	$p(x)$
5	.1
4	.4
3	.1
2	.3
1	.1
	1.0

4. x	$p(x)$
399	.20
154	.20
125	.20
100	.20
-200	.20
	1.00

5. x	$p(x)$
36	.12
30	.18
24	.20
18	.32
12	.09
6	.05
0	.04
	1.00

Consider a discrete random variable taking on only the values $x_1, x_2, \cdots, x_{10}$. Symbolize the following as expectations:

6. $x_1 p(x_1) + x_2 p(x_2) + \cdots + x_{10} p(x_{10})$

7. $x_1^2 p(x_1) + x_2^2 p(x_2) + \cdots + x_{10}^2 p(x_{10})$

8. $[x_1^2 p(x_1) + x_2^2 p(x_2) + \cdots + x_{10}^2 p(x_{10})]$
$$- [x_1 p(x_1) + x_2 p(x_2) + \cdots + x_{10} p(x_{10})]^2$$

9. $[x_1 - E(X)]^2 p(x_1) + [x_2 - E(X)]^2 p(x_2) + \cdots + [x_{10} - E(X)]^2 p(x_{10})$

10. $(x_1^2 - 4x_1 + 5)p(x_1) + (x_2^2 - 4x_2 + 5)p(x_2)$
$$+ \cdots + (x_{10}^2 - 4x_{10} + 5)p(x_{10})$$

Consider two random variables, X and Y. Simplify the following:

11. $E[(X + 35)/10]$

12. $E[X - 14Y + E(Y) + 7] - E(X + Y - 5)$

13. $E[X^2 - 2XE(X) + E^2(X)]$

14. $E[X^2 + Y^2 - 2(X + Y)^2]$

15. $E(17)$

16. $E[(X - E(X))(Y - E(Y))]$

See if you can prove the following for discrete random variables: (Hint: turn each expectation into the equivalent weighted sum.)

17. $E(aX) = aE(X)$

18. $E(aX + b) = aE(X) + b$
19. $E[X - E(X)] = 0$

For the variables defined in problems 1, 2, and 3 above, find

20. $Var(X)$, or σ_X^2
21. $E(5X - 12)$
22. $E[(X^2 + 2)/10]$
23. Suppose that the random variables given in problems 3 and 4 above were independent. Let us call the variable in problem 3, X, and that in problem 4, Y. Then, find the value of $E(XY)$.
24. For X and Y as defined in 23 above, consider two new variables that are functions of X and Y:

$$W = X^2 + 1$$
$$Z = 3Y + 2.$$

Show, for Z and W so defined, that if X and Y are independent random variables, then

$$Cov(W, Z) = E(WZ) - E(W)E(Z) = 0.$$

(Hint: start by finding the distribution of W, the distribution of Z, and the *joint* distribution of W and Z, where the probability of a particular (W, Z) combination is just the probability of all of the different (X, Y) combinations leading to that (W, Z) combination.)

Appendix C:

Tables

Table I

CUMULATIVE NORMAL PROBABILITIES

z	$F(z)$	z	$F(z)$	z	$F(z)$	z	$F(z)$
.00	.5000000	.21	.5831662	.42	.6627573	.63	.7356527
.01	.5039894	.22	.5870604	.43	.6664022	.64	.7389137
.02	.5079783	.23	.5909541	.44	.6700314	.65	.7421539
.03	.5119665	.24	.5948349	.45	.6736448	.66	.7453731
.04	.5159534	.25	.5987063	.46	.6772419	.67	.7485711
.05	.5199388	.26	.6025681	.47	.6808225	.68	.7517478
.06	.5239222	.27	.6064199	.48	.6843863	.69	.7549029
.07	.5279032	.28	.6102612	.49	.6879331	.70	.7580363
.08	.5318814	.29	.6140919	.50	.6914625	.71	.7611479
.09	.5358564	.30	.6179114	.51	.6949743	.72	.7642375
.10	.5398278	.31	.6217195	.52	.6984682	.73	.7673049
.11	.5437953	.32	.6255158	.53	.7019440	.74	.7703500
.12	.5477584	.33	.6293000	.54	.7054015	.75	.7733726
.13	.5517168	.34	.6330717	.55	.7088403	.76	.7763727
.14	.5556700	.35	.6368307	.56	.7122603	.77	.7793501
.15	.5596177	.36	.6405764	.57	.7156612	.78	.7823046
.16	.5635595	.37	.6443088	.58	.7190427	.79	.7852361
.17	.5674949	.38	.6480273	.59	.7224047	.80	.7881446
.18	.5714237	.39	.6517317	.60	.7257469	.81	.7910299
.19	.5753454	.40	.6554217	.61	.7290691	.82	.7938919
.20	.5792597	.41	.6590970	.62	.7323711	.83	.7967306

Table I   879

Table I (continued)

z	$F(z)$	z	$F(z)$	z	$F(z)$	z	$F(z)$
.84	.7995458	1.32	.9065825	1.79	.9632730	2.26	.9880894
.85	.8023375	1.33	.9082409	1.80	.9640697	2.27	.9883962
.86	.8051055	1.34	.9098773	1.81	.9648521	2.28	.9886962
.87	.8078498	1.35	.9114920	1.82	.9656205	2.29	.9889893
.88	.8105703	1.36	.9130850	1.83	.9663750	2.30	.9892759
.89	.8132671	1.37	.9146565	1.84	.9671159	2.31	.9895559
.90	.8159399	1.38	.9162067	1.85	.9678432	2.32	.9898296
.91	.8185887	1.39	.9177356	1.86	.9685572	2.33	.9900969
.92	.8212136	1.40	.9192433	1.87	.9692581	2.34	.9903581
.93	.8238145	1.41	.9207302	1.88	.9699460	2.35	.9906133
.94	.8263912	1.42	.9221962	1.89	.9706210	2.36	.9908625
.95	.8289439	1.43	.9236415	1.90	.9712834	2.37	.9911060
.96	.8314724	1.44	.9250663	1.91	.9719334	2.38	.9913437
.97	.8339768	1.45	.9264707	1.92	.9725711	2.39	.9915758
.98	.8364569	1.46	.9278550	1.93	.9731966	2.40	.9918025
.99	.8389129	1.47	.9292191	1.94	.9738102	2.41	.9920237
1.00	.8413447	1.48	.9305634	1.95	.9744119	2.42	.9922397
1.01	.8437524	1.49	.9318879	1.96	.9750021	2.43	.9924506
1.02	.8461358	1.50	.9331928	1.97	.9755808	2.44	.9926564
1.03	.8484950	1.51	.9344783	1.98	.9761482	2.45	.9928572
1.04	.8508300	1.52	.9357445	1.99	.9767045	2.46	.9930531
1.05	.8531409	1.53	.9369916	2.00	.9772499	2.47	.9932443
1.06	.8554277	1.54	.9382198	2.01	.9777844	2.48	.9934309
1.07	.8576903	1.55	.9394292	2.02	.9783083	2.49	.9936128
1.08	.8599289	1.56	.9406201	2.03	.9788217	2.50	.9937903
1.09	.8621434	1.57	.9417924	2.04	.9793248	2.51	.9939634
1.10	.8643339	1.58	.9429466	2.05	.9798178	2.52	.9941323
1.11	.8665005	1.59	.9440826	2.06	.9803007	2.53	.9942969
1.12	.8686431	1.60	.9452007	2.07	.9807738	2.54	.9944574
1.13	.8707619	1.61	.9463011	2.08	.9812372	2.55	.9946139
1.14	.8728568	1.62	.9473839	2.09	.9816911	2.56	.9947664
1.15	.8749281	1.63	.9484493	2.10	.9821356	2.57	.9949151
1.16	.8769756	1.64	.9494974	2.11	.9825708	2.58	.9950600
1.17	.8789995	1.65	.9505285	2.12	.9829970	2.59	.9952012
1.18	.8809999	1.66	.9515428	2.13	.9834142	2.60	.9953388
1.19	.8829768	1.67	.9525403	2.14	.9838226	2.70	.9965330
1.20	.8849303	1.68	.9535213	2.15	.9842224	2.80	.9974449
1.21	.8868606	1.69	.9544860	2.16	.9846137	2.90	.9981342
1.22	.8887676	1.70	.9554345	2.17	.9849966	3.00	.9986501
1.23	.8906514	1.71	.9563671	2.18	.9853713	3.20	.9993129
1.24	.8925123	1.72	.9572838	2.19	.9857379	3.40	.9996631
1.25	.8943502	1.73	.9581849	2.20	.9860966	3.60	.9998409
1.26	.8961653	1.74	.9590705	2.21	.9864474	3.80	.9999277
1.27	.8979577	1.75	.9599408	2.22	.9867906	4.00	.9999683
1.28	.8997274	1.76	.9607961	2.23	.9871263	4.50	.9999966
1.29	.9014747	1.77	.9616364	2.24	.9874545	5.00	.9999997
1.30	.9031995	1.78	.9624620	2.25	.9877755	5.50	.9999999
1.31	.9049021						

$p - P(r)$

Table II
BINOMIAL PROBABILITIES $\binom{N}{r}p^r q^{N-r}$

of successes

prob of at least
3 are

N=5
at least

							p				
N	r	.05	.10	.15	.20	.25	.30	.35	.40	.45	.50
1	0	.9500	.9000	.8500	.8000	.7500	.7000	.6500	.6000	.5500	.5000
	1	.0500	.1000	.1500	.2000	.2500	.3000	.3500	.4000	.4500	.5000
2	0	.9025	.8100	.7225	.6400	.5625	.4900	.4225	.3600	.3025	.2500
	1	.0950	.1800	.2550	.3200	.3750	.4200	.4550	.4800	.4950	.5000
	2	.0025	.0100	.0225	.0400	.0625	.0900	.1225	.1600	.2025	.2500
3	0	.8574	.7290	.6141	.5120	.4219	.3430	.2746	.2160	.1664	.1250
	1	.1354	.2430	.3251	.3840	.4219	.4410	.4436	.4320	.4084	.3750
	2	.0071	.0270	.0574	.0960	.1406	.1890	.2389	.2880	.3341	.3750
	3	.0001	.0010	.0034	.0080	.0156	.0270	.0429	.0640	.0911	.1250
4	0	.8145	.6561	.5220	.4096	.3164	.2401	.1785	.1296	.0915	.0625
	1	.1715	.2916	.3685	.4096	.4219	.4116	.3845	.3456	.2995	.2500
	2	.0135	.0486	.0975	.1536	.2109	.2646	.3105	.3456	.3675	.3750
	3	.0005	.0036	.0115	.0256	.0469	.0756	.1115	.1536	.2005	.2500
	4	.0000	.0001	.0005	.0016	.0039	.0081	.0150	.0256	.0410	.0625
5	0	.7738	.5905	.4437	.3277	.2373	.1681	.1160	.0778	.0503	.0312
	1	.2036	.3280	.3915	.4096	.3955	.3602	.3124	.2592	.2059	.1562
	2	.0214	.0729	.1382	.2048	.2637	.3087	.3364	.3456	.3369	.3125
	3	.0011	.0081	.0244	.0512	.0879	.1323	.1811	.2304	.2757	.3125
	4	.0000	.0004	.0022	.0064	.0146	.0284	.0488	.0768	.1128	.1562
	5	.0000	.0000	.0001	.0003	.0010	.0024	.0053	.0102	.0185	.0312
6	0	.7351	.5314	.3771	.2621	.1780	.1176	.0754	.0467	.0277	.0156
	1	.2321	.3543	.3993	.3932	.3560	.3025	.2437	.1866	.1359	.0938
	2	.0305	.0984	.1762	.2458	.2966	.3241	.3280	.3110	.2780	.2344
	3	.0021	.0146	.0415	.0819	.1318	.1852	.2355	.2765	.3032	.3125
	4	.0001	.0012	.0055	.0154	.0330	.0595	.0951	.1382	.1861	.2344
	5	.0000	.0001	.0004	.0015	.0044	.0102	.0205	.0369	.0609	.0938
	6	.0000	.0000	.0000	.0001	.0002	.0007	.0018	.0041	.0083	.0156
7	0	.6983	.4783	.3206	.2097	.1335	.0824	.0490	.0280	.0152	.0078
	1	.2573	.3720	.3960	.3670	.3115	.2471	.1848	.1306	.0872	.0547
	2	.0406	.1240	.2097	.2753	.3115	.3177	.2985	.2613	.2140	.1641
	3	.0036	.0230	.0617	.1147	.1730	.2269	.2679	.2903	.2918	.2734
	4	.0002	.0026	.0109	.0287	.0577	.0972	.1442	.1935	.2388	.2734
	5	.0000	.0002	.0012	.0043	.0115	.0250	.0466	.0774	.1172	.1641
	6	.0000	.0000	.0001	.0004	.0013	.0036	.0084	.0172	.0320	.0547
	7	.0000	.0000	.0000	.0000	.0001	.0002	.0006	.0016	.0037	.0078
8	0	.6634	.4305	.2725	.1678	.1001	.0576	.0319	.0168	.0084	.0039
	1	.2793	.3826	.3847	.3355	.2760	.1977	.1373	.0896	.0548	.0312
	2	.0515	.1488	.2376	.2936	.3115	.2965	.2587	.2090	.1569	.1094
	3	.0054	.0331	.0839	.1468	.2076	.2541	.2786	.2787	.2568	.2188
	4	.0004	.0046	.0185	.0459	.0865	.1361	.1875	.2322	.2627	.2734
	5	.0000	.0004	.0026	.0092	.0231	.0467	.0808	.1239	.1719	.2188
	6	.0000	.0000	.0002	.0011	.0038	.0100	.0217	.0413	.0703	.1094
	7	.0000	.0000	.0000	.0001	.0004	.0012	.0033	.0079	.0164	.0312
	8	.0000	.0000	.0000	.0000	.0000	.0001	.0002	.0007	.0017	.0039

Table II 881

Table II (continued)

N	r	.05	.10	.15	.20	.25	.30	.35	.40	.45	.50
9	0	.6302	.3874	.2316	.1342	.0751	.0404	.0277	.0101	.0046	.0020
	1	.2985	.3874	.3679	.3020	.2253	.1556	.1004	.0605	.0339	.0176
	2	.0629	.1722	.2597	.3020	.3003	.2668	.2162	.1612	.1110	.0703
	3	.0077	.0446	.1069	.1762	.2336	.2668	.2716	.2508	.2119	.1641
	4	.0006	.0074	.0283	.0661	.1168	.1715	.2194	.2508	.2600	.2461
	5	.0000	.0008	.0050	.0165	.0389	.0735	.1181	.1672	.2128	.2461
	6	.0000	.0001	.0006	.0028	.0087	.0210	.0424	.0743	.1160	.1641
	7	.0000	.0000	.0000	.0003	.0012	.0039	.0098	.0212	.0407	.0703
	8	.0000	.0000	.0000	.0000	.0001	.0004	.0013	.0035	.0083	.0176
	9	.0000	.0000	.0000	.0000	.0000	.0000	.0001	.0003	.0008	.0020
10	0	.5987	.3487	.1969	.1074	.0563	.0282	.0135	.0060	.0025	.0010
	1	.3151	.3874	.3474	.2684	.1877	.1211	.0725	.0403	.0207	.0098
	2	.0746	.1937	.2759	.3020	.2816	.2335	.1757	.1209	.0763	.0439
	3	.0105	.0574	.1298	.2013	.2503	.2668	.2522	.2150	.1665	.1172
	4	.0010	.0112	.0401	.0881	.1460	.2001	.2377	.2508	.2384	.2051
	5	.0001	.0015	.0085	.0264	.0584	.1029	.1536	.2007	.2340	.2461
	6	.0000	.0001	.0012	.0055	.0162	.0368	.0689	.1115	.1596	.2051
	7	.0000	.0000	.0001	.0008	.0031	.0090	.0212	.0425	.0746	.1172
	8	.0000	.0000	.0000	.0001	.0004	.0014	.0043	.0106	.0229	.0439
	9	.0000	.0000	.0000	.0000	.0000	.0001	.0005	.0016	.0042	.0098
	10	.0000	.0000	.0000	.0000	.0000	.0000	.0000	.0001	.0003	.0016
11	0	.5688	.3138	.1673	.0859	.0422	.0198	.0088	.0036	.0014	.0005
	1	.3293	.3835	.3248	.2362	.1549	.0932	.0518	.0266	.0125	.0054
	2	.0867	.2131	.2866	.2953	.2581	.1998	.1395	.0887	.0513	.0269
	3	.0137	.0710	.1517	.2215	.2581	.2568	.2254	.1774	.1259	.0806
	4	.0014	.0158	.0536	.1107	.1721	.2201	.2428	.2365	.2060	.1611
	5	.0001	.0025	.0132	.0388	.0803	.1231	.1830	.2207	.2360	.2256
	6	.0000	.0003	.0023	.0097	.0268	.0566	.0985	.1471	.1931	.2256
	7	.0000	.0000	.0003	.0017	.0064	.0173	.0379	.0701	.1128	.1611
	8	.0000	.0000	.0000	.0002	.0011	.0037	.0102	.0234	.0462	.0806
	9	.0000	.0000	.0000	.0000	.0001	.0005	.0018	.0052	.0126	.0269
	10	.0000	.0000	.0000	.0000	.0000	.0000	.0002	.0007	.0021	.0054
	11	.0000	.0000	.0000	.0000	.0000	.0000	.0000	.0000	.0002	.0005
12	0	.5404	.2824	.1422	.0687	.0317	.0138	.0057	.0022	.0008	.0002
	1	.3413	.3766	.3012	.2062	.1267	.0712	.0368	.0174	.0075	.0029
	2	.0988	.2301	.2924	.2835	.2323	.1678	.1088	.0639	.0339	.0161
	3	.0173	.0852	.1720	.2362	.2581	.2397	.1954	.1419	.0923	.0537
	4	.0021	.0213	.0683	.1329	.1936	.2311	.2367	.2128	.1700	.1208
	5	.0002	.0038	.0193	.0532	.1032	.1585	.2039	.2270	.2225	.1934
	6	.0000	.0005	.0040	.0155	.0401	.0792	.1281	.1766	.2124	.2256
	7	.0000	.0000	.0006	.0033	.0115	.0291	.0591	.1009	.1489	.1934
	8	.0000	.0000	.0001	.0005	.0024	.0078	.0199	.0420	.0762	.1208
	9	.0000	.0000	.0000	.0001	.0004	.0015	.0048	.0125	.0277	.0537
	10	.0000	.0000	.0000	.0000	.0000	.0002	.0008	.0025	.0068	.0161
	11	.0000	.0000	.0000	.0000	.0000	.0000	.0001	.0003	.0010	.0029
	12	.0000	.0000	.0000	.0000	.0000	.0000	.0000	.0000	.0001	.0002

Table II (continued)

N	r	.05	.10	.15	.20	.25	.30	.35	.40	.45	.50
13	0	.5133	.2542	.1209	.0550	.0238	.0097	.0037	.0013	.0004	.0001
	1	.3512	.3672	.2774	.1787	.1029	.0540	.0259	.0113	.0045	.0016
	2	.1109	.2448	.2937	.2680	.2059	.1388	.0836	.0453	.0220	.0095
	3	.0214	.0997	.1900	.2457	.2517	.2181	.1651	.1107	.0660	.0349
	4	.0028	.0277	.0838	.1535	.2097	.2337	.2222	.1845	.1350	.0873
	5	.0003	.0055	.0266	.0691	.1258	.1803	.2154	.2214	.1989	.1571
	6	.0000	.0008	.0063	.0230	.0559	.1030	.1546	.1968	.2169	.2095
	7	.0000	.0001	.0011	.0058	.0186	.0442	.0833	.1312	.1775	.2095
	8	.0000	.0000	.0001	.0011	.0047	.0142	.0336	.0656	.1089	.1571
	9	.0000	.0000	.0000	.0001	.0009	.0034	.0101	.0243	.0495	.0873
	10	.0000	.0000	.0000	.0000	.0001	.0006	.0022	.0065	.0162	.0349
	11	.0000	.0000	.9000	.0000	.0000	.0001	.0003	.0012	.0036	.0095
	12	.0000	.0000	.0000	.0000	.0000	.0000	.0000	.0001	.0005	.0016
	13	.0000	.0000	.0000	.0000	.0000	.0000	.0000	.0000	.0000	.0001
14	0	.4877	.2288	.1028	.0440	.0178	.0068	.0024	.0008	.0002	.0001
	1	.3593	.3559	.2539	.1539	.0832	.0407	.0181	.0073	.0027	.0009
	2	.1229	.2570	.2912	.2501	.1802	.1134	.0634	.0317	.0141	.0056
	3	.0259	.1142	.2056	.2501	.2402	.1943	.1366	.0845	.0462	.0222
	4	.0037	.0349	.0998	.1720	.2202	.2290	.2022	.1549	.1040	.0611
	5	.0004	.0078	.0352	.0860	.1468	.1963	.2178	.2066	.1701	.1222
	6	.0000	.0013	.0093	.0322	.0734	.1262	.1759	.2066	.2088	.1833
	7	.0000	.0002	.0019	.0092	.0280	.0618	.1082	.1574	.1952	.2095
	8	.0000	.0000	.0003	.0020	.0082	.0232	.0510	.0918	.1398	.1833
	9	.0000	.0000	.0000	.0003	.0018	.0066	.0183	.0408	.0762	.1222
	10	.0000	.0000	.0000	.0000	.0003	.0014	.0049	.0136	.0312	.0611
	11	.0000	.0000	.0000	.0000	.0000	.0002	.0010	.0033	.0093	.0222
	12	.0000	.0000	.0000	.0000	.0000	.0000	.0001	.0005	.0019	.0056
	13	.0000	.0000	.0000	.0000	.0000	.0000	.0000	.0001	.0002	.0009
	14	.0000	.0000	.0000	.0000	.0000	.0000	.0000	.0000	.0000	.0001
15	0	.4633	.2059	.0874	.0352	.0134	.0047	.0016	.0005	.0001	.0000
	1	.3658	.3432	.2312	.1319	.0668	.0305	.0126	.0047	.0016	.0005
	2	.1348	.2669	.2856	.2309	.1559	.0916	.0476	.0219	.0090	.0032
	3	.0307	.1285	.2184	.2501	.2252	.1700	.1110	.0634	.0318	.0139
	4	.0049	.0428	.1156	.1876	.2252	.2186	.1792	.1268	.0780	.0417
	5	.0006	.0105	.0449	.1032	.1651	.2061	.2123	.1859	.1404	.0916
	6	.0000	.0019	.0132	.0430	.0917	.1472	.1906	.2066	.1914	.1527
	7	.0000	.0003	.0030	.0138	.0393	.0811	.1319	.1771	.2013	.1964
	8	.0000	.0000	.0005	.0035	.0131	.0348	.0710	.1181	.1647	.1964
	9	.0000	.0000	.0001	.0007	.0034	.0116	.0298	.0612	.1048	.1527
	10	.0000	.0000	.0000	.0001	.0007	.0030	.0096	.0245	.0515	.0916
	11	.0000	.0000	.0000	.0000	.0001	.0006	.0024	.0074	.0191	.0417
	12	.0000	.0000	.0000	.0000	.0000	.0001	.0004	.0016	.0052	.0139
	13	.0000	.0000	.0000	.0000	.0000	.0000	.0001	.0003	.0010	.0032
	14	.0000	.0000	.0000	.0000	.0000	.0000	.0000	.0000	.0001	.0005
	15	.0000	.0000	.0000	.0000	.0000	.0000	.0000	.0000	.0000	.0000
16	0	.4401	.1853	.0743	.0281	.0100	.0033	.0010	.0003	.0001	.0000
	1	.3706	.3294	.2097	.1126	.0535	.0228	.0087	.0030	.0009	.0002
	2	.1463	.2745	.2775	.2111	.1336	.0732	.0353	.0150	.0056	.0018

Table II　883

Table II (continued)

N	r	.05	.10	.15	.20	.25	.30	.35	.40	.45	.50
16	3	.0359	.1423	.2285	.2463	.2079	.1465	.0888	.0468	.0215	.0085
	4	.0061	.0514	.1311	.2001	.2252	.2040	.1553	.1014	.0572	.0278
	5	.0008	.0137	.0555	.1201	.1802	.2099	.2008	.1623	.1123	.0667
	6	.0001	.0028	.0180	.0550	.1101	.1649	.1982	.1983	.1684	.1222
	7	.0000	.0004	.0045	.0197	.0524	.1010	.1524	.1889	.1969	.1746
	8	.0000	.0001	.0009	.0055	.0197	.0487	.0923	.1417	.1812	.1964
	9	.0000	.0000	.0001	.0012	.0058	.0185	.0442	.0840	.1318	.1746
	10	.0000	.0000	.0000	.0002	.0014	.0056	.0167	.0392	.0755	.1222
	11	.0000	.0000	.0000	.0000	.0002	.0013	.0049	.0142	.0337	.0667
	12	.0000	.0000	.0000	.0000	.0000	.0002	.0011	.0040	.0115	.0278
	13	.0000	.0000	.0000	.0000	.0000	.0000	.0002	.0008	.0029	.0085
	14	.0000	.0000	.0000	.0000	.0000	.0000	.0000	.0001	.0005	.0018
	15	.0000	.0000	.0000	.0000	.0000	.0000	.0000	.0000	.0001	.0002
	16	.0000	.0000	.0000	.0000	.0000	.0000	.0000	.0000	.0000	.0000
17	0	.4181	.1668	.0631	.0225	.0075	.0023	.0007	.0002	.0000	.0000
	1	.3741	.3150	.1893	.0957	.0426	.0169	.0060	.0019	.0005	.0001
	2	.1575	.2800	.2673	.1914	.1136	.0581	.0260	.0102	.0035	.0010
	3	.0415	.1556	.2359	.2393	.1893	.1245	.0701	.0341	.0144	.0052
	4	.0076	.0605	.1457	.2093	.2209	.1868	.1320	.0796	.0411	.0182
	5	.0010	.0175	.0668	.1361	.1914	.2081	.1849	.1379	.0875	.0472
	6	.0001	.0039	.0236	.0680	.1276	.1784	.1991	.1839	.1432	.0944
	7	.0000	.0007	.0065	.0267	.0668	.1201	.1685	.1927	.1841	.1484
	8	.0000	.0001	.0014	.0084	.0279	.0644	.1143	.1606	.1883	.1855
	9	.0000	.0000	.0003	.0021	.0093	.0276	.0611	.1070	.1540	.1855
	10	.0000	.0000	.0000	.0004	.0025	.0095	.0263	.0571	.1008	.1484
	11	.0000	.0000	.0000	.0001	.0005	.0026	.0090	.0242	.0525	.0944
	12	.0000	.0000	.0000	.0000	.0001	.0006	.0024	.0081	.0215	.0472
	13	.0000	.0000	.0000	.0000	.0000	.0001	.0005	.0021	.0068	.0182
	14	.0000	.0000	.0000	.0000	.0000	.0000	.0001	.0004	.0016	.0052
	15	.0000	.0000	.0000	.0000	.0000	.0000	.0000	.0001	.0003	.0010
	16	.0000	.0000	.0000	.0000	.0000	.0000	.0000	.0000	.0000	.0001
	17	.0000	.0000	.0000	.0000	.0000	.0000	.0000	.0000	.0000	.0000
18	0	.3972	.1501	.0536	.0180	.0056	.0016	.0004	.0001	.0000	.0000
	1	.3763	.3002	.1704	.0811	.0338	.0126	.0042	.0012	.0003	.0001
	2	.1683	.2835	.2556	.1723	.0958	.0458	.0190	.0069	.0022	.0006
	3	.0473	.1680	.2406	.2297	.1704	.1046	.0547	.0246	.0095	.0031
	4	.0093	.0700	.1592	.2153	.2130	.1681	.1104	.0614	.0291	.0117
	5	.0014	.0218	.0787	.1507	.1988	.2017	.1664	.1146	.0666	.0327
	6	.0002	.0052	.0310	.0816	.1436	.1873	.1941	.1655	.1181	.0708
	7	.0000	.0010	.0091	.0350	.0820	.1376	.1792	.1892	.1657	.1214
	8	.0000	.0002	.0022	.0120	.0376	.0811	.1327	.1734	.1864	.1669
	9	.0000	.0000	.0004	.0033	.0139	.0386	.0794	.1284	.1694	.1855
18	10	.0000	.0000	.0001	.0008	.0042	.0149	.0385	.0771	.1248	.1669
	11	.0000	.0000	.0000	.0001	.0010	.0046	.0151	.0374	.0742	.1214
	12	.0000	.0000	.0000	.0000	.0002	.0012	.0047	.0145	.0354	.0708
	13	.0000	.0000	.0000	.0000	.0000	.0002	.0012	.0045	.0134	.0327
	14	.0000	.0000	.0000	.0000	.0000	.0000	.0002	.0011	.0039	.0117

Table II (continued)

N	r	.05	.10	.15	.20	.25	.30	.35	.40	.45	.50
						p					
	15	.0000	.0000	.0000	.0000	.0000	.0000	.0000	.0002	.0009	.0031
	16	.0000	.0000	.0000	.0000	.0000	.0000	.0000	.0000	.0001	.0006
	17	.0000	.0000	.0000	.0000	.0000	.0000	.0000	.0000	.0000	.0001
	18	.0000	.0000	.0000	.0000	.0000	.0000	.0000	.0000	.0000	.0000
19	0	.3774	.1351	.0456	.0144	.0042	.0011	.0003	.0001	.0000	.0000
	1	.3774	.2852	.1529	.0685	.0268	.0093	.0029	.0008	.0002	.0000
	2	.1787	.2852	.2428	.1540	.0803	.0358	.0138	.0046	.0013	.0003
	3	.0533	.1796	.2428	.2182	.1517	.0869	.0422	.0175	.0062	.0018
	4	.0112	.0798	.1714	.2182	.2023	.1491	.0909	.0467	.0203	.0074
	5	.0018	.0266	.0907	.1636	.2023	.1916	.1468	.0933	.0497	.0222
	6	.0002	.0069	.0374	.0955	.1574	.1916	.1844	.1451	.0949	.0518
	7	.0000	.0014	.0122	.0443	.0974	.1525	.1844	.1797	.1443	.0961
	8	.0000	.0002	.0032	.0166	.0487	.0981	.1489	.1797	.1771	.1442
	9	.0000	.0000	.0007	.0051	.0198	.0514	.0980	.1464	.1771	.1762
	10	.0000	.0000	.0001	.0013	.0066	.0220	.0528	.0976	.1449	.1762
	11	.0000	.0000	.0000	.0003	.0018	.0077	.0233	.0532	.0970	.1442
	12	.0000	.0000	.0000	.0000	.0004	.0022	.0083	.0237	.0529	.0961
	13	.0000	.0000	.0000	.0000	.0001	.0005	.0024	.0085	.0233	.0518
	14	.0000	.0000	.0000	.0000	.0000	.0001	.0006	.0024	.0082	.0222
	15	.0000	.0000	.0000	.0000	.0000	.0000	.0001	.0005	.0022	.0074
	16	.0000	.0000	.0000	.0000	.0000	.0000	.0000	.0001	.0005	.0018
	17	.0000	.0000	.0000	.0000	.0000	.0000	.0000	.0000	.0001	.0003
	18	.0000	.0000	.0000	.0000	.0000	.0000	.0000	.0000	.0000	.0000
	19	.0000	.0000	.0000	.0000	.0000	.0000	.0000	.0000	.0000	.0000
20	0	.3585	.1216	.0388	.0115	.0032	.0008	.0002	.0000	.0000	.0000
	1	.3774	.2702	.1368	.0576	.0211	.0068	.0020	.0005	.0001	.0000
	2	.1887	.2852	.2293	.1369	.0669	.0278	.0100	.0031	.0008	.0002
	3	.0596	.1901	.2428	.2054	.1339	.0716	.0323	.0123	.0040	.0011
	4	.0133	.0898	.1821	.2182	.1897	.1304	.0738	.0350	.0139	.0046
	5	.0022	.0319	.1028	.1746	.2023	.1789	.1272	.0746	.0365	.0148
	6	.0003	.0089	.0454	.1091	.1686	.1916	.1712	.1244	.0746	.0370
	7	.0000	.0020	.0160	.0545	.1124	.1643	.1844	.1659	.1221	.0739
	8	.0000	.0004	.0046	.0222	.0609	.1144	.1614	.1797	.1623	.1201
	9	.0000	.0001	.0011	.0074	.0271	.0654	.1158	.1597	.1771	.1602
	10	.0000	.0000	.0002	.0020	.0099	.0308	.0686	.1171	.1593	.1762
	11	.0000	.0000	.0000	.0005	.0030	.0120	.0336	.0710	.1185	.1602
	12	.0000	.0000	.0000	.0001	.0008	.0039	.0136	.0355	.0727	.1201
	13	.0000	.0000	.0000	.0000	.0002	.0010	.0045	.0146	.0366	.0739
	14	.0000	.0000	.0000	.0000	.0000	.0002	.0012	.0049	.0150	.0370
	15	.0000	.0000	.0000	.0000	.0000	.0000	.0003	.0013	.0049	.0148
	16	.0000	.0000	.0000	.0000	.0000	.0000	.0000	.0003	.0013	.0046
	17	.0000	.0000	.0000	.0000	.0000	.0000	.0000	.0000	.0002	.0011
	18	.0000	.0000	.0000	.0000	.0000	.0000	.0000	.0000	.0000	.0002
	19	.0000	.0000	.0000	.0000	.0000	.0000	.0000	.0000	.0000	.0000
	20	.0000	.0000	.0000	.0000	.0000	.0000	.0000	.0000	.0000	.0000

This table is reproduced by permission from R. S. Burington and D. C. May, *Handbook of Probability and Statistics with Tables.* McGraw-Hill Book Company, (ed. 2), 1970.

Table III   885

Table III
UPPER PERCENTAGE POINTS OF THE t DISTRIBUTION

ν	$Q = 0.4$ $2Q = 0.8$	0.25 0.5	0.1 0.2	0.05 0.1	0.025 0.05	0.01 0.02	0.005 0.01	0.001 0.002
1	0.325	1.000	3.078	6.314	12.706	31.821	63.657	318.31
2	.289	0.816	1.886	2.920	4.303	6.965	9.925	22.326
3	.277	.765	1.638	2.353	3.182	4.541	5.841	10.213
4	.271	.741	1.533	2.132	2.776	3.747	4.604	7.173
5	0.267	0.727	1.476	2.015	2.571	3.365	4.032	5.893
6	.265	.718	1.440	1.943	2.447	3.143	3.707	5.208
7	.263	.711	1.415	1.895	2.365	2.998	3.499	4.785
8	.262	.706	1.397	1.860	2.306	2.896	3.355	4.501
9	.261	.703	1.383	1.833	2.262	2.821	3.250	4.297
10	0.260	0.700	1.372	1.812	2.228	2.764	3.169	4.144
11	.260	.697	1.363	1.796	2.201	2.718	3.106	4.025
12	.259	.695	1.356	1.782	2.179	2.681	3.055	3.930
13	.259	.694	1.350	1.771	2.160	2.650	3.012	3.852
14	.258	.692	1.345	1.761	2.145	2.624	2.977	3.787
15	0.258	0.691	1.341	1.753	2.131	2.602	2.947	3.733
16	.258	.690	1.337	1.746	2.120	2.583	2.921	3.686
17	.257	.689	1.333	1.740	2.110	2.567	2.898	3.646
18	.257	.688	1.330	1.734	2.101	2.552	2.878	3.610
19	.257	.688	1.328	1.729	2.093	2.539	2.861	3.579
20	0.257	0.687	1.325	1.725	2.086	2.528	2.845	3.552
21	.257	.686	1.323	1.721	2.080	2.518	2.831	3.527
22	.256	.686	1.321	1.717	2.074	2.508	2.819	3.505
23	.256	.685	1.319	1.714	2.069	2.500	2.807	3.485
24	.256	.685	1.318	1.711	2.064	2.492	2.797	3.467
25	0.256	0.684	1.316	1.708	2.060	2.485	2.787	3.450
26	.256	.684	1.315	1.706	2.056	2.479	2.779	3.435
27	.256	.684	1.314	1.703	2.052	2.473	2.771	3.421
28	.256	.683	1.313	1.701	2.048	2.467	2.763	3.408
29	.256	.683	1.311	1.699	2.045	2.462	2.756	3.396
30	0.256	0.683	1.310	1.697	2.042	2.457	2.750	3.385
40	.255	.681	1.303	1.684	2.021	2.423	2.704	3.307
60	.254	.679	1.296	1.671	2.000	2.390	2.660	3.232
120	.254	.677	1.289	1.658	1.980	2.358	2.617	3.160
∞	.253	.674	1 282	1.645	1.960	2.326	2.576	3.090

Table IV
UPPER PERCENTAGE POINTS OF THE χ^2 DISTRIBUTION

ν \ Q	0.995	0.990	0.975	0.950	0.900	0.750	0.500
1	392704.10^{-10}	157088.10^{-9}	982069.10^{-9}	393214.10^{-8}	0.0157908	0.1015308	0.454937
2	0.0100251	0.0201007	0.0506356	0.102587	0.210720	0.575364	1.38629
3	0.0717212	0.114832	0.215795	0.351846	0.584375	1.212534	2.36597
4	0.206990	0.297110	0.484419	0.710721	1.063623	1.92255	3.35670
5	0.411740	0.554300	0.831211	1.145476	1.61031	2.67460	4.35146
6	0.675727	0.872085	1.237347	1.63539	2.20413	3.45460	5.34812
7	0.989265	1.239043	1.68987	2.16735	2.83311	4.25485	6.34581
8	1.344419	1.646482	2.17973	2.73264	3.48954	5.07064	7.34412
9	1.734926	2.087912	2.70039	3.32511	4.16816	5.89883	8.34283
10	2.15585	2.55821	3.24697	3.94030	4.86518	6.73720	9.34182
11	2.60321	3.05347	3.81575	4.57481	5.57779	7.58412	10.3410
12	3.07382	3.57056	4.40379	5.22603	6.30380	8.43842	11.3403
13	3.56503	4.10691	5.00874	5.89186	7.04150	9.29906	12.3398
14	4.07468	4.66043	5.62872	6.57063	7.78953	10.1653	13.3393
15	4.60094	5.22935	6.26214	7.26094	8.54675	11.0365	14.3389
16	5.14224	5.81221	6.90766	7.96164	9.31223	11.9122	15.3385
17	5.69724	6.40776	7.56418	8.67176	10.0852	12.7919	16.3381
18	6.26481	7.01491	8.23075	9.39046	10.8649	13.6753	17.3379
19	6.84398	7.63273	8.90655	10.1170	11.6509	14.5620	18.3376
20	7.43386	8.26040	9.59083	10.8508	12.4426	15.4518	19.3374
21	8.03366	8.89720	10.28293	11.5913	13.2396	16.3444	20.3372
22	8.64272	9.54249	10.9823	12.3380	14.0415	17.2396	21.3370
23	9.26042	10.19567	11.6885	13.0905	14.8479	18.1373	22.3369
24	9.88623	10.8564	12.4011	13.8484	15.6587	19.0372	23.3367
25	10.5197	11.5240	13.1197	14.6114	16.4734	19.9393	24.3366
26	11.1603	12.1981	13.8439	15.3791	17.2919	20.8434	25.3364
27	11.8076	12.8786	14.5733	16.1513	18.1138	21.7494	26.3363
28	12.4613	13.5648	15.3079	16.9279	18.9392	22.6572	27.3363
29	13.1211	14.2565	16.0471	17.7083	19.7677	23.5666	28.3362
30	13.7867	14.9535	16.7908	18.4926	20.5992	24.4776	29.3360
40	20.7065	22.1643	24.4331	26.5093	29.0505	33.6603	39.3354
50	27.9907	29.7067	32.3574	34.7642	37.6886	42.9421	49.3349
60	35.5346	37.4848	40.4817	43.1879	46.4589	52.2938	59.3347
70	43.2752	45.4418	48.7576	51.7393	55.3290	61.6983	69.3344
80	51.1720	53.5400	57.1532	60.3915	64.2778	71.1445	79.3343
90	59.1963	61.7541	65.6466	69.1260	73.2912	80.6247	89.3342
100	67.3276	70.0648	74.2219	77.9295	82.3581	90.1332	99.3341
z_Q	-2.5758	-2.3263	-1.9600	-1.6449	-1.2816	-0.6745	0.0000

Table IV   887

Table IV (continued)

Q / ν	0.250	0.100	0.050	0.025	0.010	0.005	0.001
1	1.32330	2.70554	3.84146	5.02389	6.63490	7.87944	10.828
2	2.77259	4.60517	5.99147	7.37776	9.21034	10.5966	13.816
3	4.10835	6.25139	7.81473	9.34840	11.3449	12.8381	16.266
4	5.38527	7.77944	9.48773	11.1433	13.2767	14.8602	18.467
5	6.62568	9.23635	11.0705	12.8325	15.0863	16.7496	20.515
6	7.84080	10.6446	12.5916	14.4494	16.8119	18.5476	22.458
7	9.03715	12.0170	14.0671	16.0128	18.4753	20.2777	24.322
8	10.2188	13.3616	15.5073	17.5346	20.0902	21.9550	26.125
9	11.3887	14.6837	16.9190	19.0228	21.6660	23.5893	27.877
10	12.5489	15.9871	18.3070	20.4831	23.2093	25.1882	29.588
11	13.7007	17.2750	19.6751	21.9200	24.7250	26.7569	31.264
12	14.8454	18.5494	21.0261	23.3367	26.2170	28.2995	32.909
13	15.9839	19.8119	22.3621	24.7356	27.6883	29.8194	34.528
14	17.1170	21.0642	23.6848	26.1190	29.1413	31.3193	36.123
15	18.2451	22.3072	24.9958	27.4884	30.5779	32.8013	37.697
16	19.3688	23.5418	26.2962	28.8454	31.9999	34.2672	39.252
17	20.4887	24.7690	27.5871	30.1910	33.4087	35.7185	40.790
18	21.6049	25.9894	28.8693	31.5264	34.8053	37.1564	42.312
19	22.7178	27.2036	30.1435	32.8523	36.1908	38.5822	43.820
20	23.8277	28.4120	31.4104	34.1696	37.5662	39.9968	45.315
21	24.9348	29.6151	32.6705	35.4789	38.9321	41.4010	46.797
22	26.0393	30.8133	33.9244	36.7807	40.2894	42.7956	48.268
23	27.1413	32.0069	35.1725	38.0757	41.6384	44.1813	49.728
24	28.2412	33.1963	36.4151	39.3641	42.9798	45.5585	51.179
25	29.3389	34.3816	37.6525	40.6465	44.3141	46.9278	52.620
26	30.4345	35.5631	38.8852	41.9232	45.6417	48.2899	54.052
27	31.5284	36.7412	40.1133	43.1944	46.9630	49.6449	55.476
28	32.6205	37.9159	41.3372	44.4607	48.2782	50.9933	56.892
29	33.7109	39.0875	42.5569	45.7222	49.5879	52.3356	58.302
30	34.7998	40.2560	43.7729	46.9792	50.8922	53.6720	59.703
40	45.6160	51.8050	55.7585	59.3417	63.6907	66.7659	73.402
50	56.3336	63.1671	67.5048	71.4202	76.1539	79.4900	86.661
60	66.9814	74.3970	79.0819	83.2976	88.3794	91.9517	99.607
70	77.5766	85.5271	90.5312	95.0231	100.425	104.215	112.317
80	88.1303	96.5782	101.879	106.629	112.329	116.321	124.839
90	98.6499	107.565	113.145	118.136	124.116	128.299	137.208
100	109.141	118.498	124.342	129.561	135.807	140.169	149.449
z_Q	+0.6745	+1.2816	+1.6449	+1.9600	+2.3263	+2.5758	+3.0902

Table V
PERCENTAGE POINTS OF THE F DISTRIBUTION
UPPER 5% POINTS

v_2 \ v_1	1	2	3	4	5	6	7	8	9	10	12	15	20	24	30	40	60	120	∞
1	161.4	199.5	215.7	224.6	230.2	234.0	236.8	238.9	240.5	241.9	243.9	245.9	248.0	249.1	250.1	251.1	252.2	253.3	254.3
2	18.51	19.00	19.16	19.25	19.30	19.33	19.35	19.37	19.38	19.40	19.41	19.43	19.45	19.45	19.46	19.47	19.48	19.49	19.50
3	10.13	9.55	9.28	9.12	9.01	8.94	8.89	8.85	8.81	8.79	8.74	8.70	8.66	8.64	8.62	8.59	8.57	8.55	8.53
4	7.71	6.94	6.59	6.39	6.26	6.16	6.09	6.04	6.00	5.96	5.91	5.86	5.80	5.77	5.75	5.72	5.69	5.66	5.63
5	6.61	5.79	5.41	5.19	5.05	4.95	4.88	4.82	4.77	4.74	4.68	4.62	4.56	4.53	4.50	4.46	4.43	4.40	4.36
6	5.99	5.14	4.76	4.53	4.39	4.28	4.21	4.15	4.10	4.06	4.00	3.94	3.87	3.84	3.81	3.77	3.74	3.70	3.67
7	5.59	4.74	4.35	4.12	3.97	3.87	3.79	3.73	3.68	3.64	3.57	3.51	3.44	3.41	3.38	3.34	3.30	3.27	3.23
8	5.32	4.46	4.07	3.84	3.69	3.58	3.50	3.44	3.39	3.35	3.28	3.22	3.15	3.12	3.08	3.04	3.01	2.97	2.93
9	5.12	4.26	3.86	3.63	3.48	3.37	3.29	3.23	3.18	3.14	3.07	3.01	2.94	2.90	2.86	2.83	2.79	2.75	2.71
10	4.96	4.10	3.71	3.48	3.33	3.22	3.14	3.07	3.02	2.98	2.91	2.85	2.77	2.74	2.70	2.66	2.62	2.58	2.54
11	4.84	3.98	3.59	3.36	3.20	3.09	3.01	2.95	2.90	2.85	2.79	2.72	2.65	2.61	2.57	2.53	2.49	2.45	2.40
12	4.75	3.89	3.49	3.26	3.11	3.00	2.91	2.85	2.80	2.75	2.69	2.62	2.54	2.51	2.47	2.43	2.38	2.34	2.30
13	4.67	3.81	3.41	3.18	3.03	2.92	2.83	2.77	2.71	2.67	2.60	2.53	2.46	2.42	2.38	2.34	2.30	2.25	2.21
14	4.60	3.74	3.34	3.11	2.96	2.85	2.76	2.70	2.65	2.60	2.53	2.46	2.39	2.35	2.31	2.27	2.22	2.18	2.13
15	4.54	3.68	3.29	3.06	2.90	2.79	2.71	2.64	2.59	2.54	2.48	2.40	2.33	2.29	2.25	2.20	2.16	2.11	2.07
16	4.49	3.63	3.24	3.01	2.85	2.74	2.66	2.59	2.54	2.49	2.42	2.35	2.28	2.24	2.19	2.15	2.11	2.06	2.01
17	4.45	3.59	3.20	2.96	2.81	2.70	2.61	2.55	2.49	2.45	2.38	2.31	2.23	2.19	2.15	2.10	2.06	2.01	1.96
18	4.41	3.55	3.16	2.93	2.77	2.66	2.58	2.51	2.46	2.41	2.34	2.27	2.19	2.15	2.11	2.06	2.02	1.97	1.92
19	4.38	3.52	3.13	2.90	2.74	2.63	2.54	2.48	2.42	2.38	2.31	2.23	2.16	2.11	2.07	2.03	1.98	1.93	1.88
20	4.35	3.49	3.10	2.87	2.71	2.60	2.51	2.45	2.39	2.35	2.28	2.20	2.12	2.08	2.04	1.99	1.95	1.90	1.84
21	4.32	3.47	3.07	2.84	2.68	2.57	2.49	2.42	2.37	2.32	2.25	2.18	2.10	2.05	2.01	1.96	1.92	1.87	1.81
22	4.30	3.44	3.05	2.82	2.66	2.55	2.46	2.40	2.34	2.30	2.23	2.15	2.07	2.03	1.98	1.94	1.89	1.84	1.78
23	4.28	3.42	3.03	2.80	2.64	2.53	2.44	2.37	2.32	2.27	2.20	2.13	2.05	2.01	1.96	1.91	1.86	1.81	1.76
24	4.26	3.40	3.01	2.78	2.62	2.51	2.42	2.36	2.30	2.25	2.18	2.11	2.03	1.98	1.94	1.89	1.84	1.79	1.73
25	4.24	3.39	2.99	2.76	2.60	2.49	2.40	2.34	2.28	2.24	2.16	2.09	2.01	1.96	1.92	1.87	1.82	1.77	1.71
26	4.23	3.37	2.98	2.74	2.59	2.47	2.39	2.32	2.27	2.22	2.15	2.07	1.99	1.95	1.90	1.85	1.80	1.75	1.69
27	4.21	3.35	2.96	2.73	2.57	2.46	2.37	2.31	2.25	2.20	2.13	2.06	1.97	1.93	1.88	1.84	1.79	1.73	1.67
28	4.20	3.34	2.95	2.71	2.56	2.45	2.36	2.29	2.24	2.19	2.12	2.04	1.96	1.91	1.87	1.82	1.77	1.71	1.65
29	4.18	3.33	2.93	2.70	2.55	2.43	2.35	2.28	2.22	2.18	2.10	2.03	1.94	1.90	1.85	1.81	1.75	1.70	1.64
30	4.17	3.32	2.92	2.69	2.53	2.42	2.33	2.27	2.21	2.16	2.09	2.01	1.93	1.89	1.84	1.79	1.74	1.68	1.62
40	4.08	3.23	2.84	2.61	2.45	2.34	2.25	2.18	2.12	2.08	2.00	1.92	1.84	1.79	1.74	1.69	1.64	1.58	1.51
60	4.00	3.15	2.76	2.53	2.37	2.25	2.17	2.10	2.04	1.99	1.92	1.84	1.75	1.70	1.65	1.59	1.53	1.47	1.39
120	3.92	3.07	2.68	2.45	2.29	2.17	2.09	2.02	1.96	1.91	1.83	1.75	1.66	1.61	1.55	1.50	1.43	1.35	1.25
∞	3.84	3.00	2.60	2.37	2.21	2.10	2.01	1.94	1.88	1.83	1.75	1.67	1.57	1.52	1.46	1.39	1.32	1.22	1.00

Table V   889

Table V (continued)
UPPER 2.5% POINTS

ν_2 \ ν_1	1	2	3	4	5	6	7	8	9	10	12	15	20	24	30	40	60	120	∞
1	647.8	799.5	864.2	899.6	921.8	937.1	948.2	956.7	963.3	968.6	976.7	984.9	993.1	997.2	1001	1006	1010	1014	1018
2	38.51	39.00	39.17	39.25	39.30	39.33	39.36	39.37	39.39	39.40	39.41	39.43	39.45	39.46	39.46	39.47	39.48	39.49	39.50
3	17.44	16.04	15.44	15.10	14.88	14.73	14.62	14.54	14.47	14.42	14.34	14.25	14.17	14.12	14.08	14.04	13.99	13.95	13.90
4	12.22	10.65	9.98	9.60	9.36	9.20	9.07	8.98	8.90	8.84	8.75	8.66	8.56	8.51	8.46	8.41	8.36	8.31	8.26
5	10.01	8.43	7.76	7.39	7.15	6.98	6.85	6.76	6.68	6.62	6.52	6.43	6.33	6.28	6.23	6.18	6.12	6.07	6.02
6	8.81	7.26	6.60	6.23	5.99	5.82	5.70	5.60	5.52	5.46	5.37	5.27	5.17	5.12	5.07	5.01	4.96	4.90	4.85
7	8.07	6.54	5.89	5.52	5.29	5.12	4.99	4.90	4.82	4.76	4.67	4.57	4.47	4.42	4.36	4.31	4.25	4.20	4.14
8	7.57	6.06	5.42	5.05	4.82	4.65	4.53	4.43	4.36	4.30	4.20	4.10	4.00	3.95	3.89	3.84	3.78	3.73	3.67
9	7.21	5.71	5.08	4.72	4.48	4.32	4.20	4.10	4.03	3.96	3.87	3.77	3.67	3.61	3.56	3.51	3.45	3.39	3.33
10	6.94	5.46	4.83	4.47	4.24	4.07	3.95	3.85	3.78	3.72	3.62	3.52	3.42	3.37	3.31	3.26	3.20	3.14	3.08
11	6.72	5.26	4.63	4.28	4.04	3.88	3.76	3.66	3.59	3.53	3.43	3.33	3.23	3.17	3.12	3.06	3.00	2.94	2.88
12	6.55	5.10	4.47	4.12	3.89	3.73	3.61	3.51	3.44	3.37	3.28	3.18	3.07	3.02	2.96	2.91	2.85	2.79	2.72
13	6.41	4.97	4.35	4.00	3.77	3.60	3.48	3.39	3.31	3.25	3.15	3.05	2.95	2.89	2.84	2.78	2.72	2.66	2.60
14	6.30	4.86	4.24	3.89	3.66	3.50	3.38	3.29	3.21	3.15	3.05	2.95	2.84	2.79	2.73	2.67	2.61	2.55	2.49
15	6.20	4.77	4.15	3.80	3.58	3.41	3.29	3.20	3.12	3.06	2.96	2.86	2.76	2.70	2.64	2.59	2.52	2.46	2.40
16	6.12	4.69	4.08	3.73	3.50	3.34	3.22	3.12	3.05	2.99	2.89	2.79	2.68	2.63	2.57	2.51	2.45	2.38	2.32
17	6.04	4.62	4.01	3.66	3.44	3.28	3.16	3.06	2.98	2.92	2.82	2.72	2.62	2.56	2.50	2.44	2.38	2.32	2.25
18	5.98	4.56	3.95	3.61	3.38	3.22	3.10	3.01	2.93	2.87	2.77	2.67	2.56	2.50	2.44	2.38	2.32	2.26	2.19
19	5.92	4.51	3.90	3.56	3.33	3.17	3.05	2.96	2.88	2.82	2.72	2.62	2.51	2.45	2.39	2.33	2.27	2.20	2.13
20	5.87	4.46	3.86	3.51	3.29	3.13	3.01	2.91	2.84	2.77	2.68	2.57	2.46	2.41	2.35	2.29	2.22	2.16	2.09
21	5.83	4.42	3.82	3.48	3.25	3.09	2.97	2.87	2.80	2.73	2.64	2.53	2.42	2.37	2.31	2.25	2.18	2.11	2.04
22	5.79	4.38	3.78	3.44	3.22	3.05	2.93	2.84	2.76	2.70	2.60	2.50	2.39	2.33	2.27	2.21	2.14	2.08	2.00
23	5.75	4.35	3.75	3.41	3.18	3.02	2.90	2.81	2.73	2.67	2.57	2.47	2.36	2.30	2.24	2.18	2.11	2.04	1.97
24	5.72	4.32	3.72	3.38	3.15	2.99	2.87	2.78	2.70	2.64	2.54	2.44	2.33	2.27	2.21	2.15	2.08	2.01	1.94
25	5.69	4.29	3.69	3.35	3.13	2.97	2.85	2.75	2.68	2.61	2.51	2.41	2.30	2.24	2.18	2.12	2.05	1.98	1.91
26	5.66	4.27	3.67	3.33	3.10	2.94	2.82	2.73	2.65	2.59	2.49	2.39	2.28	2.22	2.16	2.09	2.03	1.95	1.88
27	5.63	4.24	3.65	3.31	3.08	2.92	2.80	2.71	2.63	2.57	2.47	2.36	2.25	2.19	2.13	2.07	2.00	1.93	1.85
28	5.61	4.22	3.63	3.29	3.06	2.90	2.78	2.69	2.61	2.55	2.45	2.34	2.23	2.17	2.11	2.05	1.98	1.91	1.83
29	5.59	4.20	3.61	3.27	3.04	2.88	2.76	2.67	2.59	2.53	2.43	2.32	2.21	2.15	2.09	2.03	1.96	1.89	1.81
30	5.57	4.18	3.59	3.25	3.03	2.87	2.75	2.65	2.57	2.51	2.41	2.31	2.20	2.14	2.07	2.01	1.94	1.87	1.79
40	5.42	4.05	3.46	3.13	2.90	2.74	2.62	2.53	2.45	2.39	2.29	2.18	2.07	2.01	1.94	1.88	1.80	1.72	1.64
60	5.29	3.93	3.34	3.01	2.79	2.63	2.51	2.41	2.33	2.27	2.17	2.06	1.94	1.88	1.82	1.74	1.67	1.58	1.48
120	5.15	3.80	3.23	2.89	2.67	2.52	2.39	2.30	2.22	2.16	2.05	1.94	1.82	1.76	1.69	1.61	1.53	1.43	1.31
∞	5.02	3.69	3.12	2.79	2.57	2.41	2.29	2.19	2.11	2.05	1.94	1.83	1.71	1.64	1.57	1.48	1.39	1.27	1.00

Table V (continued)
UPPER 1% POINTS

ν_2 \ ν_1	1	2	3	4	5	6	7	8	9	10	12	15	20	24	30	40	60	120	∞
1	4052	4999.5	5403	5625	5764	5859	5928	5982	6022	6056	6106	6157	6209	6235	6261	6287	6313	6339	6366
2	98.50	99.00	99.17	99.25	99.30	99.33	99.36	99.37	99.39	99.40	99.42	99.43	99.45	99.46	99.47	99.47	99.48	99.49	99.50
3	34.12	30.82	29.46	28.71	28.24	27.91	27.67	27.49	27.35	27.23	27.05	26.87	26.69	26.60	26.50	26.41	26.32	26.22	26.13
4	21.20	18.00	16.69	15.98	15.52	15.21	14.98	14.80	14.66	14.55	14.37	14.20	14.02	13.93	13.84	13.75	13.65	13.56	13.46
5	16.26	13.27	12.06	11.39	10.97	10.67	10.46	10.29	10.16	10.05	9.89	9.72	9.55	9.47	9.38	9.29	9.20	9.11	9.02
6	13.75	10.92	9.78	9.15	8.75	8.47	8.26	8.10	7.98	7.87	7.72	7.56	7.40	7.31	7.23	7.14	7.06	6.97	6.88
7	12.25	9.55	8.45	7.85	7.46	7.19	6.99	6.84	6.72	6.62	6.47	6.31	6.16	6.07	5.99	5.91	5.82	5.74	5.65
8	11.26	8.65	7.59	7.01	6.63	6.37	6.18	6.03	5.91	5.81	5.67	5.52	5.36	5.28	5.20	5.12	5.03	4.95	4.86
9	10.56	8.02	6.99	6.42	6.06	5.80	5.61	5.47	5.35	5.26	5.11	4.96	4.81	4.73	4.65	4.57	4.48	4.40	4.31
10	10.04	7.56	6.55	5.99	5.64	5.39	5.20	5.06	4.94	4.85	4.71	4.56	4.41	4.33	4.25	4.17	4.08	4.00	3.91
11	9.65	7.21	6.22	5.67	5.32	5.07	4.89	4.74	4.63	4.54	4.40	4.25	4.10	4.02	3.94	3.86	3.78	3.69	3.60
12	9.33	6.93	5.95	5.41	5.06	4.82	4.64	4.50	4.39	4.30	4.16	4.01	3.86	3.78	3.70	3.62	3.54	3.45	3.36
13	9.07	6.70	5.74	5.21	4.86	4.62	4.44	4.30	4.19	4.10	3.96	3.82	3.66	3.59	3.51	3.43	3.34	3.25	3.17
14	8.86	6.51	5.56	5.04	4.69	4.46	4.28	4.14	4.03	3.94	3.80	3.66	3.51	3.43	3.35	3.27	3.18	3.09	3.00
15	8.68	6.36	5.42	4.89	4.56	4.32	4.14	4.00	3.89	3.80	3.67	3.52	3.37	3.29	3.21	3.13	3.05	2.96	2.87
16	8.53	6.23	5.29	4.77	4.44	4.20	4.03	3.89	3.78	3.69	3.55	3.41	3.26	3.18	3.10	3.02	2.93	2.84	2.75
17	8.40	6.11	5.18	4.67	4.34	4.10	3.93	3.79	3.68	3.59	3.46	3.31	3.16	3.08	3.00	2.92	2.83	2.75	2.65
18	8.29	6.01	5.09	4.58	4.25	4.01	3.84	3.71	3.60	3.51	3.37	3.23	3.08	3.00	2.92	2.84	2.75	2.66	2.57
19	8.18	5.93	5.01	4.50	4.17	3.94	3.77	3.63	3.52	3.43	3.30	3.15	3.00	2.92	2.84	2.76	2.67	2.58	2.49
20	8.10	5.85	4.94	4.43	4.10	3.87	3.70	3.56	3.46	3.37	3.23	3.09	2.94	2.86	2.78	2.69	2.61	2.52	2.42
21	8.02	5.78	4.87	4.37	4.04	3.81	3.64	3.51	3.40	3.31	3.17	3.03	2.88	2.80	2.72	2.64	2.55	2.46	2.36
22	7.95	5.72	4.82	4.31	3.99	3.76	3.59	3.45	3.35	3.26	3.12	2.98	2.83	2.75	2.67	2.58	2.50	2.40	2.31
23	7.88	5.66	4.76	4.26	3.94	3.71	3.54	3.41	3.30	3.21	3.07	2.93	2.78	2.70	2.62	2.54	2.45	2.35	2.26
24	7.82	5.61	4.72	4.22	3.90	3.67	3.50	3.36	3.26	3.17	3.03	2.89	2.74	2.66	2.58	2.49	2.40	2.31	2.21
25	7.77	5.57	4.68	4.18	3.85	3.63	3.46	3.32	3.22	3.13	2.99	2.85	2.70	2.62	2.54	2.45	2.36	2.27	2.17
26	7.72	5.53	4.64	4.14	3.82	3.59	3.42	3.29	3.18	3.09	2.96	2.81	2.66	2.58	2.50	2.42	2.33	2.23	2.13
27	7.68	5.49	4.60	4.11	3.78	3.56	3.39	3.26	3.15	3.06	2.93	2.78	2.63	2.55	2.47	2.38	2.29	2.20	2.10
28	7.64	5.45	4.57	4.07	3.75	3.53	3.36	3.23	3.12	3.03	2.90	2.75	2.60	2.52	2.44	2.35	2.26	2.17	2.06
29	7.60	5.42	4.54	4.04	3.73	3.50	3.33	3.20	3.09	3.00	2.87	2.73	2.57	2.49	2.41	2.33	2.23	2.14	2.03
30	7.56	5.39	4.51	4.02	3.70	3.47	3.30	3.17	3.07	2.98	2.84	2.70	2.55	2.47	2.39	2.30	2.21	2.11	2.01
40	7.31	5.18	4.31	3.83	3.51	3.29	3.12	2.99	2.89	2.80	2.66	2.52	2.37	2.29	2.20	2.11	2.02	1.92	1.80
60	7.08	4.98	4.13	3.65	3.34	3.12	2.95	2.82	2.72	2.63	2.50	2.35	2.20	2.12	2.03	1.94	1.84	1.73	1.60
120	6.85	4.79	3.95	3.48	3.17	2.96	2.79	2.66	2.56	2.47	2.34	2.19	2.03	1.95	1.86	1.76	1.66	1.53	1.38
∞	6.63	4.61	3.78	3.32	3.02	2.80	2.64	2.51	2.41	2.32	2.18	2.04	1.88	1.79	1.70	1.59	1.47	1.32	1.00

Table VI   891

Table VI
THE TRANSFORMATION OF r TO Z

r	r (3rd decimal)					r	r (3rd decimal)				
	.000	.002	.004	.006	.008		.000	.002	.004	.006	.008
.00	.0000	.0020	.0040	.0060	.0080	.35	.3654	.3677	.3700	.3723	.3746
1	.0100	.0120	.0140	.0160	.0180	6	.3769	.3792	.3815	.3838	.3861
2	.0200	.0220	.0240	.0260	.0280	7	.3884	.3907	.3931	.3954	.3977
3	.0300	.0320	.0340	.0360	.0380	8	.4001	.4024	.4047	.4071	.4094
4	.0400	.0420	.0440	.0460	.0480	9	.4118	.4142	.4165	.4189	.4213
.05	.0500	.0520	.0541	.0561	.0581	.40	.4236	.4260	.4284	.4308	.4332
6	.0601	.0621	.0641	.0661	.0681	1	.4356	.4380	.4404	.4428	.4453
7	.0701	.0721	.0741	.0761	.0782	2	.4477	.4501	.4526	.4550	.4574
8	.0802	.0822	.0842	.0862	.0882	3	.4599	.4624	.4648	.4673	.4698
9	.0902	.0923	.0943	.0963	.0983	4	.4722	.4747	.4772	.4797	.4822
.10	.1003	.1024	.1044	.1064	.1084	.45	.4847	.4872	.4897	.4922	.4948
1	.1104	.1125	.1145	.1165	.1186	6	.4973	.4999	.5024	.5049	.5075
2	.1206	.1226	.1246	.1267	.1287	7	.5101	.5126	.5152	.5178	.5204
3	.1307	.1328	.1348	.1368	.1389	8	.5230	.5256	.5282	.5308	.5334
4	.1409	.1430	.1450	.1471	.1491	9	.5361	.5387	.5413	.5440	.5466
.15	.1511	.1532	.1552	.1573	.1593	.50	.5493	.5520	.5547	.5573	.5600
6	.1614	.1634	.1655	.1676	.1696	1	.5627	.5654	.5682	.5709	.5736
7	.1717	.1737	.1758	.1779	.1799	2	.5763	.5791	.5818	.5846	.5874
8	.1820	.1841	.1861	.1882	.1903	3	.5901	.5929	.5957	.5985	.6013
9	.1923	.1944	.1965	.1986	.2007	4	.6042	.6070	.6098	.6127	.6155
.20	.2027	.2048	.2069	.2090	.2111	.55	.6184	.6213	.6241	.6270	.6299
1	.2132	.2153	.2174	.2195	.2216	6	.6328	.6358	.6387	.6416	.6446
2	.2237	.2258	.2279	.2300	.2321	7	.6475	.6505	.6535	.6565	.6595
3	.2342	.2363	.2384	.2405	.2427	8	.6625	.6655	.6685	.6716	.6746
4	.2448	.2469	.2490	.2512	.2533	9	.6777	.6807	.6838	.6869	.6900
.25	.2554	.2575	.2597	.2618	.2640	.60	.6931	.6963	.6994	.7026	.7057
6	.2661	.2683	.2704	.2726	.2747	1	.7089	.7121	.7153	.7185	.7218
7	.2769	.2790	.2812	.2833	.2855	2	.7250	.7283	.7315	.7348	.7381
8	.2877	.2899	.2920	.2942	.2964	3	.7414	.7447	.7481	.7514	.7548
9	.2986	.3008	.3029	.3051	.3073	4	.7582	.7616	.7650	.7684	.7718
.30	.3095	.3117	.3139	.3161	.3183	.65	.7753	.7788	.7823	.7858	.7893
1	.3205	.3228	.3250	.3272	.3294	6	.7928	.7964	.7999	.8035	.8071
2	.3316	.3339	.3361	.3383	.3406	7	.8107	.8144	.8180	.8217	.8254
3	.3428	.3451	.3473	.3496	.3518	8	.8291	.8328	.8366	.8404	.8441
4	.3541	.3564	.3586	.3609	.3632	9	.8480	.8518	.8556	.8595	.8634

Table VI (continued)

r	r (3rd decimal)					r	r (3rd decimal)				
	.000	.002	.004	.006	.008		.000	.002	.004	.006	.008
.70	.8673	.8712	.8752	.8792	.8832	.85	1.256	1.263	1.271	1.278	1.286
1	.8872	.8912	.8953	.8994	.9035	6	1.293	1.301	1.309	1.317	1.325
2	.9076	.9118	.9160	.9202	.9245	7	1.333	1.341	1.350	1.358	1.367
3	.9287	.9330	.9373	.9417	.9461	8	1.376	1.385	1.394	1.403	1.412
4	.9505	.9549	.9594	.9639	.9684	9	1.422	1.432	1.442	1.452	1.462
.75	0.973	0.978	0.982	0.987	0.991	.90	1.472	1.483	1.494	1.505	1.516
6	0.996	1.001	1.006	1.011	1.015	1	1.528	1.539	1.551	1.564	1.576
7	1.020	1.025	1.030	1.035	1.040	2	1.589	1.602	1.616	1.630	1.644
8	1.045	1.050	1.056	1.061	1.066	3	1.658	1.673	1.689	1.705	1.721
9	1.071	1.077	1.082	1.088	1.093	4	1.738	1.756	1.774	1.792	1.812
.80	1.099	1.104	1.110	1.116	1.121	.95	1.832	1.853	1.874	1.897	1.921
1	1.127	1.133	1.139	1.145	1.151	6	1.946	1.972	2.000	2.029	2.060
2	1.157	1.163	1.169	1.175	1.182	7	2.092	2.127	2.165	2.205	2.249
3	1.188	1.195	1.201	1.208	1.214	8	2.298	2.351	2.410	2.477	2.555
4	1.221	1.228	1.235	1.242	1.249	9	2.647	2.759	2.903	3.106	3.453

Table VII   893

Table VII
COEFFICIENTS OF ORTHOGONAL POLYNOMIALS

	$J = 3$		$J = 4$			$J = 5$			
	1	2	1	2	3	1	2	3	4
X_1	−1	1	−3	1	−1	−2	2	−1	1
X_2	0	−2	−1	−1	3	−1	−1	2	−4
X_3	1	1	1	−1	−3	0	−2	0	6
X_4			3	1	1	1	−1	−2	−4
X_5						2	2	1	1
Σc_j^2	2	6	20	4	20	10	14	10	70

	$J = 6$					$J = 7$					
	1	2	3	4	5	1	2	3	4	5	6
X_1	−5	5	−5	1	−1	−3	5	−1	3	−1	1
X_2	−3	−1	7	−3	5	−2	0	1	−7	4	−6
X_3	−1	−4	4	2	−10	−1	−3	1	1	−5	15
X_4	1	−4	−4	2	10	0	−4	0	6	0	−20
X_5	3	−1	−7	−3	−5	1	−3	−1	1	5	15
X_6	5	5	5	1	1	2	0	−1	−7	−4	−6
X_7						3	5	1	3	1	1
Σc_j^2	70	84	180	28	252	28	84	6	154	84	924

	$J = 8$						$J = 9$					
	1	2	3	4	5	6	1	2	3	4	5	6
X_1	−7	7	−7	7	−7	1	−4	28	−14	14	−4	4
X_2	−5	1	5	−13	23	−5	−3	7	7	−21	11	−17
X_3	−3	−3	7	−3	−17	9	−2	−8	13	−11	−4	22
X_4	−1	−5	3	9	−15	−5	−1	−17	9	9	−9	1
X_5	1	−5	−3	9	15	−5	0	−20	0	18	0	−20
X_6	3	−3	−7	−3	17	9	1	−17	−9	9	9	1
X_7	5	1	−5	−13	−23	−5	2	−8	−13	−11	4	22
X_8	7	7	7	7	7	1	3	7	−7	−21	−11	−17
X_9							4	28	14	14	4	4
Σc_j^2	168	168	264	616	2184	264	60	2772	990	2002	468	1980

Table VII (continued)

	J = 10							J = 11				
	1	2	3	4	5	6	1	2	3	4	5	6
X_1	−9	6	−42	18	−6	3	−5	15	−30	6	−3	15
X_2	−7	2	14	−22	14	−11	−4	6	6	−6	6	−48
X_3	−5	−1	35	−17	−1	10	−3	−1	22	−6	1	29
X_4	−3	−3	31	3	−11	6	−2	−6	23	−1	−4	36
X_5	−1	−4	12	18	−6	−8	−1	−9	14	4	−4	−12
X_6	1	−4	−12	18	6	−8	0	−10	0	6	0	−40
X_7	3	−3	−31	3	11	6	1	−9	−14	4	4	−12
X_8	5	−1	−35	−17	1	10	2	−6	−23	−1	4	36
X_9	7	2	−14	−22	−14	−11	3	−1	−22	−6	−1	29
X_{10}	9	6	42	18	6	3	4	6	−6	−6	−6	−48
X_{11}							5	15	30	6	3	15
Σc_j^2	330	132	8580	2860	780	660	110	858	4290	286	156	11220

	J = 12							J = 13				
	1	2	3	4	5	6	1	2	3	4	5	6
X_1	−11	55	−33	33	−33	11	−6	22	−11	99	−22	22
X_2	−9	25	3	−27	57	−31	−5	11	0	−66	33	−55
X_3	−7	1	21	−33	21	11	−4	2	6	−96	18	8
X_4	−5	−17	25	−13	−29	25	−3	−5	8	−54	−11	43
X_5	−3	−29	19	12	−44	4	−2	−10	7	11	−26	22
X_6	−1	−35	7	28	−20	−20	−1	−13	4	64	−20	−20
X_7	1	−35	−7	28	20	−20	0	−14	0	84	0	−40
X_8	3	−29	−19	12	44	4	1	−13	−4	64	20	−20
X_9	5	−17	−25	−13	29	25	2	−10	−7	11	26	22
X_{10}	7	1	−21	−33	−21	11	3	−5	−8	−54	11	43
X_{11}	9	25	−3	−27	−57	−31	4	2	−6	−96	−18	8
X_{12}	11	55	33	33	33	11	5	11	0	−66	−33	−55
X_{13}							6	22	11	99	22	22
Σc_j^2	572	12012	5148	8008	15912	4488	182	2002	572	68068	6188	14212

Table VII 895

Table VII (continued)

	\multicolumn{6}{c}{$J = 14$}						\multicolumn{6}{c}{$J = 15$}					
	1	2	3	4	5	6	1	2	3	4	5	6
X_1	−13	13	−143	143	−143	143	−7	91	−91	1001	−1001	143
X_2	−11	7	−11	−77	187	−319	−6	52	−13	−429	1144	−286
X_3	−9	2	66	−132	132	−11	−5	19	35	−869	979	−55
X_4	−7	−2	98	−92	−28	227	−4	−8	58	−704	44	176
X_5	−5	−5	95	−13	−139	185	−3	−29	61	−249	−751	197
X_6	−3	−7	67	63	−145	−25	−2	−44	49	251	−1000	50
X_7	−1	−8	24	108	−60	−200	−1	−53	27	621	−675	−125
X_8	1	−8	−24	108	60	−200	0	−56	0	756	0	−200
X_9	3	−7	−67	63	145	−25	1	−53	−27	621	675	−125
X_{10}	5	−5	−95	−13	139	185	2	−44	−49	251	1000	50
X_{11}	7	−2	−98	−92	28	227	3	−29	−61	−249	751	197
X_{12}	9	2	−66	−132	−132	−11	4	−8	−58	−704	−44	176
X_{13}	11	7	11	−77	−187	−319	5	19	−35	−869	−979	−55
X_{14}	13	13	143	143	143	143	6	52	13	−429	−1144	−286
X_{15}							7	91	91	1001	1001	143
Σc_j^2	910	728	97240	136136	235144	497420	280	37128	39780	6446460	10581480	426360

Table VIII

FACTORIALS OF INTEGERS

n	$n!$	n	$n!$
1	1	26	4.03291×10^{26}
2	2	27	1.08889×10^{28}
3	6	28	3.04888×10^{29}
4	24	29	8.84176×10^{30}
5	120	30	2.65253×10^{32}
6	720	31	8.22284×10^{23}
7	5040	32	2.63131×10^{35}
8	40320	33	8.68332×10^{36}
9	362880	34	2.95233×10^{38}
10	3.62880×10^{6}	35	1.03331×10^{40}
11	3.99168×10^{7}	36	3.71993×10^{41}
12	4.79002×10^{8}	37	1.37638×10^{43}
13	6.22702×10^{9}	38	5.23023×10^{44}
14	8.71783×10^{10}	39	2.03979×10^{46}
15	1.30767×10^{12}	40	8.15915×10^{47}
16	2.09228×10^{13}	41	3.34525×10^{49}
17	3.55687×10^{14}	42	1.40501×10^{51}
18	6.40327×10^{15}	43	6.04153×10^{52}
19	1.21645×10^{17}	44	2.65827×10^{54}
20	2.43290×10^{18}	45	1.19622×10^{56}
21	5.10909×10^{19}	46	5.50262×10^{57}
22	1.12400×10^{21}	47	2.58623×10^{59}
23	2.58520×10^{22}	48	1.24139×10^{61}
24	6.20448×10^{23}	49	6.08282×10^{62}
25	1.55112×10^{25}	50	3.04141×10^{64}

Table X   897

Table IX

BINOMIAL COEFFICIENTS, $\binom{N}{r}$

N \ r	0	1	2	3	4	5	6	7	8	9	10
1	1	1									
2	1	2	1								
3	1	3	3	1							
4	1	4	6	4	1						
5	1	5	10	10	5	1					
6	1	6	15	20	15	6	1				
7	1	7	21	35	35	21	7	1			
8	1	8	28	56	70	56	28	8	1		
9	1	9	36	84	126	126	84	36	9	1	
10	1	10	45	120	210	252	210	120	45	10	1
11	1	11	55	165	330	462	462	330	165	55	11
12	1	12	66	220	495	792	924	792	495	220	66
13	1	13	78	286	715	1287	1716	1716	1287	715	286
14	1	14	91	364	1001	2002	3003	3432	3003	2002	1001
15	1	15	105	455	1365	3003	5005	6435	6435	5005	3003
16	1	16	120	560	1820	4368	8008	11440	12870	11440	8008
17	1	17	136	680	2380	6188	12376	19448	24310	24310	19448
18	1	18	153	816	3060	8568	18564	31824	43758	48620	43758
19	1	19	171	969	3876	11628	27132	50388	75582	92378	92378
20	1	20	190	1140	4845	15504	38760	77520	125970	167960	184756

Table X

SELECTED VALUES OF e^{-m}

m

	.0	.1	.2	.3	.4	.5	.6	.7	.8	.9
0.	1.00000	.90484	.81873	.74082	.67032	.60653	.54881	.49659	.44933	.40657
1.0	.36788	.33287	.30119	.27253	.24660	.22313	.20190	.18268	.16530	.14957
2.0	.13534	.12246	.11080	.10026	.09072	.08209	.07427	.06721	.06081	.05502
3.0	.04979	.04505	.04076	.03688	.03337	.03020	.02732	.02472	.02237	.02024
4.0	.01832	.01657	.01500	.01357	.01228	.01111	.01005	.00910	.00823	.00745
5.0	.00674	.00610	.00552	.00499	.00452	.00409	.00370	.00335	.00303	.00274
6.0	.00248	.00224	.00203	.00184	.00166	.00150	.00136	.00123	.00111	.00101
7.0	.00091	.00083	.00075	.00068	.00061	.00055	.00050	.00045	.00041	.00037
8.0	.00034	.00030	.00027	.00025	.00022	.00020	.00018	.00017	.00015	.00014
9.0	.00012	.00011	.00010	.00009	.00008	.00007	.00007	.00006	.00006	.00005

For higher or more precise values of m, note that $e^{-a-b} = e^{-a}e^{-b}$, and that $e^{-ab} = (e^{-a})^b$. Hence, within each row, the entry in any column is .9048 times the entry in the preceding column.

Table XI
POWERS AND ROOTS

N	N²	√N	√10N
1	1	1.00 000	3.16 228
2	4	1.41 421	4.47 214
3	9	1.73 205	5.47 723
4	16	2.00 000	6.32 456
5	25	2.23 607	7.07 107
6	36	2.44 949	7.74 597
7	49	2.64 575	8.36 660
8	64	2.82 843	8.94 427
9	81	3.00 000	9.48 683
10	100	3.16 228	10.00 00
11	121	3.31 662	10.48 81
12	144	3.46 410	10.95 45
13	169	3.60 555	11.40 18
14	196	3.74 166	11.83 22
15	225	3.87 298	12.24 74
16	256	4.00 000	12.64 91
17	289	4.12 311	13.03 84
18	324	4.24 264	13.41 64
19	361	4.35 890	13.78 40
20	400	4.47 214	14.14 21
21	441	4.58 258	14.49 14
22	484	4.69 042	14.83 24
23	529	4.79 583	15.16 58
24	576	4.89 898	15.49 19
25	625	5.00 000	15.81 14
26	676	5.09 902	16.12 45
27	729	5.19 615	16.43 17
28	784	5.29 150	16.73 32
29	841	5.38 516	17.02 94
30	900	5.47 723	17.32 05
31	961	5.56 776	17.60 68
32	1 024	5.65 685	17.88 85
33	1 089	5.74 456	18.16 59
34	1 156	5.83 095	18.43 91
35	1 225	5.91 608	18.70 83
36	1 296	6.00 000	18.97 37
37	1 369	6.08 276	19.23 54
38	1 444	6.16 441	19.49 36
39	1 521	6.24 500	19.74 84
40	1 600	6.32 456	20.00 00
41	1 681	6.40 312	20.24 85
42	1 764	6.48 074	20.49 39
43	1 849	6.55 744	20.73 64
44	1 936	6.63 325	20.97 62
45	2 025	6.70 820	21.21 32
46	2 116	6.78 233	21.44 76
47	2 209	6.85 565	21.67 95
48	2 304	6.92 820	21.90 89
49	2 401	7.00 000	22.13 59
50	2 500	7.07 107	22.36 07
N	N²	√N	√10N

N	N²	√N	√10N
50	2 500	7.07 107	22.36 07
51	2 601	7.14 143	22.58 32
52	2 704	7.21 110	22.80 35
53	2 809	7.28 011	23.02 17
54	2 916	7.34 847	23.23 79
55	3 025	7.41 620	23.45 21
56	3 136	7.48 331	23.66 43
57	3 249	7.54 983	23.87 47
58	3 364	7.61 577	24.08 32
59	3 481	7.68 115	24.28 99
60	3 600	7.74 597	24.49 49
61	3 721	7.81 025	24.69 82
62	3 844	7.87 401	24.89 98
63	3 969	7.93 725	25.09 98
64	4 096	8.00 000	25.29 82
65	4 225	8.06 226	25.49 51
66	4 356	8.12 404	25.69 05
67	4 489	8.18 535	25.88 44
68	4 624	8.24 621	26.07 68
69	4 761	8.30 662	26.26 79
70	4 900	8.36 660	26.45 75
71	5 041	8.42 615	26.64 58
72	5 184	8.48 528	26.83 28
73	5 329	8.54 400	27.01 85
74	5 476	8.60 233	27.20 29
75	5 625	8.66 025	27.38 61
76	5 776	8.71 780	27.56 81
77	5 929	8.77 496	27.74 89
78	6 084	8.83 176	27.92 85
79	6 241	8.88 819	28.10 69
80	6 400	8.94 427	28.28 43
81	6 561	9.00 000	28.46 05
82	6 724	9.05 539	28.63 56
83	6 889	9.11 043	28.80 97
84	7 056	9.16 515	28.98 28
85	7 225	9.21 954	29.15 48
86	7 396	9.27 362	29.32 58
87	7 569	9.32 738	29.49 58
88	7 744	9.38 083	29.66 48
89	7 921	9.43 398	29.83 29
90	8 100	9.48 683	30.00 00
91	8 281	9.53 939	30.16 62
92	8 464	9.59 166	30.33 15
93	8 649	9.64 365	30.49 59
94	8 836	9.69 536	30.65 94
95	9 025	9.74 679	30.82 21
96	9 216	9.79 796	30.98 39
97	9 409	9.84 886	31.14 48
98	9 604	9.89 949	31.30 50
99	9 801	9.94 987	31.46 43
100	10 000	10.00 000	31.62 28
N	N²	√N	√10N

Table XI   899

Table XI (continued)

N	N²	√N	√10N
100	10 000	10.00 00	31.62 28
101	10 201	10.04 99	31.78 05
102	10 404	10.09 95	31.93 74
103	10 609	10.14 89	32.09 36
104	10 816	10.19 80	32.24 90
105	11 025	10.24 70	32.40 37
106	11 236	10.29 56	32.55 76
107	11 449	10.34 41	32.71 09
108	11 664	10.39 23	32.86 34
109	11 881	10.44 03	33.01 51
110	12 100	10.48 81	33.16 62
111	12 321	10.53 57	33.31 67
112	12 544	10.58 30	33.46 64
113	12 769	10.63 01	33.61 55
114	12 996	10.67 71	33.76 39
115	13 225	10.72 38	33.91 16
116	13 456	10.77 03	34.05 88
117	13 689	10.81 67	34.20 53
118	13 924	10.86 28	34.35 11
119	14 161	10.90 87	34.49 64
120	14 400	10.95 45	34.64 10
121	14 641	11.00 00	34.78 51
122	14 884	11.04 54	34.92 85
123	15 129	11.09 05	35.07 14
124	15 376	11.13 55	35.21 36
125	15 625	11.18 03	35.35 53
126	15 876	11.22 50	35.49 65
127	16 129	11.26 94	35.63 71
128	16 384	11.31 37	35.77 71
129	16 641	11.35 78	35.91 66
130	16 900	11.40 18	36.05 55
131	17 161	11.44 55	36.19 39
132	17 424	11.48 91	36.33 18
133	17 689	11.53 26	36.46 92
134	17 956	11.57 58	36.60 60
135	18 225	11.61 90	36.74 23
136	18 496	11.66 19	36.87 82
137	18 769	11.70 47	37.01 35
138	19 044	11.74 73	37.14 84
139	19 321	11.78 98	37.28 27
140	19 600	11.83 22	37.41 66
141	19 881	11.87 43	37.55 00
142	20 164	11.91 64	37.68 29
143	20 449	11.95 83	37.81 53
144	20 736	12.00 00	37.94 73
145	21 025	12.04 16	38.07 89
146	21 316	12.08 30	38.20 99
147	21 609	12.12 44	38.34 06
148	21 904	12.16 55	38.47 08
149	22 201	12.20 66	38.60 05
150	22 500	12.24 74	38.72 98
N	N²	√N	√10N

N	N²	√N	√10N
150	22 500	12.24 74	38.72 98
151	22 801	12.28 82	38.85 87
152	23 104	12.32 88	38.98 72
153	23 409	12.36 93	39.11 52
154	23 716	12.40 97	39.24 28
155	24 025	12.44 99	39.37 00
156	24 336	12.49 00	39.49 68
157	24 649	12.53 00	39.62 32
158	24 964	12.56 98	39.74 92
159	25 281	12.60 95	39.87 48
160	25 600	12.64 91	40.00 00
161	25 921	12.68 86	40.12 48
162	26 244	12.72 79	40.24 92
163	26 569	12.76 71	40.37 33
164	26 896	12.80 62	40.49 69
165	27 225	12.84 52	40.62 02
166	27 556	12.88 41	40.74 31
167	27 889	12.92 28	40.86 56
168	28 224	12.96 15	40.98 78
169	28 561	13.00 00	41.10 96
170	28 900	13.03 84	41.23 11
171	29 241	13.07 67	41.35 21
172	29 584	13.11 49	41.47 29
173	29 929	13.15 29	41.59 33
174	30 276	13.19 09	41.71 33
175	30 625	13.22 88	41.83 30
176	30 976	13.26 65	41.95 24
177	31 329	13.30 41	42.07 14
178	31 684	13.34 17	42.19 00
179	32 041	13.37 91	42.30 84
180	32 400	13.41 64	42.42 64
181	32 761	13.45 36	42.54 41
182	33 124	13.49 07	42.66 15
183	33 489	13.52 77	42.77 85
184	33 856	13.56 47	42.89 52
185	34 225	13.60 15	43.01 16
186	34 596	13.63 82	43.12 77
187	34 969	13.67 48	43.24 35
188	35 344	13.71 13	43.35 90
189	35 721	13.74 77	43.47 41
190	36 100	13.78 40	43.58 90
191	36 481	13.82 03	43.70 35
192	36 864	13.85 64	43.81 78
193	37 249	13.89 24	43.93 18
194	37 636	13.92 84	44.04 54
195	38 025	13.96 42	44.15 88
196	38 416	14.00 00	44.27 19
197	38 809	14.03 57	44.38 47
198	39 204	14.07 12	44.49 72
199	39 601	14.10 67	44.60 94
200	40 000	14.14 21	44.72 14
N	N²	√N	√10N

Table XI (continued)

N	N^2	$\sqrt{N}$	$\sqrt{10N}$
200	40 000	14.14 21	44.72 14
201	40 401	14.17 74	44.83 30
202	40 804	14.21 27	44.94 44
203	41 209	14.24 78	45.05 55
204	41 616	14.28 29	45.16 64
205	42 025	14.31 78	45.27 69
206	42 436	14.35 27	45.38 72
207	42 849	14.38 75	45.49 73
208	43 264	14.42 22	45.60 70
209	43 681	14.45 68	45.71 65
210	44 100	14.49 14	45.82 58
211	44 521	14.52 58	45.93 47
212	44 944	14.56 02	46.04 35
213	45 369	14.59 45	46.15 19
214	45 796	14.62 87	46.26 01
215	46 225	14.66 29	46.36 81
216	46 656	14.69 69	46.47 58
217	47 089	14.73 09	46.58 33
218	47 524	14.76 48	46.69 05
219	47 961	14.79 86	46.79 74
220	48 400	14.83 24	46.90 42
221	48 841	14.86 61	47.01 06
222	49 284	14.89 97	47.11 69
223	49 729	14.93 32	47.22 29
224	50 176	14.96 66	47.32 86
225	50 625	15.00 00	47.43 42
226	51 076	15.03 33	47.53 95
227	51 529	15.06 65	47.64 45
228	51 984	15.09 97	47.74 93
229	52 441	15.13 27	47.85 39
230	52 900	15.16 58	47.95 83
231	53 361	15.19 87	48.06 25
232	53 824	15.23 15	48.16 64
233	54 289	15.26 43	48.27 01
234	54 756	15.29 71	48.37 35
235	55 225	15.32 97	48.47 68
236	55 696	15.36 23	48.57 98
237	56 169	15.39 48	48.68 26
238	56 644	15.42 72	48.78 52
239	57 121	15.45 96	48.88 76
240	57 600	15.49 19	48.98 98
241	58 081	15.52 42	49.09 18
242	58 564	15.55 63	49.19 35
243	59 049	15.58 85	49.29 50
244	59 536	15.62 05	49.39 64
245	60 025	15.65 25	49.49 75
246	60 516	15.68 44	49.59 84
247	61 009	15.71 62	49.69 91
248	61 504	15.74 80	49.79 96
249	62 001	15.77 97	49.89 99
250	62 500	15.81 14	50.00 00
N	N^2	$\sqrt{N}$	$\sqrt{10N}$

N	N^2	$\sqrt{N}$	$\sqrt{10N}$
250	62 500	15.81 14	50.00 00
251	63 001	15.84 30	50.09 99
252	63 504	15.87 45	50.19 96
253	64 009	15.90 60	50.29 91
254	64 516	15.93 74	50.39 84
255	65 025	15.96 87	50.49 75
256	65 536	16.00 00	50.59 64
257	66 049	16.03 12	50.69 52
258	66 564	16.06 24	50.79 37
259	67 081	16.09 35	50.89 20
260	67 600	16.12 45	50.99 02
261	68 121	16.15 55	51.08 82
262	68 644	16.18 64	51.18 59
263	69 169	16.21 73	51.28 35
264	69 696	16.24 81	51.38 09
265	70 225	16.27 88	51.47 82
266	70 756	16.30 95	51.57 52
267	71 289	16.34 01	51.67 20
268	71 824	16.37 07	51.76 87
269	72 361	16.40 12	51.86 52
270	72 900	16.43 17	51.96 15
271	73 441	16.46 21	52.05 77
272	73 984	16.49 24	52.15 36
273	74 529	16.52 27	52.24 94
274	75 076	16.55 29	52.34 50
275	75 625	16.58 31	52.44 04
276	76 176	16.61 32	52.53 57
277	76 729	16.64 33	52.63 08
278	77 284	16.67 33	52.72 57
279	77 841	16.70 33	52.82 05
280	78 400	16.73 32	52.91 50
281	78 961	16.76 31	53.00 94
282	79 524	16.79 29	53.10 37
283	80 089	16.82 26	53.19 77
284	80 656	16.85 23	53.29 17
285	81 225	16.88 19	53.38 54
286	81 796	16.91 15	53.47 90
287	82 369	16.94 11	53.57 24
288	82 944	16.97 06	53.66 56
289	83 521	17.00 00	53.75 87
290	84 100	17.02 94	53.85 16
291	84 681	17.05 87	53.94 44
292	85 264	17.08 80	54.03 70
293	85 849	17.11 72	54.12 95
294	86 436	17.14 64	54.22 18
295	87 025	17.17 56	54.31 39
296	87 616	17.20 47	54.40 59
297	88 209	17.23 37	54.49 77
298	88 804	17.26 27	54.58 94
299	89 401	17.29 16	54.68 09
300	90 000	17.32 05	54.77 23
N	N^2	$\sqrt{N}$	$\sqrt{10N}$

Table XI 901

Table XI (continued)

N	N²	√N	√10N
300	90 000	17.32 05	54.77 23
301	90 601	17.34 94	54.86 35
302	91 204	17.37 81	54.95 45
303	91 809	17.40 69	55.04 54
304	92 416	17.43 56	55.13 62
305	93 025	17.46 42	55.22 68
306	93 636	17.49 29	55.31 73
307	94 249	17.52 14	55.40 76
308	94 864	17.54 99	55.49 77
309	95 481	17.57 84	55.58 78
310	96 100	17.60 68	55.67 76
311	96 721	17.63 52	55.76 74
312	97 344	17.66 35	55.85 70
313	97 969	17.69 18	55.94 64
314	98 596	17.72 00	56.03 57
315	99 225	17.74 82	56.12 49
316	99 856	17.77 64	56.21 39
317	100 489	17.80 45	56.30 28
318	101 124	17.83 26	56.39 15
319	101 761	17.86 06	56.48 01
320	102 400	17.88 85	56.56 85
321	103 041	17.91 65	56.65 69
322	103 684	17.94 44	56.74 50
323	104 329	17.97 22	56.83 31
324	104 976	18.00 00	55.92 10
325	105 625	18.02 78	57.00 88
326	106 276	18.05 55	57.09 64
327	106 929	18.08 31	57.18 39
328	107 584	18.11 08	57.27 13
329	108 241	18.13 84	57.35 85
330	108 900	18.16 59	57.44 56
331	109 561	18.19 34	57.53 26
332	110 224	18.22 09	57.61 94
333	110 889	18.24 83	57.70 62
334	111 556	18.27 57	57.79 27
335	112 225	18.30 30	57.87 92
336	112 896	18.33 03	57.96 55
337	113 569	18.35 76	58.05 17
338	114 244	18.38 48	58.13 78
339	114 921	18.41 20	58.22 37
340	115 600	18 43 91	58.30 95
341	116 281	18.46 62	58.39 52
342	116 964	18.49 32	58.48 08
343	117 649	18.52 03	58.56 62
344	118 336	18.54 72	58.65 15
345	119 025	18.57 42	58.73 67
346	119 716	18.60 11	58.82 18
347	120 409	18.62 79	58.90 67
348	121 104	18.65 48	58.99 15
349	121 801	18.68 15	59.07 62
350	122 500	18.70 83	59.16 08

N	N²	√N	√10N
350	122 500	18.70 83	59.16 08
351	123 201	18.73 50	59.24 53
352	123 904	18.76 17	59.32 96
353	124 609	18.78 83	59.41 38
354	125 316	18.81 49	59.49 79
355	126 025	18.84 14	59.58 19
356	126 736	18.86 80	59.66 57
357	127 449	18.89 44	59.74 95
358	128 164	18.92 09	59.83 31
359	128 881	18.94 73	59.91 66
360	129 600	18.97 37	60.00 00
361	130 321	19.00 00	60.08 33
362	131 044	19.02 63	60.16 64
363	131 769	19.05 26	60.24 95
364	132 496	19.07 88	60.33 24
365	133 225	19.10 50	60.41 52
366	133 956	19.13 11	60.49 79
367	134 689	19.15 72	60.58 05
368	135 424	19.18 33	60.66 30
369	136 161	19.20 94	60.74 54
370	136 900	19.23 54	60.82 76
371	137 641	19.26 14	60.90 98
372	138 384	19.28 73	60.99 18
373	139 129	19.31 32	61.07 37
374	139 876	19.33 91	61.15 55
375	140 625	19.36 49	61.23 72
376	141 376	19.39 07	61.31 88
377	142 129	19.41 65	61.40 03
378	142 884	19.44 22	61.48 17
379	143 641	19.46 79	61.56 30
380	144 400	19.49 36	61.64 41
381	145 161	19.51 92	61.72 52
382	145 924	19.54 48	61.80 61
383	146 689	19.57 04	61.88 70
384	147 456	19.59 59	61.96 77
385	148 225	19.62 14	62.04 84
386	148 996	19.64 69	62.12 89
387	149 769	19.67 23	62.20 93
388	150 544	19.69 77	62.28 96
389	151 321	19.72 31	62.36 99
390	152 100	19.74 84	62.45 00
391	152 881	19.77 37	62.53 00
392	153 664	19.79 90	62.60 99
393	154 449	19.82 42	62.68 97
394	155 236	19.84 94	62.76 94
395	156 025	19.87 46	62.84 90
396	156 816	19.89 97	62.92 85
397	157 609	19.92 49	63.00 79
398	158 404	19.94 99	63.08 72
399	159 201	19.97 50	63.16 64
400	160 000	20.00 00	63.24 56

Table XI (continued)

N	N^2	$\sqrt{N}$	$\sqrt{10N}$	N	N^2	$\sqrt{N}$	$\sqrt{10N}$
400	160 000	20.00 00	63.24 56	**450**	202 500	21.21 32	67.08 20
401	160 801	20.02 50	63.32 46	451	203 401	21.23 68	67.15 65
402	161 604	20.04 99	63.40 35	452	204 304	21.26 03	67.23 09
403	162 409	20.07 49	63.48 23	453	205 209	21.28 38	67.30 53
404	163 216	20.09 98	63.56 10	454	206 116	21.30 73	67.37 95
405	164 025	20.12 46	63.63 96	455	207 025	21.33 07	67.45 37
406	164 836	20.14 94	63.71 81	**456**	207 936	21.35 42	67.52 78
407	165 649	20.17 42	63.79 66	457	208 849	21.37 76	67.60 18
408	166 464	20.19 90	63.87 49	458	209 764	21.40 09	67.67 57
409	167 281	20.22 37	63.95 31	459	210 681	21.42 43	67.74 95
410	168 100	20.24 85	64.03 12	460	211 600	21.44 76	67.82 33
411	168 921	20.27 31	64.10 93	**461**	212 521	21.47 09	67.89 70
412	169 744	20.29 78	64.18 72	462	213 444	21.49 42	67.97 06
413	170 569	20.32 24	64.26 51	463	214 369	21.51 74	68.04 41
414	171 396	20.34 70	64.34 28	464	215 296	21.54 07	68.11 75
415	172 225	20.37 15	64.42 05	465	216 225	21.56 39	68.19 09
416	173 056	20.39 61	64.49 81	**466**	217 156	21.58 70	68.26 42
417	173 889	20.42 06	64.57 55	467	218 089	21.61 02	68.33 74
418	174 724	20.44 50	64.65 29	468	219 024	21.63 33	68.41 05
419	175 561	20.46 95	64.73 02	469	219 961	21.65 64	68.48 36
420	176 400	20.49 39	64.80 74	470	220 900	21.67 95	68.55 65
421	177 241	20.51 83	64.88 45	**471**	221 841	21.70 25	68.62 94
422	178 084	20.54 26	64.96 15	472	222 784	21.72 56	68.70 23
423	178 929	20.56 70	65.03 85	473	223 729	21.74 86	68.77 50
424	179 776	20.59 13	65.11 53	474	224 676	21.77 15	68.84 77
425	180 625	20.61 55	65.19 20	475	225 625	21.79 45	68.92 02
426	181 476	20.63 98	65.26 87	**476**	226 576	21.81 74	68.99 28
427	182 329	20.66 40	65.34 52	477	227 529	21.84 03	69.06 52
428	183 184	20.68 82	65.42 17	478	228 484	21.86 32	69.13 75
429	184 041	20.71 23	65.49 81	479	229 441	21.88 61	69.20 98
430	184 900	20.73 64	65.57 44	480	230 400	21.90 89	69.28 20
431	185 761	20.76 05	65.65 06	**481**	231 361	21.93 17	69.35 42
432	186 624	20.78 46	65.72 67	482	232 324	21.95 45	69.42 62
433	187 489	20.80 87	65.80 27	483	233 289	21.97 73	69.49 82
434	188 356	20.83 27	65.87 87	484	234 256	22.00 00	69.57 01
435	189 225	20.85 67	65.95 45	485	235 225	22.02 27	69.64 19
436	190 096	20.88 06	66.03 03	**486**	236 196	22.04 54	69.71 37
437	190 969	20.90 45	66.10 60	487	237 169	22.06 81	69.78 54
438	191 844	20.92 84	66.18 16	488	238 144	22.09 07	69.85 70
439	192 721	20.95 23	66.25 71	489	239 121	22.11 33	69.92 85
440	193 600	20.97 62	66.33 25	490	240 100	22.13 59	70.00 00
441	194 481	21.00 00	66.40 78	**491**	241 081	22.15 85	70.07 14
442	195 364	21.02 38	66.48 31	492	242 064	22.18 11	70.14 27
443	196 249	21.04 76	66.55 82	493	243 049	22.20 36	70.21 40
444	197 136	21.07 13	66.63 33	494	244 036	22.22 61	70.28 51
445	198 025	21.09 50	66.70 83	495	245 025	22.24 86	70.35 62
446	198 916	21.11 87	66.78 32	**496**	246 016	22.27 11	70.42 73
447	199 809	21.14 24	66.85 81	497	247 009	22.29 35	70.49 82
448	200 704	21.16 60	66.93 28	498	248 004	22.31 59	70.56 91
449	201 601	21.18 96	67.00 75	499	249 001	22.33 83	70.63 99
450	202 500	21.21 32	67.08 20	500	250 000	22.36 07	70.71 07
N	N^2	$\sqrt{N}$	$\sqrt{10N}$	N	N^2	$\sqrt{N}$	$\sqrt{10N}$

Table XI 903

Table XI (continued)

N	N²	√N	√10N
500	250 000	22.36 07	70.71 07
501	251 001	22.38 30	70.78 14
502	252 004	22.40 54	70.85 20
503	253 009	22.42 77	70.92 25
504	254 016	22.44 99	70.99 30
505	255 025	22.47 22	71.06 34
506	256 036	22.49 44	71.13 37
507	257 049	22.51 67	71.20 39
508	258 064	22.53 89	71.27 41
509	259 081	22.56 10	71.34 42
510	260 100	22.58 32	71.41 43
511	261 121	22.60 53	71.48 43
512	262 144	22.62 74	71.55 42
513	263 169	22.64 95	71.62 40
514	264 196	22.67 16	71.69 38
515	265 225	22.69 36	71.76 35
516	266 256	22.71 56	71.83 31
517	267 289	22.73 76	71.90 27
518	268 324	22.75 96	71.97 22
519	269 361	22.78 16	72.04 17
520	270 400	22.80 35	72.11 10
521	271 441	22.82 54	72.18 03
522	272 484	22.84 73	72.24 96
523	273 529	22.86 92	72.31 87
524	274 576	22.89 10	72.38 78
525	275 625	22.91 29	72.45 69
526	276 676	22.93 47	72.52 59
527	277 729	22.95 65	72.59 48
528	278 784	22.97 83	72.66 36
529	279 841	23.00 00	72.73 24
530	280 900	23.02 17	72.80 11
531	281 961	23.04 34	72.86 97
532	283 024	23.06 51	72.93 83
533	284 089	23.08 68	73.00 68
534	285 156	23.10 84	73.07 53
535	286 225	23.13 01	73.14 37
536	287 296	23.15 17	73.21 20
537	288 369	23.17 33	73.28 03
538	289 444	23.19 48	73.34 85
539	290 521	23.21 64	73.41 66
540	291 600	23.23 79	73.48 47
541	292 681	23.25 94	73.55 27
542	293 764	23.28 09	73.62 06
543	294 849	23.30 24	73.68 85
544	295 936	23.32 38	73.75 64
545	297 025	23.34 52	73.82 41
546	298 116	23.36 66	73.89 18
547	299 209	23.38 80	73.95 94
548	300 304	23.40 94	74.02 70
549	301 401	23.43 07	74.09 45
550	302 500	23.45 21	74.16 20
N	N²	√N	√10N

N	N²	√N	√10N
550	302 500	23.45 21	74.16 20
551	303 601	23.47 34	74.22 94
552	304 704	23.49 47	74.29 67
553	305 809	23.51 60	74.36 40
554	306 916	23.53 72	74.43 12
555	308 025	23.55 84	74.49 83
556	309 136	23.57 97	74.56 54
557	310 249	23.60 08	74.63 24
558	311 364	23.62 20	74.69 94
559	312 481	23.64 32	74.76 63
560	313 600	23.66 43	74.83 31
561	314 721	23.68 54	74.89 99
562	315 844	23.70 65	74.96 67
563	316 969	23.72 76	75.03 33
564	318 096	23.74 87	75.09 99
565	319 225	23.76 97	75.16 65
566	320 356	23.79 08	75.23 30
567	321 489	23.81 18	75.29 94
568	322 624	23.83 28	75.36 58
569	323 761	23.85 37	75.43 21
570	324 900	23.87 47	75.49 83
571	326 041	23.89 56	75.56 45
572	327 184	23.91 65	75.63 07
573	328 329	23.93 74	75.69 68
574	329 476	23.95 83	75.76 28
575	330 625	23.97 92	75.82 88
576	331 776	24.00 00	75.89 47
577	332 929	24.02 08	75.96 05
578	334 084	24.04 16	76.02 63
579	335 241	24.06 24	76.09 20
580	336 400	24.08 32	76.15 77
581	337 561	24.10 39	76.22 34
582	338 724	24.12 47	76.28 89
583	339 889	24.14 54	76.35 44
584	341 056	24.16 61	76.41 99
585	342 225	24.18 68	76.48 53
586	343 396	24.20 74	76.55 06
587	344 569	24.22 81	76.61 59
588	345 744	24.24 87	76.68 12
589	346 921	24.26 93	76.74 63
590	348 100	24.28 99	76.81 15
591	349 281	24.31 05	76.87 65
592	350 464	24.33 11	76.94 15
593	351 649	24.35 16	77.00 65
594	352 836	24.37 21	77.07 14
595	354 025	24.39 26	77.13 62
596	355 216	24.41 31	77.20 10
597	356 409	24.43 36	77.26 58
598	357 604	24.45 40	77.33 05
599	358 801	24.47 45	77.39 51
600	360 000	24.49 49	77.45 97
N	N²	√N	√10N

Table XI (continued)

N	N²	√N	√10N	N	N²	√N	√10N
600	360 000	24.49 49	77.45 97	**650**	422 500	25.49 51	80.62 26
601	361 201	24.51 53	77.52 42	651	423 801	25.51 47	80.68 46
602	362 404	24.53 57	77.58 87	652	425 104	25.53 43	80.74 65
603	363 609	24.55 61	77.65 31	653	426 409	25.55 39	80.80 84
604	364 816	24.57 64	77.71 74	654	427 716	25.57 34	80.87 03
605	366 025	24.59 67	77.78 17	655	429 025	25.59 30	80.93 21
606	367 236	24.61 71	77.84 60	**656**	430 336	25.61 25	80.99 38
607	368 449	24.63 74	77.91 02	657	431 649	25.63 20	81.05 55
608	369 664	24.65 77	77.97 44	658	432 964	25.65 15	81.11 72
609	370 881	24.67 79	78.03 85	659	434 281	25.67 10	81.17 88
610	372 100	24.69 82	78.10 25	660	435 600	25.69 05	81.24 04
611	373 321	24.71 84	78.16 65	**661**	436 921	25.70 99	81.30 19
612	374 544	24.73 86	78.23 04	662	438 244	25.72 94	81.36 34
613	375 769	24.75 88	78.29 43	663	439 569	25.74 88	81.42 48
614	376 996	24.77 90	78.35 82	664	440 896	25.76 82	81.48 62
615	378 225	24.79 92	78.42 19	665	442 225	25.78 76	81.54 75
616	379 456	24.81 93	78.48 57	**666**	443 556	25.80 70	81.60 88
617	380 689	24.83 95	78.54 93	667	444 889	25.82 63	81.67 01
618	381 924	24.85 96	78.61 30	668	446 224	25.84 57	81.73 13
619	383 161	24.87 97	78.67 66	669	447 561	25.86 50	81.79 24
620	384 400	24.89 98	78.74 01	670	448 900	25.88 44	81.85 35
621	385 641	24.91 99	78.80 36	**671**	450 241	25.90 37	81.91 46
622	386 884	24.93 99	78.86 70	672	451 584	25.92 30	81.97 56
623	388 129	24.96 00	78.93 03	673	452 929	25.94 22	82.03 66
624	389 376	24.98 00	78.99 37	674	454 276	25.96 15	82.09 75
625	390 625	25.00 00	79.05 69	675	455 625	25.98 08	82.15 84
626	391 876	25.02 00	79.12 02	**676**	456 976	26.00 00	82.21 92
627	393 129	25.04 00	79.18 33	677	458 329	26.01 92	82.28 00
628	394 384	25.05 99	79.24 65	678	459 684	26.03 84	82.34 08
629	395 641	25.07 99	79.30 95	679	461 041	26.05 76	82.40 15
630	396 900	25.09 98	79.37 25	680	462 400	26.07 68	82.46 21
631	398 161	25.11 97	79.43 55	**681**	463 761	26.09 60	82.52 27
632	399 424	25.13 96	79.49 84	682	465 124	26.11 51	82.58 33
633	400 689	25.15 95	79.56 13	683	466 489	26.13 43	82.64 38
634	401 956	25.17 94	79.62 41	684	467 856	26.15 34	82.70 43
635	403 225	25.19 92	79.68 69	685	469 225	26.17 25	82.76 47
636	404 496	25.21 90	79.74 96	**686**	470 596	26.19 16	82.82 51
637	405 769	25.23 89	79.81 23	687	471 969	26.21 07	82.88 55
638	407 044	25.25 87	79.87 49	688	473 344	26.22 98	82.94 58
639	408 321	25.27 84	79.93 75	689	474 721	26.24 88	83.00 60
640	409 600	25.29 82	80.00 00	690	476 100	26.26 79	83.06 62
641	410 881	25.31 80	80.06 25	**691**	477 481	26.28 69	83.12 64
642	412 164	25.33 77	80.12 49	692	478 864	26.30 59	83.18 65
643	413 449	25.35 74	80.18 73	693	480 249	26.32 49	83.24 66
644	414 736	25.37 72	80.24 96	694	481 636	26.34 39	83.30 67
645	416 025	25.39 69	80.31 19	695	483 025	26.36 29	83.36 67
646	417 316	25.41 65	80.37 41	**696**	484 416	26.38 18	83.42 66
647	418 609	25.43 62	80.43 63	697	485 809	26.40 08	83.48 65
648	419 904	25.45 58	80.49 84	698	487 204	26.41 97	83.54 64
649	421 201	25.47 55	80.56 05	699	488 601	26.43 86	83.60 62
650	422 500	25.49 51	80.62 26	700	490 000	26.45 75	83.66 60
N	N²	√N	√10N	N	N²	√N	√10N

Table XI   905

Table XI (continued)

N	N²	√N̄	√10N̄
700	490 000	26.45 75	83.66 60
701	491 401	26.47 64	83.72 57
702	492 804	26.49 53	83.78 54
703	494 209	26.51 41	83.84 51
704	495 616	26.53 30	83.90 47
705	497 025	26.55 18	83.96 43
706	498 436	26.57 07	84.02 38
707	499 849	26.58 95	84.08 33
708	501 264	26.60 83	84.14 27
709	502 681	26.62 71	84.20 21
710	504 100	26.64 58	84.26 15
711	505 521	26.66 46	84.32 08
712	506 944	26.68 33	84.38 01
713	508 369	26.70 21	84.43 93
714	509 796	26.72 08	84.49 85
715	511 225	26.73·95	84.55 77
716	512 656	26.75 82	84.61 68
717	514 089	26.77 69	84.67 59
718	515 524	26.79 55	84.73 49
719	516 961	26.81 42	84.79 39
720	518 400	26.83 28	84.85 28
721	519 841	26.85 14	84.91 17
722	521 284	26.87 01	84.97 06
723	522 729	26.88 87	85.02 94
724	524 176	26.90 72	85.08 82
725	525 625	26.92 58	85.14 69
726	527 076	26.94 44	85.20 56
727	528 529	26.96 29	85.26 43
728	529 984	26.98 15	85.32 29
729	531 441	27.00 00	85.38 15
730	532 900	27.01 85	85.44 00
731	534 361	27.03 70	85.49 85
732	535 824	27.05 55	85.55 70
733	537 289	27.07 40	85.61 54
734	538 756	27.09 24	85.67 38
735	540 225	27.11 09	85.73 21
736	541 696	27.12 93	85.79 04
737	543 169	27.14 77	85.84 87
738	544 644	27.16 62	85.90 69
739	546 121	27.18 46	85.96 51
740	547 600	27.20 29	86.02 33
741	549 081	27.22 13	86.08 14
742	550 564	27.23 97	86.13 94
743	552 049	27.25 80	86.19 74
744	553 536	27.27 64	86.25 54
745	555 025	27.29 47	86.31 34
746	556 516	27.31 30	86.37 13
747	558 009	27.33 13	86.42 92
748	559 504	27.34 96	86.48 70
749	561 001	27.36 79	86.54 48
750	562 500	27.38 61	86.60 25

N	N²	√N̄	√10N̄
750	562 500	27.38 61	86.60 25
751	564 001	27.40 44	86.66 03
752	565 504	27.42 26	86.71 79
753	567 009	27.44 08	86.77 56
754	568 516	27.45 91	86.83 32
755	570 025	27.47 73	86.89 07
756	571 536	27.49 55	86.94 83
757	573 049	27.51 36	87.00 57
758	574 564	27.53 18	87.06 32
759	576 081	27.55 00	87.12 06
760	577 600	27.56 81	87.17 80
761	579 121	27.58 62	87.23 53
762	580 644	27.60 43	87.29 26
763	582 169	27.62 25	87.34 99
764	583 696	27.64 05	87.40 71
765	585 225	27.65 86	87.46 43
766	586 756	27.67 67	87.52 14
767	588 289	27.69 48	87.57 85
768	589 824	27.71 28	87.63 56
769	591 361	27.73 08	87.69 26
770	592 900	27.74 89	87.74 96
771	594 441	27.76 69	87.80 66
772	595 984	27.78 49	87.86 35
773	597 529	27.80 29	87.92 04
774	599 076	27.82 09	87.97 73
775	600 625	27.83 88	88.03 41
776	602 176	27.85 68	88.09 09
777	603 729	27.87 47	88.14 76
778	605 284	27.89 27	88.20 43
779	606 841	27.91 06	88.26 10
780	608 400	27.92 85	88.31 76
781	609 961	27.94 64	88.37 42
782	611 524	27.96 43	88.43 08
783	613 089	27.98 21	88.48 73
784	614 656	28.00 00	88.54 38
785	616 225	28.01 79	88.60 02
786	617 796	28.03 57	88.65 66
787	619 369	28.05 35	88.71 30
788	620 944	28.07 13	88.76 94
789	622 521	28.08 91	88.82 57
790	624 100	28.10 69	88.88 19
791	625 681	28.12 47	88.93 82
792	627 264	28.14 25	88.99 44
793	628 849	28.16 03	89.05 05
794	630 436	28.17 80	89.10 67
795	632 025	28.19 57	89.16 28
796	633 616	28.21 35	89.21 88
797	635 209	28.23 12	89.27 49
798	636 804	28.24 89	89.33 08
799	638 401	28.26 66	89.38 68
800	640 000	28.28 43	89.44 27

Table XI (continued)

N	N²	√N	√10N
800	640 000	28.28 43	89.44 27
801	641 601	28.30 19	89.49 86
802	643 204	28.31 96	89.55 45
803	644 809	28.33 73	89.61 03
804	646 416	28.35 49	89.66 60
805	648 025	28.37 25	89.72 18
806	649 636	28.39 01	89.77 75
807	651 249	28.40 77	89.83 32
808	652 864	28.42 53	89.88 88
809	654 481	28.44 29	89.94 44
810	656 100	28.46 05	90.00 00
811	657 721	28.47 81	90.05 55
812	659 344	28.49 56	90.11 10
813	660 969	28.51 32	90.16 65
814	662 596	28.53 07	90.22 19
815	664 225	28.54 82	90.27 74
816	665 856	28.56 57	90.33 27
817	667 489	28.58 32	90.38 81
818	669 124	28.60 07	90.44 34
819	670 761	28.61 82	90.49 86
820	672 400	28.63 56	90.55 39
821	674 041	28.65 31	90.60 91
822	675 684	28.67 05	90.66 42
823	677 329	28.68 80	90.71 93
824	678 976	28.70 54	90.77 44
825	680 625	28.72 28	90.82 95
826	682 276	28.74 02	90.88 45
827	683 929	28.75 76	90.93 95
828	685 584	28.77 50	90.99 45
829	687 241	28.79 24	91.04 94
830	688 900	28.80 97	91.10 43
831	690 561	28.82 71	91.15 92
832	692 224	28.84 44	91.21 40
833	693 889	28.86 17	91.26 88
834	695 556	28.87 91	91.32 36
835	697 225	28.89 64	91.37 83
836	698 896	28.91 37	91.43 30
837	700 569	28.93 10	91.48 77
838	702 244	28.94 82	91.54 23
839	703 921	28.96 55	91.59 69
840	705 600	28.98 28	91.65 15
841	707 281	29.00 00	91.70 61
842	708 964	29.01 72	91.76 06
843	710 649	29.03 45	91.81 50
844	712 336	29.05 17	91.86 95
845	714 025	29.06 89	91.92 39
846	715 716	29.08 61	91.97 83
847	717 409	29.10 33	92.03 26
848	719 104	29.12 04	92.08 69
849	720 801	29.13 76	92.14 12
850	722 500	29.15 48	92.19 54

N	N²	√N	√10N
850	722 500	29.15 48	92.19 54
851	724 201	29.17 19	92.24 97
852	725 904	29.18 90	92.30 38
853	727 609	29.20 62	92.35 80
854	729 316	29.22 33	92.41 21
855	731 025	29.24 04	92.46 62
856	732 736	29.25 75	92.52 03
857	734 449	29.27 46	92.57 43
858	736 164	29.29 16	92.62 83
859	737 881	29.30 87	92.68 23
860	739 600	29.32 58	92.73 62
861	741 321	29.34 28	92.79 01
862	743 044	29.35 98	92.84 40
863	744 769	29.37 69	92.89 78
864	746 496	29.39 39	92.95 16
865	748 225	29.41 09	93.00 54
866	749 956	29.42 79	93.05 91
867	751 689	29.44 49	93.11 28
868	753 424	29.46 18	93.16 65
869	755 161	29.47 88	93.22 02
870	756 900	29.49 58	93.27 38
871	758 641	29.51 27	93.32 74
872	760 384	29.52 96	93.38 09
873	762 129	29.54 66	93.43 45
874	763 876	29.56 35	93.48 80
875	765 625	29.58 04	93.54 14
876	767 376	29.59 73	93.59 49
877	769 129	29.61 42	93.64 83
878	770 884	29.63 11	93.70 17
879	772 641	29.64 79	93.75 50
880	774 400	29.66 48	93.80 83
881	776 161	29.68 16	93.86 16
882	777 924	29.69 85	93.91 49
883	779 689	29.71 53	93.96 81
884	781 456	29.73 21	94.02 13
885	783 225	29.74 89	94.07 44
886	784 996	29.76 58	94.12 76
887	786 769	29.78 25	94.18 07
888	788 544	29.79 93	94.23 38
889	790 321	29.81 61	94.28 68
890	792 100	29.83 29	94.33 98
891	793 881	29.84 96	94.39 28
892	795 664	29.86 64	94.44 58
893	797 449	29.88 31	94.49 87
894	799 236	29.89 98	94.55 16
895	801 025	29.91 66	94.60 44
896	802 816	29.93 33	94.65 73
897	804 609	29.95 00	94.71 01
898	806 404	29.96 66	94.76 29
899	808 201	29.98 33	94.81 56
900	810 000	30.00 00	94.86 83

N	N²	√N	√10N

Table XI   907

Table XI (continued)

N	N²	√N	√10N
900	810 000	30.00 00	94.86 83
901	811 801	30.01 67	94.92 10
902	813 604	30.03 33	94.97 37
903	815 409	30.05 00	95.02 63
904	817 216	30.06 66	95.07 89
905	819 025	30.08 32	95.13 15
906	820 836	30.09 98	95.18 40
907	822 649	30.11 64	95.23 65
908	824 464	30.13 30	95.28 90
909	826 281	30.14 96	95.34 15
910	828 100	30.16 62	95.39 39
911	829 921	30.18 28	95.44 63
912	831 744	30.19 93	95.49 87
913	833 569	30.21 59	95.55 10
914	835 396	30.23 24	95.60 33
915	837 225	30.24 90	95.65 56
916	839 056	30.26 55	95.70 79
917	840 889	30.28 20	95.76 01
918	842 724	30.29 85	95.81 23
919	844 561	30.31 50	95.86 45
920	846 400	30.33 15	95.91 66
921	848 241	30.34 80	95.96 87
922	850 084	30.36 45	96.02 08
923	851 929	30.38 09	96.07 29
924	853 776	30.39 74	96.12 49
925	855 625	30.41 38	96.17 69
926	857 476	30.43 02	96.22 89
927	859 329	30.44 67	96.28 08
928	861 184	30.46 31	96.33 28
929	863 041	30.47 95	96.38 46
930	864 900	30.49 59	96.43 65
931	866 761	30.51 23	96.48 83
932	868 624	30.52 87	96.54 01
933	870 489	30.54 50	96.59 19
934	872 356	30.56 14	96.64 37
935	874 225	30.57 78	96.69 54
936	876 096	30.59 41	96.74 71
937	877 969	30.61 05	96.79 88
938	879 844	30.62 68	96.85 04
939	881 721	30.64 31	96.90 20
940	883 600	30.65 94	96.95 36
941	885 481	30.67 57	97.00 52
942	887 364	30.69 20	97.05 67
943	889 249	30.70 83	97.10 82
944	891 136	30.72 46	97.15 97
945	893 025	30.74 09	97.21 11
946	894 916	30.75 71	97.26 25
947	896 809	30.77 34	97.31 39
948	898 704	30.78 96	97.36 53
949	900 601	30.80 58	97.41 66
950	902 500	30.82 21	97.46 79
N	N²	√N	√10N

N	N²	√N	√10N
950	902 500	30.82 21	97.46 79
951	904 401	30.83 83	97.51 92
952	906 304	30.85 45	97.57 05
953	908 209	30.87 07	97.62 17
954	910 116	30.88 69	97.67 29
955	912 025	30.90 31	97.72 41
956	913 936	30.91 92	97.77 53
957	915 849	30.93 54	97.82 64
958	917 764	30.95 16	97.87 75
959	919 681	30.96 77	97.92 85
960	921 600	30.98 39	97.97 96
961	923 521	31.00 00	98.03 06
962	925 444	31.01 61	98.08 16
963	927 369	31.03 22	98.13 26
964	929 296	31.04 83	98.18 35
965	931 225	31.06 44	98.23 44
966	933 156	31.08 05	98.28 53
967	935 089	31.09 66	98.33 62
968	937 024	31.11 27	98.38 70
969	938 961	31.12 88	98.43 78
970	940 900	31.14 48	98.48 86
971	942 841	31.16 09	98.53 93
972	944 784	31.17 69	98.59 01
973	946 729	31.19 29	98.64 08
974	948 676	31.20 90	98.69 14
975	950 625	31.22 50	98.74 21
976	952 576	31.24 10	98.79 27
977	954 529	31.25 70	98.84 33
978	956 484	31.27 30	98.89 39
979	958 441	31.28 90	98.94 44
980	960 400	31.30 50	98.99 49
981	962 361	31.32 09	99.04 54
982	964 324	31.33 69	99.09 59
983	966 289	31.35 28	99.14 64
984	968 256	31.36 88	99.19 68
985	970 225	31.38 47	99.24 72
986	972 196	31.40 06	99.29 75
987	974 169	31.41 66	99.34 79
988	976 144	31.43 25	99.39 82
989	978 121	31.44 84	99.44 85
990	980 100	31.46 43	99.49 87
991	982 081	31.48 02	99.54 90
992	984 064	31.49 60	99.59 92
993	986 049	31.51 19	99.64 94
994	988 036	31.52 78	99.69 95
995	990 025	31.54 36	99.74 97
996	992 016	31.55 95	99.79 98
997	994 009	31.57 53	99.84 99
998	996 004	31.59 11	99.89 99
999	998 001	31.60 70	99.95 00
1000	1000 000	31.62 28	100.00 00
N	N²	√N	√10N

References
and Suggestions
for Further Reading

SETS, RELATIONS, AND FUNCTIONS

Birkhoff, G., and MacLane, S.   *A survey of modern algebra* (ed. 3).   New York: Macmillan, 1965.

Kemeny, J. G., Mirkil, H., Snell, J. L., and Thompson, G. L.   *Finite mathematical structures.*   Englewood Cliffs, N.J.: Prentice-Hall, 1959.

Kershner, R. B., and Wilcox, L. R.   *The anatomy of mathematics.*   New York: Ronald Press, 1950.

MEASUREMENT AND RELATED MATTERS

Coombs, C. H., Dawes, R. M., and Tversky, A.   *Mathematical psychology, an elementary introduction.*   Englewood Cliffs, N.J.: Prentice-Hall, 1970.

Stevens, S. S.   *Handbook of experimental psychology.*   New York: Wiley, 1951.

Thrall, R., Coombs, C., and Davis, R.   *Decision processes.*   New York: Wiley, 1959.

Torgerson, W.   *Theory and methods of scaling.*   New York: Wiley, 1958.

PROBABILITY THEORY

Feller, W.   *An introduction to probability theory and its applications*, Vol. I, (ed. 3).   New York: Wiley, 1968.

Jeffreys, H.   *Theory of probability.*   Oxford: Clarendon Press, 1961.

Kyburg, H. E., and Smokler, H. E.   *Studies in subjective probability.*   New York: Wiley, 1964.

908

Mosteller F., Rourke, R. E. K., and Thomas, G. B.   *Probability with statistical applications.*   Reading, Mass.: Addison-Wesley, 1961.

Parzen, E.   *Modern probability theory and its applications.*   New York: Wiley, 1960.

STATISTICS

Alexander, H. W.   *Elements of mathematical statistics.*   New York: Wiley, 1961.

Anderson, R. L., and Bancroft, T. A.   *Statistical theory in research.*   New York: McGraw-Hill, 1952.

Bennett, C. A., and Franklin, N. L.   *Statistical analysis in chemistry and the chemical industry.*   New York: Wiley, 1954.

Box, G. E. P.   Non-normality and tests on variances.   *Biometrika 40*, 1953, 318–335.

Box, G. E. P.   Some theorems on quadratic forms applied in the study of analysis of variance problems: I. Effect of inequality of variance in the one-way classification.   *Annals Math. Stat. 25*, 1954, 290–302.

Box, G. E. P.   Some theorems on quadratic forms applied in the study of analysis of variance problems: II. Effect of inequality of variance and of correlation of errors in the two-way classification."   *Annals Math. Stat. 25*, 1954, 484–498.

Brunk, H. D.   *An introduction to mathematical statistics* (ed. 2).   Boston: Ginn, 1965.

Carroll, J. B.   The nature of the data, or how to choose a correlation coefficient.   *Psychometrika 26*, 1961, 347–372.

Cooley, W. W., and Lohnes, P. R.   *Multivariate procedures for the behavioral sciences.*   New York: Wiley, 1962.

Cramer, H.   *Mathematical methods of statistics.*   Princeton, N.J.: Princeton University Press, 1946.

Dixon, W., and Massey, F.   *Introduction to statistical analysis* (ed. 2).   New York: McGraw-Hill, 1957.

Freund, J. E., *Mathematical statistics.*   Englewood Cliffs, N.J.: Prentice-Hall, 1962.

Grant, D. A.   Testing the null hypothesis and the strategy and tactics of investigating theoretical models.   *Psych. Rev. 69*, 1962, 54–61.

Graybill, F. A.   *An introduction to linear statistical models.*   New York: McGraw-Hill, 1961.

Green, B. F., and Tukey, J. W.   Complex analyses of variance: General problems.   *Psychometrika 25*, 1960, 127–152.

Hays, W. L.   Statistical theory, in *Annual Review of Psychology*, Vol. 19, 417–436.   Palo Alto, Calif.: Annual Reviews, 1968.

Hodges, J. L., and Lehmann, E. L.   *Basic concepts of probability and statistics.*   San Francisco: Holden-Day, 1964.

Hoel, P. G.   *Introduction to mathematical statistics* (ed. 3).   New York: Wiley, 1962.

Hogg, R. V., and Craig, A. T. *Introduction to mathematical statistics* (ed. 2). New York: Macmillan, 1965.

Huff, D. *How to lie with statistics*. New York: Norton, 1954.

Kempthorne, O. The randomization theory of experimental inference. *J. Am. Stat. Ass. 50*, 1955, 946–967.

Kendall, M. G., and Stuart, A. *The advanced theory of statistics*, 3 Vols. London: Griffin, 1961–1966.

Lehmann, E. L. *Testing statistical hypotheses*. New York: Wiley, 1959.

Lewis, D. *Quantitative methods in psychology*. New York: McGraw-Hill, 1960.

McNemar, Q. *Psychological statistics* (ed. 4). New York: Wiley, 1968.

Mood, A. M., and Graybill, F. A. *Introduction to the theory of statistics* (ed. 2). New York: McGraw-Hill, 1963.

Paull, A. E. On a preliminary test for pooling mean squares in the analysis of variance. *Annals Math. Stat. 21*, 1950, 539–556.

Rao, C. R. *Advanced statistical methods in biometric research*. New York: Wiley, 1952.

Stillson, D. W. *Probability and statistics in psychological research and theory*. San Francisco: Holden-Day, 1966.

Tate, R. F., and Klett, G. W. Optimal confidence intervals for the variance of a normal distribution. *J. Amer. Stat. Ass. 54*, 1959, 674–682.

Tukey, J. W. One degree of freedom for non-additivity. *Biometrics 5*, 1949, 232–242.

Wallis, W. A., and Roberts, H. V. *Statistics: A new approach*. New York: Free Press, 1956.

Wilks, S. S. *Mathematical statistics*. New York: Wiley, 1962.

DECISION THEORY AND BAYESIAN INFERENCE

Blackwell, D., and Girshick, M. A. *Theory of games and statistical decisions*. New York: Wiley, 1954.

Chernoff, H., and Moses, L. E. *Elementary decision theory*. New York: Wiley, 1959.

Edwards, W., Lindman, H., and Savage, L. J. Bayesian statistical inference for psychological research. *Psychological Review 70*, 1963, 193–242.

Ferguson, T. S. *Mathematical statistics: A decision-theoretic approach*. New York: Academic Press, 1967.

Fishburn, P. C. *Decision and value theory*. New York: Wiley, 1964.

Good, I. J. *The estimation of probabilities*. Cambridge, Mass.: MIT Press, 1965.

Lindley, D. V. *Introduction to probability and statistics from a Bayesian viewpoint* (2 Vols.). London: Cambridge University Press, 1965.

Luce, R. D., and Raiffa, H. *Games and decisions*. New York: Wiley, 1957.

Pratt, J. W., Raiffa, H., and Schlaifer, R. *Introduction to statistical decision theory*. New York: McGraw-Hill, 1965.

Raiffa, H. *Decision analysis*. Reading, Mass.: Addison-Wesley, 1968.

Raiffa, H., and Schlaifer, R.   *Applied statistical decision theory.*   Boston: Graduate School of Business, Harvard University, 1961.

Savage, L. J.   *The foundations of statistics.*   New York: Wiley, 1954.

Schlaifer, R.   *Probability and statistics for business decisions.*   New York: McGraw-Hill, 1959.

Schlaifer, R.   *Analysis of decisions under uncertainty.*   New York: McGraw-Hill, 1969.

Schmitt, S. A.   *Measuring uncertainty: An elementary introduction to Bayesian statistics.*   Reading, Mass.: Addison-Wesley Publishing Company, Inc., 1969.

Von Neumann, J., and Morgenstern, O.   *Theory of games and economic behavior.*   Princeton, N.J.: Princeton University Press, 1944.

Wald, A.   *Statistical decision functions.*   New York: Wiley, 1950.

EXPERIMENTAL DESIGN

Cochran, W. G., and Cox, G.   *Experimental designs* (ed. 2).   New York: Wiley, 1957.

Cox, D. R.   *Planning experiments.*   New York: Wiley, 1958.

Edwards, A.   *Experimental design in psychological research* (rev. ed.).   New York: Holt, Rinehart and Winston, 1960.

Fisher, R. A.   *The design of experiments* (ed. 8).   Edinburgh: Oliver & Boyd, 1966.

Lindquist, E. F.   *Design and analysis of experiments in psychology and education.*   Boston: Houghton Mifflin, 1953.

Scheffe, H.   *The analysis of variance.*   New York: Wiley, 1959.

Snedecor, G. W., and Cochran, W. G.   *Statistical methods* (ed. 6).   Ames, Iowa: Iowa State College Press, 1967.

Winer, B. J.   *Statistical principles in experimental design.*   New York: McGraw-Hill, 1962.

METHODS FOR CATEGORICAL AND ORDER DATA

Attneave, F.   *Applications of information theory to psychology.*   New York: Holt, Rinehart and Winston, 1959.

Cochran, W. G.   The comparison of percentages in matched samples.   *Biometrika 37*, 1950, 256–266.

Garner, W. R.   *Uncertainty and structure as psychological concepts.*   New York: Wiley, 1962.

Goodman, L. A., and Kruskal, W. H.   Measures of association for cross-classifications.   *J. Amer. Stat. Ass. 49*, 1954, 732–764.

Kendall, M. G.   *Rank correlation methods* (ed. 3).   London: Griffin, 1963.

Kruskal, W. H.   Ordinal measures of association.   *J. Amer. Stat. Ass. 53*, 1958, 814–861.

Kruskal, W. H., and Wallis, W. A.   Use of ranks in one-criterion variance analysis.   *J. Amer. Stat. Ass.* *47*, 1952, 583–621.

Maxwell, A. E.   *Analysing qualitative data.*   London: Methuen, 1961.

Siegel, S.   *Nonparametric methods for the behavioral sciences.*   New York: McGraw-Hill, 1956.

Walsh, J. E.   *A handbook of nonparametric statistics* (2 Vols.).   New York: Van Nostrand, 1962.

TABLES

Burington, R. S., and May, D. C.   *Handbook of probability and statistics with tables* (ed. 2).   New York: McGraw-Hill, 1970.

Fisher, R. A., and Yates, F.   *Statistical tables for biological, agricultural and medical research* (ed. 6).   Edinburgh: Oliver & Boyd, 1963.

Owen, D. B.   *Handbook of statistical tables.*   Reading, Mass.: Addison Wesley, 1962.

Pearson, E. S., and Hartley, H. O.   *Biometrika tables for statisticians*, Vol. I (ed. 3).   London: Cambridge University Press, 1967.

The Rand Corporation.   A million random digits with 100,000 normal deviates.   New York: Free Press, 1955.

Glossary

913

∞ . . . . an infinite value (60)

Σ. . . . . sum of (see Appendix A)

$\int_a^b f(x)\,dx$ . . . . area under the curve generated by the function $y = f(x)$, as cut off by the x interval with limits a and b (122)

$N!$ . . . . factorial of the integer N (173)

$\binom{N}{r}$ . . . . number of unordered combinations of N things taken r at a time, $0 \leq r \leq N$; a binomial coefficient (176)

e . . . . mathematical constant equal approximately to 2.7182818; base of the natural system of logarithms (202)

$\exp [x]$ . . . . alternate form of e^x (31)

$\log_e m$ . . . . logarithm to the base e for the value m (699)

π . . . . mathematical constant, equal approximately to 3.14159265; ratio of the circumference of a circle to its diameter (31)

$A \propto B$ . . . . A is proportional to B (838)

STATISTICAL SYMBOLS USED IN THIS TEXT

$\mathcal{A}$ . . . . family of events defined by some sample space (51)

a_j . . . . effect of the randomly selected treatment appearing as the jth such treatment in the experiment (528)

$\boldsymbol{a}(v)$ . . . . effect associated with the randomly selected treatment v (527)

(A_j, B_k) . . . . joint event of an observation in category A_j on attribute A, and in category B_k on attribute B (728)

$A_h = \dfrac{\psi_h}{n w_h}$ . . . . weight given to the orthogonal polynomial coefficient representing a trend of degree h in estimation of a curvilinear prediction function (695)

α . . . . (small Greek alpha) probability of rejecting H_0 when it is true; probability of Type I error (354)

$\alpha_j = \mu_j - \mu$ . . . . effect associated with the jth treatment applied; a fixed effect (460)

est. α_j . . . . estimated effect for treatment j (484)

b_k . . . . random effect associated with treatment k (543)

$b_{Y \cdot X}$ . . . . sample coefficient of linear regression of Y on X (630)

$b_{1 \cdot 2}$ . . . . linear regression coefficient for prediction of z_1 from z_2 (703)

$b_{12 \cdot 3}$ . . . . multiple regression coefficient applied to variable z_2 in predictions of z_1 from z_2 and z_3 (703)

$b_{13 \cdot 2}$ . . . . multiple regression coefficient applied to variable z_3 in predictions of z_1 from z_2 and z_3 (703)

$b_{12 \cdot 3 \ldots K}$ . . . . multiple regression coefficient for weighting z_2 in the prediction of z_1 from $z_2, \cdots, z_K$ (705)

β . . . . (small Greek beta) probability of failing to reject H_0 when it is false; probability of Type II error (354)

$1 - \beta$ . . . . power of a statistical test against some given true alternative to the null hypothesis (354)

β_k . . . . fixed effect associated with the treatment or factor level k (494)

$\beta_{Y \cdot X}$ . . . . regression coefficient for prediction of Y from X in the population (638)

$\beta_1, \beta_2, \cdots, \beta_{J-1} \ldots$ population regression coefficients for linear and curvilinear trends between Y and X (676)

C................   number of columns in the data table for a two-factor experiment (499)

C_{AB}.............   coefficient of contingency (745)

c_j...............   constant applied to a sample or population mean corresponding to treatment j in a comparison (584)

c_{jk}...............   random interaction effect associated with the combination of treatment j with treatment k (543)

$(c_1, \cdots, c_n)$......   set of n constant weights figuring in some linear combustion of n random variables (312)

χ^2...............   (small Greek chi, squared) random variable chi-square; also the Pearson chi-square statistic (431, 721)

$\chi^2_{(\nu)}$...............   chi-square variable with ν degrees of freedom (435)

$\chi^2_{(N-1;\alpha/2)}; \chi^2_{(N-1;1-\alpha/2)}$ values in the distribution of χ^2 with $N-1$ degrees of freedom, cutting off the upper $\alpha/2$ and $1-\alpha/2$ proportion of cases, respectively (441)

$$\chi^2 = \sum_j \frac{(f_{oj} - f_{ej})^2}{f_{ej}}$$ Pearson χ^2 statistic in a test of goodness of fit (721)

$\text{cov.}(X,Y)$.........   covariance of the two random variables X and Y (874)

$\text{cov.}(M_1,M_2)$.......   covariance of pairs of sample means (425)

$D_i = (Y_{i1} - Y_{i2})$...   difference between the scores for members of matched pair i from among N such pairs (425)

d_i...............   deviation of the score for observation i from the grand mean (222)

δ.................   (small Greek delta) noncentrality parameter for noncentral t or noncentral F (411, 488)

$\Delta = \dfrac{|\mu_1 - \mu_2|}{\sigma_{Y|X}}$......   (capital Greek delta) absolute difference between two population means, relative to the standard deviation for either population (419)

ΔX..............   width of an arbitrarily small interval of values for random variable X (120)

$E(X)$..............   expectation of random variable X (see Appendix B)

$E(u;D_1,H_0)$........   expected loss, given decision-rule 1 and truth of hypothesis H_0 (346)

e_{ij}...............   random error associated with the ith observation made under treatment j (461)

e_{ijk}...............   random error associated with the ith observation in the treatment-combination j, k (494)

$e_i(v)$..............   error associated with observation i made under treatment v (526)

$\eta^2_{Y \cdot X}$...............   (small Greek eta, squared) correlation ratio, for the relation of Y to X (683)

ϵ.................   (small Greek epsilon) an arbitrarily small positive number (60)

F................ F ratio computed from a sample; the random variable "F" (444)

F_α................   value of F required for significance at the α level, one-tailed, for a given number of degrees of freedom (607)

$F_{(\nu_1; \nu_2)}$ .... random variable distributed as F with ν_1 and ν_2 degrees of freedom (446)

$F_{(\alpha; \nu_1, \nu_2)}$ .... value cutting off the upper α proportion of cases in an F distribution with ν_1 and ν_2 degrees of freedom (447)

$f(a) = f(X = a)$ .... probability density for random variable X at the value $X = a$ (121)

$F(a) = p(X \leq a)$ ... cumulative probability of random variable X at the value a (123)

$f'(\theta = g)$ .... prior density for parameter $\theta = g$ (819)

$f''(\theta = g)$ .... posterior density for $\theta = g$ (819)

f .... frequency of a given measurement or event class (93)

f_{ej} .... expected frequency in event-category j (720)

f_{ejk} .... expected frequency in the cell j, k of a contingency table (730)

f_{oj} .... observed frequency in event-category j (720)

f_{ojk} .... observed frequency in the cell j, k of a contingency table (730)

γ .... (small Greek gamma) coefficient of predictive association between sets of ordered classes (800)

γ_{jk} .... fixed interaction effect associated with the combination of treatments j and k (494)

H: .... indicator of a statistical hypothesis (334)

H_0: .... hypothesis actually being tested; the "null" hypothesis (335)

H_1: .... hypothesis to be entertained if H_0 is rejected; the alternative hypothesis (335)

$H(A)$ .... average uncertainty associated with the attribute A (750)

$H(B|A)$ .... average uncertainty associated with the attribute B, given the status of an observation on attribute A (751)

h .... $1/\sigma^2$, the precision of a random process (831)

i .... class interval size in some frequency distribution (95)

i .... running subscript, ordinarily indicating the ith observation among some N distinct observations (Appendix A)

J .... number of different experimental treatments or groups in an experiment (400, Appendix A)

j .... running subscript indicating the jth treatment group or factor level (Appendix A)

K .... in a two-factor experiment, number of treatment groups or levels of the second factor; number of "blocks" of observations in an experiment (565, Appendix A)

k .... an arbitrarily small positive number (253)

k .... a running subscript, ordinarily indicating the kth treatment group or "block" of observations (Appendix A)

$L(x_1, \cdots, x_N|\theta)$ .. likelihood of a particular set of N sample values, given the value of the parameter (or set of parameters) θ (270)

LR_{AB} .... likelihood ratio, A to B (846)

λ_B .... (small Greek lambda) assymetric measure of predictive association for a contingency table (747)

λ_{AB} .... symmetric measure of predictive association in a contingency table (749)

M . . . . the sample mean (219)

M_D . . . . mean difference between N matched pairs (425)

Md . . . . the median (218)

M_e . . . . mean error in a set of N experimental observations (463)

M_{ej} . . . . mean error for the n_j observations in group j (463)

$m(o)$ . . . . measured amount of some property possessed by object o (84)

MS . . . . mean square in the analysis of variance (469)

$\mu = E(X)$ . . . . (small Greek mu) mean of the probability distribution of the random variable X (232)

μ_0 . . . . value of the population mean specified by H_0 (358)

μ_1 . . . . true value of the mean of the population; a value of the mean covered by the alternative hypothesis H_1 (358)

$\mu_G = E(G)$ . . . . mean of the sampling distribution for some statistic G (268)

μ_j . . . . true mean of the potential population of observation made under treatment j (460)

μ_{jk} . . . . mean of the potential population of observations made under the treatment combination j, k (494)

$\mathbf{\mu}(v)$ . . . . mean of all potential observations made under treatment v (525)

N . . . . total number of trials in a simple experiment or of observations in a given sample (60)

n . . . . size of any one of several samples containing the same number of observations (420)

n_j . . . . number of sample observations in treatment group j (462)

n_{jk} . . . . number of sample observations in treatment combination j, k (500)

ν . . . . (small Greek nu) degrees of freedom parameter for a t or chi-square variable (393, 434)

ν_1, ν_2 . . . . number of degrees of freedom for numerator and denominator, respectively, for an F ratio; the parameters of the F distribution (444)

$\Omega'_{AB}, \Omega''_{AB}$ . . . . (capital Greek omega) prior and posterior odds ratios (845).

ω^2 . . . . (small Greek omega, squared) population index showing the relative or proportional reduction in the variance of Y given the X status or value for an observation (414)

est. ω^2 . . . . sample estimate of the proportional reduction in variance of Y given X (417)

P . . . . sample proportion of "successes" in sampling from a Bernoulli process (187)

p . . . . probability of a given event class; probability of a "success" in a single Bernoulli trial (108)

$p'(A)$ . . . . prior probability for event A (813)

$p''(A)$ . . . . posterior probability for event A (813)

$p(A)$ . . . . probability associated with a particular event A in a probability function (109)

$p(B|A) = \dfrac{p(A \cap B)}{p(A)}$ conditional probability of event B given the event A (144)

$p(X = r; N, p)$ . . . . . . binomial probability for $X = r$, given the parameters N and p (184)

$\mathcal{P}(H)$ . . . . personal probability associated by the experimenter with the hypothesis H (382)

π . . . . the population proportion, or the probability of a success in a Bernoulli process (817)

ψ . . . . (small Greek psi) value of a particular comparison among population means (584)

$\hat{\psi}$ . . . . value of a particular comparison among sample means (585)

$\hat{\psi}_g$ . . . . value of a particular comparison g on a set of sample means (596)

φ . . . . (small Greek phi) index of mean square contingency for a population or a sample contingency table (743)

φ' . . . . Cramér's statistic for association in a contingency table (745)

$\mathcal{Q}$ . . . . proportion of cases cut off by a given value on the upper tail of a sampling distribution (397)

$Q = 1 - P$ . . . . proportion of sample "failures" in sampling from a Bernoulli process (379)

$q = 1 - p$ . . . . probability of a "failure" on a single Bernoulli trial (179)

R . . . . number of rows in the data table for a two-factor experiment (499)

$R_{1 \cdot 23}$ . . . . multiple correlation of variable 1 with variables 2 and 3 (706)

$R_{1 \cdot 2 \ldots K}$ . . . . multiple correlation of variable 1 with variables 2 through K (707)

r_{XY} . . . . sample correlation coefficient between X and Y (623)

$r_{12 \cdot 3}$ . . . . partial correlation between variables 1 and 2 with 3 held constant (710)

$r_{(Y',Y)}$ . . . . correlation between the predicted and actual values of Y (706)

r_s . . . . Spearman rank-correlation coefficient (789)

ρ_I . . . . (small Greek rho, sub I) intraclass correlation coefficient for a population (535)

ρ_{XY} . . . . population correlation coefficient between variables X and Y (637)

ρ_0 . . . . value of the population correlation specified by H_0 (662)

$\mathcal{S}$ . . . . sample space for a particular simple experiment (49)

$S = \sqrt{(J-1)F_\alpha}$ . . constant determining the width of confidence intervals for post-hoc comparisons, Scheffé method (607)

$SE(u;D)$ . . . . subjective expected loss associated with a particular decision-rule D by the experimeter (348)

S_+, S_- . . . . number of agreements and disagreements, respectively, about the ordering of pairs of objects in two rank orders (795)

S_j . . . . constant value determining the width of a confidence interval for Y_{ij} given X_j, for a problem in regression (649)

SS . . . . sum of squares in the analysis of variance (466)

S . . . . standard deviation computed for a sample of data (238)

S^2 . . . . sample variance (237)

$s = \sqrt{s^2}$ . . . . corrected standard deviation for a sample (284)

$s^2 = \dfrac{N}{N-1} S^2$ . . . . . corrected variance; the unbiased estimator of the population variance from a sample (284)

s_D^2 . . . .   sample variance based on the difference for each of N matched pairs of observations (426)

$S_{Y \cdot X}$ . . . .   sample standard error of estimate for predictions of Y from X (631)

S_z^2 . . . .   variance of z or standardized scores calculated from a sample of data (252)

$S_{z_Y \cdot z_X}$ . . . .   sample standard error of estimate for predictions of z_Y from z_X (624)

$S_{z_1 \cdot z_2}$ . . . .   standard error of estimate for predictions of z_1 from z_2 (707)

$S_{z_1 \cdot z_2 \ldots z_K}^2$ . . . .   sample variance of estimate for z_1 values predicted from $z_2, \cdot\cdot\cdot,$ z_K via a multiple regression equation (707)

$S_{1 \cdot 2 \ldots K}^2$ . . . .   sample variance of estimate in the prediction of X_1 from $X_2,$ $\cdot\cdot\cdot, X_K$ (708)

σ . . . .   (small Greek sigma) standard deviation of a random variable (243)

σ^2 . . . .   variance of a random variable (243)

σ_0^2 . . . .   value of the population variance specified by the hypothesis H_0 (439)

σ_A^2 . . . .   variance of the distribution of random effects $\mathbf{a}(v)$, representing factor A (528)

σ_B^2 . . . .   variance of the random effects representing factor B (543)

σ_{AB}^2 . . . .   variance of the random interaction effects for the factors A and B (543)

σ_e^2 . . . .   variance of random errors, e (467)

$\sigma_{\text{diff.}}$ . . . .   standard error of the difference between two means (403)

est. $\sigma_{\text{diff.}}$ . . . .   estimated standard error of the difference between two means (405)

σ_G^2 . . . .   variance of the sampling distribution of some statistic G (268)

σ_M^2 . . . .   variance of the sampling distribution of the mean (278)

σ_M . . . .   true standard error of the mean, given samples of size N from some population (280)

est. σ_M . . . .   sample estimate of the standard error of the mean (390)

est. σ_{M_D} . . . .   estimated standard error of the mean difference between N matched pairs (426)

σ_P^2 . . . .   variance of a sampling distribution of sample proportions, P (245)

σ_Y^2 . . . .   variance of the random variable Y; "marginal" variance of Y in a joint distribution of (x,y) events (646)

$\sigma_Y^2 - \sigma_{Y|X}^2$ . . . .   reduction in the variance of Y afforded by specification of X (444)

$\sigma_{Y|X}^2$ . . . .   variance of the conditional distribution of Y, given the value of X (444)

$\sigma_{Y \cdot X}$ . . . .   true standard error of estimate for predictions of Y from X in some population (638)

σ_z^2 . . . .   variance of a random variable transformed to z or standardized values (253)

$\sigma_{(z_1-z_2)}$ . . . . variance of the difference between Z values for pairs of independent samples (663)

$$t = \frac{M - E(M)}{\text{est. } \sigma_M}$$ . . . . . t ratio based on the mean of a single sample (392)

$t = \frac{M_1 - M_2 - E(M_1 - M_2)}{\text{est. } \sigma_{\text{diff.}}}$ t ratio based on the difference between means of two samples (405)

t . . . . a random variable following the "student" distribution of t with γ degrees of freedom (393)

$t_{(\alpha/2; \nu)}$ . . . . value of t in a distribution with ν degrees of freedom, cutting off the upper $\alpha/2$ proportion of cases (400)

$t(o)$ . . . . true amount of some property possessed by object o (83)

τ . . . . (small Greek tau) Kendall's coefficient of rank-order agreement (792)

θ . . . . (small Greek theta) general symbol for a population parameter (272)

θ_0 . . . . value of θ specified by the null hypothesis H_0 (538)

$\theta = \sigma_A^2/\sigma_e^2$ . . . . ratio of the variance of effects due to factor A to the error variance (536)

$u(H_0; H_1)$ . . . . loss associated with the decision to accept H_0 when H_1 is in fact true (345)

$u(H_1; H_0)$ . . . . loss associated with the decision to accept H_1 when H_0 is in fact true (346)

var.(X) . . . . variance of the random variable X (see Appendix B)

var.$(M_1 - M_2)$ . . . . variance of the difference between means (403)

var.$(\hat{\psi})$ . . . . variance of a comparison among sample means (585)

$w(b|a)$ . . . . conditional density for $Y = b$ given that $X = a$ (157)

w_g . . . . weighting factor used in obtaining the sum of squares for comparison $\hat{\psi}_g$ (596)

X . . . . a random variable; the independent variable in an experiment (110)

(x,y) . . . . a joint event, consisting of a value for variable X paired with a value for variable Y (158, 619)

$|X - \mu|$ . . . . absolute deviation (disregarding sign) of the value of X from the value of μ (253)

x . . . . a particular value which random variable X can assume (110)

x . . . . midpoint of a class interval (220)

ξ_j . . . . (small Greek xi) effect associated with treatment j, in the general regression model (679)

Y . . . . a random variable; ordinarily, the dependent variable in an experiment (312)

y' . . . . raw score on Y predicted from the value of X (629)

y_{ij} . . . . score associated with the observation i in experimental group j (461)

y_{ijk} . . . . score associated with the ith observation in the treatment combination j, k (494)

Z . . . . value corresponding to r_{XY} in the Fisher r to Z transformation (662)

$z = \dfrac{X - M}{S}$ . . . . standardized score or value corresponding to a sample value X, relative to a sample distribution (251)

$z = \dfrac{X - E(X)}{\sigma}$ . . . . . standardized score or value corresponding to a particular value of X, relative to a population distribution (252)

$|z|$ . . . . absolute value of a standardized score (255)

z'_1 . . . . predicted standardized value on variable X_1 (703)

z_X, z_Y . . . . standardized scores corresponding to particular values of X and Y, respectively (620)

$z_{(1-\alpha/2)}$ . . . . standardized value cutting off the lower $1 - \alpha/2$ proportion of cases in a normal distribution (420)

$z_M = \dfrac{M - E(M)}{\sigma_M}$ . . . standardized value corresponding to a particular value of sample M in a sampling distribution of means (280)

z'_Y . . . . predicted standardized value on variable Y (620)

ζ . . . . (small Greek zeta) value in the Fisher r to Z transformation corresponding to the population correlation ρ (662)

Solutions to
Selected
Problems

CHAPTER 1

1. (a) $\{1,2,3,4,5,6,7,8,9,10,11,12,13,14,15,16\} = A$

$$A = \{x|x \text{ is an integer, } 0 < x < 17\}$$

(b) $\{4,6,8,10,12,14,16\} = B$

$$B = \left\{x \mid \frac{x}{2} \text{ is an integer and } 4 \le x \le 17\right\}$$

(c) $\{5,7,9,11,13\} = C$

$$C = \left\{x \mid \frac{x+1}{2} \text{ is an integer and } 3 < x \le 13\right\}$$

(d) $B \subset A; C \subset A$; both are proper subsets.

3. (a) $A \cup B$
   (b) $A \cap \bar{B}$
   (c) $\emptyset \cup \bar{B}$
   (d) $A \cap (B \cup C)$
   (e) $C - (A \cap B)$

5. (a) 2; $\{x\}$, $\emptyset$
   (b) 4; $\{x,y\}$, $\{x\}$, $\{y\}$, $\emptyset$
   (c) 8; $\{x,y,z\}$, $\{x,z\}$, $\{y,z\}$, $\{x,y\}$, $\{x\}$, $\{y\}$, $\{z\}$, $\emptyset$.
   (d) 2^N = number of sets possible from N objects.

7. (a) W

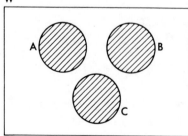

(b) W

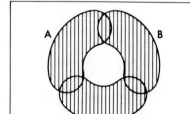

(c)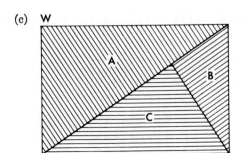

9. (a) $\{0,2,4,6\}$
   (b) $\{4\}$
   (c) $\{1\}$
   (d) $\{2,4\}$
   (e) $\{3,4,5,6\}$
   (f) $\emptyset$; no elements in common

11. (a) $A \cap B = \{1,2\}$
   $B \cap C = \{1,2\}$
   $A \cap B \cap C = \{1,2\}$
   $B \cup (A - C) = \{-3, -2, -1, 0, 1, 2\}$
   $B \cap (C - A) = \emptyset$
   (b) $A \cap B = \{0,1,2,3\}$
   $B \cap C = \{1,2,3,4,5\}$
   $A \cap B \cap C = \{1,2,3\}$
   $B \cup (A - C) = \{x | x = -3, -2 \text{ or } x \geqq -1\}$
   $B \cap (C - A) = \{x | x = 4, x = 5\}$
   (c) $A \cap B = \{-2,2\}$
   $B \cap C = \{2\}$
   $A \cap B \cap C = \{2\}$
   $B \cup (A - C) = \{x | x = -3, -1, 0, \text{ or } \sqrt{5.5} \geqq |x| \geqq \sqrt{1.5}\}$
   $B \cap (C - A) = \emptyset$
   (d) $A \cap B = \{-1,0,1,2,3\}$
   $B \cap C = C$
   $A \cap B \cap C = \{1,2,3\}$
   $B \cup (A - C) = \{x | x = -3, x = -2 \text{ or } x^3 \geqq -4.25\}$
   $B \cap (C - A) = \{x | x = 4,5\}$

13. (a) All members of the House of Representatives of the 91st congress who are above sixty years of age at its opening, or who are women members from urban districts, or both.
   (b) All members (etc.) who are women over sixty or who are Democrats from urban districts, or both.
   (c) All members (etc.) who are either women at or under sixty or are non-Democrats from non-urban districts, or both.
   (d) All members (etc.) who are female non-Democrat urban Representatives.
   (e) All members (etc.) who are neither above sixty nor from urban districts and who are male but not Democrats.
   (f) All members (etc.) who are female non-Democrats or who are females from urban districts or both.

15. (a) $m(A \cap \bar{B} \cap \bar{C}) = 8$
   (b) $m(A \cap B \cap \bar{C}) + m(A \cap \bar{B} \cap C) = 17$
   (c) $m(A \cap \bar{B} \cap \bar{C}) + m(\bar{A} \cap B \cap \bar{C}) + m(\bar{A} \cap \bar{B} \cap C) = 32$

(d) $m(A \cap B \cap \bar{C}) + m(A \cap \bar{B} \cap C) + m(\bar{A} \cap B \cap C) + m(A \cap B \cap C) = 24$

(e) 56

17.

	DOMAIN	RANGE
G:	A	$\{2\}$
H:	$\{2\}$	A
I:	A	A
J:	A	$\{3,5,6\}$
K:	A	A
$G \cap H$:	$\{2\}$	$\{2\}$
$G \cup H$:	A	A
$I \cup J$:	A	A
$J \cup K$:	A	A
$H \cap I \cap J$:	$\{2\}$	$\{6\}$
$H \cap \bar{I} \cap J \cap \bar{K}$:	$\{2\}$	$\{3,4,5\}$

Functions: $G, I, J, K, G \cap H, H \cap I \cap J$

19. $G = \{(2,2), (3,3), (4,4), (5,5), (6,6), (2,4), (2,6), (3,6)\}$

$H = \{(2,2), (3,3), (4,4), (5,5), (6,6), (3,2), (4,2), (4,3), (5,2), (5,3), (5,4), (6,2), (6,3),$
$(6,4), (6,5)\}$

$I = \{(2,3), (2,4), (2,5), (2,6), (3,4), (3,5), (3,6), (4,5), (4,6), (5,6)\}$

$J = \{(2,3), (3,4), (4,5), (5,6)\}$

$K = \{(4,2), (5,3), (6,4)\}$

$G \cap I = \{(2,4), (2,6), (3,6)\}$

$G - I = \{(2,2), (3,3), (4,4), (5,5), (6,6)\}$

$I \cup K = \{(2,3), (2,4), (2,5), (2,6), (3,4), (3,5), (3,6), (4,5), (4,6), (5,6), (4,2), (5,3),$
$(6,4)\}$

$J \cap K = \emptyset$

$\bar{G} = \{(2,3), (2,5), (3,2), (3,4), (3,5), (4,2), (4,3), (4,5), (4,6), (5,2), (5,3), (5,4), (5,6),$
$(6,2), (6,3), (6,4), (6,5)\}$

$I-J = \{(2,4), (2,5), (2,6), (3,5), (3,6), (4,6)\}$

21. Graphs not shown; see sections 1.15 and 1.19.

23. (a) $\{(a,d), (e,d), (e,a), (g,d), (g,a), (g,e), (f,d), (f,a), (f,e), (f,g), (c,d), (c,a), (c,e), (c,g),$
$(c,f), (b,d), (b,a), (b,e), (b,g), (b,f), (b,c)\}$

(b) $\{(b,f), (c,f), (c,b), (g,f), (g,b), (g,c), (a,f), (a,b), (a,c), (a,g), (d,f), (d,b), (d,c), (d,g)$
$(d,a),(e,f),(e,b),(e,c), (e,g), (e,a), (e,d)\}$

(c) $\{(b,f), (c,f), (e,a), (e,d)\}$

(d) $\{(10,500, 950), (8100, 1000), (19,200, 4000)\}$

(e) {The set of (x,y) pairs corresponding to cars a, c, d, f, and g.}

25. Let X be a well-defined set of human offspring, X a well-defined set of male human parents, Z a well-defined set of female human parents.

Let $B = X \times Y \times Z$.

$R = \{(x,y,z) \in B | x$ has biological parents $(y,z)\}$.

If the domain of R is conceived as a subset of X, and the range as some subset of $Y \times Z$, then since each element of the domain can be associated with only one element of the range, R is a function.

27. (a) $A_i - A_j = A_i \cap \bar{A}_j$

$A_i = (A_i \cap A_j) \cup (A_i \cap \bar{A}_j)$

Since

$$(A_i \cap A_j) \cap (A_i \cap \bar{A}_j) = \emptyset$$

then
$$m(A_i) = m(A_i \cap A_j) + m(A_i \cap \bar{A}_j).$$

Hence
$$m(A_i - A_j) = m(A_i) - m(A_i \cap A_j)$$

(b) If $A_i \subset A_j$, then $A_i \cap A_j = A_i$ and $\bar{A}_i \cap A_j \neq \emptyset$.
Furthermore
$$A_j = (A_i \cap A_j) \cup (\bar{A}_i \cap A_j).$$

Then
$$m(A_j) = m(A_i \cap A_j) + m(\bar{A}_i \cap A_j)$$
$$= m(A_i) + m(\bar{A}_i \cap A_j)$$

so that since
$$m(\bar{A}_i \cap A_j) > 0,$$
$$m(A_j) > m(A_i)$$

(c) $m(A_i \cup A_j) = m[(A_i - A_j) \cup (A_j - A_i)] + m(A_i \cap A_j)$
$\qquad = m(A_i - A_j) + m(A_j - A_i) + m(A_i \cap A_j)$
$\qquad = m(A_i) + m(A_j) - 2m(A_i \cap A_j) + m(A_i \cap A_j)$
$\qquad = m(A_i) + m(A_j) - m(A_i \cap A_j)$

29. (a) $m(A) + m(B) = m(A \cap \bar{B} \cap \bar{C}) + m(A \cap \bar{B} \cap C) + m(\bar{A} \cap B \cap C)$
$\qquad\qquad\qquad\quad + m(\bar{A} \cap B \cap \bar{C}) + 2m(A \cap B \cap \bar{C}) + 2m(A \cap B \cap C)$
$\quad m(C) - m(\bar{A} \cap \bar{B} \cap C) = m(A \cap \bar{B} \cap C) + m(\bar{A} \cap B \cap C) + m(A \cap B \cap C)$
$\quad m(A \cap B \cap C) = 8$

(b) $m(A \cap \bar{B} \cap C) = m(A) - m(A \cap \bar{B} \cap \bar{C}) - m(A \cap B \cap \bar{C}) - m(A \cap B \cap C)$
$\quad m(A \cap \bar{B} \cap C) = 9$

(c) 83

CHAPTER 2

1. (a) 4
   (b) 26
   (c) 12
   (d) 20
   (e) 26
   (f) 2

3. 36 pairs
   (a) 21
   (b) 6
   (c) 15
   (d) 9
   (e) 27
   (f) 22
   (g) 11
   (h) 6

5. (a) 5/22
   (b) 8/22
   (c) 7/22
   (d) 5/22
   (e) 16/22
   (f) 21/22
   (g) 16/22

7. (a) 6/26
   (b) 18/26
   (c) 10/26
   (d) 10/26
   (e) 7/26
   (f) 2/26

9. The family $\mathcal{Q}$ consists of all 2^{50} events, including $\emptyset$. We are concerned with the events L, $\bar{L}$, B, and $\bar{B}$, where
$$p(\bar{L}) = 23/50$$
$$p(L \cup B) = 37/50$$
$$p(L \cap \bar{B}) = 15/50$$

Then

(a) $p(L) = 27/50$ (b) 22/50
(c) 10/50   (d) 13/50
(e) 12/50   (f) 40/50

11. (a) .50   13. (a) 1/6   15. (a) 1/3
    (b) .46   (b) 1/18   (b) 1/6
    (c) .30   (c) 1/6   (c) 2/3
    (d) .30   (d) 5/18   (d) 5/6
    (e) .35   (e) 1/18   (e) 2/3
    (f) .06
    (g) .84
    (h) .75

17. Since the trial numbers correspond to the sequence of "natural" numbers, 1, 2, 3, 4, . . . , and the first ace does not *have* to appear in any finite number of trials, the sample space is the countably infinite set of integers.

19. (a) 5/8   21. (a) 21 to 15   23. (a) $p > \dfrac{1}{6}$
    (b) 9/11   (b) 6 to 30 or 1 to 5
    (c) .6   (c) 15 to 21 or 5 to 7   (b) $p < .375$
    (d) 17/32   (d) 9 to 27 or 1 to 3   (c) $p = .35$
    (e) 5/7   (e) 27 to 9 or 3 to 1
    (f) 4/13   (f) 22 to 14 or 11 to 7
    (g) 1/50   (g) 11 to 25
                         (h) 6 to 30 or 1 to 5

CHAPTER 3

1. (a) nominal
   (b) ratio
   (c) ordinal
   (d) ordinal
   (e) ratio
   (f) ratio
   (g) interval (zero point arbitrary)
   (h) ratio
   (i) ordinal (ordered-categories)
   (j) interval
   (k) ordinal

3. For example, with $i = 90$

	f
3124–3213	2
3034–3123	1
2944–3033	4
2854–2943	4
2764–2853	4
2674–2763	3
2584–2673	3
2494–2583	0
2404–2493	2
2314–2403	2
	25

5. (a) .211
   (b) .168
   (c) .278
   (d) .045
7. Figure not given.
9. For $i = 5.1$, middle 50 percent fall between 55.98 and 85.87. The middle (median) value is 62.99.
11. (a) .821
    (b) .587
    (c) .675
    (d) .095
    (e) .269
13.

$$p(x) = \begin{cases} \dfrac{2}{52}, & x = 1, 3, 5, 7, 9, 12, 14, 16, 18, 20 \\[2mm] \dfrac{4}{52}, & x = 2, 4, 6, 8, 10 \\[2mm] \dfrac{6}{52}, & x = 11, 22 \\[2mm] 0 & \text{elsewhere} \end{cases}$$

15.

a	EXPECTED f
20	.96
18	7.20
16	4.32
14	5.28
12	5.28
10	7.20
8	7.68
6	5.28
4	3.36
2	1.44
0	0
	48.00

17. The area under the curve must be equal to 1.00. This area is that of a rectangle with base equal to $2 - (-2) = 4$ and height equal to k. Hence $k = \dfrac{1}{4}$.
19. The density function is not continuous, as clearly shown by the cumulative distribution plot.
21. The answers should approximate the true probabilities, which are:
    (a) .124
    (b) .302
    (c) .228
    (d) .287
23. The actual values are:
    (a) .375
    (b) .250

(c) .920
(d) .125
(e) .277

CHAPTER 4

1.

		SUIT			
		SPADES	HEARTS	DIAMONDS	CLUBS
DECK	d_1	x	x	x	x
	d_2	x	x	x	x
	d_3	x	x	x	x

Each cell represents one possible pair, such as (d_1, spades) etc.

3. (a) 1/18   (b) 1/12
   (c) 2/3   (d) 1/4
   (e) 17/36   (f) 13/18
   (g) 5/9   (h) 13/18

5. (a) true
   (b) true
   (c) true
   (d) not true, unless $p(C) = 1.00$
   (e) true
   (f) true

7. Within each column, the conditional distribution is:

	1	1/6
Y	2	1/3
	3	1/3
	4	1/6

The variables are independent, since for any x and y, $p(xy) = p(x)p(y)$.

9.

		Z						
		2	3	4	5	6	7	8
	−3	0	0	0	1/48	0	0	0
	−2	0	0	1/24	0	1/16	0	0
	−1	0	1/24	0	1/8	0	1/16	0
W	0	1/48	0	1/8	0	1/8	0	1/48
	1	0	1/16	0	1/8	0	1/24	0
	2	0	0	1/16	0	1/24	0	0
	3	0	0	0	1/48	0	0	0

The variables are not independent; conditional probabilities within rows or columns do not equal the corresponding marginal probabilities.

11. (a) 1/3   13. $p(C_1|A) = .542$ 15. .30
   (b) 7/9 $p(C_2|A) = .333$
                     $p(C_3|A) = .125$

17. Let $p(A_i) = k$ for all i
       $p(B_j) = h$ for all j
then, if independent, $p(A_i, B_j) = kh$ for all i, j.

19. Since A_i and A_j are each members of a set of sets A, and A is a partition of some set W, then $p(A_i \cap A_j) = 0$. If A_i and A_j were independent, then $p(A_i \cap A_j) = p(A_i)p(A_j)$. But since $p(A_i) \neq 0$, $p(A_j) \neq 0$, $p(A_i)p(A_j) \neq 0$, and thus $p(A_i)p(A_j) \neq p(A_i \cap A_j)$. Hence A_i and A_j cannot be independent.

CHAPTER 5

1. $(3)(7!)(8!)$ *ordered* combinations or $(7!)(8!)/(3!)^4 2!$ combinations.

3. p (2 or more out of N) $= 1 - \dfrac{12!/(12-N)!}{12^N}$.

   for $N \leq 12$

   p (2 out of 2) $= \dfrac{1}{12}$

   p (2 or more out of 4) $= 1 - \dfrac{(11)(10)(9)}{12^3}$

5. (a) $\dfrac{\dbinom{10}{2}\dbinom{35}{7}}{\dbinom{45}{9}}$ (b) $\dfrac{\dbinom{15}{3}\dbinom{30}{6}}{\dbinom{45}{9}}$

   (c) $\dfrac{\dbinom{20}{4}\dbinom{25}{6}}{\dbinom{45}{9}}$ (d) $\dfrac{\dbinom{10}{0}\dbinom{35}{9}+\dbinom{10}{1}\dbinom{35}{8}}{\dbinom{45}{9}}$

   (e) $\dfrac{\dbinom{20}{9}\dbinom{25}{0}}{\dbinom{45}{9}}$

7. (a) $\dbinom{10}{0}\left(\dfrac{1}{3}\right)^0\left(\dfrac{2}{3}\right)^{10} + \dbinom{10}{1}\left(\dfrac{1}{3}\right)\left(\dfrac{2}{3}\right)^9 + \dbinom{10}{2}\left(\dfrac{1}{3}\right)^2\left(\dfrac{2}{3}\right)^8$

   (b) $\dbinom{10}{8}\left(\dfrac{1}{3}\right)^8\left(\dfrac{2}{3}\right)^2 + \dbinom{10}{9}\left(\dfrac{1}{3}\right)^9\left(\dfrac{2}{3}\right) + \dbinom{10}{10}\left(\dfrac{1}{3}\right)^{10}$

9. (a) .2501 (b) .0138
   (c) almost 0 (d) .1643
   (e) .1671 (f) .8286

11. Just under .25

13. (a) $\dfrac{3!}{1!2!}(.6)^2(.4)^2$ or .1728

    (b) $\dfrac{5!}{0!5!}(.4)^1(.6)^5$ or .0311

15. About .40

17. p(no successes out of N, process 1) $= (1-p)^N$
    p(no successes, N trials, process 2) $= (1-p)^N$
    p(no successes, N trials each process) $= [(1-p_1)(1-p_2)]^N$
    p(1 or more successes) $= 1 - [(1-p_1)(1-p_2)]^N$

19.

	AT OR BELOW MEDIAN	ABOVE MEDIAN
GROUP I	6	4
GROUP II	4	6

$$p(4,6; 10,20) = \frac{\binom{10}{4}\binom{10}{6}}{\binom{20}{10}}$$

$$= \left(\frac{10!\,10!}{4!\,6!\,6!\,4!}\right) \Big/ \left(\frac{20!}{10!\,10!}\right)$$

$$= \frac{(3.62880)^4 \times 10^{24}}{(2.4)^2(7.2)^2(2.43) \times 10^{24}}$$

This works out to a value in the neighborhood of .2. Since we are interested in a result this much *or more* deviant from a proportional split about the median, the probability we seek will be greater than .2. Hence we do not reject the hypothesis that the two groups are actually identical.

21. Imagine a stable and independent Poisson process with $m = 2$. We approximate the probability of zero defects to be slightly greater than .10 and that for 3 or fewer defects to be between .75 and .90. Hence, the probability for a number of defects between 1 and 3 should be between $.75 - .10$ or .65 and $.90 - .10$ or .80. (The exact answers are $p(0) = .1353$, $p(1 \leq x \leq 3) = .7218$.)

23. (a) $\dfrac{10!}{6!\,3!\,1!}\,(.4)^6\,(.3)^3\,(.3)^1$

(b) $\dfrac{10!}{10!\,0!\,0!}\,[(.4)^{10} + (.3)^{10} + (.3)^{10}]$

(c) $\dfrac{10!}{3!\,4!\,3!}\,(.4)^3\,(.3)^4\,(.3)^3$

CHAPTER 6

1.

	BOYS	GIRLS
MEDIAN	34.162	28.245
MODE	37.495	27.495

3. Mean = 24.44
   Median = 24.50

5. Table 3.7.1
   Mean = 101.72
   Median = 102.05
   Table 3.5.1
   Mean = 102.42
   Median = 102.08

As compared with computations based upon raw data, the more accurate figures should tend to come from distributions with smaller interval sizes.

7. Squared deviations should be smaller about the mean than for any other value.

9.

	BOYS	GIRLS
(a)	2820.9	2512
(b)	3827.6	3274
(c)	2199.5	2050.8
(d)	4249.5	3799.5

11. For $N = 5$, $E(X) = 80/31$
   For $N = 7$, $E(X) = 448/127$

13. $S^2 = 21.3314$, $S = 4.63$

15.

BOYS	GIRLS
S^2 is about 630,000	S^2 is about 518,100
S is about 793	S is about 720

17. For $N = 5$, $\text{Var}(X) = \dfrac{240}{31} - \left(\dfrac{80}{31}\right)^2$

   $N = 7$, $\text{Var}(X) = \dfrac{1792}{127} - \left(\dfrac{448}{127}\right)^2$

19. (a) $E(X) = 5.95$, $\text{Var}(X) = 3.8675$
   (b) $E(X) = 60$, $\text{Var}(X) = 12$
   (c) $E(X) = 1.25$, $\text{Var}(X) = 1.25$
   (d) $E(X) = 7.5$, $\text{Var}(X) = 11.25$
   (e) $E(X) = 5.26$, $\text{Var}(X) = 22.44$
   (f) $E(X) = 20$, $\text{Var}(X) = 19.8$
   (g) $E(X) = 20$, $\text{Var}(X) = 20$

21.

	BOYS	GIRLS
(a)	16.77 percentile rank	13.14 percentile rank
(b)	$z = -.025$	$z = .583$
(c)	$-1.413 \angle z \angle 1.172$	$-1.152 \angle z \angle 1.277$
(d)	$z = 1.11$	$z = 1.88$; more deviant for a girl

23. He should be willing to pay an amount equal to the expected value, or $6.50.

CHAPTER 7

1. $\dfrac{e^{-4}4^7}{7!} = .0595$

   $\dfrac{e^{-6}6^7}{7!} = .1377$

   $\dfrac{e^{-8}8^7}{7!} = .1396$

   $\dfrac{e^{-10}10^7}{7!} = .0901$

$m = 8$ is the maximum likelihood estimate from among this set of four possibilities.

3.

M	$p(M)$
1	1/216
4/3	6/216
5/3	21/216
2	44/216
7/3	63/216
8/3	54/216
3	27/216

5. Pooled estimate of $\mu = 102.33$

   Pooled estimate of $\sigma^2 = 30.12$

   Estimated standard error $= \sqrt{30.12/45}$

7. $\mu = 422.25$

   $\sigma^2 = 3564$

9. $M_1 - M_2 = 4.12$

   For combined samples, $Z = 1.448$

11. Under these circumstances, $M = P$, the sample proportion, and the sampling distribution of NP is binomial.

13. The z-value would be about 25.5. Even in terms of the Tchebycheff inequality, z-values this large (absolutely) should occur with probability no greater than about 1/650.

15.

$N = 2$		$N = 3$		$N = 4$	
M	$p(M)$	M	$p(M)$	M	$p(M)$
4.0	1/16	4.00	1/64	4.00	1/256
3.5	2/16	3.67	3/64	3.75	4/256
3.0	3/16	3.33	6/64	3.50	10/256
2.5	4/16	3.00	10/64	3.25	20/256
2.0	3/16	2.67	12/64	3.00	31/256
1.5	2/16	2.33	12/64	2.75	40/256
1.0	1/16	2.00	10/64	2.50	44/256
		1.67	6/64	2.25	40/256
		1.33	3/64	2.00	31/256
		1.00	1/64	1.75	20/256
				1.50	10/256
				1.25	4/256
				1.00	1/256

Each distribution is symmetric, with an increasing "flattening" of the tails of the distribution and a tendency to a "bell" form as N increases. Note that the probability of any exact value, such as $M = 2.00$ grows steadily smaller with increasing N.

17. For each distribution $E(M) = 2.5$, and $\sigma_M^2 = 1.25/N$.

CHAPTER 8

1. (a) 11.5   (b) 83.1
   (c) 97   (d) 3.8
   (e) 33.4   (f) 98.4
3. (a) .964   (b) .067
   (c) .008   (d) .309
   (e) .773
5.

	$p = .5, N = 8$ BINOMIAL	APPROX. NORMAL*	$p = .4, N = 8$ BINOMIAL	APPROX. NORMAL*
8	.0039	.007	.0007	.001
7	.0312	.032	.0079	.008
6	.1094	.106	.0413	.040
5	.2188	.218	.1239	.123
4	.2734	.274	.2322	.241
3	.2188	.218	.2787	.278
2	.1094	.106	.2090	.198
1	.0312	.032	.0896	.084
0	.0039	.007	.0168	.027

* Values read from the normal table will vary depending on rounding methods used. Note that the fit between the normal and binomial probabilities is very nearly as close for $p = .4$ as for $p = .5$, even for N as low as 8.

7. The probability of a z-value exceeding 1.259 in absolute value is about .208. One would not conclude that the mean age has changed on the basis of this evidence. The test is justified by the central limit theorem in view of the large sample.
9. The distribution of y values is normal with a mean of 0, and a variance of 5800. Ninety-nine percent of all sample values of y should fall between ± 196.467. The middle fifty percent should fall between ± 51.021.
11. A z-value greater than or equal to about .425 (absolute) should occur in a normal distribution with probability of .688. Here little doubt is cast on the supposition that the two groups do have the same tendency to react to the medication.
13. According to the gamma distribution with $m = 4$ and $r = 1$, the probability is almost zero that two or more minutes will pass before the first request. The probability is between .25 and .50 that the tenth request will occur within two minutes or less.
15. If N independent samplings are made from the same independent gamma process, with fixed parameters r and m, then the sum of the N gamma-values is also a gamma variable, with parameters Nr and m. As $N \to \infty$, the resulting gamma distribution approaches a normal distribution.
17. The probability that 90 percent or more of the population values will be covered is about .74. The probability that ten percent or fewer of the population values will be covered is nearly zero.

CHAPTER 9

1. The z-value is approximately 2.91, so that the null hypothesis, $H_0: \mu = 28.6$, can be rejected with a probability of Type I error of about .0038. Given $\alpha = .05$, the hypothesis would be rejected.

3. The null hypothesis, $H_0: p = .25$, may be rejected in favor of the alternative $H_1: p > .25$, with a probability of Type I error equal to about .04.

5. In a Poisson process with $m = 2$, six or more "successes" should occur with probability of only about .016. The null hypothesis $H_0: m = 2$ (per thousand) can be rejected in favor of $H_1: m > 2$, at beyond the $\alpha = .05$ level. (The Poisson is preferable to the normal approximation since Np is really quite small, relative to N.)

7. Yes, the hypothesis that the true mean difference is zero can be rejected at a level far beyond $\alpha = .01$. The alternative is, of course, that some change (direction unspecified) occurs.

9. The number of cases required is 52 .1 (or about 52). The critical value of M is 103.2.

11. Against $H_1: \mu = 510$, the power is only about .158. Against $H_1: \mu = 480$, the power is about .48. Against $\mu = 520$, the power is also about .48.

13. Some 43 subjects would have sufficed.

15. In a Poisson distribution, the probability is about .05 for 5 or more successes when $m = 2$. Hence, H_0 is rejected when $x \gneq 5$. However, when $m = 5$, $p(x \gneq 5)$ is between .50 and .75. Hence power is between .50 and .75.

17. The 99 percent confidence interval is $.214 \leq p \leq .711$.

19. The 95 percent confidence interval has the limits

$$(M_{\mathrm{I}} - M_{\mathrm{II}}) \pm 1.96 \sqrt{2\sigma^2/N}$$

21.

	TRUE STATE		
	$\mu = 200$	$\mu = 210$	$\mu = 220$
$\mu = 200$	.8943	.1056	.0001
DECIDE $\mu = 210$	.1056	.7888	.1056
$\mu = 220$	.0001	.1056	.8943

23.

	H_0	H_1	H_2
RULE A	1.058	1.056	1.058
RULE B	.404	1.927	.674

Both rules are admissible. The minimax-loss principle leads to the choice of Rule A, however.

Chapter 10

1. Since $z = 1.82$, H_0 cannot be rejected for $\alpha = .05$, two-tailed.

3. The z-value is about -8.72, so that $\mu = 34$ can be rejected in favor of $\mu < 34$ at far beyond the $\alpha = .01$ level.

5. The $t = -1.216$, which for 19 degrees of freedom is not significant. There is not sufficient evidence to permit the conclusion that children of smoking mothers have lower birthweight.

7. The limits of the 99 percent confidence interval are 114 ± 2.82.
9. The t value of 1.67 is not significant for 25 degrees of freedom, two-tailed. The t value of 9.24 is significant far beyond the .01 level for 25 degrees of freedom two-tailed. $H_0: \mu = 35$ is not rejected. $H_0: \mu = 30$ is rejected for $\alpha < .01$.
11. The hypothesis that the two means are identical can be rejected at a level far beyond $\alpha = .01$.
13. The difference is significant beyond the .01 level, one-tailed. The evidence supports the alternative hypothesis that recent articles have shorter sentences on the average.
15. The t-value of 1.33 is not significant. One assumes that each population is normally distributed with the same variance, and that each group may be regarded as an independent random sample.
17. About 38 (or 37.6) cases per group.
19. The t-value of 1.215 is not significant for 14 degrees of freedom. We assume a random sample of pairs, each pair independent of the others. The members within a pair are not necessarily independent, however. The population of pair-values is assumed normal.

CHAPTER 11

1. The χ^2 value is about 10.72, which, for 6 degrees of freedom, would be significant only just beyond the .10 level. Hence, for $\alpha = .05$, H_0 is not rejected.
3. The probability that $S^2 \leq 625$ is the same as the probability that $\dfrac{NS^2}{\sigma^2} \leq \dfrac{(20)(625)}{900}$, with 19 degrees of freedom. This probability is between .25 and .10.
5. The expected value is $29/30$. The most likely value is $27/30$.
7. The 95 percent confidence interval has limits 4.152 and 48.493. (Note that the χ^2 variable has six degrees of freedom in this example)
9. This is another instance of the central limit theorem. If means for samples of N independent observations approach a normal distribution as $N \to \infty$, then so must sums of independent random variables. (See section 8.12. The parent distributions must have finite mean and variance, and chi-square variables with $0 < \nu < \infty$ satisfy this condition.)
11. The F-value is 1.10, with 12 and 8 degrees of freedom. The null hypothesis is not rejected.
13. Since the test is two-tailed, in order for H_0 to be rejected at the 5 percent level, F must be greater than or equal to 2.86, or less than or equal to .349. The F-value of 2.40 does not qualify for either of these rejection regions.
15. These relations are:

If the F variable with ν_1 and ν_2 degrees of freedom is transformed to $x = \dfrac{F}{F + \dfrac{\nu_2}{\nu_1}}$, with $r = \frac{1}{2}\nu_1$ and $N' = \frac{1}{2}(\nu_1 + \nu_2)$, then the new variable follows a beta distribution with parameters r and N'. Hence, since $t^2 = F$ with 1 and ν_2 degrees of freedom, t^2 may also be transformed into a beta variable. Furthermore, $y = \chi^2$ with $m = \frac{1}{2}\nu$, $r = \frac{1}{2}$, is a gamma variable.

Since
$$F = \left(\frac{\nu_2}{\nu_1}\right)(\chi_1^2/\chi_2^2)$$

the ratio of two independent chi-square variables each divided by its degrees of freedom, then if we let

$$w = \frac{F}{F + \frac{\nu_2}{\nu_1}} = \frac{\chi_1^2}{\chi_1^2 + \chi_2^2}$$

this is a beta variable. (See section 8.14.)

CHAPTER 12

1. Estimated effects are

$$\alpha_1 = -5.3125 \quad (\text{actual } -5)$$
$$\alpha_2 = 5.9375 \quad (\text{actual } 4)$$
$$\alpha_3 = -2.3125 \quad (\text{actual } -2)$$
$$\alpha_4 = 1.6875 \quad (\text{actual } 3)$$

3.

SOURCE	S.S.	d.f.	M.S.	F
BETWEEN GROUPS	2.45	1	2.45	.422
WITHIN GROUPS	104.50	18	5.806	
TOTAL	106.95	19		

This F is not significant, so that H_0 is not rejected.

5.

SOURCE	S.S.	d.f.	M.S.	F
BETWEEN GROUPS	698.20	3	232.73	4.692
WITHIN GROUPS	793.60	16	49.60	
TOTAL	1491.80	19		

The F is significant beyond the .025 level, for 3 and 16 degrees of freedom.

7.

SOURCE	S.S.	d.f.	M.S.	F
BETWEEN GROUPS	.41	2	.205	4.051
WITHIN GROUPS	.86	17	.0506	
TOTAL	1.27	19		

For 2 and 17 degrees of freedom, this F is significant beyond the .05 level.

9.

SOURCE	S.S.	d.f.	M.S.	F
BETWEEN GROUPS	223.10	5	44.62	.663
WITHIN GROUPS	2758.22	41	67.27	
TOTAL	2981.32	46		

This F is, of course, not significant. Est. $\omega^2 = 0$.

11. Estimated $\omega^2 = .0625$
　　Est. $\alpha_1 = -.00425$
　　Est. $\alpha_2 = .08475$
　　Est. $\alpha_3 = -.03525$
　　Est. $\alpha_4 = -.04525$
13. Estimates of ω^2 are:
　　　　rows　　　.089
　　　columns　　.7436
　　interaction　　.0409
　　Estimates of effects
　　　$\alpha_1 = 12.44$　　　$\gamma_{11} = -2.10$
　　　$\alpha_2 = -11.47$　　$\gamma_{12} = -1.36$
　　　$\alpha_3 = -.97$　　　$\gamma_{13} = 3.46$
　　　　　　　　　　　　$\gamma_{21} = 2.10$
　　　$\beta_1 = 3.44$　　　$\gamma_{22} = 1.36$
　　　$\beta_2 = -3.44$　　$\gamma_{23} = -3.46$
15. The power is estimated to be between .80 and .90.
17. The power is between .30 and .40. For 10 in each group, the power rises to the vicinity of .70 to .80.
19. Yes. The hypothesis could be tested in terms of the sum of squares

$$SS \text{ mean} = N(M - k)^2$$

and the test statistic

$$F = \frac{N(M - k)^2}{MS \text{ within}}$$

with 1 and $N - J$ degrees of freedom. This is the "missing" degree of freedom in the analysis of variance, since this sort of hypothesis is seldom tested.
21. No, the statement is not generally true.

Chapter 13

1.

SOURCE	S.S.	d.f.	M.S.	F
BETWEEN STIMULI	162	7	23.14	13.23*
WITHIN STIMULI	14	8	1.75	
TOTAL	176	15		

*Significant beyond $\alpha = .01$.

3. The 95 percent confidence interval for ρ_I has limits of approximately .438 and .904.
5. The new F-value is 2.64, so that the hypothesis that $\rho_I \leq .10$ may be rejected beyond the .05 level.

7. Here, the probability desired is that for the event $[F_{(5,54)} > 2.41/(1 + 10)]$. Since the lower 5 percent of an F distribution with 5 and 54 degrees of freedom lies below a value of about .225, then the power must exceed .95 for H_1: $\rho_I = .50$.

9.

SOURCE	S.S.	d.f.	M.S.	F
BETWEEN CITIES	633.34	9	70.37	3.57
WITHIN CITIES	1000.33	20	—	
APPROACHES	685.07	2	342.54	17.39
RESIDUAL	315.26	18	19.70	
TOTAL	1632.67	29		

The F for "approaches" is significant beyond the .01 level, for 2 and 18 degrees of freedom.

11. The limits of the 95 percent confidence interval are approximately .000 and .938. Note that the confidence interval *does* include 0, reflecting the fact that F fails to reach the .05 level of significance.

13. The t-value of 1.813 is not significant for 1 and 11 degrees of freedom. Also note that $t^2 = 3.29$, approximately, as found for the F-value in problem 12.

15.

SOURCE	d.f.	M.S.	FIXED-EFFECTS F	RANDOM-EFFECTS F	MIXED F
A	5	37.20	1.436	$\frac{37.20}{23.09} = 1.611$	1.611
B	9	158.58	6.120*	$\frac{158.58}{23.09} = 6.868*$	6.868*
Interaction	45	15.55	0.600		
Error	120	25.91			

* Significant beyond $\alpha = .01$

Since Paull's criterion is satisfied by these data, we pool interaction and error to obtain pooled MS error = 3809.2/165 = 23.09 with 165 degrees of freedom.

17.

SOURCE	S.S.	d.f.	M.S.	F
Between subjects	1978.525	9		
Within subjects	254.500	30		
Lists	113.675	3	37.891	7.27*
Residual	140.575	27	5.206	
Total	2232.775	39		

* Significant beyond $\alpha = .01$

CHAPTER 14

1. One possible set (out of many) is defined by the weights:

		TREATMENTS						
		1	2	3	4	5	6	7
	1	1	1	1	-1	-1	-1	0
	2	1	-1	0	1	-1	0	0
	3	$\frac{1}{4}$	$\frac{1}{4}$	$-\frac{1}{2}$	$\frac{1}{4}$	$\frac{1}{4}$	$-\frac{1}{2}$	0
COMPARISON	4	1	-1	0	-1	1	0	0
	5	$\frac{1}{2}$	$\frac{1}{2}$	-1	$-\frac{1}{2}$	$-\frac{1}{2}$	1	0
	6	$\frac{1}{6}$	$\frac{1}{6}$	$\frac{1}{6}$	$\frac{1}{6}$	$\frac{1}{6}$	$\frac{1}{6}$	-1

3. (a) $F_{(1,40)} = \dfrac{81}{8} = 10.125^*$

   (b) $F_{(1,40)} = \dfrac{(24)^2}{8} = 72^*$

   (c) $F_{(1,40)} = \dfrac{9}{16}$

   (d) $F_{(1,40)} = \dfrac{9}{80}$

5. Since
$$SS \text{ between} = SS(\hat{\psi}_1) + SS(\hat{\psi}_2) + SS(\hat{\psi}_3) + SS(\hat{\psi}_4) = 3312,$$

$$F = \frac{(3312/4)}{40} = 20.7.$$

For 4 and 45 degrees of freedom this is significant beyond the .01 level. The results of the *post-hoc* tests between pairs of means are:

	I	II	III	IV	V
I	0	-9	-6	6	-18^*
II		0	3	15^*	-9
III			0	12^*	-12^*
IV				0	-24^*
V					0

* Significant at or beyond the .01 level.

7. For $\hat{\psi}_1$, $F < 1$ and is not significant
   For $\hat{\psi}_2$, $F = 56.83$, significant for $\alpha < .01$
   For $\hat{\psi}_3$, $F = 29.24$, significant for $\alpha < .01$
9. For $\hat{\psi}_1$, $F = 213.26$
   For $\hat{\psi}_2$, $F = 1.59$
11. Mean same-sex subjects and cues = 19.50
   Mean different-sex subjects and cues = 20.25

$$\hat{\psi} = .75$$

* Significant beyond the .01 level.

$$F = \frac{3.375}{16.07}, \text{ which is not significant}$$

This sum of squares contributes to the interaction sum of squares.

13. See section 14.13.
15. See section 14.16.
17.

	M.S.	F	SIGNIFICANCE
BETWEEN ROWS			
COMP. 1	203	6.26	N.S.
COMP. 2	25	1	N.S.
OTHER COMPS.	125	3.85	N.S.
BETWEEN COLS.			
COMP. 1	78	2.41	N.S.
OTHER COMPS.	101.4	3.12	almost at $\alpha = .01$
INTERACTION			
COMP. 1	215	6.63	almost at $\alpha = .01$
OTHER COMPS.	42.28	1.30	N.S.
ERROR	32.43		

CHAPTER 15

1. See section 15.20
3.

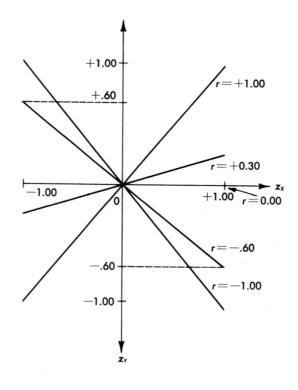

5. $\beta_{Y \cdot X} = 0$, $\rho_{XY} = 0$; no, not necessarily.

7.

SOURCE	SS	df	MS	F
LINEAR REGRESSION	1026.173	1	1026.173	3.806
ERROR AND DEVIATIONS	12941.827	48	269.621	(not significant)
TOTAL	13968.000	49		

We do not reject the H_0: $\beta_{Y \cdot X} = 0$. Estimated $\rho_{XY}^2 = .054$.

9. $r_{XY} = .25$; do not reject H_0: $\rho_{XY} = 0$.

11. $-.286 \leq \rho_{XY} \leq .666$; note that this includes $\rho_{XY} = 0$. Here one assumes bivariate normality of the joint distribution.

13. The regression equation is

$$y' = -.079(x - 5.50) + 5.35$$

with

$$S_{Y \cdot X}^2 = .409, \text{ and } S_{Y \cdot X} = .64.$$

The limits to the confidence interval are $-.1772$ and $.1614$. Note that this interval covers $\beta_{Y \cdot X} = 0$.

15. The regression equation is

$$y' = -1.28(x - 14.75) + 128.90, \text{ so that for } x = 35, y' = 102.98.$$

The limits of the confidence interval for actual y' are found from

$$102.98 \pm (1.98)(22.02) \sqrt{1.01 + \frac{(35 - 14.75)^2}{(100)(520)}}.$$

17.

SOURCE	ADJUSTED SS	df	MS	F
BETWEEN GROUPS	439.401	3	146.467	17.274*
WITHIN	161.09	19	8.479	

* Significant beyond the .01 level.

19. $U = .1789$, $V = .2934$; do not reject H_0: $\rho_1 = \rho_2 = \rho_3$.

21. For 139 degrees of freedom, $t = 5.54$ is highly significant. Hence, reject H_0: $\rho_{XY} = 0$.

CHAPTER 16

1. See section 16.1.

3. From 16.20.4 we have

$$z_{1i}' = r_{12 \cdot 3} \frac{\sqrt{1 - r_{12}^2}}{\sqrt{1 - r_{23}^2}} z_{2i} + r_{13 \cdot 2} \frac{\sqrt{1 - r_{13}^2}}{\sqrt{1 - r_{23}^2}} z_{3i}$$

so that for any i

$$z'_{1i}z_{1i} = r_{12 \cdot 3}\frac{\sqrt{1 - r_{12}^2}}{\sqrt{1 - r_{23}^2}}\,z_{2i}z_{1i} + r_{13 \cdot 2}\frac{\sqrt{1 - r_{13}^2}}{\sqrt{1 - r_{23}^2}}\,z_{3i}z_{1i}$$

Summing over all i and dividing both sides by N, and using 16.16.2, 16.16.3, and 16.20.3, we have

$$R_{1 \cdot 23}^2 = \frac{(r_{12} - r_{13}r_{23})r_{12}}{1 - r_{23}^2} + \frac{(r_{13} - r_{12}r_{23})r_{13}}{1 - r_{23}^2}$$

$$1 - R_{1 \cdot 23}^2 = \frac{1 - r_{23}^2 - r_{12}^2 - r_{13}^2 + 2r_{12}r_{13}r_{23}}{1 - r_{23}^2}.$$

But

$$(1 - r_{12}^2)(1 - r_{13 \cdot 2}^2) = (1 - r_{12}^2)\left[1 - \frac{(r_{13} - r_{12}r_{23})^2}{(1 - r_{12}^2)(1 - r_{23}^2)}\right]$$

$$= \frac{1 - r_{23}^2 - r_{12}^2 - r_{13}^2 + 2r_{12}r_{13}r_{23}}{1 - r_{23}^2}.$$

This completes the proof.

Given this result, note that it must also be true that

$$1 - R_{1 \cdot 23}^2 = (1 - r_{13}^2)(1 - r_{12 \cdot 3}^2)$$

Since $(1 - r_{12 \cdot 3}^2) \leq 1.00$, then $1 - R_{1 \cdot 23}^2 \leq (1 - r_{13}^2)$, and $R_{1 \cdot 23}^2 \geq r_{13}^2$. By the same argument, $R_{1 \cdot 23}^2 \geq r_{12}^2$.

5.

SOURCE	SS	df	MS	F
BETWEEN GROUPS	0.0791	5		
LINEAR REG.	0.0060	1	0.0060	1.875
DEV. FROM LIN.	0.0731	4	0.0183	5.719*
ERROR	0.1549	48	0.0032	
TOTAL	0.2340	53		

* Significant beyond $\alpha = .01$

7.

		GROUP					
		1	2	3	4	5	6
MEANS		.2122	.1511	.1133	.1566	.1866	.2255
WEIGHTS:							
DEGREE	1	−5	−3	−1	1	3	5
	2	5	−1	−4	−4	−1	5
	3	−5	7	4	−4	−7	5
	4	1	−3	2	2	−3	1

Comparison ψ_2 is significant beyond the .01 level. Comparisons ψ_3 and ψ_4 are not significant at the .01 level.

9.

	1	2	3	4	5
M_{Y_i}	16.3460	18.2560	18.4400	19.4020	23.7780
Y_j'	16.3442	18.2635	18.4286	19.4095	23.7762

Trends of first, second, and third degree are significant.

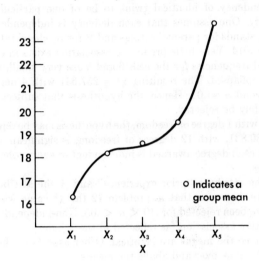

11. $r_{YW} = .675$
$r_{XY} = .425$
$r_{XW} = .431$
$r_{YW \cdot X} = .602$

There is very little effect of X on the correlation between Y and W. One cannot test this partial correlation in the usual way, since the W values were not sampled from a multivariate normal distribution, but rather were chosen in advance.

13. We solve the system of equations

$$-.17b_{12 \cdot 3} + b_{13 \cdot 2} = .38$$
$$b_{12 \cdot 3} - .17b_{13 \cdot 2} = .45$$

to obtain $b_{12 \cdot 3} = .4701$ and $b_{13 \cdot 2} = .5299$. For $z_2 = 1.9$ and $z_3 = -1.2$, the predicted value $z_1' = .257$.

15. $r_{12 \cdot 3} = .5188$; $r_{13 \cdot 2} = -.4128$.

17.

SOURCE	SS	df	MS	F
BETWEEN MEANS	1028.00	4		
LINEAR	1026.173	1	1026.173	3.568 NS
DEVIATIONS FROM LINEAR	1.827	3	.609	
ERROR	12940.00	45	287.56	

The estimated value of $\rho_{XY}^2 = .054$. The estimate of $\omega^2 - \rho^2$ is .00.

CHAPTER 17

1. For 8 degrees of freedom, the χ^2 value of 9.176 is not significant. (Note here that the expected frequencies for the extreme cells actually are very small. A preferable method would be to collapse the two extreme cells on either end of the distribution. Then $\chi^2 = 2.313$ with 6 degrees of freedom, which is, again, not significant.)

3. The χ^2 value of .5697 is not significant for 1 degree of freedom. There is little or no evidence for a tendency of identical twins to be of one particular sex, as against infants born singly. One assumes that each delivery is independent of every other.

5. If the attribute A stands for parents' ratings and B for teachers' ratings, then $\lambda_B = 0$, $\lambda_A = .028$, $\lambda_{AB} = .014$. Very little predictive association exists in either direction.

7. Since the expected frequencies for the cells 6 and 5 are very small, the first three cells (6, 5, and 4) are collapsed. The resulting $\chi^2 = 257.631$ with 4 degrees of freedom is significant far beyond $\alpha = .01$. Hence, the hypothesis that subjects were guessing on each item may safely be rejected.

9. Since $\chi^2 = 5.172$, with 1 degree of freedom, the hypothesis may be rejected for $\alpha < .025$.

11. The χ^2 value of 80.831, with 12 degrees of freedom, is significant for $\alpha < .001$. One must assume that each degree awarded is independent of every other, both within and among universities.

13. If B represents the attribute "prior experience" and A the attribute "deportment", then $\lambda_A = 0$. Note, however, that in problem 12 the χ^2 value was sufficiently large that H_0 might have been rejected for $.10 < \alpha < .05$. Non-independence is not synonymous with predictive association (as reflected in the λ index).

15. F must be zero since the means are identical. Other aspects of the distributions are reflected in the χ^2 value, over and above the means.

17. The table presented has a probability of .0389. A more extreme table in the direction of fewer problems solved by the experimental group has probability of .00108. Hence, the present table or one more extreme in the direction of H_1 has a probability of only about .04 if H_0 is true. Thus, H_0 is rejected in favor of H_1.

19. The χ^2 value of 15.624 with 9 degrees of freedom does not quite reach the value required for significance with $\alpha = .05$. Thus we do not reject the hypothesis of independence between the two sets of ratings. The value of $\varphi' = .216$.

CHAPTER 18

1. $z = \dfrac{(16/25 - .5) - \dfrac{.5}{25}}{\sqrt{.25/25}} = 1.2$

The difference is not significant.

3. The number of runs is 10. The probability of ten runs is

$$p(10; N_1 = 10, N_2 = 12) = \frac{2\dbinom{9}{4}\dbinom{11}{4}}{\dbinom{22}{10}}$$

$$= .14$$

Since the probability of 10 runs is .14, the probability of 10 *or fewer* runs must be greater than .14. Hence, the hypothesis is *not* rejected on this evidence. Note that $E(R) = 11.9$ in this instance.

5. The number of runs is eight, with a probability of more than .07. Since exactly eight runs has a probability of more than .07, eight or fewer runs has a probability in excess of .07. Hence, the hypothesis of no systematic connection between days and scores is *not* rejected.

7. $z = \dfrac{134 - 162.5}{37.16} = .767$

This is not significant.

9. $\chi^2 = .738$ approximately; for 3 d.f. not significant.

11. $Q = 13.565$, significant beyond the .01 level for 4 degrees of freedom.

13. The τ value of .412 indicates a fairly strong positive monotone relation.

15. $z = \dfrac{U' - 450}{73.54} = -.306$, which is not significant.

17. The value of τ is .733.

19. The value of τ is .792. Then $z = 76/19.67 = 3.86$, so that the value is significant beyond the .01 level.

21. $W = .263$; av. $r_S = .282$. Significant beyond the .001 level.

Chapter 19

1. The posterior probabilities are
$$p(\text{loaded}|\text{data}) = .204$$
$$p(\text{fair}|\text{data}) = .796$$

3. The posterior probabilities are
$$p(\text{rule } 1|0,2,0,0) = \frac{k^4}{k^4 + 9g^4} \qquad p(\text{rule } 2|0,2,0,0) = \frac{9g^4}{k^4 + 9g^4}$$

These are the same probabilities that would have been found if an initial sample of 4 observations had been taken, yielding the values 0, 2, 0, 0.

5. The posterior probabilities are
$$p(\pi = .2|3 \text{ of } 15) = .517$$
$$p(\pi = .3|3 \text{ of } 15) = .352$$
$$p(\pi = .4|3 \text{ of } 15) = .131$$

7. Odds of about 17 to 8 that $\mu \leq 90$, and odds of better than 25,000 to 1 that $\mu \leq 130$.

9. The posterior distribution of h would be gamma, with parameters $V'' = 22.93$ and $\nu'' = 20$. (Note that V would be based upon $\mu = 500$, so that there are 19 degrees of freedom in the sample.)

11. The 99 percent credible interval has limits given by $2.773 \pm .21$.

13. The posterior distribution is beta, with $\nu'' = 2 + 4 = 6$ and $N'' = 12 + 30 = 42$. The posterior mode is at $5/36$.

15. In the posterior distribution with $M'' = 31$ and $V'' = .823$, the probability is only about .045 that $27.6 \leq \mu \leq 29.6$. Hence H_0 is rejected in favor of H_1.

17. The limits of the credible interval are given by 506.03 ± 19.60.

19. In the posterior distribution with $M'' = 15.86$ and $N'' = 85$, the probability that $\mu \leq 15$ is only about .00004. Hence H_0 may quite safely be rejected in favor of H_1.
21. The limits of the 99 percent credible interval are $98.7 \pm .141$.
23. The limits of the 95 percent credible interval are 114 ± 2.06. Note that the posterior distribution has the Student t form, with 19 degrees of freedom.

APPENDIX A: RULES OF SUMMATION

1. $x_1 + x_2 + x_3 + x_4 + x_5.$
3. $x_1^3 + x_2^3 + x_3^3 + x_5^3 + x_6^3 + x_7^3 + x_8^3.$
5. $\displaystyle\sum_{i=1}^{3} 4 \left(\sum_{j=1}^{3} x_{ij} \right) = 4 \sum_{i=1}^{3} (x_{i1} + x_{i2}) = 4 \left(\sum_i x_{i1} + \sum_i x_{i2} \right)$
$= 4[(x_{11} + x_{21} + x_{31}) + (x_{12} + x_{22} + x_{32})].$
7. $(x_1^2 g_1 + x_2^2 g_2 + x_3^2 g_3 + x_4^2 g_4)/(g_1 + g_2 + g_3 + g_4).$
9. $\displaystyle\sum_{i=1}^{5} \left(\sum_{j=2}^{4} x_{ij}^2 + \sum_{j=2}^{4} y_i \right) = \sum_{i=1}^{5} (x_{i2}^2 + x_{i3}^2 + x_{i4}^2 + 3y_i) = 3(y_1 + \cdots + y_5)$
$+ x_{12}^2 + \cdots + x_{52}^2 + x_{13}^2 + \cdots + x_{53}^2 + x_{14}^2 + \cdots + x_{54}^2.$

11. $\displaystyle\left(\sum_{i=1}^{4} x_i \right) \Big/ N.$

13. $\displaystyle\sum_{i=2}^{4} x_i f(x_i).$

15. $\displaystyle 3 \sum_{i=1}^{N} (x_1 + y_i)^2.$

17. $\displaystyle 15 \left(\sum_{i=1}^{4} x_i \right)^2.$

19. $\displaystyle 4 \sum_{i=2}^{4} \frac{(x_i - 5)}{y_i}.$

21. $\displaystyle\left(\sum_{i=1}^{4} x_i \right)^2 \left(\sum_{i=1}^{4} y_i^2 \right).$

23. $\displaystyle\left(\sum_{i=1}^{5} x_i + 10c \right) \Big/ 21.$

25. $\displaystyle\sum_{i=1}^{N} x_i - N \left(\sum_{i=1}^{N} x_i/N \right) = 0.$

27. $N \sqrt{(a+b)c}.$

29. $\text{Var}(X) = \dfrac{\Sigma x_i^2}{N} - M_X,$ where $M_X = \left[\displaystyle\sum_{i=1}^{N} x_i \right] \Big/ N.$

31. −3.
33. 68.
35. 47/23.

APPENDIX B: THE ALGEBRA OF EXPECTATIONS

1. .75.
3. 3.1.
5. 22.29.
7. $E(X^2)$.

9. This problem should read: $\sum_{i=1}^{10} (x_i - E(X))^2 p(x_i) = \text{Var}(X)$.

11. $\dfrac{E(X)}{10} + 3.5$.

13. $E(X^2) - E^2(X) = \text{Var}(X)$.
15. 17.
17. $E(aX) = \sum_i ax_i p(x_i) = a \sum_i x_i p(x_i) = aE(X)$.

19. $E(X - E(X)) = E(X) - E(E(X)) = E(X) - E(X) = 0$.
21. $E(5X - 12) = 5E(X) - 12$; (a) −8.250; (b) −9.222; (c) 3.50.
23. $E(XY) = E(X)E(Y)$ for independent variables. Hence
$$E(XY) = (3.1)(115.60) = 358.36.$$

Index